EARLY IRISH LAW SERIES VOLUME IV
General editor: Fergus Kelly

EARLY IRISH FARMING

This calf, the symbol of Saint Luke, is from f. 124v of the seventh-century Book of Durrow in the Library of Trinity College, Dublin. It has been reproduced by kind permission of the Board of the College.

EARLY IRISH LAW SERIES VOLUME IV

EARLY IRISH FARMING

a study based mainly on the law-texts of the 7th and 8th centuries AD

[with minor revisions and corrections]

BY

FERGUS KELLY

SCHOOL OF CELTIC STUDIES
DUBLIN INSTITUTE FOR ADVANCED STUDIES
2000

This reprint 2000

ISBN 1 85500 180 2
ISSN 0790-4657

First published 1997

Printed by Dundalgan Press Ltd, Dundalk Co. Louth

EARLY IRISH LAW SERIES ISSN 0790-4657

I *Bechbretha: an Old Irish law-tract on Bee-keeping*, ed. Thomas Charles-Edwards and Fergus Kelly
ISBN 0 901282 73 1

II *Uraicecht na Ríar: the poetic grades in Early Irish law*, ed. Liam Breatnach
ISBN 0 901282 89 8

III *A guide to Early Irish law*, by Fergus Kelly
ISBN 0 901282 95 2

for

FRANK MITCHELL

CONTENTS

LIST OF FIGURES

LIST OF PLATES

Preface

I hope that this book will appeal to students of early Irish law, as well as to social historians, geographers, archaeologists, naturalists and all those with an interest in the history of Irish farming.

The intention is to present a general account of the early Irish farm, based on the written sources. By far the most detailed evidence relating to farming practice is provided by the law-texts of the seventh and eighth centuries AD. Consequently the main focus of the book is on the legal material, including later glosses and commentaries. However, other written sources must also be taken into account. It would for example be inexcusable to discuss the graded list of cereals in the eighth-century law-text *Bretha Déin Chécht* without reference to the similar list in the twelfth-century comic tale *Aislinge Meic Con Glinne* (pp. 219–20 below).

The plan of the book is straightforward. In the first six chapters I deal with early Irish domestic animals, and I then devote two chapters to the crops cultivated in fields and gardens. By way of contrast, the ninth chapter examines the exploitation of wild animals and plants. I provide in Chapter 10 a general account of the foods consumed by the early Irish, and then in Chapter 11 assemble the evidence with regard to farm lay-out. Chapter 12 is concerned with land-tenure, Chapter 13 with farm labour, and Chapter 14 with the tools and technology of the early Irish farm. Appendix A consists of editions of seven legal passages on topics connected with farming. Finally, the early Irish units of measurement – vitally important for the interpretation of the textual evidence – are treated in Appendix B.

Illustrations of relevance to early Irish farming have been reproduced photographically from various manuscripts: acknowledgements are made in the accompanying captions. There are also freehand drawings by my son Stephen of animals and human figures from high crosses at Banagher, Bealin, Castledermot and Clonmacnois, as well as from the Book of Kells. We have together devised illustrations of an early Irish onion, skirret, trench-and-bank, field-wall, fence, axe and billhook. It must be stressed that these are speculative attempts to represent the sometimes obscure descriptions of these items in the law-texts.

Perhaps some future researcher may possess the combined textual and archaeological expertise to improve on our efforts!

I have been working on the topic of early Irish farming off and on for nearly two decades. During this period I have been helped by far too many people to thank individually. I would like therefore to express my deep gratitude to everyone who replied to my queries, or who helped me with references, corrections or ideas. Many scholars read and commented upon earlier versions of all or part of this book, and I am extremely grateful to all of them for their various contributions. In particular I should mention the detailed criticisms of the final draft by Liam Breatnach, Thomas Charles-Edwards and Mick Monk. I am also very much indebted to Michelle O Riordan who finalised the whole work for the printer, and dealt most skilfully with the illustrations. The photograph on the front cover shows some of the Muckross herd of Kerry cattle (see p. 31) at the Bourn Vincent Memorial Park, Killarney, Co. Kerry, and was taken by my wife Elizabeth. The back cover is from folio 67^r of the Book of Kells, and has been reproduced by kind permission of the Board of Trinity College, Dublin.

Finally, tribute should be paid to the anonymous authors of the law-texts whose keen observation of contemporary farming practices enabled the present book to be written.

Fergus Kelly

Abbreviations

AC *Annála Connacht: the annals of Connacht* (ed. Freeman, Dublin 1944, repr. 1983)

Ann. Clon. *The annals of Clonmacnoise, translated into English A.D. 1627 by Conell Mageoghagan* (ed. D. Murphy, Dublin 1896)

ACL *Archiv für celtische Lexikographie* i–iii (ed. Stokes and Meyer, Halle 1900–1907)

AFM *Annála ríoghachta Éireann: annals of the kingdom of Ireland by the four masters* i–vii (ed. O'Donovan, Dublin 1848–51)

AI *The annals of Inisfallen* (ed. Mac Airt, Dublin 1951, repr. 1988)

AL *Ancient laws of Ireland* i–vi (ed. Hancock et al., Dublin 1865–1901)

ALC *The annals of Loch Cé* i–ii (ed. Hennessy, London 1871)

ALIW *Ancient laws and institutes of Wales* i–ii (ed. Owen, London 1841)

AU *Annála Uladh: annals of Ulster* i–iv (ed. Hennessy and MacCarthy, Dublin 1887–1901)

AU[2] *The annals of Ulster* i (ed. Mac Airt and Mac Niocaill, Dublin 1983)

BB *Bechbretha* (ed. Charles-Edwards and Kelly, Dublin 1983)

CG *Críth Gablach* (ed. Binchy, Dublin 1941, repr. 1979)

CGH *Corpus genealogiarum Hiberniae* (ed. M. O'Brien, Dublin 1962, repr. 1976)

CIH *Corpus iuris hibernici* i–vi (ed. Binchy, Dublin 1978)

CS *Chronicum Scotorum* (ed. Hennessy, London 1866)

DIL (*Contributions to a*) *Dictionary of the Irish language* (ed. Quin et al., Dublin 1913–76, repr. 1983)

GEIL *A guide to early Irish law* (Kelly, Dublin 1988, repr. 1991)

GMWL *A glossary of mediaeval Welsh law* (T. Lewis, Manchester 1913)

GPC *Geiriadur Prifysgol Cymru: a dictionary of the Welsh language* (ed. Thomas et al. Caerdydd 1950-)

IEW *Indogermanisches etymologisches Wörterbuch* i (Pokorny, Bern 1959)

IP *The Irish penitentials* (ed. Bieler, Dublin 1963, repr. 1975)

JRSAI *Journal of the Royal Society of Antiquaries of Ireland* (Dublin)

LB *Leabhar Breac.* Lithographic reproduction of transcript by O'Longan, Dublin 1872–76

LEIA *Lexique étymologique de l'irlandais ancien* (ed. Vendryes, Bachellery and Lambert, Dublin and Paris, 1959-)
LL *The Book of Leinster* i–vi (ed. Best et al. Dublin 1954–83)
LU *Lebor na hUidre* (ed. Best and Bergin, Dublin 1929, repr. 1992)
O'Dav. O'Davoren's Glossary (ed. Stokes, *ACL* ii 197–504)
OED *Oxford English dictionary* (Oxford 1933)
O'Mulc. O'Mulconry's Glossary (ed. Stokes, *ACL* i 232–324)
PRIA *Proceedings of the Royal Irish Academy* (Dublin 1836-)
RC *Revue celtique* (Paris 1870–1934)
SEIL *Studies in early Irish law* (ed. Binchy, Dublin 1936)
Thes. *Thesaurus palaeohibernicus* i–ii (ed. Stokes and Strachan, Cambridge 1901–3, repr. Dublin 1987)
TBC LL *Táin Bó Cúalnge from the Book of Leinster* (ed. C. O'Rahilly, Dublin 1967, repr. 1984)
TBC Rec.I *Táin Bó Cúailnge*, Recension I (ed. C. O'Rahilly, Dublin 1976)
TBC St *The Stowe version of Táin Bó Cuailnge* (ed. C. O'Rahilly, Dublin 1961, repr. 1978)
UM *The Book of Uí Maine.* Collotype facsimile (ed. Macalister, Dublin 1941)
VKG *Vergleichende Grammatik der keltischen Sprachen* i–ii (Pedersen, Göttingen 1908–13)
YBL *The Yellow Book of Lecan.* Photo-lithographic reproduction (ed. Atkinson, Dublin 1896)
ZCP *Zeitschrift für celtische Philologie* (Halle, Tübingen, 1897-)

Introduction

Physical description of Ireland

It has often been pointed out that Ireland is roughly saucer-shaped, with mainly flat land in the middle and mountains forming much of the perimeter. It is difficult for water to drain away from the centre, which consequently has many lakes, marshes and bogs. Much of the farm-land in this area is subject to periodic flooding and is therefore too wet for tillage, though valuable for summer-grazing. This applies, for example, to the callows (Irish *caladh*) along the River Shannon. Drainage is also a problem over large areas of Ulster and Connacht where the retreat of the last ice-sheet left a landscape of small blunt-nosed hills (drumlins), often interspersed with lakes.[1]

Some of the best land in Ireland is to be found in the well-drained plains in the drier climate of the east, and these regions feature prominently in early Irish sources. The ninth-century author of a geographical triad is no doubt thinking of their agricultural importance when he lists the three plains of Ireland as *Mag Midi* 'the plain of Meath', *Mag Lifi* 'the Liffey plain', and *Mag Line* 'Moylinny (Co. Antrim)'.[2]

In spite of its relatively small area Ireland is geologically quite varied, with a consequent diversity of land-types. The author of one Old Irish law-text makes an interesting classification of different types of land on the basis of agricultural value (p. 394). In their treatment of farming matters other legal authors take into account the different physical conditions which may obtain in different areas. For example, the main law-text on farming, *Bretha Comaithchesa*, distinguishes the type of field-boundary which would be appropriate to the deep soils of the eastern plains from that which would suit the stony terrain of much of the west and north (p. 372). The law-texts also deal with various aspects of hill-farming

[1] The term *drumlin*, which has passed into the international geological vocabulary, undoubtedly contains Irish *druimm* 'back, hill', but the second element is uncertain. See Quin and Freeman, 'Some Irish topographical terms', 85–6.

[2] Meyer, *Triads*, 4 § 41. Triad no. 52 (Yellow Book of Lecan version) refers also to the Connacht plain of *Mag Crúachan* around the royal site of Crúachu (Rathcroghan, Co. Roscommon).

(p. 43), and even take into account the special problems of keeping cattle in unusual locations such as a marine island (*muirinis mara*).[3]

Climate

Foreign observers have often commented on the mild but wet and windy climate of Ireland. Writing in the late twelfth century Giraldus Cambrensis remarks on the difficulty experienced by Irish farmers in harvesting their grain-crops on account of the unceasing rain, and makes the somewhat exaggerated claim that 'this country more than any other suffers from storms of wind and rain'.[4]

The native annalists naturally record only weather conditions which are extreme by Irish standards, but they present a similar picture of a climate which may be very difficult for the growing of cereals and pulses. For instance, the *Annals of Inisfallen* record that much of the corn-crop of the year 1012 was destroyed by heavy rain.[5] The same annals record that a gale in the autumn of 1077 greatly damaged the corn-crop.[6] According to the *Annals of Connacht* terrible thunder and lightning in the year 1328 caused corn to grow 'pale and empty' (i.e. grainless).[7] In May of 1471 hail-showers – with hailstones of two or three inches in length – destroyed beans and other crops in various parts of Ireland.[8] Drought is also occasionally mentioned. The *Annals of Ulster* record an almost total failure of bread in 773 caused by unusual drought and heat of the sun,[9] and the *Annals of Inisfallen* refer to a hot summer in 1129 which resulted in great mortality among livestock through lack of water.[10]

The winters were usually so mild that there was sufficient growth of grass and other herbage to support cattle without saving hay. However, on occasions when snow persisted for more than a few days, cattle were at risk. For example, the *Annals of Ulster* record that nearly all the cattle of Ireland died in the year 748 as a result of

[3] *CIH* i 207.3 = *AL* iv 6.10. The accompanying gloss gives *Inis Cathaig* (Scattery Island) in the Shannon Estuary as an example of such a location.

[4] O'Meara, *Giraldus: Topography*, 34 § 2 = Dimock, *Giraldi Topographia*, 27.

[5] *AI* 182 s.a. 1012 § 4.

[6] *AI* 234 s.a. 1077 § 7.

[7] *AC* 262 s.a. 1328 § 3 *tornech 7 tentich anbail . . . cur fāsadur arbanna finna fāsa.*

[8] *AC* 554 s.a. 1471 § 15.

[9] AU^2 226 s.a. 772 (recte 773) § 4.

[10] *AI* 292 s.a. 1129 § 9.

snow of unusual depth.[11] Evidently such weather-conditions were rare: otherwise the saving of hay would surely have been generally practised by early Irish farmers (see discussion on p. 47).

According to the ninth-century wisdom-text *Tecosca Cormaic*, the ideal weather is 'a fine frosty winter, a dry windy spring, a dry showery summer, and a heavy-dewed fruitful autumn'.[12] *Immacallam in dá thúarad* views the most unseasonable types of weather as being 'a leafy winter, a gloomy summer, an autumn without fruits, a spring without flowers'.[13]

The human impact in prehistoric times

Humans first came to Ireland at least as early as 7000 BC,[14] and would have found a country almost entirely covered by woods. These Mesolithic people lived by hunting, fishing and gathering, and seem generally to have stayed near the sea-shore or by the margins of lakes and rivers. It is believed that they made small clearings in the woodland, but did not sow crops or keep domestic animals.[15] Neolithic farmers brought a new way of life to Ireland about 4000 BC. In contrast to the previous inhabitants, their impact on the environment was considerable. The pollen-records from this period show a major decline in the primeval woodland, with increase of hazel-scrub. There is also a marked upsurge in grass pollen as well as that of various weeds of disturbed ground.[16] Recent archaeobotanical studies indicate that the main cereal cultivated by these farmers was emmer wheat (see p. 225), but there is also evidence of einkorn and barley.[17] Their domestic animals were cattle, pigs, sheep and possibly goats.[18] By about 3000 BC, Neolithic society was sufficiently well-organized in the fertile Boyne valley for the construction of massive passage-graves such as those of Knowth, Dowth and Newgrange. The system of

[11] *AU*^{2} 202 s.a. 747 (recte 748) § 3.

[12] Meyer, *Tecosca Cormaic*, 50 § 36.

[13] Stokes, 'Colloquy', 46 § 231.

[14] Mitchell, *The Irish landscape*, 98; *Reading the Irish landscape*, 78; Cooney and Grogan, *Irish prehistory*, 7.

[15] Mitchell, *The Irish landscape*, 101; *Reading the Irish landscape*, 84–6.

[16] Herity and Eogan, *Ireland in prehistory*, 25.

[17] Monk, 'Archaeobotanical study of samples from pipeline sites', 187.

[18] McCormick, 'The animal bones', 182.

drystone field-walls constructed by Neolithic people at Behy, Co. Mayo, shows a high level of farm-organisation in this region.[19]

The next major event in the history of Irish farming was the arrival of Bronze-users about 2300 BC.[20] Their axes were more effective than the stone axes of their Neolithic predecessors, and their bronze sickles made harvesting a much easier operation.[21] G. F. Mitchell suggests that in the later stages of the Bronze Age – about 700 BC – the first simple wooden ploughs were used in Ireland.[22] Archaeological evidence for the period between 700 and 300 BC is particularly scanty. The linguistic evidence suggests that Celtic-speaking invaders from Britain or the Continent arrived around the beginning of this period, perhaps introducing the use of iron to Ireland. However, the archaeological evidence shows little sign of a significant external intrusion.[23] The Celticization of the country may therefore have been a gradual process rather than a dramatic replacement of one culture by another.

The period between 300 BC and AD 300 shows some diminution in agricultural activity, possibly associated with worsening climatic conditions.[24] In a sample at Red Bog, Co. Louth, pollen from cereals and weeds of cultivation disappears abruptly from the pollen-diagram about 300 BC, and that of grass decreases. Hazel rises in value, followed by rises in ash, oak and elm.[25] The evidence from a number of sites indicates a significant revival of agriculture around AD 300. This marks the beginning of what has been described as the Destruction-phase of the Irish woodlands.[26] Later Irish tradition contains many references to the clearing of plains in prehistory, and may thus preserve folk-memory of these

[19]Mitchell, *The Irish landscape*, 140; *Reading the Irish landscape*, 106, 124; Caulfield, 'The neolithic settlement of North Connaught'.

[20]Mitchell, *The Irish landscape*, 145; *Reading the Irish landscape*, 129; Cooney and Grogan, *Irish prehistory*, 95.

[21]Herity and Eogan, *Ireland in prehistory*, 174–5, 188 (illustrations of bronze axes and sickles).

[22]Mitchell, *The Irish landscape*, 158; *Reading the Irish landscape*, 143.

[23]Cooney and Grogan, *Irish prehistory*, 174, 186; Champion, 'The myth of Iron Age invasions in Ireland'; Waddell, 'The question of the Celticization of Ireland'; Raftery, 'The Celtic Iron Age in Ireland: problems of origin'.

[24]Cooney and Grogan, *Irish prehistory*, 195.

[25]Mitchell, *The Irish landscape*, 159–60; *Reading the Irish landscape*, 144–5. See also the conclusions of Weir, 'A palynological study', 108.

[26]Mitchell, *The Irish landscape*, 166; *Reading the Irish landscape*, 153; cf. Weir, 'A palynological study', 107.

clearances.[27] Major road-construction seems also to be associated with this period, and it is significant that a common Irish word for road is *slige*, which literally means 'hacking, clearing'.

Archaeological information in the historical period

Until shortly before the Christian era archaeology is our only source of information on life in Ireland. Thereafter, the evidence of texts of various types (see next section) can be used to supplement the findings of archaeology. From nut-hulls, fruit-stones, cereal-grains, or chaff-fragments the archaeologist can draw conclusions about foods of plant origin. In chapters 7 and 8 of this book, an attempt is made to fit in the archaeological data on edible plants with the information found in texts of the early Christian period. Pollen-analysis can provide detailed information over a long period of time on the types of plants growing in a particular area. This information can in turn give us insight into the types of agriculture practised nearby. If enough annual rings are visible in a piece of oak, its date can be established by dendrochronology.[28] For example, it can be shown that the oak used to make one of the door-jambs for a house excavated at Glenarm, Co. Antrim, was felled in AD 648.[29] A number of watermills have likewise been dated by this method (see p. 485 below).

Being generally tougher than vegetable refuse, animal-bones survive in greater abundance and tell us much about the foods of animal origin which were consumed. A detailed study of animal-bones can provide a remarkable amount of information on stock-raising and butchering practices.[30] An expert can tell the age at which an animal was killed, and how well it was nourished prior to its death. A smashed frontal bone of the skull shows that

[27] E. J. Gwynn, *Metrical Dindshenchas* iii 198.9–16; 330.19–20; 380.5–12. According to Middle Irish tradition (Macalister, *Lebor Gabála* iii 50 §§ 18–19), the only plain which was never tree-covered was *Senmag Elta Étair* 'the old plain of (Elta) Étar', near Clontarf, Co. Dublin. See Hogan, *Onomasticon*, s.vv. *Senmag elta edair, Senmag étair, Mag elta.*

[28] At the Palaeoecology Centre of Queen's University, Belfast, Michael Baillie and David Brown have established a 7000-year chronology based largely on Irish oak remains. See Baillie, 'Dendrochronology: the prospects for dating throughout Ireland'; Baillie, *Tree-ring dating and archaeology.*

[29] Lynn, 'Deer Park farms', 14.

[30] See McCormick, 'Stock-rearing in early Christian Ireland'.

the animal was killed by a heavy blow on the forehead, and a hole in the scapula shows that the carcase was hung up for curing by salt and/or smoke. In chapters 1–3, I discuss the textual evidence for domestic livestock and attempt to correlate it with the findings of archaeology.

Another branch of archaeology – aerial photography – provides information on settlement and field-patterns, which is of particular relevance to chapter 11. In certain lights, earthworks often show up much more clearly from the air than when viewed from the ground. The remains of earthworks may also be detected by aerial photography as cropmarks in cereal-fields at an early stage of growth.[31]

THE EVIDENCE OF THE WRITTEN SOURCES

The use of writing in the Mediterranean region goes back at least as far as 3000 BC but the earliest writing in Irish (the Ogham inscriptions) dates from about the fifth century AD.[32] There are a few references to Ireland in early Greek and Latin sources, but these provide little detailed information on the country.[33] The earliest extensive document relating to Ireland is the *Confession* of Saint Patrick, written in Latin about the late fifth century. Though concerned mainly with spiritual matters, this text also includes a small amount of information on early Irish society, including a brief reference to the herding of livestock (p. 438 below). Much more detail is provided by texts – in both Old Irish and Latin – dating mainly from the seventh to the ninth centuries. For our purposes the most important of these are the Old Irish law-texts, and they form the basis of this book. Information on farming is also to be found in annals, saints' Lives, penitentials, wisdom-texts, sagas, and poetry. Purely linguistic evidence may also be of significance in the discussion on early Irish farming: see p. 15 below.

Law-texts

It will be seen that a large proportion of the footnotes in this book refer to D. A. Binchy's *Corpus Iuris Hibernici* (*CIH*), published in Dublin in 1978. This six-volume work contains a transcription

[31] Norman and St Joseph, *The early development of Irish society*, 1–15.

[32] McManus, *A guide to Ogam*, 97.

[33] T. F. O'Rahilly, *Early Irish history and mythology*, 1–2; Tacitus, *Agricola* (ed. Peterson and Hutton), 24.

of most of the extant legal material in Irish from the native tradition. Approximately fifty law-texts survive – often in an incomplete state – and there are also many short fragments from other lost texts.[34] There is solid linguistic evidence that most of the law-texts were written in the seventh or eighth centuries AD.[35] They survive, however, in manuscripts ranging in date from the twelfth to the sixteenth centuries. Successive copyists have been responsible for many misreadings, modernisations of spelling, omissions, and additions. Furthermore, some of the manuscripts are physically in poor condition and hard to read. For these reasons – and because of the many linguistic difficulties – the interpretation of the law-texts is often highly problematic.

The Old Irish law-texts deal with a wide variety of topics, such as contracts, theft, injury, suretyship, marriage, kinship, distraint, legal procedure, etc. Practically all the law-texts contain some information on farming or farm-produce (see also p. 503). For the purposes of this book the most important law-text is *Críth Gablach*, written about AD 700, which deals primarily with the legal implications of rank in early Irish society.[36] In his discussion of this topic, the author provides a detailed account of the material possessions which persons of different rank are expected to own. This gives a fascinating insight into the land-ownership, farm-buildings, livestock, tools, and farming practice of the period. Another crucial law-text is *Bretha Comaithchesa* 'judgements of neighbourhood' which deals with many aspects of farming law, including the notoriously sensitive area of animal-trespass (see Chapter 4 below). *Cáin Aicillne* 'the law of base clientship' details the annual food-rent which a client must deliver to his lord in return for a fief of stock, land or farming equipment. This text gives us a good idea of the sort of foodstuffs – meat, cheese, grain, and vegetables – which the ordinary client-farmer (*aithech*) produces from his land.

The information which the law-texts provide on farming and other aspects of daily life is of course of a very different kind to that provided by archaeology. Much archaeological evidence consists of physical matter whose identity can be determined with precision. There will always be some uncertainties, but in a typical

[34] See the list at *GEIL* 266–81.

[35] E.g. *CG* Introd. pp. xiii–xvii; *BB* 12–14; Breatnach, *Uraicecht na Ríar*, 77.

[36] See Charles-Edwards, '*Críth Gablach* and the law of status'.

excavation the great majority of plant and animal remains can be securely identified. In a law-text, by contrast, our information is at best second-hand, and we have to form a judgement as to how much trust we are entitled to place in the author's statements. In some cases we have independent means of checking them, and usually find a high degree of accuracy. I edit in Appendix A below some hitherto untranslated legal passages on livestock. One deals with defects in cows and sheep (p. 506), another with the rearing of a calf under various circumstances by a cow which is not its mother (p. 545), and another with the qualities and defects of horses (p. 555). These texts display specialized knowledge, and could only have been written by lawyers who were either personally experienced in stock-raising or else well briefed by experts; perhaps their knowledge stemmed from discussions with their professional clients.

A legal passage which displays similar expertise is the section on damage to trees in *Bretha Comaithchesa*, discussed on p. 380. Modern botanists have been impressed by the scientific manner in which the author arranges twenty-eight trees and shrubs into four groups of seven in accordance with their economic value. Modern brewers have also confirmed the account of the procedure for malting barley given in *Cáin Aicillne* (see p. 246). Any modern farmer could likewise confirm that the description of field-walls and fences in *Bretha Comaithchesa* is realistic, (see p. 372). and would recognise the six land-types distinguished in the law-text *Cis lir fodla tíre?* (see p. 394).

But in some cases the information which the law-texts provide is not so easy to reconcile with what we know from other sources. Part of the problem stems from the fact that the authors of the law-texts are often of necessity generalizing about law and society, and in doing so may distort some details. *Críth Gablach* is particularly open to this criticism. For example, in his discussion of the wealth which is expected to be possessed by persons of various ranks, the author states that an *ócaire* has seven cows, seven pigs and seven sheep. The grade above him has ten cows, ten pigs and ten sheep, and above him again the *mruigfer* has twenty cows, twenty pigs and twenty sheep.[37] Obviously, real life is never as neat and tidy as this,

[37] It is implicit that the grade below the *mruigfer* – here called the *bóaire febsa* – has twelve cows, twelve pigs, and twelve sheep, but the text only refers to cows (*CIH* ii 563.9; iii 779.28–9 = *CG* 6.157–8).

and we can assume that the author does not expect us to take him literally. The archaeological evidence in fact consistently indicates that cattle were more numerous on the average early Irish farm than pigs or sheep: see p. 27 below.

It is also important to bear in mind that the author of a law-text is trying to deal with a particular topic in accordance with established legal principles. In doing so, he may sometimes lay down rules which seem difficult or even impossible to enforce. This applies, for instance, to some of the complicated rules relating to bee-trespass (p. 145) or the unauthorized use of horses (p. 169). But such material ought not to be dismissed as being of purely academic interest and without relevance to real life. It represents the efforts – more or less successful – of a legal system trying to cope with the huge variety of disputes which can arise in a farming community.

In Appendix 1 of *A guide to early Irish law* (*GEIL*) I provide a list of law-texts with information on translations where such exist. For most texts, the reader is still dependent on the five-volume *Ancient Laws of Ireland* (*AL*), which was published between 1865 and 1901. Unfortunately, these translations are often inaccurate, especially in volumes i–ii. The great scholar Rudolf Thurneysen published corrected translations (in German) of some of these texts, and also edited and translated other law-texts which had been omitted in *AL*. The journals *Ériu*, *Celtica*, *Zeitschrift für celtische Philologie*, and *Peritia* contain English translations of a number of law-texts by D. A. Binchy and other scholars.[38] For the purposes of the present book, the most important of these is the medical law-text *Bretha Crólige*, published by Binchy in *Ériu* 12 (1938) 1–77.

The ultimate intention is for the Early Irish Law Series to provide reliable translations of all surviving Old Irish law-texts. *Bechbretha* (1983) and *Uraicecht na Ríar* (1987) have already been published, and *Bretha Comaithchesa*, *Bretha Nemed Toísech*, the *Heptads*, and *Cáin Adomnáin* are in preparation.

[38]See *GEIL* 287-95 (Bibliography).

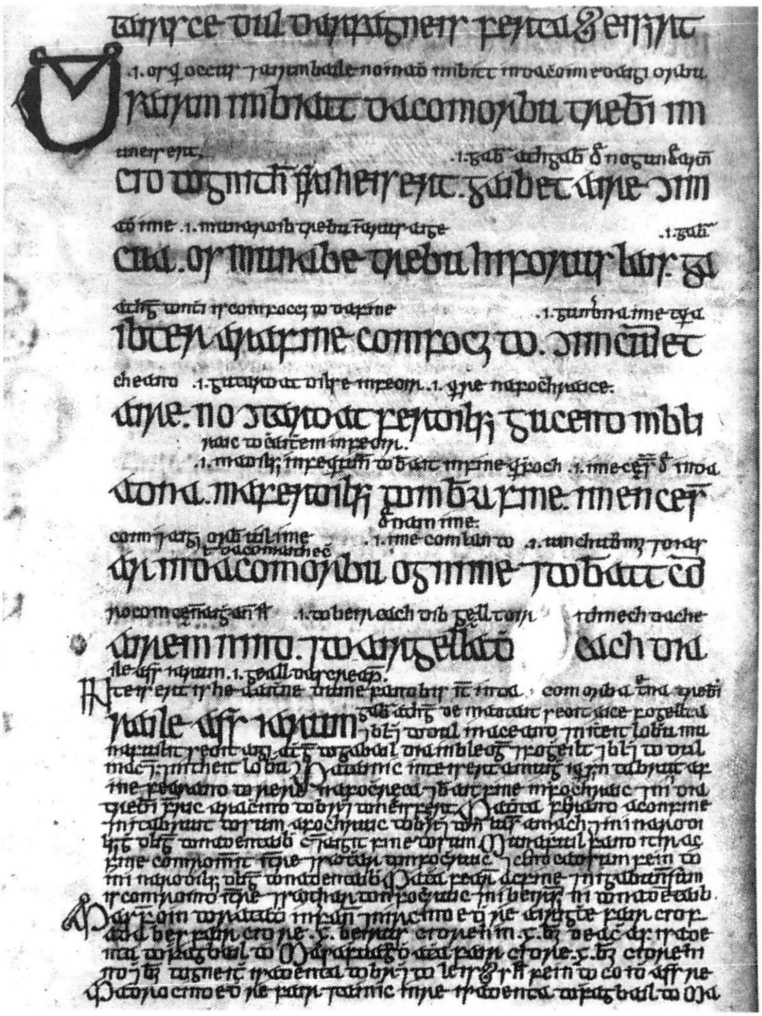

Plate 1. Part of f. 66^b of the fourteenth-century manuscript Rawlinson B 487 in the Bodleian Library, Oxford, which has been reproduced by kind permission of the Trustees. The Old Irish text (in large minuscule) is from the main law-text on farming *Bretha Comaithchesa*, 'judgements of neighbourhood', which was written in about the eighth century. This passage deals with land which has been neglected, and the responsibilities of the owner's kinsmen. The material in small minuscule consists of Middle Irish glosses and commentary: see discussion opposite.
The text is reproduced at *CIH* i 75.24–29, and the glosses and commentary are reproduced at *CIH* i 75.30–76.16. There is a translation at *AL* iv 129.

Legal glosses and commentary

Plate 1 contains a reproduction of part of the main law-text on farming, *Bretha Comaithchesa.* The manuscript was copied about the fourteenth century, but the Old Irish text (in large minuscule) was originally composed in the eighth century. The interlinear material in small minuscule consists of explanatory glosses, dating from about the twelfth century. The material at the bottom of the page is commentary from approximately the same period as the glosses. Most law-texts are accompanied by glosses and commentary (sometimes very extensive) dating from four or five hundred years after the original date of composition. However, one manuscript contains glosses and commentary which are dateable to the ninth century, at most two hundred years later than the original texts.[39]

The very bulk of the surviving glosses and commentary is intimidating. Nonetheless, this material is of immense value, and is frequently referred to in this book. For our purposes, the most important type of gloss is the simple explanation of a word or phrase in the text. Where the text is straightforward, the gloss is of limited significance. But if the text is defective or obscure, the gloss can be very helpful in elucidating the meaning. Legal glossators are not of course to be regarded as infallible,[40] but – unlike modern editors – they come from the same unbroken learned tradition as the authors of the Old Irish texts, and had access to much legal material now lost.[41]

Legal commentary takes various forms. Sometimes the commentator merely provides a paraphrase of the Old Irish text.[42] But more often the commentator expands on the content of the text. Thus the law-text on accidents *Bretha Étgid* simply states the general legal principle: 'mating is an immunity for bulls and rams'.[43] The accompanying commentary – running to almost seven hundred words – gives examples of the sort of circumstances in which

[39] *CIH* iii 877.4–924.31. See Binchy, 'A text on the forms of distraint', 72.

[40] For example, a glossator at *CIH* i 71.4–5 = *AL* iv 94.4 explains *airlim n-aībil* as being animal-trespass caused by the heat of the sun. But it is much more likely that *aíbel* here refers to sexual heat: see p. 137.

[41] For a discussion of the value of the glosses and commentary of one law-text, see *BB* 14–24.

[42] E.g. *CIH* ii 448.10–26 = *BB* 169–70 Appendix 1 (b).

[43] *CIH* i 276.3 = *AL* iii 230.z *blā tarb ⁊ reithe darmna.*

men or animals might be killed or injured by a bull or ram during the mating season, and discusses the owner's liability, if any.

Legal commentary is of particular interest when the original text is missing or fragmentary. For example, it seems that the law-text *Findśruth Fíthail* dealt with the liability of a herdsman when an animal under his charge is killed or stolen. Only the opening words of the Old Irish text survive, stating inconclusively 'any animal which dies violently'.[44] This fragment is followed by commentary which I take to be based on the rest of the Old Irish text, now lost. The circumstances are discussed in which the herdsman may be exonerated from blame in the event of animals under his charge being drowned, taken by wolves, stolen, etc. (see p. 182). Later commentary thus sometimes fills out the gaps in the Old Irish texts.

But there are difficulties with regard to the dating of such material. In general, the later glosses and commentaries seem to reflect the content of the Old Irish texts quite closely, and provide little evidence of legal or social change between the main period of text-writing (seventh–eighth centuries) and the main period of the glosses and commentaries (twelfth–fourteenth centuries). Nonetheless, there are indications of change in some areas; for example, the glosses and commentary tend to extend royal powers beyond the limits set in the Old Irish law-texts (see p. 408). In relation to farming practice, there are also a few hints of change. Commentary on co-operative ploughing (*comar*) refers to the substitution of a horse for an ox.[45] This may reflect the increased use of horses for ploughing in post-Norman Ireland (see p. 95). Another commentary mentions a sheaf of corn (*punnann*),[46] whereas the available evidence suggests that in early Irish harvesting practice corn was cut just below the ear, and probably not bound in sheaves (see p. 238). It is clear from such cases that the evidence of later glosses and commentaries must be used with caution in relation to the period of the Old Irish law-texts. On the other hand, it would be wrong to dismiss it as totally irrelevant for the early period. In this book I have indicated as far as possible whether my source on a particular topic is an Old Irish law-text or a later gloss or commentary.

[44] *CIH* vi 2139.17 *nach míl atbaill guin.*
[45] *CIH* vi 2178.23–4.
[46] *CIH* i 287.39 = *AL* iii 284.7.

Annals

The earlier annals contain legendary material relating to Ireland, as well as events from Greek and Roman history. Information of relevance to Irish farming is to be found in entries from the sixth century onwards. These deal mainly with good and bad harvests, cattle-plagues, and multiple or monstrous births by livestock. The most informative annals on such matters are the *Annals of Ulster* (up to AD 1588),[47] the *Annals of Inisfallen* (up to 1321 and irregularly to 1450), and the *Annals of Connacht* (1224 to 1544). The *Annals of Clonmacnoise* (up to 1408) survive only in an English translation made in 1627 by Conell Mageoghagan.

Saints' Lives

There are numerous saints' Lives in Irish and in Latin among our sources, ranging in date from the seventh-century *Vita Sancti Patricii* by Muirchú maccu Machtheni[48] to the sixteenth-century *Betha Colaim Chille* by Maghnas Ó Domhnaill.[49] For our purposes the chief value of the saints' Lives is in the frequent description of farming practices, often to illustrate a miracle.

Ecclesiastical legislation

The annals frequently record the promulgation of ecclesiastical law (*cáin*). Such laws were promulgated by a prominent ecclesiastic or by a king, or in some cases by an ecclesiastic and a king together.[50] For our purposes the most important is *Cáin Domnaig* 'the law of Sunday', which attempts to regulate Sunday observance (p. 455). A lost text which would undoubtedly have been very valuable is *Cáin Dar Í* 'the law of Dar Í', which is known to have dealt with offences against cattle (p. 28).[51]

Penitentials, etc.

In the early Irish Church it was customary for a sinner to expiate his offence by fasting and other austerities. Consequently, penitential texts often provide detailed information on the types of food normally eaten by the early Irish. Texts such as the *Canons*

[47]The dating for much of the earlier part of the *Annals of Ulster* is one year behind the true date: see Gearóid Mac Niocaill's Foreword to AU^2 p. xi.

[48]Bieler, *Patrician texts,* 62–123.

[49]O'Kelleher and Schoepperle, *Betha Colaim Chille.*

[50]*GEIL* 22, 281–82.

[51]*GEIL* 282 § 76. See Ó Riain, 'A misunderstood annal: a hitherto unnoticed *cáin*'.

of Adamnan deal with the subject of unclean foods.[52] The ultimate source of this material is the Old Testament, but there are some adaptations to Irish conditions. For example, Christians are expressly forbidden to accept or buy cattle which have been seized in a raid.[53]

Information on diet is also provided by monastic rules, such as *The monastery of Tallaght* (ed. Gwynn and Purton).

Wisdom-texts

The Old Irish wisdom-texts, ranging in date from the seventh to the ninth centuries, contain a great variety of material: observation of nature and human behaviour, discussion of kingship, etiquette, health, weather and many other topics.[54] The *Triads of Ireland* (*Trecheng Breth Féne*) are particularly rich in observations pertaining to farming. Triad 75, for example, takes a lyrical view of commonplace events on the farm – 'Three slender things that best support the world: the slender stream of milk from the cow's teat into the pail, the slender blade of green corn above the ground, the slender thread over the hand of a skilled woman'.[55]

Sagas

By their very nature the Old Irish sagas must be regarded as generally unreliable as sources of information on early Irish farming. For dramatic purposes, everyday reality must be exaggerated so as to fit the heroic mode. Hence, Mac Dathó's pig in the tale *Scéla Mucce Meic Dathó* is represented as having been fed on the milk of sixty cows for seven years.[56] Its bulk is so huge that forty oxen are needed to drag its carcase into the feasting-hall, and nine men are required to hold up its belly.[57]

But one later tale, *Aislinge Meic Con Glinne*, is exceptional in that it is a very valuable source of information on early Irish food. Surviving in two versions dating from the early twelfth century, it contains parodies on various forms of early Irish saga. Satire is also directed against the Church and its tenets: there is even a parody of the crucifixion. Because of the Church's concentration on the spiritual

[52] *IP* 176-81. Unclean foods are also discussed in 'Irish Canons I', *IP* 160–62.
[53] *IP* 178 § 15.
[54] A brief account of the seven surviving wisdom-texts is given in *GEIL* 284–6.
[55] Meyer, *Triads*, 10 § 75. I have made some minor alterations to his translation.
[56] Thurneysen, *Scéla Mucce Meic Dathó*, 6.12–13.
[57] Thurneysen, *Scéla Mucce Meic Dathó*, 7 (R version, cf. 23.9–11).

benefits of fasting (see under 'penitentials' above) the author of the *Aislinge* centres his tale around a vision of food dreamed by a fasting monk.[58] This account describes all sorts of dishes, sometimes in great detail, and is particularly useful as a source of information on the various milk-products consumed by the early Irish.

Poetry

A large corpus of poetry of various types has survived from the Old Irish period. Much of it is genealogical or heroic and contains little information of relevance to farming. Religious poetry is likewise rather unhelpful for our purposes. Occasionally, however, a religious poem contains information which may be significant, even when set in a foreign context. Thus Greek and Latin versions of the apocryphal *Gospel of Thomas* describe a miracle in which the youthful Jesus sowed a handful of wheat, which yielded a hundred measures.[59] In the Old Irish poetical version of this gospel, the field (*gortán*) was planted with onion (*cainnenn*) rather than wheat.[60] This substitution cannot be regarded as of great significance, but it does help to confirm the evidence of the law-texts, which treat the onion as a particularly important vegetable (see p. 251).

One might expect the extensive collection of poems on place-lore (*dindṡenchas*) to provide some general information on farming practice in particular localities. But these poems, dating in the main from about the eleventh century, are disappointingly uninformative on such matters.[61]

TERMINOLOGY

The vocabulary of farming is normally conservative. So, for example, the Old Irish words for 'cow', 'pig', 'sheep', 'hen', 'wheat' and 'oats' – *bó, mucc, cáera, cercc, cruithnecht, corcae* – survive into Modern Irish with the same meanings.[62] We are therefore generally justified in taking a particular farming term to have had the same meaning in Old Irish as it has in the modern language. However, this approach must be used with caution. For example, in the spoken Irish of Baile Bhúirne the word *fothacha* (*fathacha*) is used of

[58]Jackson, *Aislinge*, 14.430ff. = Meyer, *Aislinge*, 35.20ff.
[59]James, *The Apocryphal New Testament*, 52 (xii); 63 (x).
[60]Carney, *The poems of Blathmac, etc.* 104 § 46. The MS has *di chainniu*.
[61]E. J. Gwynn, *Metrical Dindshenchas* i–v.
[62]The same meanings are retained likewise in Scottish Gaelic and Manx.

the disease of horses called 'strangles' in English.[63] But we know from a reference in the *Annals of Connacht* that in the earlier language *fothach* was used primarily of nasal and pulmonary glanders, which is often fatal to horses and may be transmitted to humans (see p. 211). Glanders was eradicated in Ireland in the nineteenth century, so it seems that the word *fothach* was transferred to another equine disease with similar symptoms.

In this book, I sometimes give cognates from the P-Celtic group of languages (Welsh, Breton, and Cornish) particularly where there is an element of doubt about the meaning of an Irish farming term. For example, the conventional translation of Old Irish *glaisen* as 'woad' is confirmed by the the fact that its Old Cornish cognate *glesin* glosses Anglo-Saxon *wād* 'woad'. But here again it must be stressed that cognates occasionally have different meanings. Thus Welsh *banw* 'young pig' has precisely the same meaning as its Irish cognate *banb*, but the Breton cognate *banv* has acquired the meaning 'sow with litter'.[64] Similarly, Old Irish *cainnenn* (which I take to mean 'onion') is cognate with Welsh *cennin* 'leek' and Breton *kignen* 'garlic' (see p. 251).

Changes in terminology may provide an indication of development in farming practice. For example, in Old Irish the ordinary word for 'bean' is *seib*, but in Middle Irish this is replaced by *pónair*, a borrowing from Old Norse.[65] The explanation may be that Norse colonists introduced an improved type of bean to Ireland, or that beans featured prominently in their diet (see p. 249).

ILLUSTRATIONS RELATING TO FARMING, ETC.

Illustrations in manuscripts provide some information about early Irish farming, particularly in relation to livestock. By far the most useful source is the eighth-century gospel-book known as the Book of Kells. Where this book was written is a matter for debate, but its scribe was most probably Irish and may have carried out his work at the Columban monastery of Iona.[66] Some of his illustrations seem

[63] Ó Cuív, 'Fragments of Irish medieval treatises on horses', 121–2; Ó Cuív, *Cnósach*, 104 s.v. *fathacha.*

[64] Hemon, *Geriadur*, s.v. *banv.*

[65] *Seib* is itself a borrowing from Latin *faba*: see p. 249.

[66] Meehan, *The Book of Kells*, 10, 90–92. For further discussion by various scholars on the book's place of origin, see O'Mahony (ed.), *The Book of Kells*, 46–7, 269, 333, 487, etc.

likely to have been drawn from life – presumably in the monastic farmyard. For example, the cock and two hens reproduced from folio 67[r] on the back cover of this book are fairly realistic.

Other realistic illustrations of domestic animals from the Book of Kells are a female goat (reproduced on p. 78), a cat and a rat (p. 243), and a dog and a hare (p. 299). More stylized representations of animals are also found in this manuscript. On folio 290[v] the winged calf – the symbol of Saint Luke – is portrayed in bright colour with a green and purple body, red feet, blue head and a red tail. It is obvious that this illustration tells us nothing about the colours of early Irish cattle in real life.

In spite of such an auspicious start in the Books of Durrow, Kells, Dimma, etc., the tradition of illustrating manuscripts made little headway in Ireland. Most of the creativity of the later scribes seems to have gone into the perfection of the script and the designing of intricate illuminated capitals. In the hundreds of manuscripts which have survived from the twelfth to the sixteenth centuries we find remarkably few illustrations, and these are often of poor quality. For example, the representation of a feast in the *Tech Midchúarda* 'house of mead-circuit' on p. 29[a] of the twelfth-century Book of Leinster is crudely executed (p. 356). But there are occasional exceptions: I reproduce on p. 103 a lively sketch of a crowing cock from the fourteenth-century Book of Ballymote. Nicely drawn fish are also found in a number of manuscripts of various dates (see p. 285). Many of the thousands of illuminated capitals in our manuscripts contain animal elements, usually dogs in various contorted positions (e.g. p. 177). By contrast, the scribe of a Latin psalter – probably working at the monastery of Bangor, Co. Down, in the twelfth century – used birds to form some of his capitals. I reproduce on page 106 a capital **D** which was clearly inspired by a plump farmyard goose.

Domestic animals are also portrayed on stone crosses. The ninth-century high cross at Banagher, Co. Offaly, has a fine carving of a horse and rider (reproduced on p. 98), and a cross from the same period at Bealin, Co. Westmeath, shows a deer being caught by a hound (p. 276). Other crosses depict a chariot drawn by two horses (p. 498), and what is probably a hunting scene (p. 299).

Plate 2. A pig from f. 25 of the thirteenth-century Welsh legal manuscript Peniarth 28, in the National Library of Wales, Aberystwyth. It has been reproduced by kind permission of the Trustees.

The illustration is placed above a section entitled *De lege porrcorum* 'concerning the law of pigs'. The section begins: *Porrcellus quamdiu sit in* crowyn, *non exiens foras, i denarius legalis* 'a piglet as long as it may be in the pig-sty, and does not go out, is worth one legal penny'.

MEDIEVAL EUROPEAN FARMING

Early Irish farming cannot be treated in isolation from what was going on in the rest of Europe, particularly in similar climatic zones. The medieval Welsh law-texts are especially valuable for comparative purposes. Though they date in the main from the twelfth and thirteenth centuries – roughly half a millennium later than the Old Irish law-texts – they contribute much useful information on agricultural practice and terminology among a kindred people farming in similar conditions. The illustration on page 18 is taken from folio 25 of the Welsh legal manuscript Peniarth 28.[67] The text is in Latin, but many of the technical terms are in Welsh. For example, in the second line below the illustration of the pig, the Welsh word *crowyn* 'pig-sty' is used. This is cognate with Irish *cró* of the same meaning.

Anglo-Saxon sources may also be helpful. The eleventh-century *Rectitudines singularum personarum* describes the rights and duties of various farm-workers, such as the cowherd, oxherd, swineherd, shepherd, beekeeper, cheese-maker, etc.[68] A text from about the same period on the duties of an estate-manager or reeve (*gerefa*)[69] provides much detailed information on medieval English farming practice, and sometimes casts light on problems in the Irish texts.[70] Further afield, the sixth-century laws of the Franks are particularly useful in their detailed treatment of the values assigned to domestic animals and birds at various ages.[71] Writings on agriculture and livestock by Greek and Latin authors such as Xenephon, Aristotle, Herodotus, Pliny, Virgil, Varro, Columella, Cato, etc. are obviously of limited relevance to the cool damp conditions of north-west Europe. Nonetheless, this material is important as a general background to the study of early Irish farming. Occasionally, there are interesting parallels. For example, both Aristotle and Virgil refer to the feeding of salted herbage to livestock for the purpose of increasing their milk-yield. The same practice may have been carried out on a category of early Irish cow described as an *úairtnech*: see p. 525.

[67] Emanuel, *The Latin texts of the Welsh laws*, 155.

[68] Liebermann, *Die Gesetze der Angelsachsen* i 445–53.

[69] Ibid. 453–5.

[70] E.g. p. 267 below.

[71] Rivers, *Laws of the Salian and Ripuarian Franks*, 43–50, 80–81, 185–6.

Anglo-Norman farming in Ireland

In this book I concentrate on the information provided by Irish-language sources dating from the pre-Norman period. However, it is obviously impossible to draw a neat line between Irish and Anglo-Norman farming practice. The introduction of Anglo-Norman farming in fact predated the Norman invasion with the consecration of the Cistercian abbey of Mellifont in 1157. The wealth of the Cistercians was based on their efficient farming methods and their capacity to find markets for their produce.[72] After the military invasion of 1169, Anglo-Norman colonists took over much of the best land in the country, particularly in the east. Many rural boroughs were set up, primarily for agricultural purposes.[73] Manorial records and other documents in Latin, Norman French and English give us some idea of the various changes and developments in farming practice in the colonized areas over the next few centuries.[74]

A particularly important innovation was the practice of saving hay to supply fodder for livestock during the winter (p. 47). The Anglo-Normans are also likely to have been responsible for developments in ploughing technology, and may have brought in the wheeled plough and the mouldboard (p. 471). The written and archaeological evidence indicates that the Irish ox-team was yoked abreast (p. 475), whereas the Anglo-Norman team of eight was yoked in tandem.[75] It is likely that fields were generally larger in the Anglo-Norman system, with open-field cultivation of strips of winter-corn (wheat or rye), spring-corn (oats), and fallow in rotation.[76] The success of Anglo-Norman cereal-farming in Ireland is demonstrated by the large amounts of grain which were exported to England and elsewhere, particularly during the thirteenth century. For example, the mayor of London imported

[72]Mitchell, *The Irish landscape*, 183; *Reading the Irish landscape*, 173.

[73]Down, 'Colonial society and economy in the high middle ages', 457.

[74]E.g. Curtis, 'Rental of the manor of Lisronagh, 1333'; N. B. White, *The Red Book of Ormond*; Curtis, *Calendar of Ormond deeds* i–vi (1172–1603); Davies and Quinn, 'The Irish Pipe Roll of 14 John, 1211–1212'.

[75]The eight-ox team is mentioned at N. B. White, *The Red Book of Ormond*, 28; and at Davies and Quinn, 'The Irish Pipe Roll of 14 John, 1211–1212', 38–40 (and footnote 162).

[76]Otway-Ruthven, *A history of medieval Ireland*, 117; Otway-Ruthven, 'The organization of Anglo-Irish agriculture in the Middle Ages', 9.

one thousand crannocks of wheat from Ireland in 1224–5.[77] This trade declined during the fourteenth century for reasons which include the unstable political situation in Ireland, the devastating effects of the Black Death of 1348, and a marked deterioration in the climate.[78]

Giraldus Cambrensis was obviously unimpressed with Irish sheep of the late twelfth century, which he describes as black.[79] The implication is that among contemporary Anglo-Norman breeds, there was a much higher proportion of white sheep.[80] It is likely, therefore, that larger white-fleeced breeds were imported by the colonists to improve the Irish stock. Where lists have survived in manorial records, sheep are more numerous than any other domestic animal,[81] and there was a flourishing export trade in wool and sheepskins.[82] This situation contrasts with the evidence from the early Christian period, which indicates a preponderance of cattle on Irish farms.

In Ireland, as elsewhere in western Europe, a much heavier breed of horse was introduced about the thirteenth century (p. 95). It was strong enough to be used for ploughing, and was faster than the draught ox. The presence of draught horses (*affri*) is recorded on Anglo-Norman manors in Ireland, but usually in smaller numbers than the draught oxen, which continued to be the main ploughing animals.[83]

The introduction of various other mammals, fish, and birds is likewise associated with the Anglo-Normans. Of particular importance was the rabbit, which provided a regular supply of meat and skins (p. 133). The mute swan and the pheasant are also likely be have been introduced at this period, as were certain freshwater fish.[84] In addition, the fallow deer was brought in for hunting:[85] its venison is generally held to be superior in

[77] Down, 'Colonial society and economy in the high middle ages', 484–5.
[78] Mitchell, *The Irish landscape*, 184–5; *Reading the Irish landscape*, 174.
[79] O'Meara, *Giraldus: Topography*, 101 § 93 = Dimock, *Giraldi Topographia*, 150.21.
[80] For references to black, dun and white sheep in the Old Irish law-texts, see p. 70.
[81] Down, 'Colonial society and economy in the high middle ages', 478–9.
[82] Otway-Ruthven, *A history of medieval Ireland*, 117.
[83] Down, 'Colonial society and economy in the high middle ages', 474–5.
[84] The earliest reference to the pike in Ireland known to me is in the Civil Survey of Co. Tipperary of 1654-56 (Simington, *Civil Survey* ii 282).
[85] Rackham, *The history of the countryside*, 49.

flavour to that of the native red deer.[86] It is difficult to assess the degree of influence which the Anglo-Norman and Irish farming traditions exerted on each other, and there must have been much local variation. For example, those in the Gaelic areas of Laois and Offaly would have been in a favourable position to learn from their Anglo-Norman neighbours in Kilkenny. By contrast, those in north-west Ulster would have been remote from any lastingly colonized areas. Loan-words in Irish sometimes indicate the adoption of Anglo-Norman farming practices. For example, Anglo-Norman influence in field, kitchen-garden, and orchard is shown by such loan-words as *ráca* 'rake', *uinneman* 'onion', *gairleóc* 'garlic', *péire* 'pear', and *sirín* 'cherry'.[87] The most likely explanation of Irish *speal* 'scythe' is that it goes back to a Middle English verb, meaning 'to skin, peel':[88] this would fit in with the theory that hay-making was introduced by the Anglo-Normans and adopted in some areas by the Irish.[89] There is also sixteenth-century evidence of the adoption by the Irish of the three-course rotation of winter-corn, spring-corn and fallow.[90]

It can likewise be assumed that the colonists drew on the farming experience of the native Irish, particularly with regard to livestock.[91] Much of the farmwork on a typical Anglo-Norman manor was done by the *betagh*, whose tasks are carefully described in the rental of the manor of Lisronagh for the year 1333.[92] The *betagh* was an unfree tenant, almost always Irish.[93] The term itself is a borrowing from Irish *bíattach* 'one who supplies food[-rent]', and it is clear that the Anglo-Norman lords regularly retained as tenants the dependants of the displaced Irish lords.[94] There would thus have been an element of continuity as well as innovation at the start of Anglo-Norman farming in Ireland.

[86]Von Bunau, 'Fallow versus red'.

[87]Giraldus comments on the few kinds of apple-tree grown in pre-Norman Ireland (O'Meara, *Giraldus: Topography*, 102 § 93 = Dimock, *Giraldi Topographia*, 152.4–6.).

[88]O'Rahilly, '*Sp-* in Gaelic', 26; *LEIA* S-177. See p. 480.

[89]See Nicholls, 'Gaelic society and economy in the high middle ages', 413.

[90]Ibid. 411.

[91]This is indicated, for example, by the borrowing of the Irish word *badhún* (*bódhún*) 'cow-enclosure' into English (*OED* s.v. *bawn*). Irish *gerrán* 'work-pony' (anglicized *garran*, etc.) is also common in English documents: see p. 94 below.

[92]Curtis, 'Rental of the manor of Lisronagh, 1333', 47, 52–3.

[93]Down, 'Colonial society and economy in the high middle ages', 457–8.

[94]Otway-Ruthven, *A history of medieval Ireland*, 112.

Differences between Irish and Anglo-Norman farming practices also aroused mutual hostility and contempt. As the Anglo-Normans took over the best cereal-growing areas in the country, it is not surprising that the Irish cereal-growing tradition – described in considerable detail in the Old Irish law-texts – went into a decline. Writing in the late fourteenth century, the Catalan pilgrim Count John de Perilhos claims that no corn was sown in the O'Neill kingdom, though he refers to the consumption of oatcakes.[95] The lack of cereal-growing by the Irish is often lamented in the writings of later English commentators. For instance, in his *Treatise of Ireland*, written about 1600, John Dymmok links the great want of grain in the country to 'the wandering and idle life' of the people. With some justification, he attributes the underdeveloped state of Irish agriculture to the great exactions of the Irish lords upon their tenants, who enjoy no security of tenure.[96] Even when cereal-growing was practised, the unfamiliar methods used by the Irish were liable to disturb English officials. The custom of separating the grain from the stalk by burning (Irish *loiscreán*) was remarked on by Fynes Moryson in his *Description of Ireland*,[97] and forbidden by statute in 1634.[98] Repeated legislation in the early seventeenth century was directed against the practice of attaching the plough directly to the tails of horses (p. 477).[99]

Friction between settled farmers and wandering herdsmen is a worldwide phenomenon which goes back many thousands of years, and is still a source of conflict in many countries. In the Old Irish law-texts, the emphasis is very much on the rights of the settled farmer, and various categories of wanderer are denounced and discriminated against (p. 425). In post-Norman times, however, political and economic circumstances combined to inflate the relative importance of pastoralism among the Irish.[100] Consequently, the wandering Irish and their herds of cattle posed a persistent threat to the settled farmers of the colonized areas,

[95] Mahaffy, 'Two early tours in Ireland', 8.

[96] Butler, *A treatice of Ireland by John Dymmok*, 5.

[97] Morley, *Ireland under Elizabeth and James I*, 427. See discussion under 'threshing' on p. 240 below.

[98] G. O'Brien, *The economic history of Ireland in the seventeenth century*, 39.

[99] Ibid. 38–9.

[100] Nicholls, *Land, law and society in sixteenth-century Ireland*, 9–11. In 'Gaelic society and economy in the high middle ages', 397–9, the same author provides some general comments on pastoralism among the Irish in post-Norman Ireland.

and were frequently the target of disapproval and legislation. A statute of 1430 forbids Irish rymers, outlaws and felons to graze their *keryaghtes* on the borders of the Pale.[101] This refers to the common Irish practice of forming a migratory herd of livestock with attendants (*caoraigheacht*, anglicized *keryaghte*, *creaght*, etc.).[102] Such a herd was grazed on other people's land, with or without the owners' permission. Obviously this sort of behaviour was anathema to the settled farmer with his vulnerable cornfields and meadows.

Even the annual migration of people and their livestock to the hills aroused censure. In his dialogue entitled *A view of the present state of Ireland*, written in 1595, Edmund Spenser attributes to Irenaeus the opinion that 'this keeping of cows is of itself a very idle life and a fit nursery for a thief'.[103] In particular Irenaeus dislikes the 'Irish manner of keeping boolies [Irish *búaile*] in the summer upon the mountains and living after that savage sort'. Spenser also refers to the Irish custom of using the blood of living cattle for food.[104] In 1611, the Lord Deputy proposed that a ban be placed on the Irish practice of blowing into the cow's vagina to bring down her milk (p. 39), but this legislation was never enacted.[105]

Modern farming

I hope that modern farmers may be interested in at least some sections of this book. There have of course been phenomenal developments in Irish farming since the seventh century, especially during the past fifty years. Most agricultural processes have been mechanized, new crops and breeds have been introduced, and artificial insemination has largely replaced the visit to the bull. The farmer now farms mainly to sell at a profit – often for export – rather than to feed and clothe the household from the produce of the farm.

Nonetheless, much of the material in this book will be familiar to the modern farmer. The four basic farm animals are still the cow, the sheep, the pig, and the horse, as they were in the

[101] Nicholls, 'Gaelic society and economy in the high middle ages', 414.
[102] Simms, 'Nomadry in medieval Ireland'; Lucas, *Cattle in ancient Ireland*, 68–124. See p. 427 below.
[103] Morley, *Ireland under Elizabeth and James I*, 198.
[104] Ibid. 98. See p. 53 below.
[105] Lucas, *Cattle in ancient Ireland*, 55.

seventh century. Cereals are still the most widely grown crops, and of these bread-wheat still has the highest prestige. Like his or her predecessor, the modern farmer has to cope with storms, excessive rainfall, drought, and occasional heavy snow. The section on animal diseases in Chapter 6 will strike a chord with today's stock-breeder, who has to devote a great deal of effort and resources to the treatment and prevention of disease. As in modern times, sudden outbreaks of infectious disease, particularly in cattle, had devastating consequences on early Irish farming communities. The modern cereal-grower will recognise his predecessor's problems with weeds in his cornfields (p. 233) and mildew on his corn (p. 237).

When we turn to the human aspects of early Irish farming life, we also find many modern parallels. Of particular contemporary interest is the degree of control which a wife was entitled to exercise over the running of the farm. According to the eighth-century law-text *Cáin Lánamna*, if she has made a significant property-contribution to the marriage, she is entitled to make certain beneficial contracts on her own initiative, such as the rent of land or the arrangement of co-operative ploughing with kinsmen.[106] Even a wife of lower status who has brought in no property is still legally entitled to defer foolish contracts which her husband may have made in relation to food, clothing, cattle and sheep (p. 400). A grown-up son is similarly entitled to annul his father's foolish contract, especially when it could diminish the value of his own inheritance (p. 400). The Old Irish law-texts also deal with the awkward position of the landless man who has married into a farm, the *cliamhain isteach* of Modern Irish (p. 416). Another question which is still much debated, especially in Ulster, is the extent to which a farmer is morally entitled to labour on a Sunday. The ninth-century text *Cáin Domnaig* prohibits the carrying out of a wide range of farming and domestic activities on a Sunday, but allows essential tasks such as herding cattle, bringing them to water, pursuing a cow in heat, rescuing an animal stuck in a marsh, etc. (see p. 455).

Disputes with neighbours over fences are a notorious source of bitterness in any mixed farming community, and it is clear from the law-texts that early Ireland was no exception. The main text on the law of farming, *Bretha Comaithchesa*, sets out in great detail

[106] *CIH* ii 505.35-506.26 = *SEIL* 18-19.

the correct design and dimensions of the proper fence (p. 372). Finally, like its modern counterpart the early Irish farm was a dangerous place, and the law-texts devote much attention to the legal aspects of farm accidents (p. 150).

1
Livestock (i): Cattle

It is clear from both the archaeological evidence and the written sources that cattle occupied a position of central importance in early Irish society. Nearly half the domestic animal bones found in excavations of early Christian sites are bovine,[1] and cattle feature prominently throughout Old and Middle Irish literature. In the law-texts of the seventh and eighth centuries there are very frequent references to cattle, and fines, tribute, bride-prices, fosterage-fees, and other payments are commonly expressed in terms of the milch cow (p. 58). The ordinary well-off farmer is called a *bóaire*, lit. 'cow-freeman', in the law-texts; this probably refers to the fact that he pays an annual rent of one cow to his lord.[2]

The wisdom-texts accord a similar prominence to cattle. For example, a ninth-century triad gives the three renovators of the world as 'the womb of a woman, the udder of a cow, and the moulding-block of a smith',[3] and another triad lists the three best things in a household as 'oxen, men, and axes'.[4] The justice of a king ensures abundant milk-yields within his territory,[5] whereas injustice on his part results in 'dryness of milking'.[6] Similarly there is a claim in one version of the later prologue to the *Senchas Már* that a false judgement by the legendary judge Fachtna caused his cows to refuse milk to their calves.[7]

The annals also bear testimony to the social and economic importance of the cow: many entries refer to plagues and mortalities of

[1]The combined percentages from seventeen major early Christian sites are 46 per cent cattle, 31 per cent pigs, and 23 per cent sheep or goats. These figures are derived from a table given by McCormick, 'The effect of the Anglo-Norman settlement', 43. The percentages are calculated on the basis of the minimum number of individuals rather than on fragment count: see discussion at Van Wijngaarden-Bakker, 'The animal remains from Newgrange (2)', 24–5.

[2]*CG* p. 77.

[3]Meyer, *Triads*, 20 § 148.

[4]Meyer, *Triads*, 28 § 227.

[5]Kelly, *Audacht Morainn*, 6 § 18.

[6]*CIH* i 219.18 = *AL* iv 52.8.

[7]*CIH* v 1655.10 = *AL* i 24.20.

cattle.[8] Other entries record cattle-raids by kings against neighbouring territories. The purpose of such raids was doubtless to some extent economic: the simple acquisition of further wealth. But a successful cattle-raid also served to demonstrate a king's military and political power, and was frequently employed as a means whereby a recently inaugurated king could establish a reputation for himself.[9] The importance of cattle-raiding is also indicated by the number of tales which deal with such episodes. A Middle Irish list of tales gives the titles of eleven *tána* 'drivings [of cattle]', e.g. *Táin Bó Cúailnge* 'the cattle-raid of Cooley', *Táin teora n-erc nEchach* 'the driving of the three cows of Eochu'.[10] Disputes about the cattle-tribute known as the *Bóraime* (*Bórama*), which was imposed upon the Leinstermen by the kings of Tara, are described in various texts.[11]

It is clear from our sources that the relationship between cattle and people was viewed as having a significance beyond the purely economic one. Cattle are prominent in early Irish mythology, and are associated with the fairies in many texts. For example, the tale *Táin Bó Fraích* tells of twelve white red-eared cattle which came from a fairy mound.[12] Other texts refer to the supernatural grey cow *Glas Goibnenn* which belonged to the god of smithcraft Goibniu, and was renowned for her massive milk-production.[13] Bulls may likewise have magical qualities. A method of selecting a ruler by means of a 'bull-feast' (*tarbḟeis*) is described in *Serglige Con Culainn* and other tales. A white bull is killed for this purpose, and a man is chosen to eat his fill of the meat and broth. He then sleeps while four druids make an incantation of truth over him. During his sleep he is expected to see in a vision the likeness of the man who should rightfully be king.[14]

[8]See p. 194–5 A legendary king took his name from a cattle-plague: *Bressal Bódíbad* 'B. of the cow-mortalities' (*CGH* 120; E. J. Gwynn, *Metrical Dindshenchas* iii 42.48–52).

[9]See the section entitled 'The inauguration raid' in Lucas, *Cattle in ancient Ireland*, 146–8. Cf. Ó Riain ' The *crech ríg* or "royal prey" '.

[10]Mac Cana, *Learned tales*, 42.

[11]See especially *LL* v 1268.37638–1318.39321 = Stokes, 'The *Bórama*'; Ó Buachalla, 'Leinster tribute feud'; AU^2 176 s.a. 720 (recte 721) § 8. The term *bóraime* is also used with reference to other exactions of cattle-tribute, e.g. AU^2 252 s.a. 797 (recte 798) § 2. See *DIL* for further citations.

[12]Meid, *Táin Bó Fraích*, 1.5–6. For a discussion of white red-eared cattle, see p. 33.

[13]M. Byrne '*Airec Menman Uraird maic Coisse*', 59.9–13.

[14]Dillon, *Serglige Con Culainn*, 9.246–53.

Some texts refer to the existence of emotional bonds between cattle and people. For example, in the prose *Dindṡenchas*, we are told how a certain landowner reared each year's calves (*indud*) in his own house until May. As a result his cattle loved him (*coro carsat a cethre hé*), and when he died they spent three days and three nights mourning around his body.[15] A similar association of cattle with human grief is to be found in an entry in the *Annals of Ulster* which records the murder in AD 738 of Cernach son of Fogartach 'whom the calves of the cows and the women of this lower world long bewailed'.[16] The idea that calves should take part in the general mourning for important men or women is likewise attested in a number of literary texts. For example, the tale *Aided Óenḟir Aífe* describes the death of the young warrior Conlae at the hands of his father Cú Chulainn. For three days after his burial, no calf was allowed to drink milk from its mother. The bawling of the hungry calves throughout Ulster was evidently regarded as an expression of the general grief at Conlae's tragic death.[17]

An intimate relationship with cattle is also reflected in various expressions and appellations. The use of the word *lóeg* 'calf' as a term of endearment among humans is well attested, and there is even mention of a bell named *Bóbán Coímgein* 'Kevin's Calf'.[18] In a phrase very much out of tune with twentieth-century piety, the Blessed Virgin Mary is addressed as *Bó Bhithbhliocht* 'Ever-milking Cow'.[19]

Types and Breeds

Our domestic cattle (*Bos taurus*) belong to the family *Bovidae*, and are related to the European and American bison, the wild and domestic buffaloes of Asia and Africa, the yak of Tibet, the domestic humped cattle (zebu) of Asia and Africa, and various other species of Asian cattle.[20] It is believed that the progenitor of *Bos taurus* was the aurochs or wild ox (*Bos primigenius*), which was at one time widespread over most of the northern Hemisphere,

[15] Stokes, 'The prose tales in the Rennes Dindshenchas, (1)', 308–9; *LL* iii 706.21231–42 (shorter version).

[16] *AU*[2] 190 s.a. 737 (recte 738) § 3.

[17] Meyer, 'The death of Conla', 120 § 13. For a more detailed discussion, see Lucas, *Cattle in ancient Ireland*, 7–11.

[18] See *DIL* s.vv. *lóeg*, *bóbán*.

[19] McKenna, *Aithdioghluim dána* i 352 § 27.

[20] See Clutton-Brock, *Domesticated mammals*, 62–70.

apart from America. This long-horned bovine is illustrated in a number of Palaeolithic cave paintings, of which those at Lascaux in the Dordogne are particularly well executed. The bulls in these paintings are large and black, whereas the cows are reddish and considerably smaller.[21] The aurochs is mentioned by Julius Caesar as being present in the Hercynian wood in Germany,[22] but human hunting ultimately brought about its extinction: the last specimen is believed to have been killed in Poland in 1627.

Archaeological evidence indicates that the aurochs was first domesticated before 6000 BC,[23] but it is probable that domestic cattle were not brought to Ireland until about 3500 BC.[24] As in the case of other mammals, the process of domestication lead to marked reduction in size.[25] Consequently, the bones found in Neolithic sites in mainland Europe, Britain and Ireland indicate small short-horned cattle.[26] In stature, these cattle must have been similar to the small hardy breeds still kept in hilly or rough terrain in various parts of Europe. It is clear from pictorial and written records that variations in colour, size, horns, hardiness, and disposition were early established in domestic cattle through both natural selection and human interference. For example, a mural painted about 2500 BC at Abousir in Egypt shows cattle with two distinct types of horns.[27] Writing in the first century AD, the Latin author Columella refers to the many varieties of domestic cattle in Italy, as well as in Asia, Gaul and Epirus.[28]

The earliest detailed accounts of Irish cattle are to be found in the law-texts of the seventh and eighth centuries AD. Because of the extremely complex genetic make-up of all domestic cattle, it would be misleading to suggest a close identification between the cattle described in these texts and any modern breed. There is a consensus among archaeologists, however, that early Irish cattle

[21] Clutton-Brock (*Domesticated mammals*, 64) summarizes the study by Zeuner, 'The colour of the wild cattle of Lascaux' (*Man* 53 (1953) 68–9).

[22] Caesar, *The Gallic war* (ed. Edwards), 6.28.

[23] Clutton-Brock, *Domesticated mammals*, 66; Gautier, *La domestication*, 145.

[24] Van Wijngaarden-Bakker, 'The animal remains from Newgrange (1)', 333.

[25] Clutton-Brock, *Domesticated mammals*, 22; Méniel, *Chasse et élevage*, 12–13.

[26] It was formerly held that this type (*Bos longifrons* or *Bos brachyceros*) was too small to have originated from the massive *Bos primigenius*, and must therefore derive from another wild species. See, however, Clutton-Brock, *Domesticated mammals*, 65.

[27] Reproduced in Gautier, *La domestication*, 110. Cf. Clutton-Brock, *Domesticated mammals*, 27, fig. 2.1.

[28] Columella, *De re rustica* (ed. Forster and Heffner), 6.1.1–2.

were generally similar in size and build to modern Kerry cattle.[29] The law-text *Cáin Aicillne* states that a fully-grown cow suitable to be given as rent to a lord should have a girth of twenty fists, i.e. eighty inches.[30] John O'Loan points out that this is slightly larger than typical girths recorded for modern Kerry cows.[31]

It is implicit in a number of legal references that the cattle of this period were often black.[32] For example, the eighth-century law-text *Bretha Nemed Toísech* speaks of milch cows 'as black as a blackbird'.[33] However, as we have seen above, the ancestral aurochs exhibited colour-variation between black and reddish. It is noteworthy too that in spite of selection against red in the modern Kerry breed, calves of this colour are occasionally born.[34] Old and Middle Irish sources refer to cattle of various shades other than black, such as crimson-red (*flann*),[35] flame-red (*derc*),[36] and brown (*donn*).[37] Saint Cíarán, abbot of Clonmacnois, is reputed to have owned a wonderful dun-coloured cow (*in Odor Chíaráin*) which yielded twelve measures of milk at every milking – one for each of the twelve apostles of Ireland.[38] The hide of this cow (*Seiche na hUidre Cíaráin*) was famed as a relic,[39] and it was claimed that anybody who died on it was assured of eternal life with Christ.[40] In this expectation, the *Annals of Inisfallen* record the death of Tadg macc Conchobuir, king of Connachta, in AD 900 'after having renounced the world on the hide of Cíarán's

[29] By kind permission of the Office of Public Works, the front cover of this book shows some of the Muckross herd of Kerry cattle at the Bourn Vincent Memorial Park, Killarney, Co. Kerry. The photograph was taken by Elizabeth Kelly.

[30] *CIH* ii 483.12; v 1784.32 = Thurneysen, 'Aus dem irischen Recht I', 355 § 13 *fiche dorn a timcomac*. For a discussion of the length of the *dorn* 'fist', see p. 564.

[31] O'Loan, 'Livestock in the Brehon laws', 70; O'Loan, 'A history of early Irish farming (3)', 171. He gives 76 inches for a Kerry cow, 84 inches for a Dairy Shorthorn cow, and 88 inches for a Hereford cow.

[32] Varro, *Rerum rusticarum* (ed. Hooper and Ash), 2.5.8, states that the best colour for cattle is black, next red, then dun and finally white; he regards white cattle as the most delicate, and black as the hardiest.

[33] *CIH* vi 2216.7–8; cf. *CIH* iv 1299.10.

[34] Curran, *Kerry and Dexter cattle*, 21–3.

[35] *CIH* iii 885.37.

[36] *TBC* Rec. I 57.1866.

[37] *TBC* Rec. I 124.4133.

[38] Stokes, *Félire*, 202.33–5 (commentary).

[39] Stokes, *Félire*, 204.3 (commentary).

[40] Stokes, *Lismore Lives*, 123.4117–18; Plummer, *Vitae* i 205 § xv.

Dun Cow'.[41] According to a later tradition the hide of this cow was used in the making of *Leabhar na hUidhre* 'the Book of the Dun Cow', one of the most important surviving Irish manuscripts.[42]

There are also some references to cows of more than one colour. A Middle Irish tale speaks of a hornless brindled cow (*máel ríabach*),[43] and a topographical poem refers to a brindled bull (*tarb ríabach*).[44] In later tradition the last three days of March, generally cold and wet, are called *laetha na seanriabhaí* 'the days of the old brindled cow'.[45] Later sources also speak of a white-backed cow (*druimfhionn*) with a red, grey or black body: cows of this type – called drimmons – are still distinguished in Co. Kerry.[46]

White cattle seem to have had a rarity value, and were no doubt especially favoured because of the association of white with virtue and purity. Such cattle are mentioned quite frequently in our sources,[47] sometimes with special reference to outstanding milk-yields.[48] At least three islands bear the name *Inis Bó Finne* 'the island of the white cow' (anglicized Inisbofin),[49] and in later Irish the Milky Way was called *Bóthar na Bó Finne* 'the road of the white cow'.[50] A difficult legal commentary distinguishes cattle with various amounts of white in their coats, including pure white (*finn*), white-rumped (*loisinn*), and white-backed(?) (*finnisse*).[51]

[41] *AI* 140 s.a. 900.

[42] *LU* 89.2780; cf. Keating, *Foras feasa* i 78.z.

[43] Bergin, 'A story of Flann mac Lonáin', 47 § 7.

[44] E. J. Gwynn, *Metrical Dindshenchas* iv 70.25.

[45] De Bhaldraithe, *Cín Lae Amhlaoibh*, 4.105–8. See also Dinneen, *Foclóir*, s.v. *riabhach.*

[46] Curran, *Kerry and Dexter cattle*, 5–6.

[47] E.g. Bergin, 'A story of Flann mac Lonáin', 47 § 7; Dillon, *Serglige Con Culainn*, 9.246; *CIH* ii 352.26, 353.3 = *AL* i 64.1, 15.

[48] Mac Niocaill, 'Register of St. Saviour's Chantry', 205; Plummer, *Vitae* i 235 § ii.

[49] Hogan, *Onomasticon*, s.vv. *Inis bó finde.*

[50] Dinneen, *Foclóir* s.v. *bóthar.*

[51] *CIH* iii 885.34–9. The form *loisinn* seems to be a compound of *los* 'tail, end' + *finn* 'white': see *DIL* L 212.6. One can compare *loisinnán*, lit. 'the white-tipped one', which is applied to the fox because of the white tip to its tail. In relation to a cow's body, *los* might refer to the rump, the tail, or the tail-tip. *Finnisse* is otherwise unattested, but seems also to be a compound of *finn.* The other adjectives applied to cattle in this commentary are *flann* 'crimson-red', *muillech* '(?)' and *taulach*, for which Plummer suggests 'having broad foreheads' (*DIL* s.v. 2 *tulach*).

Early Irish gospel-books contain many representations of the calf (usually winged) which is the symbol of Saint Luke.[52] These illustrations are generally too stylized to give much information on what contemporary calves actually looked like. One of the most realistic is the sturdy calf on f. 124v of the seventh-century Book of Durrow, which is reproduced as the frontispiece of the present book.

White red-eared cattle

In general, the early Irish textual evidence does not justify our classifying cattle in separate breeds in the modern sense. An intriguing exception is the white red-eared cow (*bó find áuderg*) which is very often mentioned as a distinct breed. For example, a law-text refers to the payment of seven white cows with red ears (totalling two *cumal*s in value)[53] as part of the fine for satirizing Cernodon, a legendary king of Ulster.[54] Such cattle are particularly prominent in tales with magical or mythological elements: in the Book of Leinster version of *Táin Bó Cúailnge* the fury Morrígain attacks Cú Chulainn in the guise of a white red-eared heifer,[55] and in other texts these cattle are associated with the Otherworld.[56] They also appear in saints' Lives. According to the ninth-century *Bethu Brigte*, Saint Brigit spent her early years at the house of a druid, whose slave her mother was. When the time came to wean the infant Brigit, she vomited up all food which was given to her. The druid became anxious about the child and assigned to her a white red-eared cow to sustain her; she was cured as a result.[57] According to *Vita Sancti Albei*, Saint Ailbe miraculously supplied the holy Modanus with two white red-eared oxen for ploughing.[58]

Because of the obviously fabulous nature of most of these references, it might be assumed that white red-eared cattle never existed in real life, particularly as there is a similar literary

[52] E.g. Book of Armagh ff. 32v, 69v; Book of Durrow ff. 2r, 124v, Book of Kells ff. 27v, 129v, 201v, 290v.

[53] This valuation agrees approximately with the common equation of one *cumal* to three milch cows: see Appendix B, p. 592.

[54] *CIH* vi 2114.3 = Dillon, 'Stories from the law-tracts', 45 § iv; *CIH* iii 1134.18 = E. J. Gwynn, 'Privileges', 46.26–7.

[55] *TBC* LL 54.1993.

[56] E.g. Meid, *Táin Bó Fraích* 1.5–6.

[57] Ó hAodha, *Bethu Brigte*, 2 § 5.

[58] Heist, *Vitae* 127 § 40. Cf. Meyer, *Sanas Cormaic*, 48 § 585. For further references see Lucas, *Cattle in ancient Ireland*, 240–5.

reference to white red-eared horses.[59] However, white red-eared cattle are also mentioned in Welsh law-texts, which generally contain little mythology. The thirteenth-century *Llyfr Blegywryd* states that the fine for an offence against the honour of the lord of Dinefwr must be paid in white red-eared cattle (*gwarthec gwynyon clustgochyon*).[60] Also from the thirteenth century, *Llyfr Iorwerth* states that the fine for a similar offence against the king of Aberffraw should include a white bull with red ears.[61] Furthermore, as Osborn Bergin pointed out, an old British breed of white red-eared cattle has been preserved at Chillingham in Northumberland, as well as in other locations in Britain.[62] So the Irish references may after all be based on reality, perhaps originating from the tales of travellers who had seen such cattle in Britain. It is of course possible that the breed was at some stage present in Ireland also.

Shaggy or long-haired cattle

Cattle left out of doors in winter grow a thick coat to improve heat insulation. This normally falls off in spring, as Triad 105 observes: 'three living things which shed dead things: a deer shedding its antlers, a tree shedding its leaves, cattle shedding their winter coat (lit. their stinking hair)'.[63] However, in the case of certain breeds which live in cool climates, this thick coat is retained throughout the year. For example, the Galloway of southern Scotland, a hornless black or dun-coloured breed, has a dense mossy undercoat with an outer covering of long fine hair which is never shed. The Highland or Kyloe of the western Highlands and Islands of Scotland is equipped with even better insulation. This magnificently horned breed has a permanent long shaggy coat which varies greatly in colour, and may be red, yellow, dun, cream, black or brindled. M. L. Ryder suggests that the hairiness of such breeds may result from deliberate selection.[64]

[59] *LL* v 1137.33460 = Windisch, '*Tochmarc Ferbe*', 462.10–11. One can compare the reference to white red-eared dogs in the medieval Welsh tale *Pwyll Pendeuic Dyuet* (ed. R. L. Thomson, 1.20–21).

[60] Williams and Powell, *Llyfr Blegywryd*, 4.2–3 = *ALIW*, Dimetian Code, I ii § 6.

[61] Wiliam, *Llyfr Iorwerth*, 2 § 3.10 (see note on p. 105) = *ALIW*, Venedotian Code, I ii § 3.

[62] Bergin, 'White red-eared cows'.

[63] Meyer, *Triads*, 14 § 105 (translation slightly emended). Cf. Meyer, *Sanas Cormaic*, 21 § 242.

[64] Ryder, 'Livestock', 383.

The early Irish sources do not distinguish long-haired varieties of cattle, and one law-text appears to regard hairiness as a defect. A description of the perfect cow includes the stipulation that it should not be shaggy (*finchad*).[65] Possibly, however, this does not refer to genetic hairiness but rather to the fact that the retention of a shaggy coat in summer may be a sign that a cow is in poor condition.

Horned and hornless cattle

In all bovines, a small proportion never grows horns. By breeding from hornless individuals man has developed various hornless ('polled') breeds, such as the Aberdeen Angus, the Galloway, the Red Poll, and the Irish Moiled (Ir. *móel* 'bald, hornless').[66] It has sometimes been claimed that the Irish Moiled was introduced to this country by Norse settlers.[67] However, there are references to hornless cattle in pre-Norse texts,[68] and hornless bovine skulls are quite common on early Christian sites.[69]

The main advantage of hornlessness is that there is less danger of injury to other cattle or to farm workers.[70] For example, one law-text refers to a *móel machae*, 'a hornless one of the cattle-enclosure', in a context which implies that such an animal cannot injure other cattle in a confined space.[71] Hornless cattle are also less likely to damage fences and other field-boundaries. However, the absence of horns has the disadvantage that the animal is less well equipped to defend itself, especially against predators. Furthermore, there is the minor convenience that a distinctive notch may be cut in the horn to indicate ownership.[72] There also seems to have been a general preference among the early Irish for the appearance of a horned bovine, and in the sagas there is admiration for the bull with a fine pair of horns. According to *Táin Bó Cúailnge*, for example,

[65] *CIH* ii 674.39 = Appendix A, text 1 § 1 (9). The translation 'shaggy' is uncertain: see discussion on p. 509.

[66] There is a herd of Irish Moiled at Toxteth Park near Liverpool. For a brief discussion of moiled (moiley, muley, etc.) cattle, see Curran, *Kerry and Dexter cattle*, 7–10; Ó Sé, *The history of native Irish cattle*, 7.

[67] Wilson, 'The Scandinavian origin of the hornless cattle of the British Isles'.

[68] E.g. *TBC* Rec. I 57.1866.

[69] McCormick ('Stock-rearing in early Christian Ireland', 156) notes that 7 per cent of the bovine skulls at Moynagh crannóg are hornless.

[70] See p. 179 for a discussion of goring by cattle.

[71] *CIH* vi 2133.17-18.

[72] There is a reference at *CIH* iv 1253.1–5 to the cutting of notches in the horns or ears of livestock for this purpose.

one of the prize possessions of Ailill, king of Connacht, was his bull *Findbennach* 'the white-horned one'. It was jealousy on account of this bull which drove his wife Medb to ask Dáire mac Fíachnai to sell her the Brown Bull of Cúailnge (*Donn Cúailnge*): his refusal sparked off the military campaign described in *Táin Bó Cúailnge.* The law-text *Bretha Nemed Déidenach* specifies that a poet is entitled to a *bennán bó* 'horned cow' for the type of poem known as a *laíd.*[73]

A respect for horned cattle is also indicated by the later proverb *Bíonn adharca fada ar na buaibh thar lear* 'cattle across the sea have long horns'.[74] However, Triad 85 makes the point that a cow may have fine horns but her value depends on her production of milk: 'three uglinesses which hide fairness: a horned cow without milk, a fine-looking but slow horse, a handsome person without property'.[75] According to a gloss in *Cormac's Glossary*, a cow with short horns is called a *bó chrocc*, but this term is otherwise unknown.[76]

For the economic importance of horns, see p. 56 below.

Breeding of large cattle

Large cattle require more food, and are sometimes difficult to handle. However, they provide large carcases, and may be regarded as status symbols by their owners. Finbar McCormick has observed that the cattle-bones on royal sites tend to be larger than average, and suggests the practice of competitive cattle-breeding among the early Irish aristocracy.[77] In view of the stress on the size of cattle which we sometimes find in the early literature, this seems very plausible.[78]

[73] *CIH* iii 1120.34 = E. J. Gwynn, 'Privileges', 27.6; *CIH* v 1605.8 = *AL* v 62.18–19. O'Davoren (*CIH* iv 1474.32 = O'Dav. 233 § 226) explains *bennán* as *laog bó* 'a cow's calf', but this contradicts the preceding statement in *Bretha Nemed Déidenach*: *bó i lōgh laoidhe* 'a cow as payment for a *laíd*' (*CIH* iii 1120.31 = E. J. Gwynn, 'Privileges', 27.4). Compare the use of *bennán* in the meaning 'antlered one, stag' at Murphy, *Early Irish lyrics*, 122 (poem 46) § 1.

[74] Cf. T. F. O'Rahilly, *A miscellany of Irish proverbs*, 33 § 117.

[75] Meyer, *Triads* 10 § 85 (translation emended). The manuscripts have variant readings (*bó*) *bindech, binnech, beinnech, bennach,* which I take to be from *benn* 'horn'. Meyer, however, reads *bindech,* which he takes to be a derivative of *bind* 'sweet', and translates 'sweet-lowing'.

[76] Meyer, *Sanas Cormaic*, 21 § 242 *bō crocc .i. adarca gairdi fuirri* 'i.e. short horns on her'.

[77] McCormick, 'Stock-rearing in early Christian Ireland', 184; McCormick, 'Dairying and beef production in early Christian Ireland', 264.

[78] E.g. *TBC* LL 36.1323–6.

CATTLE-MANAGEMENT

Cows and Calves

In modern cattle husbandry it is normal for a cow to have her first calf at three years old, or even younger. Similarly, in medieval Welsh law-texts it is expected that a cow be brought to the bull at 2¼ and have her first calf at three, when she is called a *cynflith*, lit. 'first milk'.[79] In early Ireland, on the other hand, the written evidence indicates that the normal practice was for a cow to be bulled at 3¼ and then to produce her first calf at four. Commentary on cattle values (evidently based on a lost Old Irish law-text) states that a cow is worth 20 scruples when she bears her first calf, 'and she is then four years old'.[80] The slow maturing of early Irish cattle may be partly due to poor winter feeding (see p. 46 below).

The law-texts devote a good deal of attention to the behaviour of a cow towards her calf. In a list of the faults which detract from the value of a cow, one text includes failure to love her calf, and incapacity to provide enough milk to feed it.[81] Another text deals with the legal problems which may arise if cows belonging to different owners rear their calves in close proximity to one another.[82] If a cow adopts (*for-ling*) the calf of another cow, the calf remains the property of the owner of the cow which bore it. If a calf is born dead and another calf is given to the mother, this calf is counted as belonging to its natural mother 'because it is not done for the sake of the calf but for the sake of the cow's milk'. Similarly, if somebody places his cow on another's calf without its owner's permission, he does not acquire any share in the ownership of this calf. But if the calf-owner agrees to the arrangement, they divide between them the value of the calf.

The law-texts take into account the protective instincts of a cow towards her calf. Heptad 7 states that the owner of a cow which has newly calved (*nuídlech*) is not liable for any injury which she

[79] Williams and Powell, *Llyfr Blegywryd*, 88.8–12 = *ALIW*, Dimetian Code, II xxvi § 12. Varro, *Rerum rusticarum* (ed. Hooper and Ash), 2.5.13, states that a heifer should not be impregnated before she is two years old so that she may calve at three, and that it is all the better if she does not calve until four. Cf. Virgil, *Georgics* (ed. Fairclough), 3.60–1; Columella, *De re rustica* (ed. Forster and Heffner), 6.21.2.

[80] *CIH* iii 845.32–3 = Appendix A, text 3 § 4.

[81] *CIH* ii 674.39–41 = Appendix A, text 1 § 1 (8), (17).

[82] *CIH* iii 967.35–968.15; iv 1267.17–31 = Appendix A, text 5.

may inflict on anyone who comes near the calf.[83] According to the accompanying commentary, this immunity lasts for a period of three days. The law-text on accidents, *Bretha Étgid*, indicates a similar period by stating that the cow is not held to be liable as long as it is giving colostrum (beestings) (*nús*).[84] Because of the cow's uncertain temper – and also for the sake of the welfare of the calf – it is illegal to distrain a cow which has newly calved.[85]

A Middle Irish verse preserved in commentary on *Amrae Choluimb Chille* advices the stockman to care for his calves and their mothers (*glinnig do lóegu 7 glinnig a máithre*). If he does so, the cows will provide a lasting supply of milk (*búanblicht*).[86]

Milking

In the history of the domestication of livestock one of the most remarkable achievements has been to persuade such a large animal as a cow to surrender to a human manipulator most or all of the milk which nature has provided for her calf. This has been effected by the selective breeding of docile animals which will submit to milking, whether by hand or, in recent times, by machine. In addition, various methods have been used to stimulate the instinctive milk-yielding response from the cow. For example, a frieze from about 2500 BC at Al-'Ubaid in Mesopotamia (Iraq) shows two cows being milked with their calves tied in front of them.[87] The idea is obviously that the presence of the calves causes the cows to yield their milk – no doubt enough is left in the udder to feed the calves after the milker has taken his quota. Early Irish sources likewise stress the necessity for the calf to be present at the milking. The saints' Lives are particularly rich in accounts of the miraculous restoration or substitution of a missing calf for the purpose of getting a cow to provide milk.[88] A Latin Life of Saint Cainnech (Canice) describes how a wolf ate a calf of two cows (*vitulus vaccarum duarum*).[89] The owner bewailed the loss of the calf to Saint

[83] *CIH* i 9.35 = *AL* v 152.2–3. For a discussion of the etymology of the term, see *LEIA* N-24 s.v. *nuidlech.* Its relationship with the once-attested Welsh *nouidligi* of the same meaning is uncertain.

[84] *CIH* i 275.12 = *AL* iii 228.7.

[85] *CIH* iii 897.11 = Appendix A, text 2 § 1 (1).

[86] Stokes, 'The Bodleian Amra', 252 § 54.

[87] Lucas, *Cattle in ancient Ireland*, 50.

[88] Many examples are given by Lucas, *Cattle in ancient Ireland*, 45–51.

[89] Heist, *Vitae*, 197–8 § 59. This is the *macc dá bó* 'child of two cows' of an Old Irish law-text: see *CIH* iii 967.35; iv 1267.17 = Appendix A, text 5.

Cainnech, who told him to return to his cows and clap his hands. When he did so, the wolf returned and put its head into the calf-tie (*vinculum vituli*). Miraculously, the cows treated the wolf as their calf and licked it as they were being milked. The wolf continued to return at morning and evening milking-times for this pupose until the end of the season.

In his *Cattle in ancient Ireland* Lucas cites descriptions in sixteenth- to nineteenth-century sources of a practice which is not to my knowledge mentioned in early Irish texts.[90] This is the making of a dummy from the skin of a calf which has died or been slaughtered. The cow recognizes the smell of her own calf, and will not yield any milk unless she is able to lick the stuffed skin. Lucas also gives references to the custom of blowing into the vagina of a cow whose milk-flow has failed. He suggests that both practices are of great antiquity in Ireland, and may even have been introduced by the first cattle-keeping people in Neolithic times.[91] Another method of stimulating a cow's milk-flow is by singing or playing music during the milking.[92] Our sources contain references to music increasing cows' milk-yield, though it must be admitted that they occur in literary contexts and therefore cannot be taken as firm evidence of actual practice. According to the ninth-century tale *Longes mac nUislenn* the singing of the sons of Uisliu was so melodious that it increased the milk-yield of cows and other animals by two thirds.[93] Another tale attributes the same rate of increase to the cows, sheep and other milking animals which hear the flute of Mac Díchóeme.[94]

In natural conditions, a calf is constantly with its mother and takes milk at frequent intervals. The farmer, however, restricts access after a few weeks, so that the calf can only be suckled at the twice-daily milking. Our sources make frequent reference to the importance of keeping calves away from their mothers. For example, one law-text refers to the offence of leaving a way open

[90] Lucas, *Cattle in ancient Ireland*, 51–5.

[91] Lucas, *Cattle in ancient Ireland*, 58. For a more detailed discussion of both practices, see Amoroso and Jewell 'The exploitation of the milk-ejection reflex by primitive peoples'.

[92] Amoroso and Jewell, 'The exploitation of the milk-ejection reflex by primitive peoples', 132, 135–6.

[93] Hull, *Longes mac nUislenn* 45.102–4 § 8 *no-mbligtis dā trian blechta d'immarcraid ūadaib.*

[94] Thurneysen, 'Die Flöte von Mac Dīchoeme', 118 § 10.

between somebody else's cows and calves,[95] or breaking a fence so as to allow the calves to get to the cows.[96] The saints' Lives contain many references to the use of children as herds to keep the calves away from the cows.[97] At night, the calves are housed separately in a calf-pen (*lías lóeg*).[98]

If a farmer allows an older calf (*gamain*) continued access to its mother, the milk-yield will obviously be reduced. Furthermore, the calf is liable to develop an aggressive and unruly temperament. A ninth-century tetrad states 'there are four who are not amenable to restraint or discipline: a priest's servant, a miller's hound, a widow's son, and a stripper's calf (*gamain gamnaige*).[99]

As is general in farming communities throughout the world, our sources indicate that cows were milked early in the morning, and again in the evening.[100] The literary references imply that milking was normally carried out by women, but it is clear from some legal references that the milker (*bligre*) might also be male (see p. 451). In summer, cows were sometimes milked away from the farm at a temporary milking-place (*áirge*). At other times milking seems usually to have taken place in a milking-enclosure (*indes*) in the farmyard. Cows were restrained at milking by a spancel (*búarach*).[101]

The stream of milk was directed into a wooden pail (*derb*): for this reason a milch cow is sometimes called a *bó derba*.[102] If the milking was carried out at a distance from the dwelling-house, it could be transferred to a larger vessel strapped to the back.[103] Legal references suggest that milking pails were of a standard height of

[95] *CIH* ii 402.10 = *AL* i 232.5–6 *fācbāil ōbēle iter bū ⁊ laega.*
[96] *CIH* ii 385.22–3 = *AL* i 168.5 *aurb[a] ria laegaib do buaib.*
[97] Lucas, *Cattle in ancient Ireland,* 15–16.
[98] See p. 364 below.
[99] Meyer, *Triads,* 30 § 234. As in Modern Irish, *gamnach* 'stripper' refers to a cow which has not come into calf in a particular year, and is still accompanied by the calf of the previous year.
[100] E.g. Plummer, *Vitae* i 235 § ii.
[101] For the various methods of restraining animals, see p. 493ff.
[102] E.g. *CIH* v 1762.32 = *AL* ii 162.8 *bó derba cona deirb* 'milch cow with its milking-pail'. This corresponds to the *bó bainne* 'milch cow' later in the same commentary (*CIH* v 1766.12–13 = *AL* ii 176.16–17).
[103] A. O. and M. O. Anderson, *Adomnan's Life of Columba,* 360 (new pagination: 116) = f. 65b, refers to a milking-vessel carried on the back. Its lid was fastened with a peg (Latin *gergenna*).

about twelve inches.[104] There is little precise information on milk-yields. However, in a discussion on the proper characteristics of a cow, one law-text states that a cow is expected to yield a measure of twelve inches (*mesar da ordlach déc*) from her three teats; the milk from the fourth teat is for the calf.[105] Presumably, this refers to the yield at each of the two daily milkings.

For cattle, the natural sequence is for lactation to begin about May with the birth of the calves.[106] It continues throughout the summer, and then dries off with the approach of winter. The farmer can, however, extend his milking season by ensuring that a couple of his cows are kept from the bull at the usual time for mating (late summer), and are bulled in the autumn instead. A few calves will thus be born later than the rest, and so the milking season is extended.

If a milking cow fails to come into calf – or is deliberately kept from the bull – she may continue to give milk for a whole year, provided that she is adequately fed and looked after. Ninth-century legal commentary refers to a cow with constant milk (*bó bithblicht*) every month for a year.[107] Later commentaries likewise contain a number of references to winter-milk (*lacht geimrid*).[108] One commentator remarks – perhaps in surprise – that his source (presumably an Old Irish law-text now lost) does not refer to the value of milk in winter and spring, but only in summer and autumn.[109]

For a discussion of the importance of milk and milk-products in the early Irish diet, see p. 323.

[104] Cf. *CIH* v 1766.13 = *AL* ii 176.17; *CIH* iii 920.28.

[105] *CIH* ii 675.6–7 = Appendix A, text 1 § 5 (1). A wooden bucket found at Ballinderry is approximately 12 inches in height (Hencken, 'Ballinderry crannóg no. 1', 141).

[106] Legal commentary uses the phrase *lulgachus Beltaine* 'May calving' (e.g. *CIH* iii 999.30 = McLeod, *Early Irish contract law*, 295 [1]).

[107] *CIH* iii 894.9–10; cf. *CIH* i 161.24–5; Bergin, 'A story of Flann mac Lonáin', 46.18.

[108] E.g. *CIH* ii 533.14; v 1815.7 = *AL* iii 46.13.

[109] *CIH* ii 716.6–8.

THE DIET OF CATTLE

Cattle are herbivorous, and in natural conditions consume grasses and a wide variety of other herbs. They also eat the leaves of most tree-species: their particular partiality to the leaves of the elm is reflected in the description of this tree as 'friend of cattle' (*carae cethrae*) and 'sustenance of cattle' (*lúth cethrae*).[110] A reference in the eighth-century tale *Fled Bricrenn* shows that heather (*fráech*) and furze (*foigdech*) were regarded as inferior feeding for cattle.[111]

The Irish geographer Dícuill, writing in the early ninth century, quotes Iulius Solinus to the effect that Ireland 'is so rich in pastures as to endanger the cattle unless they are now and then removed from their feeding grounds'.[112] This seems to be a way of explaining the disorder of grass tetany, common in modern times among cattle introduced too quickly to lush spring pastures. Otherwise, our sources provide us with little detailed information on the subject of grass (*fér*). In some texts, the term *airthend* is used apparently with the meaning of some particularly valued type of grass.[113] Thurneysen suggests that it may be Timothy grass (*Phleum pratense*),[114] but it is more likely to be the same word as Modern Irish *fiorthann* (also *feorainn, iarthainn, faorthann, fiothran*, etc.) 'creeping bent-grass, *Agrostis stolonifera*', which is commonly anglicized *fiorin*. A ninth-century author argues against over-stocking on grassland in the following triad: 'three fewnesses which are better than plenty: a fewness of fine words, a fewness of cows in grass, a fewness of friends around ale'.[115]

Middle Irish topographical poems lay particular emphasis on the presence of clover (*semmar*) in the fertile plains of Ireland,

[110]See McManus, 'Irish letter-names', 136.2 (L) and 140.13-22 (glosses to L). There is disagreement among glossators as to whether *luis* refers to elm, pine or rowan. Cattle's taste for elm-leaves is noted by Mitchell, *Reading the Irish landscape*, 88, and by Rackham, *The history of the countryside*, 243. Cato, *De agri cultura* (ed. Hooper and Ash), 6.3, likewise refers to the feeding of elm-leaves to cattle and sheep. Aristotle, *Historia animalium* (ed. Thompson), 8.7, includes elm-leaves among the foods used to fatten cattle.

[111]*LU* 248.8123 = Windisch, 'Das Fest des Bricriu', 256.29 § 9. *Foigdeach* is explained by the word *aitenn* 'furze, gorse' in a gloss on this text at H 3. 18 f. 607^{b}12 (quoted *DIL* s.v. 2 *foigdech*).

[112]Tierney, *Dicuili liber de mensura orbis terrae*, 96 § 21.

[113]*DIL* s.v. *airthend* (b).

[114]Thurneysen, Review of *Togail na Tebe* (ed. Calder), 421. He refutes Calder's identification with 'oats'. See De Bhaldraithe, 'Irish names', xlix, s.v. *irthenn*.

[115]Meyer, *Triads*, 12 § 93.

and it is likely that its value as a food for cattle was recognized.[116] In maritime areas seaweed must have been useful for feeding cattle and providing them with salt and other nutrients. In the *Canones Wallici*, which are probably of Breton origin,[117] there is a reference to the feeding of seaweed (*alga maris*) to cattle as a substitute for grass.[118] A passage in *Cormac's Glossary* speaks of a hag cutting *femmain bolgach* 'bladdered seaweed', a term which presumably includes bladder wrack (*Fucus vesiculosus*), knotted wrack (*Ascophyllum nodosum*) and other wracks.[119] The text provides no information on the purpose for which she cut the seaweed, but it may have been for cattle fodder. The *úairtnech* (p. 525) may possibly be a cow which is deliberately fed salty herbage so as to increase its thirst, and hence its milk-yield.

Summer-grazing

When the main spring growth had been consumed at the farm, it seems to have been a frequent practice for cattle and other live-stock to be driven off to the hills or other rough land where they could graze during the summer. The summering of cattle in this manner is attested among many peoples throughout the world. It is still practised in Switzerland, Norway, Spain, etc. and is in harmony with the animal's natural tendency to migrate seasonally in search of fresh pasture. In his *Cattle in ancient Ireland*, Lucas provides a summary of what English and Irish sources from the sixteenth to the nineteenth centuries have to say about this practice, which was known as 'booleying', from Irish *búaile* 'cattle-enclosure'.[120]

Earlier sources provide less detail, but it is nonetheless implicit in many texts that cattle were commonly summered in the care of

[116]E. J. Gwynn, *Metrical Dindshenchas* iii 236.16; iv 148.32; 306.27.

[117]For discussion of the date and provenance of the '*Canones Wallici*', see Fleuriot, 'Un fragment en Latin de très anciennes lois bretonnes armoricaines du vi^{e} siècle'; Dumville, 'On the dating of the early Breton lawcodes'.

[118]*IP* 158 § 61. Le Duc, 'Notes sur les anciennes lois bretonnes (*Canones Wallici*)', 106, is doubtful about this passage, in part because he claims that the feeding of seaweed to a ruminant will cause it to vomit. However, the consumption of sea-weed by cattle is well attested in Ireland and Scotland, e.g. Ó hEochaidh, 'Seanchas iascaireachta', 38; M. Martin, *A description of the Western Islands*, 139; Fraser Darling, *Natural history in the Highlands and Islands*, 159. For seaweed-eating sheep, see Fenton, *The Northern Isles*, 466.

[119]Meyer, *Sanas Cormaic*, 92.18–19, 93.8 § 1059 s.v. *prūll*.

[120]Lucas, *Cattle in ancient Ireland*, 58–67; Aalen, 'A note on transhumance in the Wicklow mountains'; Aalen, 'Transhumance in the Wicklow mountains'.

young people or professional herdsmen on hills or other uninhabited lands away from the farm.[121] One law-text states that a king is entitled to requisition dry cattle in the wilderness (*sescṡlabrae i ndíthrub*) for his army when it is returning from a campaign; however, he must subsequently restore their equivalent.[122] Another law-text refers to the general entitlement of cattle to graze freely on mountain (*dind*),[123] and legal glosses make a regular distinction between the home farm (*senlis* or *senbaile*) and the summer milking-place (*áirge*).[124] Literary sources likewise contain many references to the summering of cattle on hill pasture. For example, the mythological tale *Tochmarc Étaíne* tells of the presence of Findlám, the cowherd of Tara, 'in Slíab Fúait in the midst of a wilderness',[125] and a hagiographical anecdote describes the visit of Saint Máel Rúain of Tallaght to a summer milking-place (*áirge*) in a clearing in Slíab Mairgge (Slievemargie, Co. Laois).[126] The summer-pasture might be a considerable distance from the home farm: in a Life of Saint Cóemgen it seems to be regarded as perfectly credible that a rich farmer from Mide (Meath) should bring his cattle to a valley near Glenn dá Locha (Glendalough, Co. Wicklow) on a grazing circuit (*ar cuairt bhuailltechuis*).[127]

Cattle and other livestock would obviously be exposed to many dangers in the hills, and both law-texts and saints' Lives lay special stress on attacks by wolves (p. 186). There was also the risk of falling

[121] See p. 438 for a discussion of the task of herding livestock.

[122] *CIH* ii 570.1–5 = *CG* 22.561–5.

[123] *CIH* i 291.1 = *AL* iii 296.11.

[124] E.g. *CIH* ii 370.12–14, 16 = *AL* i 132.12–14, 15. For further non-legal references to this distinction, see Lucas, *Cattle in ancient Ireland*, 65. Compare the Welsh distinction between the winter-dwelling (*hendref*, lit. 'old habitation') and the summer-house (*hafoty, hafdy*).

[125] Bergin and Best, '*Tochmarc Étaíne*', 188 § 20. The name *Slíab Fúait* was applied to the mountainous area of the Fews, Co. Armagh (Hogan, *Onomasticon*, s.v. *Slíab fúait*).

[126] Best, 'Story of Mael Ruain', 34 = *LL* v 1246.36948. *Luid Mael Ruain Tamlachta fechtas dia āirge .i. ceppán i Sléib Mairgge* 'Máel Rúain of Tallaght once went to his summer milking-place, i.e. a clearing in Slievemargie'; cf. *Thes.* ii 328.5 *āirgech airslébe* 'milkmaid of a mountainside'. For other instances of hill-grazing, see Best, 'Amairgen son of Ecet Salach', 34 § 5 = *LL* ii 436.13603–4; L. Gwynn, '*De Ṡíl Chonairi Móir*', 133.11–12.

[127] Plummer, *Bethada* i 157 § iv (8); trans. ii 153. The province of Mide extended further south than the modern Cos. Meath and Westmeath, but nonetheless was always a long way from Glendalough: see Smyth, *Celtic Leinster*, 144–8. Cattle are still driven distances of thirty miles for summer-grazing in Extremadura, Spain.

from a cliff (p. 183), drowning in a quaking bog or marsh (p. 182), or being driven off by thieves or brigands (p. 165). Another danger was the pit-fall (*cuithech*) excavated to trap deer: one legal commentary specifies that a trapper must give public warning of any trap which he has set on mountain (*slíab*) or undivided land (*dírann*).[128] To enable the cowherd to keep track of the cattle as they grazed, some or all of them had a bell (*cloc*) around their necks.[129] At night, they were confined in a pen (*búaile*), near to which the cowherd slept in a house or hut.[130] It is clear from our texts that a herd-dog was regularly in attendance.[131]

It is likely that some farmers had enough land near their homes to sustain their cattle throughout the year, and would therefore not need to expose them to the dangers of grazing on hills or other rough land in summer. Such cattle might be grazed on damp pasture (*clúain* or *calad*), i.e. land by streams or lakes which would be flooded or water-logged in winter. Grazing might also be available on a summer-dry lake or turlough (*turloch*).[132] A reference in *Fled Bricrenn* suggests that the most highly prized beef was that of cattle which had never summered on hill or moorland. The menu in an especially luxurious feast includes a fattened cow (*bó thúir*) which had never grazed on heather or furze (*fráech no foigdech*) but only on lush green grass and corn (*luigfér glasfeóir 7 arbar*).[133]

Winter-grazing

With the approach of winter, a major preoccupation in farms of all types would be the provision of grazing for cattle. The law-texts often refer to a category of land called *etham ndíguin* 'preserved grassland'.[134] This appears to have consisted of enclosed fields of

[128] *CIH* i 285.11–14 = *AL* iii 272.5–8. See p. 278 below.

[129] *CIH* ii 375.27 = *AL* i 126.3.

[130] The cowherd Findlám and his wife are represented as living in a house (*tech*) in the midst of a wilderness (Bergin and Best, '*Tochmarc Étaíne*', 188 § 20).

[131] The same tale refers to a kennel (*cró*) in the herdsman's house. See also p. 119 below.

[132] For *turloch* (*tur* 'dry' + *loch* 'lake') see Quin and Freeman, 'Some Irish topographical terms', 154.

[133] *LU* 248.8122–4 = Windisch, 'Das Fest des Bricriu', 256.28-30 § 9: see discussion by Knott, '*Bó thúir*'. For *foigdech*, see p. 42 above.

[134] E.g. *CIH* i 70.4 = *AL* iv 90.17. This text (*Bretha Comaithchesa*) makes a distinction between cattle-trespass on *etham ndíguin* and the less serious offence of trespass on *athlumpaire* (*CIH* i 66.36 = *AL* iv 78.7) or *athbronnad* (*CIH* i 70.4 = *AL* iv 90.18). The terms *athlumpaire* and *athbrondad* seem to be synonymous, and may refer to grassland which had been grazed down before the trespass took place.

grass, grazed down in the spring, and then allowed to grow during the summer. If most or all of the cattle were kept on hill, moorland or woodland pasture during the summer, the farm could have a good store of fresh grass for them on their return.[135]

Cattle would also be allowed to graze the stubble of the various cereal-crops after they had been harvested. The archaeological and linguistic evidence indicates that corn was generally cut fairly high up the stalk (see p. 238), so there would have been a good deal of edible matter left for cattle and other livestock on cornland.

It is clear from legal sources that the branches of two evergreen woody plants, holly (*Ilex aquifolium*) and ivy (*Hedera helix*), were sometimes cut for winter fodder. A gloss on the text on distraint refers to 'a hook or sickle for cutting ivy or holly'.[136] In a series of glossed extracts from a lost law-text, there is also a reference to holly and ivy as nourishment for cattle.[137] Evidence of the feeding of holly-branches to cattle is found in later sources. Count John de Perilhos, a pilgrim to Saint Patrick's Purgatory at the end of the fourteenth century, says that 'the beasts eat only grass instead of oats, and the leaves of the holly, which they roast a little on account of the prickles which are in the leaves'.[138] In Co. Kerry the relatively spine-free upper branches of holly were regularly fed to cattle a generation ago.[139]

Absence of hay-making

In medieval Europe the cutting and drying of grass to feed cattle during the winter months was clearly standard practice in most temperate regions. Hay-making was sufficiently important for the emperor Charlemagne to ordain in the late eighth century that the name of July be changed to *Heuvimanoth* 'hay-month'.[140]

[135] Legal glosses imply that the beginning of November (*Samain*) was the usual time for the return of the cattle to the home farm (*CIH* ii 370.16 = *AL* i 132.15; *CIH* ii 373.29–30 = *AL* i 138.34–5). See Plummer, 'Some passages in the Brehon laws. IV', 116–17 § 42.

[136] *CIH* ii 374.12–13; v 1684.30 = *AL* i 140.8–9 *bac no corrān buana edinn no cuilind.*

[137] *CIH* ii 657.13 *Asa [m] buar buan biata .i. in cuilenn 7 int eidind* 'from which long-living cattle have been fed(?) i.e. holly and ivy'. Cato, *De agri cultura* (ed. Hooper and Ash), 54.2, refers to the use of ivy-leaves as a substitute for hay in the diet of cattle.

[138] Mahaffy, 'Two early tours in Ireland', 8.

[139] Kelly, 'The Old Irish tree-list', 110 (I owe this information to Oliver Mooney of the Forestry and Wildlife Service); Rackham, *The history of the countryside*, 120.

[140] L. White, *Medieval technology*, 56[2].

In Ireland, on the other hand, it seems that there was hardly any hay-making before the coming of the Normans.[141] The English historian Bede, writing in the eighth century, remarks that because of the mildness of the Irish climate no one cuts hay for winter use.[142] Similar observations were made by Giraldus Cambrensis in the twelfth century.[143] Negative evidence from native sources supports these statements. The Old Irish law-texts, which provide extraordinarily detailed information on farming routine, contain no reference to hay-making, and no term for 'hay' has been identified in the language of this period.[144] There is also no archaeological evidence from pre-Norman Ireland of the standard implements of mowing: the scythe, the hay-rake, the pitchfork, and the mower's anvil. In non-legal sources of early Irish origin the only mention of hay which I have found is in the seventh-century Penitential of Cummean in Latin, but here hay (*foenum*) is for use as bedding rather than cattle fodder.[145]

It is clear from the adoption of various new types of grain (p. 222) and the spread of the water-mill (p. 482) that there was no general resistance to agricultural innovation in Ireland in the early Christian period. It is probable, therefore, that Bede's explanation is correct, and that the general mildness of Irish winters allowed sufficient growth in the grass to keep the cattle alive.[146] In addition, the wetness of the average Irish summer must always have discouraged hay-making. It is clear from the annals, however, that on the rare occasions when snow covered the ground for any length of time the Irish system of winter-grazing broke down. The *Annals of Ulster* record that in the year 748 there was 'snow of unusual depth so that nearly all the cattle of the whole of Ireland perished'.[147] This report may be exaggerated, but there is no doubt that it records a

[141] For post-Norman haymaking, see Lucas, *Cattle in ancient Ireland*, 37.

[142] Colgrave and Mynors, *Bede's ecclesiastical history of the English people*, 18 (bk. 1, ch. 1).

[143] O'Meara, *Giraldus: Topography*, 53 § 26 = Dimock, *Giraldi Topographia*, 67.1.

[144] The later language has no special word for 'hay'; the ordinary word for 'grass' is used, e.g. *ag baint an fhéir* 'cutting the hay'. In Gaelic-speaking areas of eastern Scotland, *saidhe* (a borrowing of English *hay*) is used.

[145] *IP* 112.30.

[146] Rackham, *The history of the countryside*, 334–5, notes that the Domesday Book of 1086 records a particularly low proportion of meadow in Devon and Cornwall, and suggests that this can be explained by the extensive pasture and long growing season in the south-west.

[147] *AU*^{2} 202 s.a. 747 (recte 748) § 3.

large-scale disaster. Similar associations of snow with the death of livestock are recorded in the annals on numerous occasions.[148]

Training and use of oxen

Most castrated male bovines would be slaughtered in their first or second year. However, a farmer would occasionally select one on the basis of its strength and docility, and train it as a draught ox, probably starting in its third year.[149] The ox was the tractor of the early Irish farm, and pulled the plough (*arathar*) and the heavy farm-cart (*carr*). Trained oxen were highly prized: in a ninth-century gloss on *Bechbretha* they are classed along with milch cows as 'noble dignitaries of livestock'.[150] According to Triad 76, the three best hands in the world are the hand of a good carpenter, the hand of a good smith, and the hoof of a good ox (*dorn deg-daim*).[151] Legal commentary records a tradition that Saint Patrick left a special blessing on oxen.[152]

The law-texts stress that the high value assigned to the ox only applies if it is properly trained. Such an animal is commonly referred to as a *dam ríatae* 'trained ox' or *dam cach máma* 'an ox of every yoke'.[153] An ox which will not submit to being yoked to a plough or cart is obviously of value only as a source of meat and hide, and Heptad 37 advises against accepting such an ox as a pledge.[154] Because of its crucial rôle on the farm, an ox cannot be distrained for its owner's offences while under the yoke for ploughing or for transport.[155]

When fully grown, an ox is considerably larger than a milch cow. This is illustrated by an Old Irish legal passage on trees, which makes it clear that tanning an oxhide (*damṡeiche*) requires

[148]See Lucas, *Cattle in ancient Ireland*, 33.

[149]According to medieval Welsh law, an ox is first put to the plough in its third year (Williams and Powell, *Llyfr Blegywryd* 89.8–11 = *ALIW*, Dimetian Code, II xxvii § 4).

[150]*CIH* iii 924.24 = *BB* 86 § 52[e] (B).

[151]Meyer, *Triads*, 10 § 76. Meyer reads *dorn degmna(i)* 'the hand of a skilled woman' with the Book of Lecan version. However, all the other versions have *dorn daim* or *dorn de(a)gdaim*. The glossed version in H 1. 15 (Trinity College Dublin MS no. 1289) has *dorn daimh .i. cos ag treabhadh* 'the hoof of an ox, i.e. a foot ploughing'.

[152]*CIH* ii 707.33–4.

[153]A trained ox is called a *dam airceas* at *CIH* vi 1946.29; cf. *CIH* ii 391.8 = *AL* i 184.13–14 (wrong word-division) *im damu nadbet airchesa* 'about oxen which are not trained'.

[154]*CIH* i 35.22 = *AL* v 250.5.

[155]*CIH* iii 897.14–15 = Appendix A, text 2 § 1 (12).

a good deal more oak-bark than tanning a cowhide (*bóseiche*).[156] According to the text on *comingaire* 'joint herding', an ox requires as much grazing as a milch cow and a two-year-old heifer combined.[157] A difficult passage in the same text refers to an ox which is powerful enough to defend the other cattle of the herd against wolves. Such an ox is called a *dam conchaid* 'a wolf-fighting ox'.[158] Legal commentary likewise refers to the ox as defending the cows.[159] In a personal communication, Cedric Vaughan of the National Library of Wales informs me that a flock of sheep is often defended by castrated males (wethers) rather than by rams.

According to *Críth Gablach,* the highest grade of *bóaire* is expected to have six oxen, which would enable him to do all his ploughing from his own resources.[160] Lower down the social scale, the *bóaire febsa* has two oxen[161] and the *ócaire* has only one ox.[162] Both would therefore have normally ploughed in co-operation with other farmers (see p. 445). The term *dam* is sometimes used, particularly in non-legal sources, of a castrated male reared solely for consumption: see p. 52 below.

BULLS

The law-texts give evidence of widespread ownership of bulls, even by those of relatively low rank. The author of *Críth Gablach* expects the lowest grade of adult freeman, the *ócaire,* to keep a bull even though he has only seven cows.[163] He expects the highest grade of *bóaire* to own twenty cows, six oxen and two bulls.[164] There are also references in the law-texts to the loan of bulls from one farmer

[156] *CIH* i 202.20; ii 582.7 = *AL* iv 148.19–20; cf. *CIH* vi 2184.3–4.

[157] *CIH* i 192.15–16; ii 576.35–6 = *AL* iv 100.18.

[158] *CIH* i 192.18–19; ii 577.6–7 = *AL* iv 100.z. The meaning of *dam conchaid* ('der Wölfe bekämpfende Ochse') was first identified by Thurneysen, 'Aus dem irischen Recht III', 341.

[159] *CIH* vi 1978.29 *dithne* (=*díthnid*) *na bú* 'it defends the cows'. Liam Breatnach points out to me that this passage derives from Wasserschleben, *Die irische Kanonensammlung,* 99, bk. 29 ch. 4, where the corresponding Latin text has *defendit vaccas.*

[160] *CIH* ii 564.5; iii 780.12 = *CG* 8.194–5.

[161] This is implicit in the statement (*CIH* ii 563.9; iii 779.29 = *CG* 6.158) that a person of this grade owns half a ploughing outfit (*leth n-arathair*).

[162] *CIH* iii 778.27 = *CG* 6.134.

[163] *CIH* iii 778.24 = *CG* 4.90.

[164] *CIH* ii 564.5; iii 780.12 = *CG* 8.194.

to another – no doubt the object of such transactions would often have been the avoidance of in-breeding.[165]

Naturally, the law-texts stress the necessity for a bull's presence at the time when the cows are in heat. In *Cáin Domnaig* 'the law of Sunday', bringing a bull to a cow (*tarb do boin*) is classed as work which is of lawful necessity, and is therefore permitted on a Sunday.[166] Likewise, *Di Chetharšlicht Athgabálae* states that it is illegal for a bull to be distrained while a cow is in heat.[167] According to later commentary on this text, if a bull is seized under such circumstances, the owner is entitled to compensation (*aithgein*) for the calf and milk which would have resulted if the bull had been present.[168]

There is evidence that in some cases a stud-fee is payable to a bull-owner. *Cáin Lánamna* refers to the transaction of *síl cethrae do lúaig*, which may mean 'to pay for the seed of cattle'.[169] A passage quoted in legal commentary suggests that the bull-owner's fee is one quarter of the value of the calf.[170] In other circumstances, however, it seems that no such payment is due. For example, a text describes an agreement of 'joint herding' (*comingaire*), in which cattle belonging to two or more farmers graze together. The author remarks that the bull in such a herd serves not only his own cows but also those of the other farmers, but makes no mention of any fee for this service.[171] A legal gloss likewise refers to a general principle that a calf belongs to its mother not to its father, and compares the case of a slave who impregnates a free woman but acquires no rights in respect of his child.[172]

The importance of breeding from a bull of good stock (*cinél*) is recognised.[173] According to one law-text, the qualities of an animal depend on nurture (*imchoimét*), pasturage (*scor*) and heredity

[165] *CIH* iv 1518.37 = O'Dav. 428 § 1310; *CIH* v 1578.33–6.

[166] Hull, '*Cáin Domnaig*', 162.15.

[167] *CIH* iv 1458.7 = *AL* ii 44.6; cf. *CIH* iii 897.16 = Appendix A, text 2 § 1 (14).

[168] *CIH* iv 1458.7–9 = *AL* ii 44.7–9. In addition, a fine of five *séts* must be paid for illegal distraint.

[169] The MS has *sil cethra do luad* (*CIH* ii 506.16 = *SEIL* 19.5–6 § 5). I owe this interpretaion to Donnchadh Ó Corráin. For other suggestions, see *SEIL* 23 (5).

[170] *CIH* vi 2075.37–8.

[171] *CIH* i 192.17–18; ii 577.5–6 = *AL* iv 100.x-y.

[172] *CIH* ii 547.7–9.

[173] The term *cinél* is used in connection with a bull in commentary at *CIH* vi 2075.38.

(*úadadbar cineda*). The accompanying commentary states that this applies to stallions, bulls and rams.[174]

For injury inflicted by bulls, especially at times of bulling, see p. 150.

Slaughter

The life of most bovines is ended by a human slaughterer, and our sources refer to the use of a special spear or an axe for this task (see p. 496). *Críth Gablach* states that a king's manservant (*fergniae*) should be able to kill an ox with one blow.[175]

The archaeological evidence shows that cattle might be slaughtered at any time of year, mainly between seven and twenty-four months of age.[176] The law-texts also provide evidence of all-year-round slaughter. Thus the highest grade of *bóaire* is expected to have a sack of salt in his house at every season for use when cattle are being cut up.[177] In later times in Ireland, it seems to have been regular to slaughter male calves at a few days old,[178] but the archaeological and literary evidence does not indicate that this was a widespread practice in the early Christian period. Only 1½ per cent of the cattle-bones at Moynagh crannóg, Co. Meath, are from animals under the age of six months, whereas nearly 14 per cent are from between six and twelve months.[179] One can compare the well-known passage in the eighth-century tale *Longes mac nUislenn* 'the exile of the sons of Uisliu' which describes how Deirdriu's fosterfather was skinning a calf before cooking it for her.[180] The calf is described as a *lóeg fothlai* 'calf of withdrawal', i.e. a weaned calf. The calf's blood stained the snow, so the slaughter is likely to have taken place in winter. One could guess that the author had in mind the slaughter in early winter of a calf whose mother had recently stopped milking. A poem attributed to Cíarán clearly prefers slaughter at a much later stage, and advises

[174] *CIH* vi 2076.8–10 . . . *ūadabhar cinadha .i. adbar aneich ōr chinestar .i. cullaig ⁊ tairbh ⁊ reithedha.* For *úad-* see Thurneysen, *Grammar*, 524 § 847.

[175] *CIH* ii 569.38 = *CG* 22.556–7.

[176] McCormick, 'Stock-rearing in early Christian Ireland', 87.

[177] *CIH* ii 564.1–3; iii 780.8–10 = *CG* 8.188–191.

[178] Examples from the seventeenth to nineteenth centuries are given by Lucas, *Cattle in ancient Ireland*, 224–6.

[179] McCormick, 'Stock-rearing in early Christian Ireland', 87, table 3.7.

[180] Hull, *Longes mac nUislenn*, 45.90–1 § 7.

that a calf should not be killed until it has reached the age of an ox.[181]

The slaughter of cattle or other domestic animals is well attested throughout the world as part of religious or funeral ceremonies, and was widely practised among the Celts of Gaul.[182] There are some possible references to crane-slaughter (*corrguinecht*) for magical purposes,[183] but ritual cattle-slaughter is rarely mentioned. One instance is in the account of the funeral ceremony of a king in the mythological Old Irish tale *Tochmarc Étaíne.* His funeral mound (*fert*) is dug, lamentation (*gubae*) is raised for him, and his cattle are slaughtered.[184] In Irish material of biblical origin there are of course many references to Jewish practices of animal-sacrifice.[185] For example, legal commentary includes *audbart i recht* 'sacrifice according to [Mosaic] law' as one of the four uses of cows.[186]

CATTLE-PRODUCTS

Milk

Though sheep and goats were also milked on early Irish farms, it is clear from the written sources that cows provided by far the largest proportion of milk for human consumption. Legal commentary compares the relative value of the milk-yield of a fully-grown cow (*bó mór*), a goat and a sheep during the milking-season, i.e. summer and autumn. The milk-yield of the cow is worth 6 scruples (equivalent to 18 pence), that of the goat is worth 1⅓ pence, and that of the sheep is worth only half or one third of a penny.[187]

For a discussion of the importance of milk and milk-products in the early Irish diet, see p. 323.

[181] Ó Riain, *Corpus genealogiarum sanctorum Hiberniae*, 182.1 *ni horta laeg ría n-áes daim.*
[182] Méniel, *Les sacrifices d'animaux chez les Gaulois*; Méniel, *Chasse et élevage*, 101–41.
[183] See p. 128.
[184] Bergin and Best, '*Tochmarc Étaíne*', 166 § 4 = *LU* 326.10738–9; cf. Best 'Adventures of Art', 160 § 13. See also the discussion on the name *Bóguine* on p. 158 below.
[185] See *DIL* s.v. *idbart.*
[186] *CIH* vi 1978.25–7 *ara hudbairt a recht.* Liam Breatnach points out to me that this passage derives from Wasserschleben, *Die irische Kanonensammlung*, 99, bk. 29 ch. 4.
[187] *CIH* i 306.28–9, 307.3–11; ii 716.6–10.

Beef

Seventy-five per cent of the cattle-bones found at Moynagh crannóg, Co. Meath, are from animals under three years of age.[188] It can be assumed that most of the remaining twenty-five per cent come from cows which were slaughtered because of infertility, dangerous temperament, serious injury, or old age. If possible, such cows would be fattened up before slaughter: legal commentary refers to a fat cow (*bó méth*),[189] which is equated with a cow for slaughter (*bó marta*).[190] Oxen which were too old or otherwise unsuitable for work would also be slaughtered to provide beef and hides.

Bull-flesh is very red and strong-flavoured, and generally found distasteful by the human palate. The wisdom-text *Tecosca Cormaic* regards it as actually harmful, and includes it among the things which are worst for the body.[191]

Marrow

The manner in which most cattle-bones on early Christan sites have been broken shows that the marrow (*smir*) was extracted for consumption.[192] The sagas of the Ulster cycle also refer to a marrow-mash (*smirchomairt* or *smirammair*), apparently used as a poultice or bath to cure the wounds of warriors.[193] It is uncertain whether this marrow-mash is a literary invention, or was actually used as a form of medical treatment.

Blood

Lucas devotes a chapter of his *Cattle in ancient Ireland* to the custom of drawing blood from living cattle.[194] This practice is extensively documented from Ireland of the sixteenth to nineteenth centuries, and was mainly for the purpose of supplementing people's diet during times of scarcity. He also provides evidence of a belief that the bleeding of cattle – like that of humans – had a beneficial effect on their health. In some areas,

[188] McCormick, 'Stock-rearing in early Christian Ireland', 87, table 3.7. He notes that there is a similar distribution at other sites of this period.
[189] *CIH* i 229.37 = *AL* v 444.29.
[190] *CIH* i 229.28 = *AL* v 444.18.
[191] Meyer, *Tecosca Cormaic*, 38 § 21. For the ritual consumption of bull-flesh, see p. 28 above.
[192] Hencken, 'Ballinderry crannóg no. 1', 231–2.
[193] E.g. *TBC* Rec. I 100.3299–3300; *TBC* LL 105.3780–5; cf. Meid, *Táin Bó Fraích*, 9.225–7.
[194] Lucas, *Cattle in ancient Ireland*, 200–22.

the bleeding of cattle clearly had a ritual significance and the blood was not consumed. This practice was traditionally carried out on Mayday (*Lá Bealtaine*), and was evidently felt to provide magical protection for the cattle.

There are a few literary references to the bleeding of cattle for food in earlier Irish sources. For example, a fragmentary Early Modern Irish recension of *Táin Bó Cúailnge* refers to the provision of gore and blood (*crú 7 fuil*) to the men of Ireland by the cattle of Conaille Muirthemne.[195] Another text describes how the Munstermen drew blood from their cattle.[196]

Hides

It is clear from the written sources that the hide (*gemen* or *seiche*) of a bovine served many purposes.[197] It was used as a surface on which food could be placed,[198] a bed-covering,[199] a covering for the inside of a chariot (*forgemen*),[200] and a surface on which ears of corn could be flailed.[201] The skins of young or fetal calves were particularly valued as a source of vellum for manuscripts.[202] Oak-bark (see p. 382) supplied the tannin used to make leather (*lethar*) from bovine hides. The leather items most frequently mentioned in our sources are shoes,[203] but there are also references to leather belts,[204] bags,[205] halters,[206] dog-muzzles,[207] etc.

On account of its size and thickness, the bovine hide is more suited than that of other livestock for covering boats of wood or

[195] Thurneysen, 'Táin Bó Cúailghni nach H.2.17', 544. For a full discussion of the literary references, see C. O'Rahilly, 'The bleeding of living cattle'.
[196] Sjoestedt, '*Forbuis Droma Damhghaire*', 52 § 53.
[197] It is included in a list of domestic utensils in *CIH* ii 472.4 = *AL* v 406.17. For the uses of sheepskin and goatskin, see p. 72 and p. 79 respectively.
[198] E.g. Jackson, *Aislinge*, 19.583 = Meyer, *Aislinge*, 47.11.
[199] E.g. *CIH* ii 567.23 = *CG* 16.403.
[200] E.g. *TBC* Rec. I 22.718.
[201] E.g. *CIH* i 283.23 = *AL* iii 264.14.
[202] Bischoff, *Latin palaeography*, 9; K. Ryan, 'Holes and flaws', 245.
[203] E.g. *CIH* i 202.20; ii 582.7; vi 2184.3–4 = *AL* iv 148.19–20; Jackson, *Aislinge*, 4.101–2 = Meyer, *Aislinge* 9.21–2 *dí chuarán . . . do dondlethar* 'two shoes . . . of brown leather'. The word *cúarán* lit. 'curved one' can be applied to shoes of both tanned or untanned cowhide. It was borrowed into Welsh in the form *cuaran* (see *GPC* s.v. *curan, cuaran*; *GMWL* 82 s.v. *kuaran*). For a summary of the archaeological evidence, see Edwards, *The archaeology of early medieval Ireland*, 79–80.
[204] E.g. Jackson, *Aislinge* 34.1067 = Meyer, *Aislinge*, 89.9.
[205] E.g. Meyer, 'Zwiegespräch zwischen Gúaire und Marbán', 455 § 6.
[206] *CIH* iii 972.7 *coirtegh adastair* 'tanning a halter'.
[207] *CIH* iv 1389.13.

wattle. The tenth-century *Navigatio Sancti Brendani* describes the manufacture by Saint Brendan and his monks of a light wooden boat. They covered it with bovine hides (*coriis bouinis*) which had been tanned in oak-bark, and smeared all the joints of the hides on the outside with butter (*ex butyro*).[208] Other texts provide evidence of the use of hide-covered boats of various sizes. Thus there are references to a boat made with a single hide (*curach óenṡeichi*),[209] and to a larger boat with three hides (*náu trechodlide*).[210]

Writing in the late twelfth century, Giraldus Cambrensis refers to the export of the hides of wild and domestic animals from Ireland.[211] Legal commentary from about the same period discusses salvage rights relating to furs, hides, and other cargoes.[212]

Tallow

The bovine carcase yields large quantities of tallow. The main use of this substance in the early Irish economy was evidently in the production of candles. According to *Críth Gablach*, a candelabrum is one of the normal fittings in the house of a prosperous farmer, and it should always have a fresh candle in it (*caindel for caindelbrai cen meth*).[213]

The law-text on base clientship, *Cáin Aicillne*, describes the correct dimensions and manufacture of the three handfuls of candles which a client must give to his lord as part of his annual rent.[214] Each candle is made by dipping a stripped rush (*simin*) in the tallow (*geir*) and grease (*úsca*) of a slaughtered beast.[215] The length of the candle must be 8 fists (i.e. 2½ feet), and the rush from which it is made must be the thickness of one grain at its base. A grain (*gráinne*) is normally one third of an inch (*ordlach*).

For liturgical purposes, candles made of beeswax were used: see p. 114.

[208]Selmer, *Navigatio Sancti Brendani*, 10–11 ch. 4.
[209]Meyer, *Sanas Cormaic*, 21 § 229 *ōn nōi ōenṡeiched*; Mulchrone, *Bethu Phátraic*, 132.2633 *hi curuch ōenṡeichi.*
[210]Van Hamel, *Immrama*, 28.94 = Stokes, 'The voyage of Máel Dúin, (1)', 458.z *dogní . . . nói trechodlidi*; Van Hamel, *Immrama*, 100.168 = Stokes, 'The voyage of the Húi Corra', 38 § 33 *curach trechodlaide.*
[211]O'Meara, *Giraldus: Topography*, 35 § 2 = Dimock, *Giraldi Topographia*, 28.18–19.
[212]*CIH* vi 2155.20.
[213]*CIH* ii 563.24–5; iii 780.3–4 = *CG* 7.181; cf. *CIH* ii 376.10 = *AL* i 126.5.
[214]*CIH* ii 483.16–19 = Thurneysen, 'Aus dem irischen Recht I', 355 § 13.
[215]See *DIL* s.v. *ítharnae*; Russell, 'The sounds of a silence', 18.

Bones and horns

The law-text *Bretha Nemed Toísech* contains a reference – probably humorous or sarcastic – to the combmaker's ability to chant a spell over a dunghill so that all the antlers and bones and horns come to the top.[216] This passage makes it clear that these were the main raw materials used by the combmaker (*círmaire*).[217] The archaeological finds in early Christian sites in Ireland commonly include bone combs, as well as bone pins, needles, spindle-whorls, etc.[218] Decorated bovine bones have also been found.[219]

As we have seen above (p. 30) the ancestral wild ox (*Bos primigenius*) had long horns. However, during the process of domestication a short-horned type evolved in Europe and western Asia. This type is called *Bos longifrons* (or *Bos brachyceros*),[220] and because of its particular association with Celtic sites on the Continent as well in Britain and Ireland, it is sometimes described as the 'Celtic Shorthorn'.[221] On account of the generally small size of the horns of early Irish cattle, there was evidently a trade in imported wild ox horns from the Continent.[222] A gloss on a legal heptad refers to foreign valuables, including ox horns (*cuirn búabaill*).[223] The Welsh law-texts likewise refer to the use of ox horns to provide drinking-goblets for kings.[224] According to Julius Caesar, the large horns of the wild ox (*urus*) were much sought after by the Germani, who trapped these animals in pits in the

[216] *CIH* vi 2220.5–7 = Meyer, *Triads*, 16 § 117. For a general discussion, see Dunlevy, 'Early Irish combs'.

[217] An Iron Age weaver's comb of horn is illustrated at Raftery, *La Tène in Ireland*, 237.

[218] E.g. Hencken, 'Ballinderry crannóg no. 1', 163–5; Hencken, 'Lagore crannóg', 184–97; cf. Mytum, *The origins of early Christian Ireland*, 242–5.

[219] Hencken, 'Lagore crannóg', 181–3.

[220] The designation *Bos longifrons* refers to the long frontal bone of this type. It is also called *Bos brachyceros* 'short-horned ox'. See Clutton-Brock, *Domesticated mammals*, 65.

[221] E.g. Cunliffe, *Iron Age communities in Britain*, 379; Ryder, 'Livestock', 383.

[222] It is important to distinguish the extinct wild ox (*Bos primigenius*, German *Aurochs*) from the still surviving European bison (*Bison bonasus*, German *Wisent*). The horns of the latter are considerably smaller.

[223] *CIH* i 26.39 = *AL* v 220.23; *CIH* iv 1507.34 = O'Dav. 380 § 1051.

[224] Emanuel, *The Latin texts of the Welsh laws*, 127.31–4. For a detailed discussion, see Mac Cana, 'Ir. *buaball*, W. *bual* "drinking horn"'.

Hercynian wood. The horns were edged with silver, and used as drinking cups at their banquets.[225]

Dung

Legal commentators stress the value of the dung of cattle as a fertilizer: see p. 229. I have found no early references to the use of dung as building material or fuel.[226]

Cattle prices and values

No economic system can guarantee absolute stability in cattle-prices. A series of harsh winters or an outbreak of bovine disease or of war may cause the death of large numbers of cattle with a consequent increase in the price which can be asked by sellers of the surviving cattle. The loss of the cereal harvest may likewise inflate the price of cattle because of people's increased dependence on milk-products and meat.

Conversely, excessive numbers of cattle may depress prices. This could happen if the human population were severely depleted by plague, or if a series of good harvests caused an abundance of grain. Success in cattle-raiding might have the same effect. The *Annals of Connacht* record that in the year 1536 Ó Domhnaill amassed so many cattle in his campaign against Ó Conchobhair and Ó Dubhda that one or even two bullocks for slaughter could be bought for a single groat (*bonn*) in his camp.[227]

In spite of such inevitable fluctuation, the authors of the law-texts work on the assumption that cattle have a constant value. To do otherwise would obviously entail practical difficulties for judges, as the fixed payments for many services and offences are given in terms of cattle. If each such payment had to be adjusted in line with current market values, the whole system would become chaotic. There is, however, some flexibility in the system in that certain payments can be made in either cattle or silver,[228] or in a combination of these two currencies.[229]

[225] Caesar, *The Gallic war* (ed. Edwards), 6.28; cf. Isidore, *Etymologiarum* (ed. Lindsay), 12.1.34.

[226] In later Irish cow-dung dried for fuel is called *bóithreán*: see Dinneen, *Foclóir* s.v.; R. B. Breatnach, *Seana-chaint na nDéise II*, 52; Ó Cuív, *Cnósach*, 42.

[227] *AC* 694 s.a. 1536 § 18.

[228] E.g. *CIH* vi 2307.10–11 = Binchy, '*Bretha Déin Chécht*', 26 § 7.

[229] E.g. *CIH* vi 2312.33 = Binchy, '*Bretha Déin Chécht*', 40 § 30. See p. 588 below for payment consisting of one third cattle, one third horses, and one third silver.

The currency system described in the law-texts deals mainly in cows and young cattle, though there are occasional references to fines payable in oxen.[230] The basic unit of value is the milch cow, for which the Old Irish terms are *laulgach* or *bó mlicht* (often simply *bó*). The milch cow is taken to be equivalent in value to one ounce (*ungae*) of silver. By modern standards, this seems a very low value to assign to a cow. However, the equivalence of an ounce of silver to a milch cow is clearly stated in both the Old Irish law-texts and in later legal commentary. For example, the eighth-century law-text *Bretha im Ḟuillemu Gell* refers to the purchase of a cow for an ounce [of silver].[231] A similar equation is attested in a non-legal source. An ounce [of silver] is given as equivalent to one and a third of the value of a milch cow in an entry for the year 1106 in the *Annals of Ulster*.[232]

The relationship of the milch cow to other forms of currency is fairly consistent in the law-texts. The largest unit of value in the system is the *cumal*, a term which basically means 'female slave' (see p. 591) and is often equivalent to three milch cows. Another very common unit of currency is the *sét*, originally a standard object of value. It is generally fixed at half the value of a milch cow.

In the remainder of this section I discuss the value of male and female cattle at various stages of their lives. The main source is a section of commentary on animal values, which is clearly based on a lost Old Irish law-text.[233] The degree of detail in this commentary (and in other sources) testifies to the crucial importance of cattle in the early Irish economic system. The unit of value used in this commentary is the scruple (*screpul*) of silver, which is one twenty-fourth of an ounce of silver.

It is important to distinguish between the the value (*lóg*) of a bovine or other domestic animal and its *díre*, i.e. the fine paid for illegally killing it. The latter is normally about fifteen or twenty times greater, see p. 76 (sheep) and p. 87 (pigs).

Young calf (*lóeg*)

It is clear that a milch cow is normally accompanied by her calf when she is given in a payment or other transaction. For example, an illegal injury to a person's shin entails a fine of three *sét*s, which

[230] E.g. *CIH* iv 1263.16–17.
[231] *CIH* ii 467.11–12 = *AL* v 392.1–2.
[232] *AU*^{2} 544 s.a. 1106 § 4. The figure given at *GEIL* 116^{112} should be corrected.
[233] *CIH* iii 845.22–846.14 = Appendix A, text 3.

includes a milch cow with a calf (*laulgach co lóeg*).[234] Similarly, a passage on clientship states that the best fief which a lord can give his client is a milch cow with her calf.[235] The value of the calf in such circumstances is clearly included in that of the milch cow.

However, in cases such as the theft or illegal killing of a calf belonging to somebody else, it is obviously necessary in law for a separate value to be assigned to a calf. According to the legal commentary on animal-values, a calf of either sex has a value of 2 scruples from the time of its birth in the spring (*errach*) until the following November (*Samain*).

The main Old Irish text on base clientship, *Cáin Aicillne*, distinguishes four categories of castrated male bovines which may be included in the food-rent which a client gives to his lord.[236] The smallest is a male calf worth a bushel of grain (*ag lóige méich*).[237] The calf is only acceptable if it has a minimum girth of eight fists (probably 32 inches), and is sufficiently plump that its haunches cover its kidneys, except for the space of three fingers.[238] It is expected to have grazed on grass with the milch cows, and its castration wound must have healed (*slán ó chull*). It must not have died of sickness or disease, but have been slaughtered by its owner. It is suitable for cooking in summer.[239]

Older calf (*gamain*)

According to the commentary on animal-values, a female calf is worth 3 scruples from its first November i.e. after weaning, until the following May (*Beltaine*) when it is approximately one year old. The value of a male calf, however, remains at 2 scruples. This is in line with the general principle that a male bovine has only two thirds the value of a female from the time of its weaning until its maturity (*forbart*) i.e. until it is fully trained as an ox.[240]

[234] *CIH* vi 2315.38–9 = Binchy, '*Bretha Déin Chécht*', 46 § 36.
[235] *CIH* iii 920.6.
[236] For a discussion of the relationship between lord and client, see p. 448.
[237] *CIH* ii 480.21–3; v 1782.39 = Thurneysen, 'Aus dem irischen Recht I', 347 § 8. In the law-texts, the bushel of grain seems to have been worth between 1 and 2 scruples: see Appendix B, p. 588.
[238] The 32-inch girth is a little smaller than that of modern Hereford or Aberdeen Angus calves of three months (K. Ryan, 'Holes and flaws', 252). For a discussion of the *dorn* 'fist' as a measure, see Appendix B, p. 564.
[239] *CIH* ii 483.28 = Thurneysen, 'Aus dem irischen Recht I', 355 § 13 *ag lōige mēich for fulucht a samrad.*
[240] *CIH* iii 846.9–10 = Appendix A, text 3 § 11.

The term *gamain* is not used in this commentary. However, a distinction between a young calf (*lóeg*) and an older calf (*gamain*) is well-attested in the law-texts.[241] Though there is some local variation, this distinction is generally maintained in Modern Irish and Scottish Gaelic. For example, in the Gaelic of South Uist a *laogh* becomes a *gamhain* at the beginning of November (*Samhain*).[242] A cow whose calf is allowed to stay with her beyond the normal time of weaning is described as a *gamnach* 'stripper', a term which means literally 'one accompanied by a *gamain*' (see p. 40 above).

In *Cáin Aicillne,* the second smallest category of male bovine is given as an *ag lóige dá míach* 'male calf worth two bushels of grain'.[243] It must have a minimum girth of ten fists (probably 40 inches), and have grazed with milch cows since the beginning of the summer. I suggest that this is a calf at the *gamain* stage.

Yearling bullock (*dartaid*)[244]

According to the commentary on animal-values, a male bovine attains the value of 2⅔ scruples in May. This goes up to 4 scruples in November, when it is aged 1½ years. This commentary does not give a name to the male bovine at this stage, but it is likely that the term *dartaid* covers the period of a castrated male's life between its first and second birthdays. The *dartaid* is the smallest unit used in fines or payments which are expressed in terms of cattle. For example, a person who pulls a handful of thatch from another's roof must pay him a *dartaid.*[245] In a gloss on *Cáin Aicillne* the *dartaid* is equated with the *ag lóige trí míach* 'bullock worth three bushels'.[246] This bullock must have a minimum girth of twelve fists (probably 48 inches), and have grazed with milch cows from the beginning of summer until shown to the lord (before slaughter) in the winter.[247]

[241] E.g. *CIH* iii 893.29–30 = Appendix A, text 4 § 1 (5).

[242] McDonald, *Gaelic words and expressions*, 85 s.v. *crodh.*

[243] *CIH* ii 482.1–4; v 1784.11 = Thurneysen, 'Aus dem irischen Recht I', 351 § 9.

[244] In general, the textual evidence indicates that the *dartaid* is male. For example, the 'ox-ogham' (*damogam*) in *Auraicept na nÉces* arranges male bovines in the order (1) *tarb,* (2) *dam,* (3) *colpthach firenn,* (4) *dartaid* (Calder, *Auraicept*, 294.5750–8). The Yellow Book of Lecan version of *Cormac's Glossary* refers to a *dartaid boinenn* 'female *dartaid*' (Meyer, *Sanas Cormaic*, 19 § 209), but this is probably a scribal error for *dairt.* One can compare the same gloss at *CIH* iii 922.21 which has *dairt baininn* 'female *dairt*'. For a discussion of the etymology of *dartaid,* see below.

[245] *CIH* ii 564.19; iii 780.25 = *CG* 8.214.

[246] *CIH* ii 485.14 = *AL* ii 258.17.

[247] *CIH* ii 482.21–4 = Thurneysen, 'Aus dem irischen Recht I', 353 § 10.

A difficulty with the identification of *dartaid* as a yearling bullock is the fact that legal glossators and commentators often take it to have a value of only 2 scruples – the same as a new-born calf.[248]

Yearling heifer (*dairt*)[249]

The commentary on animal-values assigns a value of 4 scruples to the heifer from May, when she is approximately one year old. This goes up to 6 scruples in November. It is likely that the term *dairt* – very common as a unit of value in the law-texts – generally refers to a heifer of between one and two years.[250]

As in the case of *dartaid* dicussed above, there are difficulties with this identification. From a reference in the law-text *Bretha Déin Chécht* it would seem that the author took the *dairt* to be worth one third of a milch cow, i.e. 8 scruples.[251] On the other hand, later glossators and commentators usually take the *dairt* to be worth only 4 scruples.[252]

[248]E.g. *CIH* i 203.21 = *AL* iv 150.22; *CIH* i 203.30–1 = *AL* iv 152.12. Scholars have had great difficulty in deciding whether the *dartaid* and *dairt* should be regarded as calves in their first year, or as yearlings. For example, in the glossary to *CG*, Binchy took *dartaid* to be a male calf in its first year, *dairt* to be a female calf in its first year, and *colpthach* (*colpdach*) to be a yearling calf. In his edition of *Bretha Déin Chécht*, on the other hand, he took *dartaid* and *dairt* to be male and female yearlings respectively, and *colpthach* to be a two-year-old heifer (Binchy, '*Bretha Déin Chécht*', 19).

The evidence of Modern Irish and Scottish Gaelic is not helpful, as the words *dartaid* and *dairt* seem to have become confused. For example, in his *Foclóir* Dinneen translates *dairt* as 'heifer, young cow, young bullock'. In his *Faclair*, Dwelly translates both *dairt* and *dartaidh* as 'heifer'. In both dictionaries, *dartán* (*dartach*) is explained as a two-year-old bull. So far as I have been able to find, neither *dartaidh* nor *dairt* is recorded from speech in any Irish or Scottish Gaelic dialect-survey.

[249]The etymology of *dairt* is also a problem. It looks like a passive formation from the root *dar-* 'bulls' (*VGK* ii 504 § 701), but a yearling heifer is obviously too young to be bulled. Similarly, *dartaid* seems to contain the agent suffix *-id*, so one might expect it to mean 'one which bulls'. However, the male yearling is too young for bulling, and is generally castrated. The explanation may be that these terms were originally applied to older bovines, but were at some stage transferred to yearlings. *Dartaid* is attested as both a male and a female personal name, e.g. Knott, *Togail Bruidne Da Derga*, 22.729; 45.1495; *DIL* s.v. *Dartaid.*

[250]Commentary at *CIH* vi 2105.29–34 refers to a calf (*láeg*) becoming a *dairt* at the end of the same year (*i cinn na bliadna-sin badēn*), i.e. when it is one year old.

[251]*CIH* vi 2309.27–9 = Binchy, '*Bretha Déin Chécht*', 32 § 16. The glossator is clearly surprised by this figure, and suggests that *dairt* here stands for a *colpthach* worth eight scruples.

[252]E.g. *CIH* i 68.2 = *AL* iv 82.22; *CIH* ii 463.32 = *AL* v 380.19. At *CIH* v 1778.29 = *AL* ii 220.19–20 it has a value of only 3 scruples.

Two-year-old bullock (*colpthach ḟirenn*)

According to the commentary on animal-values, the value of a male bovine goes up to 5⅓ scruples in May, when it is approximately two years old. It retains this value until the following May. I would identify this stage with the *colpthach ḟirenn* 'male *colpthach*' of the Old Irish law-texts.

The author of *Cáin Aicillne* explicitly equates the *colpthach ḟirenn* with the bullock worth four bushels (*ag lóige .iiii. míach*) which is slaughtered as part of a client's rent to his lord.[253] Its girth must be 14 fists (probably 56 inches) and it must have recovered from its castration while a *dartaid*.[254] Its haunches must cover its kidneys, and it must not have died of disease. It must have grazed with milch cows until shown to the lord.

Two-year-old heifer (*colpthach*)

The commentary on animal-values gives 8 scruples as the value of a heifer from May, when she is approximately two years old, until the following May. I suggest that this is the *colpthach*, regularly given in the law-texts as a unit of value between the *dairt* and the *samaisc*.[255] In the glosses the *colpthach* is generally taken to be worth 8 scruples i.e. one third of a milch cow.[256]

The glossators also refer occasionally to a *colpthach* (later spelling *colpach*) worth 6 scruples.[257] This is presumably to be identified with the *colpthach ḟirenn* 'male *colpthach*' discussed above. In the Old Irish law-texts the unqualified term *colpthach* seems always to refer to a two-year-old heifer. In Modern Irish dialects, *colpach* (*colpa*, *colan*) generally refers to a year-old heifer, but is also well attested of a two-year-old.[258]

[253] *CIH* ii 482.34–6 = Thurneysen, 'Aus dem irischen Recht I', 354 § 11.
[254] *CIH* ii 482.35 = Thurneysen, 'Aus dem irischen Recht I', 354 § 11 *doērna slān ina dartadas* 'it has come safely through [its castration] while a *dartaid*'.
[255] E.g. *CIH* vi 2308.32–3 = Binchy, '*Bretha Déin Chécht*', 30 § 13; cf. *CIH* vi 2104.33.
[256] E.g. *CIH* i 70.28 = *AL* iv 92.14–15; *CIH* ii 470.18 = *AL* v 402.16; *CIH* v 1778.19 = *AL* ii 220.5.
[257] E.g. *CIH* i 203.11 = *AL* iv 148.9; *CIH* v 1778.29 = *AL* ii 220.20.
[258] Wagner, *Linguistic Atlas* ii–iii, questionnaire nos. 19–20, e.g. points 7, 12, 17, 22, 29, 37, 51; Ó Cuív, *Cnósach*, 195 s.v. *samhaisg*. The Scots form *colpindach*, *colpnoch*, etc. was applied to 'ane young beast or kow, of the age of an or twa yeires (Skene)' (Craigie et al., *A dictionary of the older Scottish tongue* s.v. *colpindach*).

Three-year-old male bovine

The commentary on animal-values does not specify a name or value for the three-year-old male, presumably because the majority of males would already have been slaughtered at this stage. Those still alive would mostly be strong castrated males suitable for training as oxen, as well as an occasional young bull (*tairbíne*) for breeding purposes.

Another section of this commentary states that castrated males continue to have two-thirds the value of a female of the same age until they have reached full development (*co hind a forbarta*).[259]

Three-year-old dry heifer (*samaisc*)

According to the commentary on animal-values, a heifer at the age of three becomes a *samaisc*, a term which means 'summer-dry' (*sam* + *sesc*). She has not yet been brought to the bull, and is rated as having a value of 12 scruples.[260] This agrees with her normal value in the law-texts. For example, in an Old Irish passage on theft a *samaisc* is equated with half an ounce [of silver], i.e. half the value of a fully grown milch cow (24 scruples).[261] A similar equation is found in the *Annals of Ulster* for the year 1106, where the *ag ndára* (bulling heifer) is given half the value of a milch cow.[262]

From non-legal sources it is clear that the term *samaisc* may be used of an older heifer which would be expected to have come into calf but has not yet done so. This usage is illustrated in a passage in the saga *Longes mac nUislenn*.[263] This saga describes the tragic career of the beautiful Deirdriu, who was reared in secret to be the wife of the old king Conchobar. One day, however, she was attracted by the song of the young warrior Noísiu mac Uislenn, and slipped out to see him. To compliment her, he remarked 'fair is the heifer (*samaisc*) which goes past us'. Adopting his metaphor, she replied: *dlegtair samaisci móra bale na bít tairb* 'heifers are bound to be big where there are no bulls'. He then pointed out that she had the bull of the whole province (i.e. king Conchobar), but she pleaded

[259] *CIH* iii 846.9–11 = Appendix A, text 3 § 11.

[260] *CIH* iii 845.30–1 = Appendix A, text 3 § 4 *Samaisc ō sin co rodairter 7 dā screpul .x. uirre.*

[261] *CIH* vi 1975.4. Cf. *CIH* ii 512.11 = *AL* ii 380.8 *agh lethlōige bō* 'a heifer half the value of a cow'.

[262] *AU*[2] 544 s.a. 1106 § 4.

[263] Hull, *Longes mac nUislenn*, 46.109–24 § 9; Tymoczko, 'Animal imagery in *Loinges Mac nUislenn*', 149-51.

that she would prefer a young bull (*tairbín*) like Noísiu, and obliged him to elope with her. As a result both met violent and untimely deaths.

The use made of the term *samaisc* by the author of this tale is significant, as it implies that a heifer can still be described as a *samaisc* even if she is old enough to have had a calf. In Deirdriu's metaphor the infertility of the *samaisc* is simply due to the fact that she has not yet been brought to a bull. In the modern language *samhaisc* has a similarly broad meaning; for example a *samhaisc* is defined as 'a beast of three or maybe four years which has not yet born a calf'.[264] In Breton the term *hañvesk* – the exact cognate of *samaisc* – is used of a cow of any age which passes a year without calving.[265] So far as I have been able to find out, Irish *samhaisc* is never used in this way of a cow of proven fertility which has failed to calve in a particular year. In the law-texts such a cow would be simply described as *bó ṡesc* 'dry cow', probably also worth half a milch cow (for a discussion of the values of dry cows, see below).

In-calf heifer (*bó inlóeg*)

In the scheme set out in the commentary on animal-values, the term *samaisc* is applied only from when the heifer is three years old until she is impregnated by a bull. Her value then increases to 16 scruples i.e. two thirds of the value of a fully-grown milch cow.[266] This agrees with the Old Irish law-texts which regularly give the *bó inlóeg* a value intermediate between a *samaisc* and a milch cow.[267]

The author of this commentary is here dealing with the value of a heifer expecting her first calf. Eugene O'Curry held that the term *bó inlóeg* applies only to this category.[268] It seems more likely, however, that in the law-texts *bó inlóeg* refers to an in-calf cow of any age. Her value would be less than that of a milch cow, because of the chance that she may abort or that her calf may die at birth.

Milch cow (*laulgach, bó mlicht, bó mór*)

The commentary on animal-values gives the four-year-old cow a value of 20 scruples on the birth of her first calf. This increases to 22 scruples on the birth of her second calf. She reaches her maximum

[264] Ó Dubhda, 'Foclóir agus téarmaí feirmeoirechta', 29.
[265] Hemon, *Geriadur* s.vv. *hañvesk, hañveskenn.*
[266] *CIH* iii 845.31–2 = Appendix A, text 3 § 4; cf. *CIH* vi 1978.20–1.
[267] E.g. *CIH* vi 2308.31–2 = Binchy, '*Bretha Déin Chécht*', 30 § 13; *CIH* iii 1098.31–3; *CIH* iii 1119.29–30 = E. J. Gwynn, 'Privileges', 25.19–21; cf. *CIH* vi 2104.32–3.
[268] Quoted *AL* v 49 n. and *DIL* s.v. *indláeg.*

value of 24 scruples at the age of six with the birth of her third calf. Later legal commentary contains many references to the *bó threlóeg*, which O'Curry took to mean 'a three-calf cow', i.e. a six-year-old milch cow in her full maturity.[269]

According to medieval Welsh law, a cow is in her prime from her second to her fifth calf, after which her value declines.[270] The Old Irish law-texts provide no specific information on the values assigned to cows which have passed their prime. However, according to later commentary a cow's value is divided into three: a third (i.e. 8 scruples) for her flesh, a third for her milk and calf, and a third for her potential (*saíltinche*).[271] In the case of an aging cow, the value of her potential would naturally decline. On this commentary's assessment, a cow no longer capable of producing a calf would only be worth 8 scruples for her flesh.[272]

Dry cow (*bó ṡesc*)

After being bulled in July or August, it would be normal for milch cows to be dried off before winter. Most of these cows would produce calves the following spring. Each year, however, it would be expected that some of them would fail to conceive, and therefore miss a year's milk-production. Such a cow is termed a *bó ṡesc* 'dry cow', and her value is reduced during her period of infertility. From one law-text it seems that a dry cow has only half the value of a milch cow, i.e. 12 scruples.[273]

Older male bovines

It is clear from *Cormac's Glossary* that an ox which pulls the plough (*dam timchill arathair*) can attain the same value as a milch cow, i.e. 24 scruples.[274] But according to the commentary on animal-values, it takes an ox eight years to reach this value, whereas a milch cow does so after six years, provided that she has borne

[269] Quoted *AL* v 49 n. Binchy, however, suggests in 'Varia. III', 231–2, that *trelóeg* means 'having had a calf, after calving'. A *bó threlóeg* would thus refer to any cow with a calf, in contrast to a *bó inlóeg* 'in-calf cow'. But the three-calf *trelóeg* seems to fit in well with the commentary's view that a cow reaches her maximum value on the birth of her third calf (*CIH* iii 845.34–5 = Appendix A, text 3 § 6).

[270] Williams and Powell, *Llyfr Blegywryd* 89.30–1 = *ALIW*, Dimetian Code, II xxvii § 16.

[271] *CIH* i 306.26–9.

[272] Cf. *CIH* vi 2230.18 *.8. screbuill lōgh bōcarna* 'the value of a cow's flesh is eight scruples'.

[273] *CIH* ii 675.34–7 = Mac Niocaill, '*Tír cumaile*', 82.22–6.

[274] Meyer, *Sanas Cormaic*, 19 § 209; cf. Williams and Powell, *Llyfr Blegywryd*, 89.13–16 = *ALIW*, Dimetian Code, II xxvii § 8.

three calves.[275] According to another legal commentary, one third of an ox's value is for its flesh, one third for its work (*gnímrad*), and one third for its potential (*saíltinche*).[276] An ox incapable of work through age, injury or unsuitable temperament would thus be worth only 8 scruples for its flesh (and hide).

According to the commentary on animal-values, a bull normally has only half the value of a female of the same age.[277] By this reckoning the value of a bull does not exceed 12 scruples.

[275] *CIH* iii 845.36–8 = Appendix A, text 3 § 6.
[276] *CIH* i 306.30–1.
[277] *CIH* iii 846.10–11 = Appendix A, text 3 § 11.

2
Livestock (ii)

Sheep

After dogs, it is thought that sheep and goats were the first animals to be brought under human control, probably around 8000 BC.[1] Sheep must originally have been kept primarily for their flesh. Early on, however, man learned to breed long-fleeced sheep which provided the enormously valuable fibre, wool (Old Irish *olann*).[2] In early Irish sources, much more emphasis is placed on sheep as wool-producers than as providers of meat. The seventh-century wisdom-text *Audacht Morainn* refers to sheep being valued on account of their fleeces 'for the garments of the people'.[3] Similarly, a short legal passage on the proper qualities of a sheep makes no mention of the flesh of the sheep, and is concerned mainly with the qualities of the fleece and skin.[4]

Presumably because of their rôle as processors of wool and cloth, women have a particular association with sheep in our texts.[5] For example, a woman who has a legal claim to land must enter it accompanied by ewes, whereas a man in the same situation is accompanied by horses.[6]

Sheep-management

The domestic sheep (*Ovis aries*, Old Irish *cáera*[7]) probably originated from the Asiatic mouflon (*Ovis orientalis*), which is still

[1] Clutton-Brock, *Domesticated mammals*, 56; Gautier, *La domestication*, 126, 128; Ryder, *Sheep and man*, 3.

[2] This word has cognates in most Indo-European languages, e.g. Greek λῆνος, Latin *lana*, Gothic *wulla*, etc. (*VGK* i 179; *LEIA* O-19 s.v. *olann*).

[3] Kelly, *Audacht Morainn*, 12 § 44.

[4] *CIH* ii 675.13–17 = Appendix A, text 1 §§ 7–8.

[5] Cf. Plummer, *Vitae* i p. cxxi[5] *ubi enim ouis, ibi mulier* 'for where there is a sheep, there is a woman'.

[6] See p. 432. There is a more detailed discussion of *tellach* 'legal entry' at *GEIL* 186–9.

[7] *Cáera* is probably cognate with the Gaulish tribal name *Caeracates* 'the sheep people', but its other connections are uncertain: see *LEIA* C-8–9 s.v. *cáera*. The word for 'sheep' inherited from Indo-European is *oí*, cognate with Latin *ovis*, Greek ὄις, etc. (*LEIA* O-14). It is rather rare in our sources, e.g. *CIH* iii 1121.7 = E. J. Gwynn, 'Privileges', 27.22. Another term used almost exclusively in poetry is *cetnat*: see p. 68.

found in mountainous regions from Asia Minor to southern Iran. Its natural habitat is on foothills and open mountainside at lower altitudes than the goat or ibex.[8] Domestic sheep were first brought to Britain and Ireland by Neolithic farmers, and it has been shown that the Soay sheep of Saint Kilda in the Outer Hebrides retain many features of this early type.[9]

Texts of the Old and Middle Irish period contain numerous references to sheep. In general, they do not seem to be viewed as animals of the hills: according to a twelfth-century version of *Táin Bó Cúailnge* they grazed on greens, lawns and plains.[10] The early eighth-century law-text on status, *Críth Gablach,* refers to the presence all the year round of sheep on the green (*faithche*) of a prosperous farmer,[11] and an eleventh-century poem speaks of the call of a young sheep from its green (*gáir chetnata dá ḟaithche*).[12] Other texts likewise assume that a farm may be expected to have sheep at the entrance causeway (*airdrochet*).[13] At night such sheep must be brought for safety into a sheep-pen (*lías caírech*).[14] This is emphasized in a passage in *Bretha Nemed Déidenach*: 'it has been enjoined on the young sheep in front of his house to go into its pen; full penalty-fine (*lándíre*) for a sheep which sleeps in its pen above its hurdle'.[15] The implication is that after dark the householder cannot claim full penalty-fine for loose sheep if they are stolen, killed by wandering dogs, etc.

The law-texts often refer to the danger which wolves present to sheep. However, in spite of wolves and other predators (see p. 186) it is clear that sheep were sometimes brought to graze on common

[8]Clutton-Brock, *Domesticated mammals*, 52.

[9]Clutton-Brock, *Domesticated mammals*, 54; Ryder, *Sheep and man*, 493, 511, 549; Fraser Darling, *Natural history in the Highlands and Islands*, 211–12.

[10]*TBC* LL 2.60–61 *murthréta caírech d'[ḟ]aichthib ⁊ d'urlannaib ⁊ rédib.*

[11]*CIH* iii 780.14–15 = *CG* 8.198.

[12]E. J. Gwynn, *Metrical Dindshenchas* iv 368.11. The word *cetnat* (also *cetnait, cethnait*) is mainly used in poetry, and is a diminutive of *cit* 'sheep', which is attested only in glossaries. For a discussion of the various etymologies which have been proposed for this word, see *LEIA* C-108 s.v. *cit.*

[13]E.g. *CIH* iv 1231.2; O'Keeffe, '*Dál Caladbuig*', 19 § 5.

[14]*CIH* ii 564.5; iii 780.12 = *CG* 8.194; Connolly and Picard, 'Cogitosus: Life of Saint Brigit', 15 § 7.

[15]*CIH* iii 1121.6–7 = E. J. Gwynn, 'Privileges', 27.21–2 *Roherbadh do cethnait comhair a tighe techt ina lias; láindíre do oí ina lias uasa a chleith codla.* A version of the first sentence is to be found at *CIH* iv 1487.30 = O'Dav. 290 § 576. For a discussion of the second sentence see Watkins, 'The syntax of the Old Irish verb', 49.

land away from the farm. To lessen the risks, the law-text on joint-herding (*comingaire*) specifies that lambs should not go until the beginning of August, by which time they would have passed their period of greatest vulnerability.[16] The social organisation of the sheep is such that leadership of the flock is assumed by a single individual.[17] This rôle can be taken over by a shepherd, whose call the flock will learn to follow. This is reflected in the Old Irish term for shepherd, *áugaire*, the basic meaning of which is 'sheep-caller'.[18] As far as one can tell from the written sources, shepherds led rather than drove their flocks, and their dogs' function was to protect rather than to round up the sheep (see p. 119). The occupation of shepherd was regarded as being of low status, like that of cowherd.[19]

The life-cycle of the sheep normally begins in March or April with the ewe giving birth to a lamb (*úan*) or lambs, and it is clear from legal commentary that their care is regarded as a job for children (see p. 452). Twin lambs seem to have been fairly common, and are mentioned in a legal gloss.[20] Apart from the occasional male selected to grow into a ram (*reithe*), all male lambs are castrated after weaning. A castrated male, i.e. a wether (*molt*), seems normally to have been slaughtered in its first summer or autumn: see p. 72 below. In October or November, the ewes come into heat (*broth*), and are tupped by a ram. According to an Old Irish law-text it is illegal to distrain[21] either ewes or rams at the period of mating.[22] A young ewe which has not yet had a lamb is called an *oisc*, a term which literally means 'dry sheep' (*oí* + *sesc*).[23] Though no information survives from our period with regard to the age at which a ewe was normally expected to lamb for the first time, it is likely to have been three years.[24]

[16] *CIH* i 192.2–3; ii 576.26 = *AL* iv 100.6–7.

[17] Clutton-Brock, *Domesticated mammals*, 55.

[18] L. Breatnach, 'On the agent suffix *-e* in Irish', 194. For a discussion of the social position of the professional shepherd, see p. 442 below.

[19] *CIH* ii 363.23 = *AL* i 104.x.

[20] *CIH* ii 378.4 = *AL* i 146.14–15. In the *Annals of Connacht* (*AC* 312 s.a. 1355 § 18), an unlikely claim is made for the year 1355: *X. n-uain do breith d'oenchairig in hoc anno* 'one sheep gave birth to ten lambs in this year'.

[21] For a discussion of the legal process of distraint, see p. 170.

[22] *CIH* iii 897.19–20 = Appendix A, text 2, § 1 (20), (22).

[23] It survives as Modern Irish *fóisc*, Scottish Gaelic *othaisg*.

[24] At the present time ewes bred under poor conditions are often mated so as to lamb at about three years.

Fleece

The Asiatic mouflon, probably the ancestor of all domestic sheep,[25] is dun-coloured with a black throat ruff, a white rump patch and white underparts. This colour-scheme provides it with an effective camouflage in its natural environment.[26] Its outer coat is stiff and hairy, and in winter it grows a short woolly undercoat which is shed each spring. By selective breeding over many millennia man has produced breeds in which the undercoat grows to a considerable length and is not shed. It must therefore be periodically plucked by hand or cut with shears.

In modern times the majority of the world's sheep has white or cream-coloured fleeces, though occasionally black or brown individuals occur. It is clear from the Old Irish texts, however, that white-fleeced strains had not yet achieved their present dominance. Naturally, white fleeces were particularly prized, as this wool is of attractive appearance, and most suitable for dyeing. In the law-text *Bretha Nemed Toísech* white sheep are represented as being twice or three times as valuable as those of other colours.[27] The penalty-fine (*díre*) for killing a sheep which is white (*find*) is fixed at fifteen *séts*.[28] For a sheep which is dun-coloured (*odar*) ten *séts* must be paid, and a sheep which is black (*dub*) merits only five *séts*. It seems likely that at this period white sheep were relatively uncommon. Writing approximately four centuries later, Giraldus Cambrensis describes Irish sheep as being dark-coloured (*niger*).[29]

As we have seen above, the ancestral mouflon is predominantly dark but has white underparts. The author of an Old Irish law-text on the purchase of cattle and sheep advises that sheep with such two-coloured fleeces be avoided.[30] In his view a proper sheep should be uniformly white, black or dun-coloured (*lachtnae*)[31]

[25] Clutton-Brock, *Domesticated mammals*, 53.

[26] Ryder, *Sheep and man*, 16.

[27] Columella, *De re rustica* (ed. Forster and Heffner), 7.2.4, states that white (*albus*) is the best colour for a sheep, preferable to black (*pullus*) or brown (*fuscus*).

[28] *CIH* vi 2227.34–5. For a discussion of the *díre* of sheep, see p. 76 below.

[29] O'Meara, *Giraldus: Topography*, 101 § 93 = Dimock, *Giraldi Topographia*, 150.21.

[30] *CIH* ii 675.14–15 = see Appendix A, text 1 § 8 (2). For discussion of my emendation of MS *forofinn* to *fofind*, see p. 518. In his *De re rustica* (ed. Forster and Heffner), 7.3.7, Columella warns against buying a ewe which is piebald (*varia*).

[31] Liam Breatnach points out to me that the Manx form of this adjective is preserved in the four-horned *Loghtan* (or *Loaghtyn*) breed still surviving in the Isle of Man. For this breed, see Ryder, *Sheep and man*, 512.

Hence, if it is black or dun-coloured, it should not be white-backed (*forfind*) or white-bellied (*fofind*).

In the same passage the buyer is warned against other undesirable qualities in a sheep. It should not be cluster-haired (*congalfinnach*), which presumably means that its fleece should be long and straight rather than growing in tight curls which would be difficult to shear.[32] On the other hand, it should not be *letheirlach*, probably meaning 'with uneven tufts'.[33] Irish glossaries refer to a type of rough wool called *cintecal* which is not fine enough for clothing, but can be used to make a rug (*pell*).[34]

The buyer is reminded that the ideal sheep should be easy to shear (*solomrad*).[35] A difficult passage in the law-text *Bretha Déin Chécht* distinguishes two types of fleece of equal value: the *cnaí errginn* (or *eirginn*) and the *cnaí fogumraid*.[36] Binchy suggests that these are fleeces shorn in the spring and autumn respectively.[37] Even if this interpretation is correct, it does not prove that it was customary to shear the same sheep twice in the one year. Though shearing twice a year is practicable in some countries,[38] it is normal in Irish conditions to shear a sheep only once, i.e. after the 'rise' in the wool, which generally occurs about June.[39] A single shearing is indicated by the gloss in *O'Mulconry's Glossary* which explains the word *sartan* as meaning 'a year's wool which is on sheep'.[40] The

[32]Compare the phrase *casithir rethe copad* 'as curly as a tufted(?) ram' (Knott, *Togail Bruidne Da Derga*, 26.881–2). See *DIL* s.v. *copp*.

[33]*CIH* ii 675.16 = Appendix A, text 1, § 8 (9).

[34]e.g. *CIH* ii 611.1. The tenth-century glossator Cormac mac Cuilennáin (Meyer, *Sanas Cormaic*, 21 § 239) correctly identifies this as a borrowing from a Welsh term referring to some sort of rough blanket: see *GMWL* 66 s.v. *kenhughel*, and Russell, 'Brittonic words in Irish glossaries', 175.

[35]*CIH* ii 675.13 = Appendix A, text 1 § 7 (1). The west face of the Tall Cross at Monasterboice, Co. Louth, which dates from the ninth or tenth century, has a carving which is rather indistinct, but generally taken to represent the shearing of a sheep (Harbison, *High Crosses* i 150 (discussion); i 367–8 (dating); ii fig. 496 (photograph)).

[36]*CIH* vi 2308.36 = Binchy, '*Bretha Déin Chécht*', 30 § 13. However, in commentary on *Uraicecht Becc* (*CIH* v 1609.29–30; vi 2325.37–8 = *AL* v 80.20), the *cnaí erginn* is only given half the value of the other fleece, here called the *cnaí fogebrind* (*fogemrind*).

[37]Binchy, '*Bretha Déin Chécht*', 56, suggests that the word *eirgenn* is a compound of *errach* 'spring' + *cenn* 'head, end'. The form *fogumrad* must be from *fogamar* 'autumn', but is otherwise unattested.

[38]Ryder, *Sheep and man*, 524, 694.

[39]For the sheep-shears (*deimes*), see p. 500.

[40]O'Mulc. 273 § 845; *CIH* iii 1076.39–41 *Sartan . . . oland bliadhna bís for caīribh.*

author of *Bretha Déin Chécht* may be distinguishing the main shearing of the older sheep from the autumn shearing of the previous year's lambs.

The archaeological evidence indicates an increase in sheep-numbers in the tenth century in the Meath area, and raises the possibility of the export of wool at this period.[41] For a brief discussion of domestic wool-processing, see p. 449 below.

Sheep-skins

With the wool left on, sheepskins can be used to make warm outer clothing, rugs, bed-covers, etc.[42] The skin can also be used to make leather. For example, a Latin poem in the seventh-century *Hisperica Famina* describes the preparation of a sheep's skin for a book-satchel. The skin is scraped, stretched on a wall between thick stakes, and dried in a smoky fire.[43] The poet makes no reference to the tanning of the skin, but it is likely that oak-bark was used for this purpose (see p. 382). Untanned sheepskin can also be used to make parchment.[44]

Meat

Though our sources concentrate on the importance of sheep's wool, there are also some references to the consumption of mutton. According to an Old Irish passage on clientship, the food-rent which a client gives to his lord includes a wether for a feast (*molt feisse*) with its fleece (*lí*). This is decribed as a wether of summer food (*molt sambíd*), so it is clear that it is eaten in the summer.[45] There is also a reference to an autumn wether (*molt fogmair*) being given to a lord as part of the autumn food-rent (*fáer*).[46] A reference in the lament of the Old Woman of Beare shows that it was a usual practice for wethers to be killed for a wedding feast.[47] In excavations from our period some of the sheep scapulas have holes

[41] McCormick, 'Stock-rearing in early Christian Ireland', 110–11. For the export of wool in post-Norman times, see Timothy O'Neill, *Merchants and mariners*, 58–65.

[42] There are some literary references to the skin of a wether (*moltchroicenn*): see *DIL* s.v. *molt*.

[43] Herren, *Hisperica Famina* 104.513–106.530.

[44] K. Ryan, 'Holes and flaws', 245.

[45] *CIH* iii 920.10–12.

[46] *CIH* iv 1504.25 = O'Dav. 365 § 977; *CIH* v 1910.27 = Crigger, 'A man is better than his birth', 340.

[47] Murphy, *Early Irish lyrics*, 76 § 11b.

which were obviously made to facilitate hanging. This suggests that mutton was sometimes cured for later consumption.[48]

It is likely that tallow from sheep – like that from cattle – was used in candle-making: see p. 55 above.

Milk

Some modern breeds of sheep produce fair yields of milk for human consumption, much of which is made into cheese. Early Irish sources contain a number of direct and indirect references to the milking of sheep.[49] A poem in the comic tale *Aislinge Meic Con Glinne* lists a large number of foodstuffs, including 'fair white porridge made with pure sheep's milk'.[50]

In general, however, the milk of a sheep was not held in much esteem, and was regarded as inferior to that of a goat. This view is clearly expressed in an eleventh- or twelfth-century prophetical poem ascribed to Bécán mac Dé, which gives one of the signs of the End of the World as *blicht cáerach ag gobor glas* 'the grey goat will have the milk-yield of a sheep' i.e. the milk-yield of goats will be reduced in quality and quantity.[51] Legal commentary from approximately the same period takes the milk-yield of a goat to be worth one and a third pence, whereas that of a sheep is worth only half or a third of a penny.[52]

Horns

In the ancestral mouflon, both rams and ewes have horns. Over the millennia, however, horns have to a large extent been bred out so that many modern breeds of domestic sheep are hornless. Even in horned breeds, most ewes are hornless.

By selective breeding in the opposite direction, man has also managed to develop four-horned breeds of sheep. I have found no reference to such sheep in Old Irish sources, but the archaeological evidence shows that they were fairly common in early Christian Ireland. For example, in the excavations from

[48] McCormick, 'Stock-rearing in early Christian Ireland', 115–16.

[49] E.g. Mulchrone, *Bethu Phátraic*, 7.134–7; Plummer, *Bethada* i 57 (66); Ó hAodha, *Bethu Brigte*, 16.527 § 44; Meyer, 'The guesting of Athirne', 6 § 7.

[50] Jackson, *Aislinge*, 13.402–3 = Meyer, *Aislinge*, 33.29.

[51] Knott, 'Poem of prophecies', 64 § 30. It makes better sense to translate *gobor* as 'goat' rather than 'pony'. Knott gives this alternative in her notes.

[52] *CIH* i 307.3–11; ii 716.8–9. On the other hand, commentary at *CIH* iv 1207.22–6 values the milk of a two-scruple sheep at 1½ pence, and that of a three-scruple sheep at 2¼ pence.

about AD 800 at Moynagh crannóg in Co. Meath, 12 per cent of the sheep-skulls were four-horned.[53] The surviving Manx breed known as the *Loghtan* or *Loaghtyn* (p. 70) is four-horned.

The main use of sheep's horns seems to have been to make combs. It is clear from a triad in the law-text *Bretha Nemed Toísech* that these horns were normally straightened after being softened – by boiling or steam – over a fire.[54] They were then cut into shape by the combmaker (*círmaire*). Four-horned sheep were presumably bred because they provided greater amounts of horn.

Sheep types

Fig. 1. A sheep with David the shepherd from f.4v of the 'Southampton Psalter' in the Library of St. John's College, Cambridge. It is reproduced by kind permission of the Master and Fellows of the College.

The Old Irish law-texts provide a good deal of information on the different colours of sheep, but there is no evidence in the written sources that distinct breeds had been developed by this period. Finbar McCormick has pointed out that the sheep-bones from Midland excavations such as Lagore and Moynagh are generally bigger than those of Cahercommaun, Co. Clare, but attributes this difference to richer grazing rather than genetic factors.[55]

The earliest representation of a sheep in a manuscript believed to be of Irish origin is the cartoon-like illustration of David the shepherd in f. 4^v of the 'Southampton Psalter' in the Library of St. John's College, Cambridge.[56] The sheep's dappled coat is of no special significance, as David is similarly ornamented. However, the Roman nose characteristic of modern breeds such as the Leicester is clearly true to life, and the thick curled horns are also noteworthy. The

[53] McCormick, 'Stock-rearing in early Christian Ireland', 164.

[54] *CIH* vi 2220.5–6 = Meyer, *Triads*, 16 § 117. The statement here that the combmaker must straighten the ram's horn by his breath without fire (*adarc reithi do dírgud dia anáil cen tenid*) is presumably intended to be humorous.

[55] McCormick, 'Stock-rearing in early Christian Ireland', 163.

[56] This psalter has been dated to the second half of the tenth or the first half of the eleventh century (F. Henry, 'Remarks on the decoration of three Irish psalters', 34).

sheep depicted here seems further removed from the primitive mouflon type than the Soay of Saint Kilda (see p. 68 above), as the Soay has a concave rather than a Roman-nosed profile.[57]

Sheep as currency

Fines and other payments are often expressed in terms of cattle, but for smaller amounts the sheep is a common unit of currency. This usage is illustrated in a law-text on illegal injury, *Bretha Déin Chécht*, which lists the fines due for a small facial wound.[58] In this list the highest grade of noble is paid a milch cow, whereas the lowest grade of adult freeman, the *ócaire*, gets a yearling bullock (*dartaid*). Below him, the *fer midboth* – a semi-independent youth on his father's land – gets a perfect sheep (*cáera inraic*).[59] Below him again, a *bruider*[60] gets a sheep of lesser value (*cáera forcsen*).[61] Finally, the lowest grade recognised, the apprentice (*inol*), gets two fleeces if he is with a master, and one fleece if he is not.

A slightly different treatment of sheep-values is found in the text on status *Uraicecht Becc*.[62] Here the *fer midboth* has an honour-price of a yearling heifer (*dairt*). A youth called a *gairid* has an honour-price of one sheep (simply *cáera*). Below him the *flescach* has an honour-price of a lamb worth a bushel of grain (*úan méich*). Again at the lowest rung of the ladder, the apprentice has an honour-price of a fleece of wool (*cnaí olla*) or a ball of wool (*ceirtle*) or a non-broody hen (see p. 102).

In another law-text, a distinction is made between the value of a male and a female lamb.[63] If a person removes one stake from

[57] Fraser Darling, *Natural history in the Highlands and Islands*, 211–2.

[58] *CIH* vi 2308.30–37 = Binchy, '*Bretha Déin Chécht*', 30 § 13.

[59] Compare the Old Irish passage at *CIH* ii 715.11 which forbids the distraint of *caīrig indīre* 'sheep suitable for payment'. The gloss explains: *.i. na caīrigh as ferr bīs asin buathligh* (= *isin buailidh*) 'i.e. the best sheep which are in the pen'.

[60] For the *bruider* (= *mruigfer*?), see p. 424.

[61] The meaning of *forcsen* is uncertain here. It seems to be gen. sing. of *forcsiu* vb.n. of *for-acci* 'surveys, examines; is equivalent to'. The context indicates that a *cáera forcsen* is of less value than a *cáera inraic*, and Binchy suggests 'a young(?) sheep'; cf. *CIH* vi 2307.15 = Binchy, '*Bretha Déin Chécht*', 26 § 8 *caura forcsen*. However, there is a clear instance in *Bretha Nemed Toísech* of *laulgach forcsen* with the meaning 'milch cow of equivalent value', i.e. of the same value as a *laulgach inraic* 'perfect milch cow': *lulaicce indruicce tar ēisi inngit(h)er .iiii.orai lulaca forcsin* 'in place of a perfect milch cow, four equivalent milch cows are given' (*CIH* vi 2231.15. I owe this translation to Liam Breatnach).

[62] *CIH* v 1609.14–1610.21 = *AL* v 80.12–84.21.

[63] *CIH* ii 580.34–5.

another's fence, he must pay a wether lamb (*úanmolt*),[64] whereas if he removes two stakes he must pay a female lamb (*úan bainenn*). If he removes three or more stakes the fine must be paid in cattle.

Penalty-fine *(díre)*

It is important to distinguish between the ordinary commercial value of a sheep or other domestic animal and its *díre* i.e. the fine paid for illegally killing it. As we have seen, a law-text fixes the *díre* of a white sheep at fifteen *séts* (= 180 scruples), of a dun-coloured sheep at ten *séts* (= 120 scruples), and of a black sheep at five *séts* (= 60 scruples).[65] The *díre* of a black sheep is thus twenty times greater than the value (*lóg*) which is assigned to it in legal commentary.[66] In addition, the culprit would have to make restitution (*aithgein*), either by replacing the sheep or by paying its commercial value.

Value *(lóg)*

The value of a sheep in ordinary circumstances is discussed in a legal commentary which also deals with cattle- and pig-values.[67] It may be based on a lost Old Irish law-text.

At its birth in the spring a lamb of either sex is worth one penny, i.e one third of a scruple. At some stage (unspecified in the surviving versions)[68] the value of the female lamb rises to one and a half pence, whereas that of the male remains at one penny. In August the female's value doubles to one scruple (three pence), and the male's value doubles to two pence.[69] In August of the following year the female's value doubles again to two scruples. At the same time, the male's value is increased to two thirds that of the female. Then, at the age of two the female attains her maximum value of

[64]Compare the reference in a quotation at *CIH* iv 1491.17 = O'Dav. 306 § 680 to a vessel worth a wether lamb (*lestar úanmuilt*): see p. 580.

[65] *CIH* vi 2227.34–5; *CIH* iv 1531.22 = O'Dav. 485 § 1616.

[66]It should be pointed out that the maximum value of three scruples assigned to a sheep in legal commentary may be one scruple less than that of the Old Irish law-texts. There is evidence in the law-texts of a pig with a value of four scruples: see p. 87. As the values of pigs and sheep are generally parallel (p. 86), there is the possibility of an adult (black) sheep with a value of four scruples. The penalty-fine for this sheep would be fifteen times greater than its value, as in the case of an adult pig.

[67] *CIH* iii 845.22–846.14 = Appendix A, text 3.

[68]In footnote[n] to *CIH* iii 845.39 Binchy suggests adding *co beltaine* 'to May'. The version at *CIH* vi 2106.28 does not have complete readings at this point.

[69]An alternative ('or one and a half pence') is given at *CIH* iii 846.3–4 = Appendix A, text 3 § 8.

three scruples, and the male's value is raised to two scruples.[70] No distinction is made between the value of a ram (*reithe*) and of a castrated male, i.e a wether (*molt*).

In other legal commentaries the emphasis is on the value of adult females, which are classed either as 'sheep of two scruples' or 'sheep of three scruples'. The distinction seems to be one of quality rather than age or fertility, and is to be compared to the distinction noted above between the *cáera inraic* and the *cáera forccsen*. As in the case of other livestock, the sheep's value is viewed by commentators[71] as consisting of three elements: one third for its flesh, one third for its lamb, wool and milk, and one third for its potential (*saíltinche*).[72] The commentators make no reference to wool colour.

The figures given in these commentaries are to be regarded as representing a general average, and obviously cannot always have been maintained at a constant level. Under exceptional circumstances the value of a sheep might be very much greater. In 1315 the *Annals of Inisfallen* record that in the besieged camp of Edward Bruce at Coleraine, the four quarters of a sheep were worth two shillings sterling.[73]

In texts of a literary nature, the values of individual domestic animals are sometimes hugely inflated for dramatic purposes. This applies particularly in the case of horses, dogs or bulls of outstanding quality. Sheep are hardly mentioned at all in such material. An exception is the Book of Leinster version of *Táin Bó Cúailnge*, which refers to an especially fine ram belonging to Queen Medb; it was worth one *cumal*, i.e. at least seventy-two scruples.[74]

GOATS

The goat (*Capra hircus*) was first domesticated at around the same time as the sheep, i.e. about 8000 BC.[75] It derived from the bezoar goat (*Capra aegagrus*), native to the mountain regions of Asia Minor

[70] According to one version (*CIH* iii 846.7–8 = Appendix A, text 3 § 10), the male may attain the value of three scruples after a further year.

[71] E.g. *CIH* iv 1207.18–19.

[72] For a discussion of this term, see p. 65.

[73] *AI* 418 s.a. 1315. For the value of the shilling (*solidus*), see p. 598.

[74] *TBC* LL 2.62–3. For the value of the *cumal*, see p. 591.

[75] Clutton-Brock, *Domesticated mammals*, 59.

and the Middle East. Its bones – apart from the horns – are difficult to distinguish from those of sheep.[76] Consequently, excavation reports generally list the sheep and goat bones together.[77] The evidence of the written sources is therefore of special significance.

Fig. 2. This illustration of a female goat is from f. 41v of the eighth-century Book of Kells. The artist has not attempted to represent both horns in side view.

The Old Irish law-texts indicate that the goat (Old Irish *gabor*)[78] was not particularly common or important. It is not mentioned in the eighth-century law-text *Críth Gablach*, which gives a fairly detailed account of early Irish farming. It is included, however, in lists of livestock in two other law-texts.[79] It is also mentioned in an unedited text on clientship, which states that both the female goat (*gabor bainenn*) and the buck-goat (*bocc*)[80] may be given in fief by a lord to his client.[81] In return for the female goat, the client must give small quantities of products made from goat's milk. In the first year he must give his lord a tub (*cúad*) twelve inches high full of sweet cheese (*milsén*), and in the second year he gives the same quantity of butter. In the third year he gives an *ian*-measure five inches high of 'man-butter' (*ferimb*) and four inches high of 'woman-butter' (*banimb*).[82] In return for a buck-goat, the client must in the first year supply three day's labour at road-clearing (*slige*) for the lord. The text makes no reference to duties for the second and third years.

[76]Van Wijngaarden-Bakker, 'The animal remains from Newgrange (1)', 338.

[77]E.g. Hencken, 'Ballinderry crannóg no. 1', 234–5; McCormick, 'The effect of the Anglo-Norman settlement', 43.

[78]It is cognate with Welsh *gafr*, Breton *gavr, gaor* (*VKG* i 117). A rare poetic word for goat in Irish is *cadla*, of uncertain etymology: see *LEIA* C-5 s.v. 3 *cadla*.

[79]*CIH* ii 422.35 = *AL* i 268.21; *CIH* ii 573.24–5. Cf. also *CIH* i 13.21 = *AL* v 166.5.

[80]Mod.Ir. *boc(án), poc(án)*. This word is found in both Celtic (Welsh *bwch*, Breton *bouch*) and in Germanic (Old English *bucca*, Old Norse *bukkr*). It is uncertain whether it is a borrowing from Celtic to Germanic, or vice versa. See *LEIA* B-62 s.v. 2 *boc*.

[81]*CIH* iii 920.28–32.

[82]For the association of the smaller measure of butter with a woman, compare the *bairgen banḟuini* 'loaf baked for a woman', which is half the size of a standard loaf (see p. 330 below).

According to legal commentary, dating from about the twelfth century, a goat is valued at a lower rate than a sheep (see p. 76). Thus a female goat can never exceed a value of two scruples i.e. one twelfth of a milch cow.[83] The commentary states that one third of the female goat's value is for her flesh, one third for her potential (*saíltinche*) and the remaining third for her milk and her kid (*mennán*).[84] Of this third, one and a third pence are for the milk, and two thirds of a penny are for the kid. The value of a buck-goat is not discussed.

The goat was also valued for its skin, used to make shoes. A quotation from an undentified source in *O'Davoren's Glossary* refers to *bróc chóel do chorrgubul gabair* 'a slender shoe of goat's skin'.[85]

Pigs

The archaeological evidence indicates that the wild pig (*Sus scrofa*)[86] had already been domesticated by about 7000 BC in places as far apart as Jericho and New Guinea.[87] In Ireland, bones from the domestic pig have been found at Neolithic sites such as Ringneill Quay (Co. Down) and Tankardstown South (Co. Limerick).[88] They comprise nearly one third of the animal bones found at early Christian sites.[89]

The pig (Old Irish *mucc*)[90] is particularly prominent in the law-texts, and also has a fairly high profile in the wisdom-texts[91] and in the annals.[92] It is mentioned occasionally in narrative literature. For example, the tale *Scéla Mucce Meic Dathó* 'the story of Mac

[83] *CIH* i 307.8; iv 1208.1–2; v 1628.11 = *AL* iii 378.6–7.

[84] *Menn(án)* is cognate with Welsh *myn*, Breton *menn* (*LEIA* M-38 s.v. *menn*).

[85] *CIH* iv 1479.33 = O'Dav. 255 § 366. O'Davoren identifies *corrgubul* with *craicenn* 'skin', but no other attestations of this word are known.

[86] For the hunting of the wild pig in Ireland, see p. 281.

[87] Clutton-Brock, *Domesticated mammals*, 72, 76.

[88] Mitchell, *The Irish landscape*, 123; *Reading the Irish landscape*, 105; McCormick, 'The animal bones', 183.

[89] The combined percentages from seventeen major early Christian sites are 46 per cent cattle, 31 per cent pigs and 23 per cent sheep or goats: see p. 27 above.

[90] *Mucc* is cognate with Welsh *moch*, Breton *moc'h* 'pigs' (*LEIA* M-68 s.v. *mucc*). A common poetical word for pig is *mát* (*máta*), e.g. *CIH* vi 2113.6 = Dillon, 'Stories from the law-tracts', 44 § ii (wrong word-division at *CIH* vi 2217.8); *CIH* iv 1515.39 = O'Dav. 415 § 1241. Its etymology is uncertain : see discussion at *LEIA* M-23 s.v. *mát*.

[91] E.g. Kelly, *Audacht Morainn*, 12 § 45; Meyer, *Triads*, 8 § 68; 12 § 92.

[92] The many references to the acorn-crop reflect the annalists' concern for the fattening of pigs: see p. 84 below.

Dathó's Pig' deals humorously with the question of which warrior is to be given the honour of carving this massive pig at a feast.[93] Another tale (surviving only in a single extract) makes mention of a pig named Babgither, which is described as 'the noblest pig in Ireland'.[94] In the tale *Aígidecht Aithirni* 'the guesting of Aithirne', the flesh of a pig is preferred to that of a calf, sheep or bull.[95]

The Irish domestic pig of this period was a small long-legged hairy animal, and doubtless bore a general resemblance to the pig reproduced from a medieval Welsh legal manuscript on p. 18.[96] It would have been quite unlike its modern counterpart, which has been transformed over the last couple of centuries by crossing with quick-fattening oriental types.[97] A Middle Irish *Bríatharogam* refers to pigs of various colours: *find* 'white, light-coloured', *líath* 'grey', *loch* 'black', *crón* 'reddish brown', and *forglas* 'blue-black(?)'.[98]

Pig-management

According to *Críth Gablach*, a prosperous farmer of *mruigḟer* rank would be expected to own two breeding sows (*birit*).[99] In the wild the pregnant sow chooses a sheltered spot in which to excavate a large nest, lined with grass and leaves, and covered with branches. In the case of the domestic pig, man provides the shelter of a pig-sty (*cró* or *foil*), but the sow (*cráin*) retains the instinct to construct a nest from any available material.[100] The term *fóir* seems to be applied to a pig's nest in Heptad 7, which states that there is no liability in the case of *fuil muc i fóir* 'blood shed by pigs in a nest'.[101] *Fóir* is listed along with *cró* 'sty' among words connected with pigs in *O'Davoren's Glossary*.[102]

[93] Thurneysen, *Scéla Mucce Meic Dathó*, 7.5–7.

[94] Meyer, *Sanas Cormaic* 12 § 118; *CIH* ii 607.34–5.

[95] Meyer, 'The guesting of Athirne', 6 § 8.

[96] Pigs are described as 'hairy' in Herren, *Hisperica Famina*, 76.163; 88.314.

[97] Clutton-Brock, *Domesticated mammals*, 75.

[98] Calder, *Auraicept*, 288.5668–79.

[99] *CIH* ii 564.6; iii 780.13 = *CG* 8.196. The original meaning of this word is 'one which bears': see *LEIA* B-52 s.v. *birit*.

[100] Cf. Greene, 'OIr. *sopp*, etc.', 179.

[101] *CIH* i 10.18 = *AL* v 152.4. The glossator explains: *.i. oc breith muc* 'i.e. giving birth to pigs'. Cf. *CIH* iv 1506.13 = O'Dav. 373 § 1010 *diam orc a foir foserbthar* 'if it be a piglet which is stolen out of the nest'.

[102] *CIH* iv 1479.28 = O'Dav. 254 § 360. Proinsias Mac Cana suggests that *fóir* is the same word as Modern Irish *fáir* 'hen's nest'; cf. *cearc fáire* 'broody hen'. The meanings given for 1 *fóir* in *DIL* include 'burrow (of fox)', 'lair (of deer)'. It is probable that the enties listed in *DIL* under *foir* belong under 1 *fóir*.

According to a version of Triad 148, the uterus of a sow (*brú birite*) is classed as one of the three renovators of the world.[103] Farrowing (*doth* or *trog*)[104] takes place in the spring, and legal commentary refers to a litter of up to nine piglets.[105] Compared with the young of other domestic animals, the piglet (*orc* or *orcán*) is weakly developed and very dependent in the first few weeks of its life.[106] When away from the nest it communicates with its mother by squealing (*éigem*). A passage in the law-text *Bretha Étgid* deals with the legal consequences if a person maliciously disturbs another's sow by imitating the squealing of her piglets.[107] In the wild, the weakest piglet often does not survive. In the case of domesticated pigs, however, the runt of the litter may be taken from its mother and hand-reared on milk. It is clear from *Cáin Lánamna* that this task is regarded as belonging to the farmer's wife, as she gets two thirds of the meat from such a pig in the event of a divorce.[108] This piglet may be reared as a pet (*orcpheta*), in which case it is liable to become troublesome as it grows older.[109] The main law-text on farming, *Bretha Comaithchesa*, refers specifically to the trespasses of a pet pig,[110] and early commentary likewise mentions the offences of 'pet pigs which follow everybody'.[111]

It is clear from the law-text on joint-herding (*comingaire*) that young pigs were normally kept in the vicinity of the farm until the beginning of August.[112] In one passage in *Bretha Comaithchesa*, a

[103] Meyer, *Triads* 20 § 148 (v.l.).

[104] Farrowing is also called *andud*, e.g. *CIH* ii 715.9. See p. 532.

[105] *CIH* i 307.1 = *AL* iii 372.13. Varro, *Rerum rusticarum* (ed. Hooper and Ash), 2.4.19, regards eight piglets as the largest litter which a sow can feed.

[106] Clutton-Brock, *Domesticated mammals*, 73.

[107] *CIH* iii 943.36–7 *acht nīb neach dorōna ēigem n-orcāin doib namā* 'provided only it is not somebody who may have made the squealing of a piglet to them', cf. RIA MS 670 (D v 2) f. 22[a]28–9 (not in *CIH*) *acht nīp nech dorōnu ēigem n-orcāin doib namā*. The version at *CIH* i 278.24ff. = *AL* iii 242.13ff. omits this clause.

[108] *CIH* ii 509.2–3 = *SEIL* 33; *CIH* ii 509.18 = *AL* ii 368.15. At *SEIL* 35 (and in the Index) Thurneysen follows *AL* ii 368.15 in reading *in torc peta* 'the pet boar', but it makes better sense to read *int orcpheta* 'the pet piglet'; cf. *CIH* vi 2181.19 *int oircpeta*.

[109] T. F. O'Rahilly, *A miscellany of Irish proverbs*, 72 § 251, quotes the modern Irish triad: *Na trí peataí is measa – peata sagairt, peata bacaigh, peata muice* 'the three worst pets – a pet priest, a pet beggar, a pet pig'.

[110] *CIH* i 72.18 = *AL* iv 108.9.

[111] *CIH* iii 891.23–4 *orccpeta muc tiagdae i ndiaidh caich*; cf. *CIH* ii 391.15 = *AL* i 190.16.

[112] *CIH* i 192.2–3 = *AL* iv 100.6–7.

pig at this stage is called a *scuichid*.[113] By August the young pigs were regarded as hardy enough to survive the rigours and dangers of life in the woods. It is implicit in the text on joint-herding that it was a common practice for pigs belonging to a number of owners to be joined together in a single herd (*trét*) under the control of a swineherd (*muccid*). If the pigs were feeding in nearby woods, it is to be assumed that the swineherd would bring them back to their sties each night.[114] Often, however, they would be far from home: a version of *Táin Bó Cúailnge* refers to pigs foraging 'in woods and valley-slopes and remote places'.[115] In such cases the swineherd would obviously have to camp out with the pigs for long periods.

The evidence of the saints' Lives suggests that young slaves were often given the job of herding the pigs, and protecting them from robbers. *Bethu Brigte* represents the nobleman Dubthach as being angry with Brigit (his daughter by a slave woman) because two boars had been stolen from his herd.[116] The swineherd would no doubt also be expected to guard against wolves, though our sources regard lambs and calves as their main prey among domestic animals (see p. 186).

Diet

Pigs eat a wide variety of foods of animal and vegetable origin, and thus will consume practically all household scraps. For instance, an Old Irish tale refers to the leavings (*tresc*) of a meal being fed to a herd of pigs.[117] Pigs will even eat carrion; according to the *Canones Adomnani* pigs which have grown fat on carrion are unsuitable for human consumption until they have returned to their original thinness.[118]

Cereals are a major constituent in modern pig-feeding systems, and our sources likewise contain references to the fattening of pigs on cereals. Legal commentary on *Cáin Lánamna* speaks of a pig

[113] *CIH* i 73.11; 195.27 = *AL* iv 112.7 (emending *scuithe* to *scuichid*). Cf. *CIH* iv 1523.39 = O'Dav. 451 § 1428 *Scuithid .i. ainm do orc*; *CIH* vi 2227.39 *scuichid*. *Scuichid* is an agent noun from *scu(i)ch-* 'moves': see Binchy, 'Varia hibernica', 39.
[114] *CIH* i 72.12; 192.35 = *AL* iv 96.24–5; *CIH* ii 572.26–7.
[115] *TBC* LL 2.66–7.
[116] Ó hAodha, *Bethu Brigte* 2.58–65 § 9; cf. *CIH* i 166.34; vi 2139.17.
[117] Dillon, 'The story of the finding of Cashel', 66.71 § 4.
[118] *IP* 176 §§ 6–7.

fattened on corn.[119] The same commentary refers also to the fattening of pigs on milk,[120] and the fabulous pig of *Scéla Mucce Meic Dathó* is said to have been reared for seven years on the milk of sixty cows.[121]

When kept in a farmyard, pigs are fed from a wooden trough (*oircél*);[122] hence a boar specially fattened up for cooking is called a *torc oircéla*.[123] Pigs also eat grass and other leaves, and are very fond of roots for which they plough up the ground with their teeth. For this reason, the Old Irish law-texts regarded trespass by pigs as a more serious offence than that of other livestock (see p. 142). The root of the dandelion (*Taraxacum officinale*) seems to have been especially relished by pigs, and was consequently called *serbán mucc*, i.e. 'the bitter one of pigs'.[124] In *Hisperica Famina* there is a reference to pigs eating the roots of ferns.[125]

A food frequently associated with pigs in our sources is the acorn crop (*daurmess* 'oak-fruit', often simply referred to as *mess*).[126] This crop comes in September and October and serves to fatten up the young pigs for immediate killing or to provide them with reserves to survive the hungry days of the coming winter.[127] Pigs and acorns were so closely linked in the early Irish mind that the phrase *mucc remi-thuit mess* 'a pig which dies before the acorn-crop' is used as a proverbial expression to describe a lost opportunity.[128] A legal commentator claims that a single oak can provide enough acorns to fatten one pig in a good year, but also points out that there are

[119] *CIH* ii 509.13 = *AL* ii 368.3 *mucc . . . mēthtar for arbar.* Aristotle (*Historia animalium* (ed. Thompson), 8.6) refers to the fattening of pigs on barley, millet, acorns, and fruit. He also mentions their fondness for roots.

[120] *CIH* ii 509.14 = *AL* ii 368.4 *mucc mēthar for lacht.* Cf. *CIH* iii 914.23 *torc . . . for ass.*

[121] Thurneysen, *Scēla Mucce Meic Dathó*, 6.12–13 § 5.

[122] E.g. *CIH* i 278.24; iii 943.30 = *AL* iii 242.13. In the accompanying commentary, *oircél* is explained by the term *omar* (*ammar*). The word *oircél* is also used of the similarly shaped chute of a mill, e.g. *CIH* ii 374.20 = *AL* i 124.18.

[123] *CIH* ii 367.33 = *AL* i 122.6.

[124] E.g. Stokes, '*Materia medica*', 230 § 26.

[125] Herren, *Hisperica Famina*, 76.165.

[126] *Mess* is cognate with Welsh *mes*, Breton *mez* 'acorns' (*LEIA* M-43).

[127] Note the distinction in legal commentary at *CIH* vi 2096.20–4 between the fat pig (*muc mēth*) and the lean pig (*muc cael*). Cf. *LL* v 1121.32940–60 = Windisch, '*De chophur in dá muccida*', 235.19–236.45 = Roider, *De chophur in dá muccida*, 26.12–32.41.

[128] E.g. Knott, *Togail Bruidne Da Derga*, 18.611; 34.1137–8.

years in which the acorn-crop fails completely.[129] The annals sometimes record particularly good years for acorns.[130] The *Annals of Clonmacnoise* state that in the year 1038 acorns were so abundant that even the runts of the litters were fattened.[131]

A law-text on distraint distinguishes pigs feeding on acorns (*muca for mesruth*) from another category described as *muca denma*.[132] The meaning of the latter term is uncertain: a glossator suggests that it refers to pigs fattening on any food other than acorns.[133] It is illegal to distrain either category of pig.

Mention should also be made of the account in the comic tale *Fled Bricrenn* of the diet of a boar specially reared for a great feast.[134] The quality of the foodstuffs has obviously been much exaggerated for dramatic effect, but the passage at least gives us some idea of the seasonal variation in a pig's diet. In spring this special boar gets porridge made with fresh milk (*littiu lemnachta*) and gruel (*menadach*), in summer curds (*croth*) and fresh milk, in autumn nut-kernels[135] and wheat, and in winter flesh and stewed meat (*enbruithe*).

Meat

The flesh of the pig was particularly appreciated by the early Irish: a ninth-century triad includes the death of a fat pig (*bás muicce méithe*) among the 'three deaths which are better than life'.[136] Another triad – in the early eighth-century law-text *Críth Gablach* – refers to the boar 'which removes dishonour at every season' i.e. provides a feast for high-ranking guests whenever they visit.[137] The same idea is expressed in the seventh-century wisdom text *Audacht Morainn*, where the pig's fat side is described as 'the freeing from shame of every face'.[138]

129 *CIH* vi 2183.19–21.

130 *AU*^{2} 222 s.a. 768 (recte 769) § 6 *Habundantia dairmesa*; *ALC* i 404.28 s.a. 1254 *imad dairmhessa.*

131 *Ann. Clon.* 176 s.a. 1038. Conell Mageoghagan's 1627 translation of the lost Irish original has 'The[re] was such an abundance of ackorns this yeare that it fattened the pigges of pigges'.

132 *CIH* iii 897.20 = Appendix A, text 2 § 1 (23)–(24).

133 *CIH* ii 715.10–11.

134 *LU* 248.8118–22 = Windisch, 'Das Fest des Bricriu', 256.24–8 § 9.

135 Compare the description of a pig as *siúr na gcnó gcuill* 'sister of the hazel-nuts' in a lament for a pig (Quin, '*Truagh truagh an mhuc*', 30 § 9).

136 Meyer, *Triads*, 12 § 92.

137 *CIH* ii 563.27–8; iii 780.6–7 = *CG* 7.185–6.

138 Kelly, *Audacht Morainn*, 12 § 45.

Pigs were sometimes killed as sucklings (*comlachtaid*) for immediate consumption.[139] More commonly, however, they were well fattened up before slaughter. Thus the *lupait*[140] – apparently a female pig of six to eight months – is said to have been regularly killed at Martinmas, i.e. November 11th.[141] It is defined in an early legal gloss as a *banb samna* 'young pig of November'.[142]

Many pigs were not slaughtered in their first autumn or winter. In his study of animal bones on the early Christian site at Moynagh crannóg, Finbar McCormick has found that sixty-seven per cent of pig-bones came from animals slaughtered between their second August and the following spring.[143] This finding tallies with the evidence of the law-texts, as there are many references to the slaughter of older pigs. For example, *Críth Gablach* states that a prosperous farmer of *bóaire* rank should have a bronze cauldron large enough to cook a boar (*torc*).[144] This *bóaire* is expected to own four *tuirc forais* 'household boars' (as well as two sows and twenty other pigs).[145] These boars were presumably barrows (castrated males) reared for killing in their second year. Another law-text includes the belly of a pen-reared(?) boar (*tarr tuirc airrbe*) in the food-rent paid by a client to his lord.[146]

There is evidence too of the killing of sows after they had produced two or three litters.[147] *Cáin Aicillne* refers to food-rents which include a flitch of bacon of fixed dimensions from a *deiling* or *trichem*.[148] These terms are taken by legal glossators to refer to sows of two and three litters respectively.[149] The glossators assume that the *deiling* is two years old, and the *trichem* three years old. A

[139] Hencken, 'Ballinderry crannóg no. 2', 71, reports the bones of suckling pigs.

[140] The etymology of this word is unknown, but the final *-it* may be a feminine suffix, cf. Thurneysen, 'Altirisch *canait*'.

[141] *CIH* iv 1513.41 = O'Dav. 407 § 1194.

[142] *CIH* iii 918.38; cf. *CIH* iv 1504.26 = O'Dav. 366 § 978.

[143] McCormick, 'Stock-rearing in early Christian Ireland', 104 (and table 3.9).

[144] *CIH* ii 564.7; iii 780.14 = *CG* 8.197.

[145] *CIH* ii 564.6; iii 780.13 = *CG* 8.195–6.

[146] *CIH* iii 920.20. *DIL* A 166.60 is probably correct to take *airrbe* here to be for *airbe* 'fence, pen'.

[147] Today sows are normally killed at four or five years of age.

[148] *CIH* ii 482.7; 482.25 = Thurneysen, 'Aus dem irischen Recht I', 351 § 9; 353 § 10; cf. Thurneysen, 'Aus dem irischen Recht IV', 208.

[149] E.g. *CIH* ii 482.17–18, discussed by Thurneysen, 'Aus dem irischen Recht I', 352; cf. *CIH* iv 1492.27–8 = O'Dav. 311 § 706. There is a reference to a male *trichem* (*torc trichem* = *torc-t*[*h*]*richem*?) at Bergin and Best, '*Tochmarc Étaíne*', 176 § 4.

barren or milkless sow is only good for the pot: a verse preserved in a legal gloss refers to the cooking of milkless sows (*feis cen lacht*).[150]

Finbar McCormick notes that in cases where it was possible to determine, approximately half the pig-scapulas at Moynagh crannóg were holed.[151] This shows that these pigs must have hung up for curing with salt and/or smoke. The written sources contain many references to salted pork (see p. 336).

Lard

The lard (*blonac*[152]) of the pig is much more palatable than that of cattle or sheep, and there are some references to its consumption. For instance, the author of *Aislinge Meic Con Glinne* refers enthusiastically to the 'rich juicy lard of a well-fed choice boar'.[153] It is clear from a verse preserved in legal commentary that lard might be eaten with bread:

> Give the bread,
> and a piece of the big lard;
> good were your mother and father,
> give the buttermilk after it.[154]

Pig-values

In general, values in early Irish society were not expressed in terms of pigs, nor were these animals used as a normal medium of currency, as were cattle and sheep (see Appendix B, p. 587).[155]

In the main part of the legal commentary on the values of domestic animals, it is simply stated that a pig is of the same value (*comlóg*) as a sheep of the same age and sex.[156] But an additional paragraph gives a somewhat different arrangement.[157] It assigns a value of one

[150] *CIH* ii 375.12 = *AL* i 140.37; *CIH* iv 1506.36 = O'Dav. 376 § 1026; Calder, *Auraicept*, 241.4543.
[151] McCormick, 'Stock-rearing in early Christian Ireland', 109.
[152] Pedersen took this word to be a borrowing from Welsh *bloneg* of the same meaning (*VKG* i 24; *LEIA* B-60 s.v. *blonac*).
[153] Jackson, *Aislinge*, 34.1049 (and note on p. 66) = Meyer, *Aislinge*, 87.15.
[154] *CIH* v 1605.26–8 = *AL* v 64.25–8.
[155] However, the male piglet (*torcbanb*) is used as a unit of value at *CIH* iii 1033.27–31 (Thurneysen, *Cóic Conara Fugill*, 41 § 61), as are a young sow (*céis*) and a piglet (*banb*) at *CIH* vi 2215.30–31 (trans. L. Breatnach, *Uraicecht na Ríar*, 31).
[156] *CIH* iii 846.8 = Appendix A, text 3 § 10; cf. *CIH* vi 2106.33.
[157] *CIH* iii 846.12–14 = Appendix A, text 3 § 12.

penny[158] – i.e. one third of a scruple – to the male or female piglet (*lúrcc*)[159] from birth until May. From May until August it is called a *banb*,[160] and is valued at two pence. In November its value increases to one scruple,[161] and in May of the next year this goes up to four and a half pence. In the following spring it attains its maximum value of three scruples.[162] The same sum is given as the value of a mature sow in another commentary.[163] However, the Old Irish law-text *Bretha Nemed Toísech* provides evidence of slightly higher pig-values. The twenty-first part (i.e. *ind érca*) of the honour-price of a *cano* poet is given as a young sow (*céis*).[164] As the honour-price of this grade of poet is seven *séts*, the value of the *céis* can be taken as four scruples.[165] Another passage in *Bretha Nemed Toísech* also indicates a higher rate of value: the render (*frithgnam*) which a client pays to his lord includes a pig (*mucc*) worth three scruples.[166] No information is given as to the age of the pig, but the evidence of another text on clientship, *Cáin Aicillne*, suggests that it is a young pig which has not yet reached its maximum value. According to this text the client's render includes a pig which is nine fists in length (*náe nduirn a fot*).[167] It is given to the lord in winter, presumably for immediate consumption.[168]

As in the case of other domestic animals (see p. 76 above), there is an important distinction between a pig's ordinary commercial value (*lóg*) and its *díre*, i.e. the fine which must be paid for killing it illegally. For example, *Bretha Nemed Toísech* states that five *séts* must be paid as the *díre* of a young sow (*céis*)[169] or a young boar

[158]Cf. *CIH* iv 1208.8 *pinginn ar gach n-orc* 'a penny for every piglet'.

[159]*Lúrcc* is from *lú* + *orcc* 'small piglet', as at *CIH* iii 1109.22 *lulaig* no *luuan* no *luorc* 'a small calf or a small lamb or a small piglet', and *CIH* iii 924.25 = *BB* 86 § 53e *luain 7 luoirc 7 luluígh 7 mendain* 'small lambs and small piglets and small calves and kids'.

[160]The term *banb(h)* survives in Hiberno-English as *bonav, bannoo, bonham*, etc. It is cognate with Welsh *banw* 'piglet'.

[161]According to commentary at *CIH* iii 816.36, a *deil* (= *deiling*?) is worth a scruple. See Thurneysen, 'Aus dem irischen Recht IV', 208.

[162]Contrast the commentary at *CIH* iii 816.38, which gives the *trechem* (*trichem*) 'three-year-old sow' a value of only 2 scruples.

[163]*CIH* iv 1208.9. The sow is valued at 9 pence = 3 scruples.

[164]*CIH* vi 2215.29–30 = L. Breatnach, *Uraicecht na Ríar*, 29.33–7.

[165]See L. Breatnach, *Uraicecht na Ríar*, 32 (note to ll. 32–44); 184 (table 9).

[166]*CIH* vi 2230.19–20.

[167]The length of the *dorn* is probably four inches, see p. 564.

[168]*CIH* ii 483.35–6 = Thurneysen, 'Aus dem irischen Recht I', 356.11.

[169]Commentary at *CIH* i 191.15–17; v 1856.38–40 distinguishes three categories of pig: *mucc mór* 'full-grown pig', *céis* 'young sow', *banb* 'piglet'.

(*céistorc*).[170] This is fifteen times the value assigned to it in the same text. In addition, the culprit would have to give restitution (*aithgein*), either by replacing the pig or by paying its commercial value.

Horses

The domestication of the wild horse (*Equus ferus*) was accomplished much later than that of the four animals already discussed. It is probable that horses were first brought under human control as a food source in the plains of what is now the Ukraine around 4000 BC.[171] At an early stage in the domestication of the horse, its potential as a means of rapid transport was realised: abnormal wear on the premolars of some of the Ukrainian horse-skulls indicates the use of bit and bridle. The archaeological evidence for the domesticated horse (*Equus caballus*) in Ireland goes back to about 2000 BC, but its bones are never found in abundance.[172] Crannógs (lake-dwellings) have proved the richest source of animal-bones in excavations from the early Christian period: it can be assumed that the vast majority of these bones are from animals which were eaten by the occupants of the crannógs. Horse-bones are found on all such sites, but in fairly low percentages. For example, of the 7278 fragments of animal-bones found in the Moynagh crannóg from about AD 800, only 85 (i.e. 1.2 per cent) belong to the horse.[173] The religious taboo on the eating of horseflesh (see p. 352) may be partly responsible for the low proportion of horse-bones.

The evidence of the Old Irish law-texts fits in with the findings of archaeology. Thus the text on status, *Críth Gablach*, lists the approximate numbers of livestock (excluding young animals) which would be expected on the farms of various grades of freeman. The most prosperous type of *bóaire* is said to own twenty cows, two bulls and six oxen, but only two horses – one for riding

[170] *CIH* vi 2228.3.

[171] Clutton-Brock, *Horse power*, 55; Gautier, *La domestication*, 150–1.

[172] Van Wijngaarden-Bakker, 'The animal remains from Newgrange (1)', 346; Clutton-Brock, *Horse power*, 58. The existence in the early Christian period of feral horses in Ireland is suggested by a literary reference to a *fíadgraig* 'herd of wild horses' (Ó Cathasaigh, *Heroic biography*, 121.73).

[173] McCormick, 'Stock-rearing in early Christian Ireland', 5 (table 1.1).

and one for farmwork.[174] Throughout the written sources, the horse is associated in particular with men of high rank, and much stress is laid on their possession of ornate riding equipment. The lowest grade of lord is expected to own a riding horse (*ech slíasta*) with a silver bridle (*srían arggait*), as well as four other horses with unornamented bridles.[175] A higher grade of lord has a bridle of silver and a bridle of gold.[176] Because of their high value, such bridles are often given as pledges.[177]

Our sources concentrate to a large extent on the rôle of the horse in racing and warfare. Horse-racing was clearly a regular entertainment for kings (see p. 99 below), and there are many references to the warrior's chariot (*carpat*) pulled by two horses.[178] The sagas contain numerous descriptions of such horses, mostly couched in flowery rhetorical language. The most famed horses of Irish literary tradition were *Líath Machae* 'the Grey of Machae' and *Dub Saiglenn* 'the Black of Saigliu'. They were twins of about the same age as the hero Cú Chulainn, and grew up along with him.[179] Driven by the charioteer Lóeg, they pulled Cú Chulainn's chariot. In *Aided Chon Culainn,* the tale of the death of Cú Chulainn, we are told how both horses were injured during the fatal attack on their master. After Cú Chulainn's death, *Dub Saiglenn* rode away with Lóeg, but *Líath Machae* stayed by the body for three days and three nights, allowing no one to approach. Finally, *Líath Machae* jumped into a lake and drowned.[180]

The horse also has a special place in the legal system, in that a man who lays claim to land occupied by another must formally enter this land accompanied by a fixed number of horses.[181]

Types and breeds

Two main types of horse are distinguished in our sources. As we have seen, the author of *Críth Gablach* expects a farmer of

174 *CIH* ii 564.6; iii 780.13 = *CG* 8.196. The text at this point actually mentions only the *ech slíasta* 'horse of thigh' i.e. riding horse. However, the lower grade of *bóaire* has both a riding horse and a work-horse, *CIH* ii 563.9; iii 779.29 = *CG* 6.158.

175 *CIH* ii 566.25–6; iii 782.31–2 = *CG* 14.346 *co nglassrianaib* lit. 'with grey bridles'.

176 *CIH* ii 567.25; iii 783.32–3 = *CG* 16.407.

177 E.g. *CIH* ii 475.4 = *AL* v 414.24–5.

178 See Greene, 'The chariot as described in Irish literature', 63.

179 Van Hamel, *Compert Chon Culainn,* 4 § 3; *LU* 321.10587–90.

180 Van Hamel, *Compert Chon Culainn,* 117 § 46.

181 See p. 432. For a more detailed discussion of legal entry (*tellach*), see *GEIL* 186–8.

bóaire rank to own a work-pony (*capall fognamo*) and a horse for riding (*ech immrimme*).[182] Some six centuries later, the *Annals of Connacht*[183] make a similar distinction between the work-pony (*capall*) and the more prestigious horse for riding (*ech*).[184] The early Irish work-pony was small and sturdy, similar in build to the modern Connemara pony.[185] In archaeological writings such ponies are often described as being of 'Celtic' type because of their particular association with the early Celtic-speaking peoples of the continent.[186] In Old Irish sources, there are a number of references which suggest that imported stock was used to improve the speed and size of the early Irish horse. For example, the law-texts assign especial value to the horse from overseas (*ech allmuir*).[187] Britain is often given as the source of this type of horse, which is consequently termed *ech bretnach* 'British horse'.[188] It seems likely that such horses derive from stock brought to Britain during the period of Roman occupation. Juliet Clutton-Brock points out that bones of the large Roman military horse – ultimately, it is thought, of Scythian origin – have been found in archaeological excavations all over the Roman world, including Britain.[189] One might expect it to be more economic to import stallions rather than mares to improve the quality of Irish horses in the early Christian period. However, Irish mares at this

[182] *CIH* ii 563.9; iii 779.29 = *CG* 6.158

[183] *AC* 276 s.a. 1336 § 8.

[184] It should be stressed, however, that *ech* is also used as a generic term to include all kinds of horses. It is cognate with the words for 'horse' in many Indo-European languages, e.g. Latin *equus*, Sanskrit *aśva-*, etc. In one law-text (*CIH* ii 422.35; iv 1455.24; vi 1958.41 = *AL* i 268.20) a distinction is made between *ech* and *marc* (cognate with Welsh *march* 'horse'). In legal glosses, *ech* is taken to refer to the male, and *marc* to the female (*CIH* ii 422.36; iii 1079.42–3; v 1722.32 = *AL* i 304.y). But elsewhere *marc* is used as a non-specific poetic term for 'horse': see P. Kelly, 'Dialekte im Altirischen?'. Its derivative *marcach* is well-attested in the meaning 'rider', and is not confined to poetic language. Another word for 'horse', used mainly in poetry and glossaries, is *pell* or *fell*.

[185] Hencken, 'Ballinderry crannóg no. 1', 233.

[186] Clutton-Brock, *Domesticated mammals*, 86. The Latin word for work-horse *caballus* (which ultimately gave Italian *caballo*, French *cheval*, etc.) is almost certainly a borrowing from Celtic, and cognate with Irish *capall*, Welsh *ceffyl*. See *LEIA* C-33 s.v. *capall*.

[187] E.g. *CIH* vi 2312.31 = Binchy, '*Bretha Déin Chécht*', 40 § 30; *CIH* iii 897.18 = Appendix A, text 2 § 1 (18).

[188] E.g. *CIH* iv 1484.30 = O'Dav. 277 § 501; Jackson, *Aislinge*, 43.1328 = Meyer, *Aislinge*, 111.14; AU^2 466 s.a. 1029 § 6.

[189] Clutton-Brock, *Domesticated mammals*, 88.

period may have been too small to bear the larger foals of foreign stallions. Certainly, the importation of foreign mares is indicated by the use of the term *gaillit*, a derivative of *gall* 'Gaulish, foreign', to refer to a good quality mare.[190] Glossators take such mares to be of British origin,[191] and in one glossary a male foal from a *gaillit* is termed a *gaillire*.[192]

In our sources, the small Irish work-pony is often called the *gerrán*.[193]

It is regularly identified as a pack-animal. For example, in a version of *Compert Mongáin*, a *gerrán* has a pack-saddle (*srathar*) on its back bringing corn and flour from a mill,[194] and in an entry in *O'Mulconry's Glossary* it is equated with the pack-horse (*suma*) which carries loads of wheat.[195] The term *gerrán* passed into the English of Ireland in post-Norman times in the form *garran*, and is frequent in official documents, travellers' records, etc.[196] In Scotland the word is now applied to the Garron, the mainland type of Highland pony (as distinct from the lighter Western Isles pony).

Colour

In a discussion on the value of domestic animals and other goods, the law-text *Bretha Nemed Déidenach* states that they should be judged for their size, shape, colour and speed.[197] The most prestigious

[190] E.g. *CIH* iii 918.10 (*giallit* of the MS must be emended to *gaillit*); *CIH* iv 1507.34–6 = O'Dav. 380 § 1051. The fem. suffix *-it* is also found in *birit* 'sow', cf. Thurneysen, 'Altirisch *canait*'.

[191] E.g. *CIH* i 26.39 = *AL* v 220.22.

[192] *CIH* iv 1507.36 = O'Dav. 308 § 1051. For the suffix *-(a)ire*, see Thurneysen, *Grammar*, 172 § 269.

[193] *Gerrán* may simply be a derivative of the adjective *gerr* 'short, low-sized'; cf. Welsh *gerran* 'dwarf'. There is also a possibility that *gerrán* means 'that which has been castrated' (from *gerraid* 'cuts, castrates'), and was originally applied to geldings; cf. *meile* 'gelding', perhaps from **mel-* 'to crush' (*LEIA* M-29). For the agent-suffix *-án*, see Kelly, 'An Old-Irish text on court-procedure', 103 § 7(h). Like other agent-suffixes, *-án* can be used with a passive sense; cf. *crochaire* (a) 'hangman', (b) 'one who is hanged'.

[194] Meyer, *The Voyage of Bran* i (Appendix) 68.1 *gerrán bān maircech ⁊ sensrathar fair* 'a white galled work-pony with an old pack-saddle on it'. Cf. O'Grady, *Silva Gadelica* i 234.8–10 (trans. ii 265–6) = *LU* 95.2938–40.

[195] O'Mulc. 273 § 856 *Suma .i. gerrān.* The word *suma(d)* is from Late Latin *suma*, *sauma* 'load, pack, beast of burden': see *LEIA* S-204 s.v. *suma.*

[196] E.g. Hardiman, *Ancient Irish deeds*, 51. See *OED* s.v. *garron.*

[197] *CIH* iii 1131.6 = E. J. Gwynn, 'Privileges', 41.32–3 *Mesir eocha ar a méd sgéo crotha, dathaibh sgēo rethaibh* 'may you judge horses by their size and shape, by their colours and runnings'.

colour for a horse seems to have been white, and the poetic term *gabor* is used especially of a white or partly white horse.[198] There are also references to horses which are black (*dub*), grey (*líath*), dark grey (*dubglas*), dun (*odar*), brown (*donn*), and yellow (*buide*). Colour combinations include roan (*crón*) with a white head,[199] and grey (*glas*) with speckled mane.[200]

Qualities

An Old Irish legal passage gives some information on the qualities which the buyer of a horse should look for.[201] In physique the horse should be large and broad-chested, but not too tall. It should hold its head high, and not have too large a mouth. In temperament it should be mettlesome, but also easy to handle, and not given to kicking or jumping around. It should be healthy, with no trace of lameness, glanders, or shivering. Similar descriptions of chariot horses are found in the sagas.[202] As one would expect, the names given to outstanding horses often celebrate their speed, e.g. *Grip* 'swift',[203] *Gáeth* 'wind', *Athach* 'gust', *Side* 'whirlwind'.[204]

Care of mares and foals

The need for careful treatment of a mare after foaling (*láir íar trog*) is taken into account in a legal passage on distraint. Here it is stated that no mare may be brought away on distraint until nine days after giving birth.[205] A quotation from a lost law-text of the *Senchas Már* collection emphasizes that a foal (*lurchaire*)[206] should not be separated from its mother.[207] Calves, by contrast, are generally kept from their mothers so that most of the milk can be taken for human consumption (see p. 38). Though mares are milked in many parts of the world, the early Irish records refer only to the milking of cows, goats and sheep.

[198] Meyer, *Sanas Cormaic*, 55 § 675.
[199] *LU* 261.8625–6 = Windisch, 'Das Fest des Bricriu', 277.10 § 47.
[200] Dillon, *Serglige Con Culainn*, 17.492.
[201] *CIH* iv 1235.19–27 = Appendix A, text 7.
[202] E.g. *LU* 260.8590–4 = Windisch, 'Das Fest des Bricriu', 276.1–5 § 45; *LU* 279.9254–60.
[203] *LL* v 1214.36042; Binchy, *Scéla Cano*, 9.238.
[204] *LL* i 124.3910–11; Thurneysen, 'Mittelirische Verslehren', 58 § 111.
[205] *CIH* ii 715.5–7.
[206] *Lurán*, a diminutive of *lurchaire*, is also well-attested: see *DIL* s.v. *lurán*. For the formation, compare *doburchú* 'otter' with diminutive *dobrán*.
[207] *CIH* iv 1513.28–9 = O'Dav. 406 § 1189.

Housing and restraint

Our sources stress that unsupervised horses should be kept in confinement or under restraint so that they cannot wander away or cause damage to crops. There are also numerous references to the horse-thief (*echtháid*), who seems to have been regarded with particular abhorrence in our texts – perhaps because of the special association of the horse with the higher-ranking members of society. According to the main law-text on farming, *Bretha Comaithchesa*, horses should be properly tied up at night or else confined in a stable (*inne*).[208] An entry in the *Annals of Inisfallen* for the year 1068 similarly implies that it was normal for horses to be housed at night for their own safety.[209] Legal glossators also refer to confinement in a shelter in the corner of a field (*bacc n-achaid*).[210] When grazing – particularly in unfenced pasture – a horse should be under some form of restraint (*cuimrech*). According to a glossator this normally consisted of a tie between the horse's head and its legs known as a *langpheitir*.[211] It is illegal to put an excessively severe restraint on a horse belonging to somebody else.[212]

Training

The various tasks which man has assigned to the horse are of course highly unnatural to it. Much effort must therefore be expended in teaching the young horse to submit to pulling a cart or carrying a pack or rider. The basic means of controlling a horse is by persuading it to keep a bit (*glomar*) in its mouth.[213] Attached to the rings of the bit is the bridle (*srían*), which fits around the horse's muzzle. In a list of domestic tools and accoutrements in a law-text, the *srían* is distinguished from the *all*.[214] It seems from the accompanying glosses that the former is a smaller bridle designed for a pack-horse or riding horse, whereas the latter is for a chariot-horse.[215] The same list also includes the halter (*adastar*), which is stated in a legal gloss to be made of

[208] *CIH* i 72.12–13 = *AL* iv 96.25–6; cf. *CIH* i 6.1 = *AL* v 136.9–10.

[209] *AI* 224 s.a. 1068 § 4.

[210] E.g. *CIH* iv 1518.20 = O'Dav. 426 § 1300; *CIH* ii 715.14.

[211] *CIH* i 240.10 = *AL* v 478.17–18. The word *langpheitir* is probably an early loan from English: see p. 494 below.

[212] *CIH* i 239.36 = *AL* v 476.23 *sārcuibreach for eocha* 'excessive binding on horses'.

[213] At *CIH* v 1537.25, the *díre* ('penalty-fine') for a bit is given as 5 *séts*.

[214] *CIH* ii 373.28; v 1901.26 = *AL* i 124.14.

[215] *CIH* ii 373.30–1 = *AL* i 138.35–6. Cf. Greene, 'The chariot as described in Irish literature', 69 § 9.

leather.[216] The training of a horse normally starts in its third year. In one law-text untrained horses (*anindle ech*) are described by the glossator as 'stallions and colts still with the mares'.[217] Young stallions not required for breeding are castrated so as to make them more docile. According to a legal passage on distraint, a recently castrated gelding cannot be removed from its owner until twenty-seven days after the operation.[218]

Load-carrying

The simplest task required of a horse is that of carrying a load (*aire*) on its back.[219] The evidence of our texts indicates that the work-horse (*capall fognamo*) spent much of its life carrying loads, and therefore filled a similar rôle to that of the donkey in nineteenth-century Ireland. These loads might be balanced on a pack-saddle (*srathar*),[220] which was secured onto the horse's back. Only a trained animal could be employed for this purpose. Hence the phrase 'a pack-saddle on a colt' (*srathar for serrach*) is used as a proverbial expression for an impossible undertaking.[221] *Bethu Phátraic* refers to a load (*marclach*) carried by a work-pony (*gerrán*).[222] This load is said to have consisted of two bags (*bolcc*) of wheat, which were presumably balanced on either side of the horse's back.

A legal quotation dating from about the eighth century states that a horse should not carry a full load on boggy or rocky terrain.[223] As the horseshoe had not been introduced to Ireland at this period, the hoofs of a heavily laden horse would be particularly liable to injury on sharp rocks. There is evidence that the Romans used both nailed iron horseshoes and hipposandals tied with leather straps.[224]

[216] *CIH* vi 2184.12–14. Here the glossator equates the word *adastar* with *bráigdech* 'halter', a derivative of *brágae* 'neck'.

[217] *CIH* ii 471.15–17 *cullaigh ⁊ searraig lāra.*

[218] *CIH* ii 715.7–8.

[219] E.g. *LL* v 1247.36986 *aire .uii. n-ech do braich ⁊ biud* 'the load of seven horses of malt and food'.

[220] E.g. *CIH* vi 2137.5; *Thes.* ii 290.14. The word *srathar* is a loan from Latin *stratura* 'pack-saddle', which also gave Welsh *ystrodur* of the same meaning. See *LEIA* S-182 s.v. *srathar.* What is possibly part of a pack-saddle from the early Christian period was found at Ballinderry (Hencken, 'Ballinderry crannóg no. 1', 133, 137–8).

[221] Meyer, *Aislinge*, 125.19 (H 3. 18 version).

[222] Mulchrone, *Bethu Phátraic*, 142.2840–7. *Marclach* is a derivative of *marc* 'horse': see *LEIA* M-20 s.v. *marc.*

[223] *CIH* iii 759.39; iv 1290.24.

[224] Langdon, *Horses, oxen and technological innovation*, 10.

However, the horseshoe seems to have gone out of use throughout Europe at the end of the Roman era, and does not reappear until the late ninth century.[225] The horseshoes found in excavations at Dublin have been dated to the eleventh and twelfth centuries.[226]

Draught

Early Irish sources contain more references to the use of horses for draught than for riding. The sagas in particular provide many descriptions of the horses which pull the warrior's chariot (*carpat*).[227] There is a close association between ownership of a chariot and high social status; thus the law-text *Bretha Crólige* refers to the person who is *cairptech*, i.e. of chariot-owning rank.[228] It would seem from the detailed lists of possessions in *Críth Gablach* that even the most prosperous grade of commoner is not expected to own a chariot.[229] In early Christian Ireland, pulling the plough was a task for oxen rather than horses.[230] It was not until the late thirteenth century that Ireland saw the introduction of the great plough-horse which had been developed – originally for military purposes – on the Continent.[231] As a plough-horse is faster, cheaper, and more enduring than an ox, its use gradually spread throughout western Europe.[232] References in Giolla na Naomh mac Duinn Shléibhe Mhic Aodhagáin's fourteenth-century legal manual suggest that both oxen and horses were used for

[225] L. White, *Medieval technology*, 58–9; Clutton-Brock, *Horse power*, 73.

[226] Kavanagh, 'The horse in Viking Ireland', 115–17. Greene, '*Cró, crú* and similar words', 2, points out that the basic meaning of OIr. *crue* is 'hoof', and that 'horseshoe' is a later development. In *DIL* s.v. 1 *cró* (h) the secondary meaning 'horseshoe' is erroneously placed first.

[227] See p. 92 above.

[228] *CIH* vi 2302.10 = Binchy, '*Bretha Crólige*', 46 § 58.

[229] *CIH* ii 563.18–564.43; iii 779.37–781.8 = *CG* 7–10 §§ 14–18.

[230] Confusion has been caused by the mistranslation of *ōin eich agar for blā* at *AL* v 279.6–7 (= *CIH* i 44.1–2; vi 1993.21; 2247.15). It means 'the loan of a horse which is driven on a race-course' and not 'the loan of a horse to plough upon land'.

[231] Lucas, 'Irish ploughing practices (2)', 67; Davis, *The medieval war-horse*, 21.

[232] L. White, *Medieval technology*, 62; Langdon, *Horses, oxen and technological innovation*, 61, 164.

ploughing in Ireland at this time,[233] but by the end of the fifteenth century it seems that oxen had been totally superseded.[234]

In the early period, work-horses pulled the harrow (p. 478) and the farm-cart (p. 498). An eighth-century law-text mentions a 'horse of cart and load' (*capall cairr 7 airi*).[235]

Riding

According to *Críth Gablach*, persons of the rank of *bóaire febsa* and above are expected to own a horse specifically for riding (*ech immrimme*),[236] and the Middle Irish commentary on the text on fosterage states that the son of a lord or king should be taught riding (*marcaigecht*).[237] A gloss on the same text emphasizes that a fosterfather is not obliged to teach riding to the sons of commoners.[238] References to riding by women are rare in our sources. The clearest case is in the eighth-century voyage-tale *Immram Curaig Maíle Dúin.*[239] Here the voyagers are met by a female rider seated on an ornamented horse-cloth (*echdíllat*) on a prize horse (*ech búada*). But this incident took place on a mysterious island ruled by a widowed queen with seventeen daughters. The author may have intended to draw attention to the unusual nature of the island by introducing a female rider. Generally, high-ranking women in early Irish literature are represented as travelling in chariots.[240] The law-texts stress the importance of the horse as a means of relaying warnings or other messages. It is thus illegal to distrain a 'horse of warning' (*ech robaid*) or 'the horse of a man who strives for the honour

[233] *CIH* ii 698.3–4. It is noteworthy that there is a reference to 'fifty trained horses for ploughing and tilling the earth' (*cæoca each riata . . . fri har 7 fri trebad in talman*) in commentary on *Félire Óengusso* (Stokes, *Félire*, 88.24). It seems surprising that as late as the thirteenth century, the Welsh law-text *Llyfr Iorwerth* forbids the use of either male or female horses in ploughing (Wiliam, *Llyfr Iorwerth*, 98 § 152.12–15 = *ALIW*, Venedotian Code, III xxiv § 23).

[234] Lucas, 'Irish ploughing practices (2)', 68.

[235] *CIH* ii 471.18 = *AL* v 406.1; cf. *CIH* i 106.19; vi 1977.41.

[236] *CIH* ii 563.9; iii 779.29 = *CG* 6.158. The lower-ranking *ócaire* has a horse which is for both riding and work (*CIH* iii 778.24–5 = *CG* 4.91).

[237] *CIH* v 1760.33; 1761.5 = *AL* ii 154.6; 156.1–2.

[238] *CIH* v 1762.22–3 = *AL* ii 160.21–2.

[239] Van Hamel, *Immrama*, 43.606–7 = Stokes, 'The voyage of Máel Dúin, (2)', 62 § xxviii.

[240] E.g. *TBC* Rec. I 2.39. The female messenger (*baineachlach*) in the Middle Irish tale *Cath Bóinde* seems to have travelled on horseback (J. O'Neill, '*Cath Boinde*', 178.20).

of the territory'.[241] A king is expected to have in his service a professional messenger (*techt* or *techtaire*). He is given a fairly high status, with an honour-price half that of his employer.[242] Because the messenger's tasks are often urgent or dangerous, he is not held liable for the injury or loss of a horse entrusted to him on an errand.[243] In non-legal sources there are also references to a courier of lower rank, known as an *echlach*.[244] In the law-texts this term is only used in the secondary meaning 'prostitute'.[245]

Early Irish sources provide little evidence of the use of mounted warriors. In the method of warfare described in the Old Irish sagas, the combatants come to battle in horse-drawn chariots, but then dismount and fight on foot. One version of *Togail Bruidne Da Derga* describes the three horsemen (*marcaig*) of King Conaire, but gives no indication that these men have a military function.[246] On the other hand, the Market Cross at Kells, Co. Meath, dating from about the middle of the ninth century, shows four horsemen clearly armed with swords and shields.[247] Another cross from around the same period (at Bealin, Co. Westmeath) shows a huntsman riding a horse as his dog catches a stag.[248] The high cross from Banagher, Co. Offaly, contains a detailed carving in relief of a horse and rider (reproduced on p. 98).[249] The rider has long hair or some form of head-dress extending down to his shoulders and is apparently carrying a crozier. The horse is nicely proportioned, and has a long well-groomed mane and tail. The bridle and reins are clearly delineated, but there seems to be no saddle, and there are certainly no stirrups, spurs or girth-strap. This horse exemplifies very well the riding techniques and equipment of pre-Norman Ireland.

[241] *CIH* iii 897.17–18 = Appendix A, text 2 § 1 (16)–(17).
[242] *CIH* ii 568.31–2 = *CG* 19.484.
[243] *CIH* i 19.29–20.3 = *AL* v 190.17–18.
[244] E.g. Meyer, *Fianaigecht*, 78 § 25ff.
[245] E.g. *CIH* i 42.14 = *AL* v 272.4 *echlach oides a corp do cāch* 'a prostitute who lends her body to everyone'; cf. *CIH* i 221.18 = *AL* iv 56.29; *CIH* iii 913.17-18.
[246] Stokes, 'The destruction of Dá Derga's hostel', 291 § 117 = *LU* 233.7619–24. In the margin, *marcach* is equated with *ritire*, probably a borrowing from Old English *rídere* 'knight' (*LEIA* R-34).
[247] Harbison, *High Crosses* i 105 (discussion), i 372 (dating); ii fig. 334 (photograph).
[248] Harbison, *High Crosses* i 27 (discussion); i 377–9 (dating); ii fig. 73 (photograph).
[249] Harbison, *High Crosses* i 26 (discussion); i 377–9 (dating); ii fig. 66 (photograph).

Fig. 3. A horse and rider from the high cross of Banagher, Co. Offaly, which dates from about the ninth century.

The Old Irish law-texts refer only to the horse-cloth (*echdíllat*) under the rider, not to the saddle. Saddles seem to have originally been brought to Ireland by the Norsemen, as the Irish word *sadall* is from Old Norse *söðull* 'saddle'.[250] But they do not seem to have been generally used by the Irish until much later. Writing in the late twelfth century, Giraldus Cambrensis remarks on the fact that the Irish rode without saddles,[251] and the same observation is made in a French record of King John's visit to Ireland in 1200.[252] Many scholars have traced the close connection between military success and the adoption of the stirrup, which allowed a horseman much greater freedom to use his weapon – whether sword, lance or bow.[253] The stirrup probably originated in China, and reached western Europe about the eighth century AD. It was used to great effect by the Normans, and no doubt contributed to the extraordinary success of their small bands of knights in Ireland. The Irish word *stiróip* is a borrowing from Middle English *stirope*, and *spor* is likewise from Middle English *spore* 'spur'.

[250]Marstrander, *Bidrag*, 74, 115, 127.
[251]O'Meara, *Giraldus: Topography*, 101 § 93 = Dimock, *Giraldi Topographia*, 150.28–9. See Kavanagh, 'The horse in Viking Ireland', 93.
[252]Duffy, 'King John's expedition to Ireland', 22–3. Marie Therese Flanagan, 'Irish and Anglo-Norman warfare', 72, draws attention however to the references to saddles in the twelfth-century *Lebor na Cert* (ed. Dillon), 106. 1571; 146. 2185.
[253]L. White, *Medieval technology*, 14–38; Clutton-Brock, *Horse power*, 73–7.

Horse-racing

A vivid description of early Irish horse-racing is found in *Immram Curaig Maíle Dúin*: 'When [the voyagers] had gone a short distance from land they saw many people by the island shore, who held a horse-race (*grafand*) after reaching the green (*faithche*) of the island. Each horse was swifter than the wind, and great was their noise and their shouting and their tumult. Máel Dúin then heard the blows of the horse-whips (*echlasc*) at the meeting. He also heard what each of them was saying: "Bring the grey horse". "Drive the dun horse over there". "Bring the white horse". "My horse is faster". "The jumping of my horse is better"'.[254] Such races were a feature of the regular public assembly (*óenach*),[255] and seem to have been held particularly at the feast of Lugnasad in early August.[256] Kings were clearly able to enjoy horse-racing more frequently: *Críth Gablach*'s account of a king's weekly schedule sets aside Friday for horse-races.[257] The term *ech búada* 'horse of victory, prize horse' is used of a swift riding horse, often in the context of horse-racing. For example, a law-text states that such a horse cannot be distrained – even for its own offences – at a time when races are being held.[258] Another term, *ech aige*, is used specifically of a horse kept for racing.[259] There is evidence that chariot-races also took place: *Bretha Nemed Déidenach* refers to 'the chariot which gains the victory at an assembly'.[260]

Horse-values

The main categories of livestock used in the early Irish currency system are female cattle and sheep (see Appendix B, p. 587). There are, however, a few instances in our sources where a fine or value is given in terms of horses. *Bretha Étgid* states that a payment of more than one *cumal* should consist of one third cattle, one third horses, and one third silver. One third of the horses should be mares.[261] The horse also features in the payments due to poets. According to *Bretha Nemed Déidenach* a poem in the *emain* metre

[254] Van Hamel, *Immrama*, 31.183–91 = Stokes, 'The voyage of Máel Dúin, (1)', 466.
[255] *CIH* i 242.6–7 = *AL* v 484.29–30; Hull, '*Noínden Ulad*', 28.24–6.
[256] *CIH* ii 471.22–6 = *AL* v 406.6–12.
[257] *CIH* ii 569.32 = *CG* 21.546–7.
[258] *CIH* iii 897.16–17 = Appendix A, text 2 § 1 (15).
[259] *CIH* ii 471.22 = *AL* v 406.6.
[260] *CIH* iii 1123.7 = E. J. Gwynn, 'Privileges', 30.15 *carbad beireas buaidh aonaigh.*
[261] *CIH* iv 1263.15–18 = *AL* iii 150.12–16. See p. 588.

requires the payment of a horse worth two cows (*ech dá bó*).[262] The suggestion is made in commentary on this passage in *Uraicecht Becc* that the horse itself is only worth one cow, but that it is accompanied by a fine bridle also worth a cow.[263] Other law-texts show that the value of a horse can be much greater than this. *Bretha im Ḟuillemu Gell* states that if a lord pledges his bridle on behalf of another, he is entitled to receive a horse worth ten *sét*s as interest (*fuillem*).[264] The *sét* is normally equivalent to half a milch cow, so this horse can be taken to be worth approximately five milch cows.[265] The horse which must be handed over as interest on a king's bridle is valued at thirty *sét*s i.e. fifteen milch cows. A sixteenth-century glossator on this text remarks: 'this is the horse of the greatest price which the [law-]book speaks of'.[266]

Middle Irish legal commentary gives standard values for all types of livestock except horses. This is no doubt in recognition of the fact that the value of horses varies much more than that of any other domestic animal. The commentators are however able to lay down some general principles of valuation. At birth, a foal (*serrach*) of either sex is said to be worth one twelfth of its mother's value if it resembles her. If it resembles its father it is worth one twelfth of his value. If it resembles neither of them, it has one twenty-fourth of the value of each parent.[267] Before birth, a foal is said to be worth more than it is after birth, having one ninth of its mother's value. The commentator explains this anomaly as being due to the fact that the death of the foal in its mother's womb is more likely to harm her than its death afterwards.[268] As with other livestock, legal commentary takes one third of an adult horse's value to be for its body (*colainn*), one third for its potential (*saíltinche*) and one third for its work (*gnímrad*). If a horse has no potential and is incapable of work, it is said to have no value because its carcase (*mart*) is worthless.[269] The commentator thus shares the view – widely expressed in our sources – that horse-meat is unsuitable for human

[262] *CIH* iii 1119.31 = E. J. Gwynn, 'Privileges', 25.21–2.
[263] *CIH* v 1605.10–11 (*laitnemach* is a misprint for *taitnemach*) = *AL* v 62.21–2.
[264] *CIH* ii 475.1 = *AL* v 414.22.
[265] The glossator in fact takes this horse to be worth *six* cows (*CIH* ii 475.3 = *AL* v 416.1).
[266] *CIH* ii 475.7 = *AL* v 416.2.
[267] *CIH* iv 1208.25–8.
[268] *CIH* iv 1208.19–24.
[269] *CIH* i 306.40–41 = *AL* iii 372.6–8.

consumption. Nevertheless, there is some evidence of the eating of horse-meat in early Christian Ireland: see p. 352.

3
Livestock (iii)

Hens

The domestic hen has been developed over many millennia from the red junglefowl (*Gallus gallus*), which still exists as a wild bird in the rain-forests of India, Burma, Malaysia, etc. This bird (Old Irish *cerc*) was obviously of considerable importance in the early Irish economy, as it is frequently mentioned in the law-texts, and sometimes portrayed in manuscripts – as in folio 67ʳ of the Book of Kells reproduced on the back cover of this book. It features much more prominently than the domestic goose or duck, and a tenth-century poem quotes the saying *fó cía beith cerc i trebad* 'it is good that there should be a hen on a farm'.[1]

Hens are included along with cattle and sheep in the early Irish currency system (Appendix B, p. 588). For example, the price which must be paid for a serious offence against an apprentice (*inol*) is fixed at a fleece, a ball of yarn or a non-broody hen (*cerc cen rún*).[2] The distance within which a cock-crow can be heard is used as an approximate measurement of area (Appendix B, p. 571).

In legal commentary a laying hen (*cerc céin dothas*) is valued at two bushels of grain, whereas a sexually active cock (*cailech céin íunas*) has only the value of one bushel. When a hen no longer lays and a cock is no longer capable of sexual activity, their value is reduced to half a bushel, as they are fit only for the cooking-pot.[3] A chicken (*eirín*) has one ninth of its mother's value while at her feet.[4] When no longer dependent a female chicken attains the value of one bushel, which is doubled when it starts to lay.[5]

Legal commentary provides no information on the value of a male chicken from the time when it is able to fend for itself until sexual maturity. Presumably most males would be killed during

[1] *LL* i 211.6322 = Dobbs, 'On the graves of Leinster men', 141 § 42.

[2] *CIH* v 1609.32 = *AL* v 80.12–13. The literal meaning of the phrase is 'a hen without a secret'; see discussion under 'eggs' below.

[3] *CIH* v 1609.38–9 = *AL* v 82.16; cf. *CIH* vi 1979.26–7.

[4] The Old Irish form is *eréne*: see p. 218.

[5] *CIH* iv 1209.4–7; v 1628.34–8 = *AL* iii 380.2–5.

this period, or else neutered for fattening as capons.[6] There are a few references to the capon in our sources, and the term used is *gaillín* (or *gaillén*), evidently a diminutive derived from Latin *gallus* 'cock'.[7] The victim of a Middle Irish satire is addressed 'O comb of a capon, smoky-coloured, crooked and ragged' (*a ulcha gaillín detbutánaig cuarlupánaig*).[8]

The domestic hen is very vulnerable to predators as it is a poor flyer and cannot run fast. This vulnerability is obviously increased by the frequent practice in both early and modern times of clipping hens' wings to prevent their flying up into trees or onto roofs.[9] Legal sources also refer to the temporary restriction of their mobility by the use of a hood: see p. 494.

Fig. 4. A crowing cock from p. 125 of the fourteenth-century Book of Ballymote (MS no. 536 = 23 P 12) in the Library of the Royal Irish Academy. It has been reproduced by kind permission of the Council of the Royal Irish Academy.

Our only information on the housing of hens is to be found in a short passage in a law-text on distraint, probably dating from the eighth century.[10] This passage indicates that when hens were not loose in or near the farmyard, they were normally housed in a movable hen-coop (*árus*), presumably made of wickerwork. This coop was either kept in an outhouse, or fixed up in a tree where

[6]Fattened fowl (Latin *altilia*) are mentioned in Irish Canon I (*IP* 160 § 9). This term may be applied to either male or female fowl fattened up for the pot.

[7]In his discussion of the various words *gall* in his *Glossary* the tenth-century scholar Cormac mac Cuilennáin includes the otherwise unattested borrowing from Latin *gallus* 'cock' (Meyer, *Sanas Cormaic*, 56 § 683). *Gaillín* and *gaillén* are regular diminutives of *gall*.

[8]Thurneysen, 'Mittelirische Verslehren', 98 § 160. The reading is that of McLaughlin, 'Early Irish satire', 15 § 15. Compare the English expression 'cockscomb'.

[9]*CIH* iii 897.35 = Binchy, 'A text on the forms of distraint', 78 § 3; *CIH* ii 578.36 = *AL* iv 118.z.

[10]*CIH* iii 897.34–7 = Binchy, 'A text on the forms of distraint', 78 § 3.

the hens would be out of the reach of their main predator, the fox (p. 188).[11]

Eggs

Many species of wild bird will continue to lay for a considerable period if their eggs are regularly removed from the nest. For example, up to thirty eggs can be taken in one season from the nest of a pair of gulls without causing them to abandon their nesting-site. These eggs were evidently prized by the early Irish: the twelfth-century *Lebor na Cert* includes twenty gulls' eggs among the stipend (*túarastal*) payable to the king of the Ulaid.[12]

At some remote period in Indian prehistory, it was discovered that the Red Junglefowl was particularly easy to exploit as a source of eggs. The wild ancestral type normally has a clutch of five to six eggs twice a year. By selective breeding man has now produced a hen which can lay 300 eggs annually.[13] 'Broodiness' – the instinctive urge to sit on and hatch the eggs – has been bred to a minimum. The egg-production attained by the early Irish hen was considerably lower than that of modern times. A reference in legal commentary suggests that a hen was expected to lay fifty eggs each year.[14] According to a quotation in *O'Davoren's Glossary* from a lost law-text, the egg of a full-grown hen (*og rerchirce*) should be four inches in circumference and five inches around the long axis.[15] These measurements are about the same as a small pullet or bantam egg.

Hen-eggs were clearly of importance in the early Irish diet, and are quite frequently mentioned in the texts. In the early twelfth-century tale *Aislinge Meic Con Glinne* they are described as 'pearls of the household'.[16] They feature prominently in penitential literature, as they served to provide protein for meatless diets. Thus a clerical penitent could only eat bread, gruel, fresh milk, garden

[11] I take *mad i mbarr* to mean 'if it (the coop) is in the branches of a tree' rather than Binchy's 'if they (the hens) are [roosting] in the branches of a tree'.

[12] Dillon, *Lebor na Cert*, 84.1238; 86.1272 *Fichi ug failind*; cf. 128.1899.

[13] In large-scale poultry-farming, a hen is normally killed for eating after a year or two. A hen can however live for ten years or more, though with declining egg-production.

[14] *CIH* iii 982.22; *CIH* v 1610.16 = *AL* v 84.18.

[15] *CIH* iv 1521.21 = O'Dav. 440 § 1375; *CIH* iv 1526.9–10 = O'Dav. 461 § 1484.

[16] Jackson, *Aislinge*, 39.1203–4 = Meyer, *Aislinge*, 101.3.

herbs and a few hard-boiled eggs.[17] In a gloss on the law-text *Bretha Crólige*, an egg-yolk (*buidecán*) is included along with butter, curds, and porridge in the 'soft fare of fosterage'.[18] This is food regarded as being suitable for injured children between seven and ten years of age.

Because of its fairly constant size, the hen-egg (halved) was used as a unit of measurement for liquids. See Appendix B, p. 576.

Geese

Our sources contain a number of references to the farmyard goose (Old Irish *géd*) which is a domesticated form of the greylag goose (*Anser anser*), native to the British Isles, Scandinavia, and much of Eastern Europe and Central Asia.

The goose is included in lists of livestock in an Old Irish law-text[19] and in commentary to *Uraicecht Becc*.[20] The law-text on joint-grazing refers to the large amount of grass which geese consume, and claims that the grazing of two geese is equivalent to one sheep.[21] The grazing of geese is also mentioned in the Middle Irish tale *Erchoitmed ingine Gulide*.[22]

Legal commentary on the value of various domestic animals says that one third of a goose's value is for its flesh, one third for its brood and one third for its potential (*saíltinche*).[23] However, no figure is given for the total value of the goose. One would expect a goose to be worth considerably more than a hen on account of its larger size. It is certainly clear from the Middle Irish tale *Fled Dúin na nGéd* ('The Feast of the Fort of the Geese') that a goose-egg was more prized than a hen-egg. The king of Ulster, Congal Cláen, was deeply insulted because he found the egg of a red-feathered hen on his plate at a feast given by the king of Ireland. There were goose-eggs on the plates of all the other kings.[24]

For the hunting of wild geese, see p. 299.

[17] E. J. Gwynn, 'An Irish penitential', 142 § 7 = *IP* 263 § 7 (trans. only). The Irish term is *ug tírimm* lit. 'dry egg'. A similar diet is described for a presbyter or deacon who has committed fornication, *IP* 112 II § 2.

[18] *CIH* vi 2300.39 = Binchy, '*Bretha Crólige*', 42 § 52[5].

[19] *CIH* ii 573.25.

[20] *CIH* v 1608.17 = *AL* v 76.19.

[21] *CIH* i 192.14; ii 577.2 = *AL* iv 100.22–3.

[22] Meyer, *Hibernica minora*, 66.11 § 6 *geilt ængeóid* 'the grazing of one goose'.

[23] *CIH* v 1628.29–32.

[24] Lehmann, *Fled Dúin na nGéd*, 8.256–60.

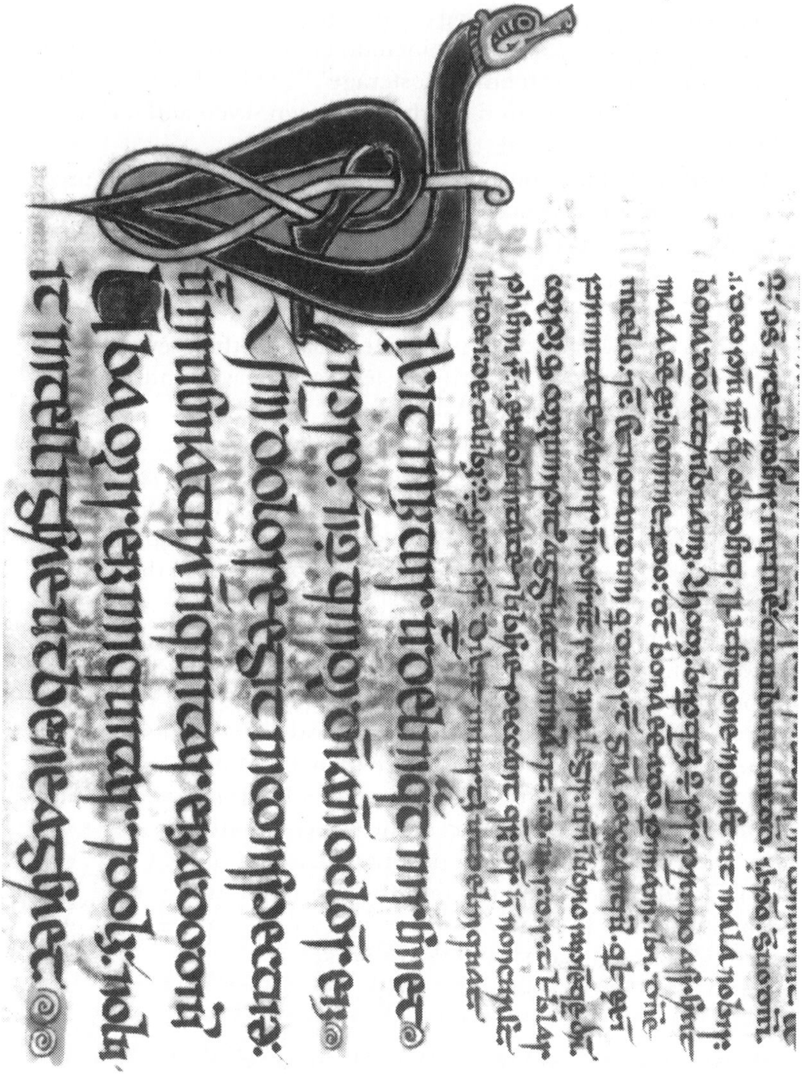

Plate 3. An initial **D** in the form of a goose from the 'Coupar Angus Psalter' (MS Pal. Lat. 65 in the Vatican Library). This psalter was written in the north of Ireland, probably at the monastery of Bangor, in the twelfth century.[25] It has been reproduced by kind permission of the Trustees and Librarian of the Vatican Library.

[25]See Henry and Marsh-Micheli, 'A century of Irish illumination (1070–1170)', 157–9.

Ducks

The farmyard duck is a domesticated form of the mallard (*Anas platyrhynchos*), native to most of the northern hemisphere. The Old Irish term *lachu* is used for both the wild and domesticated types.

The duck is rarely mentioned in our sources, and seems to have been of minor importance in early Irish farms. I have found no references to it in legal material. However, the typical association of duck and pond is mentioned in a proverbial expression in a late Life of Saint Máedóc of Ferns. Here the parting of saints Máedóc and Dallán is compared with 'the parting of a woman from her son, or a cow from her calf, or a bitch from her pups, or a duck from its pond'.[26]

For the hunting of wild ducks, see p. 298

Doves

Early Irish literature contains many references to the dove (*colum*), mainly in religious contexts.[27] Saint Columba or *Columb Cille* took his name from this bird, a symbol of peace. Saint Patrick is likewise described as 'a dove for gentleness and simplicity' (*colum ar chennsai 7 diúiti*).[28]

The domestic dove is descended from the rock dove (*Columba livia*), which nests in cliffs and caves throughout much of Europe, north Africa and Asia. It still occurs in the wild on the coasts of Scotland and western Ireland. Medieval monasteries in Britain and the continent are known to have commonly had a *columbar* (*columbarium*) 'dove-cote', in which doves were reared for eating. I have found no mention of the rearing of doves in early Irish monasteries. After the Norman invasion, however, dove-cotes are often mentioned in official records,[29] and a verse preserved in the *Irish grammatical tracts* refers to a pointed dove-cote (*teagh corr colam*).[30]

[26] Plummer, *Bethada* i 267 (236), trans. ii 259 (236) *fa dedhail mná rea mac, nó oighe rea láogh, no con fri cuaine, no lachan re linn.*

[27] See *DIL* s.v. 1 *colum*; *LEIA* C-160 s.v. 1 *colum.*

[28] Mulchrone, *Bethu Phátraic*, 37.648.

[29] E.g. Curtis, *Calendar of Ormond Deeds* iii 48; 123 § 138; 138 § 156.

[30] Bergin, 'Irish grammatical tracts: II Declension', 106.1343 § 53.

Peafowl

The gorgeous feathers of the peacock (*Pavo cristatus*) are illustrated very accurately on folio 32ᵛ of the Book of Kells, and it is possible that these birds – ultimately of Indian origin – may have sometimes been kept in early Irish monasteries. They were prized by the ancient Romans both for their beauty and their flesh,[31] and they featured on the menu of many medieval English banquets.[32] They were commonly kept on Anglo-Norman manors in Ireland.[33]

An argument in favour of the occasional presence of peafowl in early Christian Ireland is the fact that it has a native Irish name: *gésachtach* 'the screecher'. The ninth-century author of the St. Gall glosses employs this word to explain Latin *pavo*,[34] and it may also be present in a list of birds in verse in the Book of the Dean of Lismore.[35]

Bees

The honeybee was clearly of considerable importance in the early Irish economy. This is demonstrated by the existence of a special law-text on beekeeping, *Bechbretha*, which dates from the seventh century.[36] There are also many references to honeybees in other law-texts, in saints' Lives, in sagas and in poetry. Bee-plagues in AD 951 and 993 were regarded as sufficiently important to be mentioned in the *Annals of Ulster* (see p. 214 below).

The western honeybee (*Apis mellifera*) is native to Europe, Africa and the Middle East. Its stocks of honey have been raided by man for many thousands of years. For example, a rock-painting in Spain, dating from about 6000 BC, depicts a man taking honey from a nest high up a tree.[37] The earliest evidence of the keeping of bees in man-made structures comes from a relief in the Sun Temple at Abu Ghorab, Egypt, from about 2400 BC. It shows honey being transferred from hives to large storage vessels.[38]

[31] Varro, *Rerum rusticarum* (ed. Hooper and Ash), 3.6.1–6, devotes a whole chapter to the rearing of peafowl for the table.

[32] E.g. Rackham, *The history of the countryside*, 119.

[33] Down, 'Colonial society and economy in the high middle ages', 478.

[34] *Thes.* ii 60.28.

[35] Cameron, *Reliquiae Celticae* i 74.5 *ij zassidi* MS (= *dā ghēasadich* 'two peafowl(?)'). I am grateful to Pádraig Ó Macháin for this reference.

[36] *CIH* ii 444.12–457.10 = Charles-Edwards and Kelly, *Bechbretha*, 50–88.

[37] Crane, *The archaeology of beekeeping*, 21–2.

[38] Ibid. 36.

References to beekeeping occur in the Hittite laws of about 1300 BC, and are frequent in Greek and Latin sources.

Expert opinion is divided as to whether the honeybee spread naturally to Britain and Ireland when there was still a land connection with the continent, or whether it was brought over by man. The entomologist Michael Chinery regards it as an introduced species whereas Eva Crane believes it to be native to both islands.[39]

The Irish evidence is inconclusive. One tradition holds that the first swarm of honeybees was brought from Britain by the saint Mo Domnóc in the early seventh century.[40] However, early saints were credited with the introduction of a number of agricultural and other techniques, some of which are known from archaeological evidence to be of far greater antiquity.[41] The truth may be that the monasteries practised beekeeping on a larger scale than was known in pre-Christian Ireland, and so acquired the reputation for introducing honeybees.

The linguistic evidence suggests that honeybees have been in Ireland since long before the coming of Christianity. The Irish language has native words for 'bee' (*bech*), 'honey' (*mil*), and 'mead' (*mid*), all with cognates in other Indo-European languages.[42] It might be argued that these Irish words were applied originally to the native species of bumblebee (*Bombus*) whose honey-stores have no doubt been raided ever since the arrival of man.[43] However, bumblebees store honey in such small quantities that it seems unlikely that their honey was ever of much economic importance, or obtainable in sufficient quantity to make mead.[44]

There is also some linguistic evidence that Irish beekeeping goes back to the Common Celtic period, and that the practice of keeping bees may have been introduced to this country by Celtic-speaking colonists in pre-Christian times. An Old Irish term for the second swarm which leaves a bee-hive annually is *tarbṡaithe* lit. 'bull swarm'.

[39] Ibid. 14–5. See also the discussion at *BB* 39–40.

[40] Stokes, *Félire*, 60 (February 13).

[41] *BB* 40.

[42] *IEW* 116, 723, 707; *LEIA* B-24 s.v. *bech*, M-50 s.v. *mil*, M-48 s.v. *mid*.

[43] The raiding of bumblebee nests (Mod.Ir. *cúasnóg*, *prúais(n)eog*, *talmhóg*) used to be a common pastime of children, particularly at hay-making time.

[44] For further linguistic arguments indicating pre-Christian beekeeping in Ireland, see *BB* 41.

Exact cognates are found in medieval Welsh (*taruheit*) and in Breton (*tarvhed*), which indicate a Common Celtic **tarwo-satios*.[45] The fact that a differentiation was made between the first and second swarms strongly suggests that there was human supervision of the bee-colonies, most probably in artificial hives.

Hives

Man first exploited honeybees by hunting for their nests in trees or cliffs, and climbing up to take the honey. A later development – particularly suited to well-wooded countryside – was the making of cavities in accessible trees in the hope that honeybees would nest there.[46] The honey could then be removed at intervals with relative ease.

Both these methods are still practised in various parts of the world, but by far the greatest amount of honey is taken from man-made structures to which a colony of honeybees has been introduced. As we have seen above, bee-hives (probably of unbaked mud) were already in use in Egypt at least as early as 2400 BC. In Roman writings there are references to hives made from hollowed-out logs, wickerwork, bark, wooden boards, earthenware and cow-dung.[47]

In the seventh-century *Bechbretha* and its ninth-century glosses, the term used for bee-hive is *lestar*.[48] I have found no written evidence as to the type of materials commonly used in the construction of Irish hives at this period, and no archaeological evidence has come to light. Because *lestar* is used in other contexts of watertight vessels for liquids, I would guess that bee-hives made from hollowed logs were the commonest type. Support for this theory is provided by a legal quotation, probably Old Irish, in which the word *crann* 'tree' seems to be used in the meaning 'bee-hive'.[49] From about the eleventh century, *lestar* is replaced by the terms *ceis*[50] and *clíab*,[51] both basically meaning 'basket'. This

[45] *CIH* iii 924.1 = *BB* 62 § 20[b] (B). *Tarbṡaithe* could conceivably be an Old Irish calque on the Welsh term, but is more likely to go back to Common Celtic: see *BB* 41.

[46] Crane, *The archaeology of beekeeping*, 77–88.

[47] Ibid. 52.

[48] E.g. *CIH* ii 449.21 = *BB* 68 § 30; *CIH* iii 923.41 = *BB* 60 § 18[a].

[49] *CIH* iii 959.25; 982.19 = *BB* 184, Appendix 4 (c). Compare the use of the derivative *crannóg* to mean 'bee-hive' in Baile Bhúirne (Ó Cuív, *Cnósach*, 69).

[50] E.g. *CIH* ii 449.23 = *BB* 68 § 30[b].

[51] E.g. *CIH* ii 384.35 = *AL* i 170.19; Dillon, *Lebor na Cert*, 128.1900.

indicates the adoption of a wickerwork type of hive at this period. There is no mention of the straw hive (*corcóg*) in sources in the Irish language before about the seventeenth century.[52]

It is clear from *Bechbretha* that the hives were normally kept within the farmyard (*les*) or in the orchard (*lubgort*) of the early Irish farm.[53] A reference in another law-text[54] to a *bechdín* 'bee-shelter' suggests that a number of hives may sometimes have been placed together under some form of roofing.[55]

Hives could also be placed in the infield (*faithche*) near the dwelling-house, or even in an outfield location (*sechtar ḟaithchi*).[56] In the latter case it seems likely that the hives would only be kept there for short periods so as to be near a particularly rich source of nectar, such as the heather in August or September.[57]

Swarms

Early Irish lawyers devised an ingenious set of rules on beekeeping which must have helped to spread this asset throughout the land-owning community (see p. 145). The basic principle was one which might seem rather pedantic, i.e. that the bee which takes nectar from a neighbour's flowers is guilty of 'grazing-trespass' (*tairsce*)[58] like any other domestic animal. However, this assumption enabled swarms from a beekeeper's hives to be gradually distributed among his neighbours until they all owned bees; at this stage mutual trespasses would be expected to cancel each other out.

When a landowner starts beekeeping, he is normally entitled to three years of freedom, in which his bees can forage on his neighbours' land without liability. In the fourth year, however, the first swarm (*cétṡaithe*)[59] to issue from his hive has to be given to the nearest landowner on whose property his bees have done most 'grazing'.[60] Subsequent swarms are distributed to other neighbours, depending on their distance from the hive and the

[52] *BB* 44.
[53] *CIH* ii 455.31 = *BB* 84 § 50; cf. Lehmann, *Fled Dúin na nGéd*, 10.316.
[54] *CIH* ii 384.20 = *AL* i 166.27.
[55] For a discussion of 'bee-shelters', see Crane, *The archaeology of beekeeping*, 163–95.
[56] *CIH* ii 456.11–20 =*BB* 86 §§ 52–3.
[57] *BB* 45.
[58] *CIH* ii 444.14 = *BB* 50 § 3.
[59] Cognate with medieval Welsh *kyn*[*t*]*heit* 'first swarm', *BB* 192 § 2.
[60] *CIH* ii 447.4–6 = *BB* 60–2 §§ 18–19.

amount of 'grazing' on their land.[61] The neighbour whose turn it is to receive a swarm is responsible for keeping a watch on the hive at times when swarming is likely – 'any period of brightness', as the text correctly points out.[62] If he misses his swarm he must wait until the following year.

Much of *Bechbretha* is concerned with the legal intricacies connected with the swarming of bees into somebody else's property, a subject obviously difficult to legislate for. The text distinguishes two situations in which a swarm, tracked by its owner, enters a neighbour's land.[63] In one case the tracked swarm settles temporarily in the branches of a tree or on open land (e.g. on a fence).[64] The owner can recover this swarm without damaging the neighbour's property. This is done by means of a spread cloth (*brat scarthae*) onto which the bees are tipped.[65] The landowner is entitled to one third of the bees' honey for a year.[66]

But if the bees have made a permanent home in a cavity in a tree from which they cannot be moved without damage to the tree, the procedure is quite different. The bees are left in the tree, and the beekeeper gets one third of the honey for three years. After this period the bees and all their produce become the property of the landowner.[67]

If a person finds a stray swarm of unidentified origin in a great wood or unshared land or inaccessible country, it becomes his property – except for a ninth of the honey which goes to the head of his kin-group and to his church.[68] But if he finds a stray swarm on another's land, he only gets a proportion of the honey (varying between a half and a quarter) for a year.[69]

[61] *CIH* ii 447.13–28 = *BB* 62 §§ 20–22.

[62] *CIH* ii 447.28–31 = *BB* 62 § 23. For a discussion of swarming and the bees' life-cycle, see *BB* 45–9.

[63] I ignore the added complexity caused by taking the rank of the land-owner into consideration.

[64] *CIH* ii 451.17–452.26 = *BB* 74–8 §§ 39–42.

[65] For an account of this method of recovering a swarm, see *BB* 47–8.

[66] *CIH* ii 451.35 = *BB* 76 § 41.

[67] *CIH* ii 450.13–24 = *BB* 72–4 §§ 36–8.

[68] *CIH* ii 455.1–4 = *BB* 84 § 49.

[69] *CIH* ii 454.13–28 = *BB* 82 §§ 46–8.

Honey

Honey is mentioned in the medical law-text *Bretha Crólige* as a *tarsunn* or 'relish',[70] and the Old Irish tale *Fled Bricrenn* speaks of 'wheaten bread baked with honey'.[71] In addition to providing flavour, honey must have been a useful foodstuff in its own right, as it contains energy-giving carbohydrates, as well as proteins, minerals and vitamins (though only in small quantities). Being a supersaturated sugar syrup, honey keeps well and must have been particularly appreciated during the winter. Its medical properties were highly rated,[72] though *Bretha Crólige* warns against its use where there is infection of the stomach.[73]

A law-text refers to a cup filled to a certain measure with *mellit* as part of the return which a person gives for the grant of a colony of bees.[74] *Mellit* may be some honey-based drink distinct from mead or bragget, and is perhaps to be identified with hydromel, an unfermented mixture of honey and water.[75] The word seems to be a borrowing from Latin *mellitus* 'honeyed', which is well attested in the phrase *panis mellitus* 'honey-bread'.[76]

Mead

Mead (Old Irish *mid*) is made by fermenting honey with water, and may be the earliest intoxicant used by man. It is mentioned frequently in our sources, and seems to have been regarded as a more prestigious drink than beer. In *Aislinge Meic Con Glinne* mead is described as 'the relish of noble stock'.[77] The legendary banqueting hall associated with the royal site of Tara was known as the *Tech Midchúarda* 'house of the mead-circuit'. A poorly preserved illustration in the Book of Leinster (reproduced on p. 356) depicts what may be a vat of mead in the centre of the *Tech Midchúarda.*

Another alcoholic drink mentioned in our sources is *brocóit* 'bragget', made from malt and honey: see p. 335.

[70] *CIH* vi 2298.14–15 = Binchy, '*Bretha Crólige*', 36 § 45.

[71] *LU* 248.8125 = Windisch, 'Das Fest des Bricriu', 256.31 § 9.

[72] See brief discussion at *BB* 100.

[73] *CIH* vi 2292.27–32 = Binchy, '*Bretha Crólige*', 20 § 25.

[74] *CIH* iii 920.32–3 *Somuīne bech . . . ian ōil lān di mellit* 'profit from bees . . . an *ól*-measure cup full of *mellit*'. For the *ól* as a liquid measure, see Appendix B, p. 578.

[75] Greek ὑδρόμελι; Latin *hydromeli.*

[76] Toussaint-Samat, *A history of food*, 31–2.

[77] Jackson, *Aislinge*, 38.1192 = Meyer, *Aislinge*, 99.16 *sercoll socheneóil.*

Beeswax

Beeswax (Old Irish *céir*) is an important product of the honeybee, and is used to make candles, writing tablets, seals, adhesives, etc.[78] Surprisingly, it is not mentioned in Irish legal material, though it must have been an essential substance in every monastery.

Dogs

The relationship of the domestic dog with man goes back to at least 10,000 BC.[79] At the end of the last Ice Age, both man and the wolf (*Canis lupus*) would have been in competition for the same types of prey. Gradually, associations built up between the two species. These may have originated from the capture of wolf-cubs by hunters, or from their attraction to the easier pickings at human encampments. Because the social structure and behaviour-patterns of a wolf-pack are similar to those of a group of people, wolves were able to fit relatively easily into human society.

Over the millennia, many different types of dog have been bred from the ancestral wolf, with huge variations in size, temperament and physical attributes. Roman authors broadly distinguish guard-dogs, hunting dogs, herd dogs and lap-dogs. All these types are well documented from early Irish sources. A particularly important source is the fragmentary law-text on dogs, *Conṡlechtae*.[80]

Guard-dogs

One of a dog's most common functions is to guard its owner's house or encampment from intruders, whether human or animal. It has two attributes for this purpose – its bark and its bite. The ancestral wolf does not normally bark, but man has bred this characteristic into practically all domestic dogs.[81] The bark serves to alert the household, so even a small dog can be of importance in guarding a home.

[78] Crane, *The archaeology of beekeeping*, 240–6. *Céir* is a borrowing from Latin *cera* 'wax' (*LEIA* C-56).

[79] Clutton-Brock, *Domesticated mammals*, 34.

[80] Quotations with glosses and commentary are to be found at *CIH* i 111.3–114.7; 311.36–314.16; iv 1395.15–1396.6; v 1550.24–1551.12. See also *GEIL* 275 § 48. For a discussion of *Conṡlechtae*, see L. Breatnach, 'On the glossing of early Irish law-texts', 15–20.

[81] Clutton-Brock, *Domesticated mammals* 39–40.

In early Irish sources, however, more stress is laid on the large fierce guard-dog whose function is to attack the intruder. This type of dog is known as an *árchú* 'slaughter-hound' (cognate with Welsh *aergi*), as it is bred and trained to kill.[82] These dogs are particularly associated with the homes of nobles and other powerful men. The best-known literary account of such a dog is to be found in *Táin Bó Cúailnge*.[83] Here the author describes how the slaughter-hound of the smith Culann was so strong that it had to be held by three chains with three men on each chain. According to one version, this dog was brought from Spain. On the night of a feast, this dog was let loose when Culann thought that all his guests had arrived. The young Cú Chulainn (then known as Sétantae) came late, and was immediately attacked by the dog. Those in the house assumed that the youth would be killed, but he managed to smash the dog to pieces against a pillar. Culann bewailed the death of his dog, whom he described as 'a man of his household' who protected his goods and livestock. However, the youth volunteered to act as his guard-dog until another pup had been reared from the same breed. He thus acquired the name *Cú Chulainn* 'hound of Culann'.

The Old Irish law-texts also contain many references to the *árchú*. For example, an unedited text on clientship describes how a young *árchú* is expected to gradually extend its guarding duties in a fortified dwelling (*dún*).[84] In the first year it guards the doorway (*dorus*) and the cleared area (*airscartad*) in front. As it gains in size and confidence in its second year it also guards the plank (*comlae*) leading up to the door. In its third year it guards the bridge of access (*drochet*) to the dwelling.

Other law-texts similarly refer to the area around the house for which the guard-dog is responsible. A section of Irish canon law in Latin states that the most valuable type of dog is the 'dog of four doors' i.e. of the house where its master lives, of the sheep-fold, of the byres of the calves and of the oxen.[85] Anyone who illegally kills such a dog must pay the owner ten cows and substitute a dog of

[82]In his glossary to *Stories from the Táin*, Strachan makes a distinction (repeated at *DIL* s.v. *árchú* and *LEIA* A-81 s.v. *ar-*, A-82 s.v. *ár*) between *árchú* 'slaughter-hound' and *archú* 'watch-dog'. It is probable, however, that *archú* is a ghost-word.

[83]*TBC* Rec. I 17.540–19.607. Another famous guard-dog in early Irish literature is Ailbe, owned by Mac Dathó, king of Leinster. He guarded the whole of Leinster (Thurneysen, *Scéla Mucce Meic Dathó* 1.1–2).

[84]*CIH* iii 920.41–2.

[85]*IP* 174 § 4 = *GEIL* 354 (5) § 4.

the same breed. The dog of four doors (*cú chethardoruis*) is also referred to in the law-text *Bretha Nemed Toísech.*[86]

In some circumstances the duties of the *árchú* are not merely defensive. The 'lord of blood-vengeance' (*aire échta*) is expected to own a dog of this type.[87] He is a prominent member of the kingdom (*túath*), whose duty is to avenge any serious crimes – ranging from rape to murder – committed by a member of another *túath.* If no reparation is made, he must organize an avenging party to cross the border and kill the offender. A fierce dog with good tracking ability would obviously be of help in hunting down such a person if he attempted to escape.

The law-texts envisage a similar rôle for the dog of the local champion (*ánruth*) – in fact it seems likely that the champion would often act as a 'lord of blood-vengeance'. His *árchú* should be a 'dog of three accomplishments': an ability to track a trail of blood (*derglorgairecht*), an ability to seize (*gabáltaige*) and an ability to protect a man under attack from two warrior-bands (*dingbáil fir ḟorráin iter dá ḟiallach*).[88] The hound of a champion (*árchú ánruith*)[89] is worth the same as a man, i.e. seven *cumal*s.[90]

Not all guard-dogs were fierce mastiffs like the highly-valued *árchú.* The law-texts also refer to dogs of lower prestige which might also carry out this function, particularly in the farmyards of commoners. They are described in contemptuous terms such as *cú otraig* 'dunghill dog',[91] or *cú chrumdumai,* 'dog of the maggot-heap'.[92] In *Táin Bó Cúailnge* there is a similar contrast

[86] *CIH* vi 2216.37.

[87] *CIH* vi 2128.28–33. For a discussion of the office of *aire échta,* see *CG* 70–2; McCone, 'Werewolves, cyclopes, *díberga,* and *fíanna*', 7–9.

[88] *CIH* iv 1390.24–35; cf. *CIH* i 186.37, where the text has simply *lorgairecht* 'tracking ability' rather than *derglorgairecht* 'tracking a red (i.e. bloody) trail'.

[89] *CIH* iv 1390.24–7; 1522.25–6 = O'Dav. 445 § 1397); cf. *CIH* i 320.17 = *AL* iii 446.27. *DIL* s.v. *anrath* takes *anra(i)th* in these examples to be an 'adj. used of a fierce or vicious dog', but it makes better sense to read *ánruith,* gen. sing. of *ánruth* 'champion'.

[90] The gloss at *CIH* v 1550.31 which takes the *díre* of a champion's hound (*árchú ánruith*) to be only three scruples must be mistaken.

[91] *CIH* iii 897.38 = Binchy, 'A text on the forms of distraint', 78 § 4; cf. *CIH* ii 376.24; v 1902.8 = *AL* i 126.8.

[92] *CIH* i 111.6; iv 1386.25. At *CIH* iii 1107.35, a glossator suggests that *cú chrumdum[ai]* refers to a young pup (*culén beg*) which has not yet started to work. However, this does not fit in with the other attestations, which evidently refer to an adult dog.

between a *cú mór* 'great hound' and an *aithechmatad* 'peasant cur'.[93]

Hunting dogs

Man learned early on to use the dog's remarkable hunting skills to his own advantage. An Egyptian tomb illustration of *c.*2600 BC shows a team of leashed hunting dogs,[94] and a rock painting from the central Sahara, perhaps as early as 4000 BC, shows a pack of dogs bringing down a wild sheep already wounded by spears.[95] From early Irish sources, it is clear that dogs were used mainly in the hunting of deer, wild pigs and hares (see Chapter 9). The general strategy of hunting with dogs is illustrated by the names of three legendary dogs in a Middle Irish text. They are *Aig* 'drive hither!', *Taig* 'drive thither!', and *Tairchell* 'hem in!'.[96]

The usual term for a hunting dog is *mílchú*, literally 'animal-hound'. Unlike the *árchú*, which was bred for size and aggression, the *mílchú* was bred for speed and ability to detect the prey. Such dogs are associated particularly with the aristocracy. One law-text states that it is right that a lord (*flaith*) should have a hunting dog,[97] and in another text the king's weekly routine is said to include a day spent watching his dogs hunting.[98] The hunting dog does not, however, have the high value of the trained guard-dog. According to a quotation in *O'Davoren's Glossary*, the hunting dog of a lord has a *díre* of only two *séts*, i.e the equivalent of one milch cow.[99]

Fig. 5. This illustration of a hunting dog catching a hare is from f. 48r of the eighth-century Book of Kells in the Library of Trinity College, Dublin.

[93] *TBC* Rec. I 43.1383–4.
[94] Gautier, *La domestication*, 124.
[95] Ibid. 111–13.
[96] *LL* i 124.3911 = Thurneysen,'Mittelirische Verslehren', 59 § 111.
[97] *CIH* iv 1268.16; vi 2128.26.
[98] *CIH* ii 569.31 = *CG* 21.545–6. A marginal entry on p. 161 of the Trinity College Dublin MS H 4. 22 (no. 1363) attributes special knowledge about dogs to the kings of Connacht (Abbott and Gwynn, *Catalogue*, 210).
[99] *CIH* iv 1516.12 = O'Dav. 417 § 1248.

Other words for hunting dog are also attested in Old Irish; this suggests that more than one breed may have already been in existence in early Christian Ireland. A legal fragment distinguishes the *gadar* from the *mílchú*,[100] but provides no evidence as to how they differed. In an article on the origin of Irish breeds of dogs, R. F. Scharff suggests that the function of the *gadar* was to track the position of the game by scent; the larger and swifter *mílchú* was then unleashed to hunt it down.[101] One can compare the distinction in Welsh between the *bytheiad* 'scenting hound' and the *milgi* 'greyhound'.[102]

On linguistic grounds Carl Marstrander argued that the word *gadar* (also *gagar*) was borrowed from Old Norse *gagarr* 'dog'.[103] However, its presence in an Old Irish law-text suggests that the word was in the language before the Viking incursions of the ninth and tenth centuries – no other Norse loan is found in a law-text. There is also literary and historical evidence that dogs were exported from Ireland to the Norse countries. In *Njáls Saga*, the great dog Sámr is represented as being from Ireland,[104] and in Snorri Sturlason's *Heimskringla* there is praise for the skill of an Irish dog which was able to divide a mixed herd of cattle into two groups on the basis of their markings. This dog was given by its Irish owner to the Norse king Óláfr Tryggvason, and named Vígi.[105]

Another term used in Old Irish of the hunting dog is *archocaid*, a word of unknown derivation.[106] It occurs in a list of types of dog in a law-text on distraint,[107] and is defined by a ninth-century glossator on this text as a *cú selga* 'hound of the chase'.[108] Later glossators – probably writing around the twelfth century – identify it as

[100] *CIH* iv 1268.14. A Middle Irish tale (Stokes, 'The death of Crimthann, etc.', 174.7) distinguishes a *mílchú*, a *gadar*, and a *matud* ('cur'). In other sources *mílchú* and *gadar* are regarded as synonymous, e.g. *CIH* i 113.3; iv 1516.12.

[101] Scharff, 'On the breeds of dogs peculiar to Ireland', 81.

[102] Linnard, 'The nine huntings', 130; Jenkins, *The law of Hywel Dda*, 291 (note to 182.13). Welsh *milgi* is the exact cognate of Irish *mílchú*.

[103] Marstrander, *Bidrag*, 158.

[104] Sveinsson, *Brennu-Njáls Saga*, 171–4 ch. 70.

[105] Monsen, *Heimskringla*, 139.

[106] The abstract noun *archogacht* glosses *uenatio* 'hunting' at *Thes.* i 466.38.

[107] *CIH* ii 377.6; v 1902.9 = *AL* i 126.9; cf. *CIH* i 312.13 = *AL* iii 412.18.

[108] *CIH* iii 889.8.

a hound which can be used both for deer-hunting and for tracking robbers.[109]

I have found no references in the written sources of the early period to the hunting of wolves by dogs (though the tale *Táin Bó Fraích* refers to the hunting of foxes).[110] It seems therefore that the massive Irish wolfhound is a later breed. The earliest descriptions of such a dog are from the sixteenth century,[111] and the archaeological evidence of the early Christian period shows no dog-bones larger than those of a modern Alsatian.[112]

Herd dogs

Herd dogs were obviously of importance to the early Irish farmer as they are mentioned quite frequently in the law-texts.[113] A list of types of dog in a text on distraint[114] includes the 'herd dog of every livestock' (*conbúachaill*[115] *cacha cethrae*). Another legal passage distinguishes three types of herd dog.[116] It begins with the cryptic statement *Ní dlig búachail beithir*, which seems to mean 'a herdsman does not require a monster' i.e. a mastiff or very large dog.[117] It then goes on to distinguish the *conbúachaill mórchethrae* 'herd dog of large livestock', the *conbúachaill láeg* 'herd dog of calves',[118] and the *conbúachaill cáerach* 'herd dog of sheep'. The penalty-fine (*díre*) for the first is 2½ cows i.e. 5 *séts*, for the second 8 scruples, and for the third 6½ scruples. Irish canon law in Latin likewise puts a high value on the herd dog, and states that if somebody kills a dog which guards the flocks, he must pay five cows and supply a dog of

[109] E.g. *CIH* iv 1469.27 = O'Dav. 212 § 100; *CIH* ii 377.7–8 = *AL* i 144.14–15.

[110] Meid, *Táin Bó Fraích*, 3.54.

[111] Scharff, 'On the breeds of dogs peculiar to Ireland', 86–7.

[112] McCormick, 'Stock-rearing in early Christian Ireland', 172.

[113] E.g. *CIH* vi 2195.15 = Thurneysen, 'Aus dem irischen Recht III', 331 § 22; *CIH* iii 897.38 = Binchy, 'A text on the forms of distraint', 78 § 4.

[114] *CIH* ii 376.26–377.6 = *AL* i 126.8–9.

[115] The two same elements, in reverse order, make up the Welsh cognate *bugeilgi* 'herd dog'.

[116] *CIH* i 171.18–20; cf. *CIH* iii 1107.33–5.

[117] The original meaning of *beithir* seems to have been 'bear'. However, as the brown bear became extinct in Ireland in prehistoric times (p. 190), the word is used in a vague sense of any fierce animal or warrior (*LEIA* B-28 s.v. *beithir*). One can compare the classification of the lion (Irish *leoman*) as a type of large dog at Stokes, 'The death of Crimthann, etc.' 174.6.

[118] Cf. *caú ingaire gamna* 'a dog for looking after calves', *LU* 281.9328 (*Síaburcharpat Con Culaind*).

the same breed. He must also pay for whatever wild animals eat of the flocks until the end of the year.[119]

Pet dogs

The breeding of miniature or toy dogs has a long history. It is likely, for example, that the Pekinese has been bred in a fairly pure form for at least two thousand years.[120] Such dogs are mainly for people's amusement, and an Old Irish triad refers to a pet dog (*orcae* or *oirce*)[121] as one of the three entertainments of a gathering.[122] Pet dogs are associated particularly with high-ranking women in our sources: the law-text on status *Críth Gablach* states that a lord of precedence (*aire tuíseo*) should have a hunting hound whereas his wife should have a pet dog.[123] In a Middle Irish topographical poem the noble Bóand, wife of Nechtain, is said to be the possessor of a pet dog (*messán*)[124] named Dabilla.[125]

Pet dogs are also associated with certain professions: the legal passage on dogs states that it is proper for a pet dog to be owned by a physician, a harpist, a queen, and a hospitaller (*briugu*).[126] According to a quotation in *O'Davoren's Glossary* the *briugu*'s pet dog is worth one third of his honour-price, provided it carries out its duties of accompanying him and watching by his pillow.[127] Elsewhere the pet dog is given the same value as a hunting hound i.e. two *séts*.[128]

Legal commentary also refers to a special supernatural function of a woman's pet dog. This is to protect her from the fairies (*túaithgeinti*) while she is giving birth.[129] If somebody kills a pet dog at this time he must not only pay a fine of three *séts*, but must also provide a priest to protect her by reading from scripture day

[119] *IP* 174 § 3 = *GEIL* 354 (5) § 3.
[120] Clutton-Brock, *Domesticated mammals*, 39.
[121] The diminutive *oircne* is also common. See *LEIA* O-28 s.v. 1 *orc.*
[122] Meyer, *Triads*, 32 § 241.
[123] *CIH* ii 567.26; iii 783.33 = *CG* 16.408.
[124] This common term for pet dog is a diminutive of *mess* 'pet, favourite': see *LEIA* M-43 s.v. *mess.* The compound *meschú* is also well attested, e.g. Mulchrone, *Bethu Phátraic*, 139.2759.
[125] E. J. Gwynn, *Metrical Dindshenchas* iii 32.81–4.
[126] *CIH* iv 1268.16–17; vi 2128.26–7.
[127] *CIH* iv 1512.10–11 = O'Dav. 400 § 1153.
[128] *CIH* iii 806.34; *CIH* iv 1531.2 = O'Dav. 482 § 1599.
[129] *CIH* iii 806.33–807.10; cf. *CIH* iv 1531.3–4 = O'Dav. 482 § 1600.

and night while she is in labour. This case is described as *in sagart i n-aithgein in mesáin* 'the priest in compensation for the pet dog'.

Cats

The domestic cat is believed to be of much later origin than the dog. Though cat bones have been found in Neolithic sites as early as 7000 BC, it is probable that these are from wild animals killed primarily for their skins. The earliest definite evidence of the domestication of the cat comes from Egyptian illustrations of *c.*1600 BC.[130] The Egyptians came to regard cats as the most sacred of all animals and carefully embalmed their bodies after death. It is thought that the earliest domestic cats originated from the African wild cat (*Felis sylvestris libyca*). When they were introduced to the countries north of the Mediterranean Sea, they interbred with the slightly larger European wild cat (*Felis sylvestris sylvestris*).

Recent finds suggest that the wild cat was once native to Ireland, but became extinct before the early Christian period.[131] The domestic cat was probably introduced from Roman Britain in the first centuries AD.[132] By the time of our main records of the seventh and eighth centuries, the domestic cat (Old Irish *catt*)[133] was obviously well established and of considerable economic importance. It is mentioned quite frequently in legal material, and a few fragments have survived of a law-text entitled *Catślechtae* 'cat-sections', which deals with the payments which must be made for killing, injuring or stealing another person's cat.[134] Cats also feature in early Irish literature: perhaps the best known Old Irish lyric poem concerns a monk and his pet cat *Pangur bán.*[135] In the sagas, cats are sometimes accredited with supernatural qualities reminiscent of their sacred status in Ancient Egypt. In *Imram Curaig Maíle Dúin* 'the voyage of Maíl Dúin's boat' the jewels in a

[130] Clutton-Brock, *Domesticated mammals*, 111.

[131] Van Wijngaarden-Bakker, 'The animal remains from Newgrange (1)', 349; 'The animal remains from Newgrange (2)', 91–2. Margaret McCarthy of the Department of Archaeology, University College Cork, has recently identified bones of wild cat from well sealed Bronze Age deposits at Chancellorland, Co. Limerick (pers. comm.).

[132] McCormick, 'The domesticated cat in early Christian and medieval Ireland', 218.

[133] *Catt* may be a direct borrowing from Latin *cattus* (*VGK* i 234), or it may be of earlier provenance (*LEIA* C-50 s.v. *catt*).

[134] *CIH* v 1550.15–23, cf. *CIH* i 110.14–21.

[135] Murphy, *Early Irish lyrics*, 2.

fort on a deserted island were guarded by a little cat. When one of Maíl Dúin's companions attempted to steal a necklace, the cat leaped on him and immediately he was reduced to ashes.[136] In *Fled Bricrenn* 'the feast of Bricriu' the warriors Lóegaire, Conall and Cú Chulainn were attacked in the cave of Crúachu by three kittens, which are described as 'monsters of sorcery' (*bíasta druídechta*).[137]

The main function of the cat was of course to keep its owner's house and grain-store free of mice (and possibly also of rats).[138] The cat was also valued because of its attractiveness as a pet. Legal commentary based on *Catślechtae* states that a cat is worth three cows if it is able to purr (*crónán*) and to guard the barn, mill and corn-drying kiln against mice. If it is only able to purr, it is worth 1½ cows.[139] Another legal commentary states that a kitten (*caitín*) has one ninth the value (*lóg*) of its mother until it is independent; its subsequent value depends on its abilities.[140]

In legal material cats tend to be associated with women, children,[141] and the kitchen, whereas the dog – particularly the hunting dog – is closer to the master of the household. Some especially favoured cats seem to have slept indoors in a basket,[142] or even on a pillow beside their mistresses.[143] The only adult male explicitly associated with the cat in legal material is the cowherd (*búachaill*). According to a fragment from *Catślechtae*, he is entitled to one third of the fine (*díre*) of three *sét*s which must be paid for killing or injuring a cat.[144] This is obviously because cats would be attracted to the warmth and fresh milk of the cowshed.[145] The cat's love for milk is reflected in the proverbial expression quoted

[136] Van Hamel, *Immrama*, 34.279–35.318 = Stokes, 'The voyage of Máel Dúin, (1)', 476–8.

[137] *LU* 265.8761–72 = Windisch, 'Das Fest des Bricriu', 282.16–283.2 § 57.

[138] For the rat, see p. 243 below.

[139] *CIH* i 110.14–15; cf. *CIH* v 1550.17–18 *cat sabaill ⁊ muilinn ⁊ ātha .i. bīs oca nimcomēt a triur* 'a cat of barn and mill and drying kiln, i.e. which guards all three'.

[140] *CIH* v 1628.26–8. The same applies to a puppy.

[141] A glossator at *CIH* v 1901.32 gives the kitten (*catín*) as an example of children's toys (*esrechta macraide*); cf. *CIH* ii 373.27 = *AL* i 138.33.

[142] I suggest that *baircne cat ban* at *CIH* ii 379.11; v 1903.3 = *AL* i 150.11 means 'a basket for women's cats', taking *báircne* to be a diminutive of *bárc* 'boat, vessel, container'. On the other hand, *Baircne* is explained as a cat's name at *CIH* iii 889.23–4. At *CIH* iv 1475.17–18 = O'Dav. 235 § 242, *Baircne* seems to be taken as the name of a particular cat, or of a type of cat.

[143] *CIH* iii 889.24.

[144] *CIH* v 1550.19–22.

[145] *CIH* iv 1501.16–17 = O'Dav. 351 § 915.

in a law-text: *cuirm lemm, lemlacht la cat* 'beer with me, fresh milk with a cat' i.e. I find beer irresistible just as a cat cannot resist fresh milk.[146]

Colour and cat-names

Both ancestral subspecies are striped, and this is still the predominant pattern in domestic cats. However, many variations have arisen by mutation, and modern cats exhibit a wide variety of fur-patterns and colours.

The names applied in Old Irish sources to individual cats or cat-types demonstrate that many different variations were known at this period. The anonymous monk who composed the poem mentioned above had a white cat *Pangur bán*. It has been suggested that this name is Old Welsh **pangur* (later *pannwr*) 'fuller (of cloth)' as he is a person who is likely to be covered in the white dust of fuller's clay.[147] Another cat-name *Bréone* quoted in a fragment of *Catṡlechtae* means 'little flame', and could refer to an orange or marmalade-coloured animal.[148] *O'Davoren's Glossary* quotes a reference, presumably also from *Catṡlechtae*, to a cat called *Glas nenta* 'nettle-grey'. This cat was evidently of little value, as its *díre* was only one *sét.*[149] It is not clear whether *Glas nenta* refers to an individual cat – perhaps in some leading case used to establish cat-values – or whether it refers to a particular type of cat, such as the common greyish striped tabby. A later legal gloss refers to white-breasted black cats.[150]

Other cat-names preserved in the fragments of *Catṡlechtae* seem to be typical pet-names referring to special characteristics of the animal. Thus the name *Méone* is a diminutive meaning 'little meow', and refers to a kitchen cat (*cat cuile*).[151] A glossator explains the name as 'a strong cat which mews'.[152] The name *Cruibne* 'little

[146] *CIH* iv 1176.9 = *AL* iii 532.4. For the use of *lemlacht la cat* and similar expressions in poetry, see Ó Cuív, 'Three Middle Irish poems', 9–11.

[147] Gruffydd, '*Pangur*'; Murphy, *Early Irish lyrics*, 172.

[148] *CIH* i 110.14; *CIH* iv 1475.15 = O'Dav. 235 § 241. See *DIL* s.vv. *breó, breóán, breo(in)ne*.

[149] *CIH* iv 1507.22 = O'Dav. 379 § 1045.

[150] *CIH* ii 379.32; v 1903.17 = *AL* i 152.33–4 *cait broinnfinda dubha.*

[151] *CIH* i 110.18; *CIH* iv 1516.8–9 = O'Dav. 417 § 1246.

[152] *CIH* v 1550.15 *trénchat dogní meighligh.*

paws', applied to a cat of barn, mill and drying kiln, is explained as being from the cat's distinctive paws.[153]

O'Davoren's Glossary quotes further cat-names which are more difficult to interpret. *Rincne*, possibly meaning 'spear',[154] is said to be a children's cat.[155] *Folum* is the name of a cat 'which is kept with the cows in the yard'.[156] I have no explanation for another name, *Iach*, used of a cat which is only entitled to half *díre*.[157]

OTHER PETS AND CAPTIVES

The animals discussed above are all of some practical use to their owners. But it is clear from our sources that other types of wild animal were occasionally kept for amusement, or possibly for religious purposes. Some animals came from abroad as gifts to powerful men. The *Annals of Inisfallen* record that the king of Scotland sent a camel (Irish *camall*) to Muirchertach Ua Briain in the year 1105.[158] The skull and jaw-bones of another exotic animal, the Barbary ape (*Macaca sylvanus*), have been recovered from late Bronze Age or early Iron Age levels at Navan fort (Emain Machae), Co. Armagh.[159]

The law-texts show that many landowners had pets of native origin. *Bretha Comaithchesa*, the law-text on the mutual obligations of neighbouring farmers, deals with offences by pets such as wolves, foxes, deer and herons.[160] Later commentary provides further detailed information on the types of native mammals which were kept as pets.[161]

The prominence given to pets in legal material is matched in the sagas and saints' Lives. In *Táin Bó Cúailnge*, for instance, king Ailill is represented as having a pet bird (*peta éoin*) on his shoulder, while queen Medb had a pine marten (*togmall*). Cú Chulainn killed both

[153] *CIH* v 1550.17. The name is spelled *Cruipne* here, but O'Davoren has the better reading *Cruibne* at *CIH* iv 1481.21 = O'Dav. 263 § 422.
[154] See *DIL* s.vv. 1 *rincne*, 3 *rincne*.
[155] *CIH* iv 1520.42 = O'Dav. 438 § 1365.
[156] *CIH* iv 1501.16–17 = O'Dav. 351 § 915.
[157] *CIH* iv 1510.9 = O'Dav. 391 § 1109.
[158] *AI* 262 s.a. 1105 § 7. Camels were also sent to Ireland by the king of England in AD 1472 (*AC* 560 s.a. 1472 § 13, which provides quite a detailed account of these animals).
[159] Lynn, 'Navan fort', 16.
[160] *CIH* i 73.19–20 = *AL* iv 114.10–12.
[161] *CIH* i 74.15–25 = *AL* iv 120.1–14; *CIH* i 98.31–5; vi 2187.14–17.

pets with stones from his sling.[162] Many saints are also recorded as having pets.[163] Saint Mo Ling had a number of animals which he kept in honour of their Creator, and fed from his own hand. His pet fox once stole and ate a hen belonging to his fellow monks, but through a miracle the bird was restored to life.[164] In his *Life of Saint Brigit,* Cogitosus gives an account of a fox who belonged to a king, and was able to perform tricks.[165]

The usual term for a pet in Old Irish is *pet(t)a,* which may be the ultimate origin of the English word *pet.* The word *treitell* (*drettel*) is also commonly used of a pet animal or a household favourite (particularly a child).[166]

Corr

Apart from cats and dogs, the pet most frequently mentioned in our sources is the *corr,* which is often identified with the heron (*Ardea cinerea*), still a common bird of rivers and lakes. A complicating factor is the occasional application of the word *corr* to another large bird, the crane (*Grus grus*).[167] For example, the Latin Life of Saint Abbán explains the Irish lake-name *Loch na Corr* as *stagnum gruum,* i.e. 'the lake of the cranes'.[168] The crane is now a rare passage migrant in the British Isles, but was formerly quite common as a breeding bird. In a chapter entitled *De grue, ejus natura* 'on the crane and its nature' in his *Topography of Ireland* Giraldus Cambrensis refers to its abundance in this country.[169] A manuscript written about 1200 illustrates this chapter of the *Topography* with a fine drawing of a crane: this is reproduced on p. 126.[170]

[162] *TBC* Rec. I 29.920–5.
[163] See Plummer, *Vitae* i Introd. p. cxlvi for further details.
[164] Plummer, *Vitae* ii 201 § xxiii.
[165] Connolly and Picard, 'Cogitosus: Life of Saint Brigit', 19 § 20.
[166] *DIL* s. vv. *drettel, treitell;* J. E. Caerwyn Williams, 'Welsh *drythyll, trythyll;* Irish *drettel, treitell*'.
[167] To this day the heron is called a 'crane' in parts of Britain and Ireland.
[168] Plummer, *Vitae* i 27 § xliii.
[169] O'Meara, *Giraldus: Topography,* 40 § 10 = Dimock, *Giraldi Topographia,* 46.1–3.
[170] For a discussion of this manuscript, see Dimock, *Giraldi Topographia,* pp. xx–xxi; Scott and Martin, *Expugnatio hibernica,* p. xxxv.

Plate 4. This crane (*grus*) is from the bottom margin of f. 8^{v} of the British Library MS Royal 13 B VIII. This manuscript, written about AD 1200, contains an illustrated version of Giraldus Cambrensis's *Topography of Ireland*. The artist has portrayed very clearly the characteristic black and white neck-stripes of the crane. The absence of a crest and the drooping feathers over the tail also distinguish it from the heron. The bottom of the page has been evened off with loss of the crane's right claw.

This illustration has been reproduced by kind permission of the Trustees of the British Library.

The heron and the crane are similar in appearance, though very different in behaviour. The heron feeds on fish, rodents, insects, etc. whereas the diet of the crane is mainly vegetarian. The bones of both birds have been found in excavations from the early Christian period.[171] There are many references to the pet *corr* in legal material, but it is difficult to be sure which bird is intended. In modern times, there is abundant information on the domestication of both birds. The heron is a difficult pet; it is strongly territorial and tends to attack other domesticated birds, as well as dogs, cats and children.[172] Even when taken as an egg, it is hard to get it to bond with a human. By contrast, the crane is easily bonded with a human.[173] In some societies the flesh of the crane was particularly relished: paintings inside ancient Egyptian tombs show cranes being fattened up by force-feeding.[174] However, Giraldus Cambrensis notes an abhorrence of crane-flesh among the Irish.[175] In the laws of the sixth-century Franks, there is a reference to the domesticated crane,[176] and evidence that this bird was used in the training of hawks.[177] In early Christian Ireland, however, hawking does not seem to have been practised: see p. 303.

The pet heron or crane (*peta cuirre*) is included among the pets listed in *Bretha Comaithchesa*,[178] and in another law-text it is stated that a fine (*díre*) of five *sét*s must be paid for killing this pet. This is the same sum as must be paid for a kitten, puppy, cock or goose.[179] A later commentator on this text expresses surprise at the high value assigned here to the *corr*, on account of the general legal principle that a wild animal which has been tamed is only entitled to a

[171] E.g. CRANE: Hencken, 'Ballinderry crannóg no. 2', 73; 'Lagore crannóg', 229. HERON: 'Ballinderry crannóg no. 1', 234; 'Ballinderry crannóg no. 2', 73; 'Lagore crannóg', 229.

[172] For this information I am grateful to an expert on the behaviour of the heron, James B. Fox of Athy, Co. Kildare.

[173] Note, however, that the crane may also be a pest of crops, as it forms large flocks when migrating and wintering: see p. 236.

[174] Varro, *Rerum rusticarum* (ed. Hooper and Ash), 3.2.14, refers also to the rearing of cranes for the table.

[175] Scott and Martin, *Expugnatio hibernica*, 96.30–1 (bk. 1, ch. 33).

[176] Rivers, *Laws of the Salian and Ripuarian Franks*, 50.

[177] Ibid. 186.

[178] *CIH* i 73.19; 196.13 = *AL* iv 114.10.

[179] *CIH* iii 1043.1; Cf. *CIH* i 238.26–7 = *AL* v 472.29–31.

half fine (*lethdíre*).[180] He explains away the discrepancy by assuming that the bird had either been originally taken from the wild as an egg, or that both its parents were themselves tame (*ar petacht*). The breeding of this bird in captivity is also suggested by a commentator's reference to *corr co haimsir gnímraid* 'a crane up to the time of activity'.[181] Here *gnímrad* may refer to the bird's sexual maturity.

Another problem connected with the word *corr* is its apparent compound *corrguine* 'sorcerer, magician'. Just as *sedguine* means 'deer-killer, hunter' (see p. 273) and *bóguine* means 'cow-killer' (see p. 158), one would expect *corrguine* to mean 'crane-killer' or 'heron-killer'. The slaughter of the bird could thus have been part of some ritual, just as a cock is killed in voodoo cults. On the other hand, a description of the ritual of the sorcerer in *O'Davoren's Glossary* refers to his delivering a satire or malediction 'standing on one leg, with one arm outstretched(?), and with one eye shut'.[182] Here he would seem to be imitating the stance of the heron, perhaps with the implication that his satire has destructive power like the beak of a heron. In a review,[183] Brian Ó Cuív discusses briefly an article by Anne Ross on the Gaulish god Esus who is sometimes depicted with crane-like birds.[184] He suggests that the druidical practice of *corrguinecht* may be connected with the crane-cult for which Ross has collected evidence among the Celtic peoples, both continental and insular.

A quotation in *O'Davoren's Glossary* from a lost law-text states that there is no penalty for taking a *corr* or a *séig* (hawk) on another's land.[185] The glossator seems to assume that these birds are hunted or trapped because they are regarded as a nuisance rather than for purposes of domestication. Thus he refers to the fact that the hawk carries off young pigs and hens (see p. 189 below). Here *corr* could apply equally to the heron or to the crane; the former could be regarded as a nuisance because of its depredations on fish, whereas the latter could inflict damage on crops.[186] Another mysterious reference to the *corr* is found in a series of legal quotations on trapping

[180] *CIH* iii 1043.1–5; v 1873.25–36.
[181] *CIH* ii 709.37.
[182] *CIH* iv 1480.12–13 = O'Dav. 257 § 383. Cf. L. Breatnach, *Uraicecht na Ríar*, 140.
[183] Ó Cuív, 'Review of *Études celtiques*, vol. IX, fascicule 2', 337.
[184] Ross, 'Esus et les trois "grues"'.
[185] *CIH* iv 1526.1–2 = O'Dav. 460 § 1480.
[186] See p. 236.

and other topics.[187] This quotation reads *aurchur la cuirr*, which seems to mean 'a cast (of a spear) at a heron or crane'.[188]

Senén

The other bird included in the *Bretha Comaithchesa* list is the *senén* (*sinén*), which seems to mean 'old bird'.[189] Its identity is difficult to establish, and the term may in fact have been applied to a number of different species. In *O'Davoren's Glossary* it is equated with a word meaning 'eagle'.[190] Similarly, in an Old Irish penitential it corresponds to *aquila* 'eagle' in the Latin text.[191] This explanation would suit the etymology 'old bird', as the eagle can live to a great age: cases have been recorded of captive eagles living for over a hundred years.

On the other hand, a glossator on *Bretha Comaithchesa* identifies it as a hawk (*sebacc*).[192] The same identification is given in another entry in *O'Davoren's Glossary*, but here it is also equated with the *préchán*.[193] In Modern Irish *préachán* is confined to the crow family (especially the rook), but in the earlier language it was used of a variety of predators and carrion-feeders, including kites, hawks, and crows.

In the absence of conclusive evidence it seems best to take the word *senén* as one which could be applied to various birds of prey.

Crows

Birds of the crow family (*Corvidae*) are commonly kept as pets in many parts of the world. Fledgling ravens (*Corvus corax*), rooks (*Corvus frugilegus*) and jackdaws (*Corvus monedula*) which have fallen from their nests can be reared as pets with relative ease. They are particularly intelligent birds and adapt readily to domestication. The Old Irish law-text *Bretha Nemed Toísech* deals

[187] In the quotations at *CIH* iii 786.4–24, trapping is mentioned at 786.14 *bir airdil* 'hidden spike' and 786.19 *cuithech no bir* 'pit or spike'.

[188] *CIH* iii 786.15–16. A difficulty is that *aurchur* (*airchor*) is usually followed by *for* or *do*. The scribe adds the correction *no iarchor* above the line: this could hardly be for *iarcor* 'subsequent contract'. The quotation is glossed *.i. mīr do chuirr* 'i.e. a piece of (or for) a heron or crane'.

[189] *CIH* i 73.19; 197.1 = *AL* iv 114.11.

[190] *CIH* iv 1480.2 = O'Dav. 256 § 374 *Cufir .i. sinēoin.*

[191] E. J. Gwynn, 'An Irish penitential', 146 Cap. i § 3 = *IP* 260 § 3 (compare the Latin text at *IP* 216 § 5.6). I follow Binchy's emendation of *sinain* to *sinéoin*, *IP* 275 note[1].

[192] *CIH* i 197.3.

[193] *CIH* iv 1523.18 = O'Dav. 449 § 1414 *Senēn .i. prēchān no seabac.*

with offences by a pet crow (*teithre*)[194] which is expected to be kept on a string.[195] Legal commentary also refers to a pet jackdaw (*cáóc*).[196]

I have found no references to the caging of smaller birds.

Wild mammals

Bretha Comaithchesa lists only three pet mammals taken from the wild: the fox, the wolf and the deer.[197] The later commentator on this text provides a much more comprehensive list,[198] and ingeniously fits in the offences which they would be likely to commit with those of ordinary domestic animals.[199] The wild pig is thus regarded as likely to commit the same sort of offences – such as rooting up the ground – as the domestic pig. If it does so on a neighbour's land, the usual fines for pig-trespass apply.

The commentator equates the offences of the three largest wild carnivores – the wolf (*Canis lupus*, Irish *cú allaid*), the fox (*Vulpes vulpes*, Irish *sinnach*), and the otter (*Lutra lutra*, Irish *dobrán*[200]) – with those of the domestic dog. The offences of the red squirrel (*Sciurus vulgaris*, Irish *íara*) and of the stoat[201] (*Mustela erminea*, Irish *ess*) are equated with those of the domestic cat. The position of the pine marten (*Martes martes*, Irish *togán*[202]) is uncertain, either through scribal confusion, or because the lawyers found it difficult to decide whether it should be classed with the cat or the dog. In one version it is listed with the dog (cf. the name *crannchú* 'tree dog' also applied to this animal),[203] but in another version it is placed

[194] See *DIL* s.v. 1 *tethra*.

[195] *CIH* vi 2216.34.

[196] *CIH* v 1873.25 *isin cāōigh*.

[197] *CIH* i 73.19–20; 197.1–2 = *AL* iv 114.10–12.

[198] There are versions at *CIH* i 74.15–18 = *AL* iv 120.1–5; *CIH* i 98.31–5; *CIH* vi 2187.14–17. In the latter case the commentator is dealing with the sale of wild animals rather than their offences; however, the basic list is the same.

[199] Compare the commentary on the offences of a wild animal (*fíad*) which has been tamed (*CIH* iv 1152.6–15; v 1642.21–30).

[200] *Dobrán* = *doburchú*, lit. 'water-dog'. In the Rawlinson B 487 version of this list (*CIH* i 74.15 = *AL* iv 120.1) the scribe has omitted the initial *do-* of *dobrán*. This entry under *bran* 'raven' in *DIL* B 157.51 should therefore be transferred to *dobrán*.

[201] The weasel (*Mustela nivalis*) is not native to Ireland.

[202] This is a diminutive of the older term *togmall*. Legal commentary at *CIH* v 1585.28–30 gives the fox (*sinnach*) and pine marten (*toghān*) as examples of captive wild animals which are not eaten.

[203] See *DIL* C 509.55–6.

along with the cat.[204] In the third version it is listed both with the cat *and* the dog.[205]

The only native Irish ruminant, the red deer (*Cervus elaphus*, Irish *ag allaid*) is logically placed along with another ruminant, the domestic bovine, and the badger (*Meles meles*, Irish *brocc*) and wild pig (*Sus scrofa*, Irish *mucc allaid*) are fitted in with the domestic pig. Rather surprisingly, the hare (*Lepus timidus*, Irish *míl maige*) is classed in one version with the domestic hen.[206]

Later introductions

As we have seen in the Introduction, most of the written information on early Irish farming comes from the law-texts of the seventh and eighth centuries. These texts were copied, explained and expanded by later jurists, working mainly between the twelfth and the sixteenth centuries. In general, these later scribes stuck closely to the original texts, and ignored farming innovations brought in by the Anglo-Normans. For example, it is practically certain that the Anglo-Normans introduced rabbits (see p. 133 below), which are mentioned as early as *c.*1185.[207] Yet there are no references to rabbits in the post-Norman legal material in the Irish language.

Ass

The domestic ass or donkey (*Equus asinus*) originated in the semi-desert of north Africa, and was probably domesticated in Egypt by about 3500 BC.[208] It is frequently mentioned in the Old and New Testaments, and must therefore have been well known to the Irish in religious contexts. For example, the tenth-century poem *Saltair na Rann* refers to Christ's entry into Jerusalem on an ass (*assan*).[209]

Although the ass was of enormous importance as a beast of burden on Irish farms of the eighteenth and nineteenth centuries,

[204] *CIH* vi 2187.16–17.

[205] *CIH* i 98.32–4.

[206] *CIH* vi 2187.17. Paul Russell suggests to me that the commentator may be thinking of the 'form' – comparable to a hen's nest – which a hare makes in the open for its young.

[207] Curtis, *Ormond deeds* i 4 § 7.

[208] Clutton-Brock, *Domesticated mammals*, 91; Gautier, *La domestication*, 155.

[209] Stokes, *Saltair na Rann* 113.7688. *As(s)an* is a borrowing from Latin *asinus*. In later Irish this word is replaced by *as(s)al*, from the Latin diminutive *assellus*. See *LEIA* A-93 s.v. *asan*.

it hardly features at all in the earlier records of this country. There is no mention of it in the law-texts, nor in their associated glosses or commentary. In one Old Irish version of *Cáin Domnaig*, it is stated that a person should not ride on a horse or ass (*imrim for ech nó assan*) on Sundays.[210] However, this portion of text may owe more to Old Testament tradition than to current Irish practice.[211] In another Old Irish version, there is no mention of riding on an ass; possibly the redactor omitted it as being irrelevant in an Irish context.[212] To my knowledge, the earliest explicit reference to an ass on Irish soil is in an anecdote in a legal manuscript, which records that a cardinal came from Rome to instruct certain ecclesiastics, by name Ua Annócc, Ua Ceillchín of Cell Mór, Ua Slu[a]isti of Cúil Ó Sluaisti and Ua Glésáin.[213] They, however, stole the cardinal's horses, mules and asses (*eich 7 múil 7 asain*). As a result the Pope sold the tribute and dues of Ireland – which had formerly gone to him – to the English. This incident is alleged to have taken place in the reign of Domnall Mór Ua Briain, king of Munster, who died in 1194.

Mule

A mule is the offspring of a jackass (a male *Equus asinus*) and a mare (a female *Equus caballus*). It has the advantage of being larger and stronger than an ass, and having greater stamina and endurance than a horse.[214] Mules are frequently depicted in Mesopotamian art of the first millennium BC, and are known to have been of great importance in the Roman economy, being used for riding, ploughing, carrying baggage and drawing carts.

As in the case of the ass, there is little mention of the mule in early Irish sources (see previous section). It seems to have been an animal which was generally unfamiliar, so a ninth-century glossator

[210] O'Keeffe, '*Cáin Domnaig*', 202 § 17.

[211] Schmitz, *Die Bussbücher und die Bussdisciplin der Kirche* i 786 § 9 (Poenitentiale Laurentianum) . . . *et non mittat in opera non servum, non ancillam, non bovem, non asinum* 'and let him not put to work his man-servant nor his maidservant nor his ox nor his ass'. Cf. Exodus 20:17 (tenth commandment).

[212] Hull, '*Cáin Domnaig*', 160 § 1. In most respects the activities forbidden in § 1 of this text agree closely with those of O'Keeffe, '*Cáin Domnaig*', 200–2 § 17.

[213] *CIH* vi 1950.18–23; translated by Bergin, 'What brought the Saxons to Ireland'.

[214] Clutton-Brock, *Domesticated mammals*, 95.

on Priscian defines the Latin word *mulio* 'mule-keeper' in terms of the familiar horse.[215]

The offspring of a stallion and a female ass is known as a 'hinny' (usually 'jennet' in Ireland). It is of much less use to man than a mule and does not seem to be mentioned in early Irish sources.

Rabbit

The domestic rabbit (*Oryctolagus cuniculus*) is of southern European origin and was probably introduced to Ireland by the Anglo-Normans in the late twelfth century.[216] Rabbits were kept for food in specially constructed rabbit-warrens,[217] which are often mentioned in the records of medieval Ireland.[218] For example, the Rental of the Manor of Lisronagh, 1333, refers to the rabbit-warrens (*cunicularia*) on this estate which are valued at ten shillings. They are expected to provide an annual crop of twenty-four rabbits, worth one penny each.[219] Similarly, a deed of 1416 records a transfer of lands by Alice Archer and her son Simon MacCarrowyll (*Mac Cearbhaill*) to Richard O'Hedyane (*Ó hÉideáin*), archbishop of Cashel.[220] This includes a rabbit-warren in the tenement of Rathsax in the cantred of Elyoffogyrthe (*Éile Uí Fhógartaigh*).

Fish

The *Annals of Clonmacnoise* describe the theft in 1061 of two salmon kept in 'the kings ffountaine or fishpond' at Kincora,[221] and legal commentary contains a number of references to fishweirs (p. 287). So far as we know, the construction of special ponds for introduced fish such as pike, perch, carp, etc. belongs exclusively to the post-Norman period in Ireland.[222]

[215] *Thes.* ii 87.26–7 **mulio** *.i. custos mulorum .i. echaire* 'i.e. keeper of mules i.e. horse-keeper'. In the next gloss he gaelicizes Latin *mulus* as *múl.* The mule is also mentioned in Plummer, *Bethada* i 76 § xliv (142); trans. ii 74.

[216] Irish *coinín* 'rabbit' is a borrowing from Middle English *cunin, conyng,* (later *cony*).

[217] Irish *coinigéar* is a borrowing from Middle English *conynger* 'rabbit-warren'. This word is common in placenames (Joyce, *Irish names of places* i 481).

[218] Curtis, *Ormond deeds* i 34 § 74; 201 § 504.

[219] Curtis, 'Rental of the manor of Lisronagh, 1333', 45, 50.

[220] Curtis, *Ormond deeds* iii 14 § 22. Cf. 18 § 28; 47 § 66 (2).

[221] *Ann. Clon.* 178 s.a. 1061. See p. 291 below.

[222] For medieval English fishponds and their fish, see Rackham, *The history of the countryside,* 366. Cf. Mitchell, *The Irish landscape,* 184; *Reading the Irish landscape,* 174.

4

Offences by domestic animals

There have doubtless been disputes about injury or damage perpetrated by animals ever since people first learned how to domesticate them.[1] Most full-grown domestic animals are capable of causing injury to other animals or to humans. Furthermore, most domestic animals are herbivores, and therefore pose a constant threat to crops. Already in the earliest surviving law-codes, we find legislation dealing with such problems. The Code of King Hammurabi of Babylon (who reigned about 2000 BC) considers the case of an ox which has shown signs of aggressiveness but has not had its horns blunted by its owner; if it subsequently gores a freeman to death the owner must pay half a *mina* of silver.[2] This code also lays down penalties for the theft of cattle, sheep, asses or pigs,[3] and deals with the case of a shepherd who allows his sheep to graze on young corn belonging to another.[4]

The Book of Exodus deals with the killing of a man or woman by an ox under various circumstances (21:28–32), the accidental killing of an ox or ass in a pit (22:33–4), the killing of one ox by another (21:35–6), the theft of livestock (22:1–4), damage to a field or vineyard by livestock (22:5), and other issues involving livestock (22:9–13). Early Irish law on the theft of livestock clearly derives from Exodus 22:1, and it is likely that Exodus 21:35–6 is the model for a passage in the text on joint-grazing which deals with the killing of one bullock by another.[5]

The offences of livestock described in the Old-Irish law-texts range from trifling nuisance – such as defecation on a neighbour's land by a dog or cat – to attacks resulting in loss of human life. When dealing with damage or injury by livestock, the law-texts make a basic distinction between an 'animal-offence' (*ropchin*) and a 'human offence' (*duinechin*). In the former case, there is no malice or negligence on the part of the owner, and consequently

[1] For a general discussion on offences by things or animals, see MacCormack, 'On thing-liability (*Sachhaftung*) in early law'.

[2] Johns, *The oldest code of laws in the world*, 52 § 251.

[3] Ibid. 3 § 8.

[4] Ibid. 15 § 57.

[5] *CIH* i 192.24–33 = *AL* iv 102.7–19.

the fine is smaller. For example, if a horse which has never before behaved aggressively bites a passer-by without provocation, the owner's liability is merely for an 'animal-offence'. But if the horse is known to be a biter (*ech daintech*) the owner's guilt is greater, because he should have taken steps to prevent such an accident. His liability is therefore for a 'human offence'.[6]

In certain circumstances no liability at all is attached to injury or even death caused by livestock. This applies in the case of illegal or negligent behaviour on the part of the victim. For example, if a person is stung while robbing a bee-hive, he has no redress against the owner of the bees. There is likewise no liability if he is stung while carelessly watching the bee-hives at swarming time.[7]

GRAZING-TRESPASS

Most of the law-text on neighbourhood, *Bretha Comaithchesa,* deals with grazing-trespass. It is a composite text, and consequently shows some variation in the way in which such offences are treated, and in the legal terminology employed.

In anticipation of possible grazing-trespass by his livestock, each farmer must lodge a fore-pledge (*tairgille*) with his immediate neighbours.[8] In the event of trespass, the fore-pledge guarantees payment of the appropriate fine.[9] In some cases the fore-pledge itself is forfeit as payment for the trespass.[10] The text stresses the landowner's obligation to maintain properly constructed fences around his land.[11]

Much attention is devoted to the circumstances of the grazing-trespass and the amount of damage done. *Bretha Comaithchesa* makes the obvious point that there is regrowth of grass in summer but not in winter.[12] Consequently the

[6] Meyer, *Triads*, 22 § 168. The vice of biting is commonest among stallions: see West, *Black's veterinary dictionary*, 591.

[7] *CIH* ii 449.4–6 = *BB* 66 § 27.

[8] *CIH* ii 412.1–3 = *AL* i 260.9–11.

[9] *CIH* i 64.28 = *AL* iv 70.2; *CIH* ii 444.12 = *BB* 50 § 1 (see discussion at *BB* 90, 113–15).

[10] *CIH* i 197.2 = *AL* iv 114.12.

[11] *CIH* i 73.7–18 = *AL* iv 112.

[12] *CIH* i 69.32–3 = *AL* iv 90.1–2. Commentary (*CIH* ii 684.14–15; v 1579.33–4; 1865.32–41) states that if the victim of grazing-trespass does not take legal action before the grass has grown again, he has no entitlement to restitution (*aithgein*), but only to a penalty-fine (*smacht*).

fines for winter-trespass (*gamḟuacht*) are heavier than those for summer-trespass (*samḟuacht*).[13] Winter-trespass is deemed to apply during the three months of winter (*gaimred*) and the first two months of spring (*errach*). Summer-trespass applies during the last month of spring, the three months of summer (*samrad*), and the three months of autumn (*fogmar*).[14] It is clear that 1st April was regarded for legal purposes as being the date at which grass-growth was fully under way, and 1st November the date at which it ceased.[15]

The quality of the land which has been grazed is also taken into account. Thus the fine (*smacht*) when livestock trespass on preserved pasture (*etham ndíguin*) is twice as much as when they trespass on aftergrass (*athbronnad*) or moorland (*móin*).[16] Another factor which is taken into consideration is the hour at which the offence is committed. If there is grazing-trespass by night, it entails a full penalty (*ógcaithig*).[17] The text explains that this is because cattle should be locked up in an enclosure (*búaile*) at night.[18] Similarly, pigs should spend the night in a sty (*foil*), sheep should be in a pen (*lías*), and horses should either be properly tied up, or else housed in a stable (*inne*). According to *Bretha Nemed Toísech*, a farmer whose cattle are guilty of lying-down trespass (*feis*) must hand over a bushel of wheat and a milch cow for each period of twenty-four hours in which the trespass continues.[19]

[13] *CIH* i 70.3–6 = *AL* iv 90.16–20. According to *Bretha Comaithchesa* commentary at *CIH* ii 576.10–11, eight bushels of grain are paid for winter-trespass, and four bushels for summer-trespass. In both cases a heifer (*ag*) is the fine (*díre*) for the grass.

[14] *CIH* i 69.19–29 = *AL* iv 88.18–26. Summer was regarded as comprising the months of May, June and July. Thus in the late seventh century Adomnán took 15th June to be the middle of Summer (A. O. and M. O. Anderson, *Adomnán's Life of Columba*, 328–30 (new pagination: 96–8) = ff. 54b–55a). The summer solstice was actually on 18th–19th June at this date.

[15] By the sun (i.e. length of day) 1st April in the seventh century corresponds to about 4th April in the twentieth century; 1st November likewise corresponds to about 4th November. For a brief discussion of the mechanism, see p. 460 (footnote).

[16] *CIH* i 70.3–5 = *AL* iv 90.16–19.

[17] *CIH* i 72.4 = *AL* iv 96.21. The reading *ocaichach* of the MS shows coalescence of the lenited *g* and *c* (Thurneysen, *Grammar*, 87 § 137).

[18] *CIH* i 72.10–11 = *AL* iv 96.23–4. By day they should be under the control of a cowherd (*búachaill*); cf. *CIH* vi 2134.4–5.

[19] *CIH* vi 2228.8–10. *Bretha Comaithchesa* defines *feis* as 'every lying and remaining', *CIH* i 66.37–8 = *AL* iv 78.8–9.

The distance which the trespassing livestock have travelled has a bearing on the severity of the penalty. If they have only crossed one or two holdings, or a road, or a shallow stream, it is regarded as a lesser form of grazing-trespass (*tairsce*).[20] But a more serious view is taken if they have crossed three or four holdings. In this case the trespass is classed as *ruiriud* – which literally means 'great running' – and entails a full penalty. The text holds that to allow cattle to stray so far from their own fields betrays complete neglect.[21]

As every livestock farmer knows, some individuals in his herd are particularly likely to break out – in Modern Irish a cow of this disposition is decribed as a *bó bhradach*. The Old Irish law-texts similarly recognise that a single individual may act as ringleader and initiate a general break-out. This may be of special legal significance in cases where livestock belonging to more than one owner are grazing together. *Bretha Comaithchesa* decribes a bovine of this nature as a *foirrgid* 'raider'. If this animal causes the rest of the herd (*éit*) to trespass, it is held responsible for half the fine payable to the landowner.[22] The same text deals with the case of a pet piglet (*orcpheta*) which is guilty of leaping-trespass (*airlim*) into a cornfield four times in one day. The other pigs in the herd (*trét*) have only trespassed once. In this situation half the liability devolves on the pet piglet and the other half on the rest of the herd.[23]

In certain circumstances, no liability is attached to grazing-trespass. For instance, if cattle commit leaping-trespass when pursued by men or dogs, no penalty (*smacht*) is due from their owner.[24] There is likewise no penalty if they trespass when in heat (*oíbel*)[25] or in any situation of panic or emergency. According to a glossator, cattle frightened by thunder or lightning would fall into

[20] *CIH* i 75.15–17 = *AL* iv 126.5–7. *Tairsce* seems generally to be used for trespass over unfenced land: see discussion at *BB* 94.

[21] *CIH* i 77.38–9 = *AL* iv 136.19–20 *ōgcathach insin, ar is ōg a follugud* 'that is full penalty, because its neglect is complete'.

[22] *CIH* i 73.1–2; 195.10–19 = *AL* iv 108.13–14.

[23] *CIH* i 72.14–30 = *AL* iv 108.8–12.

[24] *CIH* i 71.1–2 = *AL* iv 92.25–7.

[25] A glossator (*CIH* i 71.5 = *AL* iv 94.4) suggests that this refers to the heat of the sun, but *oíbel(l)* (*aíbel(l)*) is well-attested in the sense of sexual excitement in animals. It also is used of the frenzy of cattle tormented by gadflies: see T. F. O'Rahilly, 'Ir. *aobh, aoibheall,* etc.'. See p. 203 for a discussion on warble-flies (gadflies) in cattle.

this category.[26] There is also no penalty for any form of trespass by a bovine which has lost its senses, or which returns to its familiar surroundings (presumably after having been moved to another place).[27]

A tradition is recorded in a gloss and commentary on Heptad 43 that livestock which are found trespassing on a king's land (*mruig ríg*) on the day of his inauguration become his property. This may have been a widespread royal prerogative, although the reference is only to the entitlement of the king of Cashel to any livestock found in Fíad Mugain in Éile (Ely O'Carroll) on the day of his inauguration. It is also claimed that livestock found on the road of Adamar (Rót Adamair) in Uí Chonaill Gabra are forfeit to the abbot of Lismore on the day he takes office.[28]

Aircsiu

If a farmer observes his own livestock on his neighbour's land but does nothing about it, he is guilty of the offence of *aircsiu* 'looking on'.[29] Because there is malice or negligence on his part, this offence is treated more seriously than most other forms of trespass, and is classed as a 'human offence' (p. 134). It consequently entails a heavier fine, described as a 'human penalty' (*duinechaithig*).[30]

Án 7 tán

A fine of five *sét*s is to be paid for *án 7 tán* 'driving to and fro'.[31] This is the offence of driving cattle across a neighbour's land without permission. The law requires that cattle should only be driven across a neighbour's land with his prior consent.[32] To minimise the risk of damage to his land, the operation must be carried out

[26] *CIH* i 71.6 = *AL* iv 94.6; cf. *CIH* ii 575.35; 678.4 *airlim ria torainn no aībaill* 'leaping-trespass before thunder or heat'.

[27] *CIH* i 41.2 = *AL* v 268.18.

[28] *CIH* i 40.2–9; v 1844.33–40 = *AL* v 266.1–7; *CIH* iii 908.15–22. Fíad Mugain (also Tír Mugain (Mumain)) has been identified with Fithmoone, near Thurles, Co. Tipperary (Hogan, *Onomasticon* s.vv. *fiadh mughaine, fid mogain, tír mughain*).

[29] *CIH* i 235.29 = *AL* v 464.7–8; cf. *CIH* iii 913.5–6; *CIH* vi 2228.7.

[30] *CIH* ii 580.24–5 = *AL* iv 146.16–17.

[31] *CIH* i 205.15 = *AL* iv 156.20; cf. *CIH* ii 571.8 = Binchy, 'An archaic legal poem', 158.72.

[32] The neighbour cannot refuse the cattle-owner's request if the land has no other access. But according to commentary at *CIH* v 1858.29–36 = *AL* iv 158.14–22 a fee must be paid.

by six men – three representatives of the landowner, and three of the cattle-owner.[33]

If a cattle-owner puts his cattle into the enclosure (*búaile*) of another farmer, he must pay one *sét*. The author of the text concedes that his act does no more harm than good – the glossator is no doubt correct in explaining that the value of the grass consumed by the cattle is no greater than the value of the dung which they leave behind in the enclosure. Nonetheless, the cattle-owner must pay a fine because of the stated legal principle that 'nothing done without permission is good'.[34]

If someone drives the cattle of his neighbour into his (i.e. the neighbour's) pasture, he must pay a fine of three *séts*. This is the same fine as he would pay if he were to drive his own cattle into the neighbour's pasture.[35] If someone breaks open a fence so as to allow cattle (his or another's) into grassland, he must pay a fine of five *séts*.[36] Further penalties may also be due for the actual damage to the fence: see p. 378.

As with other forms of trespass, the rules are relaxed in cases of emergency, described as a 'driving of necessity' (*án éicne*). For example, no fine is due for driving a cow in heat to a bull across a neighbour's land.[37]

Fothlae 7 tothlae

These offences are mentioned together in a number of law-texts.[38] Both terms are compounds of the verb *tlenaid* 'steals', which suggests that these trespasses are carried out in a secretive or underhand manner. The only detailed descriptions are found in Old Irish commentary on *Bretha Comaithchesa*.[39] In the case of *fothlae*, the commentator envisages a party of guests arriving at a person's house and unyoking their horses on his neighbour's land.

[33] *CIH* i 205.11 = *AL* iv 156.18–19.

[34] *CIH* i 236.13–14 = *AL* v 466.1–3.

[35] *CIH* i 236.21–2 = *AL* v 466.4–5.

[36] *CIH* ii 385.22 = *AL* i 168.4 (*airba* omitted). The figure of five *séts* is supplied by a glossator, *CIH* i 386.7–8 = *AL* i 174.11. See Binchy, 'An archaic legal poem', 157.31.

[37] *CIH* i 205.15–16 = *AL* iv 156.20–21.

[38] E.g. *CIH* ii 571.8 = Binchy, 'An archaic legal poem', 158.71; *CIH* ii 675.5 = Appendix A, text 1 § 4 (2).

[39] *CIH* i 194.7–14; ii 578.10–18 = *AL* iv 106.20–108.3. There is a a corrected translation at Binchy, 'An archaic legal poem', 164. He expresses doubt as to whether the commentator's examples of *fothlae* and *tothlae* reflect the true distinction between these terms, but makes no further suggestions.

If the host, noticing his guests' bridles, fails to ensure that they have unyoked their horses on his own land, he is guilty of *fothlae*, and must pay for the damage done by the horses. In another passage, *fothlae* is listed as an offence liable for a 'human penalty' (*duinechaithig*).[40]

In the case of *tothlae* the host plays a more active rôle in the horses' trespass. According to the commentator, he makes a remark such as 'this is fine land to unyoke in', and thus encourages his guests to unyoke their horses on a neighbour's land. He must therefore pay for any damage which they do.

The above examples illustrate the offences of *fothlae* and *tothlae* in relation to horses, but we can assume that other livestock could also be guilty of this form of trespass.[41]

Damage to crops

The destruction of a crop by trespassing livestock could cause severe hardship in the affected household. Predictably, the main emphasis in legal material is on corn (*ith*), with special mention of young corn (*fochenn*)[42] and corn in a rick (*dais*).[43] In spite of the seriousness of such offences, legal commentary enjoins that domestic animals found in corn or other crops must not be killed. Instead, they should be seized and their owner should pay a fine appropriate to the amount of damage done.[44]

Hens are identified as a particular menace to corn.[45] Commentary on *Bretha Comaithchesa* lists the three hen-offences outside a yard as 'depredations on drying-kilns, barns and corn-ricks'.[46] Another version refers also to their depredations in cornfields and mills.[47] The fine is fixed at one bushel (*míach*) of grain for

[40] *CIH* ii 580.25 = *AL* iv 146.17.

[41] The law-text *Di Astud Chirt 7 Dligid* refers to the man who secretly introduces (*do-tlen*) his cattle onto a neighbour's pasture (*CIH* i 236.27–8 = *AL* v 466.6–7). For this he must pay a fine of three *sêts* with restoration of equivalent grass. This is probably a case of *tothlae* (*to-tlen*), though at *DIL* D ii, 383.38 it is listed under *di-tlen*.

[42] E.g. *CIH* ii 385.9 = *AL* i 166.28; *CIH* i 266.43 = *AL* iii 186.2–3.

[43] *CIH* ii 384.21 = *AL* i 166.28.

[44] *CIH* ii 698.31–2.

[45] E.g. Wasserschleben, *Die irische Kanonensammlung*, 215, bk. 53 ch. 9.

[46] *CIH* i 73.23–4 (omitted *AL*) *a trī cercfogla a sechtarlis .i. fogal re hāthaib 7 [ṡ]ablaib 7 daisib arbhu.*

[47] *CIH* i 74.9–11 = *AL* iv 118.21–3; cf. *CIH* i 98.28–9.

each hen, up to seven hens.[48] A cock, however, is only liable for a fine of half a bushel. If hens have been guilty of habitual trespass (*bithbinche*) on a neighbour's land, the law requires that each of them should wear a hood (*cochall*) to restrain them. If their owner fails to comply with this requirement, he is liable to pay a heavier 'human penalty' (*duinechaithig*).[49] Damage done by other domestic fowl or by pet birds (such as herons or ravens) is assessed at the same rate as that of hens.[50] Horses, too, are represented as a threat to growing corn,[51] and *Bretha Nemed Toísech* refers to the destruction of a rick of corn by horses.[52]

Other crops may also be damaged by domestic animals. Again, hens seem to have been a major nuisance. Commentary on *Bretha Comaithchesa* gives the three hen-offences in another's enclosure (*lis*) as 'swallowing bees,[53] and destroying madder and onions'.[54] It seems likely that the damage to these crops is caused mainly by their scratching at the roots. For offences by hens in an enclosure or garden (*lubgort*) the owner must pay a half scruple of silver.[55]

Damage by sheep to a crop of woad (*glaisen*) is mentioned in the Old Irish tale *Scéla Éogain 7 Cormaic*.[56] As a young man Cormac mac Airt was approaching the gates of Tara for the first time, when he saw the steward of the king Mac Con telling a woman something which caused her to weep. On enquiring what was wrong, he was told that the woman's sheep had broken into the queen's woad-garden (*glaisengort*) and had eaten the leaves off the plants. Mac Con had passed judgement that the woman's sheep be forfeit for

[48] *CIH* i 74.12–13 = *AL* iv 118.25–6. This passage implies that the maximum payable for hen-trespass *i sechtarlis* was seven bushels, even if more than seven hens were involved.

[49] *CIH* i 73.26–9 = *AL* iv 116.21–5.

[50] *CIH* i 73.30–31 = *AL* iv 116.26–7.

[51] Meyer, 'Conall Corc and the Corco Luigde', 60.2; Hull, 'Conall Corc and the Corco Luigde', 899.

[52] CIH vi 2216.38.

[53] See p. 180 below.

[54] *CIH* i 73.22–3 = *AL* iv 116.18–19 *a tri cercfogla a lis: maethslucud bech 7 lot roidh 7 cainninne*; cf. *CIH* iii 802.21. For a discussion of these crops, see Chapter 8. Another version (*CIH* i 74.9) refers more generally to hen-offences 'against bees and plant(s) and corn-ricks' (*fri beochu 7 lus 7 cruacha*).

[55] *CIH* i 73.25 = *AL* iv 116.20–21.

[56] Ó Cathasaigh, *Heroic biography*, 122.85–97; O Daly, *Cath Maige Mucrama* 70 §§ 17–18, cf. 58 §§ 63–4.

their offence, and this was the cause of her distress. Cormac immediately pointed out to the steward that the judgement should have been 'one shearing for another' i.e. the woman should only have to forfeit the shearings of her sheep in recompense for their shearing of the woad-plants. When Mac Con was told of this judgement he realised that he had been guilty of injustice, and handed over the kingship of Tara to Cormac.

This tale is obviously apocryphal, but it serves to illustrate two contrasting views on how the law should deal with offences by domestic animals. The principle that an animal be forfeit to its victim is cited in a number of texts. For example, the law-text on beekeeping *Bechbretha* refers to the general principle that an animal is forfeit for its offence (p. 156), and a story in *Cormac's Glossary* illustrates this principle in relation to damage to a knife-handle by a pet dog (p. 149). *Bretha Nemed Toísech* likewise states that every type of animal is forfeit for its first offence (p. 154). However, the author of *Scéla Éogain 7 Cormaic* clearly felt that a fine was a more appropriate punishment than forfeiture of the offending animals in a case of grazing-trespass. For the payment of fines in fleeces of wool, see p. 75 above.

Rooting-trespass

The authors of the law-texts take into account the fact that the various types of livestock employ different methods of grazing, and consequently damage grassland to varying degrees. Cattle, for example, have a bony pad instead of top teeth, and therefore cannot crop grass as low as horses, which – as one legal quotation points out – 'graze to the soil'.[57] Our sources are in agreement that the most destructive grazing is the rooting of pigs, as they 'dig down to the gravel (subsoil)'.[58] Pig-trespass of this type is therefore specially penalized. According to *Bretha Comaithchesa*, if pigs root up a neighbour's land, their owner must provide alternative land for the neighbour's use until the sward has grown again on the damaged land. Presumably he must also fill in the

[57] *CIH* iii 788.11 *geilit eich co hūir.* The same observation is made in *TBC* LL 13.444. Another tale likewise contrasts the grazing of horses and cows (M. E. Byrne, '*Airec Menman Uraird Maic Coisse*', 59.10–13).

[58] *CIH* vi 788.11 *claidhit muca co grian.*

holes made by his pigs, though this is not specified in the text.[59] An ingenious test is used to decide when this land is ready for re-use. Two horses are yoked together (*i córait*), and are loosed onto the field. If no earth adheres to their teeth as they graze, the land is declared to have recovered from the rooting-trespass.[60]

In addition to the above, commentary on *Bretha Comaithchesa* specifies fines which must be paid by the pig-owner. For every fully grown pig (*muc mór*) which has trespassed, he must pay a bushel of grain, for every young sow (*céis*) he must pay half a bushel, and for every piglet (*banb*) he must pay four scoops (*máim*) of grain; in another version these fines are halved.[61]

In cases where pigs are simply guilty of grazing-trespass (*gleth*) rather than rooting-trespass (*fochlaid*) their owner pays the same fines as for other livestock.[62] Another law-text makes the correct observation that sows do not root up the ground at the time of farrowing (*ré duith*).[63]

Heptad 45 considers the case of a domestic animal which gives birth while trespassing on another's land. No fine is payable for any damage or soiling of the ground at the birth.[64]

Old-Irish commentary on *Bretha Comaithchesa* includes digging under a house (*fobach n-aitreibe*) among the offences which dogs are liable to commit.[65]

Defecation

The dung of grazing animals is generally regarded as beneficial because of its use as a fertilizer.[66] The only reference which I

[59] Commentary at *CIH* i 69.1–3 = *AL* iv 86.19–21 refers to holes made by trespassing livestock (*cethra*). The owner of these animals must fill up the holes with the same type of soil, which he must press and stamp down (*a sonnad ⁊ a sálad*). There is a similar passage at *CIH* i 291.1–5 = *AL* iii 296.11-18, which is discussed under 'defecation' below.

[60] *CIH* i 72.1–4; 191.8–10 = *AL* iv 96.17–20; cf. *CIH* ii 576.18–19.

[61] *CIH* i 191.15–19 = *AL* iv 98.19–24. Commentary at *CIH* i 191.3–5 = *AL* iv 98.13–16 refers to a procedure alleged to be 'in the old judgements' but which seems quite impractical. This is that the owner of the rooting pigs should fill up each hole alternately with corn and butter.

[62] *CIH* i 72.1 = *AL* iv 96.16–17.

[63] *CIH* iii 897.26 = Appendix A, text 2 § 3.

[64] *CIH* i 41.20 = *AL* v 268.21 *fourt cethra ac touth.* The glossator takes *fourt* (otherwise unattested) to refer to the afterbirth.

[65] *CIH* i 197.9–10, ii 579.2 = *AL* iv 122.5. Cf. *CIH* iv 1501.25–6 = O'Dav. 352 § 919.

[66] *CIH* i 236.13–14 = *AL* v 466.1–3.

have noted where defecation by all domestic animals is treated as an offence is when this occurs on paved roads – presumably near dwellings. Legal commentary counts 'fouling a street and pavements' (*salchad sráite 7 chlachán*) as one form of the general offence of *fodraime* 'damage to crops, fences and roads'.[67] Another commentary deals with the special case of cattle which have been grazed without permission on the hill (*dind*) where a public gathering is to take place, and have cut up and soiled the ground. As well as paying fines, their owner must press down (*sonnad*) and stamp (*sálad*) the ground, and cover over the affected area with fine clay of the same type. On the day of the meeting, he must provide rugs for the kings to sit on, and rushes for those of lesser rank.[68]

Bretha Comaithchesa takes a severe line towards dog-owners who allow their dogs to defecate on other people's land. The dog-owner is required to remove the faeces and to replace the contaminated soil. He must then give the land-owner the same quantities of butter, curds and dough.[69] Commentary on this text specifies that all soil which is affected even by the stench of the faeces must be removed.[70] Cat-excrement is particularly vile, so it is not surprising to find that the offences of domestic animals listed in *Bretha Nemed Toísech* include the fouling of floor-strewings of rushes by a kitten (*caitíne*).[71]

As we have seen above, legal commentary pays special attention to the damage done by hens to corn and other crops. Hens may also be a nuisance in a neighbour's dwelling-house: the three hen-offences inside a house are given as 'theft, pollution, and squirting'.[72] Of these, theft (*foxal*) clearly refers to the taking of food by the hen within the house. Pollution (*corbad*) presumably includes drinking liquids from a container and scratching or pecking at foodstuffs, thereby making them unfit for human

[67] *CIH* iii 802.20–22.

[68] *CIH* i 291.1–5; v 1647.10–14 = *AL* iii 296.11–18.

[69] *CIH* i 74.26–30; 197.6–17 = *AL* iv 120.15–22. Commentary at *CIH* i 197.10–11; ii 579.2–3 = *AL* iv 122.6–7 advises a slightly different procedure: the faeces is removed and replaced by a sod which is covered by cow-dung (*bóchor*) for a month.

[70] *CIH* i 98.21–6. It seems that if the dog-owner (*fer bunaid*) witnesses the offence, but does nothing about it, he is liable for a 'human penalty' (*duinechaithig*).

[71] *CIH* vi 2216.37–8. Liam Breatnach points out that one should read *cía ro corba sretha naís* 'in the event that it pollutes strewings of rushes (*nóes*)'.

[72] *CIH* i 73.21–2 = *AL* iv 116.17–18.

consumption.[73] Squirting (*dórtad*) must refer to defecation by the hen within the house.[74] For any of these offences the owner of the hens must pay three loaves of bread.[75]

Theft of food

The main domestic animal likely to steal food from a neighbour's house is the cat. In relation to such offences, the law-text *Bretha Étgid* lays down the general principle that 'the kitchen is an immunity for cats' (*blaí catt cuile*).[76] The accompanying commentary explains that the owner of the cat is not liable if food has been carelessly left in the kitchen without proper supervision. On the other hand, if the cat takes food from a secure place of storage (*daingen tige*) or a vessel (*lestar*), the owner must replace the food which has been eaten or damaged. If the cat has been previously guilty of the same offence, a fine must also be paid.

Flying-trespass

The Old-Irish law-texts devote an extraordinary amount of attention to the legal problems associated with trespass by the flying animals of the farm, particularly honeybees. One law-text states that the three most difficult trespasses are those of bees, horses and pigs.[77] It describes a method of identifying the trespassing bees, but unfortunately the text is difficult to understand. Possibly, flour was sprinkled on the bees as they trespassed so that a witness could recognise them when they returned to their hives.[78]

The basic problem for the early Irish lawyers in the case of bees was how to fit their foraging into the general framework of the law of animal-trespass. Bees forage for nectar and pollen within a radius of about a mile from their hives, and are therefore almost certain to trespass on land which does not belong to the beekeeper. In all other legal systems – so far as I am aware – bee-trespass is ignored, except in cases of swarming or where there is danger to people or

[73] One can compare Irish Canons I § 17 (*IP* 162) which sets a severe penance for drinking what has been contaminated by a hen, cock or other birds.

[74] For this use of *dórtad* see *DIL* D ii, 367.14.

[75] *CIH* i 73.25 = *AL* iv 116.20.

[76] *CIH* i 290.32; v 1647.1 = *AL* iii 296.1.

[77] *CIH* iii 898.20 = Binchy, 'A text on the forms of distraint', 80 § 11 = *BB* 189(c). Cf. *CIH* ii 573.25.

[78] See discussion at *BB* 190.

livestock. The early Irish lawyers, however, viewed the matter differently, and felt that the owners of the lands on which the bees foraged must be compensated. The author of the main law-text on beekeeping, *Bechbretha,* invokes the principle that 'no one should support a gratuity for another in Irish law'.[79] This approach is echoed in later commentary which alleges that bees take away 'taste and produce' by foraging on a neighbour's land.[80] This seems to be only half true: if a beekeeper owned many hives his bees could reduce the quantity but not the quality of his neighbour's honey. The solution proposed in *Bechbretha* to the problem of bee-trespass is complex, but it has the advantage of spreading the asset of beekeeping throughout the community. After a three-year period of grace the beekeeper is required to give a swarm of bees to his immediate neighbours, starting with the one whose land is nearest.[81] After a few years all his neighbours will therefore be beekeepers themselves, and the trespasses of their bees should cancel out.[82]

There are differences of opinion in the law-texts as to the correct terminology to be used in relation to bee-trespass. In *Bechbretha* the trespass of a bee is classed as *tairsce,* a word used in other texts to describe the less serious forms of cattle-trespass.[83] But in a passage on the trespass-penalties of flying animals, bees are regarded as being guilty of *airlim* 'leaping-trespass' because they cannot be restrained by fences.[84]

The same passage also refers briefly to the offences of geese, hens and pet herons. In the case of *airlim* by a hen, the owner must pay a fine (*smacht*) consisting of a twelve-inch tub full of grain. He must also ensure that its wings are clipped and spancels put on its legs.[85] The same topic is also touched on in a legal poem on land-law, which refers to 'flying-trespass of birds, of hens, of herons, if they are pets'.[86]

[79] *CIH* ii 445.3 = *BB* 54 § 10.

[80] *CIH* iii 788.11.

[81] *CIH* ii 447.5–6 = *BB* 62 § 19.

[82] However, at *CIH* ii 450.24 = BB 74 § 38 it is implied that one beekeeper might in certain unspecified circumstances demand fines for trespass from another beekeeper.

[83] *CIH* ii 444.14 = *BB* 50 § 3.

[84] *CIH* i 196.20, ii 578.26 = *BB* 186(b). For further discussion on the terminology relating to bee-offences, see *BB* 187.

[85] *CIH* ii 578.34–6 = *AL* iv 118.27–30.

[86] *CIH* ii 571.9 = Binchy, 'An archaic legal poem', 158.76. The MS has *aurlimm en cerrcc corr mad beth pettai.* Following Thurneysen's suggestion ('Aus dem irischen

Fig. 6. This illustration in the bottom margin of p. 281 of the fourteenth-century Book of Ballymote (MS no. 536 = 23 P 12) in the Library of the Royal Irish Academy shows a naked man pursued by a large dog. It has been reproduced by kind permission of the Council of the Royal Irish Academy.

Offences against persons

The larger domestic animals are stronger and heavier than men, and are therefore capable of inflicting severe injuries, which may be fatal. Smaller domestic animals can also cause injury by tooth, hoof, horn, beak, claw, talon or sting.

On early Irish farms, it is likely that most injuries by domestic animals were inflicted on the farmer or on a member of his immediate family. In such cases there would be no need for the law to become involved. But if the victim were a neighbour, guest, passer-by or employee, he – or his kin in the event of his death – might have a legal claim against the owner. The law-texts devote much attention to the circumstances of the injury or death. If it can be demonstrated that the victim was wholly to blame, the owner is free from liability.

Injury by dogs

The law-texts deal more frequently with attacks on people by dogs than by any other domestic animal. This is not surprising: as we have seen in Chapter 3, the guard-dog (*árchú*) was specially bred for its bulk and ferocity. Attacks by dogs also feature in other early Irish sources. For example, Cú Chulainn's narrow escape from a guard-dog is one of the best-known episodes in the sagas (see p. 115). Another dog-attack – possibly based on an actual incident – is described in the genealogies of the saints.[87] A priest of Tír Dá

Recht IV', 204), Binchy takes from *cerrc* to *pettai* to be a gloss and omits these words in his edition ('An archaic legal poem', 165). But I believe that they should be retained in spite of the lack of connective alliteration: for some discussion on the rules of alliteration in Old Irish, see Kelly, 'A poem in praise of Columb Cille', 5; Kelly, '*Tiughraind Bhécáin*', 70.

[87] Ó Riain, *Corpus genealogiarum sanctorum Hiberniae*, 111 § 668.

Glas named Semplán went with his attendants on an errand to the house of Diarmait of Lecc na Sinnach. They found Diarmait clearing the entrance-path to his house with a shovel. Diarmait's dog attacked the party and bit Semplán, who then struck the dog. Diarmait retaliated by hitting Semplán with his shovel, thereby smashing the service-set (*meinistir*) of Columba, which was on the priest's back. The case went for arbitration to Ruidén, king of Uí Dróna, and Diarmait had to pay a fine of seven *cumal*s to the Columban community.

To reduce the incidence of attacks on legitimate visitors, the law-texts specify various forms of warning or restraint. The fragmentary text on dog-law, *Conṡlechtae*, states that a dog which is known to be aggressive (*cú foilmnech*) should have a bell (*cloc*) or rattle (*crothal*) round its neck.[88] The owner's obligation to keep the dog chained up or on a leash during the day is also stressed.[89] According to legal commentary on *Conṡlechtae*, a guard-dog should be tied to the ninth stake (*cleth*) from the doorway so that it cannot bite people in the middle of the house or out in the pathway.[90] Anyone who is foolish enough to get within the range of a leashed guard-dog has no claim against the owner.[91] The owner of a herd-dog (*canis pecorum*) is immune from liability for any injury which it may inflict within the cow-enclosure (*in bovello*) or in the fields where the cattle are. But he is liable for any injury which may be inflicted outside.[92]

A guard-dog may only be loosed on its owner's property from the time the cows come into the milking-place (*indes*) until sunrise. Any injury which it inflicts between these times is free from liability.[93] From a brief quotation in a legal glossary it would also seem that a daytime attack is free from liability if there is no tearing of the person's skin or clothing.[94]

An injury inflicted during a fight between two dogs (*congal*) is generally free from liability.[95] One law-text goes so far as to say that

[88] *CIH* i 111.22; iv 1483.13–15 = O'Dav. 271 § 471.
[89] Wasserschleben, *Die irische Kanonensammlung*, 215, bk. 53 ch. 5.
[90] *CIH* iv 1389.26–7 = *AL* iii 412.19–20; *CIH* iii 807.11–12. The commentator presumably envisages a circular house of wattling woven around vertical stakes: see p. 363.
[91] *CIH* vi 2216.33.
[92] Wasserschleben, *Die irische Kanonensammlung*, 214, bk. 53 ch. 5.
[93] *CIH* iv 1395.19–20; cf. *CIH* i 313.36–42; v 1550.31–2.
[94] *CIH* iv 1513.9–10 = O'Dav. 404 § 1179.
[95] *CIH* i 268.12 = *AL* iii 192.9.

even if it is a king who attempts to separate fighting dogs, he has no claim against the owners if he is injured.[96]

The severity with which the law regards an attack by a dog may depend on whether its aggressive nature was previously known to the owner. According to legal commentary, if it is a 'dog of first offence' (*cú chétchintach*), the owner is only responsible for medical care (*othras*) or recompense (*aithgein*): no further fines are due.[97] Other law-texts treat a dog's first offence quite differently, and invoke the principle that it becomes the property of its victim.[98] He can then decide whether to kill the offending animal or keep it himself. The same principle underlies a legendary legal case quoted in *Cormac's Glossary*.[99] Cairpre Músc coveted a pet dog (*oirce*) which belonged to his relatives in Britain. He devised a trick to secure it: he rubbed fat and beef lard on the ornamental handle of his knife, and left it in front of the dog at night. On the following morning, he found that the dog had thoroughly gnawed the handle, so he pretended to be greatly enraged. His hosts offered to pay compensation, but he demanded that the dog itself be given as recompense for the damage. He argued that it was a principle of British law that every guilty one is forfeit for its offence (*cach bibdu ina chinaid*).[100] His view prevailed, and he brought the dog back with him to Ireland.

If a person is injured by a dog, he must give public notice (*airfócrae* or *apad*) of the offence. According to a law-text on distraint, he is required to place a plank of wood across the dog's feeding-trough with notice that it is not to be fed.[101] Presumably, if the owner has already admitted his liability – or at least given a pledge to indicate

[96] *CIH* iv 1395.24–5.

[97] *CIH* vi 2151.31 (other legal commentary refers to circumstances in which full fines are to be paid even for a first bite (*CIH* i 312.35–42 = *AL* iii 414.24–416.5)). *Othras* involves taking responsibility for the nursing of the injured person, paying all medical fees, providing a substitute to do his or her normal work, and compensating for any permanent disability. For a general account of the law relating to illegal injury, see *GEIL* 129–32.

[98] *CIH* ii 449.11–19 = *BB* 66 § 29 (especially gloss[f]); cf. Meyer, *Triads*, 22 § 169; *IP* 174 § 2 = *GEIL* 354 (5) § 2.

[99] Meyer, *Sanas Cormaic*, 75–6 § 883.

[100] In fact this seems rather to be a principle of Irish than of British law: I have found no trace of it in relation to animal-offences in medieval Welsh law. See further discussion on p. 142 above.

[101] *CIH* iii 897.37–9 = Binchy, 'A text on the forms of distraint', 78 § 4.

his willingness to submit to arbitration – this formality is unnecessary. If the owner ignores the notice and continues to feed the dog, his guilt is thereby increased and he must pay the fine for a 'human offence' (*duinechin*) rather than for an 'animal offence' (*ropchin*).[102]

An offence committed by a dog straying during the day is also classed as a 'human offence'. The explanation given for this rule is that it should be either tied up or on a leash at this time.[103] However, legal commentary on *Conṡlechtae* takes a less severe view of the straying of a bitch in heat (*cú ṡodaig*). Nonetheless, her owner is required to warn his four immediate neighbours of her state.[104]

In some cases, a person other than the owner may be held liable for the dog's offences. This applies if someone is guilty of inciting (*muilliud* or *inmuilliud*) another's dog against people or livestock.[105] It may also apply in the case of a wandering dog (*cú ṡeichill*)[106] which leaves its owner and attaches itself to another household. If the householder feeds it for six days, he becomes liable for its offences rather than the original owner. To avoid responsibility for 'last sighting' (*tigradus*), he must drive off the dog three times, and give public notice (*escaire*) of its presence.[107] This category of dog is perhaps to be identified with the *táidchú ḟoíndil* 'wandering thieving dog', mentioned in *Bechbretha*.[108] The text emphasizes that it should not be harboured by anybody, no matter how high-ranking.

Injury by cattle

As one would expect, the law-texts devote more attention to injuries caused by male than by female bovines. The dangerous temperament of the bull (*tarb*) is stressed, and it is made clear that in some circumstances a person injured by a bull has no legal redress against the owner. The text on accidents, *Bretha Étgid*, gives

[102] *CIH* vi 2195.8–9 = Thurneysen, 'Aus dem irischen Recht III', 329 § 20. In a triad at *CIH* vi 2099.21 it is implied that *any* offence by a dog is treated as a 'human offence', but this is an oversimplification. The same claim is made with regard to the offences of a chained dog (*cú áraig*) in Meyer, *Triads*, 22 § 168.
[103] *CIH* ii 572.25–7.
[104] *CIH* i 285.16–17 = *AL* iii 272.12–13; cf. *CIH* i 112.29.
[105] *CIH* i 334.25; iv 1177.9; 1391.1 = *AL* iii 518.20.
[106] Cf. *CIH* i 312.17 *cú faeindil* 'wandering dog'.
[107] *CIH* iii 857.4–17; 971.17–29.
[108] *CIH* ii 451.24 = *BB* 74 § 39; *CIH* i 55.2 = *AL* v 318.14 (Heptad 63). The glossators, on the other hand, take this to be a metaphorical phrase referring to a human thief.

it as a general legal principle that a bull is immune from liability while the cows are in heat.[109] The owner also has no responsibility for any injury caused by his bull when it is out on loan, provided he has warned the borrower of its innate viciousness (*bithbinche*). If the bull has been borrowed without the owner's permission, the borrower is liable for any injury inflicted on a third party.[110] In other law-texts it is implied that a fierce bull should be under restraint at times when it might pose a threat to neighbours or passers-by. Thus *Bretha Nemed Toísech* refers to a bull which is tied up (*cuimrechtach*).[111]

An ox (*dam*) is generally more docile than a bull, but can sometimes be dangerous. From the law-texts it would appear that most injuries caused by oxen took place on the ploughing-field. As ploughing regularly involved co-operation by up to four neighbours (see p. 445), it is easy to imagine how complex legal disputes could arise. An ox belonging to neighbour X might provoke an ox belonging to neighbour Y, which might injure neighbour Z. To avoid litigation in such cases, the law-text on accidents makes a general rule that no one is held accountable for injuries caused by 'ox-rage' (*damgal*) during ploughing.[112] Oxen are similarly exempt when they are being caught for yoking,[113] or when pulling a cart on a journey (*immitecht*).[114]

Female bovines are generally placid, and therefore pose little danger to neighbours or passers-by. However, a cow which has just calved (*nuídlech*) is liable to be nervous and unpredictable. For this reason, the law-text on accidents, *Bretha Étgid*, states that if a newly-calved cow injures somebody, no liability is incurred by the

[109] *CIH* i 276.3 = *AL* iii 230.z (quoted in part at *CIH* iv 1494.7 = O'Dav. 318 § 741) *blā tarb 7 reithe darmna* 'heat is the immunity of bulls and rams'. This is the only attestation of the term *darmna* 'heat (in cows or ewes)'. The glossator equates it with *dáir* of the same meaning.
[110] *CIH* v 1578.33–4.
[111] *CIH* vi 2216.33–4.
[112] *CIH* i 284.1–4 = *AL* iii 266.13–14. It is similarly implied at *CIH* vi 2306.15 = Binchy, '*Bretha Déin Chécht*', 24 § 4 that no one is held responsible if a person is injured when oxen in a plough-team stumble (*fuil arat*[*h*]*air oc* [*c*]*uicilga*). See discussion in Binchy's notes.
[113] *CIH* i 10.17 = *AL* v 152.3–4.
[114] *CIH* i 284.3 = *AL* iii 266.17.

owner.[115] This only applies when her calf is tied up in a pen (*lías*) or enclosure (*airbe*); under such circumstances the cow would naturally be particularly distressed.[116]

Among low-ranking farmers, most injuries caused by a newly-calved cow would be inflicted on the owner or a member of his family and could not therefore give rise to litigation. Legal commentary shows that in the case of more affluent farmers such injuries would be likely to be suffered by employees, particularly the cowherd (*búachaill*) and milker (*bligre*). It is emphasized that they have no legal entitlement to compensation for any injuries by a newly-calved cow, except in the case of a cow of innate viciousness (*bó bithbinche*).[117]

It is well known that a previously docile cow may go mad for no apparent reason. As the owner could not be expected to anticipate such an occurrence, Heptad 7 absolves him from responsibilty for injuries which might be caused at the immediate outbreak of the madness.[118] Glosses and commentary make it clear, however, that he must attempt to slaughter the mad cow,[119] and must also warn his four nearest neighbours as soon as possible.[120]

Injury by horses

In early Irish sources we find many references to injuries caused directly or indirectly by horses. There are numerous records in the annals of death or injury resulting from a fall (*escar*) from a horse.[121] It is clear from Heptad 7 that if somebody is thrown from a borrowed horse, the owner is generally not held to be legally responsible. The accompanying commentary does however stress that the owner must give a warning if the horse is known to be in the habit of bucking (*prepad*) or shying (*scáth*).[122]

As in the case of other male animals, a stallion (*echcullach* or *cullach eich*) is more likely to be vicious than a mare. A stallion is

[115] *CIH* i 275.12–13 = *AL* iii 228.7–8. The same rule is given in Heptad 7 (*CIH* i 9.35; iii 1046.33; v 1886.15 = *AL* v 152.2–3); the accompanying gloss fixes three days as the period in which a cow is regarded as a *nuídlech* (*nuithlech, nuthlech* MSS).

[116] *CIH* i 275.13 = *AL* iii 228.7–8 *acht bid ō lias nō airbe adriastar a laegh.* Another version (not printed in *CIH*) in RIA MS 670 (D v 2) f. 19^b15 has *do lias.*

[117] *CIH* i 10.5–10 = *AL* v 152.12–24.

[118] *CIH* i 10.17 = *AL* v 152.3.

[119] *CIH* i 10.20–21 = *AL* v 154.10.

[120] *CIH* i 285.16–17 = *AL* iii 272.12–13.

[121] See *DIL* s.v. 1 *escar*.

[122] *CIH* i 10.24 = *AL* v 154.25–6.

generally free from liability for offences committed while the mares are in heat.[123] If the owner of a horse is aware that it is a biter (*daintech*), he must take special steps to ensure that it is kept under restraint. If he fails to do so, and the horse bites someone, the injury is classed as a 'human offence' (*duinechin*).[124]

Injuries involving horses are particularly likely to happen on roads or where many people or vehicles are crowded together. The law-text on accidents makes the general rule that 'an assembly is an exemption for horses'.[125] According to the accompanying commentary, this means that a horse-owner is not normally liable for injuries inflicted by his horse at a fair, nor is he entitled to compensation if his horse is itself injured. Exceptions to this general principle occur where the horse is known to possess a vicious temperament (*bithbinche*): if the owner lends such a horse to another he may be wholly or partly liable for its offences at a fair.[126] He is also liable if his horse causes an injury through shying (*bedg*), leaping (*redg*), a kick (*lúa*), an awkward jump (*sáebléim*), a twist while under [the rider's] hand (*cor fo láim*), bucking (*prep*) or putting its head in the fork of a tree (*cenn i ngabal*).[127] If the borrower damages the horse through roughness (*borblachas*) or through stretching it beyond its strength (*rige tar tracht*), he must pay a fine to the owner.[128]

Accidents involving chariots are similarly immune from liability when they occur at a fair.[129] To reduce the risk of chariot-accidents on the roads, the law-text on this topic specifies that a high road (*slige*) should be wide enough for two chariots to pass. A second-class road (*rout*) should be able to accommodate a chariot and two horsemen.[130] The law-texts also refer to the problem of persons being injured by stones or clods of earth thrown up by horses' hoofs. In general, the rider is not held responsible. Thus Heptad 8

[123] *CIH* iii 1046.34–5 = *AL* v 152.5–6 (glossed at 154.28–30); cf. *CIH* ii 376.18 = *AL* i 126.6–7.

[124] Meyer, *Triads*, 22 § 168; cf. *CIH* iv 1494.41–2 = O'Dav. 321 § 758.

[125] *CIH* i 265.41 = *AL* iii 180.6 *blā ech aenach*. Cf. Kelly, *Audacht Morainn*, 10.71 § 28.

[126] *CIH* i 266.28–267.1 = *AL* iii 184.4–186.5.

[127] *CIH* i 266.6–8; iii 934.34–7 = *AL* iii 180.16–20.

[128] *CIH* i 266.4–6; iii 934.29–31 = *AL* iii 180.12–15.

[129] *CIH* i 283.28 = *AL* iii 264.22.

[130] *CIH* iii 893.23–5 = Appendix A, text 4 § 1 (1)–(2).

includes 'a rebounding from the hoofs of horses' among the seven reboundings which may cause injury, but do not entail liability.[131]

Injury by pigs

Pigs – particularly boars and sows with young – may be very aggressive towards humans, and have extremely sharp teeth. *Cáin Adomnáin* speaks of women being killed by pigs,[132] and in *Bethu Phátraic* we find a reference to the eating of a child by pigs.[133] In response to the entreaties of the child's parents, Patrick ordered that his bones be collected, and through prayer restored him to life. The law-text *Bretha Nemed Toísech* describes two cases in which a child is attacked by pigs. In the first of these, the trouble was started by a hunting dog (*mílchú*), apparently belonging to a boy named Béimnech, great-grandson of Conlae.[134] The dog attacked a herd of pigs, which turned on it and gave chase. The dog escaped from them, but they then attacked Béimnech, and tore him with their teeth from sole to forehead, inflicting fifty wounds. A case for compensation and sick-maintenance was submitted to Coirpre Lifechair, who judged that the pigs be forfeit – a common penalty for a first offence by a domestic animal (see p. 142).[135] But when Coirpre's father, King Cormac mac Airt, heard of this judgement, he overturned it. He ruled that it was not the pigs but the dog which was responsible in law for the boy's injuries. (In the second case described in this text, pigs killed a calf and a child: see p. 180 below.)

Other law-texts also discuss various circumstances in which injury inflicted by pigs is free from liability. One text states that pigs are totally exempt if they injure someone while being driven from their digging (*clas*), or while being moved from their trough (*omar*), or when disturbed at an accustomed feeding-place in a wood, or after

[131] *CIH* i 11.18 = *AL* v 156.15 *athsceinm a cruaib* [*e*]*ach*. Cf. *CIH* i 283.9 = *AL* iii 262.21 *blā rob caebad* 'throwing up clods is the immunity of animals'. At *AL* v 157.21 *a cruaib* is translated 'from horses' shoes', but there is no evidence of horseshoes in Ireland at this period: see p. 94 above.

[132] Meyer, *Cáin Adamnáin*, 28 § 42.

[133] Mulchrone, *Bethu Phátraic*, 119.2319–120.2333.

[134] *CIH* vi 2217.8–23. A later version at *CIH* vi 2113.6–15 is edited by Dillon, 'Stories from the law-tracts', 44 § ii (trans. 52–3 § ii).

[135] *CIH* vi 2217.12 *dīubruiter cach cethra 'na .c.cinadh* 'all livestock are taken away for their first offence'.

being wounded out in the fields.[136] Similarly, the law-text on accidents gives it as a general rule that 'the feeding-trough or the sty is the exemption of pigs'.[137]

Heptad 7 lists two types of situation in which there is no liability for injury by pigs. One is the case of 'bloodshed by pigs being seized in a sty': this implies that if a person – for whatever reason – catches or manhandles pigs in their sty, the owner is not liable for any injuries which may result.[138] The other case concerns a sow with young; her owner is exempt for her offences.[139] A reference in another law-text suggests that a boar (*muccullach*)[140] is immune from liability during the mating season.[141]

Injury by sheep

Sheep seldom cause injury to humans, although rams can be aggressive, particularly when the ewes are in heat. According to the law-text on accidents, the owner of a ram is free from liability for any injuries which it may inflict during the period of mating.[142] But if a ram of known fierceness (*reithe lond*) causes injury at other times, its owner must pay as for a 'human offence' (*duinechin*).[143]

Injury by bees

As we have seen in Chapter 3, honeybees have a remarkably high profile in early Irish law, partly because of the legal problems connected with bee-trespass. The main text on bee-law, *Bechbretha*, also devotes much attention to the legal consequences of stings from bees.[144]

[136] *CIH* iv 1268.11–13. Versions are also found in later commentary at *CIH* i 10.28–9; v 1828.16–19 = *AL* v 154.21–4. I take *oc suiguid feada rognathaith-* of *CIH* iv 1268.12–13 to be for *oc suidiu ḟedo rognáthaigetar* 'at a situation (feeding-place?) in a wood which they frequent'. The later version at *CIH* v 1828.19 has simply *ag ithe mesa na caille* 'eating the mast of the wood'.

[137] *CIH* iii 943.30 = *AL* iii 242.13 *blā oircēl no chrō*; cf. *CIH* i 278.24. The unpublished version in RIA MS 670 (D v 2) f. 22ª21 has *blaī mucc orcēl ⁊ crō.*

[138] *CIH* i 10.17–18 = *AL* v 152.4 *fuil muc oca ngabāil a crū.*

[139] *CIH* i 10.18 = *AL* v 152.4. The text has *fuil muc a foir*, which the glossator explains as *oc breith muc* 'giving birth to pigs'. I take *fóir* to refer to the nest in which the sow gives birth to her piglets, and where they spend the first weeks of their life: see p. 80.

[140] Cf. *CIH* iii 897.21 = Appendix A, text 2 § 1 (25) *cullaig* (sic leg.) *muc.*

[141] *CIH* ii 376.20 = *AL* i 126.7. The glossator has *i n-aimsir lāith* 'in the time of mating'.

[142] *CIH* i 276.3 = *AL* iii 230.25; cf. *CIH* iii 1046.35 = *AL* v 154.29.

[143] Meyer, *Triads*, 22 § 168.

[144] *CIH* ii 449.4–32 = *BB* 66–72 §§ 27–35.

If a person is stung while robbing or moving the bee-hives, or even while looking over the hives at swarming time, the beekeeper is not liable. But if a bee stings a neighbour or passer-by who is not interfering with the bees in any way, the beekeeper must provide him with a meal of honey. The victim must, however, swear an oath that he did not kill the bee which stung him. This is because of the general principle in early Irish law that the life of an animal is forfeit for its offence.[145] Consequently, if the victim has killed the offending bee, he is felt to have already been requited for the sting. A ninth-century glossator on *Bechbretha* raises the point that a honeybee normally dies anyway after stinging a human.[146] This is because its sting becomes imbedded in the skin and it ruptures its abdomen in trying to withdraw. The glossator takes the view that the victim is still entitled to receive honey from the beekeeper because he has not deliberately caused the death of the bee.

After a general account of the law relating to bee-stings, the author of *Bechbretha* moves on to deal with the case of Congal Cáech, king of Tara, allegedly blinded in one eye by bees. According to early Irish custom, this meant that he had to abdicate, as a blemished man could not remain in the kingship.[147] Congal, we are told, sued the beekeeper, and the judgement was that lots be cast on all the hives in the apiary. The hive on which the lot fell became his property. The author of *Bechbretha* points out that in such a case, all the bees of this hive are forfeit for the offence of a single bee.[148]

In spite of being documented in a nearly contemporary law-text,[149] the story of Congal Cáech contains some dubious features. Though he was undoubtedly a king of the Cruithni, there is no early evidence outside *Bechbretha* that he ever actually held the kingship of Tara.[150] The claim that his eye was blinded by a bee-sting is also suspect: the human eye-closing reflex is so fast

[145] *CIH* ii 449.13 = *BB* 66 § 29.
[146] *CIH* iii 924.5–6 = *BB* 66 § 29^{e}.
[147] *GEIL* 19.
[148] *CIH* ii 449.28 = *BB* 70 § 34. For the application of the same principle to cases where one animal kills another, see p. 176 below.
[149] Congal is known to have died at the battle of Mag Rath in AD 637 (*BB* 123), and *Bechbretha* can be dated on linguistic grounds to the seventh century, possibly as early as AD 650 (*BB* 13).
[150] See discussion at *BB* 126–31.

that a bee's sting would be unlikely to penetrate to the cornea.[151] It is possible, therefore, that the leading case of Congal's bee-sting is fictitious.

Bechbretha does not deal with cases in which a person dies as a result of a sting inflicted by someone else's bees.[152] In later commentary, however, it is stated that the beekeeper must pay a fine of two hives. The rule is given in the formula 'one hive for blinding, two hives for killing' (*ceis isin cáechad, dá chis isin marbad*).[153]

[151] See discussion at *BB* 121–22.
[152] Death from a bee-sting is treated in medieval Welsh law: see *BB* 198 § 6.
[153] *CIH* i 316.37 = *AL* iii 432.14.

5
Offences against domestic animals

The legal concept of cruelty to animals is not recognised in early Irish law. Consequently, all the offences listed below (except bestiality) relate only to livestock belonging to others. For example, the offence of 'excessive fettering on horses' (*sárchuimrech for echu*) might occur while a person restrained a borrowed horse which was being troublesome.[1] It could not apply to his own horse.

Killing

There are many situations in which a person illegally causes the death of a domestic animal belonging to somebody else. It may simply be for food: legal commentary describes the case where an animal is stolen and then eaten. The finding of its entrails, bones and gore near a suspect's house can be used as evidence against him.[2] On the other hand, malice may lie behind the killing of an animal: legal commentary refers also to the case of a wife who deliberately starves a household dog to death.[3] Large-scale slaughter of cattle was a tactic widely used in early Irish warfare, with the intention of humiliating an enemy people and damaging their economy.[4] The epithet *Bóguine* 'cow-killer' was attached to the sixth-century king Énnae, son of Conall Gulban, from whom the Cenél mBóguine were descended.[5] Possibly, Énnae Bóguine was famous – or notorious – for the number of his enemies' cattle which he had slaughtered.[6]

[1] *CIH* i 239.36 = *AL* v 476.23.

[2] *CIH* ii 692.17–18.

[3] *CIH* i 149.10–11. This short commentary is attached to the words *denugudh con* 'hastening(?) [the death] of a dog'. It is likely that these two words belong to an Old Irish text on legal disputes within marriage (usually leading to divorce). At the beginning of this section (*CIH* i 144.5), there is a quotation from the main text on the law of marriage, *Cáin Lánamna*, but the rest of the material in this section seems to be independent of that text.

[4] E.g. AU^2 574 § 6 s.a. 1128.

[5] *CGH* 163.

[6] Cf. Stokes, 'The prose tales in the Rennes Dindṡenchas, (2)', 153 § 142. On the other hand, it is possible that *bóguine* refers to the sacrificial slaughter of cattle (see p. 52). Meyer ('Zur keltischen Wortkunde. VII', 629) draws attention to the Greek cognate βουφόνος 'ox-killer, priest'. A further possibility is that *bóguine* means 'one who slaughters cattle [for guests]', i.e. a generous host.

The death of livestock may result from carelessness or rough treatment rather than from an intention to kill: a farmer may find his neighbour's animals on his land and drive them off in a manner which causes their death or serious injury. Heptad 5 lists seven such illegal drivings and specifies that fines must be paid by the person responsible, who may of course counter-claim for the damage done to his crops or grassland. The seven drivings are: 'driving them into the sea, driving them into a mire, driving them into a marsh, driving them into a place where wolves are congregated, driving them with spite or carelessness through which an animal dies, startling them with angry fierceness through which bones are broken, driving them into a cow-house infected with the disease of staggers(?) or cattle-plague (unless they have come out of it)'.[7] Non-legal sources provide instances of fatal injury as a result of careless driving of trespassing livestock. For example, the Life of Saint Colmán mac Lúacháin recounts the tale of a legendary bull named Grogín, which had the unusual habit of mating with mares as readily as with cows. One day this bull strayed into a cornfield belonging to a man named Mac Coisemnaig, who chased it out and in doing so broke one of its legs, thereby causing its death. In recompense, Mac Coisemnaig was required to hand over the field in which the accident had occurred to the owner of the bull. He in turn gave it 'to God and to Colmán' and it was thereafter known as Grogín's field (*Gortín Grogín*).[8] A similar tale describes the fatal impaling of a valuable horse as it was being driven from a neighbour's cornfield.[9]

Heptad 5 also refers to the legal concept of 'last sighting' (*tigradus*) in relation to the illegal driving of livestock. This means that although a farmer who drove his neighbour's cattle into the sea may not have actually witnessed their drowning, he is nonetheless legally responsible for it as he saw them in a situation of potential danger.

Livestock may die as a result of passive negligence. For example, if a man leaves a fencing stake in a dangerous position, he is responsible if a neighbour's animal dies through becoming impaled upon it.[10] Naturally, he is not responsible in the unlikely event of an

[7] *CIH* i 6.23–6 = *AL* v 138.10–15.
[8] Meyer, *Betha Colmáin*, 38–40 § 41.
[9] Meyer, 'König Guaire', 2.6–9; O'Grady, *Silva Gadelica* i 401.23–4 (trans. ii 437).
[10] *CIH* iii 907.31–2; cf. *CIH* i 201.3 = *AL* iv 140.26; *CIH* vi 2099.22.

animal being spiked by a stake in a properly-constructed fence.[11] Similar provisions apply if a domestic animal dies as a result of negligence on common land. If a man digs a deer-pit (*cuithech*) and neglects to notify the king and people of the area, he must pay a fine if any livestock fall into it.[12]

Early Irish law places a responsibility on an uninvolved passer-by to intervene and report if he sees domestic animals in trouble. If he fails to do so, he becomes guilty of the offence of *aircsiu* 'looking on' (see p. 138 above). According to commentary on a short quotation from an Old Irish law-text, this applies if he sees any animal near a pit or wolves or a thief.[13] The concept of 'last sighting' (*tigradus*) applies here also, so the passer-by still commits an offence even if he does not actually witness the death or theft of the animal.

In certain circumstances, a person is free to deliberately kill a domestic animal belonging to another. He may, for example, kill his neighbour's cat if he finds it in his larder (*daingen tige*) or in a food-vessel (*lestar*).[14] A stray dog may likewise be killed if it becomes a danger or a nuisance. According to commentary based on the lost law-text on dogs, *Conṡlechtae*, it is permissible to kill a straying bitch in heat (*cú ṡodaig*) if it is impossible to keep her confined.[15]

There are also circumstances in which there is no penalty if a person accidentally causes the death of another's animal. For example, if a dog-owner allows another person to take his dog deer-hunting, he has no claim if it is killed during the hunt.[16] The same applies if a person employs a messenger to carry out some business for him, and entrusts him with his horse for this purpose. The messenger is not held responsible if the horse is killed or injured on the journey.[17]

[11] *CIH* i 289.22 = *AL* iii 290.10; cf. *CIH* iv 1329.9 = Thurneysen, 'Aus dem irischen Recht V', 390.

[12] *CIH* i 321.20–35 = *AL* iii 452.25–454.5; *CIH* i 285.13–14 = *AL* iii 272.8.

[13] *CIH* ii 708.24; iii 787.39; iv 1182.35. Compare also commentary on the text on distraint at *CIH* v 1738.39–1740.6 = *AL* ii 56.19–60.17.

[14] *CIH* i 290.32–4 = *AL* iii 296.1–6.

[15] *CIH* i 112.38–9.

[16] *CIH* i 20.3 = *AL* v 190.18–19. The accompanying commentary gives some exceptions to this general rule, concentrating on the nature of the agreement between the borrower and the dog-owner.

[17] *CIH* i 19.29–20.3 = *AL* v 190.17–18.

The payments due for the illegal killing of livestock vary according to the value of the animal and the circumstances of the offence. Where there are no mitigating factors, the culprit must pay restitution (*aithgein*) as well as a penalty-fine (*díre*). Restitution consists of the normal value of the animal, or its replacement by another animal of the same type and quality. The penalty-fine, on the other hand, is much greater than the animal's commercial value. The two types of payment are clearly illustrated in a short law-text on dogs entitled *De canibus sinodus sapientium.* Here it is laid down that a person who illegally kills a dog which guards the house must pay ten cows, and provide a substitute of the same breed. In the case of a dog which guards the livestock, he must pay five cows and provide a substitute. In the latter case there may be additional restitution: he must compensate for any livestock eaten by wild animals until the end of the year.[18]

In some cases, a person who kills or causes the death of an animal need only pay restitution. The reasoning behind such cases is illustrated in the commentary on the use of somebody else's dog for hunting.[19] Because of the general principle that such a loan is at the owner's risk, the borrower is not obliged to pay a penalty-fine (*díre*) for the dog's death. On the other hand, if the borrower has not kept to the terms of the hunting agreement, he is to some degree culpable. Thus if he agreed to hunt only deer, and caused the dog's death by hunting other animals, he is liable for restitution (*aithgein*). If he did this without being specifically forbidden, he is liable for half restitution (*lethaithgein*).

The penalty-fine for a domestic animal seems generally to be fifteen or twenty times its normal commercial value: see p. 76 (sheep) and p. 87 (pigs). In the case of a dog, the penalty-fine may be linked to its owner's rank. A legal quotation in *O'Davoren's Glossary* states that the penalty-fine for a hunting dog belonging to one of the higher grades of lord is one seventh of his honour-price (*eneclann*) if it is killed through inadvertence (*anfót*).[20] As the honour-price of the higher grades of lord ranges from fifteen to thirty *séts*, the fine is therefore two or three *séts*. It would obviously be greater if the killing were deliberate. Legal commentary gives a general rule that the penalty-fine for killing a lapdog (*mesán*) is one third of its

[18] *IP* 174 §§ 3–4 = *GEIL* 354 (5) §§ 3–4.
[19] *CIH* i 20.5–12 =*AL* v 194.20–29.
[20] *CIH* iv 1516.12–14 = O'Dav. 417 § 1248.

owner's honour-price in the case of deliberate killing, and one sixth in the case of inadvertence.[21]

INJURY

To inflict serious injury on another's animal is often tantamount to killing it. Thus an ox with a broken leg is of no use for ploughing or haulage, and fit only to be slaughtered.[22] In such cases it seems likely that the culprit must pay the full penalty-fine and restitution for the animal.[23] But if the injury is less serious, the fines are correspondingly lighter. So, if a dog is subjected to shaving (*berrad*), cutting off of its ears (*máelad*),[24] or other mutilation (*gerrad*), the culprit must pay half the fine for killing it (*lethfíach báis*).[25] Just as cattle were sometimes slaughtered with the intention of humiliating their owners (see previous section), non-fatal injury or abuse might also serve this purpose. For example, a Life of Saint Finnchua describes how livestock belonging to his community were mutilated (*gerrtar a n-indile*) and his shepherds beaten by local people who objected to his farming in their area.[26] A purely symbolic assault on an animal serves to challenge a warrior in the tale *Aided Cheitt maic Mágach*. Conall Cernach pulled tufts out of the mane of the horses of Cett mac Mágach, leaving a wisp in front of his chariot.[27]

Commentary on *béoathchummae cethrae* ('living injury to livestock') contains a very detailed account of the fines payable for various injuries to animals.[28] Some of the distinctions drawn are so

[21] *CIH* i 113.23–4; cf. *CIH* iv 1512.10–11 = O'Dav. 400 § 1153.

[22] Note the emphasis in legal commentary on leg-injury to livestock, e.g. *CIH* v 1723.34, 1724.16 = *AL* ii 4.5, 29, etc.

[23] The main law-text on distraint refers to *athcuma do con lomnai* 'mutilating your leashed dog' (*CIH* ii 402.9 = *AL* i 232.3–4). According to the glossator, penalty-fine (*eneclann*) and restitution (*aithgein*) are due for this offence.

[24] I have no other examples of *máelaid* in the meaning 'cuts off the ears'. The adjective *máel* has a wide variety of applications, including 'bald, shaved, hornless (of cattle), headless (of a corpse)'.

[25] *CIH* iv 1268.18.

[26] Stokes, *Lismore Lives*, 87.2899.

[27] Meyer, *Death-tales*, 36 § 3. Cett was at first disposed to see this as a friendly gesture, but his charioteer convinced him that it was an insult. A Welsh example of the mutilation of horses as an insult is to be found at D. S. Thomson, *Branwen uerch Lyr*, 3.73–7.

[28] *CIH* i 304.27–307.11; iv 1204.19–1209.15; v 1624.14–1628.42 = *AL* iii 356.19–380.24.

fine that it is difficult to imagine them being applied in practice. Nonetheless, this commentary is valuable as a general guide to later legal theory on this topic. As one commentator points out, the law relating to animal injury is based on that of human injury.[29] Consequently we find that three categories of blow against an animal are distinguished: a white [i.e. bloodless] blow (*bánbéim*), a lump-blow (*cnocbéim*), and blood-letting (*fuiliugud*). These distinctions broadly follow those of the Old Irish law-text on human injury, *Bretha Déin Chécht.*[30]

The payments specified in this commentary for the illegal injury of an animal are generally fractions of the penalty-fine (*díre*) due for killing it. For example, if a person inflicts a lump-blow on an animal he must pay one twenty-fourth of its *díre,*[31] and if he inflicts a white blow he must pay one forty-eighth.[32] This commentary takes a much less serious view of the shaving (*giunad*) of another person's animal than does the law-text quoted above. If the animal is shaved bare, the culprit must pay one twenty-fourth of its *díre,*[33] but if it is only partially shaved he is liable for one forty-eighth.[34]

Naturally, a human parallel cannot be found for all animal injuries. In the case of a cow milk-production is of paramount importance, so an injury to the udder is regarded as particularly serious. The general principle is expressed in the phrase *bó ar úth* 'a cow for an udder'.[35] A person who causes the destruction of the whole udder must therefore give another cow to the owner. There are also substantial fines for damage to a teat (*sine*) or milk-duct (*sreb*).[36] Another commentary deals with the offence of stealing milk from another person's cow, and refers to the danger of damaging the udder in the process.[37]

When a person is illegally injured, early Irish law normally requires that the culprit organizes his removal from home for

[29] *CIH* iv 1204.30; v 1624.23–4 = *AL* iii 358.1–2.
[30] *CIH* vi 2311.17–21 = Binchy, '*Bretha Déin Chécht*', 36 § 24.
[31] *CIH* i 305.17 = *AL* iii 360.16.
[32] *CIH* i 305.22–3 (supplying *.xl.*) = *AL* iii 360.25–6.
[33] *CIH* i 305.17 = *AL* iii 360.16–17. For a discussion of such offences against persons, see commentary at *CIH* i 303.36–304.15 = *AL* iii 352.9–354.11.
[34] *CIH* i 305.23 = *AL* iii 360.25–7.
[35] *CIH* v 1626.37; 1636.9 = *AL* iii 364.12.
[36] *CIH* iv 1206.14–38 = *AL* iii 364.13–366.7.
[37] *CIH* i 144.30–33.

nursing (*othras*) until he is cured.[38] The culprit must ensure that the injured person is provided with suitable food and accommodation, and pays all medical expenses. It seems that this system was extended to illegally injured animals. However, one legal fragment states that it does not apply in the case of a herd dog (*conbúachaill*).[39] The glossator puts forward the plausible explanation that the dog might learn to trust those who looked after it, and subsequently allow them to steal its owner's livestock. Consequently, the dog must be treated in its own home at the culprit's expense.

For a general discussion of the treatment of injured livestock, see pp. (216–18).

CECHRAD 7 SALCHAD

Legal commentary refers to the offence of *cechrad* 'driving another person's animals into a marsh (*cechair*)' and *salchad* 'causing them to become dirty (*salach*)'. If the intention is to bring about the death of the animals, the culprit must pay a fine of five *séts*, even if no injury has resulted from his action.[40]

ARSON

The offence of arson (*forloscud*) often affected livestock. The law-text on arson *Bretha Forloiscthe* has not survived, but it is clear from quotations and from commentary based on this text that it dealt both with the burning of the pens and enclosures of domestic animals,[41] and with the burning of the animals themselves.[42] According to one quotation, the three things which Saint Patrick forbade most strenuously were the killing of trained oxen, offences against milch cows and arson.[43] In a poetical version of this triad, the third offence is given as the burning of animal-enclosures.[44] It is clear from the annals that much arson

[38] See *GEIL* 130–3; Binchy, 'Sick-maintenance in Irish law'.

[39] *CIH* iv 1242.18 = Binchy, 'Sick-maintenance in Irish law', 87 § 6.

[40] *CIH* iv 1181.22, 26; cf. *CIH* iii 827.9; iv 1327.18.

[41] *CIH* i 103.10; vi 1945.38.

[42] *CIH* vi 1946.41; 1947.1.

[43] *CIH* vi 1946.28–30.

[44] E. J. Gwynn, *Metrical Dindshenchas* iv 156.136. Gwynn discusses the difficulties of this line in his note on p. 417. He gives the reading *loscud lés fás* 'burning empty byres', following Bergin's suggestion that *les, leis, lios* of the MSS should be read as gen. plur. *lés*, an early spelling (unexpected in a Middle Irish poem) of *lías*. He

– like the killing or stealing of cattle – took place during warfare, and therefore entailed special legal difficulties. The invocation of Patrick's name demonstrates the Church's desire to put an end to these traditional practices: see further discussion under 'theft' below.

In cases where the culprit can be brought to justice, arson is treated like other crimes of violence: as a commentator observes, 'a firebrand is like a naked weapon' (*amal arm nocht int aithinne*).[45] The arsonist must therefore pay the full penalty-fine (*díre*) for any livestock whose death he has caused.[46] It is further specified (in what is probably a quotation from the original law-text) that he must pay the *díre* of an adult cow or sheep for any calf or lamb which is burnt.

THEFT

Early Irish law relating to the theft of livestock is particularly complex. As with other serious crimes, it is an offence against a person's honour to steal his property, or property belonging to another which happens to be on his land. A thief must therefore pay the landowner's honour-price.[47] The fragmentary law-text on theft, *Bretha im Gata*, gives some examples of how this system worked.[48] For example, if an article belonging to a commoner is stolen from the house of a king, the thief must pay the full honour-price of the king.[49] Two thirds of this sum go to the king, and one third to the commoner. In addition, the commoner receives restitution (*aithgein*) of the stolen article.[50]

Bretha im Gata provides details on the rates of restitution due for the theft of various animals. If a sheep has been stolen, the thief

---

comments that the burning of empty byres seems pointless. However, it is probable that the poet is merely following the distinction made in the law-text between two offences: burning the enclosures and burning the animals. If the enclosures are burnt when empty, the fine is only one *sét*; e.g. *CIH* vi 1945.38 *.s. i loscud lias laogh ɫ caorach ɫ foil muc* 'a *sét* for burning enclosures of calves or of sheep or sties of pigs'. Cf. *CIH* i 103.9–10.

[45] *CIH* vi 1947.9–10.

[46] *CIH* vi 1947.1.

[47] For a discussion of honour-price (*lóg n-enech* or *eneclann*), see *GEIL* 8–9.

[48] *CIH* ii 477.31–479.22 = Hull, '*Bretha im Gatta*', 215–23. Other unedited material on theft at *CIH* vi 1974.11–1980.39 may come from the missing part of this text.

[49] The honour-price of the petty king (*rí túaithe*) is seven *cumal*s, often equivalent to twenty-one milch cows (see p. 592).

[50] *CIH* ii 478.26–9 = Hull, '*Bretha im Gatta*', 220–21. See also discussion at *BB* 163–4.

must give back four sheep to the owner. If he has stolen a cow or ox, he must give back five cows or five oxen. If he has stolen a horse, pig, or inanimate object (*marbdil*) he must make double restitution. These rates are taken in part from Canon Law, based ultimately on Exodus 22:1.[51]

The disproportionately low value assigned to horses in *Bretha im Gata* seems out of line with the special esteem for them which is found throughout early Irish literature. As we have seen in Chapter 2, the horse was associated particularly with kings and lords, and horse-racing was clearly a very important entertainment. It is not surprising, therefore, that later commentary gives a higher rate of compensation for horse-theft: the culprit must restore **four** horses for a single horse, whether it be draught-horse, pack-horse or riding-horse.[52] Presumably the additional horses must be of the same type and quality as the stolen horse. This is implicit in the general legal principle on restitution which states that 'everything must be paid for by what is appropriate for it'.[53]

Condemnation of the horse-thief (*echtháid*) is voiced in a number of texts. The wisdom-text *Tecosca Cormaic* gives the 'oath of a horse-thief' as an example of the worst type of oath.[54] Another wisdom-text, *Audacht Morainn*, advises that a king 'should not sell his honour or soul for the horses of a horse-thief'.[55] According to a legal glossator, the client of a lord has three public duties of attack (*fubae*): hunting down pirates, horse-thieves, and wolves.[56] Horse-theft might be carried out on a large scale. A difficult Old Irish legal passage deals with the theft of a herd (*graig*) of twelve horses belonging to a number of owners.[57] If one of the owners is of higher rank than the others, he gets restitution (*aithgein*) as well as payment of his honour-price (*díre*). The owner next in rank

[51] See discussion by Hull, '*Bretha im Gatta*', 218; Ó Corráin, Breatnach, and Breen, 'The laws of the Irish', 413–15; Charles-Edwards, 'The construction of the *Hibernensis*', v (forthcoming).

[52] *CIH* i 106.19 = Hull, '*Bretha im Gatta*', 224.

[53] *CIH* iv 1479.8 = O'Dav. 251 § 346 *dorenar gach dūil dia cuinnfi fadeisin*. Cf. *CIH* i 261.34–5; iii 1079.3; vi 1980.27–8 (the principle is expanded upon in the passage which follows at *CIH* vi 1980.29–39).

[54] Meyer, *Tecosca Cormaic*, 44 § 27.15.

[55] Thurneysen, 'Morands Fürstenspiegel', 86 § 43 (trans. 105 § 43); Kelly, *Audacht Morainn*, 68 § 43 (Recensions A, L, N). The reference to the soul (*anam*) implies that dealing in stolen horses may affect the king's chance of entering Heaven.

[56] *CIH* ii 487.2; v 1788.8–9 = *AL* ii 270.4–5.

[57] *CIH* i 32.12–16; ii 582.23–7 = *AL* v 242.9–14.

gets double restitution, whereas the other owners only get simple restitution, i.e. the return of their own horses or substitutes of the same value.

As we have seen in Chapter 1 (p. 28), the large-scale theft of cattle is referred to frequently in early Irish writings, especially in sagas and annals.[58] The Church was strongly opposed to cattle-raiding – partly no doubt because of its own extensive farming interests. The penitential text *Canones Adomnani* states that 'cattle seized in a raid are not to be taken by Christians whether in trade or as gifts: for what Christ rejects, how shall the soldier of Christ receive?'.[59] A lost canon law-text *Cáin Dar Í* 'the law of Dar Í' seems to have dealt mainly with cattle-raiding, as its contents are summarized in a gloss on Colmán's Hymn by the words *cen bú do gait* 'not to steal cows'.[60] This text is also called *Cáin Bóslechtae* 'the law of cow-sections'[61] or simply *Bóslechtae* 'cow-sections'.[62] It is by the latter title that its promulgation in AD 810 is recorded in the *Annals of Inisfallen*.[63] A quotation from the law-text *Bretha Forloiscthe* (see p. 164) accredits Saint Patrick with having most strenuously forbidden offences against milch cows (glossed 'i.e. stealing them').[64]

Most cattle-raids were carried out across political boundaries, and therefore not readily amenable to normal legal processes, even under clerical pressure. However, early Irish law does provide a framework by which justice could be done in such cases. This is by means of the *cairde* 'treaty between kingdoms',[65] which allowed for the prosecution of murderers, thieves, arsonists, rapists, and satirists across a territorial boundary.[66] *Cáin Bóslechtae* seems to have been promulgated with extensive cattle-raiding in mind. This is implied in the short text *Rith na Cánann* 'the extent of the laws', which states that *Cáin Bóslechtae* is applied in cases involving up

[58] Lucas, *Cattle in ancient Ireland*, 125–99.

[59] *IP* 178 § 15. However, legal commentary at *CIH* v 1642.33 is less scrupulous: a lord is entitled to receive from his base client (*dáerchéile*) one third of what he gets in a raid or plundering expedition (*trian a chota creiche 7 ruathair*).

[60] *Thes.* ii 306⁵.

[61] *CIH* i 254.22; iii 823.1; iv 1413.25; v 1539n; vi 1972k.

[62] *CIH* iv 1367.4 = Thurneysen, 'Aus dem irischen Recht V', 388. See *GEIL* 275–6.

[63] *AI* 122 s.a. 810. See Ó Riain, 'A misunderstood annal: a hitherto unnoticed *cáin*'.

[64] *CIH* vi 1946.28–30 *fogail fri bū blichta .i. i ngait.*

[65] See *GEIL* 279.

[66] *CIH* iii 791.5–6.

to a hundred warriors. Each of them must pay 3½ *cumal*s for an offence against cattle.[67]

The law-texts are also concerned with cattle-theft on a smaller scale within a community. One text refers to the marking of the ears or horns of livestock; presumably this practice would have been partly to discourage local thieves.[68] The discovery of stolen cattle after a period of time has elapsed brings special legal problems, which are treated in a commentary on theft.[69] For example, if a heifer is stolen as a two-year-old (*colpthach*) or a three-year-old (*samaisc*), and the thief is not sued until she is a full-grown cow (*bó mór*), he must provide restitution (*aithgein*) appropriate to a full-grown cow, but a penalty-fine (*díre*) appropriate to a heifer. The same logic is applied to the case of a cow which is stolen in her prime, but whose rightful ownership is not established until she is old, blind or lame. The thief must pay a penalty-fine appropriate to a healthy cow, but restitution appropriate to a defective cow.

The law-text on beekeeping, *Bechbretha*, contains a section on the theft of bee-hives.[70] The principles applied in this text differ markedly from those of *Bretha im Gata*. There is no mention of the restitution (*aithgein*) of bee-hives as in the case of other domestic animals. Instead, the culprit is required to pay a penalty-fine (*díre*) which varies according to the location of the hives. If the hives are stolen from a courtyard (*les*) or garden (*lubgort*), the fine is the same as for household goods. This amounts to the full honour-price of the beekeeper.[71] If the hives are stolen from a field near the house (*faithche*), the fine is the same as for large animals, such as milch cows or trained oxen. According to later commentators, this amounts to half the beekeeper's honour-price.[72] If the hives are stolen from outside this area (*sechtar faithchi*), the fine is the same as for small livestock. According to later commentators, this amounts to one seventh of the beekeeper's honour-price. It is clear, therefore, that the location of the hives in relation to their owner's house is of great significance when assessing the fine payable by a thief.

67 *CIH* v 1587.4 = Thurneysen, 'Aus dem irischen Recht V', 383.
68 *CIH* iv 1253.1–2.
69 *CIH* i 173.1–15; cf. *CIH* ii 715.32–41.
70 *CIH* ii 455.31–456.29 = *BB* 84–88 §§ 50–54.
71 See discussion of this difficult passage at *BB* 161–3.
72 *CIH* ii 456.2–3 = *BB* 171 (h); *CIH* iv 1287.27–8 = *BB* 178 (k).

Glossators and commentators seem surprised by the absence of references to restitution in *Bechbretha*'s treatment of bee-theft. A sixteenth-century glossator suggests that for the theft of hives from a field near the house, the culprit should pay half the beekeeper's honour-price, and also restore four hives.[73] Here he follows the 'four for one' rate of restitution laid down for sheep-theft in *Bretha im Gata.* Other glossators take the same view.[74]

UNAUTHORIZED USE

To borrow somebody else's property without permission is obviously a lesser offence than theft, and in certain circumstances may not be an offence at all. Thus Triad 163 states that unauthorized use (*foimrimm*) does not entail a fine (*díre*) if carried out in fear or to warn of danger.[75] A gloss on this triad suggests that a horse can legally be taken in these situations.[76]

Normally, however, the unauthorized use of a horse, boat, cart, chariot, pot or vessel is illegal.[77] An Old Irish commentary on *Di Chetharṡlicht Athgabálae* deals with the unauthorized use of horses belonging to persons ranging in rank from king to *bóaire* (prosperous farmer).[78] The culprit must pay a fine of one cow for the mere act of mounting (*forlaim*) and another cow for alighting (*tairlim*).[79] These fines are applied irrespective of the rank of the horse-owner. However, if the culprit rides off with the horse, the commentator specifies that he must pay additional fines based on the owner's rank and on the distance travelled.[80] So if he borrows a *bóaire*'s

[73] *CIH* ii 456.15–16 = *BB* 86 § 52^{d} (III).

[74] E.g. *CIH* ii 456.22 = *BB* 86 § 53^{d} (IV). See also commentary at *CIH* iii 959.25–7 = *BB* 184 (c).

[75] Meyer, *Triads*, 22 § 163.

[76] Meyer, *Triads*, 40 § 163.

[77] *CIH* ii 384.15–16 = *AL* i 166.21–3; *CIH* ii 401.16–17 = *AL* i 230.z. Cf. *CIH* iii 915.16, which refers to the unauthorized taking of a neighbour's livestock (*cethra*) for the purpose of legal entry into a holding of land. Here, *cethra* presumably refers to the horses required by law for making a formal claim to land: see p. 432.

[78] *CIH* iii 894.35–41.

[79] The law-text *Bretha Nemed Déidenach* also refers to the offence of mounting a dignitary's horse, and advises *ní ḟorlais* 'do not mount [it]', 2 sing. pres. subj. of *for-ling* (*CIH* iii 1121.13 = E. J. Gwynn, 'Privileges', 27.27; *CIH* iv 1512.30 = O'Dav. 402 § 1167).

[80] A later commentator views the unauthorized use of the horse of a king, bishop, poet or learned man as a much more serious offence than the unauthorized use of a horse belonging to a person of lower rank (*CIH* ii 709.40–710.4). In the latter case,

horse, he must pay an extra yearling heifer for every crossroads (*bélot*), and a two-year-old heifer for each area of rough ground (*drobél*). If he crosses the boundary of another territory he must pay a further cow. In the case of a king's horse, the fines are much greater: five *séts* for every crossroads, half a *cumal* for each area of rough ground, and a *cumal* for crossing a boundary.[81] It must be admitted that these additional fines seem very difficult to assess, and may have had little relevance in practical law.

Opening sheds, etc.

Another illegal act of great destructive potential is to open a farmer's shed, pen, or enclosure, thereby allowing livestock to escape.[82] The culprit must provide restitution (*aithgein*) for any animals which are lost.

A similar offence is to leave an opening so that a farmer's calves can gain access to the cows.[83] The culprit must provide restitution of the shortfall in the milk-yield. This amount must be sworn by the woman who has milked the cows at the previous milking-time (*etrud*). If he fails to make restitution, he must pay the honour-price (*eneclann*) of the cattle-owner.[84]

Illegal distraint

The practice of distraint (*athgabál*) was central to the enforcement of law in early Irish society, and is described in detail in *Di Chetharšlicht Athgabálae* and associated material.[85] If a person has

a fine (*fiach foimrime*) need only be paid if the horse is ridden away for some distance (*tar naī n-imaire* 'over nine ridges').

[81] The sense demands the emendation of *bo* to *cumal* at *CIH* iii 894.37, as Binchy suggests. This would fit in proportionately with the fines given for riding off with the horse of an *aire tuíseo* 'lord of precedence' and an *aire déso* 'lord of vassalry'.

[82] *CIH* vi 2191.12–21.

[83] *CIH* ii 402.10; iii 895.12 = *AL* i 232.5–6.

[84] *CIH* iii 895.12–14.

[85] This text is found at *CIH* ii 352.25–422.36; iv 1438.36–1465.27; v 1723.11–1755.16 = *AL* i 64–304; ii 2–118.6. Another Old Irish text follows the glosses and commentary on *Di Chetharšlicht Athgabálae* in the legal manuscript H 3. 18. It is printed at *CIH* iii 896.9–901.13, but has never been fully edited and translated. *CIH* iii 897.29–898.31 is edited and translated by Binchy, 'A text on the forms of distraint', 78–81. *CIH* iii 896.19–41 is translated by L. Breatnach, 'Lawyers in early Ireland', 11–12. *CIH* iii 897.10–29 is edited and translated by me in Appendix A (text 2) of the present book. For a general account of the topic, see Binchy, 'Distraint in Irish law'. There is a shorter treatment in *GEIL* 177–86.

a claim against another, he must first give formal notice of his intention to distrain his livestock. There is then a period of a few day's delay, the length of which varies according to the nature of the offence. If the defendant does not pay the appropriate fine or give a pledge to indicate his willingness to submit to arbitration, the plaintiff can then enter his land and remove livestock to the value of the amount due.

The livestock distrained are normally cattle. One text explains that horses and sheep are generally avoided because of their vulnerability when being driven, and because of the danger of their being attacked by wolves or stolen by thieves.[86] Pigs are likewise regarded as unsuitable because of their habit of digging up the ground.[87] But if the defendant does not have cattle, the plaintiff has no alternative but to distrain other animals.[88]

In normal circumstances, the plaintiff brings the distrained animals to a private pound (*forus*),[89] either on his own or on a third party's land. They remain there for a further period of delay. If the defendant fails to meet his legal obligations to the plaintiff by the end of this period, the progressive forfeiture of his livestock begins. On the first day, livestock to the value of five *sét*s are forfeit. On each subsequent day livestock to the value of three *sét*s are forfeit until none remain.

There are many restrictions on the practice of distraint, and if the plaintiff fails to observe these he himself becomes guilty of the offence of illegal distraint (*indliged n-athgabálae*). This normally entails a fine of five *sét*s, but it may be reduced in cases of ignorance, inadvertence, carelessness or difficulty.[90] These restrictions are mainly designed to protect the interests of the defendant, and to cause as little disruption as possible to the normal farming routine. Hence it is forbidden to distrain any animal classed as *nemed*

[86] *CIH* iii 897.22–3 = Appendix A, text 2 § 2.

[87] *CIH* iii 897.24–6 = Appendix A, text 2 § 3.

[88] *CIH* iii 897.22–6 = Appendix A, text 2 §§ 2–3.

[89] The pound is also sometimes termed *comann*, e.g. *CIH* iii 897.33 = Binchy, 'A text on the forms of distraint', 78 § 2; *CIH* iv 1455.25 = *AL* i 268.21. In glosses and commentary *mainder* is generally used, e.g. *CIH* v 1723.16, 34 = *AL* ii 2.9, 4.6. There are Old Irish instances of *mainder* in quotations from *Antéchtae Breth* (*CIH* vi 2211.26 = L. Breatnach, 'The first third of *Bretha Nemed Toísech*,' 10 § 5) and from *Bretha Nemed Dédenach* (*CIH* ii 725.11 = L. Breatnach, 'The first third of *Bretha Nemed Toísech*', 30).

[90] *CIH* iv 1459.22–33 = *AL* ii 48.11–54.13.

i.e. 'immune, privileged'.[91] These are mostly animals at a crucial stage of their reproductive cycle, such as a cow just after calving, a bull while the cows are in heat, a ewe in season or a boar, ram or stallion at mating time. If the plaintiff distrains at such times he must compensate for any loss which his action has caused. According to commentary, if he distrains a bull at mating time he must not only pay the fine of five *sét*s, but also compensate for any consequent loss of calves and milk-yield.[92]

Other animals are excluded in the interests of the defendant's superiors. It is therefore not permissible to distrain a cow due as rent for a lord or Church.[93] Immune livestock can sometimes be distrained for their own offences. For instance, it is normally illegal to distrain oxen during the ploughing season.[94] However, if the oxen themselves have committed an offence – such as goring or grazing-trespass – it is permitted to distrain them, even at such a crucial time.[95] On the other hand, it is specified that a champion horse (*ech búada*) cannot be distrained in the racing season, even for its own offences.[96] The same applies to pigs being fattened on mast.[97] It is illegal for a poet to distrain any type of animal: he is expected to pursue a claim by means of satire.[98]

Some animals are clearly regarded as particularly awkward to move to a pound. The plaintiff must therefore carry out certain formal acts as a substitute for ordinary distraint. In the case of oxen, he must remove their tackle (*amlach*) and give public notice (*apad*) that they should do no ploughing. If the defendant defies this prohibition, he is guilty of a 'human offence' (*duinechin*) bringing a heavy fine.[99]

[91] Heptad 41 (*CIH* i 38.19–21 = *AL* v 260.1–4) lists seven *nemid chethrae* 'immune livestock'. A list of twenty-five categories of animal described as *nemid athgabálae* 'those immune from distraint' is found at *CIH* iii 897.10–21 = Appendix A, text 2 § 1. The term *deorad* 'beyond legal process, privileged' is used interchangeably with *nemed* in this text (*CIH* iii 897.16 = § 1 (14); 897.21 = § 1 (25)).

[92] *CIH* i 181.25–6.

[93] *CIH* i 38.21 = *AL* v 260.3 *bō cīss flatha no eclasa.*

[94] *CIH* iii 897.15 = Appendix A, text 2 § 1 (12).

[95] *CIH* iii 898.17–18 = Binchy, 'A text on the forms of distraint', 80 § 10.

[96] *CIH* iii 897.16–17 = Appendix A, text 2 § 1 (15).

[97] *CIH* iii 897.27–9 = Appendix A, text 2 § 3.

[98] *CIH* vi 2226.33.

[99] *CIH* iii 898.17–18 = Binchy, 'A text on the forms of distraint', 80 § 10 (and his note at p. 85). Similar procedures are laid down for the distraint of dogs (see p. 149 above).

In the case of normal distraint, the law-texts devote a great deal of attention to the manner in which the livestock are brought to the pound and the conditions in which they are kept there. Firstly, the plaintiff must carry out the distraint early in the morning. If the cattle have been brought to the milking-place (*indes*) before he arrives, he cannot distrain them on that day.[100] It is also stressed that the distrained animals must not be driven too fast:[101] if any are injured double restitution must be given.[102]

For their own safety, distrained livestock should not be driven to a cave, or into the wilderness or a wood, or in secret or in the dark.[103] They should not be brought onto the land of lawless persons, such as a thief or satirist, or to an empty house.[104] Animals of different types must not be kept together, and stallions (*eich*) and mares (*mairc*) must be housed separately.[105] When distrained livestock escape from their pound and suffer death or injury, the person who distrains them obviously bears the major responsibility if it results from his carelessness. He is guilty of what the commentator describes as 'neglectful custody' (*faill choiméta*).[106] But the main law-text on distraint, *Di Chetharšlicht Athgabálae*, also applies the principle of 'culpable onlooking' (*aircsiu*) in such cases. Consequently, if a third party sees the distrained livestock in a dangerous situation, and leaves them there without reporting it, he too is guilty of an offence, defined as 'bad leaving' (*mífácbáil*). He must pay one third restitution of any livestock lost.[107]

In certain exceptional circumstances the defendant can obtain a postponement (*taurbaid*) of the distraint of his livestock. This applies in domestic crises such as sickness, childbirth, death, moving house, or the arrival of guests.[108] To distrain in defiance of a postponement is classed as an illegal distraint.[109]

[100] *CIH* iii 897.1–3.
[101] Commentary distinguishes *luathimáin* 'fast driving' (e.g. *CIH* v 1724.16 = *AL* ii 4.29) from *certimáin* 'proper driving' (e.g. *CIH* v 1724.1 = *AL* ii 4.10).
[102] *CIH* v 1742.1 = *AL* ii 68.11 (read *-briathar*).
[103] *CIH* iii 900.12–13.
[104] *CIH* i 40.20–3 = *AL* v 266.19–23.
[105] *CIH* ii 422.14–35; iv 1455.24–5 = *AL* i 268.17–22.
[106] *CIH* v 1724.17 = *AL* ii 4.30–31.
[107] *CIH* v 1738.39 = *AL* ii 56.19. The following commentary provides a detailed application of this general principle.
[108] *GEIL* 183–4.
[109] *CIH* i 37.34–5 = *AL* v 256.31.

BEWITCHING

In Irish folklore of the nineteenth and early twentieth centuries we find many references to sickness in cattle caused by sorcery, the 'evil eye',[110] or the fairies.[111] In early Irish sources there is evidence that the bewitching of livestock was counted as a legal offence. One law-text refers to the crime of *fubae do grega*, which literally means 'an attack on one's horses'.[112] The glossators take this to include such physical assaults as beating them with a stick, scaring them with a rag, or causing injury through sorcery (*corrguinecht*) – a fine of five *séts* is due for such offences.[113] Another legal reference to the bewitching of livestock occurs in a passage on drowning. Here a glossator suggests that an animal may have been drowned as a result of witchcraft (*geinntlecht*).[114]

The law-text on clientship, *Cáin Aicillne*, also refers to the death of livestock through supernatural agency in a passage which details the size and condition of a bullock (*ag*) which a client gives as rent to his lord.[115] The author stipulates that the animal must have been properly slaughtered by its owner, and not have died as a result of magical pestilence (*teidm fithnaisi*) or disease (*galar*).[116] *Teidm fithnaisi* is explained in *O'Davoren's Glossary* as 'a short disease' (*galar gairet*), but it is clear from other attestations of the word *fithnaise* that the supernatural is usually involved.[117] It is probable that the 'destructions' (*milliuda*) included among the afflictions of cattle in *Immacallam in dá thúarad* were regarded as being the result of sorcery.[118] In later medical manuscripts there are many references

[110]In legal commentary, the 'evil eye' is called *drochrusc* (*CIH* i 144.34), *drochsūil* (*CIH* ii 673.3), or *bēim sūla* (*CIH* iii 955.1).

[111]E.g. Logan, *Making the cure* 163–4; Ó hEochaidh, 'Seanchas na caorach', 142–3.

[112]*CIH* ii 383.5; v 1689.13 (*fub-* MS) = *AL* i 162.22.

[113]For *corrguinecht*, see p. 128 above. *Fubae* is also used of casting malign spells (*pisóca*) in other contexts, e.g. *CIH* ii 387.31–2 = *AL* i 176.3. See also *DIL* s.v. *fuba* (b).

[114]*CIH* v 1546.24–5. Binchy emends *gleinntlecht* of the MS.

[115]*CIH* ii 480.23 = Thurneysen, 'Aus dem irischen Recht I', 347; *CIH* iv 1503.1 = O'Dav. 358 § 946.

[116]It must be admitted, however, that the glossator at *CIH* ii 481.2 = *AL* ii 238.y does not interpret the text in this way. He refers to *bristi* 'fractures' and *galair* (= *galar*) *nairni* 'disease of the kidney': see *DIL* A 420.4–34.

[117]For example, it is recorded in the *Chronicum Scotorum* (*CS* 230 s.a. 985) that there was a magic colic (*treagad fithnaisi*) in the east of Ireland caused by demons.

[118]Stokes, 'Colloquy', 46 § 233. See *DIL* s.v. *milliud* (c); cf. Meyer, *Sanas Cormaic*, 73 § 858.

to the bewitching of cattle (*mille ba*), which may be caused by an elf-shot (*urchar millte*).[119]

The most obscure offence relating to the bewitching of animals is described in the law-text on distraint as *mímír do choin*, which means literally 'a bad morsel to a dog'.[120] From the accompanying gloss it does not appear to refer to the deliberate poisoning of a dog, but rather to trying out on it a charm intended to bewitch a person. The culprit is only required to pay half the dog's penalty-fine (*díre*) as he was merely 'trying out the spell to see if it was magic'[121] with no intention to kill.

Bestiality

Sexual relations between animals and humans feature occasionally in early Irish literature. For example, Triad 236 recounts the story of the Beast of Leittir Dalláin, which was one of the wonders of Glenn Dalláin.[122] It had a human head, but the rest of its body was in the shape of a blacksmith's bellows. It was the result of a union between a water-horse (*ech uisci*) and a daughter of the local priest. In his *Topography of Ireland* Giraldus Cambrensis includes many similar tales. In some cases such acts may have had ritual significance.[123]

I have found no references to bestiality in the Old-Irish law-texts or in later legal glosses and commentary. In penitential literature, however, such practices are regularly proscribed. The penances seem relatively light; for example, the *Penitential of Cummean* states that a cleric who sins with an animal must do one year's penance.[124] In other penitentials the penance is for three years[125] or for two and a half years.[126]

[119] RIA MSS 461 (24 B 2), pp. 172, 181; 462 (24 M 34), p. 191; 465 (23 N 20), pp. 40, 61. For these MSS, see p. 218 below.
[120] *CIH* ii 387.32; v 1692.35 = *AL* i 176.4.
[121] *CIH* ii 388.13–14; v 1692.35–6 = *AL* i 180.30 *froma uptha dūs inbud amainsi.*
[122] Meyer, *Triads*, 30 § 236. *Glenn Dalláin* is Glencar, Co. Sligo (Hogan, *Onomasticon*).
[123] See p. 353.
[124] *IP* 114 § 6.
[125] *IP* 100 § 10; cf. 264 § 24.
[126] *IP* 68 § 7.

Offences against animals by other animals

A farmer's animals may break into his neighbour's land, and cause damage not only to grass or crops, but also to livestock. For example, his bull may attack and kill or injure his neighbour's bull, or his dog may worry his neighbour's sheep. The Old Irish law-texts devote a good deal of attention to such offences, particularly by cattle and dogs.

It is clear from our sources that the practice of joint-herding (*comingaire*) was a major factor in the incidence of offences by livestock against other livestock. With animals belonging to different owners herded together under the control of a single herdsman, it is easy to imagine how an attack by an animal belonging to farmer A on an animal belonging to farmer B could lead to a legal dispute. In cases where the culprit can be identified, the general rule is that its owner must give compensation (*aithgein*) to the owner of the victim. This compensation typically consists of an animal of the same quality as the victim.[127]

But if the culprit cannot be identified, our legal sources present two alternatives. *Bechbretha* describes the case of a dead animal (*mart*) being found among pigs or cattle or dogs. There is no witness to swear that one particular animal was responsible; consequently guilt falls on the whole herd or pack.[128] One alternative is for the owners of all the suspect animals to contribute to the compensation – this seems to be the solution proposed by the author of *Bechbretha*.[129] Another option, proposed by an early *Bechbretha* glossator, is to fix guilt on one animal by casting lots: the owner of this animal then becomes liable to pay the full fine.[130]

Attacks by dogs

The hunting instincts of the domestic dog are easily aroused, and can lead to attacks on livestock. Thus a law-text includes 'attacking livestock' (*aurbach cethrae*) among the four offences of dogs.[131]

[127] *CIH* iii 788.12–14.
[128] *CIH* ii 449.28–32 = *BB* 70–2 §§ 34–5.
[129] *CIH* ii 449.31 = *BB* 72 § 35.
[130] *CIH* iii 924.7–9 = *BB* 70 § 34^{e}B. Cf. *CIH* i 318.35–6 = *AL* iii 440.13–16; *CIH* v 1724.38–41.
[131] *CIH* ii 579.2. The other three are: *duinechaithchi* 'human offences' (i.e. dog-offences caused by human malice or carelessness), *fobach n-aitreibe* 'digging under a house', and *conlón i tír* 'dog-excrement on [another's] land'.

Naturally, the dog-owner is not liable if it can be shown that the dog acted in retaliation.[132]

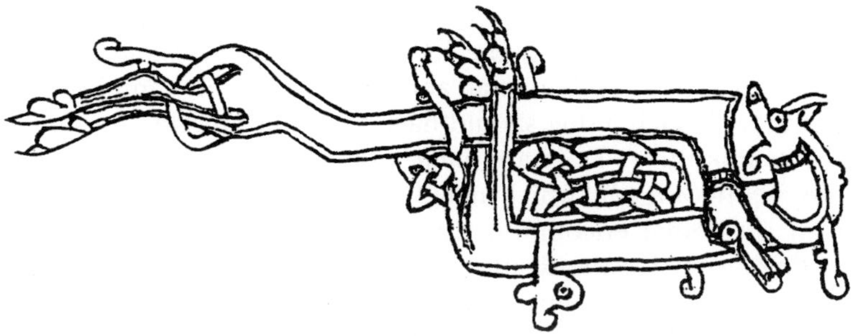

Fig. 7. This initial on p. 11 of the fourteenth-century Book of Ballymote (MS no. 536 = 23 P 12) in the Library of the Royal Irish Academy shows a dog-fight. Both dogs are collared. It has been reproduced by kind permission of the Council of the Royal Irish Academy.

Even well-trained dogs may savage livestock. *Conṡlechtae* refers to the killing of a domestic animal by a dog, while being urged on by its owner, who did not however intend that the animal be harmed.[133] The commentator explains: 'driving off [trespassing livestock] without killing was taught to the dog, but driving off with killing was what the dog did'.[134] Such attacks may also be due to a mistake by the dog-owner. Commentary on the inciting of dogs (*muilliud con*) deals with the case of the dog-owner who sets his dog on another's livestock instead of his own.[135] He has merely to provide restitution (*aithgein*) for any damage done.[136] The same applies if a neighbour's herd approaches the herd which a dog is guarding, and it bites one of the neighbour's animals to death: this is classed as a 'carcase of proper jealousy' (*mart étaig théchtai*).[137]

[132] *CIH* vi 2195.15–16 = Thurneysen, 'Aus dem irischen Recht III', 331 § 22; cf. *CIH* v 1551.2–3 (excerpt from *Conṡlechtae*).
[133] *CIH* v 1577.28 *mart con gabāla dia muillter* lit. 'a dead animal of a dog of seizure when it (the dog) is incited'. Cf. *CIH* v 1550.38–9.
[134] *CIH* v 1577.28–1578.3. The owner's liability depends on the circumstances of the attack.
[135] *CIH* i 334.25; iv 1177.9 = *AL* iii 518.20. The terms *muilliud* and *inmuilliud* are clearly synonymous, cf. *CIH* iv 1391.1.
[136] *CIH* iv 1177.22 = *AL* iii 518.25–6.
[137] *CIH* v 1550.41.

Commentary on a short quotation from an Old Irish text discusses the legal complications which arise when a domestic dog (*cú chennaid*) and a wolf (*fáel allaid, mac tíre*) together kill a young sheep.[138] Another commentary deals with the case of the herd-dog which is in the habit of snapping (*minaigecht*) at poultry, lambs and other small animals or pets which might be in a farmyard. Such a dog should have a leather muzzle (*srúblaing*) over its mouth.[139] Dogs are liable to kill or injure each other in a dog-fight (*congal*), which may be an organized entertainment or an accidental encounter. In general, no liability attaches to the dog-owner for the death or injury of the other dog. Hence, the law-text on accidents *Bretha Étgid* states the principle that 'a dog-fight is an immunity for dogs'.[140] In the accompanying commentary, however, various circumstances are outlined in which the law recognizes some degree of liability on the part of one or both owners or on those who incited the dogs to fight.

Attacks by cattle

The law-texts and later commentary devote special attention to the killing or injury of other animals by bulls. As we have seen above (p. 150), a bull-owner is generally free from liability if his bull attacks a person while the cows are in heat. He is likewise fully or partially exempt if his bull kills or injures other livestock at these times.[141] Commentary deals with the various circumstances of such incidents,[142] sometimes giving regulations which conflict in points of detail.[143] Fights between bulls are also mentioned in text[144] and commentary.[145] Fines in the case of the death of one of the bulls are less severe during the mating season. For example, if

[138] *CIH* ii 662.15–25; iv 1390.14–23.

[139] *CIH* i 313.1–3; iv 1389.11–14 = *AL* iii 416.8–10. The translation of *minaigecht* is uncertain, see *DIL* s.vv. *minach* and *minaigecht*.

[140] *CIH* i 268.12 = *AL* iii 192.9.

[141] Cf. *IP* ('Welsh Canons') 142 § 35.

[142] *CIH* i 10.30–7; iii 1047.1–5; v 1828.20–9; 1886.33–7 = *AL* v 156.1–10; *CIH* i 276.9–14 = *AL* iii 232.10–18.

[143] For example, at *CIH* i 10.32–3 = *AL* v 156.4–5 it is stated that a bull is exempt (*slán*) for any damage inflicted on a neighbour's cow in heat; but in the next line we are told that half-restitution (*lethaithgein*) may be due. I am also uncertain as to what is meant by the bull's 'own cattle' (*a cethra* (or *indile*) *budéin*) in these commentaries. As injury to these cattle may entail a fine, I am assuming that although they are in the same herd as the bull, they may belong to a different farmer.

[144] *CIH* v 1551.3.

[145] *CIH* i 276.15–31 = *AL* iii 232.19–234.18.

one bull kills another in an unprovoked attack when the cows are in heat, its owner must pay half-fine (*lethfíach*); but if the attack takes place when the cows are not in heat, the full fine (*lánfíach*) must be paid.[146] Fights between rams are treated likewise.[147]

Other cattle are much less aggressive than bulls, but may occasionally develop the habit of goring other animals. The *Comingaire*-text deals with cases of goring in a herd where cattle belonging to different owners are kept together.[148] If a bovine (*ag*) was known previously to be a gorer (*guinid*), its owner is specially penalized.[149] Where another animal has been killed, the owner of the gorer must give a substitute to the owner of the dead animal, who also takes the carcase. But in cases where one bovine kills another without having previously shown a tendency to gore, the culprit may itself be slaughtered, and the carcases of the two animals divided equally between the owners.[150] Where the culprit cannot be identified, it seems that lots are cast on the whole herd: the animal on which the lot falls – and consequently its owner – is then held to be responsible.[151]

The above regulations do not seem to apply to oxen. According to the law-text on accidents, *Bretha Étgid*, one of the exemptions of oxen is 'mutual attack' (*imoirgnechas*).[152]

Attacks by horses

Most of our legal information on the subject of horse-attacks comes from the brief statement in *Bretha Étgid* that 'a horse-fight is the exemption of horses' (*blaí ech echthres*). The accompanying commentary explains that no liability is attached to the owner of a horse which damages another horse in a fight.[153] In commentary on Heptad 7, we are told that the owner of a stallion (*cullach*) is

146 *CIH* i 276.17–19 = *AL* iii 232.22–5.

147 The commentator generally uses the term *míl* 'animal' so as to include both bulls and rams, e.g. *CIH* i 276.20 = *AL* iii 234.1.

148 *CIH* i 192.23–33, ii 577.13–24 = *AL* iv 102.6–19 (partially retranslated *BB* 134).

149 Cf. Wasserschleben, *Die irische Kanonensammlung*, 214–15, bk. 53 ch. 7.

150 *CIH* i 192.24–5, ii 577.15–16, iv 1268.8–10 = *AL* iv 102.7–9. Cf. *IP* 142 § 34; 156 § 39, though here the solution is not to kill the gorer but to make it and its victim the joint property of both owners.

151 *CIH* ii 449.28–32 = *BB* 70–72 §§ 34–5 (see discussion in notes to § 35).

152 *CIH* i 284.3 = *AL* iii 266.13. The term *imoirgnechas* (*imairgnechus* MS) is otherwise unattested: see *DIL* s.vv. *oirgnech*; 1 *imm-oirg*.

153 *CIH* i 290.24–6; v 1646.29–33 = *AL* iii 294.15–19.

not liable for any injury which it may inflict on a mare at mating time.[154]

Attacks by pigs

Our information on pig-attacks seems contradictory. *Bretha Étgid* classes fights between pigs as an exemption (*blaí*) for which neither owner would be liable in the case of death or injury.[155] On the other hand, *Bechbretha* treats the killing of a pig (or other animal) in a jointly-owned herd as an offence for which retribution must be paid by the pig-owner.[156]

Bretha Nemed Toísech describes a case brought before the legendary judge Senchae which involved the killing by pigs of both a child and a calf.[157] A herd of pigs belonging to a rich Ulster landowner named Mugnae killed and ate the son of Maine. At the time the pigs were not under the supervision of a swineherd. Maine locked the guilty pigs into a shed to initiate the legal procedure of distraint (*athgabál*).[158] However, a calf belonging to Mugnae had previously been put into the same shed, and it was also eaten by the pigs. Maine went to the judge Senchae complaining of the loss of his son, and Mugnae put in a counter-claim for the loss of his calf. Senchae was uncertain about the case, and consulted King Conchobar, who fixed damages in favour of Maine for the loss of his son, and against him for causing the death of Mugnae's calf. Because the pigs were not being supervised, Mugnae had to pay the full fine for homicide in connection with the boy's death, rather than merely forfeiting the pigs.

Attacks by hens

Modern beekeepers whom I have consulted are invariably surprised by the claim in legal commentary that 'soft swallowing of bees' (*máethślucud bech*) is one of the three hen-offences in the farmyard.[159] Hens might occasionally eat dying bees from the

[154] *CIH* i 10.30 = *AL* v 156.1. The word *cullach* can mean both 'stallion' and 'boar'.
[155] *CIH* i 290.24–5 = *AL* iii 294.15.
[156] *CIH* ii 449.29 = *BB* 70 § 34.
[157] *CIH* vi 2217.24–35. A later version at *CIH* vi 2113.16–25 is edited by Dillon, 'Stories from the law-tracts', 44–5 § iii (trans. 53 § iii).
[158] For distraint, see p. 170.
[159] *CIH* i 73.22 = *AL* iv 116.10.

ground in front of a hive, but they are not generally regarded as a threat to bees in flight or to their larvae within their hive.[160]

Another commentary deals in detail with attacks by bees on other animals and by other animals on bees, but makes no mention of their being eaten by hens.[161]

Attacks by cats

In one version of the commentary on *Bretha Étgid*, it is stated that no liability is attached to the owner of a cat which inflicts injury (*fogal*) on pets (*esrechta*)[162] during the night, but that there is liability if the offence takes place during the day.[163] The implication is that the onus is on the owner to ensure that vulnerable pets are safely locked in during the night, and perhaps also that the cat must be allowed freedom to carry out its nocturnal mouse-hunting duties (see p. 243). In a more general application, the same principle is to be found in the section on cats (*De pilacibus*) in Irish canon law: 'The *Hibernenses* say: if a cat has committed some offence by night, let not the owner pay; if it offends by day, let him pay'.[164]

160 The only comparable case known to Dr Eva Crane (author of *The archaeology of beekeeping*) is of the Denizli chickens, trained to eat wasps which prey on honeybees in western Anatolian apiaries.

161 *CIH* i 317.19–318.29 = *AL* iii 434.18–440.5.

162 In legal material, the term *esrecht* generally refers to pet animals or toys.

163 *CIH* v 1647.6–7.

164 Wasserschleben, *Die irische Kanonensammlung*, 215, bk. 53 ch. 8.

6
Accidents, diseases, etc.

ACCIDENTS TO LIVESTOCK

As we have seen in Chapter 5, there are some circumstances in which a person may be liable for the accidental death or injury of another's cattle, e.g. by digging a deer-pit on common land without proper warning or by leaving a spiked fencing-post in a dangerous position. But there are also references in the law-texts and other sources to accidents involving livestock for which no person can be held responsible.

Drowning

It is clear from our sources that the two main natural hazards for livestock were wolves (see next section) and water. The texts refer occasionally to the accidental drowning of livestock in open water (*bádud i n-uisciu*),[1] but devote more attention to the dangers of marshes or mires where an animal cannot swim to safety. In *O'Davoren's Glossary* a distinction is made between a marsh (*cechair*) which consists of a semi-solid layer over water, and a mire (*lathach*) in which water covers a layer of mud.[2] The former is regarded as particularly treacherous, and is classed as a hazard in a number of law-texts.[3] According to *Cáin Domnaig* 'the law of Sunday', the Church does not regard it as a breach of this law if a man goes to rescue his cow from a marsh on a Sunday.[4]

Legal commentary includes quaking bogs (*tonna crithaig*) among the possible causes of accidental death of livestock in a jointly-owned herd.[5] The herdsman is immune from responsibility if one of the animals under his care is drowned in a quaking bog, provided that he is immediately aware of the accident and is

[1] E.g. *CIH* v 1739.36 = *AL* ii 60.5; *CIH* vi 2139.18–19.

[2] *CIH* iv 1485.10 = O'Dav. 279 § 514. Heptad 5 (*CIH* i 6.24, v 1884.30 = *AL* v 138.11–12) likewise distinguishes between *cechair* and *lath* (from which *lathach* is a derivative).

[3] E.g. *CIH* i 192.3 = *AL* iv 100.7; *CIH* iv 1181.22. The term *féith* is used in a similar sense, e.g. *CIH* v 1739.36 = *AL* ii 60.5.

[4] Hull, '*Cáin Domnaig*', 160.14 § 1.

[5] *CIH* i 166.38–167.2; *CIH* vi 2139.18–21. The form of the second element is variable in the MSS: I take it to be gen. sing. of the substantive *crithach* 'trembling', cf. *fót crithaig* 'trembling sod'. *Tonna* is presumably nom. plur. of *tonn* 'bog, swamp'.

afterwards able to show the owner where the dead body (*mart*) lies. This provision is obviously to ensure that the herdsman is not guilty of dishonesty or negligence towards the animals under his care. If he is unable to show where the body lies, it is considered as a 'failure of guardianship' and he must restore the value of the animal to the owner.

According to the seventh-century law-text *Coibnes Uisci Thairidne*, the landowner is held responsible if a person or animal drowns in a flooded ditch (*clad*).[6] However, there is immunity in the case of the ditch of a fort, of a church, of a grave-mound, of a mill-race, of a mill-pond, of a turf-bog, or in the ditch which is around a bridge. The glossator explains that 'these ditches are so common that it is not feasible to sue their owners for liability'.

Accidents on fences

Legal commentary on joint herding of livestock gives another cause of accidental death as *brissiud i crannchaib*, which seems to mean 'breaking [of a limb] on wooden fencing'.[7] The owner of the fence is only liable if it has been constructed in a dangerous manner.[8]

Falls

As livestock were regularly brought to the hills in the summer (p. 43) there would have been a danger of falls from cliffs in many areas. According to the *Irish Penitentials*, an animal which has died in a fall from a cliff can be eaten provided that its blood has been shed. But if its bones are broken, and its blood is not shed, it must be rejected as carrion.[9]

Poisoning

There are a number of plants which may cause serious or fatal poisoning to livestock, e.g. hemlock (*Conium maculatum*), yew (*Taxus baccata*), corn cockle (*Agrostemma githago*), cowbane (*Cicuta virosa*),

[6] *CIH* ii 461.27–462.2 = Binchy, '*Coibnes Uisci Thairidne*', 70–72 §§ 13–14. Cf. *CIH* iii 907.31 *mad romarbtar cethra .i. hi claud . . . adrenar sin* 'if livestock are killed i.e. in a ditch . . . that is to be paid for'. The law-text on distraint (*CIH* ii 383.3–4 = *AL* i 162.21) likewise refers to the 'liability of one's ditch' (*im chinaid do chlaid*).

[7] *CIH* i 166.38; vi 2139.18.

[8] *CIH* ii 580.29 = *AL* iv 140.26–7 *cin . . . do crainnche* 'the liability . . . of your fencing'. Compare the paragraph (*CIH* ii 462.1 = Binchy, '*Coibnes Uisci Thairidne*', 72 § 14) which refers to injury caused by an improperly constructed fence (*anime*).

[9] *IP* 176 § 2.

and hemlock water-dropwort (*Oenanthe crocata*).[10] Ragwort (*Senecio jacobaea*) causes cirrhosis of the liver in cattle and horses. Nicholas Williams suggests that the Irish name of this plant *búa(th)fallán* (later *buachalán, bóthalán,* etc.) means 'cow-shriveller'.[11] A difficulty about this attractive etymology is that the poisoning of cattle by ragwort occurs mainly when this plant is dried in hay. Cattle will not eat living ragwort, though horses may do so when it is young.[12] The proposed etymology makes most sense if the plant acquired its name in a society in which hay-making was generally practised. However, as we have seen above (p. 47), the available evidence suggests that there was little hay-making in pre-Norman Ireland.[13]

As its name implies, henbane (*Hyoscyamus niger*) is notorious as a poisoner of hens and other poultry; in French it is called *mort aux poules* 'death to hens' (also *mort aux oies* 'death to geese'). The Irish word for this plant is *gafann* (*gaphann*), and there are references in early texts to its ill-effects on people. A tenth-century(?) poem implies that drinking *gafann* causes madness,[14] and in *O'Mulconry's Glosssary* it is defined as *bíad gonus amal gaí* 'food which wounds like a spear'.[15] Whitley Stokes notes that *gafann* 'henbane' is to be identified with Cornish *gahen* of the same meaning.[16] Nicholas Williams puts forward the theory that the Irish word is a Brittonic borrowing, and that the basic meaning of the word is 'goose-killer'.[17]

[10]West, *Black's veterinary dictionary*, 120, 258, 603, 616. I suggest on p. 310 below that *áth aba* refers to hemlock water-dropwort in the context of the accidental poisoning of humans.

[11]N. Williams, 'Some Irish plant names', 449–50. He derives it from an Indo-European compound **g^{w}ou-swolos*, which would regularly give the Irish form (with addition of the suffix *-án*). Unfortunately, no cognates are attested in P-Celtic or in other Indo-European language-groups to confirm this etymology. He traces all the current dialectal variants in Irish and Hiberno-English back to a single form *buachalán*. However, *buafalán* is attested from south Donegal (Uí Bheirn, *Cnuasach focal as Teileann*, 36): it presumably goes back to the earlier *búa(th)fallán*.

[12]West, *Black's veterinary dictionary*, 455.

[13]The plant-name could of course have been brought to Ireland by Celtic-speaking colonists. But there is no evidence that the vocabulary of hay-making goes back to Common Celtic.

[14]*LL* vi 1356.40526; *CGH* 339; Meyer, 'Aed Dub mac Colmáin', 459; F. J. Byrne, *Irish kings and high-kings*, 152.

[15]O'Mulc. 263 § 613.

[16]Stokes, '*Materia medica*', 229.

[17]N. Williams, 'Some Irish plant names', 460; N. Williams, *Díolaim luibheanna*, 90–4.

For a discussion of the legal implications of selling land infested with weeds – some of which may be poisonous to humans and/or livestock – see p. 234 below.

Other accidents

Legal commentary on joint herding refers also to the type of accident which can be categorised as an act of God (*dífoichid Dé*), for which a herdsman obviously cannot be held responsible.[18] This would include such events as the death of an animal after being struck by lightning (*saignén teintide*), an accident which is mentioned in the fourteenth-century legal manual of Giolla na Naomh mac Duinn Shléibhe Mhic Aodhagáin.[19] The *Chronicum Scotorum* records a paricularly devastating flash of lightning which struck the yew-tree of Saint Cíarán in January 1149, and killed one hundred and thirteen sheep which were sheltering beneath it.[20]

Another act of God involving the death of livestock at Sliabh Gamh (Ox mountains, Co. Sligo) is recorded in the *Annals of the Four Masters* for the year 1490. An earthquake (*maidhm talmhan*)[21] killed a hundred people, as well as many horses and cattle.[22]

Predators

Domestic animals tend to attract predators. This is partly because man has for his own convenience bred out protective characteristics such as aggression, wariness, and speed. Domestic animals are therefore often ill-equipped to fight off or escape from predators. Another cause of vulnerability is the fact that many domestic animals must spend their lives in environments which are unnatural to them. For example, the original home of the sheep (*Ovis aries*) was open mountainside where they could escape from predators by their ability to negotiate steep terrain. In the woods, thickets and marshes of early Ireland, they would have been at a great disadvantage when under pursuit. Similarly, the domestic hen derives from the red junglefowl (*Gallus gallus*) which escapes from predators by

[18] *CIH* vi 2139.19.

[19] *CIH* ii 698.9. Cf. *CIH* i 19.5–6 = *AL* v 190.16–17.

[20] *CS* 346 s.a. 1149 *Torand ocus soighnēn do tiachtain i nEnāir gur gabh an teni i n-iubhar Cíarain . . . ocus goro marbadh trí caoírigh .x. ar cēd fon niubhar.*

[21] The more usual term for an earthquake is *talamchumscugud* 'earth-movement', e.g. *CIH* vi 2220.15; *LU* 293.9759; Stokes, 'The Life of S. Féchín of Fore', 348 § 41.

[22] *AFM* iv 1184 s.a. 1490.

flying up into the trees. This instinctive response is of no avail when the hen is confronted in a low shed by a fox or stoat.

Wolves

Our sources are unanimous in pointing to the wolf (*Canis lupus*)[23] as the principal predator on livestock, particularly on lambs and calves.[24] Wolf-hunting is consequently a public duty: according to a ninth-century glossator on the law-text *Cáin Aicillne* a client must carry out a foray (*fubae*) against wolves once each week.[25] If it is known that a pack of wolves is present in a particular place, it is illegal for a person to drive his neighbour's livestock near them, and he must pay compensation for any animals which they kill.[26]

Fig. 8. This wolf is reproduced from f. 76v of the eighth-century Book of Kells in the Library of Trinity College, Dublin. The artist has caught very skilfully the lolling tongue and loping gait of a prowling wolf.

A section of Middle Irish commentary on accidents to livestock deals also with a herdsman's duties with regard to wolves. In one version of the text, a herdsman who is tending livestock which is owned by another is free from liability for 'the first seizure by the wolf'.[27] This seems to imply that his care of the herd is only held to be defective if he allows wolves to take a second animal. Another version puts more responsibility on the herdsman by stating that he is free from liability for 'a single tooth of the wolf

[23]In legal sources, the wolf is usually called *cú allaid*, lit. 'wild dog'. Other words commonly employed in Irish are *macc tíre*, lit. 'son of the land', and *fáel* (or *fáelchú*).

[24]E.g. *CIH* iii 897.23 = Appendix A, text 2 § 2; *CIH* iv 1496.3–5 = O'Dav. 326 § 786; *CIH* iv 1390.14; Meyer, *Betha Colmáin*, 54 § 51; *CIH* ii 585.14 = *AL* iv 352.22; Ó hAodha, *Bethu Brigte*, 8 § 25.

[25] *CIH* iii 890.8; cf. *CIH* ii 525.31 = Thurneysen, 'Aus dem irischen Recht I', 364–5 (*AL* ii 268.16–17).

[26] *CIH* i 6.24 = *AL* v 138.12.

[27] *CIH* vi 2139.17 *cētgabāil in mic tīri*; cf. *CIH* ii 698.9, where *pēsti* 'beasts' presumably refers to wolves.

with noticing immediately'.[28] Evidently, he must be so alert that the wolf has only time to make a single bite at its intended prey. If this causes a fatal injury, the herdsman is free from liability, but he must show the dead animal to its owner.

In some non-legal texts, there seems to be an acceptance of limited predation by wolves. The eighth-century tale *Togail Bruidne Da Derga* gives a vivid picture of the peace and prosperity of Ireland during the reign of King Conaire. The king kept seven wolves as hostages in his house to ensure that the wolves of Ireland only took one male calf from each herd during the year.[29] The saints' Lives likewise display tolerance of the consumption by wolves of calves and wethers.[30] Saint Mo Lua is even credited with providing a cooked calf in the monastic guest-house for a pack of hungry wolves on which he had taken pity. This became an annual event, and in gratitude the wolf-pack took on the responsibility of protecting the monastic livestock from other wolves and from robbers.[31]

Bulls were clearly expected to provide some protection for the cattle in their herds: a Middle Irish poem refers to a brindled bull which protected its herd from wolves and other dangers.[32] An obscure legal passage suggests that an especially powerful and aggressive ox might also be capable of defending the herd: this is the *dam conchaid* 'wolf-fighting ox' (see p. 49).[33] The law-texts do not appear to regard the wolf as being of direct danger to man, though an entry in the *Annals of Connacht* for the year 1420 states that 'many persons were killed by wolves in that year'.[34] The general consensus among zoologists is that attacks by wolves on humans are rare, and are almost invariably carried out by sick, old or rabid animals. The destruction of the remaining woods of Ireland in the sixteenth and seventeenth centuries deprived the wolf of its natural habitat, and the last wolves were killed in the early eighteenth century.[35]

[28] *CIH* i 166.37 *aenfiacail in meic diri (= thíre) cuna rāthugudh fo .c.oir.*

[29] Knott, *Togail Bruidne Da Derga*, 18.601–4 *ag fireand cacha indise* lit. 'a male calf from every cattle-enclosure'.

[30] E.g. Plummer, *Vitae* i 238 § ix; ii 142 § v; 148 § xxii.

[31] Ibid. ii 217–18 § 33.

[32] E. J. Gwynn, *Metrical Dindshenchas* iv 70.25–8.

[33] *CIH* i 192.18–19; ii 577.6–7 = *AL* iv 100.z.

[34] *AC* 452 s.a. 1420 § 7.

[35] Rackham, *The history of the countryside*, 36.

Foxes

At the present time the fox (*Vulpes vulpes*) impinges on the farmer mainly as a predator on poultry, but it may also kill young or sickly lambs. In early Irish sources, it is occasionally mentioned as a predator, with emphasis on its cunning. Thus *Aislinge Meic Con Glinne* speaks of the smooth slinking approach of a young fox (*sinchán*) past the shepherd,[36] and *Mesca Ulad* refers to the difficulty of tracking a fox in bank, thicket or woodland.[37] In a text on the wonders of Ireland, the three harmful animals found in this country are given as the wolf (*fáel*), the fox (*sinnach*), and the mouse (*luch*).[38]

Birds of prey

Fig. 9. This magnificent eagle, the symbol of Saint John, is reproduced from f. 1 of the eighth-century gospel-book MS 197B in the Library of Corpus Christi College, Cambridge. By kind permission of the Master and Fellows of the College, this illustration has been adapted from a photograph.

All the domestic fowl kept by the early Irish farmer particularly the fledglings, would have been at risk from various birds of prey, such as the peregrine falcon (*Falco peregrinus*), the sparrow hawk (*Accipiter nisus*), the goshawk (*Accipiter gentilis*), the kestrel (*Falco tinnunculus*) and the hen harrier (*Circus cyaneus*). The extent to which the larger birds of prey killed the young of domestic mammals in the early period is uncertain. There is no doubt that the golden eagle (*Aquila chrysaëtos*) will sometimes take young lambs, kids and

[36] Jackson, *Aislinge*, 33.1012 = Meyer, *Aislinge*, 85.2.

[37] Watson, *Mesca Ulad*, 15.335.

[38] Todd, *The Irish version of Nennius*, Appendix II, 218 § xxxiv; cf. Meyer, 'Irish mirabilia', 4, and O'Meara, *Giraldus: Topography*, 49–50 § 20 = Dimock, *Giraldi Topographia*, 61.24–6.

piglets. However, modern studies on eagles in Scotland indicate that their main food consists of grouse, hares and carrion.[39] Early Irish sources contain some possible references to the eagle as a carrion-feeder,[40] but I have found no evidence that it was regarded as a predator on livestock remotely comparable to the wolf. According to the law-text on 'joint-herding' (*comingaire*), lambs and piglets were not brought from the farm to commonage – often mountain pasture – until the beginning of August.[41] They would therefore have grown past the stage at which they would be most vulnerable to eagles. I reproduce on p. 188 an eagle from the gospel-book MS 197B in Corpus Christi College, Cambridge.[42]

A legal quotation in *O'Davoren's Glossary* states that three birds may be taken without any share going to the landowner.[43] They are a very small bird (*minnta*),[44] a heron (*corr*), and a hawk (*séig*). The glossator provides no explanation as to why the first two are included. Presumably a small bird was too insignificant to be of legal consequence, and the heron may have been regarded as a pest because of its depredations on fish.[45] In the case of the *séig* the glossator remarks that 'it carries off young pigs and hens' (*foxlaid orcu 7 cerca*). He explicitly identifies the *séig* with the later word for 'sparrow hawk' (*seabhac*), but I have been unable to find confirmation for his view that a sparrow hawk would take an animal as large as a young pig.[46] Possibly the author of the Old Irish law-text from which the quotation was taken used the word *séig* to include bigger birds of prey, such as buzzards, kites, harriers, etc.[47]

[39]Fraser Darling, *Natural history in the Highlands and Islands*, 100; Lockie and Stephen, 'Eagles, lambs, and lamb management in Lewis'.

[40]E.g. *CIH* ii 603.33–4 = E. J. Gwynn, 'Privileges', 54 § 12. Here *cubhair* is explained *.i. prēcháin ł hilair* 'i.e. kites or eagles'.

[41]*CIH* i 192.2–3 = *AL* iv 100.6–7.

[42]For a discussion of the manuscript, see George Henderson, *From Durrow to Kells*, 68-71. Isabel Henderson (*The Picts*, 124–5, 217, plates 37–38) compares this eagle with that inscribed on a Pictish stone at Knowe of Burrian, Birsay, Orkney.

[43]*CIH* iv 1526.1–2 = O'Dav. 460 § 1480.

[44]This word is used of any small bird, but particularly tits (*Paridae*).

[45]On the other hand, herons may also have been kept as pets: see p. 127.

[46]Recent Donegal tradition holds that a *seabhac* would take a new-born lamb (Ó hEochaidh, 'Seanchas éanlaithe Iar-Uladh', 253–4). However, in an addendum to Ó hEochaidh's article (p. 337), Breandán Ó Madagáin suggests that *seabhac* here refers to the buzzard. See next note.

[47]In an Old Irish gloss (*Thes.* ii 119.27), *séig* is used as a convenient word to explain Latin *uultur* 'vulture', a bird which would obviously have been unfamiliar to the early Irish. Similarly, in the Old Irish tale *Scél Tuáin meic Chairill* (ed. Carey, 102.65 = *LU*

Other predators

Domestic livestock or poultry may sometimes be taken by carnivores such as the stoat (*Mustela erminea*), the otter (*Lutra lutra*) or the pine marten (*Martes martes*). In legal material the only reference to these animals is as pets: see p. 130.

There is also an intriguing references to the bear as a predator on livestock. In the ninth-century tale *Cath Maige Mucrama*, the slaughter wrought by two great warriors is likened to that of bears among young pigs (*mathgamna eter banbraid*).[48] This comparison is especially interesting because there is no archaeological evidence for the brown bear (*Ursus arctos*) in Ireland after the Neolithic or Early Bronze Age.[49] Nonetheless, early Irish writers show a good deal of acquaintance with this animal, and there are native words in Irish which can be identified as meaning 'bear'.[50] It is clear that the bear featured prominently in the mythology of the continental Celts, and was admired for its strength and ferocity. In early Irish texts, words for 'bear' are often applied in a complimentary sense to warriors,[51] and the elements *Art* and *Mathgamain* – both

---

46.1354) *séig* evidently refers to the osprey or fish-hawk (*Pandion haliaëtus*). It seems, therefore, that *séig* could be used as a general term for large birds of predatory or carrion-eating habits.

[48]O Daly, *Cath Maige Mucrama*, 54 § 53 = Stokes, 'The battle of Mag Mucrime', 456 § 53.

[49]Ó Ríordáin, 'Lough Gur excavations', 354, 367; cf. Van Wijngaarden-Bakker, 'The animal remains from Newgrange (1)', 317–18; 'The animal remains from Newgrange (2)', 91.

[50]The Indo-European term for 'bear' survives in Irish in the form *art*, which is cognate with Welsh *arth*, Greek ἄρκτος, Latin *ursus*, etc. (*LEIA* A-91). Another word, probably related to *maith* 'good', is *math*, usually compounded with *gamain* 'calf' (*LEIA* M-24). This gives the personal name *Mathgamain*, surviving in surnames such as Ó Mathghamhna (O'Mahoney) and Mac Mathghamhna (MacMahon). The original meaning of *beithir* seems likewise to have been 'bear' (*LEIA* B-28–9), but in one legal reference it is apparently used of a large dog: see p. 119 above.

There are other instances of native Irish words for fauna which are absent from this island. Irish *nathir* preserves the Indo-European word for snake (*LEIA* N-4) although there are no snakes in Ireland. A similar case is that of the word *loscann*, which seems generally to refer in Irish to the common toad (*Bufo bufo*). It is however absent from Ireland, our only native toad being the rare natterjack (*Bufo calamita*) in south Kerry. After the introduction of the common frog (*Rana temporaria*) in post-Norman times (O'Meara, *Giraldus: Topography*, 52 § 25 = Dimock, *Giraldi Topographia*, 65–6), the word *loscann* was also applied to this amphibian, though the loan-word *frog*(*a*) is more common. For a discussion of another Irish word for 'frog', *lafan*, see Greene, '*Lafan: llyffant*'.

[51]See *DIL* s.v. 1 *art*.

originally meaning 'bear' – are common in personal names.[52] The brown bear is thought to have survived in Scotland until the ninth or tenth centuries AD,[53] so some general idea of its appearance and behaviour could have been maintained in the neighbouring island, especially in view of the fact that a common language was spoken in both areas. It is noteworthy that the bear's special liking for honey is mentioned in a Middle Irish text: Calvert Watkins has pointed out that the term *milchobur* 'honey-desirer' in a rhetoric in *Talland Etair* can only refer to the bear.[54]

The brown bear is omnivorous, but relies mainly on roots, leaves, fruit, honey, bees and ants. In countries where the brown bear lives near farmland, it does not seem to be regarded as a special threat to livestock. On the other hand, cases where bears have killed domestic animals are well documented, and there is no doubt that a hungry bear could make short work of a litter of young pigs, as described in *Cath Maige Mucrama*.[55]

Carrion-feeders

Two carrion-feeders of the crow family feature prominently in early Irish literature and mythology: the raven (*Corvus corax*) and the hooded crow (*Corvus cornix*).[56] No doubt because of its harsh gloating cry, which seems to express a delight in death, the hooded crow is identified with the battle-fury or war-goddess Morrígain.[57] For example, in *Táin Bó Cúailnge* she appears on slain corpses in the guise of a hooded crow (*badb*), and mocks the fighting prowess of Cú Chulainn.[58] In later Irish this bird is known as a *fennóc* (*feannóg*),[59] and we find similar instances of its association with death

[52]See *CGH* s.vv. *Artt, Artán, Artacán, Artgal, Artrí, Mathgamain.*

[53]Fraser Darling, *Natural history in the Highlands and Islands*, 65. On the other hand, Rackham (*The history of the countryside*, 33–4) believes that it became extinct in Roman times or earlier.

[54]Watkins, 'Irish *milchobur*'. He compares Old Church Slavonic *medvědъ* 'honey-eater', medieval Welsh *melfochyn* lit. 'honey-pig'.

[55]The twelfth-century Norwegian law of the Gulathing refers to attacks on farm animals by bears or wolves (Larson, *The earliest Norwegian laws*, 69 § 43).

[56]The black and grey hooded crow (scald crow) is the Irish and northern Scottish form of the entirely black carrion crow (*Corvus corone*) of England, Wales, and southern Scotland. The distinction was noted by Giraldus Cambrensis (O'Meara, *Giraldus: Topography*, 46 § 16 = Dimock, *Giraldi Topographia*, 54.y-z).

[57]She is the sinister Morgan la Fée of Arthurian legend.

[58]*TBC* Rec. I 16.498–9; cf. Nettlau, '*Táin Bó Cuailnge*, (2)', 72 §§ 111–12.

[59]*DIL* s.v. *fennóc*. The word *fennóc* looks as if it might be a derivative of *fennaid* 'tears, flays [flesh]' with *-óc* as an agent suffix. On the other hand, it may be a later

on the battlefield. Thus the O'Tooles of Wicklow chose this bird as their emblem, and are recorded as going into battle in the fourteenth century with the cry *fennockabo!* (= *feannóg abú*).[60] The raven (*bran, fiach*) is likewise often used to symbolize violent death in Irish literature of all periods.[61]

The early Irish farmer would mainly have come into conflict with carrion-feeders when he wished to preserve the skin of a dead animal. Legal commentary deals with the rights of a person who comes across the recently dead carcase (*mart*) of an animal belonging to somebody else.[62] He is entitled to one quarter of the value of the carcase for his trouble in skinning it and for preserving it from wolves and birds (*ar chonaib ⁊ énaib*).

Diseases and defects of livestock

The annals contain a number of references to animal diseases, mainly of cattle, but our most important source of information on this topic is a legal commentary giving the diseases and defects which may appear after purchase in cattle, horses, goats, sheep, dogs and poultry (and also in slaves).[63] In the list of diseases of livestock set out on p. 196 below, those which are included in this commentary are marked °. Generally a buyer is entitled to compensation if the symptoms appear within a fixed period after the sale, the length of which varies in accordance with the type of livestock involved. There is a limitation period (*iubaile*) of one year for a horse, nine months for a cow, four months for a pig, sheep or goat, and six weeks for a dog or cat.[64]

This commentary also deals with the legal implications of whether an animal has contracted an infectious disease (*galar tecmaisech*) or whether it suffers from an inherent ailment (*galar bunaid*). The distinction may be difficult to establish, so the category of a 'disease

---

form of *ennach*, well-attested in Old Irish in the meaning 'hooded crow': see *DIL* s.v. 1 *ennach*; E. J. Gwynn, '*Miscellanea Celtica*', 71–2.

[60] Greene, 'The Irish war-cry', 169.

[61] For examples, see *DIL* s.v. *bran*.

[62] *CIH* ii 661.8–9; iii 1106.29–30.

[63] *CIH* iii 1000.26–32; 1000.41–1001.2; vi 2048.24–8. These passages are edited in McLeod, *Early Irish contract law*, 288 [p], 299 [f], 300 [i], with discussion in Appendix 2. For a brief account of the place of slaves in the early Irish economy, see p. 438 below.

[64] *CIH* ii 696.6–8. For similar regulations in Welsh law, see Wiliam, *Llyfr Iorwerth*, 86 § 128.19–23 = *ALIW*, Venedotian Code, III vii § 6.

of uncertain origin' (*galar cunntabartach*) is also recognized.[65] The commentary correctly allows for the fact that similar symptoms may be caused by different types of disease. For example, stomach-pain (*idu*) is included both among the inherent ailments and infectious diseases of cattle.[66]

Diseases of livestock are also mentioned in two texts which contain prophecies of disasters to befall the people of Ireland. Thus the tenth-century *Immacallam in dá thúarad* 'colloquy of the two sages' lists seven diseases of cattle (*tedmann for cethraib*).[67] An eleventh- or twelfth-century poem of prophecies ascribed to Bécán mac Dé contains a stanza naming five afflictions of cattle from the same list.[68]

Some of the diseases mentioned in legal commentary, annals, and prophecy-texts would be readily identifiable to a modern Irish farmer or veterinary surgeon. For example, *dallṡinche* 'blind-teatedness' in cows obviously refers to blockage and loss of function of one of the teats. Similarly, *claime* 'mange' in goats raises no problems, as psoroptic mange still afflicts this animal (see p. 207 below).

Other diseases, however, are more difficult to identify. Our sources rarely provide any indication as to the nature of the symptoms. Consequently, if the Old Irish name for a disease has not survived in Modern Irish or Scottish Gaelic – and there is no other textual reference with information on the symptoms – then its identification is largely guesswork.

Another difficulty is the fact that diseases which afflict animals, like those of humans, change over a period of time. Many of the diseases which would be found among early Irish livestock have since been eradicated. For instance, *conach* 'rabies', listed among the diseases of dogs, cattle, pigs, and poultry, is no longer present in

[65] E.g. *CIH* iii 998.26.

[66] *CIH* iii 998.20, 22.

[67] Stokes, 'Colloquy', 46 § 233 *bedcacha, scamacha, bōāra, comalla, milliuda, cnuicc, crithcha.* I follow Stokes in translating *cethrai* as 'cattle' rather than 'livestock', as the inclusion of *bóára* 'cow-mortalities' suggests that the author is thinking primarily of cattle-disease. For a discussion of the identity of these diseases, see next section.

[68] Knott, 'Poem of prophecies', 72 § 65 *sgamach, beadhgach, bōār buar,/ comillid, crith-galar cuar.* Knott notes the superior readings *bōār buan* 'lasting cow-mortality' and *comaille* 'dropsy' in the YBL version of this poem. The poet departs from the arrangement of the *Immacallam* by taking *cnuicc* as a disease of humans: *cnuic a mbraīgdibh na ndaīne* 'lumps in people's throats'.

Ireland or Britain. *Fothach* 'glanders' has been eradicated from Ireland, Britain, and most of the rest of Europe. On the other hand, many present-day diseases – such as 'mad cow disease' – would have been absent from the early Irish farm.

CATTLE-DISEASES AND DEFECTS

References to cattle-mortalities are frequent in the annals and other sources. Sometimes these were serious enough to cause starvation among the human population.

Many cattle-mortalities are associated with exceptional weather conditions, and a consequent shortage of grass. The *Annals of Inisfallen* record that in 1172 bad weather resulted in the death of 'most of the livestock (*crod*) of the men of Ireland'.[69] Such losses were especially likely to occur during the hungry months of spring: the *Annals of Ulster* refer to a great starvation (*ascolt*) of livestock in the spring of 879.[70] A summer drought could also be devastating: the *Annals of Inisfallen* record that in the hot summer of 1129 the waters of Ireland dried up with consequent mortality (*ár*) of cattle.[71]

In cases such as the above it is clear that the animals were weakened through lack of nourishment, and therefore fell victim to a variety of ailments which would not normally be fatal. In other cases it is clear that a specific epidemic infection was responsible. For example, the *Annals of Ulster* record that an outbreak of cattle-disease occurred in England in 699.[72] The following year it was reported on the calends of February in Mag Trega in the district of Tethbae (in the modern Co. Longford). According to the *Fragmentary Annals of Ireland,*[73] this outbreak was heralded by an aurora borealis which was likened to three shields 'as if fighting in the sky, from east to west, like tossing waves, on the tranquil night of the Ascension of the Lord. The first was snowy, the second fiery, the third bloody, which it is thought prefigured three evils to follow: for in the same year herds of cattle throughout Ireland were almost destroyed, not only in Ireland, but indeed thoughout Europe. In the next year there was a human plague for three consecutive years.

69 *AI* 304 s.a. 1172 § 2.
70 *AU*² 334 s.a. 878 (recte 879) § 7.
71 *AI* 292 s.a. 1129 § 9.
72 *AU*² 158 s.a. 698 (recte 699) § 1.
73 Radner, *Fragmentary annals*, 46 s.a. 700 § 143.

Afterwards came the greatest famine, in which men were reduced to unmentionable foods'.[74]

Some epidemics mentioned in the annals struck both humans and their livestock. In 987 the *Annals of Ulster* record the outbreak of a disease called *bedgdíbad,* which literally means 'death from fits'. It caused many deaths among people and animals in England, Wales and Ireland, and has been identified by Mac Airt and Mac Niocaill with St Vitus' Dance.[75] In 1077 the *Annals of Inisfallen* record an outbreak of *sinech* affecting livestock which could also be contracted by humans: see p. 203 below.[76] The same annals refer to a cough called *galar na placodi* which was widespread among horses and humans in 1259.[77] The most extraordinary case of illness transmitted from animals to man is that described in the *Annals of Connacht* in 1224. 'A heavy and terrible shower fell in part of Connacht this year, that is in Tír Maine and in Sodain and in Uí Diarmata and in Clann Taidc, which brought about disease and very great sickness among the cows and beasts of those regions after they had eaten grass and leaves; and when men drank of the milk of these cattle and ate of their flesh, they suffered internal pains and various diseases'.[78] This description strangely prefigures the contamination of pasture and hence meat and milk by rain-borne radioactive fall-out, as happened in many parts of Europe after the explosion at the Chernobyl nuclear plant in 1986. I have not been able to find any explanation of the 1224 episode. Obviously, the heavy shower and the subsequent outbreak of disease may have been totally unconnected. On the other hand, it is noteworthy that the annalist gives the precise locations of the affected areas – all in what is now east Galway[79] – so his information should be taken seriously: see p. 752.

° *Máelgarb*

The epidemic most frequently mentioned in the annals is *máelgarb,* which means 'bald rough'. Presumably the symptoms

[74]I.e. cannibalism; cf. *AU*[2] 158 s.a. 699 (recte 700) § 6.

[75]*AU*[2] 420 s.a. 986 (recte 987) § 2.

[76]*AI* 234 s.a. 1077 § 6.

[77]*AI* 358 s.a. 1259 § 3. See p. 212 below.

[78]*AC* 2 s.a. 1224 § 2. Compare the pestiferous cloud described by Adomnán which caused sores on human bodies and on the udders of animals (A. O. and M. O. Anderson, *Adomnan's Life of Columba,* 330–4 (new pagination: 98–100) = ff. 55b–57a.

[79]See discussion by A. Martin Freeman, *AC*[2] footnote 1.

included skin-sores and loss of coat in the affected cattle.[80] This explanation is supported by a legal glossator who identifies it with *claime* 'skin-disorder, mange'.[81] In legal commentary it is listed as an infectious disease (*galar tecmaisech*).[82]

The *Annals of Clonmacnoise* record the appearance of this disease in the year 770.[83] According to another annalistic tradition, however, the disease did not come to Ireland until 985. Thus the *Chronicum Scotorum* states that this year saw 'the beginning of the cow-mortality (*bó-ár*) i.e. the *maílgarb*, which had not come before'.[84] The same source records another severe epidemic of *máelgarb* in 1129, which caused the death of almost all the cattle and pigs of Ireland.[85] It persisted in the northern half of the country in the following year.[86] Possibly *máelgarb* is to be identified with foot-and-mouth disease which affects cattle, sheep, goats and pigs; its symptoms include blistering on the tongue, lips, udder, teats and feet.[87]

A cattle-plague called the *máel domnaig* is attested in the *Annals of Ulster* and the *Annals of Loch Cé* for the year 1324. The first element *máel* 'bald' suggests that its symptoms were similar to those of *máelgarb*. However, I have no explanation for the second element *domnaig*, which seems to be genitive singular of *domnach* 'Sunday'.

° *Bedgach*

A *bedg* is a 'start, jump, leap', so any disease characterized by unusual jerky movements could be described as *bedgach*. It is included in legal commentary among the infectious diseases of

[80] Columella, *De re rustica* (ed. Forster and Heffner), 6.13.1, refers to *scabies* 'scab, skin-disease' in cattle.

[81] *CIH* vi 2049.37–8 = McLeod, *Early Irish contract law*, 288 [p].

[82] *CIH* iii 998.21; 1000.41.

[83] *Ann. Clon.* 123 s.a. 770. Conell Mageoghagan, the seventeenth-century translator into English of these lost Irish annals gives the anglicized spelling *moylegarow* here. At *Ann. Clon.* 160 s.a. 981 he gives *moylegarie* as the name of the same disease.

[84] *CS* 230.16–17 s.a. 985 = Stokes, 'The annals of Tigernach (2)', 345 = *AFM* ii 720 s.a. 986 (recte 987). This is to be distinguished from the *bedgdíbad* 'death from fits' affecting humans and livestock in 987 (AU^2 420 s.a. 986 (recte 987) § 2): see above.

[85] *CS* 334.4–5 s.a. 1129.

[86] *CS* 334.9 s.a. 1130. The *Annals of Loch Cé* record a cow-mortality (*bódíbath*) throughout Ireland in 1133, and state that no comparable outbreak had occurred since 699: see discussion at *ALC* i 132, footnote[4].

[87] West, *Black's veterinary dictionary*, 215.

cattle.[88] The infectious nature of *bedgach* is also stressed by the eighth-century author of Heptad 5.[89] He states that a person who drives another's livestock into an enclosure infected by *bedgach*[90] is responsible for any loss which may result. This does not of course apply if these animals have previously come out of the infected enclosure.

Bedgach is also included among the diseases of livestock in *Immacallam in dá thúarad*[91] and in Bécán's *Prophecies.*[92] For *bedgdíbad* 'jerking mortality', see p. 195 above.

° ***Scamach***

This term is used of disease of the lungs (*scam*), and is perhaps to be identified with bovine pleuropneumonia or 'lung-plague'.[93]

It is possible that *scamach* also refers to bovine tuberculosis. In legal commentary it is included among the infectious diseases of cattle,[94] and it also appears among the diseases of livestock listed in *Immacallam in dá thúarad,*[95] in *Betha Farannáin,*[96] and in Bécán's *Prophecies.*[97] The miserable hospitality described in *Erchoitmed Ingine Gulidi* includes part of a joint from a lung-infected calf (*gamain scamach*).[98]

Scamach is also recorded as a epidemic among humans: the *Annals of Ulster* refer to outbreaks in 783 and 786.[99]

[88] *CIH* iii 998.21; 1001.1. The Welsh law-texts refer to a disease of cattle, horses and sheep called *dere* (*dera*), usually translated 'staggers' (e.g. *GMWL* 114 s.v. *dere*).

[89] *CIH* i 6.25–6; v 1884.32–3 = *AL* v 138.14–15.

[90] The readings of the MSS suggest an original *imán i cathaig sáetho bedgaige* 'driving into an enclosure of the disease of *bedgach*'. There is no other attestation of *cathach* in the meaning 'enclosure'; however, commentary takes it to be equivalent to *búaile* 'cow-pen' (*CIH* i 6.33; iv 1326.12).

[91] Stokes, 'Colloquy', 46 § 233.

[92] Knott, 'Poem of prophecies', 72 § 65.

[93] This is to be compared with the horse-disease *ysgyfaint*, mentioned in medieval Welsh law (*GMWL* 120 s.v. *dueskeynt*, 302-3 s.v. *eskeueynt*; Williams and Powell, *Llyfr Blegywryd*, 92.6–7 = *ALIW*, Dimetian Code, II xxviii § 22; Wiliam, *Llyfr Iorwerth*, 83 § 123.2 = *ALIW*, Venedotian Code, III iv § 13). This word is cognate with Irish *scam*: see *LEIA* S-31 s.v. *scam*. Aristotle, *Historia animalium* (ed. Thompson), 8.23, refers to a fatal lung-disease of cattle called κραῦρος, which the editor identifies with pleuropneumonia.

[94] *CIH* iii 998.21; 1000.41.

[95] Stokes, 'Colloquy', 46 § 233.

[96] Plummer, '*Betha Farannáin*', 6.5.

[97] Knott, 'Poem of prophecies', 72 § 65.

[98] Meyer, *Hibernica minora*, 66 § 9.

[99] *AU*[2] 238 s.a. 782 (recte 783) § 8; 242 s.a. 785 (recte 786) § 9.

°***Bó-ár***

This term simply means 'cow-mortality', and probably refers to a number of different diseases. Our sources generally reserve the term for diseases which are infectious. The author of Heptad 5 refers to a cattle-enclosure infected by *bó-ár*,[100] and legal commentary includes *bó-ár* in its list of the infectious diseases of cattle.[101] According to the law-text on legal entry, cattle should not be brought into a place infected by cow-mortality (*baislec bó-áir*).[102] In *O'Davoren's Glossary*,[103] it is treated as a general term for infectious cattle-disease, and is explained 'i.e. *scamach* (pulmonary disease) or *máelgarb* (mange) or *galar rigin* (stiff sickness[104])'.

The only other indications known to me of the nature of *bó-ár* are found in glosses on the legal commentary list. Here *bó-ár* is explained *.i. int ascalt no caíle* 'i.e. the dearth or emaciation'[105] or simply *.i. caíle* 'i.e. emaciation'.[106] But it is not clear whether the glossator is taking emaciation to be a symptom of the disease(s), or whether he is making the obvious link between malnutrition and epidemic disease (see p. 194 above).

Bó-ár is included among the diseases of livestock in *Immacallam in dá thúarad*[107] and in Bécán's *Prophecies.*[108] According to a genealogy of the Ossairge, the first *bó-ár* came to Ireland in the time of Bressal Brecc, traditionally held to be the ancestor of this people.[109] Mícheál Ó Dochartaigh MRCVS suggests to me that *bó-ár* may refer to the cattle-plague known as rinderpest, which ravaged Europe on many occasions during the Middle Ages.[110]

[100] *CIH* i 6.25; v 1884.32 = *AL* v 138.15. Legal commentary at *CIH* v 1565.25 likewise refers to an infected cattle-enclosure (*buaile galair*).
[101] *CIH* iii 998.21; 1000.41.
[102] *CIH* i 207.2 = *AL* iv 6.9. *Baislec* may mean 'venemous or poisoned place', from Latin *basiliscus* 'venemous reptile, etc.': see Binchy, '*Bretha Crólige*', 69. On the other hand, it could be from Latin *basilica* in the sense of 'graveyard, place of death': see Doherty, 'The basilica in early Ireland', 313–15.
[103] *CIH* iv 1477.13 = O'Dav. 243 § 291.
[104] This presumably refers to some disease characterized by stiffness in the joints.
[105] *CIH* vi 2049.38.
[106] *CIH* iii 1001.1.
[107] Stokes, 'Colloquy', 46 § 233.
[108] Knott, 'Poem of prophecies', 72 § 65.
[109] *LL* vi 1493.45384.
[110] West, *Black's veterinary dictionary*, 468–70.

° ***Conach***

Conach is included in legal commentary among the infectious diseases of cattle.[111] The basic meaning of the term is rabies, a fatal disease of dogs which may be transmitted to other mammals and to birds (p. 215 below). In the later language *conach* is also applied to other ailments of cattle, and is often loosely translated 'murrain'.[112] An Ó Longáin medical manuscript describes a decoction to be drunk by cattle afflicted with *conach*, but gives no symptoms of the disease.[113] The term *conach* may also be used of the caterpillar of the elephant hawk-moth (*Deilephila elpenor*), which was believed to sting cattle in the muzzle.[114] The hairs (setae) of the caterpillars of certain moths and butterflies can indeed cause rashes or lesions in the muzzles of domestic animals, but the damage is never serious.[115] There was also a widespread folk-belief that the hawk-moth caterpillar ('conach worm') causes disease if eaten by cattle or pigs, but this idea has no scientific basis.[116]

° ***Opann***

Opann (*obann*) is included among the infectious diseases of cattle in legal commentary.[117] Another commentary states that if a person lends an ox to another for ploughing and it dies of *opann*, the borrower is not obliged to compensate the owner.[118]

The name of this disease is presumably the same word as *opann* (*opunn*) 'swift, sudden', referring to the swiftness with which it strikes its victims. This would suit anthrax, an infectious disease of cattle, pigs, sheep and goats. Its acute form is characterized by a sudden onset and a rapidly fatal course.[119]

[111] CIH iii 1000.41; vi 2049.38.
[112] Dinneen, *Foclóir*, s.v. 1 *conach*.
[113] RIA MS 462 (24 M 34), p. 194; cf. MSS 461 (24 B 2), p. 180; 465 (23 N 20), p. 61. For these MSS, see p. 218.
[114] Dinneen, *Foclóir*, s.v. 2 *conach*; Ó Cuív, *Cnósach*, 66. It is possible that *conach* was also used of the larger and even more impressive caterpillar of the death's head hawk-moth (*Acherontia atropos*).
[115] West, *Black's veterinary dictionary*, 96. It is doubtless for this reason that the term *spiorad neannta* 'venemous spirit' is applied in Modern Irish to hairy caterpillars.
[116] Logan, *Making the cure*, 171.
[117] *CIH* iii 998.22; 1001.1.
[118] *CIH* ii 698.9.
[119] West, *Black's veterinary dictionary*, 20.

°*Idu*

Idu means 'pain, pang, cramp, ache', and is used particularly of stomach-pains and the pains of childbirth. In legal commentary *idu* is listed as an inherent illness (*galar bunaid*) to which cattle, horses, sheep, goats, humans (in this case, slaves) and poultry are susceptible.[120] It is also listed in the same commentary as an infectious disease (*galar tecmaisech*) of cattle.[121]

°*Adbach*

Adbach[122] is listed in legal commentary both as an inherent and as an infectious disease of cattle, and in both manifestations it is explained as 'dead-calf' (*marblaogh*).[123] The abortion of calves due to infectious disease can be attributed to brucellosis.

Legal commentary also refers to abortion (*forláeg*) for which the buyer has no claim against the seller of the cow. This abortion is treated as an act of God (*díḟoichid Dé*), on the grounds that the cow's pregnancy (*indláegus*) 'is made with too much wetness'.[124]

°*Odbach*

Odbach is a derivative of *odb* 'lump, knot, protuberance', so it must be a disease characterized by swellings. It is included in legal commentary as an inherent disease of cattle, and is glossed *.i. dubuar*, which seems to be a compound of *dub* 'black' and *búar* 'diarrhoea, flux, scour'.[125] Neil McLeod makes the plausible suggestion that *odbach* is 'black-quarter' (also called 'black-leg'), a disease of cattle – and occasionally of sheep and pigs – which is characterized by rapidly increasing gas-filled swellings in the shoulder, neck, thigh, etc.[126] There is often blood-stained discharge from nostrils and rectum.[127]

120 *CIH* iii 998.20; 1000.26–32.

121 *CIH* iii 998.22.

122 As Old Irish *odb* 'lump, knot' became *adb* (*fadb*) in later Irish, it is tempting to see *odbach* and *adbach* as variant spellings of the same disease. However, they are kept distinct in all the manuscripts.

123 *CIH* iii 998.22; 1000.28; vi 2049.39. Cf. Ó Cuív, *Cnósach*, 153 s.v. *mara-ghamhain.*

124 *CIH* ii 678.22–3; iii 853.10–11; 1002.24–5 *ō imarcraid fliuchaidechta donīthar he.*

125 *CIH* iii 998.20; 1000.28. Cf. Ó Cuív, *Cnósach*, 90 s.v. *dobuar.*

126 McLeod, *Early Irish contract law*, 323. He explains *dubuar* as 'black plague', taking *búar* to be from *bó-ár* 'cow-mortality' (see p. 198 above): but there are no other attestations of *búar* in this meaning.

127 West, *Black's veterinary dictionary*, 48.

A disease called *cnuicc* 'lumps', which is mentioned in *Immacallam in dá thúarad*, could also refer to black-quarter.[128] According to a gloss in one of the manuscripts of this poem, the lumps are on the necks of the animals. Other diseases of cattle characterized by lumps are tuberculosis and ringworm. For a reference to a lump (*cnoc*) on the back of a horse, see Appendix A, text 7 § 2 (1).

°***Radrachus***

This is listed in legal commentary as an inherent defect of cattle.[129] It is defined thus: 'she (the cow) is bulled and it does not stay in her'.[130] This implies that the cow comes into heat in the normal manner, but does not become pregnant after being bulled or else aborts at an early stage. If she fails to calve in the first year, her value is reduced to that of a carcase (*mart*), i.e. one third that of a fertile cow (see p. 65). The seller must pay half of this sum to the buyer.[131]

°***Ingenas búaile***

This phrase means 'virginity of the cattle-enclosure', and is included in legal commentary among the inherent defects of cattle.[132] A cow with this defect does not allow herself to be bulled; in another text she is described as *tarbchad* 'bull-hostile'.[133]

If the defect appears in a cow which has been sold, the seller must pay the buyer one half of the value of a year's milking. If the condition persists into the second year, the cow's value is reduced to that of a carcase. The seller must then pay half the difference between this sum and the value of a normal cow (*bó inich*).[134]

°***Dallṡinche***, °***Trifnecht***

The literal meaning of *dallṡinche* (also *dallṡinechus*) is 'blind-teatedness' i.e. a teat (*sine*) on the udder is non-functional. This commonly occurs as a result of mastitis, with consequent reduction of the cow's milk-yield. Occasionally two teats may be

[128] Stokes, 'Colloquy', 46 § 233.
[129] *CIH* iii 1000.28.
[130] *CIH* iii 998.37.
[131] *CIH* iii 839.8–9; 977.3–4; 999.8–9 = McLeod, *Early Irish contract law*, 294 [e]. This is the general rule: the commentary goes on to deal in detail with the various circumstances in which a bought cow fails to calve, and their legal implications.
[132] *CIH* iii 998.20–21; 1000.29.
[133] *CIH* ii 674.39 = Appendix A, text 1 § 1 (10).
[134] *CIH* iii 839.39–840.2; 976.40–977.2; 999.1–4 = McLeod, *Early Irish contract law*, 294 [c].

affected. Blind-teatedness is included in legal commentary among the inherent defects of cattle.[135] This condition is not apparent when the cow is dry, so a buyer will not become aware of it until she starts to give milk. In general, the buyer is entitled to recoup half the cow's value from the seller.[136]

Trifnecht (also *trifnechus, treṡinechas,* etc.) is formed from the compound *trifne* (*triphne*) 'three-teated'.[137] In legal commentary it seems to be used with much the same meaning as *dallṡinche,* i.e. of a cow whose udder looks healthy but in fact has one non-functioning teat.[138] In other contexts, it is probable that *trifne* has a wider meaning, and is used of any three-teated cow, whether the defect is visible or not. For example, in *Táin Bó Cúailnge* we are told how Cú Chulainn begged a drink of milk from a lame half-blind hag – the battle-fury Morrígain in disguise – who was milking a three-teated cow (*bó thriphne*).[139] As the hag's defects are stressed by the author, it seems likely that the cow's fourth teat is visibly damaged or missing.

Comalne

Comalne (later *comalla, comaille*) is not mentioned in legal material, but is included among the diseases of cattle in *Immacallam in dá thúarad.*[140] In the Old Irish Glosses *comaln(a)e* is twice used to explain Latin *intercus* 'dropsy',[141] so in *Immacallam in dá thúarad* it may refer to some complaint such as bloat.

Crithach

Immacallam in dá thúarad also refers to *crithach* 'trembling, shaking, shivering' as a disease of cattle.[142] In Bécán's *Prophecies* the corresponding ailment is *crithgalur cúar,* probably meaning 'crooked shaking disease'.[143] As many diseases of cattle and other animals are characterized by shaking, it is not possible to provide a precise translation for *crithach* here. In a legal description of a

135 *CIH* iii 998.20; 1000.29.
136 *CIH* iii 839.12; 977.5; 999.10 = McLeod, *Early Irish contract law,* 294 [f].
137 *Trí* 'three' + *sine* 'teat': see *LEIA* T-145 s.v. *triphne.*
138 *CIH* iii 839.12; 977.5; 999.10.
139 *TBC* Rec. I 62.2041.
140 Stokes, 'Colloquy', 46 § 233. In Knott, 'Poem of prophecies', 72 § 65, the Book of Uí Maine version has *comillid,* but the YBL version has the superior reading *comaill–.*
141 *Thes.* ii 143.31; 227.21.
142 Stokes, 'Colloquy', 46 § 233 nom. plur. *crithcha.*
143 Knott, 'Poem of prophecies', 72 § 65.

sound horse, it is stipulated that it should not suffer from shivering (*crithach*).[144]

Milliud

Immacallam in dá thúarad also refers to *milliuda* 'destructions' of cattle.[145] This is probably to be taken as a reference to cattle-disease caused by bewitching: see p. 174 above.

Sinech

The *Annals of Inisfallen* record an outbreak of *sinech* among livestock in 1077, which also spread to humans.[146] The word *sine* means 'teat (of an animal)', so we can assume that this disease affected the teat. The annalist describes the symptoms as 'many lumps' (*cnuicc imda*). This suggests cow-pox (*Vaccinia*), which is characterized by lesions on the teat and udder, and can be transmitted to humans.[147] *Sinech* is also included among the diseases of livestock in *Betha Farannáin*.[148]

Warble-fly

The warble-fly or gadfly (*Hypoderma bovis*) lays its eggs on the legs of cattle.[149] After hatching, the larvae burrow under the skin, and gradually move up until they reach the animal's back. Here they remain and cause a swelling – the warble – with an opening to the air. When fully fed they drop to the ground and pupate: the life-cycle is then repeated after the emergence of a new generation of adults. The whine of this fly is notorious in that it drives cattle into a frenzy, often described as 'gadding'.[150] In Modern Irish the term *ar aoibheall* (*ag aoibheall*) is used both of the excitement of cows in heat and of cattle which are gadding.[151]

There may be a reference to the warble-fly in a legal commentary which lists various things which frighten cattle, such as thunder, lightning, banging on a shield or rattling a stone in a vessel.[152] Another cause of fright is given as *lésán foiche*, which the *Dictionary*

[144] *CIH* iv 1235.26 = Appendix A, text 7 § 2 (14).
[145] Stokes, 'Colloquy', 46 § 233.
[146] *AI* 234 s.a. 1077 § 6.
[147] West, *Black's veterinary dictionary*, 434.
[148] Plummer, '*Betha Farannáin*', 6.5.
[149] The term gadfly is sometimes also applied to various species of horsefly (Family *Tabanidae*): the adults suck the blood of horses, cattle, and other mammals.
[150] Cf. Virgil, *Georgics* (ed. Fairclough), 3.146–51.
[151] T. F. O'Rahilly, 'Ir. *aobh, aoibheall*, etc.', 1–2.
[152] *CIH* iv 1455.29–31; v 1723.6–8 (read *cloch i ndabaig* in the latter).

of the Irish language suggests may mean 'a wasp sting'.[153] *Foich* is certainly well-attested in the meaning 'wasp' (*vespa*) at all stages of the language,[154] but wasps are not recognized as particularly threatening to cattle. So perhaps *foich* should here be translated 'warble-fly'. One can compare the use of the compound *marccoich* (= *marcc-foich*) in the meaning 'horse-fly'.[155]

It has often been claimed that the holes found in early Irish manuscripts were caused by warble-fly larvae. Recent research by Kathleen Ryan has however shown that these holes are much more likely to be the result of damage during the processing of the skins for parchment-making.[156]

Other defects and disorders

In practice most litigation about cattle-defects is likely to relate to defects which would not be obvious to a competent buyer. Consequently, legal commentary deals in most detail with defects such as *ingenas búaile* 'virginity of the cattle-enclosure' or *dallṡinche* 'blind teatedness' which could not be detected at the time of sale. These have been discussed above.

Other defects are generally dealt with in a more superficial manner. Here our principal source is a short legal passage on cow-defects (*bó-ainmea*) which is edited and translated in Appendix A.[157] It specifies that a cow being put up for sale should not be lame (*losc*),[158] half-shanked (*lethcholpthach*) or blind in one eye (*cáech*).[159] Her temperament should be equable, and she should not be a kicker (*luach*), a gorer (*guinid*), aggressive (*lond*) or lacking in affection towards her calf.

[153] *DIL* L 119.6.
[154] T. F. O'Rahilly, '*Spoch. Foich.*'.
[155] Lambert, 'Les gloses bibliques de Jean Scot', 210.
[156] K. Ryan, 'Holes and flaws'.
[157] *CIH* ii 674.37–675.12 = Appendix A, text 1 §§ 1–6.
[158] Mícheál Ó Dochartaigh MRCVS has kindly drawn my attention to the use in south Donegal and Fermanagh of the phrase *loisce buaile* for 'foul-in-the-foot', acute lameness in cattle associated with grazing in muddy paddocks during the summer months.
[159] Cf. *CIH* i 173.6 (and footnote); ii 715.38–9; vi 2072.10. Legal commentary at *CIH* iii 1001.19–20 refers also to *sūile ruamanna ⁊ sūile cen imcisin* 'red eyes and eyes without sight' (McLeod, *Early Irish contract law*, 300 [o]). The former is presumably 'pink-eye', i.e. infectious keratitis.

As well as the bovine with an innately aggressive temperament, Irish law also recognised that a hitherto normal animal might suddenly go berserk for no apparent reason. Heptad 45 states that 'a bovine which has parted from its sense' is not held responsible for any damage in land-law, i.e. to crops or other livestock.[160] The glossator adds the proviso that the owner must give a public warning about the animal's condition and also must have attempted to kill it. Another legal commentator states that the owner of a mad bovine (*ag mer*) must warn the four neighbouring farmers about it.[161]

Non-legal sources contain a number of references to the sudden onset of madness in cattle. For example, *Bethu Phátraic* describes how a cow went mad (*dásachtach*) in a cattle-enclosure and killed five other cows.[162] The glossator attributes her madness to the entry of a demon into her. The young Patrick cured her and brought the dead cows back to life.

Sheep-diseases and defects

Our sources give little information on sheep-disease. The *Annals of Connacht* record that in the year 1338 almost all the sheep of Ireland died, but provide no information on the nature of the disease.[163] Most versions of the legal commentary on animal-diseases list only two inherent diseases (*galair bunaid*) of sheep: *líath* and *úthud*.[164]

° *Líath*

This is presumably a specialized use of the word *líath* 'grey', and is to be compared with Modern Irish *liathadh* 'rot in sheep'.[165] It may be foot-rot or liver-fluke.[166] *Líath* seems to be a different

[160] *CIH* i 41.2 = *AL* v 268.19.
[161] *CIH* i 285.16–17 = *AL* iii 272.12–13.
[162] Mulchrone, *Bethu Phátraic* 7.154–8.158.
[163] *AC* 280 s.a. 1338 § 7. There is one attestation of *oíba* 'mortality among sheep' (*DIL* O 114.46-8.), but it may only be a glossary word formed on the analogy of *duineba* 'human mortality'.
[164] *CIH* ii 696.12; iii 1000.29. The commentary at *CIH* vi 2048.26 adds the general term *idu* (*idha*) 'pain, colic': see under cattle-diseases on p. 200 above.
[165] Dinneen, *Foclóir*, s.v. *liathadh*.
[166] See *DIL* L 148.4–7. The medieval Welsh law-texts (e.g. Wade-Evans, *Welsh medieval law*, 75.8–9 = *ALIW*, Gwentian Code, II xii § 6) refer to a sheep-disease *llederw*, which is generally translated 'liver-fluke'.

disease from *léithuisce* (*léathuisce, líathuisce*) 'grey water', which is mentioned as an ailment of sheep in later medical manuscripts.[167]

° *Úthud*

This term (spelt *uthud, uthfaigh* and *ufadh*) is probably a compound containing the element *úth* 'udder'.[168] *Úthud* may therefore be a disease affecting the udder, such as mastitis ('blue bag').

Claime

Sheep suffer from a number of skin-afflictions, such as scab (caused by psoroptic mites), pustular dermatitis and wool-rot. None of these appear to be included in the legal commentary on animal-diseases.[169] However, the law-text *Di Chetharślicht Athgabálae* lists the offence of *foxal camtire*.[170] According to the glossator, this refers to the removal of the covering which has been placed on a sheep with skin-disease (*methir bis im caírig claim*).[171] Presumably, the covering is intended as a defence against blowfly which lays eggs on raw, dirty or scabby flesh. These hatch out as maggots and cause severe damage to the skin and fleece, sometimes with fatal results.

Claime is included in the legal commentary among the diseases of goats: I suggest below that it refers mainly to psoroptic mange.

Defects

A short Old Irish legal passage deals with defects in sheep. Much of this text is difficult to understand, but some of the defects

[167]E.g. RIA MSS 461 (24 B 2) p. 174; 462 (24 M 34) p. 191; 465 (23 N 20) p. 43. See Dinneen, *Foclóir,* s.v. *liath-uisce.*

[168]*LEIA* U-31 s.v. *uth* (recte *úth*: cf. Dinneen, *Foclóir,* s.v. *úth*; Dwelly, *Faclair,* s.v. *ùth*).

[169]The medieval Welsh law-texts include *clafri,* the first element of which is cognate with Irish *clam,* among the diseases of sheep, e.g. Wade-Evans, *Welsh medieval law,* 75.8 = *ALIW,* Gwentian Code, II xii § 6. It is generally translated 'mange, scab'. An Anglo-Saxon leechbook preserves a charm against pocks and scab in sheep (Bonser, *The medical background of Anglo-Saxon England,* 426–7). Virgil, *Georgics* (ed. Fairclough), 3.299, refers to scab (*scabies*) and foot-rot (*podagra*) in sheep, and Columella, *De re rustica* (ed. Forster and Heffner), 7.5.5, states that sheep more often than any other animal are susceptible to scab (*scabies*).

[170]*CIH* ii 390.6; v 1694.14 = *AL* i 184.10. The glossator is doubtless correct in taking *camtire* (read *camthaire*) to be from *cammaid* 'folds' + suffix *-aire,* i.e. 'that which is folded, wrap'. See *DIL* s.v. *camthir*.

[171]*CIH* ii 390.20; v 1694.14 = *AL* i 188.17. This gloss also refers to protective covering (*forbrat*) for a calf. Varro, *Rerum rusticarum* (ed. Hooper and Ash), 2.2.18, refers to jacketed sheep (*oves pelliti*), but in this case the jacket is to prevent the wool from getting dirty rather than as a protection for a sheep with skin-disease.

are straightforward. See edition, translation and discussion in Appendix A.[172]

GOAT-DISEASES

Legal commentary generally distinguishes two inherent diseases (*galair bunaid*) of goats.[173]

° ***Claime***

This term is used of skin afflictions of man and various animals (see under *máelgarb* in 'diseases of cattle' on p. 196 above). Mange affecting goats is commonly caused by psoroptic mites, and occasionally by sarcoptic mites.[174]

° ***Idu***

This general term for 'pain, colic' may perhaps refer here to goat scour.[175]

Other diseases

One version of the legal commentary adds three further diseases of goats. *Srannán* 'snorting' must refer to some respiratory ailment, such as parasitic bronchitis ('hoose').[176] *Bedgach* 'starting, leaping' (also the name of infectious disease of cattle) suits a number of goat diseases. It could, for example, be gid or staggers, a serious disease caused by the larval stage of the dog tapeworm *Taenia multiceps* entering the goat's brain.[177] *Bedgach* could also refer to louping ill (encephalitis), a tick-transmitted viral disease characterized by convulsions.[178] *Scamach* is also included among the diseases of cattle (p. 197), and is possibly to be identified here with caprine pneumonia.

[172] *CIH* ii 675.13–17 = Appendix A, text 1 § 8.

[173] *CIH* ii 696.12–13; iii 1000.29–30.

[174] Dunn, *The goatkeeper's veterinary book*, 171; West, *Black's veterinary dictionary*, 355–8.

[175] Dunn, *The goatkeeper's veterinary book*, 72.

[176] West, *Black's veterinary dictionary*, 398.

[177] Dunn, *The goatkeeper's veterinary book*, 106.

[178] This is mainly a disease of sheep but there is also evidence of its presence in goats: see Dunn, *The goatkeeper's veterinary book*, 103.

Pig-diseases

I have noted only one reference in the annals to a plague specifically affecting pigs. The *Annals of Inisfallen* record that in the year 1057 there was 'a sickness on cattle and pigs, etc.'.[179] Starvation is unlikely to have been a factor in the sickness of the pigs, as the same annals tells that there was in that year an abundance of acorns (*mess*). Legal commentary lists four inherent diseases (*galair bunaid*) which may appear in pigs.[180]

° ***Crapán (crupán)***

This term is used also of afflictions of cattle and humans.[181] It is formed from *crap* 'spasm, contraction', and seems generally to refer to lameness caused by contraction of the tendons.[182]

° ***Lésán***

Lés means 'blister, bag', so *lésán* must be a condition characterized by pustules or blisters.[183] This would suit swine pox, caused by the virus *Suipoxvirus.* Pustules, which develop into blisters, may cover the whole body of infected sucklings. In the case of older pigs, these symptoms are generally confined to hairless areas such as the snout, udder, ears and vulva.[184]

In later Irish swine pox is called *bolgach léasach.*[185] In Giolla na Naomh mac Duinn Shléibhe Mhic Aodhagáin's fourteenth-century legal manual, there is a reference to *bolg* 'blister' in the context of diseases of oxen or horses.[186] There is also mention of blisters (*dona bolgaib*) in a sixteenth-century treatise on horses, probably based on an English original.[187]

[179] *AI* 216 s.a. 1057 § 6 *teid*[*m*] *ar innilib ⁊ ar mucca ⁊rl.*
[180] *CIH* ii 696.13; iii 1000.30; vi 2048.27–8.
[181] See *DIL* s.v. *crapán* and McLeod, *Early Irish contract law*, 317.
[182] Possibly, this is to be identified with *hualog*, lit. 'the fettered [disease]', given as an ailment of pigs in the medieval Welsh law-texts, e.g. Williams and Powell, *Llyfr Blegywryd*, 93.26 = *ALIW*, Dimetian Code, II xxxi § 10.
[183] E.g. *CIH* ii 617.11.
[184] Leman, *Diseases of swine*, 361.
[185] McLeod, *Early Irish contract law*, 322.
[186] *CIH* ii 698.9. The MS definitely has *bolg* rather than *bolg-* (for *bolgach*).
[187] Ó Cuív, 'Fragments of two mediæval treatises', 52 § 17.

° ***Rúad otraig***

The meaning of this disease is 'red dung', so there is no doubt about its symptoms. Blood-stained faeces is found in a number of diseases of pigs, especially swine dysentery ('blood scour').[188]

° ***Conach***[189]

Rabies is primarily a disease of dogs (see p. 215 below) but it may also be transmitted to other mammals. Rabies in pigs has been widely recorded, with similar symptoms to canine rabies. Affected pigs may become excitable, squeal, and show muscular spasms, before suffering paralysis and death.[190]

Horse-diseases and defects

The legal commentary on animal-diseases lists six – in one version seven – inherent diseases (*galair bunaid*) of horses.[191] Some of these are also mentioned in four fragments of treatises on horses which have been edited and translated by Brian Ó Cuív.[192] These texts, transcribed in the fifteenth and sixteenth centuries, are probably translations of an untraced English original. They use many borrowed technical terms, such as *glondras* from English *glanders* and *splinda* from English *splint.* However, they also contain native terminology relating to veterinary medicine, and are therefore of assistance in identifying the diseases of horses listed in legal commentary.

° ***Seirthech***[193]

The identity of *seirthech* (also spelled *serthach*) is uncertain. It may contain the element *seir* 'heel, hock', and refer to some affliction of the hock. Treatise I on horses distinguishes three ailments of the hock: *cnáimspetan* 'bone spavin', *crupus* 'curb(s)', and *spadumus,* of unknown meaning.[194]

[188] Leman, *Diseases of swine,* 599–616.

[189] *DIL* D ii, 179.6 gives *doach* as a disease of pigs. However, in the MS (Trinity College Dublin no. 1337 = H 3. 18, p. 163) *do-* seems to be a dittography from the line above, which the scribe has overwritten with the compendium for *con.* Binchy reads *conach* here (*CIH* ii 696.13).

[190] West, *Black's veterinary dictionary,* 451–2.

[191] *CIH* ii 696.11–12; iii 1000.26–7; vi 2048.25–6.

[192] Ó Cuív, 'Fragments of two mediæval treatises on horses'; Ó Cuív, 'Fragments of Irish medieval treatises on horses'.

[193] *CIH* ii 696.11; iii 976.30; 1000.26; vi 2048.25.

[194] Ó Cuív, 'Fragments of two mediæval treatises', 38–40 §§ 20–2; cf. Ó Cuív, 'Fragments of Irish medieval treatises', 115 §§ 1–2.

° ***Deilgnech***[195]

In two versions of the legal commentary this ailment is explained *.i. na péiste gaile* 'i.e. the stomach worms'. As internal parasites are particularly serious in horses, this explanation seems very probable. The word *deilgnech* is clearly a derivative of *delg* 'thorn', and is likely to refer to the thorn-like appearance of some parasitic worms.[196] In one fragment of the treatise on horses, two types of internal parasite are mentioned: long worms and narrow worms. The latter are described as *maílchinnáin*, which means 'bald-headed ones', and are said to be the worst kind of worm.[197]

° ***Echmaig***[198]

The disease glanders is caused by the bacillus *Pseudomonas mallei.* Veterinary science distinguishes nasal, pulmonary and cutaneous glanders, though the boundaries between these forms may be indistinct. The cutaneous form is known as 'farcy', and is characterized by pustules, nodules and ulcers. On recovery these form star-shaped scars.

It is very probable that *echmaig* refers to 'farcy'. This is the interpretation given in Plunket's Latin-Irish dictionary of 1662.[199] Furthermore, one fragment of the treatise on horses recommends that a mild case of *echmaig* (*ecmuigh*) be treated by breaking with a tongs, and a severe case by firing with a red-hot iron. These treatments indicate skin-eruptions typical of farcy.[200]

[195] *CIH* ii 696.11; iii 976.30; 1000.26; vi 2048.25 = McLeod, *Early Irish contract law*, 299 [f].

[196] McLeod (*Early Irish contract law*, 318) points out that the larvae of the parasite *Oxyuris equi* are today called 'pinworms'. Other common parasitic worms of horses are roundworms, redworms, threadworms, and bots.

[197] Ó Cuív, 'Fragments of Irish medieval treatises', 117 § 13 *piasda fada 7 piasta caela darab ainm mailcinnāin.*

[198] *CIH* ii 696.11; iii 1000.27; vi 2048.25. The spelling with *-m-* seems likely to be earlier: it is used at *CIH* iii 1000.27, as well as in Plunket's *Dictionary* and in 'Treatise on horses I' (Ó Cuív, 'Fragments of two mediæval treatises', 42 § 29). The spelling *eachb–* at *CIH* 696.11 and 2048.25 is probably due to scribal confusion with *eachba*(*d*), a compound of *ech* 'horse' + *ba*(*d*) 'death', cf. *duineba*(*d*) 'human-death, plague'. A charm against *echbad* is printed by Best, 'Some Irish charms', 28 (I).

[199] Quoted by Ó Cuív, 'Fragments of two mediæval treatises', 59.

[200] Ó Cuív, 'Fragments of two mediæval treatises', 42 § 29.

° ***Fothach***[201]

This term is used of nasal and pulmonary glanders, which is often fatal to horses. This disease has been eradicated from Ireland, Britain, western Europe and north America. In Modern Irish the term *fothach* has been transferred to the disease 'strangles', which is caused by *Streptococcus equi* and characterized by coughing and swollen lymph glands in the neck.[202]

Glanders of all types may be transmitted to humans.[203] The *Annals of Connacht* record cases of fatal glanders in the year 1464. Muircheartach mac Airt Uí Mhaoilsheachlainn and his wife, along with three others, are said to have contracted the disease through going to see a horse which died of the same glanders (*dona cnapaib cétna*). The term *cnap* is here used of the nodules of cutaneous glanders. All five died within twenty-four hours of each other.[204]

° ***Ech-idu***[205]

Idu is a general term for pains, especially those of the abdomen, and is best translated 'colic' (see p. 200 above). Colic in horses may be caused by intestinal parasites (especially bots, strongyles, and pinworms), anthrax, incorrect diet, etc.

° ***Lec ós crú***[206]

The literal meaning of this complaint is 'stone above the hoof'. I would suggest that it refers to ringbone, a calcium deposit encircling the posterior bones of the front legs. It may also occur on the hind legs. McLeod suggests bone spavin, a bony enlargement on

[201] *CIH* ii 696.11; iii 976.29; 1000.27; vi 2048.25. The medieval Welsh law-texts (e.g. Williams and Powell, *Llyfr Blegywryd*, 92.7 = *ALIW*, Dimetian Code, II xxviii § 22; Wiliam, *Llyfr Iorwerth*, 83 § 123.2 = *ALIW*, Venedotian Code, III iv § 13) refer to the disease *llynmeirch* lit. 'moisture of horses': this presumably refers to the nasal discharge characteristic of glanders. The compound *bilfothach* is applied to a horse in the law-text *Bretha Nemed Déidenach* (*CIH* iii 1119.31 = E. J. Gwynn, 'Privileges', 25.21). Glossators take this to be a derivative of *fothach* 'glanders'. Thus O'Davoren has *.i. slān cen fotac* 'i.e. healthy without glanders' (*CIH* iv 1474.4 = O'Dav. 230 § 206) and another glossator has *slán hí cin foth– innti* 'she is healthy without glanders in her' (*CIH* ii 556.30). This explanation is followed at *DIL* F 393.44–6 and at McLeod, *Early Irish contract law*, 326.

[202] Ó Cuív, 'Fragments of Irish medieval treatises', 121 § 12; Fiachra Éilgeach, *Scéalta andeas*, 51.

[203] West, *Black's veterinary dictionary*, 240.

[204] *AC* 518 s.a. 1464 § 13.

[205] *CIH* ii 696.11; iii 1000.27; vi 2048.25.

[206] *CIH* ii 696.11–2; iii 976.30; 1000.27; vi 2048.25–6.

the inside of the hock, which may cause lameness.[207] However, in one of the later treatises on horses a distinction is made between *lec ós crua* and *cnáimspedain* 'bone spavin'.[208]

°*Aife seralach*

Only one version of the list of animal-diseases includes *aife seralach.*[209] There is no connective between *aife* and *seralach* in the manuscript, so it is possible that a single disease is intended. On the other hand, the *Dictionary of the Irish Language* lists them as separate diseases, though without proposing any identifications.[210] McLeod likewise distinguishes two diseases, suggesting that *aife* is tetanus (lockjaw), and that *seralach* is an ailment affecting the hock (*seir*).[211]

Tetnais (merechduin)

Tetanus or lockjaw is a generally fatal disease of horses. It is caused by the bacterium *Clostridium tetani,* the spores of which are infectious for humans.

This disease is mentioned in one version of the commentary on the law-text on accidents, *Bretha Étgid.* The commentator discusses the complex forms of liability in the case where a person borrows somebody else's horse to bring to a fair. Among other liabilities, the owner of the horse must pay for liability with regard to tetanus.[212] A glossator on this passage explains *tetnais* as *merechduin,* which seems to be a compound of *mer* 'mad, wild' and *ech* 'horse'.[213] This term presumably refers to the violent symptoms of the disease, caused by overreaction of the reflex and motor stimuli. Death is ultimately by asphyxiation.

Galar na placodi

An entry in the *Annals of Inisfallen* records that in the year 1259 there was a widespread cough (*cossachgtagh cotkenn*) on people and horses, which was called *galar na placodi.*[214] The second element may be *plocóid* 'stopper, bung', referring to the suffocating nature

[207] McLeod, *Early Irish contract law,* 321.
[208] Ó Cuív, 'Fragments of two mediæval treatises', 52 § 20.
[209] *CIH* iii 1000.27.
[210] *DIL* s.vv. 1 *aife, seralach.*
[211] McLeod, *Early Irish contract law,* 315, 324.
[212] *CIH* i 266.3; iii 935.40–1 = *AL* iii 180.11 *íc a cinaid isin tetnais fair.*
[213] *CIH* iii 1078.21–4. See *DIL* s.v. *merechduin.*
[214] *AI* 358 s.a. 1259 § 3.

of the cough.[215] Possibly, this was an epidemic of strangles (see p. 211), as human infection by *Streptococcus equi* has been recorded.[216]

Defects

The second half of a short legal passage edited in Appendix A of this book lists various defects of horses.[217]

POULTRY-DISEASES

Only one version of the legal commentary on animal diseases refers to inherent diseases (*galair bunaid*) of domestic fowl.[218] The text has *donaib huilib énaib* 'of all birds', but as hens were undoubtedly the most important type of poultry on the early Irish farm (p. 102), it can be assumed that the author was thinking primarily of hen-diseases.

I follow Binchy's punctuation of the text,[219] and take it that two types of disease are distinguished here: *confad* and *idu*. *Confad* (synonymous with *conach*) normally means 'rabies', which is primarily a disease of dogs, but can be transmitted to any warm-blooded animal.[220] Living in the farmyard, hens and other poultry would be particularly liable to be bitten by a household dog which had contracted rabies (for further discussion of this disease, see p. 215 below). Alternatively, *confad* may here refer to some other disease of poultry which affects the nervous system, such as encephalitis in pheasants.[221] *Idu* 'pain, colic', being a general term for serious disease, could refer to many of the poultry diseases distinguished by modern veterinary science.

The only annalistic reference to widespread mortality among birds is in the *Annals of Inisfallen*, where it is recorded that in the year 917 there was a great cow-mortality on livestock and

[215] Dinneen, *Foclóir*, s.v. *plocóid*. I am grateful to Seán Ua Súilleabháin for drawing my attention to this word.

[216] West, *Black's veterinary dictionary*, 520–1.

[217] *CIH* iv 1235.22-7 = Appendix A, text 7 § 2.

[218] *CIH* iii 1000.31–2.

[219] McLeod (*Early Irish contract law*, 299 [f]) takes *confuidh* with the preceding *conuibh*, reading *conach do conuibh confuidh*. However, his translation 'for mad dogs, *conach* [rabies]' seems tautological. I suggest that *do huilib énaib* has been omitted by the scribe after *conf–* at *CIH* vi 2048.28.

[220] For rabies in poultry, see Bisseru, *Rabies*, 325.

[221] I am indebted for this suggestion to Pat McCarthy of the Faculty of Veterinary Science, University of Sydney, New South Wales.

birds (*bó-ár for cethraib 7 énaib*).[222] This mortality affected wild birds, as the annalist states that the sound of blackbird (*lon*) or songthrush (*smólach*) was scarcely to be heard in that year.

Bee-diseases

The *Annals of Ulster* refer to the occurrence of a bee-mortality (*bech-díbad*) in 951.[223] Again in 993, the same annals record a great mortality of people, cattle, and bees.[224]

In Ireland the honeybee is near – or possibly even beyond – the limit of its natural range (see p. 109 above) and therefore suffers severely in cool or wet summers. The bee-mortalities mentioned in the annals may have been mainly the result of starvation caused by bad weather, though parasitic or other infections may also have contributed. It has been calculated that between 1909 and 1917 Britain lost approximately 90 per cent of its bees, with similar mortalities in Ireland. This loss has generally been attributed to the spread among bee-stocks of the parasitic mite *Acarapis woodi,* which enters the respiratory system of young bees and feeds there by piercing the tracheal wall and sucking the bee's blood. However, the researches of the bee-expert Dr L. Bailey now suggest that starvation rather than parasitic infection was the major cause of death during this period.[225] The Irish record from 993 would tend to support his view, as the cattle-mortalities in the same year suggest bad weather conditions; for a discussion of the relationship between weather and epidemics, see pp. 194–5 above.

In his *Topography,* Giraldus Cambrensis states that Ireland has honeybees, but claims that their swarms would be much more plentiful if they were not frightened off by the bitter and poisonous yew with which the woods abound. Yew branches are of course poisonous if eaten by cattle and other livestock, but the idea that bees are affected in any way by this tree is erroneous. It seems that Giraldus has based his claim on a misapplication of a line in Virgil's *Eclogues.*[226]

[222] *AI* 146 s.a. 917.
[223] *AU*² 396 s.a. 950 (recte 951) § 6.
[224] *AU*² 424 s.a. 992 (recte 993) § 7.
[225] C. Butler, *The world of the honeybee*, 20–21.
[226] O'Meara, *Giraldus: Topography*, 35 § 2 (and note) = Dimock, *Giraldi Topographia*, 28 dist. I, cap. vi. The reference is Virgil, *Eclogues* (ed. Fairclough), 9.30.

Dog-disease

° *Conach (Confad)*

By far the most terrible disease of dogs is rabies, known in Irish as *conach* or *confad* (also *confaid*).[227] In the legal commentary on animal-diseases it is listed as affecting cattle, pigs and poultry as well as dogs.[228] In dogs, the initial symptoms include lassitude, uneasiness, and a desire to lick anything cold. After a few days the furious phase of the disease sets in, and the dog will then attack with intense fury any living thing which it encounters. The rabies virus is in the dog's saliva and can be transmitted by even a minor bite. Any warm-blooded animal can be infected with rabies; in man it is called Hydrophobia because dread of water is one of the symptoms of the disease. It is generally fatal in all animals.

Another legal commentary states that the owner of a rabid dog (*cú chonfaid*) must put out a public warning (*escaire*). But his responsibility does not end here: he must ensure that the dog is killed, its body burnt and the ashes thrown into a stream.[229] These precautions suggest awareness of the fact that the rabies virus can survive for some time in the dead body of the infected animal.[230] In modern times, a case has been recorded of a man who contracted rabies after skinning a cow which had died from the disease.[231]

The words *conach* and *confad* are also used in a more general sense of any violent fury or madness.[232]

Cat-disease

There are no references to diseases of cats in the legal commentary on animal-diseases. However, another commentary on sickness in livestock (*cethrae*) states that a sick or injured cat can legally be

[227] Both words contain the element *cú* (*con-*) 'dog', *LEIA* C-187, C-192 s.vv. *conach, confa(i)d.*

[228] *CIH* ii 696.13; iii 1000.30–31, 41; vi 2048.27–8, 2049.38. *Conach* is also listed among the diseases of livestock in Plummer, '*Betha Farannáin*', 6.5.

[229] *CIH* i 285.21–2.

[230] However, in a paper delivered at the School of Celtic Studies *Tionól* in November 1994, Máirtín Ó Briain demonstrated that the motif of burning the body of a person or animal and throwing the ashes into water is quite common in Irish literature, e.g. Van Hamel, *Immrama* 35.312–16 = *LU* 60.1721–5; Ní Shéaghdha, *Agallamh na Seanórach* i 8.12–13.

[231] Baer, *The natural history of rabies* i 9.

[232] *DIL* s.vv. *conach, confad.*

regarded as fully recovered when it starts to catch mice as well as it did before.[233]

TREATMENT OF INJURY AND DISEASE IN LIVESTOCK

In stock-raising communities injury or disease is commonly treated by the animal's owner. However, from the earliest times it must have been apparent that certain individuals had a particular flair for healing animals. In recent Irish folk tradition, for example, there are many records of particular 'animal-doctors' who cured with the aid of herbs, touch, or incantations. In earlier Irish sources the curing of animals is sometimes effected by articles associated with saints. The twelfth-century Life of Saint Colmán mac Lúacháin records a tradition that the gapped bell of Saint Mo Chuta (*Bernán Mo Chuta*) could cure both humans and livestock of many ailments and diseases.[234] The sick person or animal had to drink water from the bell, which was then struck three times.[235] In his English translation of the *Annals of Clonmacnoise* – completed in 1627 – Conell Mageoghagan gives a firsthand account of a similar practice applied to a gospel-book. He recounts how he himself had seen the custodian of the Book of Durrow, whom he describes as an 'ignorant man', infuse that book in water which he then gave to sick cattle. The cattle were miraculously restored to health and the book itself suffered no ill effects. Conell believed that the Book of Durrow was one of three hundred gospel-books transcribed by Saint Columba, and attributes its properties to the power of the saint.[236]

Old Irish charms for the relief of human ailments survive from the eighth century,[237] but there are no early records in Irish sources of charms for curing animals. However, an Old Irish charm for expelling worms from a man or beast survives in a slightly garbled form in an Anglo-Saxon leech-book. Part of it reads: *gono míl,*

[233] *CIH* vi 2098.30–31.
[234] Meyer, *Betha Colmáin*, 26 § 27. The author of this text is unsure whether the bell is associated with Mo Chuta or his disciple Motura.
[235] It seems better to take *dínnech* to mean 'healing draught' rather than 'washing' as suggested by Meyer. See *DIL* s.v. 2 *dínech.*
[236] *Ann. Clon.* 96 s.a. 590. The Book of Durrow has been dated to around the middle of the seventh century (Luce, et al., *Codex Durmachensis* ii 65), whereas Columba died in 597.
[237] *Thes.* ii 248–9.

orgo míl, marbu míl 'I wound the worm, I strike the worm, I kill the worm'.[238]

In Old Irish lists of professions there is no specific mention of the animal doctor or veterinary surgeon. However, later Irish legal commentary contains a few references to the fees and duties of the animal-doctor (*liaig*).[239] A commentary on non-fatal injury to livestock (*béoathchummae cethrae*) credits the god of healing Dían Cécht with establishing the principle that the person who cures a wounded animal is entitled to one quarter of the price of the wound (i.e of the sum due for the wound if inflicted illegally).[240]

A more complex system for assessing veterinary fees is given in another commentary.[241] Here we are told that the animal-doctor is entitled to whichever is less: one quarter of the price of the wound or one quarter of the replacement-value (*aithgein*) of the animal. This commentary refers also to the necessity for a formal ruling (*derosc*) before it is legally established that the convalescence (*othras*) of an illegally injured animal has been completed. In the law of human injury – from which the law of animal injury has clearly been adapted – this ruling seems to have been made by a physician.[242] It is likely, therefore, that an animal doctor was required to make the formal ruling with regard to an animal's recovery. Legal commentary further specifies how it can be ascertained that sick or injured animals of various types are fully recovered. In the case of a dog, for example, we are told that if it chews a bone as well as it did before its illness, it is to be regarded as cured. A horse is held to have recovered if it runs as well after being taken from the water as it did beforehand.[243] A pig is held to be cured if it roots up the ground as well as formerly and a cat's recovery is gauged by the restoration of its mouse-catching ability. If the animals cannot do these things, no formal ruling of recovery can be pronounced.[244]

[238] Thurneysen, 'Ir. *marbu* "Ich töte"', 106. For further discussion see Meroney, 'Irish in Old English charms', 177–8, and Bonser, *The medical background of Anglo-Saxon England*, 245–6.

[239] The term *liaig* applies primarily to the physician who treats human disease and injury, but may also be used of the animal-doctor, e.g. *CIH* i 170.15; v 1626.4; vi 2174.33.

[240] *CIH* v 1626.3–8 = *AL* iii 362.1–4.

[241] *CIH* vi 2174.30–33.

[242] For discussion, see *GEIL* 131, and Binchy, '*Bretha Déin Chécht*', 16–18.

[243] I do not understand this reference.

[244] *CIH* vi 2098.27–32.

The tenth-century tale *Orgain Denna Ríg* refers to the accidental breaking of the leg of a chicken (*eréne circe*). The injury is treated by the application of a bandage (*cuimrech*).[245]

There is evidence that the post-Norman medical families were also involved in veterinary practice. For example, an Ó Longáin medical manuscript prescribes herbal remedies and charms for ailments of livestock such as grey water (*léithuisce*) in sheep,[246] the bewitching of cattle (*mille ba*), bleeding from the teats of cows while being milked, *conach* in cattle, etc.[247] A colophon on p. 122 of this manuscript records that it derived from an original which was in the possession of the physician Eóin Ó Callannáin in the year 1692. The family of Ó Callannáin (Callanan, Callinan) were physicians to Mac Cárthaigh (MacCarthy) of west Cork.

One fragment of a treatise on horses (see p. 209 above) was interpolated in a manuscript written in 1469 by Donnchadh Óg Ó hÍceadha.[248] The family of Ó hÍceadha (O'Hickey, Hickey) were physicians to Ó Briain (O'Brien) of Thomond. Another fragment is found in a medical manuscript written between 1516 and 1527 by An Giolla Glas Ó Caiside. He was a member of the important medical family Ó Caiside (O'Cassidy) of Fermanagh. In his book on Irish folk medicine, Patrick Logan records a recent tradition in the Fermanagh area concerning the use of 'Cassidy's rag' to cure livestock. A piece of cloth from the house of any person named Cassidy was put into water, which was then given to the sick animal to drink.[249]

[245] Greene, *Fingal Rónáin and other stories*, 18.316.
[246] See p. 206 above.
[247] RIA MS 461 (24 B 2), pp. 172–81. With slight variation the same ailments are listed in the Ó Longáin RIA MSS 462 (24 M 34), pp. 191–4, and 465 (23 N 20), pp. 40–61. For *conach* in cattle see p. 199 above.
[248] Ó Cuív, 'Fragments of Irish medieval treatises', 113–14.
[249] Logan, *Making the cure*, 170–1.

7
Crops (i): Cereals

Types of cereals

Though cereals do not feature in our sources as prominently as livestock, there is nonetheless a good deal of information relating to the types of cereals grown, the ways in which they were processed, and the forms in which they were consumed. Cereals were sufficiently important in the early Irish economy for a bushel (*míach*) of grain to be used as a form of currency.[1] There are also references in the annals showing that the destruction of cereal-crops, whether by bad weather or human agency, could bring about severe food-shortage or even famine (see p. 2).

An important list of seven cereal-grains is given in the eighth-century law-text *Bretha Déin Chécht*.[2] They are placed in the following order: *cruithnecht* (bread-wheat), *secal* (rye), *suillech* (spelt wheat?), *ibdach* (two-row barley?), *rúadán* (emmer wheat?), *éornae* (six-row barley), and *corcae* (common oat). This order is based on the relative prestige of each type of grain, which is correlated with a particular grade in human society. Thus a wheat-grain is equated with the rank of superior king, bishop or chief poet, whereas at the bottom of the scale the oat-grain is equated with the commoner of *bóaire* rank.[3]

Other texts agree in attaching more value to bread-wheat than to other cereals. For example, in a Life of Saint Finnian we are told that on weekdays the saint had a piece of barley-bread and a drink of water, but on Sundays and holy days he ate a piece of wheaten

[1] See Appendix B, p. 588.

[2] *CIH* vi 2305.6–13 = Binchy, '*Bretha Déin Chécht*', 22 § 2. I have normalized the spelling.

[3] The glossator gives two reasons why the grain of bread-wheat is at the top of the list: *ara loiget 7 ara uaisle* 'because of its small size and its dignity'. In his edition, Binchy tentatively accepts the glossator's view that the list is arranged on the basis of the relative size of each grain ('*Bretha Déin Chécht*', 49). He quotes modern measurements of a grain of wheat (0.25 in.), rye (0.29 in.), barley (0.33 in.) and oats (0.50 in.). But, as he remarks, this leaves us with the problem of fitting in the grains of *suillech, ibdach,* and *rúadán* between 0.29 in. and 0.33 in. For this reason, it seems to me likely that the author of the text was thinking primarily of the relative value and prestige of each grain.

bread with a morsel of boiled salmon, and drank a cup of mead or ale.[4] Similarly, legal commentary on fosterage states that the child of a lord is entitled to porridge made with barleymeal, whereas the child of a king is entitled to porridge made with wheatmeal.[5] In another legal commentary the value of a bushel of wheat is given as a scruple,[6] that of a bushel of barley as two pence (i.e. two thirds of a scruple), and that of a bushel of oats as one penny.[7]

The main problem in the *Bretha Déin Chécht* list is the identity of *suillech, ibdach,* and *rúadán*: see discussion on pp. 222–3. There is also a list of eight cereals in the satirical tale *Aislinge Meic Con Glinne,* which dates from the early twelfth century: *secal, serbán, máelán, rúadán, cruithnecht, éornae, fidbach,* and *corcae.*[8] No significance seems to be attached to the order in which they are placed. Of these, *serbán* 'the bitter one' and *máelán* 'the bald or awnless one' do not appear in the *Bretha Déin Chécht* list. *Fidbach* must be a later spelling of *ibdach.*

Cruithnecht (Bread-wheat)

Bread-wheat (*Triticum aestivum*, subspecies *vulgare*) is one of the world's most important food-plants. It is believed to have originated some time between 5000 and 3000 BC in the Iran/Iraq area. Unlike other cereals, it has no wild counterpart in nature.[9] It is a hybridization-and-fusion product, containing two sets of chromosomes present in emmer and durum wheats, and a third set found in a wild grass *Aegilops squarrosa.* It thus originated as a 'weed' of emmer or durum wheatfields, which proved to have outstandingly valuable characteristics.

The rather cool damp climate of Ireland is not favourable to the cultivation of bread-wheat. Consequently, although clearly the most prized cereal, it was less frequently grown than others. In recent discussions on plant-remains in early Christian sites

[4]Stokes, *Lismore Lives*, 81.2734–7.

[5]*CIH* v 1759.41–1760.2 = *AL* ii 150.2–5.

[6]This equation is also given in the Old Irish law-text *Bretha Nemed Toísech* (*CIH* vi 2230.20).

[7]*CIH* iii 806.1–3. Similar values are given for these three cereals in commentary on *Uraicecht Becc* (*CIH* v 1609.40–1610.5 = *AL* v 82.18–z).

[8]Jackson, *Aislinge* 38.1183–4 = Meyer, *Aislinge*, 99.4–5.

[9]Ucko and Dimbleby, *The domestication and exploitation of plants and animals*, 60; Zohary and Hopf, *Domestication of plants*, 47–53.

in Ireland, Mick Monk has noted that there are relatively few instances of wheat from this period.[10]

The common Old Irish term for wheat, *cruithnecht*, is a native word of uncertain etymology.[11] A less frequent term is *tuirenn*, also of obscure origin.[12] For wheat-beer – said to be 'from the juices of *tuirenn*' – see p. 334.

Secal (Rye)

Rye (*Secale cereale*) has a more northerly range than other cereals, and can grow in poorer soil. Archaeological evidence of rye is rare in the Neolithic and Bronze Age settlements in the Near East. The earliest definite evidence that it was grown as a crop in its own right – and not merely as a tolerated weed among other cereals – is from about 3000 BC in Anatolia. By around 1500 BC, it was being grown in several sites in Czechoslovakia.[13]

Rye is the most winter-hardy cereal, and therefore is particularly suitable for autumn-sowing. That it was regularly autumn-sown in Ireland in the twelfth century is suggested by a reference in *Aislinge Meic Con Glinne* to 'a wild deer cropping a field of winter-rye (*gemṡecol*) in the month of June'.[14] The position of rye next to bread-wheat in the *Bretha Déin Chécht* list is of particular interest, as it indicates that this crop was of considerable importance in the early Irish economy. This is confirmed by the archaeological evidence: Mick Monk points out that rye has been found quite widely on early Christian sites in Ireland, though forming only a small percentage (4.34 per cent) of the total cereal sample of the twenty-three sites investigated.[15]

The only word attested for rye in Old Irish is *secal*, a borrowing from post-Classical Latin *secale*. This does not of course prove that rye was introduced to Ireland from Roman Britain. In fact, there is evidence of rye from a Bronze Age site at Carrowmore,

[10] Monk, 'The archaeobotanical evidence for field crop plants', 318–21; Monk, 'Evidence from macroscopic plant remains', 33.

[11] *LEIA* C-254–5 s.v. *cruithnecht*.

[12] *LEIA* T-174 s.v. 1 *tuirenn*.

[13] Zohary and Hopf, *Domestication of plants*, 64–73.

[14] Jackson, *Aislinge*, 33.1013–14 = Meyer, *Aislinge*, 85.4.

[15] Monk, 'The archaeobotanical evidence for field crop plants', 318 (table 25.2); Monk, 'Evidence from macroscopic plant remains', 33–4.

Co. Sligo, so its presence long predates Roman influence in north-west Europe.[16] However, it is quite likely that new strains of rye came to Ireland with Christianity, and that the Latin word entered the language at this stage.[17] In this connection it is worth noting that there is a fairly high proportion of Latin loan-words in the Old Irish vocabulary connected with crops,[18] whereas the vocabulary connected with livestock is practically devoid of Latin loans. This indicates that the introduction of new strains and techniques was more a feature of arable than of pastoral farming.

Suillech (Spelt wheat?)

The identity of the third cereal in the *Bretha Déin Chécht* list is uncertain. In the accompanying gloss it is equated with barley (*éornae*).[19] The same identification is made in *O'Davoren's Glossary*,[20] where it is described as *éorna cethardruimnech* 'four-ridged barley'. This is presumably four-row barley: see discussion on p. 226 below.

Another suggestion, which seems more likely, is that of Gearóid Mac Niocaill, who holds that *suillech* is a borrowing from Latin *siligo*, a type of winter-wheat.[21] *Siligo* is distinguished from *triticum* (bread-wheat) in Insular Latin sources.[22] Giraldus Cambrensis records the story of a field of *siligo* which was miraculously turned into *triticum* through the intervention of the bishop of Cork.[23] This shows that *siligo* was regarded as inferior to *triticum*. In an Irish context, therefore, *suillech* may refer to one or other of the less prized (but hardier) wheats, such as emmer (*Triticum turgidum*, subspecies *dicoccum*), spelt (*Triticum aestivum*, subspecies *spelta*) or rivet

[16] Monk, 'Evidence from macroscopic plant remains', 32. Monk points out that rye here may possibly have been a weed in another cereal-crop.

[17] Middle Irish commentary on *Félire Óengusso* attributes the introduction of wheat and rye to saints Fínán and Déclán respectively (Stokes, *Félire*, 112). However, such claims are common in Irish learned tradition, and are of little significance.

[18] Other examples are *pis* 'pea' (Lat. *pisum*), *seib* 'bean' (Lat. *faba*), *connall* 'stubble' (Lat. *cannula*), *súst* 'flail' (Lat. *fustis*), *saball* 'barn' (Lat. *stabulum*), *sorn* 'oven (of grain-drying kiln)' (Lat. *furnus*), *muilenn* 'mill' (Lat. *molina*), *coltar* 'coulter' (Lat. *culter*).

[19] *CIH* vi 2305.23 = Binchy, '*Bretha Déin Chécht*', 22 § 2[7].

[20] *CIH* iv 1525.34–5 = O'Dav. 459 § 1475.

[21] He is quoted by Binchy, '*Bretha Déin Chécht*', 48. The main difficulty of this suggestion is that *siligo* should give Old Irish **silech*.

[22] E.g. Emanuel, *The Latin texts of the Welsh laws*, 157 *in tritico vel in siligine*.

[23] O'Meara, *Giraldus: Topography*, 89 § 78 = Dimock, *Giraldi Topographia*, 131–2. Giraldus's account gives two alternative miracles, but the point of the story is unaffected.

wheat (*Triticum turgidum*, subspecies *turgidum*). Of these, Regina Sexton believes that spelt is the most likely identification as this wheat contains a high proportion of gluten-forming proteins, and consequently produces a light fine-textured loaf of good flavour.[24] Isidore refers to the special use of *siligo* in bread-making.[25] Columella likewise praises the appearance of bread made with *siligo*, but remarks on its lack of weight.[26]

Ibdach (Two-row barley?)

The glossator takes *ibdach*, the fourth cereal on the list, to be a type of barley.[27] He observes that 'a drink is drunk from its juice', i.e. beer (see p. 332). He identifies this cereal as *éorna na n-én* 'barley of the birds' or *maothéorna na n-innsi* 'succulent barley of the islands'.[28] The latter explanation seems to be based on the interpretation of *ibdach* as meaning 'Hebridean'.[29] Much of the Hebrides consists of hilly or boggy terrain unsuitable for cultivation. However, the coastal plains or *machair*s can be brought to a fair level of fertility; indeed, the island of Tiree in the Inner Hebrides was famed for its cereal-production in the nineteenth century, being known as *Eilean ìosal an eòrna* 'low island of barley'.[30] It is therefore quite possible that *ibdach* refers to some type of barley associated with the Hebrides.

If this is so, the problem then is to establish what the distinction is between *ibdach* (= *fidbach*) and *éornae* in the *Bretha Déin Chécht* and *Aislinge Meic Con Glinne* lists. I follow Regina Sexton's suggestion that *ibdach* refers to two-row barley (*Hordeum vulgare*, subspecies

[24]Sexton, 'Cereals and cereal foodstuffs in early historic Ireland', 8–9.

[25]Isidore, *Etymologiarum* (ed. Lindsay), 17.3.7.

[26]Columella, *De re rustica* (ed. Ash), 2.6.2; 2.9.13.

[27]It appears in the *Aislinge Meic Con Glinne* list in the form *fidbach*, with inorganic *f-* and transposition of *b* and *d*.

[28]*CIH* vi 2305.27–8 = Binchy, '*Bretha Déin Chécht*', 22 § 2[11]. The same gloss is found at *CIH* iv 1514.29–30 = O'Dav. 410 § 1213. Here *midbach* has been erroneously substituted for *ibdach*, but the correct reading – though with transposition of *b* and *d* – has been added at the end of the gloss. There is a slight possibility that we should take *na nen* to be gen. plur. of *ian* 'drinking-vessel' – a reference to the use of barley for brewing. For variation between forms with *e-* and *ia-* in legal glosses and commentary, compare nom. plur. *eana* (*CIH* ii 467.18 = *AL* v 392.12) and *iana* (*CIH* vi 2329.21 = *AL* v 94.9).

[29]See discussion by Binchy, '*Bretha Déin Chécht*', 49–50.

[30]Dwelly, *Faclair* s.v. *eòrna*; cf. Fraser Darling, *Natural history in the Highlands and Islands*, 15.

distichum) and that *éornae* lower down in the list refers to six-row barley (*Hordeum vulgare*, subspecies *hexastichum*).[31] Both two-row and six-row barley are known to have been in cultivation in the Near East about 7500 BC,[32] and they are distinguished by the Roman author Columella, who wrote on farming in the first century BC.[33] He refers to two-row barley as *distichum* or *Galaticum* 'Galatian', and comments on its extraordinary weight and whiteness, and on its suitability as food for the household when mixed with wheat. He regards six-row barley (which he calls *hexastichum*) as a wholesome food for animals, and useful for humans in times of scarcity. In excavations of early Christian sites in Ireland, barley is the dominant cereal: more samples are of the six-row than the two-row type.[34]

In later Irish, a distinction is sometimes made between *eorna bheag* 'small barley' i.e. two-row barley, and *eorna mhór* 'great barley' i.e. four-row (= six-row) barley.[35] In a passage on cereals, the nineteenth-century writer Aodh Mac Domhnaill states that *eorna bheag* has special medicinal properties, and is used particularly in the manufacture of beer and spirits. *Eorna mhór* is coarser, hardier and less trouble to grow, and is suitable for bread-making.[36] In Scotland and northern England, four-row (= six-row) barley is called bere, bear-barley or bigg: see discussion on p. 226 below.

Rúadán (Emmer wheat?)

The fifth *Bretha Déin Chécht* cereal is also included in the *Aislinge Meic Con Glinne* list. *Rúadán* means 'the red one', so it must refer to a cereal with a reddish tinge in its grains or stalk. The glossator explains it as *cruithnecht rúad* 'red bread-wheat', and I believe that

[31] Sexton, 'Cereals and cereal foodstuffs in early historic Ireland', 9–11.

[32] Zohary and Hopf, *Domestication of plants*, 62–4.

[33] Columella, *De re rustica* (ed. Ash), 2.9.14–16.

[34] Monk, 'The archaeobotanical evidence for field crop plants', 317; Monk, 'Evidence from macroscopic plant remains', 33.

[35] E.g. De Bhaldraithe, *Cín Lae Amhlaoibh*, 1.26. For a discussion on four- and six-row barley, see p. 226 below.

[36] Beckett, *Fealsúnacht Aodha Mhic Dhomhnaill*, 206–9 §§ 105–6.

he is correct in identifying it as some sort of wheat.[37] One can compare the Latin wheat-name *robus*, which means 'reddish'.[38] Emmer seems a likely candidate, as Mick Monk tells me that its stalk can have a red colour.[39] Regina Sexton points out also that emmer flour is unsuitable for bread-making as it produces a rather heavy loaf: this would account for its position below the other wheats in the *Bretha Déin Chécht* list.[40]

In his edition of *Aislinge Meic Con Glinne*, Meyer took *rúadán* to refer to buckwheat (*Fagopyrum esculentum*).[41] However, this food-plant was not introduced into Europe until the thirteenth century, so it could not be present in either of our lists. It has never been a regular crop in Britain or Ireland.

Éornae (Six-row barley)

Éornae is the generic name for barley (*Hordeum*) in Irish. In excavations of the early Christian period in Ireland, both two-row barley (*Hordeum vulgare*, subspecies *distichum*) and six-row barley (*Hordeum vulgare*, subspecies *hexastichum*) are well represented, though the latter is more frequent.[42] It has been suggested on p. 223 that in the *Bretha Déin Chécht* and *Aislinge Meic Con Glinne* grain-lists *ibdach* (*fidbach*) refers to the more prized two-row barley, while *éornae* refers to the hardier but coarser six-row barley.[43]

I have referred above (p. 224) to the distinction in later Irish between *eorna bheag* 'small barley' i.e. two-row barley, and *eorna mhór* 'great barley' i.e. four-row (= six-row) barley. Mick Monk points

[37] *CIH* vi 2305.31 = Binchy, '*Bretha Déin Chécht*', 22 § 2[14]. A similar gloss is found at *CIH* iv 1521.7 = O'Dav. 439 § 1369. For the further identification of *rúadán* with *máelchruithnecht* 'bald wheat', see discussion under *máelán* on p. 228 below.

[38] Columella, *De re rustica* (ed. Ash), 2.6.1.

[39] According to Columella (*De re rustica* (ed. Ash), 2.6.3), one type of emmer (*far*) is reddish (*rutilum*). Alexander Falileyev informs me that wheat-names containing an element meaning 'red' occur in several Iranian languages, e.g. Tadjik *surxak* 'a type of wheat', from *surx* 'red'. Cf. Loth, 'Les noms et les variétés du froment', 196–7.

[40] Sexton, 'Cereals and cereal foodstuffs in early historic Ireland', 13. She does, however, point out some difficulties with this identification.

[41] Meyer, *Aislinge*, 98.5. In the glossary to his edition, Jackson follows suit.

[42] Monk, 'The archaeobotanical evidence for field crop plants', 317; Monk, 'Evidence from macroscopic plant remains', 33.

[43] It must be stressed, however, that in other contexts *éornae* may refer to two-row barley. For example, *éornae* is used in the law-text *Cáin Aicillne* (*CIH* ii 480.26 = Thurneysen, 'Aus dem irischen Recht I', 348.4) of barley grown in best-quality land to make malt fit to be given as part of a client's food-rent to his lord. This is likely to be the superior two-row barley.

out to me that both six-row and four-row barley have in fact six rows of grains along the ear. In the four-row type, the rows are arranged densely, so only four rows are visible. In the lax-eared six-row type, on the other hand, all six rows can be seen. In his gloss on the cereal-name *suillech* (see p. 222 above), O'Davoren makes an identification with *éorna cethardruimnech*, lit. 'four-ridged barley'.[44] This is presumably a term for four-row barley, though it seems more likely that *suillech* is actually a wheat. In the dialects of Scotland and northern England, there are many records of a distinction between the two-row type and the four- or six-row type. Four- or six-row barley is called bere, bear-barley or bigg,[45] and is still cultivated to some extent in Orkney.[46] In Welsh, the two-row type is called *haidd rhywiog* 'fine barley' whereas the four- or six-row type is called *haidd garw* 'rough barley'.[47]

A reference in the Old Irish tale *Fled Bricrenn* suggests that barley may sometimes have been used as a horse-feed. The three heroes Conall, Lóegaire and Cú Chulainn were asked to choose the type of food which was to be given to their horses. Conall and Lóegaire chose 'grass of two years' (*airthend dā bliadan*), i.e. grass which had not been grazed for two years.[48] Cú Chulainn, on the other hand, chose barley-grain.[49]

Corcae (Common oat)

It is probable that the oat originated as a weed of wheat and barley, and was then taken into cultivation because of its suitablity for the moist cool climates of the temperate latitudes. Two types of oats (*Avena sativa* and *A. strigosa*) have been cultivated in central Europe since about the beginning of the first millennium BC.[50] Both types are found among the plant remains on early Christian sites in Ireland.[51] After barley, the most frequent cereal on these sites is common oat.

The author of *Bretha Déin Chécht* classes the oat as the least valued cereal. A similar assessment is found in legal commentary where a

[44] *CIH* iv 1525.34–5 = O'Dav. 459 § 1475.

[45] Craigie et al., *A dictionary of the older Scottish tongue* s.v. *bere*, *OED* s.vv. *bear*, *bigg*.

[46] Fenton, *The Northern Isles*, 335.

[47] *GPC* s.vv. *haidd*, *heiddyd*.

[48] For a discussion of the identity of *airthend*, see p. 42.

[49] *LU* 267.8833–5 = Windisch, 'Das Fest des Bricriu', 285.15–17 § 63.

[50] Ucko and Dimbleby, *The domestication and exploitation of plants and animals*, 166; Zohary and Hopf, *Domestication of plants*, 77–8. For *Avena strigosa*, see under '*serbán*'.

[51] Monk, Evidence from macroscopic plant remains', 33.

bushel of barley is given twice the value of a bushel of oats.[52] *Aislinge Meic Con Glinne* likewise contains contemptuous references to oaten bread.[53] A quotation in the *Irish grammatical tracts* refers to a mixture of oats and barley (*coirceórna*, i.e. *corca* + *eórna*).[54] In Welsh this mixture is known as *siprys*,[55] and in English the term dredge (dradge) is applied to a mixture of grains, especially oats and barley, which have been sown together.[56]

Serbán (Pilcorn?)

Serbán 'the sour one' is included in the *Aislinge Meic Con Glinne* list, but not in *Bretha Déin Chécht*.[57] Glossators take it to be a form of oats,[58] a suggestion which prompted Meyer to identify it with the wild oat (*Avena fatua*).[59] However, I find this identification difficult to accept. The grains of the wild oat are small and would hardly be worth collecting except in famine conditions. I suspect that the early Irish farmer – like his modern counterpart – regarded wild oats as a persistent nuisance which took up valuable space in his fields. Its status as a pest is illustrated in the satirical verse which refers to 'a sack in which there are wild oats' (*píanán i mbí corca fásaig*) with the implication that its contents are worthless.[60] Another text regards *serbán* (here spelled *serbainn*) as a crop which can actually be milled: this would hardly apply to the wild oat.[61]

Possibly, *serbán* in our list refers to bristle-pointed oat (*Avena strigosa*) which is still sometimes grown on land too poor for *Avena sativa*, and is popularly called pilcorn, black oat, or grey oat.[62] In Modern Irish the names *coirce dubh* 'black oat' and *coirce bocht*

[52] *CIH* v 1610.1–2 = *AL* v 82.24.

[53] Jackson, *Aislinge*, 7.185–6; 35.1074 = Meyer, *Aislinge*, 17.5–6; 89.16.

[54] Bergin, 'Irish grammatical tracts: I Introductory', 23 § 94.

[55] I am indebted to Nicholas Jacobs for this information.

[56] *OED* s.v. *dredge*.

[57] Old Irish *serb* 'sour, bitter'.

[58] *CIH* ii 620.1 and *CIH* iii 1077.5–7. In both glossaries the spelling is given as *serpan* with *p* for lenited *b*, whereas the *Aislinge* has *seruan* with *u* for lenited *b* (Jackson, *Aislinge*, 38.1183 = Meyer, *Aislinge*, 99.4).

[59] Meyer, *Aislinge*, 98.5 = Jackson, *Aislinge*, 194 (Glossary).

[60] Thurneysen, 'Mittelirische Verslehren', 105 § 210 = Meroney, 'Studies in early Irish satire', 202 § 14.

[61] See Greene, *Fingal Rónáin and other stories*, 48.886 and note.

[62] Cf. Fenton, *The Northern Isles*, 335.

'poor oat' have been recorded,[63] and in Welsh it is *ceirch llwyd* 'grey oat', *blewgeirch* 'hairy oat', or *ceirch coliog* 'prickly oat'.[64]

Máelán

As in the case of *serbán*, this cereal is included in the *Aislinge Meic Con Glinne* list, but not in *Bretha Déin Chécht*. The word *máelán* is a derivative of *máel* 'bare, bald', which suggests that it was applied to an awnless cereal, i.e. one in which there is no long stiff spike or 'beard' above the grain. Wheats such as *Triticum aestivum* and *Triticum turgidum* have awnless varieties, so it is possible that *máelán* refers to a type of wheat. It is therefore tempting to identify *máelán* with *máelchruithnecht* 'bald (awnless) wheat', which is mentioned in two legal glosses.[65]

However, there are difficulties in both cases. In a gloss on *Bretha Déin Chécht*, the cereal *rúadán* is equated with *máelchruithnecht*,[66] but the difficulty here is that the *Aislinge Meic Con Glinne* list distinguishes *rúadán* and *máelán* as separate cereals. This glossator's explanation therefore argues against the identification of *máelán* with *máelchruithnecht*. The other reference to 'bald wheat' in a legal gloss is equally inconclusive. A gloss on Heptad 29 refers to 'land from which corn grows without labour, i.e. land from which only bald wheat grows'.[67] I find this gloss difficult to understand: the implication of the second part seems to be that the land is so fertile that it is only suitable for the most prized type of cereal, i.e. awnless wheat.

Meyer put forward a quite different explanation for *máelán*, suggesting that it is bere, a type of barley (see p. 226 above).[68] But he provides no evidence to support this suggestion, and there is the difficulty that all barley is awned, and so can hardly be *máel* 'bald'.

Máelán muilche is mentioned as a weed of corn-fields in legal commentary and in an Old Irish gloss on Priscian. I argue below (p. 234) that this is the wild oat (*Avena fatua*).

[63] Dinneen, *Foclóir* s.v. *coirce*; cf. Dwelly, *Faclair* s.v. *coirce-dubh*.

[64] *GPC* s.vv. *ceirch, blewgeirch*.

[65] Mick Monk (pers. comm.) makes the alternative suggestion that *máelchruithnecht* means 'free-threshing wheat', i.e. wheat in which the grains are not hulled, and so are easier to thresh: see Zohary and Hopf, *Domestication of plants*, 25.

[66] *CIH* vi 2305.32 = Binchy, '*Bretha Déin Chécht*', 22[15]. The same gloss is found at *CIH* iv 1521.7 = O'Dav. 439 § 1369.

[67] *CIH* i 27.26 = *AL* v 222.22–3 *.i. ferann asa n-ā́sann arbur can frichnam .i. asnā fāsann acht cruithnecht maol.*

[68] Meyer, *Aislinge*, 186 (Glossary).

CEREAL-CULTIVATION

Land

The law-text *Cáin Aicillne* gives a description of the type of land which is suitable for barley.[69] It should be level (*mín*). It should be *ardtreichem*, an obscure term which probably refers to deep free-draining tilth.[70] It should also be 'land of three roots' (*tír trí mecon*). This phrase is used elsewhere in legal commentary of land in which wheat is grown.[71] It may refer to the depth of the soil, or to the root-crops which can be grown in it, or to weeds whose presence indicates fertility.[72]

Cáin Aicillne specifies that the soil in which the barley is grown should be properly manured.[73] Our sources contain many references to the manuring of land. Legal commentary refers to the increase in the corn-crop brought about by manure (*fúal*, *tuar* or *aílech*).[74] The dung of cattle is held to be the best fertilizer. Commentary on *Cáin Ṡóerraith* states that a year's dung (*míaslach*) of a milch cow or a trained ox is worth one scruple.[75] The dung of an in-calf cow or an untrained ox is valued at three quarters of a scruple.[76] That of other cattle is valued at half a scruple. The same rate applies to other domestic animals whose dung is useful as a

[69] *CIH* ii 480.26–7 = Thurneysen, 'Aus dem irischen Recht I', 348 § 8 *ēorna a ardtreichem talman do mīn tīre trī mecon cona tu[a]r tēchta* 'barley from well-drained tilth(?) of level land of three roots with its proper manuring'.

[70] See discussion by Thurneysen, 'Aus dem irischen Recht I', 350; 'Aus dem irischen Recht IV', 208. Another occurrence of the phrase *ardtr(e)ichem talman* is to be found at *CIH* iv 1492.24 = O'Dav. 311 § 705.

[71] *CIH* v 1594.2 = *AL* v 18.6.

[72] Petrie's suggestion ('Ecclesiastical architecture', 218, footnote) that the three weeds are thistle, ragwort and wild carrot seems unlikely.

[73] The text has *cona tur techta*. I follow the glossator in reading *tuar*, which is well-attested in the meaning 'manure, manuring': see *DIL* s.v. 2 *túar*. The spelling *tur* of the text is probably for *túr*: see *DIL* s.v. 2 *túr*; cf. *fúar*, later *fúr*.

[74] *CIH* ii 533.27 = *AL* iii 48.27; *CIH* iv 1250.14; vi 2180.28.

[75] *CIH* v 1772.19–20 = *AL* ii 200.3–4.

[76] The extra dung produced by the calf would account for the greater value of the milch cow's dung. A trained ox (*dam ríatae*) would presumably be larger and eat more than an untrained ox (*dam anríatae*), and therefore produce more dung. In another section of commentary on the same text, a distinction is simply made between the dung of a large cow (*bó mór*), which is valued at one scruple, and the dung of smaller cattle, which is valued at half a scruple (*CIH* v 1778.15–16 = *AL* ii 220.1).

fertilizer.[77] In Old Irish law-texts, the term *gert* is often used to include the two vital products of living cattle: milk and dung.[78]

As in later farming systems, manure from the sheds of cattle and other livestock is piled in a dunghill (*otrach*) in the farmyard. Household waste may also be thrown onto this heap, as the law-text *Bretha Nemed Toísech* refers to the presence there of bones and antlers.[79] In autumn,[80] the manure is brought out to the fields in a cart.[81]

To my knowledge, early Irish sources contain no references to other methods of fertilising land, such as the application of ground limestone or shells.[82] The application of sand to improve the quality of land is well attested in post-Norman sources, such as the accounts of Old Ross in 1283–4. As Kevin Down points out, this was probably nutrient-rich sand from nearby tidal estuaries.[83] There are also records in these accounts of the burning of the turf-sward to provide a temporary increase in soil fertility. This destructive method is recorded as being carried out *more patrie* 'in the manner of the country', so it seems to have been traditional in the Old Ross area.

Ploughing

There are occasional references to autumn-sown cereals in later sources. For example, the twelfth-century *Aislinge Meic Con Glinne* speaks of winter rye (*gemṡecol*),[84] and a fourteenth-century entry in the *Annals of Connacht* refers to the damage wrought by cold

[77] *CIH* v 1772.21–22 = *AL* ii 200.6 *cach mīl is tarb*[*ach*] *tuar* 'every animal whose dung is beneficial'.

[78] E.g. *CIH* ii 514.26; 515.1 = *AL* ii 388.17, 29–30 *gert .i. in lacht ⁊ in tuar* 'i.e. the milk and the dung'. *Gert* contains the Indo-European root *g^{w}*er-* 'is warm', and refers to the warmth of fresh milk and dung. See discussions on *gert* by Thurneysen, 'Aus dem irischen Recht II', 241–42; 'Aus dem irischen Recht IV', 211; *SEIL* 40.

[79] *CIH* vi 2220.5–7 = Meyer, *Triads*, 16 § 117.

[80] *CIH* ii 481.8 = *AL* ii 240.6 *a foghmur.*

[81] *CIH* ii 384.15 = *AL* i 170.1–2 *carr .i. aīlich* no *arba* 'a cart i.e. for [transporting] manure or corn'.

[82] At *AL* iv 277.11–12, the translation 'land . . . which does not require the application of manure or shells' is given for *tīr . . . nāch ēicin dō frichnam tuair nā slige* (*CIH* ii 675.21–2). The editors took *slige* here to be a later spelling of Old Irish *slice* 'shells'. However, I believe that Mac Niocaill is correct in his re-edition of this passage ('*Tír cumaile*', 84) to take *slige* to mean 'clearing, felling'. See also Lucas, 'Sea sand and shells as manure', 184.

[83] Down, 'Colonial society and economy in the high middle ages', 473–4.

[84] E.g. Jackson, *Aislinge*, 33.1014 = Meyer, *Aislinge*, 85.4.

weather on the fields of young corn from the the end of the first fortnight of winter until well into the spring.[85] However, in the Old Irish texts spring ploughing and sowing are regarded as the norm. March is the principal month for ploughing, and in *O'Davoren's Glossary* is defined as *mí air* 'month of ploughing',[86] but in favourable weather conditions there was an expectation that some of this work might be done before March. This is implicit in a complaint in the *Annals of Inisfallen* that the snow and frost were so bad in the early part of the year 1282 that not a single day's ploughing or harrowing was done between Christmas and a week before the feast of Saint Brigit (1st February).[87] For a discussion on early Irish ploughing equipment, see pp. 468–77.

A seedbed for corn may also be prepared by digging rather than by ploughing. This method gives a higher yield, but is of course much more labour-intensive.

Harrowing

The harrow levels, compacts and breaks up the soil which has been exposed and loosened by the plough, and also helps to get rid of weeds. A widespread practice in medieval Europe was for the ox-drawn plough to be followed by a horse-drawn harrow. A reference in the law-text *Críth Gablach* suggests the same procedure in eighth-century Ireland. According to this text, the high-ranking *aire tuíseo* is expected to own a complete ploughing-outfit, and also two work-horses for harrowing (*dá chapall do foirtsiud*).[88] The form *foirtsed* is otherwise unknown, but Binchy is no doubt correct in identifying it with *foirsed* 'harrowing'.

For a discussion on harrowing equipment see p. 478.

Sowing

Our sources stress the importance of sowing the seed in good, well-prepared soil.[89] 'Sowing seed in bad soil' (*cor síl i ndrochithir*) is a proverbial expression used to describe a pointless undertaking.[90] It seems to have been general to sow corn on wide earthen ridges

[85] *AC* 282 s.a. 1339 § 4. The editor (A. Martin Freeman) translates *ar gortaib gemair* as 'winter grass', but *gemar* seems always to mean 'young corn': see p. 233 below.

[86] *CIH* iv 1516.16 = O'Dav. 417 § 1249.

[87] *AI* 380 s.a. 1282 § 2.

[88] *CIH* ii 567.27 (*do fortsed*); iii 783.34 (*do foirtsid*) = *CG* 16.410.

[89] E.g. *CIH* iv 1249.9.

[90] Jackson, *Aislinge*, 28.870 (= Meyer, *Aislinge*, 73.25, where *drochith-* is incorrectly expanded as *drochith*lainn rather than *drochith*ir).

or raised beds.[91] For example, a Middle Irish Life of Saint Lasair speaks of tribute consisting of '[the produce of] a ridge of every type of corn' (*iomaire as gach arbar*).[92] According to *Críth Gablach*, the most prosperous grade of *bóaire* (the *mruigfer*) is expected to sow sixteen *míach*s of seed annually.[93]

I have found no information on the storage of seed-corn in the earlier texts. An entry in the *Annals of Loch Cé* for the year 1236 describes a raid on the churches of Mag Éo, and complains that there was not left so much as a stack of seed or corn (*crúach ṡíl no arba*) in the great churchyard of Mag Éo or in the churchyard of the Church of Michael the Archangel.[94] It seems, therefore, that there were separate stacks for seed-corn (*síl*) and for corn which was to be eaten (*arbor*).

Young corn

The term *fochann* (*fochenn*) is used of the first growth of young corn. A ninth-century triad describes in striking language the promise of a field of young corn: 'three slender things which best support the world: the slender stream of milk from the cow's teat into the milking-pail, the slender blade of green corn above the ground, the slender thread over the hand of a skilled woman'.[95]

The Old Irish law-text on distraint contains two references to *fochann* (*fochenn*).[96] In both cases the editors of the *Ancient laws of Ireland* translate it as 'ripe corn', following a glossator.[97] But other occurrences confirm that the correct meaning is 'young corn'; for example a poem in praise of ale in *Bretha Nemed Déidenach* refers to

[91] Such cultivation-ridges from the Bronze Age are illustrated in Mitchell, *The Irish landscape*, plate 27; *Reading the Irish landscape*, 197. Compare the legal description of a raised bed – probably for vegetables – at p. 367 below.

[92] L. Gwynn, 'The Life of St. Lasair', 94.12. Cf. Meyer, *Sanas Cormaic*, 43 § 516.

[93] *CIH* ii 564.6–7; iii 780.13–14 = *CG* 8.196–7 *sé méich deec i talmain* 'sixteen bushels in the ground'. According to commentary at *CIH* iv 1230.34, the lowest grade of lord – the *aire désa* – sows only seven *míach*s. Cereal-growing may have been of relatively less significance on the farms of lords than of commoners: see p. 422. For a discussion on the capacity of the early Irish *míach* see p. 582.

[94] *ALC* i 338.5 s.a. 1236. The version at *AC* 62 s.a. 1236 § 11 has simply *cruach tsíl* 'a stack of seed[-corn]'.

[95] Meyer, *Triads*, 10 § 75. The text has *coíl foichne for tuinn*, glossed in H 1. 15 (Trinity College Dublin, no. 1289) *.i. fochan an gheamhair. Foichne* is singulative of *fochann*, i.e. a single blade of young corn; cf. Ó Máille, *An béal beo*, 44 *foichnín*.

[96] *CIH* ii 371.17; v 1683.22 = *AL* i 124.6; *CIH* ii 385.9; v 1690.28 = *AL* i 166.28.

[97] *CIH* ii 371.18; v 1683.22 = *AL* i 134.14.

corn being 'very green when it is *fochenn*'.[98] Likewise, a reference in *Auraicept na nÉces* states that corn is 'sweeter than any grass' – presumably from the point of view of a trespassing animal – when it is *fuachonn* (= *fochann*).[99]

Another well-attested word for unripe corn is *gemar*.[100]

Weeding

Weeds can smother a crop of corn, particularly in its early stages. A legal glossator refers to the task of *gortglanad* 'cleaning of the cornfield', which doubtless refers to the removal of weeds.[101] A section of commentary on defective contracts lists the faults which may emerge after sale in metal articles, vessels, domestic animals, clothing, wool, and land. The three faults of land are the presence of the weeds *máelán muilchi, dithan*, or *ithloinges*.[102] The seller is not held to have any liability if they are found on the land after three years from the time of purchase.

There has been much discussion on the identity of *máelán muilchi, dithan* (*dithen*), and *ithloinges*. The three-year period suggests that they are weeds which may lie dormant on land (as roots or seeds) and only become apparent after it is ploughed up for a crop. One problem with their identification is that the number of weeds which are a significant nuisance in modern Irish cornfields is considerable, and varies from one locality to another. A recent report lists twenty-three problem weeds of arable land in Ireland.[103] Furthermore, due to selective herbicides and improved screening of seedcorn, some of the weeds which were formerly a serious problem have been largely extirpated. For example, cornflower (*Centaurea cyanus*) used to take up considerable areas of cornland, but is now believed to be extinct everywhere in Ireland except for the Aran Islands.

[98] *CIH* iii 1130.16 = E. J. Gwynn, 'Privileges', 40.26 *roghlas imbí* [*ḟ*] *oichenn*; cf. Stokes, *Félire*, Preface p. xxv § 7.

[99] Calder, *Auraicept*, 278.5558–280.5559. A variant spelling in *fua-* is also attested in legal commentary at *CIH* iii 935.18 which refers to grazing-trespass by horses on grass, [ripe] corn and green corn (*re fér 7 re harbar 7 re fuaicend*). The other version (*CIH* i 266.43 = *AL* iii 186.2–3) has *foich*enn.

[100] *DIL* s.v. *gemar*; Dwelly, *Faclair* s.v. *geamhar*.

[101] *CIH* ii 371.18; v 1683.23 = *AL* i 134.15.

[102] *CIH* iii 1000.34–5; 1034.17 = Thurneysen, *Coic Conara Fugill*, 42 § 66 (with discussion p. 69) = McLeod, *Early Irish contract law*, 299 [g] (with discussion pp. 312–13).

[103] Doyle, 'Weeds of arable land'.

Máelán muilchi is mentioned in a number of texts,[104] and in the Old Irish tale *Tochmarc Emire* it is identified as a weed which might maliciously be sown in an enemy's corn-field (*gort*).[105] In Classical Latin sources, two main weeds of the corn-field are identified: the wild oat (*Avena fatua*) and darnel (*Lolium temulentum*). For example, in his *Eclogues*, Virgil refers to *infelix lolium et steriles nascuntur avenae* 'unhappy darnel and sterile oats grow'.[106] In commentary on this line, the fifth-century scholar Philargyrius explains that Virgil calls darnel 'unhappy' because those who eat it suffer blindness.[107] Old Irish glosses on this commentary are difficult to interpret, but it seems that *maílán muilchi* refers to the wild oat (*Avena fatua*), still a serious pest in cornfields.[108]

In the Irish glosses on a medieval tract on Latin declension, *dithen* is identified as *lolium* (darnel).[109] This grass was formerly a common weed of cornfields, but is now practically extinct in Ireland and Britain. It is particularly undesirable because its seeds are poisonous and can easily contaminate bread. Many instances have been recorded of harmful results to humans and animals from eating meal containing ground-up darnel seeds. The symptoms are giddiness, staggering, and interference with vision. There may be permanent eye-damage, or – if a large quantity has been eaten – death may result.[110] Another suggested identification given in the *Dictionary of the Irish language* is with corn marigold (*Chrysanthemum segetum*), a widespread weed of cornfields in the present day.[111] Seeds of corn marigold have been found in fair abundance in pits

[104]See *DIL* s.v. *muilche*.

[105]Van Hamel, *Compert Chon Culainn, etc.*, 37 § 37.

[106]Virgil, *Eclogues* (ed. Fairclough), 5.37.

[107]*Infelix lolium idest quia, qui lolium manducant, caecitatem patiuntur atque ideo 'infelix'* (quoted by De Vries-Edel, '*Máeláin Muilchi*'). For poisoning by darnel, see below.

[108]One version (*Thes.* ii 48.26) has *Auenae .i. mail molchi uel cuintbecha idest genus zizaniæ* 'oats, that is *maíl[án] molchi* or empty [glossing *steriles*?], that is a type of tare(?)'. Another version (*Thes.* ii 46.16) has *Auenae .i. maila muilchi* (see v.l.) *uel cuinfec uel zizaniae* 'oats, that is *maílán muilchi* or empty or tare(?)'. See also *Thes.* ii 361.v.37; 363.v.37. *Cuintbech* (*cuinfec*) is probably the same word as *cuindbech* (*cuinnbech*), glossed *fás* 'empty' at *CIH* iv 1479.11 = O'Dav. 252 § 349. McLeod says (*Early Irish contract law*, 313) that *máelán muilche* glosses *lolium* 'darnel' at *Thes.* ii 48.26, but this is not so.

[109]Stokes, *Irish glosses*, 22 no. 718.

[110]West, *Black's veterinary dictionary*, 131; A. Gwynn, *The writings of Bishop Patrick*, 94.170–7; cf. 88.74–80.

[111]*DIL* s.v. *dithen*.

at Fishamble Street, Dublin, which have been dated to the early eleventh century.[112]

Thurneysen takes the name of the third weed, *ithloinges*, to mean 'corn-invasion', but he has no comments on its identity.[113] McLeod suggests hairy tare (*Vicia hirsuta*), an occasional weed of arable land.[114] Another possibility is corn cockle (*Agrostemma githago*).[115] Its seeds, like those of darnel, are poisonous.[116]

Corn-damage

Cereals – particularly wheat – are vulnerable to adverse weather conditions. For example, the *Annals of Inisfallen* record that much of the corn-crop of the year 1012 was destroyed by heavy rain,[117] and that a gale in the autumn of 1077 greatly damaged the corn-crop.[118] A cold winter, on the other hand, seems to have been regarded as beneficial to the subsequent corn-crop: the ninth-century wisdom-text *Tecosca Cormaic* states that 'ice is the mother of corn'.[119] Drought is rarely mentioned in the context of the cereal-crop. The only example which I have noted in our sources is a reference in the *Annals of Ulster* of the year 773 to an almost total failure of bread caused by 'unaccustomed drought and heat of the sun'.[120]

Corn is liable to damage by domestic herbivores: commentary on the law-text *Bretha Étgid* refers to horse-trespass on grass (*fér*), young corn (*fochenn*) and fully-grown corn (*arbar*).[121] Wild deer may also be a problem to the cereal-grower,[122] and a Latin Life of Saint Ailbe (Albeus) contains a reference to the destruction of grass

[112] Geraghty, *Viking Dublin: botanical evidence from Fishamble Street*, 31, 41, 104. Martha Hannon and John Tierney have identified seeds of this plant in pre-Norman deposits from Waterford town (pers. comm.).

[113] Thurneysen, *Cóic Conara Fugill*, 69 § 24.

[114] McLeod, *Early Irish contract law*, 313.

[115] Cf. Mitchell, *Archaeology and environment in early Dublin*, 24–5; Geraghty, *Viking Dublin: botanical evidence*, 29, 37.

[116] West, *Black's veterinary dictionary*, 120.

[117] *AI* 182 s.a. 1012 § 4.

[118] *AI* 234 s.a. 1077 § 7.

[119] Meyer, *Tecosca Cormaic*, 36 § 17.3 *máthair etha aig.* However, autumn-sown corn might be damaged by persistent snow and frost, e.g. *AU* ii 462 s.a. 1339.

[120] *AU*^{2} 226 s.a. 772 (recte 773) § 4.

[121] *CIH* i 266.42–3 = *AL* iii 186.1–3. *Arbar* is translated 'green corn' at *AL* iii 187.2, but there is no evidence of the use of the word in this meaning.

[122] Jackson, *Aislinge*, 33.1013–14 = Meyer, *Aislinge*, 85.4–5.

and cereals by a great flock of cranes (*grues*).[123] Mice – and perhaps rats – feature mainly as a menace in grain-stores (p. 243), but there is also a reference to their destruction of standing crops : *Lochaid ag ithe na ngort uile i n-arailibh tíribh i nÉrinn* 'Mice (or rats) eating all the corn-fields in some territories in Ireland'.[124]

Modern cereal-farmers have to devote a great deal of effort and expense to protecting their crops from such insect pests as frit-flies, wireworms, leather-jackets, wheat-midges, etc. No doubt early Irish cereal-crops were attacked by at least some of these pests, but no information on this topic has come down to us in native sources. Strangely, Welsh annals relate a tradition that in or about the year 895, the crops of Ireland were destroyed by 'mole-shaped vermin with two teeth which fell from the air'.[125] As David Thornton points out, this sounds very like a visitation from the locust (*Locusta migratoria*), resident in southern Europe.[126] Swarms of locusts are recorded as having reached Britain and Ireland on rare occasions, but there are no records of any significant damage to crops. It seems likely that the Welsh annals have misattributed information relating to the well-documented Continental locust-swarms of 873.[127] Contemporary Irish annals contain no references to any such event.

Finally, growing corn may be damaged by human agency. A legal fragment states that two ounces of silver must be paid for cutting off an ear (*días*) of corn from its stem,[128] or for gathering a fistful of ears of corn while walking on a path by a cornfield.[129] Legal commentary also deals with the liability of a person who negligently or maliciously looks on as another person's corn is damaged – presumably by trespassing cattle. He is described as being *sellach in arba* 'a looker-on at corn'.[130]

[123] Plummer, *Vitae* i 62 § xliii; Heist, *Vitae*, 130 § 51. For a discussion of the *corr*, usually 'heron', but also sometimes 'crane', see p. 125.

[124] *AFM* ii 988 s.a. 1109.

[125] E.g. Williams ab Ithel, *Annales Cambriæ* 16 s.a. [896] *Panis in Hibernia defecit. Vermes de aere ceciderunt, talpæ similes, cum duobus dentibus qui totam comederunt.*

[126] Thornton, 'Locusts in Ireland?', 46. Locusts other than *Locusta migratoria* are also known in southern Europe, but this species is the most likely to reach the British Isles.

[127] Thornton, 'Locusts in Ireland?', 47–9.

[128] *CIH* iii 1105.18; 1107.5.

[129] *CIH* iii 1105.19–20.

[130] *CIH* iii 977.39; 1065.17.

Diseases of corn

In the generally damp Irish climate, cereal-crops – whose natural home is in much drier parts of the world – are prey to fungal infections, such as smut, rust (of various types), bunt, mildew, etc. An Old Irish term for such an infestation is *glasar*, clearly a derivative of *glas* 'grey, green'; it may refer to rust or mildew.[131]

CEREAL-HARVESTING

When the corn is ripe (*apaig*), the ear may droop heavily from the stalk (*connal*). The ninth-century author of a triad provides a vivid image of a field of ripe corn, which gives an impression of sorrow – presumably he is thinking of the weighed-down appearance of the corn-stalks – but in fact brings joy by providing food. He compares it with the 'sorrow' of a tree weighed down with fruit and the 'sorrow' of a feeding herd of pigs. In the latter case he may be thinking of the unusual quiet of the pigs as they concentrate on the acorns. His triad reads: *trí bróin ata ferr fáilti: brón tréoit oc ithe messa, brón guirt apaig, brón feda fo mess* 'three sorrows which are better than joy: the sorrow of a herd of pigs eating acorns, the sorrow of a ripe cornfield, the sorrow of a tree in fruit'.[132] Another triad from about the same period refers to *trí bairr for Érinn: barr dés 7 barr scoth 7 barr mesa* 'three abundances in Ireland: an abundance of ears of corn, an abundance of flowers, and an abundance of fruit'.[133]

Reaping

Corn is usually cut in the middle month of autumn (i.e. September), which is described in *O'Davoren's Glossary* as *mí búana* 'the month of reaping'.[134] Harvesting may take place earlier if the summer is exceptionally good. The *Annals of Connacht* record that in the year 1252 the summer was so hot and dry that wheat and all other cereals were harvested three weeks before the festival of *Lugnasad* at the beginning of August.[135] The same annals contain an unbelievable reference to a very late harvest in 1225: 'corn was being cut after Saint Brigit's Day (1st February) and there was ploughing at the same time'.[136]

[131] *Thes.* i 5.10.
[132] Meyer, *Triads*, 8 § 68. I have slightly rephrased his translation.
[133] Knott, *Togail Bruidne Da Derga*, 18.609–10.
[134] *CIH* iv 1516.17 = O'Dav. 417 § 1249.
[135] *AC* 106 s.a. 1252 § 9. For *Lugnasad* see p. 459 below.
[136] *AC* 20 s.a. 1225 § 37.

It is clear that a reaping-party (*meithel*)[137] might contain a large number of people. For example, a saint's Life refers to a party of one hundred and fifty reapers.[138] One of a base client's obligations is to provide labour for his lord's harvest (p. 446), and failure to do so is classed as the offence of *meth methle gíallnae* 'failure to join the reaping-party of base clientship'.[139] According to *Críth Gablach*, the highest grade of lord has twenty base clients.[140] This implies a reaping-party of at least twenty persons. Obviously, the labour of such large numbers needs direction and supervision. In the later language, the individual who leads the reaping-party is called *ceann an mheithil* 'the head of the reapers'. The annals record the death in 1311 of Maolsheachlainn mac Conchobair Ruaidh of the Clann Muircheartaigh.[141] He bore the nickname *Ceann in mheithil*, perhaps as a metaphorical reference to his military prowess.

The corn is cut with a sickle (p. 480). Michael Duignan held that the shape and size of the majority of sickles from the early Christian period indicate that 'the straw must have been cut high up the stalk, often in fact quite near the ear'.[142] There is some linguistic and textual evidence to support his claim. Thus there are references which suggest that the ears of corn were put straight into a reaping-basket (*clíab búana*) after being cut.[143] Furthermore, the word for 'sheaf of corn' is *punnann*, first attested in the tenth-century religious poem *Saltair na Rann*,[144] and generally taken to be a Norse loan-word.[145] It could be argued, therefore, that the practice of cutting corn-stalks near the ground, and binding them

[137]The Welsh cognate is *medel* of the same meaning (*GPC* s.v.).

[138]Plummer, *Bethada* i 226 § xlv (125) (trans. ii 219); cf. Van Hamel, *Immrama*, 84.86–7 § 6 = Stokes, 'The voyage of Mael Dúin, (1)', 20 § 20 *meithle mōra acco ic buain inn arba* 'they had large reaping-parties cutting the corn'. See also O'Dowd, *Meitheal*.

[139]*CIH* ii 382.20–21 = *AL* i 156.y; cf. *CIH* iii 918.22. Giolla na Naomh mac Duinn Shléibhe Mhic Aodhagáin describes such a failure as *brisid lá ar thigerna* 'he breaks a day on a lord' (*CIH* ii 698.34–5). A free client is also obliged to contribute labour to his lord (p. 447), some of which may be required at harvest-time.

[140]*CIH* ii 567.33–4 = *CG* 17.420–21.

[141]*AU* ii 420 s.a. 1311; *AC* 226 s.a. 1311 § 14.

[142]Duignan, 'Irish agriculture in early historic times', 140.

[143]Knott, *Togail Bruidne Da Derga* 16.519 = *LU* 213.6952–3; cf. E. J. Gwynn, *Metrical Dindshenchas* iv 138.81 *clíab co cendaib dés* 'a basket with ears of corn' (*dés* = *días*).

[144]Stokes, *Saltair na Rann*, 48.3329–30.

[145]Marstrander, *Bidrag*, 81. Flax, on the other hand, was harvested in sheaves in the early Christian period, e.g. *CIH* ii 510.16 = *AL* ii 372.16 *a scuapaib lin* 'in sheaves of flax'. As far as I know, *scúap* (later *scuab*) is never used of a sheaf of corn in Irish: it normally retains the meaning 'brush' of the Latin word *scopa* from which it derives.

into sheaves was associated particularly with the Norse settlers of the coastal regions.[146] On the other hand, there is a reference to a corn-rick (*dais*) in an Old Irish law-text from about the eighth century. Corn can only be ricked when in sheaves: see next section.

Corn-rick

The corn-rick (*dais*)[147] is presumably situated in a sheltered spot near the farmyard. The Old Irish law-text *Di Chetharšlicht Athgabálae* contains a reference to the payment due for damage to a farmer's corn-rick.[148] According to a Middle Irish commentary on *Bretha Comaithchesa*, the three main offences of hens outside the farmyard are depredations in drying-kilns, barns and corn-ricks.[149]

Legal glossators use the word *crúach* to explain Old Irish *dais*.[150] *Crúach* is the general term for 'corn-rick' in Middle and Modern Irish, and also in Scottish Gaelic. For example, the *Annals of Inisfallen* record the destruction of many houses and corn-ricks (*crúacha*) by the violent winds of the year 1282,[151] and the *Metrical Dindshenchas* refers to the top – presumably the thatch – of a corn-rick (*cond crúaiche*).[152] A Middle Irish tale describes how corn was cut and stacked in the field and then brought into the farmyard (*les*) and placed in a single thatched rick (*ina œnchruaich thuighthi*).[153]

A glossator on the law-text *Bretha Étgid*, writing about the twelfth century, discusses some of the legal aspects of the making of a corn-rick.[154] To be classed as a proper structure, two thirds of the sheaves (*punnann*) should be placed at the bottom, and one third on top.

However, in Scottish Gaelic *sguab* is used both of a brush and a sheaf of corn. *Ysgub*, the Welsh borrowing of Latin *scopa*, is likewise applied to both.

[146]For other evidence of Norse influence on Irish farming, see p. 249.

[147]*Dais* is cognate with Welsh *das* of the same meaning.

[148]*CIH* ii 384.20–21 = *AL* i 166.x *im dīre do daise arba* 'concerning the penalty-fine of your rick of corn'.

[149]*CIH* i 73.23–4 = *AL* iv 116.12–13 (quoted incompletely) *a trī cercfogla a sechtarlis .i. fogal re hāthaib ⁊ [s]ablaib ⁊ daisib arbhu.*

[150]E.g. *CIH* ii 385.1 = *AL* i 170.27.

[151]*AI* 380 s.a. 1282 § 2. Mac Airt reads *cruocha*, a possible variant of *cruacha*, but his *-o-* is perhaps a smudged *-a-*: see Best and Mac Neill, *The annals of Inisfallen: facsimile*, f. 50^{b}13.

[152]E. J. Gwynn, *Metrical Dindshenchas* iv 160.215.

[153]Stokes, '*Scél na Fír Flatha*, etc.' 196–7 § 46.

[154]*CIH* i 287.36–288.6 = *AL* iii 284.1–18.

Corn-straw

In many farming communities – medieval and modern – corn-straw is an important by-product of the harvest, being used for bedding, floor-covering, thatch, rope-making and basketry, as well as for fodder. But in the written sources of the Old Irish period the emphasis is on reeds for thatch, rushes for floor-covering, and withes for rope-making and basketry. This absence of early references to the uses of corn-straw would support Duignan's view that corn was generally cut near the ear, the stalks being left in the ground to be eaten off by livestock after the harvest.

There seems to have been increased use of corn-straw from the Middle Irish period onwards. The word *tugae* (*tuige*), which in Old Irish refers to any form of covering or roofing, developed the narrower meaning of corn-straw. This implies that corn-straw had become a standard material for roofing. Legal commentary also refers to the use of corn-straw as a fire-lighter, fodder for livestock and bedding for people.[155] A legal gloss mentions the use of corn-straw as a filling for boots.[156]

Cereal-processing

Threshing

The next stage is for the ears of corn to be threshed with a stick (*súst*).[157] It is clear that this operation takes place at the drying-kiln (*áith*), as one law-text refers to accidental injury by a threshing-stick at a drying-kiln.[158] The ears of corn are laid on a threshing floor (*ithland*),[159] and the action of the stick detaches the grain from the stem.

The early sources contain no references to an alternative method of separating the grain from the stem which is called *loiscreán* (*loistreán*) in later Irish.[160] This involves the careful burning of each ear of corn so that the glumes and chaff are burnt

[155] *CIH* iv 1190.9–11; 1263.12–14 = *AL* iii 150.8–11.
[156] *CIH* v 1721.14 = *AL* i 300.22–3.
[157] For a discussion of threshing sticks and flails, see p. 481.
[158] *CIH* i 273.29 = *AL* iii 220.15; cf. *CIH* i 11.12 = *AL* v 156.14–15.
[159] Mac Eoin, 'A poem by Airbertach mac Cosse', 120 § 7. In a gloss at *Thes.* ii 123.25, *ithland* glosses Latin *area* 'threshing-floor'. The word *léibend* 'flat surface' is also applied to the threshing-floor in Old Irish, e.g. *CIH* ii 242.18–19 = *AL* v 486.14.
[160] Joyce, *A social history* ii 342. This practice was forbidden by statute in 1634: see p. 23.

away, leaving the grains. This method is particularly suitable for small-scale cereal-growing, and has the advantage that the grains are partially dried by the flames.[161]

Drying

Because of the damp climate, Irish grain usually needs artificial drying (*tírad*) before it can be stored. The drying-kiln is consequently of great importance in the early Irish economy, and is frequently mentioned in our sources. According to *Críth Gablach* a farmer of *bóaire* rank is expected to own a drying-kiln.[162] Lower down the social scale, the *ócaire* merely has a share in a drying-kiln.[163]

In spite of the many references to it, our sources provide us with little information on the structure or operation of a drying-kiln.[164] A ninth-century text on the reciprocal services of the kings of Cashel refers to a kiln of thirty feet (in diameter?), which seems remarkably large.[165] However, Mick Monk suggests to me that the author may be thinking of a structure which contains a number of separate kilns. According to a legal glossator, the stone wall of a kiln is known as the *caisel*.[166] From a reference in *O'Mulconry's Glossary* it seems that the grain is placed on a wickerwork tray (*clíath*) which slots into the wall.[167] Presumably, a number of such trays could be fitted one above the other.

The operator lights a fire at the base of the kiln, and must make sure that the fire dries but does not burn the grain. The author of Triad 140 pours scorn on the incompetent operator by including 'kiln-drying with scorching' (*tírad co n-aurgorad*) among the three black husbandries.[168] Burnt ears of corn (*díasa loscthi*) are included

[161]For a brief discussion of *loiscreán*, see Ó Danachair, 'The flail and other threshing methods', 6–8. Mick Monk remarks (pers. comm.) that this technique is still practised in Turkey, and has been studied by Gordon Hillman.
[162]*CIH* ii 563.7; iii 779.27 = *CG* 6.156.
[163]*CIH* iii 778.28 = *CG* 4.96.
[164]See discussion by Monk, 'Post-Roman drying kilns'. Elizabeth O'Brien has excavated a sixth-century drying-kiln at Ballyman, Co. Dublin (pers. comm.).
[165]O'Keeffe, '*Dál Caladbuig*', 19 § 2.
[166]*CIH* i 242.36 = *AL* v 488.13.
[167]O'Mulc. 270 § 806 *Láem dono ainm don clēith bīs fon arbar oca tīrad* '*Láem* is the name of the tray which is under the corn while it is drying'. There is a similar gloss at *CIH* iii 1074.39–40.
[168]Meyer, *Triads*, 18 § 140.

along with red blackberries and green nuts among the unappetizing food favoured by the rough and misshapen Amargein son of Ecet Salach.[169]

After the grain has been dried, the ashes are scraped away from the base and sides of the kiln. These kiln-scrapings (*fobach n-átha*) may contain grains which have fallen through the trays, and are consequently sought after by hungry people. According to one law-text, it is not regarded as an offence for a person to forage among kiln-scrapings belonging to another.[170]

Winnowing

The next stage is winnowing, i.e. the separation of the grain from the chaff (*cáith*). This is effected by throwing the mixture of grain and chaff into the air when there is a light wind. The husks, awns, and bristles which make up the chaff are blown to one side, leaving the grains behind on the riddle (*críathar átho*).[171] According to *Bretha Comaithchesa*, the grain which is to be paid for animal-trespass must be checked for hardness (*cruas*) and bareness (*lomdatu*). This means that it must have been properly dried in a kiln, and properly winnowed so that only the bare grain remains.[172]

Chaff is generally held to be worthless and fit only to be burnt up.[173] There is some evidence, however, of it being stored for use as a firelighter. The early twelfth-century tale *Aislinge Meic Con Glinne* refers to material for a cooking-fire consisting of two sparks in a wisp of oat-chaff (*sopp sílchátha corcca*) and two sods of wet turf.[174] Admittedly, the passage is satirical, but the reference to chaff as a firelighter may reflect actual practice.

Storage

The dried corn (*tarae*) must be stored in a dry place safe from domestic animals. For this purpose the prosperous farmer (*bóaire*)

[169] Best, 'Amairgen son of Ecet Salach', 33 § 1. This hardly refers to the custom of separating the grain from the stalk by burning (*loiscreán*). The point of the passage seems to be that Amargein consumes only damaged or unripe food.
[170] *CIH* i 242.19 = *AL* v 486.14.
[171] *Crīathar ātho* glosses Latin *cribrum areale* 'riddle of the threshing-floor' at *Thes.* ii 235.7.
[172] *CIH* i 71.7 = *AL* iv 94.7–8. The glossator explains . . . *7 hē lom .i. cin colg* '. . . and it is bare, i.e. without awn'.
[173] E.g. O'Keeffe, 'A poem on the day of judgement', 31 § 24; cf. *CIH* iii 1028.36–6 = Thurneysen, *Cóic Conara Fuigill*, 29 § 12.
[174] Jackson, *Aislinge*, 6.159 = Meyer, *Aislinge*, 15.2.

has a barn (*saball*) near his house,[175] whereas the less prosperous farmer (*ócaire*) merely has a share in a barn.[176] In his *Life of Columba*, Adomnán describes how the saint blessed two heaps of grain in the barn at Iona.[177]

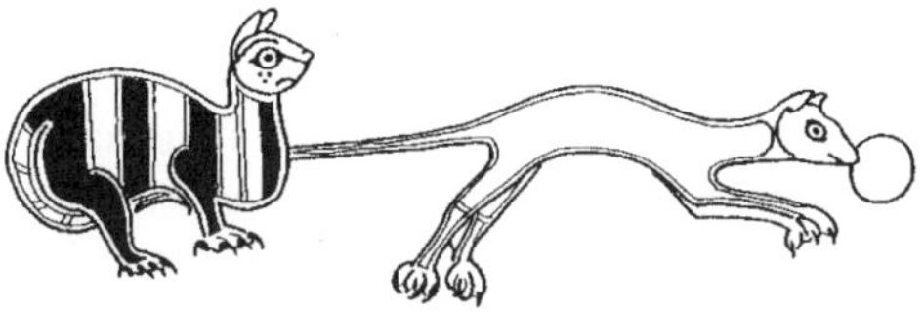

Fig. 10. This illustration from f. 48r of the eighth-century Book of Kells in the Library of Trinity College, Dublin, shows a cat with what appears to be a rat.

The area around the barn is patrolled by the cat: the fragmentary law-text on cats, *Catšlechtae*, stresses the importance of the cat in keeping mice away from the barn, the mill, and the drying-kiln.[178] The house mouse (*Mus musculus*, Irish *luch*) is classed in an Old Irish law-text as a significant destroyer of food,[179] and Giraldus Cambrensis – writing in the late twelfth century – refers to the mouse as a particular pest in Ireland.[180]

There is uncertainty about the significance at this period of an even more serious pest of grain-stores: the rat. It was widely assumed until recently that the black rat (*Rattus rattus*), originally from southern Asia, did not reach Britain or Ireland until about the twelfth century.[181] However, in a publication of 1979, James Rackham identified rat-bones from a late Roman site in York,[182] and in 1984 further rat-bones were identified from a third-century

[175] *CIH* ii 563.7; iii 779.27 = *CG* 6.156. Though *saball* is a borrowing from Latin *stabulum* 'stable', it has acquired the meaning 'barn' in Old Irish. In the later language the normal word is *scioból*, a borrowing from Latin **scoparium* 'barn' via Welsh (*y*)*sgubor* (*LEIA* S-44 s.v. *sciból*). The Old Irish compound *itech* (*ithtech*) lit. 'corn-house' is used in the Milan Glosses to explain Latin *horreum* 'barn' (*Thes.* i 332.27–8).

[176] *CIH* iii 778.28 = *CG* 4.96.

[177] A. O and M. O Anderson, *Adomnan's Life of Columba*, 520 (new pagination: 220) = f. 126a.

[178] *CIH* v 1550.17; cf. *CIH* iv 1481.21 = O'Dav. 263 § 422.

[179] *CIH* vi 2004.14; 2196.27 (trans. Thurneysen, 'Aus dem irischen Recht III', 342 § 32[1]); cf. Plummer, *Vitae* ii 11 § xxi.

[180] O'Meara, *Giraldus: Topography*, 49 § 20 = Dimock, *Giraldi Topographia*, 61.22–4.

[181] On the other hand, MacArthur, 'The identification of pestilences', 170, argued for the earlier presence of the rat.

[182] J. Rackham, '*Rattus rattus*, the introduction of the black rat into Britain'.

site in London.[183] The only occurrence of the rat in an early Christian site in Ireland is at Rathmullan, Co. Down.[184] There is also the evidence of the Book of Kells, written in the eighth century at some monastery in Ireland or Scotland (possibly Iona). On folio 48ʳ a scribe has drawn a cat with a large rodent which seems very rat-like. It is certainly quite different from the four mice illustrated on folio 34ᵛ.

The Irish word *luch* seems originally to have been applied to rats as well as mice, so the *plág lochad* recorded by the *Chronicum Scotorum* in the year 1013 may have been a plague of rats.[185] It affected the Leinstermen and Norse settlers, and so presumably originated from ships arriving at ports on the east coast. The distinction between *luch* 'mouse' (sometimes *luch beag* 'small mouse') and *luch frangcach* 'French mouse' (i.e. 'rat') only emerges in post-Norman sources, and probably indicates a dramatic increase in the rat population after the Norman invasion.

For short-term use, a measure of grain is kept in the living-house. According to legal commentary, a *bóaire* keeps in his house a bushel of corn for food (*míach arba bíd*), a bushel of malt (*míach mracha*), and a bushel of salt.[186] A similar measure is brought by a client to his lord as part of his annual food-rent. The law-text on base clientship, *Cáin Aicillne*, refers to a 'bushel of hard (i.e. ripe and dry) wheat for food'.[187]

The written evidence suggests that a tightly woven wicker basket was a common container for the transport or temporary storage of grain. For example, an Old Irish story about Saint Máel Rúain of Tallaght refers to 'a basket (*cess*) or two baskets full of malt and of corn for food'.[188] *Aislinge Meic Con Glinne* likewise refers to a basket (*clíab*) of corn.[189] Grain might also be transported in a bag (*bolc*), perhaps of leather. An episode in the Life of Saint Cíarán describes

[183] Armitage, et al., 'New evidence of the black rat'.

[184] Lynn, 'The excavation of Rathmullan', 154; McCormick, 'Stock-rearing in early Christian Ireland', 252.

[185] *CS* 256 s.a. 1013. In footnote[2] Hennessy wrongly rejects Wilde's interpretation of *plag lochad* as 'a plague of rats (or mice)' and translates 'a plague of putrefaction'.

[186] *CIH* iii 1058.3–4 = *AL* v 90.15–16; cf. *CIH* iii 888.6–7.

[187] *CIH* ii 483.16 = Thurneysen, 'Aus dem irischen Recht I', 355 § 13 *miach cruithnechta cruaid inbid.*

[188] *LL* v 1247.36984–5 *cess nō dā chess lána do braich ⁊ do arbur biid.* A misprint in *LL* has been repeated in *DIL* C 147.39; the correct reference is to line 36984, not 39984.

[189] Jackson, *Aislinge*, 28.867 = Meyer, *Aislinge* 73.18.

how a bag of oats (*bolc corcai*) which the saint was carrrying to a mill was miraculously changed into wheat, a much more valuable cereal.[190]

Grinding

Grain must be ground into meal (*men*) before it can be used to make bread or porridge. The use of the rotary quern (*bráo*, later *bró*) for this purpose goes back many thousands of years.[191] In the Old Irish law-texts the task of grinding (*mleth*) is associated particularly with the female slave: see p. 439.

In the early Christian period, the introduction of the horizontal water-mill enabled large quantities of grain to be ground quickly, though the use of the hand-quern continued. It is clear from both archaeological and written evidence that mills were already quite common in Ireland by the seventh or eighth centuries AD. According to *Críth Gablach*, farmers of *ócaire* and *bóaire* rank are expected to own a share in a mill.[192] The higher grade of *bóaire* – called a *mruigḟer* in this text – may have his own mill.[193] The existence of a special law-text on milling and mill-races, probably dating from the middle of the seventh century, also underlines the importance of the water-mill. It is entitled *Coibnes Uisci Thairidne* 'kinship of conducted water'.[194]

For a discussion of the written evidence on the technology of corn-grinding, see p. 482 below.

Malting

It is clear from our sources that most beer (*cuirm*) was brewed from barley: see p. 334. The malting process brings the value of barley up to that of wheat. Thus the law-text *Bretha Nemed Toísech* assigns the same value to a bushel of malt as it does to a bushel of wheat.[195]

[190] Stokes, *Lismore Lives*, 124.4163–5; cf. Plummer, *Bethada* i 26 § viii (*bolcc corca*).
[191] Note the cognates in other Indo-European languages, e.g. Welsh *breuan*, Old English *cweorn*, Sanskrit *grāvan*, Lithuanian *gìrnos*, etc. (*LEIA* B-92 s.v. 1 *bró*).
[192] *CIH* iii 778.28; 779.27 = *CG* 4.96; 6.155.
[193] *CIH* iii 780.11; 781.2 = *CG* 8.192; 9.238.
[194] *CIH* ii 457.11–462.18 = Binchy, '*Coibnes Uisci Thairidne*', 64–76.
[195] *CIH* vi 2230.19–20.

A quantity of malt (*mraich*)[196] is included in the annual food-rent which a base client must give to his lord. The law-text on base-clientship, *Cáin Aicillne,* gives an account of how this malt should be prepared. In general, it follows the same procedures as those practised by the modern brewer.[197] The main problem is in identifying the timescale which the author had in mind. The text refers to *mraich trí cóicthiges* 'malt of three fortnights',[198] but no interpretation of his figures brings us to a period as long as this. Furthermore, according to modern maltsters, the process generally takes only twelve to fifteen days. It is probably best, therefore, to follow the glossator in taking the period to be fifteen days.[199] This requires that we interpret the word *lá* to mean 'a twelve-hour period (of day or night)'.

The process of malting is initiated by soaking the grain for twenty-four hours. It then drains for a day and a half (*trí lá*), after which it is covered for four and a half days (*nómad*).[200] The text provides no information on the nature of the covering, but the gloss suggests that corn-straw may be used.[201] It is then exposed for a further three days. After this, it is raised into little heaps (*ina fótaib*) and kept in this state for five days. It is then subjected to raking (*círad*) and is left in ridges (*ina imairib*). The whole process has taken fifteen days: it is now ready for kiln-drying (*tírad*).

When dry the malt is brought to the lord. It is not acceptable as food-rent until it has been subjected to a test (*focull*) on three

[196] *Mraich* (later *braich*) is cognate with Welsh *brag* of the same meaning. See *LEIA* M-66 s.v. *mraich*.

[197] See Binchy 'Brewing in eighth-century Ireland'; O'Loan, 'A history of early Irish farming (3)', 180–2.

[198] *CIH* ii 481.9; v 1783.7 = Thurneysen, 'Aus dem irischen Recht I', 348.

[199] *CIH* ii 481.20 *tri cūicti*; v 1783.7 *tri cōicte* = *AL* ii 240.23. In his edition of *Cáin Aicillne,* Thurneysen raised the possibility of emending *cóicthiges* to *cóicthe* ('Aus dem irischen Recht I', 348[1]), but this would conflict with the readings of both manuscripts. See his further discussion on p. 350, and also the remarks by Binchy in 'Brewing in eighth-century Ireland', 5.

[200] *CIH* ii 481.10 = Thurneysen, 'Aus dem irischen Recht I', 348.7. *Nómad* sometimes means 'nine 24-hour days', but here seems to refer to half this time: see *DIL* N 63.14–29.

[201] *CIH* ii 481.23 = *AL* ii 240.28 *cid tuige nō arbar* lit. 'whether straw or corn'. I am unclear as to the distinction between *tuige* and *arbar* here; possibly *tuige* (*tugae*) retains its earlier meaning of any form of covering: see p. 240.

occasions.[202] It is first tested on delivery to the lord's house, probably – as Binchy suggests – by his brewer (*scóaire*). According to a gloss, the tester makes sure that it has the correct smell and checks the consistency of a grain under his tooth.[203] The malt is then ground in the lord's mill, and tested again. This test is effected by making a small cake (*toirtín*) from it, and tasting this for flavour and wholesomeness. Binchy suggests that only enough malt is ground to make a sample cake. He believes that the rest of the malt is put to ferment without being ground. However, it seems to me that the most straightforward reading of the text implies that **all** the malt is ground before fermentation, as is the practice in modern brewing. There is also evidence of grinding before fermentation in other texts. The eighth-century tale *Fled Bricrenn* refers to a ten-chute mill grinding very hard malt,[204] and a Middle Irish fragment on Saint Brigit mentions a bag of meal (*menbolg*), half of which is meal for beer (*min chorma*).[205]

The third test is carried out on the wort (*braichles*) before it is begins to ferment. If a client's malt passes all three tests, he is not held responsible if faults develop subsequently in the brewing process.[206]

For brewing, see p. 333.

[202] *CIH* ii 481.26; v 1783.28; 1784.1 = Thurneysen, 'Aus dem irischen Recht I', 348.10.
[203] *CIH* ii 481.31 = *AL* ii 242.17–18.
[204] *LU* 263.8699–8700 = Windisch, 'Das Fest des Bricriu', 279.22–3 § 52. The Middle Irish version of *Táin Bó Cúailnge* in the Book of Leinster (*TBC* LL 91.3306) also refers to malt being ground in a mill.
[205] Ó hAodha, *Bethu Brigte*, Appendix, 19.70.
[206] *CIH* ii 481.28–9 = Thurneysen, 'Aus dem irischen Recht I', 348.12–14.

8
Crops (ii)

Pulses

After the *Gramineae* (grass family), the *Papilionaceae* (pea family)[1] are the most important source of plant food for humans. The seeds of many species are rich in protein, and keep well when dried; there is also the added advantage that the roots have nitrogen-fixing nodules which enhance the fertility of the soil.

The archaeological evidence indicates that the cultivation of pulses in the Near East goes back to about 7000 BC.[2] The main type grown was *Lens culinaris* 'lentil', derived from the wild *Lens orientalis*.[3] Other important pulses in the Near East at an early period were the broad bean (*Vicia faba*), the garden pea (*Pisum sativum*) and the chickpea (*Cicer arietinum*).

All four were also cultivated by the Romans, but only the broad bean and garden pea had sufficient tolerance of cold to be introduced successfully to early Irish gardens. It is significant that the ninth-century Irish glossator on Priscian clearly did not know the correct meaning of Latin *lens* 'lentil', and explained it inaccurately as 'a kind of corn'.[4]

Pis (Pea)

As we have seen in the previous chapter (p. 219), the author of the law-text *Bretha Déin Chécht* arranges seven cereals in order of value. The eighth place in his list is assigned to *pis*, which is the garden pea (*Pisum sativum*).[5] It is probable that peas were generally allowed to ripen like cereals, and then kiln-dried prior to storage.

Seib (Bean)

The broad bean (*Vicia faba*) is placed at the bottom of the *Bretha Déin Chécht* list. As in the case of the garden pea, it was probably more often kiln-dried and stored than eaten fresh.

[1] This family was formerly called *Leguminosae*: hence plants in this family are often referred to as 'legumes'. Pulses are the seeds of *Papilionaceae*.

[2] Zohary and Hopf, *Domestication of plants*, 228–9.

[3] Ibid. 88–9.

[4] *Thes.* ii 101.22.

[5] *LEIA* P-9 s.v. *pis*.

The history of the Irish words for 'bean' is particularly interesting, as it provides evidence as to how this crop developed in Ireland. No native word for 'bean' is attested in the language, but this does not necessarily prove that beans were unknown in Ireland before Christian times. Archaeologists have identified in pre-Roman sites in Britain and the continent a small primitive type of bean, which has been termed the 'Celtic bean'.[6] It is quite likely that it was also grown in Ireland at this period, though no traces have been found.

It seems that in Roman times the Celtic bean was superseded in Britain by a larger type, closer in size and shape to the modern broad bean.[7] The Latin word *faba* 'bean' was borrowed into the Brittonic language, ultimately giving Welsh *ffa*, Cornish *fāv*, Breton *fav*.[8] This bean was also introduced to Ireland, no doubt along with the many other horticultural and agricultural innovations associated with monasticism. In Ireland the word *faba* took the form *seib*, with the regular change of Latin initial *f* to Irish *s*.[9]

About the tenth century the Irish word *seib* was almost completely ousted by *pónair*, a borrowing from Old Norse *baunir* 'beans'.[10] This replacement was carried out through the whole Gaelic-speaking area, with the result that Modern Irish has *pónaire*, Scottish Gaelic has *pònair*, and Manx has *poanrey*. The explanation may be that the Norse colonists introduced an improved type of bean, or that beans were an especially important crop in their economy. The Norsemen controlled good agricultural land around the east-coast towns of Ireland for a number of centuries.[11] It is not surprising, therefore, that their farming activities have left traces in the Irish language. As well as *pónair*, we have noted at p. 238 that the Irish word *punnann* 'sheaf' is probably of Norse origin. Another Norse horticultural term, *garðr* 'garth, enclosure,

[6] *Faba celtica nana* (Bois, *Les plantes alimentaires*, 100); cf. Cunliffe, *Iron Age communities in Britain*, 373.

[7] The bean of Roman times is nonetheless small enough to be classed as *Vicia faba* var. *minor* rather than var. *major* (Zohary and Hopf, *Domestication of plants*, 106). Monk, 'Charred seed and plant remains', 98, refers to the *Vicia faba* found in a late thirteenth-century site at Kilferagh, Co. Kilkenny, as a small 'celtic' bean.

[8] *VKG* i 192.

[9] The *-a-* to *-e-* vowel-change probably reflects Brittonic influence: see Thurneysen, *Grammar*, 571 § 921.

[10] Marstrander, *Bidrag*, 59, 96, 117.

[11] Edwards, *Archaeology of early medieval Ireland*, 189; Bradley, 'The interpretation of Scandinavian settlement in Ireland', 53–65.

garden', was borrowed into Irish (*garrda*, later *garraidhe*) and Scottish Gaelic (*gàradh*).[12]

VEGETABLES AND HERBS

It is clear from the Old Irish law-texts that a prosperous farm often had an enclosed garden (*lubgort*) near the farmhouse.[13] In Old Irish we do not find a distinction between 'vegetable' and 'herb': the term *lub* (*luib*) includes plants eaten as part of the normal diet, as well as plants used for medicinal or flavouring purposes.

There is special emphasis in our sources on the growing of plants with medicinal properties. The medical law-text *Bretha Crólige* speaks of 'the great service given by garden plants in nursing', and mentions in particular *cainnenn* and *imus.*[14] The same text makes the claim that the primary purpose of gardens was the care of the sick.[15] Another medical law-text, *Bretha Déin Chécht*, refers to 'three foreign herbs' (*tri lubai gall*) used in the treatment of a king's facial injury: see p. 257 below.[16] According to a glossator of legal and other material, a physician's bag (*línchur*) contains various small compartments so that his medicinal herbs are not mixed up.[17]

It seems from our sources that plant-cultivation was associated particularly with the monasteries. The gardener (*lubgortóir*) is listed among the seven officers of the church,[18] and Adomnán refers to a holy man named Laisrán mocu Moie who seems to have been a gardener (*hortulanus*) at the monastery of Iona.[19] There is no mention of a gardener among the servants normally employed by a king or lord.

[12] Marstrander, *Bidrag*, 78, 94. Vocabulary connected with farming was also borrowed in the opposite direction. The Irish term for a summer milking-place *áirge* occurs in many place-names of Norse origin in northern England : see Reaney, *The origin of English place-names*, 186–7.

[13] See p. 368.

[14] *CIH* vi 2299.38–40 = Binchy, '*Bretha Crólige*', 38–40 § 49.

[15] *CIH* vi 2293.25–6 = Binchy, '*Bretha Crólige*', 22 § 27.

[16] *CIH* vi 2307.21 = Binchy, '*Bretha Déin Chécht*', 26 § 9.

[17] *CIH* ii 616.35–6; iii 1075.3–6.

[18] *CIH* vi 2213.32.

[19] A. O and M. O Anderson, *Adomnan's Life of Columba*, 242 (new pagination: 42) = ff. 23b–24a.

Cainnenn (Onion?)

By far the most prominent vegetable in our sources is *cainnenn* (also *cainenn*). I believe it to be onion (*Allium cepa*), but other scholars have sometimes identified it with garlic (*Allium sativum*) or with leek (*Allium ampeloprasum* var. *porrum*). The issue cannot be resolved by comparison with other Celtic languages, because the Welsh cognate *cennin* means 'leek', whereas the Breton cognate *kignen* means 'garlic'. It might therefore be argued that Irish *cainnenn* is a generic term which includes onion, leek and garlic. But the legal references in particular suggest that *cainnenn* had a more precise meaning. This is underlined by the frequent prefixing of the adjective *fír* 'true' to *cainnenn*.[20] Binchy suggests that this refers to the freshness of the vegetable,[21] but it seems more likely that the intention is to stress that it must be a cultivated onion of a recognized type and that no wild species of *Allium* (see p. 308) can be substituted. The word *cainnenn* dropped out of general use in the Early Modern Irish period, and was replaced by *uinneman* (*uinniún*), most probably a borrowing from Norman French *oignun* 'onion'.[22]

It is clear from references in the text on status, *Críth Gablach*, that *cainnenn* was a regular element in the early Irish diet.[23] *Bretha Crólige*, which deals with illegal injury, stresses a high-ranking invalid's entitlement to this vegetable, unless the physician forbids it on medical grounds.[24] The most detailed description of *cainnenn* is in another text on status, *Uraicecht Becc*, which specifies the food which a client must provide for the annual visit of his lord. The amount depends on the rank of the lord. The lowest-ranking lord is entitled to four loaves of bread for each member of his visiting party of four men. The bread must be accompanied by a relish (*annlann*) or condiment (*tarsunn*), which

[20] In law-texts, the compound *fírchainnenn* is attested at *CIH* ii 644.6–7 = *AL* v 40.11–12; *CIH* vi 2298.15 = Binchy, '*Bretha Crólige*', 36 § 45; *CIH* ii 563.16; iii 779.36 = *CG* 7.168. It also occurs In non-legal material, e.g. Jackson, *Aislinge* 34.1069 = Meyer, *Aislinge*, 89.11; Murphy, *Early Irish lyrics*, 30 § 10.

[21] *CG* (Glossary) 53 s.v. *fír-*.

[22] Risk, 'French loan-words in Irish, (2)', 91. Norman French *oignun* (Mod.Fr. *oignon*) goes back to Latin *unio* '(single) onion'. Vendryes (*LEIA* U-20) proposed that Irish *uinniún* came from Latin *unio* by way of Welsh *wynwyn*.

[23] *CIH* iii 778.13 = *CG* 3.74; *CIH* ii 563.4; iii 779.23 = *CG* 6.150; *CIH* ii 563.16; iii 779.36 = *CG* 7.168.

[24] *CIH* vi 2298.14–15 = Binchy, '*Bretha Crólige*', 36 § 45.

may consist of *cainnenn*, honey, fish, cheese or salted meat. The text states that there must be 'sixteen cloves of true *cainnenn* for each loaf, or four plants of true *cainnenn* for each loaf'.[25] This implies that a typical *cainnenn* was expected to have about four cloves growing from the base of each plant. I suggest, therefore, that *cainnenn* refers to a bunching or shallot type of onion, as illustrated below.

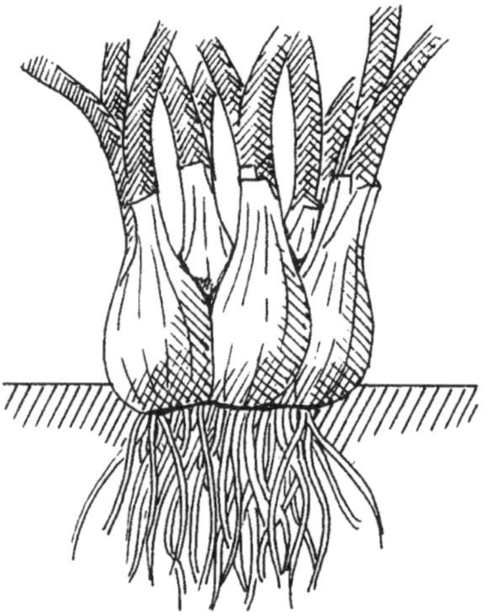

Fig. 11. A suggested illustration of the type of onion described in the eighth-century law-text Uraicecht Becc.

It might be argued that the reference to more than one 'clove' (*ingen*) in connection with this vegetable indicates garlic rather than onion.[26] However, there are various objections to this identification. In the law-text on base clientship, *Cáin Aicillne,* the food-rent which a client must give to his lord includes 'a handful of green *cainnenn* with their tops'.[27] It is further specified that each plant should be four fists in length, i.e. about sixteen inches. In the *Ancient laws of Ireland,* the translation given for *cainnenn* in this text is 'garlic'.[28] But garlic is not ready for lifting until the leaves wither, usually in July. Its cloves develop entirely underground, and so have no trace of green. Consequently, it is difficult to see how 'green *cainnenn*' (*glaschainnenn*) could apply to garlic. It suits the onion much

[25] *CIH* ii 644.6–7; v 1599.34–5; vi 2266.34–5 = *AL* v 40.11–12 *sē hingni dēc gacha bairgine nō cetre buindi fīrcaindinne cacha bairghine. Ingni* is nom. plur. of *ingen* 'nail, hoof, clove, bulb', and *buindi* is nom. plur. of *buinne* 'stalk, sprout, branch, etc.'.

[26] Compare Welsh *cennin ewinog* 'garlic', where *ewin* 'clove' is cognate with Irish *ingen.*

[27] *CIH* ii 479.23–4 = Thurneysen, 'Aus dem irischen Recht III', 371 *imglaici glascainne cona cennaib.* It seems unlikely that *cenn* refers here to the bulb of the onion. But cf. French *tête d'ail* 'bulb of garlic'.

[28] *AL* ii 255.11 = *CIH* ii 483.35. Thurneysen, 'Aus dem irischen Recht I', 357.13, translates *glaschainnenn* as 'Grün-Lauch'.

better, as the green leaves of the onion are utilised as well as the base. *Críth Gablach* also contains a reference to 'a handful of *cainnenn* with tops'.[29]

Another argument against the identification of *cainnenn* with garlic is the evidence that fairly large quantities were grown. Legal commentary expects a *bóaire* to grow six ridges of *cainnenn*,[30] and the comic tale *Aislinge Meic Con Glinne* refers to seven ridges of true *cainnenn*.[31] We do not know the length of the ridge (*indra, immaire*) which either author had in mind, but both references indicate a considerable crop. The same conclusion can be drawn from an Old Irish religious poem, probably composed in the eighth century, which refers to the planting of a small garden (*gortán*) with *cainnenn*.[32] Because of its remarkably powerful flavour, it seems unlikely that garlic would be cultivated on the scale indicated by these references. Furthermore, it is clear that the early Irish made much use of wild garlic (*crim*), as it is often mentioned in the sources (see p. 308). There would therefore be less need for a garden variety of garlic.

Finally, a Middle Irish glossator states that *cainnenn* brings tears from the eyes.[33] This is true of onion, but not of leek or garlic.

Imus (Celery?)

In the law-texts *imus* (*umus*) is always mentioned in conjunction with *cainnenn*. For example, *Bretha Crólige* includes *cainnenn* and *imus* among the three condiments which may be substituted for beer in an invalid's diet.[34] The same text states that illegally injured persons of all ranks – from high lord down to *fer midboth* – are entitled to *imus* because 'it prevents sickness and does not stir it up, [and] prevents thirst and does not infect wounds'.[35]

[29] *CIH* iii 778.13 = *CG* 3.74 *imglaice cainne co cennaibh.*

[30] *CIH* v 1611.42 = *AL* v 90.14 *sē hinnra caindi*; cf. *CIH* iii 1058.2. Lower down the social scale, the *ócaire* grows three ridges (*CIH* v 1611.12 = *AL* v 88.6; cf. *CIH* iii 1058.14). Widespread cultivation of *cainnenn* is also indicated by a legal glossator at *CIH* iii 808.25 who gives *buain na mell cainde asin talmain* 'taking bulbs of *cainnenn* out of the ground' as an example of the offence of *turorgain úire* 'digging up the soil'.

[31] Jackson, *Aislinge* 34.1069 = Meyer, *Aislinge*, 89.11.

[32] Carney, *The poems of Blathmac, etc.*, 104 § 46. He translates *cainnenn* as 'leek'.

[33] Meyer, *Hibernica minora*, 47.4.

[34] *CIH* vi 2299.39 = Binchy, *Bretha Crólige*, 38 § 49.

[35] *CIH* vi 2298.14–15 = Binchy, *Bretha Crólige*, 36 § 45.

It is clear that *imus* was widely grown. According to legal commentary on *Uraicecht Becc*, farmers of *ócaire* and *bóaire* rank are expected to grow one ridge of *imus*.[36] Binchy suggests that *imus* is to be identified with celery (*Apium graveolens*).[37] This umbellifer is native to Ireland and it is quite possible that it was cultivated in early Irish gardens.[38] It would not, however, have been nearly as tender as the modern variety of celery (*Apium graveolens* var. *dulce*) which was introduced to Britain from the continent in the seventeenth century.[39]

Another possibility is that *imus* refers to alexanders (*Smyrnium olusatrum*), a vigorous umbellifer which puts out shoots very early in the spring.[40] It was formerly a popular vegetable in Britain, but is now seldom eaten. The *Dictionary of the Irish language* also equates *imus* with parsley (*Petroselinum crispum*), but there is no firm evidence to support this identification.

Borrlus (Leek?)

In *Cáin Aicillne* this vegetable is included in the annual food-rent due to a lord. A client must give him two handfuls of *borrlus*, of which each plant must be four fists in length, i.e. sixteen inches.[41] This vegetable is also mentioned in legal commentary which discusses the proper food to be supplied by an employer to a builder and his assistants.[42]

Borrlus seems literally to mean 'the swelling plant',[43] which could apply to a wide variety of vegetables. In the translation of this text

[36] *CIH* v 1611.12 = *AL* v 88.7; *CIH* v 1611.42 = *AL* v 90.15.

[37] Binchy, '*Bretha Crólige*', 70.

[38] Traces of celery have been found in excavations of early Dublin (Mitchell, *Archaeology and environment in early Dublin*, 26).

[39] Harrison et al., *Oxford book of food plants*, 148; Bois, *Les plantes alimentaires*, 232–4.

[40] Bois, *Les plantes alimentaires*, 231; Mabey, *Food for free*, 110. This vegetable (called *hipposelinum* or *olusatrum*) is mentioned by Pliny, *Natural History* (ed. H. Rackham), 19.37.124; (ed. W. H. S. Jones), 20.46.117; cf. Columella, *De re rustica* (ed. Forster and Heffner), 11.3.36.

[41] *CIH* ii 479.24–5 = Thurneysen, 'Aus dem irischen Recht III', 371; *CIH* iv 1477.11 = O'Dav. 242 § 288; cf. *CIH* ii 483.35 = Thurneysen, 'Aus dem irischen Recht I', 356.10.

[42] *CIH* vi 2108.5.

[43] See *LEIA* B-73 s.v. *borr*. The later version of *Aislinge Meic Con Glinne* in H 3. 18 (Meyer, *Aislinge*, 123.32–3) refers to *secht n-imairib dēc do borraig firlosæ*, but the meaning is uncertain.

in the *Ancient laws of Ireland*, it is taken to be leek (*Allium ampeloprasum* var. *porrum*).[44] This seems a likely identification, as the leek is known to have been widely cultivated in Europe in the medieval period.[45] Its long-established importance in Ireland is indicated by the use of a general term for plant, *lus*, to refer specifically to the leek.[46] For example, in the fifteenth-century medical text *Rosa Anglica* the three main cultivated members of the genus *Allium* are given as *lus 7 uindemuin 7 gairleog* 'leek and onion and garlic'.[47] The same development took place in Manx, where 'leek' is one of the meanings assigned to *luss*.[48]

Braisech (Cabbage)

Wild cabbage (*Brassica oleracea*) is native to the coasts of southern and western Europe, including southern England and Wales.[49] It was already well established as a garden vegetable in Greek and Roman times.[50] It is not native to Ireland, and it is likely that it was introduced from Britain in the early Christian period, as the Irish word *braisech* is a borrowing from Latin *brassica*.[51] Another Irish word for cabbage, *cál*, is also of Latin origin, being a borrowing from *caulis* of the same meaning.[52]

Cabbage features particularly in monastic diets. For example, in *Ríagail na Céle nDé* it is stated that if cabbage is available, penitents are allowed to take some with their bread. Because cabbage is regarded merely as a relish (*tarsunn*), there is no diminution in the bread-ration.[53] In another ninth-century monastic rule, clerics are forbidden to eat cabbage which has been cooked on a Sunday.[54]

[44] *AL* ii 255.11. On the other hand, Thurneysen, 'Aus dem irischen Recht III', 372, took it to be garlic (Knoblauch).

[45] Bois, *Les plantes alimentaires*, 507–10. It was one of the most popular vegetables in the ancient Near East (Zohary and Hopf, *Domestication of plants*, 183).

[46] *DIL* s.v. 1 *lus* (b); Stokes, *Irish glosses*, 24 no. 810 hoc porrum. *lus.*

[47] Wulff, *Rosa Anglica*, 226.8–9; Carney, *Regimen na sláinte* ii 113–14.

[48] Cregeen, *A dictionary of the Manks language* s.v. *lhuss.*

[49] Bois, *Les plantes alimentaires*, 30–1.

[50] Pliny, *Natural History* (ed. W. H. S. Jones), 20.33.78–83; Zohary and Hopf, *Domestication of plants*, 186.

[51] *LEIA* B-77.

[52] The basic meaning of *caulis* is 'stalk, stem, tendril', but it has been applied specifically to the cabbage since Classical times.

[53] E. J. Gwynn, *The rule of Tallaght*, 64 § 4. Gwynn takes *Ríagail na Céle nDé* to go back to a ninth-century original.

[54] Gwynn and Purton, 'The monastery of Tallaght', 132.20 § 13.

Cabbage is also mentioned frequently in the early twelfth-century comic tale *Aislinge Meic Con Glinne.* In a fanciful genealogy of the abbot Manchín, one of his ancestors is named as Barr Braisce Bíthe, which means 'head of tender cabbage'.[55] In another passage, juicy cabbage is described as being particularly sought after by the priests.[56]

For a discussion on the cooking of cabbage, see p. 340.

Root-crops

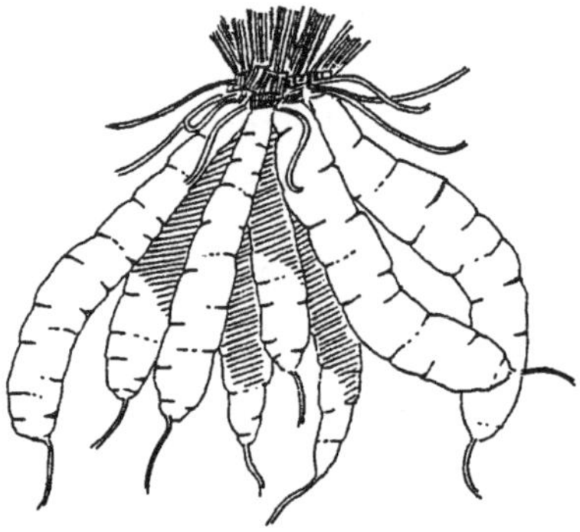

Fig. 12. A modern illustration of the roots of the skirret (*Sium sisarum*), which is perhaps to be identified with Old Irish *cerrbacán*.

There is archaeological evidence that the carrot (*Daucus carota*), the turnip (*Brassica rapa*) and the parsnip (*Pastinaca sativa*) were cultivated in Roman Britain.[57] The Irish written evidence with regard to root-crops is difficult to interpret. There are some general references to edible roots in our sources,[58] but the only specific term used is *cerrbacán,* generally translated 'carrot'. This vegetable was clearly well thought of, as it is associated in *Aislinge Meic Con Glinne* with the diet of a queen. Both editors of this text take it to refer to the carrot.[59] However, the etymology of *cerrbacán* is difficult to reconcile with this identification. Its basic meaning is 'the bent crooked one',[60] but this does not suit either the cultivated carrot or the wild carrot from which it derives. Another possibility is suggested by the meanings attested for the modern forms of the word. Modern Irish *cearrachán* may refer to skirret (*Sium sisarum*), an umbellifer with edible roots

[55]Jackson, *Aislinge,* 13.410 = Meyer, *Aislinge* 35.4.

[56]Jackson, *Aislinge,* 39.1199 = Meyer, *Aislinge* 99.27 *mian na sacart .i. braisech bélaide.*

[57]Applebaum, 'Roman Britain', 120, 212, 220.

[58]E.g. Atkinson, *Passions and homilies,* 55.519.

[59]Jackson, *Aislinge,* 38.1195 = Meyer, *Aislinge,* 99.20 *briscén mbanrígna .i. cerrbacán.* Meyer takes *briscén* to refer to a mash, cf. *brisc* 'brittle, fragile'. Jackson, on the other hand, takes it to refer to the wild plant tansy (*Chrysanthemum vulgare*): see his discussion in the notes.

[60]*LEIA* C-75 s.v. *cerr*; B-2 s.v. *bacc.*

growing at an angle from the base of the stem: see illustration on p. 256. The Scottish Gaelic form *cearacan* has the same meaning.[61] The history of this vegetable is obscure, but it is known to have been popular in Britain and the Continent from the late Middle Ages until the nineteenth century.[62] It is possible that it is to be identified with *siser*, a vegetable which Pliny records as being a favourite of the Emperor Tiberius.[63]

Foltchép (Chives)

This relative of the onion (*Allium schoenoprasum*) is commonly grown in modern herb-gardens. The new growth is regularly cut to provide a garnish for a wide variety of dishes. It seems to have been used in a similar way in early Christian Ireland. The eighth-century saga *Fled Bricrenn* contains a reference to the cutting of chives down to the ground with a sharp blade.[64]

The name *foltchép* is a compound of *folt* 'hair' and *cép* (from Latin *cepa* 'onion'), and refers to its bushy mode of growth. A ninth-century Life of Saint Patrick claims a miraculous origin for this plant.[65] While Patrick was in Óchtar Cuillend, the pregnant wife of Aillill mac Cathbath suddenly developed a craving for a plant which she had seen in a vision. She told Patrick that if she did not eat this plant, she or her unborn baby – or both of them – would die. Patrick asked her to describe the plant, and she said that it was like rushes (*lúachair*). Patrick then blessed some rushes, which immediately turned into chives. The woman ate and was cured, and afterwards gave birth to a son. Patrick then announced that any woman who ate of this plant would be cured.

'Three foreign herbs'

The eighth-century medical law-text *Bretha Déin Chécht* refers to three foreign herbs (*trí lubai gall*) which must be sought if a king is wounded in the face. If the person responsible for the wounding

[61] Dinneen, *Foclóir*; Dwelly *Faclair*. It must be admitted, however, that both Dinneen and Dwelly also record the translation 'wild carrot' for this word. Another term for skirret in Scottish Gaelic is *cromag*, lit. 'the bent one'. See also Stokes, '*Materia medica*', 232[114].

[62] Bois, *Les plantes alimentaires* 240–1; Sherman and Bowen, *The green gardening and cooking guide*, 83–4.

[63] Pliny, *Natural History* (ed. H. Rackham), 19.27.90; cf. Columella, *De re rustica* (ed. Forster and Heffner), 11.3.35.

[64] *LU* 261.8619–20 = Windisch, 'Das Fest des Bricriu', 277.2–3 § 46 . . . *amal bentair foltchíb fri lár talman co n-altain áith.*

[65] Mulchrone, *Bethu Phátraic*, 120.2345–121.2355.

fails to provide these herbs, he must pay an additional fine of one and a half ounces of silver.[66]

The identity of these herbs is uncertain. The author of the text lists them as *sraif*, *lungait*, and *arcetluim*. He claims that *sraif* is for healing the wound, *lungait* is for treating the colour of the wound, and *arcetluim* is for the skin. However, *arcetluim* is elsewhere explained as the mineral orpiment.[67] Furthermore, in his article 'Irish letter-names and their kennings', Damian McManus points out that *sraif* is well-attested with the meaning sulphur.[68] He draws attention to commentary on *Uraicecht Becc* in which the 'three foreign herbs' are treated as distinct from *sraif*, *lungait* and *arcetluim*.[69] It seems, therefore, that the author of *Bretha Déin Chécht* has confused herbal and mineral remedies. Or possibly the term *lub*, though normally meaning 'herb', may be used here in a wider sense to include minerals used for medicinal purposes.[70]

Other medicinal herbs

Apart from the three mysterious 'foreign herbs' discussed above, I have found no references in Old or Middle Irish sources to the cultivation of specific herbs for use in medicine. Similarly, there are many references to the use of herbs in the making of poisons and love-potions, but no evidence of their being specially cultivated for these purposes. Legal commentary describes the cutting of three types of herb, apparently as a defence against the 'evil eye'.[71] These are the royal herb (*ríglus*) for kings and those of royal rank, the bull-herb (*tarblus*) for lords, and the plebeian herb (*aithechlus*) for commoners. But I know of no clues as to their identity.

In the later medical material – mainly from the fourteenth to sixteenth centuries – there is a massive body of information on the use of plants (wild and cultivated) in medicine.[72]

[66] *CIH* vi 2307.21–4 = Binchy, '*Bretha Déin Chécht*', 26 § 9.

[67] Stokes, '*Materia medica*', 225.

[68] McManus, 'Irish letter-names', 159–60.

[69] *CIH* v 1610.9–10 = *AL* v 84.9–10 *dīri uigi circi, a lān do luibib gall ⁊ sraiff ⁊ luingit ⁊ airgetlaim* 'the payment for a hen's egg: its fill of foreign herbs and sulphur and [?] and orpiment'. Binchy (*CIH* v 1610[d]) suggests omitting the first ⁊ or emending it to *.i.*

[70] See *DIL* L 237.80–6.

[71] *CIH* i 145.5–9.

[72] E.g. Wulff, *Rosa Anglica*; Carney, *Regimen na sláinte* i–iii; O'Grady, *Catalogue*, 171–285.

Chenopodium

Pre-Norman sites in Ireland often contain abundant traces of fat-hen (*Chenopodium album*).[73] Seeds occur in faecal material, and so must have been of significance in the human diet.[74] I have not been able to identify any term for *Chenopodium* in Old Irish texts.

FRUIT

Apple

The wild apple (*Malus sylvestris*) is a common native tree, which bears heavy crops of small sour apples. It has no doubt been used as a food-source since man's first arrival in this country: see p. 306. In Old Irish the tree is termed *aball*,[75] and its fruit is an *uball*.[76]

The early history of the cultivation of the apple is obscure. Zohary and Hopf point out that strains of *Malus sylvestris* with unusually large or sweet fruit could have been taken into cultivation anywhere in the temperate areas of Europe and of western and central Asia.[77] It is probable that this was done many times in different areas, so it is pointless to look for a centre of diffusion. As cultivated varieties do not breed true from seed, the maintenance of a particularly palatable strain is dependent on grafting, a technique which was well known to the Greeks and Romans.[78]

It is clear from Old Irish texts that already in the early Christian period there was a generally recognised distinction between the sour wild apple and the sweeter cultivated types. For example, the ninth-century *Bethu Brigte* refers to an abundant crop of sweet apples (*ubla cumræ*) in a churchyard.[79] An eighth-century law-text

[73] Monk, 'Evidence from macroscopic remains', 34; Geraghty, *Viking Dublin: botanical evidence from Fishamble Street*, 29, 37.

[74] Mitchell, *Archaeology and environment in early Dublin*, 23. The seeds of various species of knotgrass, especially meld (*Polygonum aviculare*), also occur.

[75] *LEIA* A-6 s.v. *aball*. Another rare term for 'apple-tree' is *oblann*: see *LEIA* O-5 s.v.

[76] Another word for 'apple' is *forcán*, e.g. *CIH* iv 1506.20 = O'Dav. 374 § 1015; Meyer, *Hibernica minora*, 47.23.

[77] Zohary and Hopf, *Domestication of plants*, 166.

[78] Virgil, *Georgics* (ed. Fairclough), 2.69–82; Columella, *De re rustica* (ed. Forster and Heffner), 5.11.1–12; K. D. White, *Agricultural implements of the Roman world*, 248, 257–8. Medieval Welsh law assigns a separate value to a fruit-tree graft (*ymp*, *imp*), e.g. Wiliam, *Llyfr Iorwerth*, 90 § 138.11 = *ALIW*, Venodotian Code, III xx §§ 9–10.

[79] Ó hAodha, *Bethu Brigte*, 12.377 § 32. Cf. A. O and M. O Anderson, *Adomnan's Life of Columba*, 326 (new pagination: 96) = f. 54a.

refers to a wild apple (*fíaduball*): this implies the existence of cultivated apples.[80] A later legal passage states that the penalty (*díre*) for the destruction of an apple-tree belonging to a high-ranking person (*nemed*) is twenty *séts*. If it belongs to a commoner, however, the fine is only five *séts*. In either case, the culprit must also restore a tree of the same variety (*crann comchenóil*).[81] This passage apparently recognises the existence of different varieties of apple.[82]

It is impossible to say whether the cultivated sweet apple of early Christian Ireland derived solely from native wild stock or whether seeds or grafts were introduced from Britain or the Continent. In his *Glossary* Cormac mac Cuilennáin refers to the Italian town of Abella (now Avella) as a source of apple-seeds (*síl n-aball*).[83] But this entry is based on a passage in Isidore's *Etymologies*, and cannot be taken as evidence of a Mediterranean element in the early Irish cultivated apple.[84] Cormac's main concern is to make an etymological connection between the words *aball* and *Abella*.

Apples feature very prominently in the sagas, and there are many references to apples with various magic properties. Apples were clearly a valued foodstuff, as they provided vitamins during the winter when few other fruits or vegetables were available. For example, a monastic rule of ninth-century origin states that a penitent could eat apples with his ration of bread. If the apples were small, he could have up to five or six, but if they were big he could only have three or four.[85] The value attached to the apple is also demonstrated by a legal passage which fixes the very heavy fine of two ounces of silver for stealing even one apple from an apple-tree.[86]

The Irish term for orchard is *aballgort*, literally 'apple-field', which occurs as an element in a number of place-names.[87] A verse in the early twelfth-century *Aislinge Meic Con Glinne* gives a description of an orchard in flower: 'an array of truly-fragrant apple-trees, a

[80] *CIH* i 238.31 = *AL* v 474.1.

[81] *CIH* v 1537.21.

[82] I suggest that *comchen[e]óil* here means 'of the same variety' rather than 'of the same species'.

[83] Meyer, *Sanas Cormaic*, 111 § 1272 (cf. 10 § 94); *LL* iv 781.23338.

[84] Isidore, *Etymologiarum* (ed. Lindsay), 17.7.24.

[85] E. J. Gwynn, *The rule of Tallaght*, 2–4 § 2.

[86] *CIH* iii 1107.5.

[87] Hogan, *Onomasticon*, s.v. *Aballgort*. It is generally anglicized Oulart (Joyce, *Irish names of places* i 516–17).

wood with its pink-tipped bloom'.[88] *O'Davoren's Glossary* contains a possible reference to cider: one of the explanations of the obscure word *nenadmim* takes it to be 'the delicate juice of wild apples' (*sugh caol na n-uball fiad*).[89]

Plum

The blackthorn (*Prunus spinosa,* Old Irish *draigen*) is native to Ireland, and is placed in the third class of trees in the Old Irish tree-list.[90] It bears small sour sloes, which can only have been of use as a source of flavour, or as an emergency food in times of hunger. It has been suggested by some authorities that the garden plum (*Prunus domestica*) is a cross between the blackthorn and the wild cherry-plum (*Prunus cerasifera*) of southern Europe and neighbouring regions.[91] However, Zohary and Hopf have argued against this theory and hold that the blackthorn is unlikely to have made any contribution to the ancestry of the garden plum.[92] In their view, the garden plum was evolved from the cherry-plum by human selection of sweeter varieties, and their maintenance by grafting.[93] The grafting of plums was already practised in Roman times.

The Old Irish sources provide evidence of the cultivation of the garden plum. Thus a ninth-century legal commentary refers to *draigen cumra* 'sweet blackthorn', which I take to be some form of cultivated plum-tree.[94] From the same period, *Bethu Brigte* contains two references to *árni cumræ* 'sweet sloes' i.e. plums.[95] In Middle Irish legal commentary from about the twelfth century, a distinction is made between the plum (*áirne cumra*) and the sloe (*áirne fiadain*).[96] There is some archaeological evidence of cultivated plums in Dublin at this period. A structure (probably a fruit-loft) at High Street – which has been dated to about AD 1200 – was found to contain the well-preserved remains of various fruit,

[88] Jackson, *Aislinge*, 15.473–4 = Meyer, *Aislinge*, 39.1–2 *ecor d'ablaib fírchumra, fid cona bláth barrchorccra.*

[89] *CIH* iv 1518.17–18 = O'Dav. 426 § 1299. For a suggested emendation, see *DIL* s.v. *nenadmin.*

[90] See p. 380.

[91] E.g. Singer et al., *A history of technology* i 357.

[92] Zohary and Hopf, *Domestication of plants*, 169–71.

[93] Within the last couple of centuries, there has in addition been much hybridization with Japanese and American species of plum (ibid. 169).

[94] *CIH* ii 582.20 = *AL* iv 150.11. See discussion at Kelly, 'The Old Irish tree-list', 117.

[95] Ó hAodha, *Bethu Brigte*, 12 §§ 32, 33.

[96] *CIH* v 1874.15.

including the plum *Prunus domestica*.[97] An early eleventh-century pit at Fishamble Street contained stones of a bullace-type plum *Prunus insititia*, along with shell-fragments of walnuts (*Juglans regia*); the latter were certainly imported, and it is possible that the plums were likewise part of a consignment of exotic fruit.[98]

Other cultivated fruit

There is little evidence in Old Irish sources for the cultivation of fruit other than the apple and the plum. The wild pear (*Pyrus communis*) is not native to Ireland, and its cultivated forms do not seem to have been introduced until after the Norman invasion. The Irish word for 'pear', *péire*, is a borrowing from Norman French or Middle English.[99]

The climate of Britain and Ireland is generally not suitable for the outdoor cultivation of the grape-vine (*Vitis vinifera*). Even in the warmer parts of southern England the summer is often not hot enough to ripen grapes properly. However, because of the great prestige of wine and its importance in Christian ritual, it is probable that there has been some vine-growing in Britain since Roman times. An eleventh-century Anglo-Saxon text includes planting a vineyard among the tasks which should be undertaken in spring by an estate manager (*gerefa*).[100]

Modern experience in Ireland has shown that hardy varieties of grape can ripen outdoors in good summers, particularly in the south of the country.[101] It is therefore quite possible that monks successfully introduced vines to Ireland on a small scale along with the many other agricultural innovations of the early Christian period. However, the claim by Bede in the eighth century that Ireland had 'no lack of vines' must be greatly exaggerated.[102] Giraldus Cambrensis – who had firsthand acquaintance with the country in the late twelfth century – takes the contrary view, and

[97]Mitchell, *Archaeology and environment in early Dublin*, 27.
[98]Geraghty, *Viking Dublin: botanical evidence*, 30–1, 70.
[99]Risk, 'French loan-words in Irish (2)', 85.
[100]Liebermann, *Die Gesetze der Angelsachsen* i 454 [12] *wingeard settan*.
[101]Michael O'Callaghan makes wine commercially from his own vineyard at Longueville House Hotel, Mallow, Co. Cork.
[102]Colgrave and Mynors, *Bede's ecclesiastical history of the English people*, 20 (bk. 1, ch. 1).

remarks on their absence.[103] For an account of the importation of wine from France, see p. 319.

The Irish word for the cultivated cherry is *sirín*, a borrowing from Middle English *cherrie*.[104] This suggests that the cultivation of this fruit belongs mainly to the post-Norman era. However, stones which have been identified as belonging to the introduced *Prunus cerasus* have been found in an eleventh-century pit at Winetavern Street, Dublin, so it seems that some cultivation of the cherry preceded the Norman invasion.[105]

Dye-plants

Legal commentary on fosterage lays down strict regulations concerning the colours which may be worn by various ranks.[106] The sons of commoners may only wear clothing which is dun-coloured (*lachtnae*), yellow (*buide*), black (*dub*) or white (*find*). The sons of lords may wear clothing which is red (*derg*), grey (*glas*) or brown (*donn*), and the sons of kings may wear purple (*corcra*) or blue (*gorm*). Obviously such regulations could never have been universally obeyed to the letter. Nonetheless, they must reflect a general truth that the higher ranks wore more colourful clothing than those of lower rank. The association of purple with kingship is widely disseminated in Mediterranean civilizations, and the Roman emperor wore 'Tyrian purple' obtained from shellfish of the genera *Murex* and *Purpura*.

There is evidence that the early Irish also obtained purple from shellfish,[107] but most of their dyes were undoubtedly of vegetable origin. Many native Irish plants yield effective dyes. For example, bracken (*Pteridium aquilinum*) gives a yellow-green colour, lady's bedstraw (*Galium verum*) dyes grey, and juniper (*Juniperus communis*) dyes brown. One of the best-known native dye-plants is the lichen *Parmelia saxatilis*, which gives a very fine

[103] O'Meara, *Giraldus: Topography*, 35 § 2 = Dimock, *Giraldi Topographia*, 28.5.

[104] *DIL* s.vv. *sirín, seiríne*. See p. 308 below for a discussion of possible references to the wild cherry (*Prunus avium*).

[105] Mitchell, *Archaeology and environment in early Dublin*, 27.

[106] *CIH* v 1759.12–15 = *AL* ii 146.10–13.

[107] See Edwards, *The archaeology of early medieval Ireland*, 82–3; Mahon, 'Traditional dyestuffs in Ireland', 116-17; McCormick, 'Stock-rearing in early Christian Ireland', 207–8.

brownish red colour.[108] Its Irish and Scottish Gaelic name *crotal* (earlier *crottball*) has passed into English as 'crottle'.

Other dye-plants are introduced species. For example, saffron (*Crocus sativus*) probably originated in Asia Minor, and has long been cultivated in many parts of Europe for use as a dye, cosmetic, spice and medicine.[109] The word *cróch* (from Latin *crocus*) is well-attested in Irish, and forms an adjective *cróchdae* 'saffron-coloured'. Though saffron gives an orange-yellow colour (as in the traditional Irish saffron kilt), *cróch* is sometimes used of reddish hues, extending even to blood-red.[110]

Another introduced plant whose uses include dyeing is the onion (*Allium cepa*), which also gives a yellow colour. In a passage in *Cáin Adomnáin*, the word *sep* seems to refer to a plant used in dyeing.[111] In his edition Meyer took it to be for *seib* 'bean', but there is no other evidence of the bean being used as a dye-plant. Possibly, therefore, *sep* is here a variant of *cép*, from Latin *cepa* 'onion', cf. *foltchép* 'chives' (p. 257 above).[112]

Two introduced plants were grown exclusively for their dyeing properties: woad and madder.

Glaisen (Woad)

Woad (*Isatis tinctoria*) is a biennial plant native to the Aegean area and south-west Asia, which yields a rich lasting blue dye.[113] It has long had a particular association with the Celtic peoples; for example, Pliny states that in Gaul there is a plant called *glastum*, with which the wives and daughters-in-law of the Britons stain their bodies.[114] The word *glastum* means 'woad', and is cognate with

[108] So does the closely related *Parmelia omphalodes*. It is probable that the dye-plant mentioned in the Old Irish Glosses on St. Mark (*Thes.* i 492.28–9) is a lichen.

[109] Zohary and Hopf, *Domestication of plants*, 189–90.

[110] See *DIL* s.vv. *cróch*, *cróchdae*; *LEIA* C-242 s.v. *cróch*. I do not understand the Old Irish gloss at *Thes.* ii 361.8 where Latin *croceus* 'saffron-coloured' is explained *.i. glas* no *tinctura*. The adj. *glas* does not seem to be used elsewhere to mean yellow or orange-yellow.

[111] Meyer, *Cáin Adamnáin*, 32 § 52.

[112] Latin *c-* is normally retained in Old Irish loanwords, e.g. *cell* (Latin *cella*), *cert* (Latin *certus*). However, Middle Irish *silic* is attested as a variant (or scribal error) of *cilic* 'haircloth' (Latin *cilicium*).

[113] Zohary and Hopf, *Domestication of plants*, 191–2.

[114] Pliny, *Natural History* (ed. W. H. S. Jones), 22.2.2.

Irish *glas* 'grey, green, blue', of which Old Irish *glaisen* 'woad' is a derivative.[115]

There are many references in Old Irish sources to woad, and it is clear that it was a widely-grown crop.[116] Its prestige is reflected in the eighth-century tale *Scéla Éogain 7 Cormaic*, where the queen of Tara is represented as owning a private woad-garden (*glaisengort*).[117] There is possible evidence of a later decline in its importance: a Latin Life of Saint Boecius (Buite) refers to woad as having been cultivated for dyeing by the people of long ago.[118] The author seems to be implying that it would be an unfamiliar plant to his readers. On the other hand, legal commentary from about the twelfth century treats woad-growing as an ordinary contemporary activity.

There is considerable variation in the methods used to dye with woad-leaves.[119] Some information on the processes employed by the early Irish can be obtained from the eighth-century law-text *Cáin Lánamna*, a section of which deals with the division of a married couple's property in the event of their divorce.[120] Further details are provided by its associated glosses and commentary.[121]

According to the text, the wife gets only a third of the woad which is in containers (*a cruib*).[122] However, she gets half of any woad which has been dried (*coitethe*). Commentary gives a more detailed account of the preparation of the dye, though there is some minor variation between the different versions. The first stage is the harvesting of the woad (*búain na glaisine*). If the marriage property is divided just after the woad-harvest, the wife gets a ninth of the crop.[123] Next comes the pulping (*minugud*) of the woad-leaves in

[115] Thurneysen, *Grammar*, 96 § 155(e).

[116] E.g. *CIH* ii 675.21 = Mac Niocaill, '*Tír cumaile*', 82.4; *CIH* ii 510.13 = *SEIL* 36 § 15; *CIH* ii 657.22.

[117] O Daly, *Cath Maige Mucrama*, 70 § 17; Ó Cathasaigh, *Heroic biography*, 122.92–3.

[118] Plummer, *Vitae* i 95 § xxvii *Quandam enim herbam ortensem antiqui habebant, nomine glassen, ex cuius succo tincturam pannorum suorum faciebant* 'For the ancients had a certain garden herb, called *glassen*, from whose juice they made a dye for their clothes'.

[119] Hurry, *The woad plant and its dye*, 22–50.

[120] *CIH* ii 510.16–17 = *SEIL* 36 § 15.

[121] *CIH* ii 510.21 (read *ina dath urlam*) = *AL* ii 372.22–4; *CIH* ii 510.23–5 = *AL* ii 374.1–3; cf. *CIH* i 176.36–42 = *AL* ii 418.10–18; *CIH* v 1807.15–18; vi 2105.18–19.

[122] For examples of *cráo* (*cró*) in the meaning 'container', see Greene, '*Cró, crú* and similar words', 4–5.

[123] *CIH* v 1807.15–16 = *AL* ii 374.1–2; *CIH* vi 2105.18; cf. *CIH* i 176.36 = *AL* ii 418.10.

the first container (*cétchró*).[124] If the property is divided then, the wife gets a sixth of the pulp.[125] One version of the commentary refers to the leaves being made into cakes (*ar tuirtínib do dénumh de*) at this stage.[126] A similar process is described in nineteenth-century accounts of English woad-production.[127]

The next stage is the first drying (*cétchotud*) after which the wife is entitled to a third share.[128] The woad is then wetted and fermented in a second container (*cró tánaise*).[129] In nineteenth-century England this process, called 'couching', is said to have taken about nine weeks.[130] The final stage is the second drying, after which the dyestuff is ready for use. The commentary repeats the text's assertion that the wife is entitled to half the dried woad.[131] The gradual increase in the proportion due to the wife indicates that she is responsible for all stages in the preparation of the woad after harvesting. The actual dyeing process is explicitly stated to be a woman's task (p. 449), and involves steeping the cloth for a short while in a tub containing the dyestuff.[132]

Though the Irish word for woad, *glaisen*, is a derivative of the adjective *glas* 'grey, green, blue', our sources generally describe the colour which is obtained as *gorm*, which refers to a more limited range of the spectrum, approximating to English 'blue'. For example, in a Middle Irish Life of Saint Cíarán it is claimed of woad blessed by the saint: 'woad as good as that was never made before or after, for even if all the clothing of the Cenél Fiachrach were put into its dyestuff (*íarcain*), it would make it blue (*gorm*)'.[133] In the list quoted above of colours appropriate to boys of various ranks, it is stated that the clothing of the sons of kings should be purple (*corcra*) or blue (*gorm*). This blue is presumably the colour obtained

[124] *CIH* v 1807.16–17 = *AL* ii 374.2–3.
[125] *CIH* ii 510.24 = *AL* ii 374.2.
[126] *CIH* i 176.37 = *AL* ii 418.11.
[127] Hurry, *The woad plant and its dye*, 24.
[128] *CIH* ii 510.24–5 = *AL* ii 374.3.
[129] *CIH* ii 510.21 = *AL* ii 372.24; *CIH* i 176.38 = *AL* ii 418.13; cf. *CIH* v 1807.17.
[130] Hurry, *The woad plant and its dye*, 26.
[131] *CIH* ii 510.25 = *AL* ii 374.3; *CIH* v 1807.17–18.
[132] The term *íarcain*, of unknown etymology, seems to be used of the actual dyeing solution. Cf. *CIH* iv 1518.13 = O'Dav. 425 § 1295 *fasta iarcæ in datha .i. corcair nó glaisin* 'the fastening of the solution of the dye i.e. purple or woad'.
[133] Stokes, *Lismore Lives*, 121.4079–81.

from woad.[134] The adjective *glas* is, however, used to describe the colour of woad tattooed into the human skin. A gloss on the eighth-century *Caldron of poesy* refers to the poet Amargen's blue tattooed shank.[135]

One of the remarkable properties of woad is that the pulped leaves can be re-used to dye cloth a strong pink colour after the blue dye-stuff has been squeezed out.[136] This could explain the glossing of Latin *sandyx* 'vermilion' by *glasen* in the ninth-century Saint Gall glosses on Priscian.[137] The *Dictionary of the Irish language* may therefore not be correct in taking this to be an extension of the meaning of *glasen* to 'dye (in general)':[138] it is possible that the glossator knew of woad's capacity to dye pink.[139]

Roid (Madder)

There are some references to the cultivation of *roid*, which I believe to be the dye-plant madder (*Rubia tinctorum*), native to south-west Asia.[140] The law-text on land-types states that land which is to be categorised as 'best arable' must be suitable for flax, woad and madder (*roid*), as well as for corn, vegetables and grazing.[141] Similarly, legal commentary speaks of damage by livestock to madder and onion (*loit roide 7 cainde*),[142] and a legal gloss gives madder, onion and apple-trees as examples of cultivated plants.[143]

From these and other references it seems that this plant must have been fairly widely grown in Ireland.[144] So far, however, the only

[134]A strong blue is also obtained from the indigo plant (*Indigofera tinctoria*), which originated in India. However, large-scale importation of indigo to Britain and Ireland did not start until the nineteenth century.

[135]L. Breatnach, 'Caldron of poesy', 62 § 1[8] . . . *ica tá in colpa glas iarna crechadh*. See Meyer, 'Tätowierung bei den Iren'.

[136]I have tried this, and it actually works.

[137]*Thes.* ii 125.22.

[138]*DIL* s.v. *glasen* (b).

[139]The Old Cornish cognate *glesin* (equated with Old English *wād* 'woad') is similarly glossed by *sandyx* (*sandix*) (Graves, *The Old Cornish Vocabulary*, 283 § 654).

[140]Zohary and Hopf, *Domestication of plants*, 192–3.

[141]*CIH* ii 675.21 = Mac Niocaill, '*Tír cumaile*', 82.4. He suggests emending *roid* of the MS to *rú*, but the form *roid* is well-attested in a number of texts: see p. 268 below.

[142]*CIH* iii 802.21; cf. *CIH* i 73.22 = *AL* iv 116.10.

[143]*CIH* ii 424.6 *do roidh ł do choindind ł do ablaibh*. There is a different version of this gloss at *CIH* i 245.35 = *AL* v 500.24; cf. *CIH* v 1876.6 = *AL* v 500.29–30.

[144]E.g. Meyer, *Cáin Adamnáin*, 32 § 52 *roid* (*roig* MS) *7 glaisine 7 sep* 'madder and woad and onion(?)'; *CIH* iv 1501.7 = O'Dav. 350 § 909 *roidh 7 cainnenn 7 arbhar*

early Christian site which has produced traces of madder is Boho, Co. Fermanagh.[145] The cultivation of madder is well-documented from Britain; for example, an eleventh-century Anglo-Saxon text advises the estate manager (*gerefa*) to plant madder, and to sow flax-seed (linseed) and woad-seed.[146]

A brilliant red dye is obtained from the roots of madder. Legal commentary on the division of property after a divorce provides some information on the preparation of this dye.[147] The divorced wife is entitled to a ninth of the madder after it has been harvested (*búain*). When it has been gathered into sheaves (*trillsín*), she is entitled to a sixth, and when it has been gathered into bundles (*scriplín*) she is entitled to a third. She is due half when it has been ground into a powder (*min*) or has been made up as a dyestuff ready for use (*dath urlamh*). An obscure quotation in *O'Davoren's Glossary* seems to distinguish between a reddish brown dye (*donn-roid*) which increases the value of cloth by a quarter, and a stronger red (*coic*) which increases it by a third.[148]

Thurneysen took the original nominative singular of this plant-name to be *rú*, and is followed by the *Dictionary of the Irish language*.[149] This form occurs in a legal gloss,[150] but *roid* is very much better attested. Another Old Irish word which must be compared is *rúam* which seems to have the meaning 'red dye' in *Cáin Adomnáin*.[151] In *Cormac's Glossary* the same word is used of a plant, perhaps madder, which provides a red cosmetic for the face.[152] This word, like *rondid* 'reddens, dyes red', is clearly

'madder and and onion and corn'; *CIH* iv 1434.2 *lōg scribuil do corcair ł do roid* 'a scruple's worth of purple [dye] or madder'.

[145] Edwards, *The archaeology of early medieval Ireland*, 83. Note that a will of the year 1502 (Curtis, *Ormond deeds* iii 301 § 305) distinguishes between old madder (*antiquus sandyx*) and new madder (*novus sandyx*). The latter may have been a superior new strain.

[146] Liebermann, *Die Gesetze der Angelsachsen* i 454 [12] *mederan settan, linsed sawan, wadsæd eac swa.*

[147] *CIH* i 177.12–15 = *AL* ii 420.3–7. See discussion by Thurneysen at *SEIL* 38.

[148] *CIH* iv 1485.42 = O'Dav. 283 § 533. See *LEIA* C-142 s.v. 3 *coic.*

[149] Thurneysen, *Irisches Recht*, 75; *SEIL* 38.

[150] *CIH* ii 428.26 = *AL* v 362.12. In another gloss (*CIH* ii 390.37 = *AL* i 188.z), the glossator seems to have confused the dye-plant with the word for iron ore (Old Irish *rud*): see *SEIL* 38[2].

[151] Meyer, *Cáin Adamnáin*, 32 § 52 *mad ruam in bruit* 'if it be the red dye for a cloak(?)'.

[152] Meyer, *Sanas Cormaic*, 98 § 1094 *Ruain* (*ruam* v.l.) *.i. lus dobeir cucht* [*in faciem*] *commbi derg* 'i.e. a plant which gives a red appearance to the face'.

cognate with Old Irish *rúad* 'red'.[153] Vernam Hull suggests that the *rodgabáil* recorded in an eighth-century text as being imposed on the Aran Islands was a madder-tax.[154] This is uncertain, but it is noteworthy that a twelfth-century text on the tributes due to the kings of Ireland includes wool, thread, and various dyeing agents – possibly including madder.[155]

FLAX

The archaeological evidence indicates that cultivated flax (*Linum usitatissimum*) was developed from a wild flax, *Linum bienne*, in the Near East before 6000 BC.[156] It was grown both for its fibre, and for the oil extracted from its seeds (linseed oil). This oil can be used for cooking, lighting, and other purposes. The Old Irish term for flax is *lín* (from Latin *linum*), and it is likely that the plant was introduced to Ireland from Roman Britain. It is clear from our sources that wool was by far the most important fibre of the early Irish. Nonetheless, flax is also mentioned quite frequently, and was used particularly to make the linen tunic (*léine*) worn under the cloak (*brat*). Flax is well represented in archaeobotanical samples from the early Christian period.[157]

The law-text *Cáin Lánamna* and its associated glosses and commentary describe the methods of dividing property, including flax, when a married couple are divorced.[158] The text itself contains only a single reference: this is that a sixth of the flax which is in sheaves (*a scúapaib*) goes to the wife. However, the commentary provides much more information on the harvesting and processing of flax. If the divorce takes place while the flax is still growing, we are told that the wife is entitled to a cup of linseed (*escrae ruis*).[159] If the stalks of flax have been pulled and bound in sheaves, the wife

[153]See *LEIA* s.vv. *rúad, rúam, rúan, rondid.*

[154]Hull, 'Conall Corc and the Corco Luigde', 900; Meyer, 'Conall Corc and the Corco Luigde', 60.24.

[155]Dillon, *Lebor na Cert*, 110.1623; 114.1683. He suggests that *blaán* may be madder.

[156]Zohary and Hopf, *Domestication of plants*, 119–26.

[157]Monk, 'The archaeobotanical evidence for field crop plants', 320.

[158]*CIH* ii 510.16 = *SEIL* 36 § 15; *CIH* ii 509.27–9 = *AL* ii 368.26–30; *CIH* ii 510.20–21 = *AL* ii 372.21–2; *CIH* ii 510.25–7; v 1807.18–19; vi 2105.20–22 = *AL* ii 374.4–6; *CIH* i 177.1–11 = *AL* ii 418.19–420.2.

[159]The word *ros* (gen. sing. *ruis*) is used of small seeds, particularly linseed.

is entitled to a ninth share.[160] When the sheaves are dry,[161] the next stage is for the seeds to be dislodged by beating. This is presumably the *túargain* 'beating' referred to in the commentary: it is evidently a woman's job, as the twelfth-century tale *Buile Ṡuibne* speaks of 'women beating flax' (*mná ag túargain lín*).[162] If the couple are divorced after the flax has been beaten, the wife is entitled to a sixth share.[163]

The legal sources make no mention of the next stage in the preparation of flax: its immersion or 'retting' in water for up to twenty days. This is for the purpose of loosening the fibres from the woody core of the stem.[164] The next stage mentioned in the commentary is the bringing of the flax to the *clár*, which is no doubt to be identified with the scutching-board.[165] There the fibres are detached from the cores in readiness for twisting into yarn. If the divorce takes place when the flax has been scutched, half of it goes to the wife. She is likewise entitled to half of what has been woven into cloth (*étach*) or made up into a finished article (*lámthorud*).[166]

Ornamental flowers

The dividing line between a purely functional herb-garden and an ornamental flower-garden is very hard to draw, and it would be wrong to state dogmatically that only the former was to be found in early Christian Ireland. Nonetheless, it must be admitted that the tradition of cultivating flowers for ornament, though of high

[160] Thurneysen points out at *SEIL* 37 that one would expect the proportion at this stage to be a sixth, as in the text.

[161] The Life of Saint Cíarán of Saigir refers to the drying of flax (*lín*) on the inside wall of a house (Plummer, *Bethada* i 120 § xxii, trans. ii 116). This seems likely to refer to the drying of newly pulled flax, rather than the drying of 'retted' flax in preparation for scutching.

[162] O'Keeffe, *Buile Ṡuibne*, 17.478.

[163] In the version at *CIH* ii 509.28 = *AL* ii 368.29 the proportion is a sixth. Instead of *túargain*, the otherwise unattested term *ailgubad* is used. Thurneysen suggests (*SEIL* 37) that the first element is *ail* 'stone'.

[164] In her edition of *TBC* Rec I (123.4099, trans. 236.15), Cecile O'Rahilly took *notninus amail negarforcor hi lunggu* to mean 'I shall drub you as flax-heads(?) are beaten in a pool'. But there is no evidence for this meaning of *forcor*, and flax is not in fact beaten in water. The corresponding phrase in *TBC* LL 91.3305 (trans. 227.39–40) is also obscure.

[165] *CIH* ii 510.27 *ō dochoi ō chlār*; v 1807.19 *ō docae for clār*; vi 2105.21–2 *ō docai fo clār.*

[166] *CIH* ii 510.16 = *SEIL* 36 § 15.

antiquity in the Middle East and Mediterranean areas,[167] seems to have taken a long time to become established in northern Europe and the British Isles. I know of no literary or archaeological evidence that early Irish houses, even those of kings or nobles, had flower-gardens or flower-beds.

The cultivated rose would of course have been well known to Irish monks from their reading and in many cases from visits to the Continent. The word used in Irish for this plant is *ros* (later *rós*),[168] a borrowing from Latin *rosa.* From his gloss on Latin *rosarium,* it is clear that the ninth-century author of the Old Irish glosses on Priscian was aware of the existence of red and white roses.[169] This is not surprising: the Irish foundation at Saint Gall in Switzerland is known to have had a particularly fine garden.

The Romans also cultivated various types of lily (*lilium*), of which there is much mention in Latin literature. This plant-name was borrowed into Irish as *lile.*

[167]The hanging gardens of Babylon were one of the seven wonders of the ancient world.

[168]The *-o-* seems to be short in Old Irish *rostan* and *roschaill,* both of which are used to gloss Latin *rosetum* 'rose-garden' in the St. Gall Glosses (*Thes.* ii 104.22; 107.21). See discussion at *VKG* i 195 and *LEIA* R-44 s.v. 3 *ros.*

[169]*Thes.* ii 89.19–20.

9
Hunting and gathering

Humans first came to Ireland about 7000 BC and lived entirely by hunting, fishing and gathering.[1] This way of life continued until about 4000 BC when Neolithic colonists brought domestic animals and crops.[2] Thereafter, hunting, fishing and gathering supplied a decreasing proportion of the food eaten, though obviously there must have been fluctuation in the relative importance of wild and farm-produced food. Crop-failure in a particular year would have forced farmers to look to wild plants for sustenance, and a cattle-plague would have acted as a spur to the hunting of wild game.

It is clear from the seventh- and eighth-century Old Irish law-texts that the basic food consumed at this period came from cattle and cereals. Nonetheless, there is some mention of deer-hunting, fishing, and the collecting of nuts, wild fruit, wild herbs, and seaweed.

Hunting of deer

The red deer (*Cervus elaphus*) is our only native deer,[3] and is frequently mentioned in the sources. In Old Irish, the same terminology is often applied to deer and cattle: they were clearly viewed as similar animals. The commonest word for deer is *ag allaid*, which literally means 'wild bovine'. *Dam allaid* ('wild ox') is regularly used to mean 'stag', and *lóeg allaid* ('wild calf') is well attested as an equivalent of *oisín* (*oisén*) 'fawn'. Another general word for deer, *os*, is cognate with the words for 'ox' in many Indo-European languages.[4] *Os* occasionally retains the meaning 'bovine' in Irish,[5] so a deer is often referred to as *os allaid* 'wild deer' to make the distinction plain. This perceived close

[1] Mitchell, *The Irish landscape*, 100; *Reading the Irish landscape*, 78.

[2] Mitchell, *The Irish landscape*, 114; *Reading the Irish landscape*, 97.

[3] The Irish giant deer (*Megaloceros giganteus*) became extinct *c.*8000 BC (Mitchell, *The Irish landscape*, 74). The roe deer (*Capreolus capreolus*, Welsh *iwrch*) is native to Britain but not to Ireland. The fallow deer (*Dama dama*) is native to the Near East, but has been naturalized in Britain since the twelfth century and in Ireland since the thirteenth century (Rackham, *The history of the countryside*, 49–50).

[4] *LEIA* O-34 s.v. *oss*.

[5] See *DIL* s.v. 1 *os*(*s*).

relationship between deer and cattle is also reflected in saints' Lives. On a number of occasions, hinds miraculously supply milk in place of a cow,[6] and a stag may likewise miraculously take the place of an ox at the plough.[7] In the genealogies it is claimed of Nia Segamain, king of the Éoganacht of Caisel, that he milked 'both types of cattle' (*díabulbúar*) i.e. cows and hinds.[8]

In Irish archaeological remains of the early Christian period, there is generally only a small percentage of deer-bones.[9] For example, in the excavations of crannóg no. 1 at Ballinderry, Co. Westmeath, deer-bones comprised only about one per cent of the total.[10] Similarly, at the crannóg at Moynagh, Co. Meath, there were only eight deer-bones compared with 4,737 cattle-bones.[11] In the written sources of the period there is a general association of deer-hunting (*sedguinecht*)[12] with royalty and aristocracy. For such persons deer-hunting served more as a pastime and as military training than as a means to secure food for the pot. In fact, in the many descriptions of feasts in the Old Irish sagas, there is remarkably little reference to venison:[13] pork and beef were much preferred.

Many authors have discussed the fundamental incompatibility between crop-cultivation and hunting.[14] From the cultivator's point of view, game such as deer or wild pigs are a menace, and must be discouraged. Medieval English kings got over this problem by setting aside large expanses of land as 'deer-forest', generally away from cultivated areas. So far as I have been able to discover, there is no mention of royal deer-forests or

[6]Plummer, *Vitae* i 67 § iv; 99 § iv; 163 § xxx; 251 § xxxi; Plummer, *Bethada* i 129 § xiv (28) (trans. ii 125).

[7]E.g. Stokes, *Félire*, 72.28; Plummer, *Vitae* i 179 § xxv.

[8]*CGH* 362.53–5 = *LL* vi 1378.41295 *lasa mbligtis diabulbuar .i. baī ⁊ elti* 'by whom double-cattle were milked i.e. cows and hinds'. This power came from his fairy mother Flidais, and was shared by all his subjects during his reign (Stokes, *Cóir Anmann*, 294 § 26).

[9]Edwards, *The archaeology of early medieval Ireland*, 64.

[10]Hencken, 'Ballinderry crannóg no. 1', 230, 233. A higher percentage is found in crannóg no. 2 (Hencken, 'Ballinderry crannóg no. 2', 71).

[11]McCormick, 'Stock-rearing in early Christian Ireland' 5, table 1.1.

[12]In Old Irish a deer-hunter is *sedguine*, a compound of *sed* 'deer' (cognate with Welsh *hydd* of the same meaning) and the agent noun *-guine* 'killer'. *Sedguinecht* is the abstract noun.

[13]An otherwise unattested word, *sidhin*, is defined in a gloss as *ossfeoil* 'deer-meat' (*CIH* ii 619.37).

[14]E.g. Clutton-Brock, *Domesticated mammals*, 189.

hunting preserves in the Old Irish texts. However, there is good evidence that a typical Irish king of our period had at least one professional hunter in his employment. An Old Irish text on the personnel in a king's household refers to hunters (*selcithi*) and trappers (*cuthguiri*).[15] A genealogical text explains how the royal swineherd (*rígmuccid*) of king Conn Cétchathach also acted as his deer-hunter (*fer selgga oss n-alta*).[16] Hounds were looked after by a kennel-lad, called in legal commentary *gilla con.*[17] In the Gaelic Ireland of post-Norman times, a lord was entitled to exact a 'black rent' (*dubhdholadh*, anglicized *dowgollo*) from his subjects which supported his huntsmen and also provided food for his hounds.[18]

In medieval England and Scotland royal power was able to conserve deer by enforcing a closed season for hunting.[19] Legislation also protected hinds and fawns, and forbade any deer-hunting in snowy weather. There are no records of such conservation measures in Irish legal material. There is however evidence that deer – and perhaps other wild animals – could be privately owned.[20] Such an animal was given a special mark (*fíad co comartha*). If somebody killed it within its owner's infield (*faithche*), he was required to pay half the penalty-fine (*díre*) of a heifer. One quarter was due it it was killed in the outfield, and one eighth if it was killed on mountain or undivided land.[21]

Chasing with hounds

Of the various techniques used to catch deer, early Irish literature devotes most attention to chasing with hounds.[22] First, the hounds must locate their quarry by sight or scent, and then pursue it until it is cornered or exhausted. The huntsmen then dispatch it with spears[23] or other weapons. In medieval forms of hunting in England, Scotland and France, much importance was

[15] O Daly, '*Lānellach tigi rīch*', 83 § 16.

[16] *LL* iii 759.35 = Stokes, 'The prose tales in the Rennes Dindṡenchas, (2)', 55 § 103.

[17] *CIH* v 1607.25 = *AL* v 72.y.

[18] Carew MS 625 ff. 28–41. See Simms, *From kings to warlords*, 144–5; *DIL* D ii, 332.79–333.9.

[19] Gilbert, *Hunting*, 67–8.

[20] See also the discussion on pet animals taken from the wild at p. 130.

[21] *CIH* iv 1152.6–15; v 1642.21–30 = O'Grady, *Catalogue*, 80–81.

[22] I illustrate on p. 276 a hound catching a deer from a ninth-century high cross at Bealin, Co. Westmeath (Harbison, *High Crosses*, i 27 (discussion); i 377–9 (dating); ii fig. 73 (photograph).

[23] E.g. Plummer, *Bethada* i 38 § xxiii (70), trans. ii 37.

attached to the cureé, the ceremonial cutting up of the deer.[24] Tradition prescribed that different parts of the deer were given to the owner of the first dog, second dog, etc.[25] Our sources indicate the existence of similar conventions in Ireland. An Old Irish law-text states that the first judgement ever given in this country related to the division of a deer's carcase.[26] The background to this judgement is purely mythological, but the details probably bear some relation to actual practice in seventh- or eighth-century Ireland. According to the preamble, the judgement was delivered by the legendary judge Amairgein to resolve a dispute between the invaders Éber and Éremón, who were brothers. Éber and his followers had gone hunting (*selg*) in the mountains, and had killed twelve deer. Éremón and his followers had in the meantime prepared a habitation. Éber's followers argued that they were not obliged to share any of the venison with the others. The matter was then submitted to Amargein.

His judgement as preserved in the law-text[27] (and also in a later poem)[28] is difficult to translate, and does not seem to relate closely to the dispute described in the preamble. It deals rather with the division of the deer's carcase between the various persons involved in a hunt, as well as the owner of the land where the killing takes place. The first person who wounds the deer is entitled to the *classach*, which presumably refers to some part of its body.[29] The person who flays the deer gets its shoulder (*lethe*), and the owner of the hounds gets the haunch (*cés*).[30] Another person – perhaps he who

[24] Gilbert, *Hunting*, 67.

[25] Compare the regulations in medieval Welsh law, e.g. Wiliam, *Llyfr Iorwerth*, 89–90 §§ 136–137 = *ALIW*, Venedotian Code, III xviii §§ 1–7.

[26] *CIH* vi 2127.6–18.

[27] *CIH* vi 2127.16–18. There is also a shorter version in the Bodleian MS Rawlinson B 512 f. 97^{b}1 lines 30–5.

[28] *LL* i 55.1755–56.1778.

[29] The poem (*LL* i 55.1763–4) has the spelling *clossach*, with a marginal gloss stating that '*sét* (= *sed*) and *clossach* are two words for deer'. This poem says that the *clossach* is due to the first man or hound who breaks the deer's skin. It is difficult to be sure what *classach* (*clossach*) means in this context. In Modern Scottish Gaelic *closach* refers to the whole carcase of an animal (e.g. Dwelly, *Faclair* s.v. *closach*), but it is quite clear in the law-text that the first person to wound the deer does **not** get the whole carcase. Perhaps it means that he has responsibility for dividing up the carcase among the others, and is entitled to the remainder. But this seems to leave him with rather a raw deal.

[30] In Scotland, this was the most prized part (Gilbert, *Hunting*, 67).

actually kills the deer – gets the neck (*muinél*),[31] and the hounds themselves get the legs (*cossa*). The last man on the scene gets the intestines (*inathar*) and the rest of the hunting-party get the liver (*áe*).[32] Finally, the landowner gets the belly (*tarr*).

It is illegal to prevent another's hound from hunting.[33]

Fig. 13. This illustration of a hound catching a deer is from a ninth-century high cross at Bealin, Co. Westmeath.

Deer-ambush

A different approach is for the hunters to lie in wait for their quarry. They may conceal themselves near a watering-place or by a route known to be frequented by deer. If large numbers of men are at the hunters' disposal, the deer may be driven towards the site of the ambush. Kings and nobles were thus able to kill deer with little risk or effort. In his *Hunting and hunting reserves in medieval Scotland*, John Gilbert provides a number of examples of this type of hunt being set up for Scottish kings.[34] He also points out that the descriptions in sixteenth-century Gaelic poetry of hunting by Finn and his Fiana fall into this category.[35] Presumably the poets are attributing contemporary practices to this legendary hunter and warrior.

[31]The law-text has *fir iarn muinel*, with word-division between *iarn* and *muinél*. Perhaps read *fiur iairn muinél* 'to the man of the iron [i.e. the weapon which kills the animal] is the neck' (the corresponding poem has *d'iurn munel* 'the neck is to the iron'). Liam Breatnach points out that one could also read gen. sing. *fir* with the MS.

[32]*CIH* vi 2127.17 *æi la fiallach nurgadach*. I do not understand (*n*)*urgadach*.

[33]*CIH* i 335.1; iv 1178.1, 1393.30 *A coscc do fiad* 'preventing it (the hound) from wild game'.

[34]Gilbert, *Hunting*, 54–5.

[35]Ibid. 55.

The Old Irish term for such an ambush is *erelc*, which I take to be from *air-* 'before, in front' + *selc* 'hunting'.[36] This word occurs in a metathesized form *elerc* in a twelfth-century Gaelic record of land-grants to the monastery of Deer in Aberdeenshire.[37] The word passed into Scots as *elrick*, which was used at least until the eighteenth century of the spot where the hunters await the driven deer.[38] Another hunting term of Gaelic origin which passed into Scots is *tinchell*, used of the beaters who drove the deer towards the place of ambush.[39] This is from Gaelic *timchioll*, Old Irish *timchell*, the basic meaning of which is 'going around, surrounding'.[40] This reflects the beaters' task of outflanking the deer and driving them to the *elrick*.

The use of the Gaelic terms *elrick* and *tinchell* in medieval Scots documents shows the extent to which Scottish hunting practice drew on Gaelic as well as Anglo-Norman tradition.[41]

Driving to a barrier

A refinement of the previous method is for a barrier of some type to be erected to trap or confine the driven deer. They can then be more easily killed by the waiting hunters. In *O'Davoren's Glossary*, there is a reference to a 'deer-fence' (*osairbe*), which may be a barrier of this type.[42] The Old Irish tale *Scéla Cano meic Gartnáin* refers to the catching of deer in nets, but does not explain the hunting techniques involved.[43]

In medieval Scottish tradition, it was held that the Picts formerly hunted by driving deer into nets and killing them there. This method was regarded as ignoble by the Scots.[44]

[36]In an Old Irish gloss at *Thes.* i 63.32, it is used of ambushes prepared by Saul against David.

[37]Jackson, *The Gaelic notes in the Book of Deer*, 31 (text II 12). See his discussion on p. 52.

[38]Gilbert, *Hunting*, 52. The word is not included in Craigie, et al., *A dictionary of the older Scottish tongue*. Elrick is quite common as a place-name in Scotland (Watson, *The history of the Celtic place-names of Scotland*, 137, 184, 489).

[39]Gilbert, *Hunting*, 52.

[40]*DIL* s.v. *timchell*.

[41]For further discussion, see Gilbert, *Hunting*, 6, 8, 52, 55, 226.

[42]*CIH* iv 1521.1–2 = O'Dav. 438 § 1366.

[43]Binchy, *Scéla Cano*, 1.9 *.l. lín fri haige altai* 'fifty nets for deer'.

[44]Gilbert, *Hunting*, 57.

Hidden spikes

Irish legal sources devote more attention to the trapping of deer on spikes or in pits than to other ways of killing deer. This is obviously because these methods, if carelessly used, constitute a general hazard to livestock and people. The fixed spike (*bir airndil*) seems usually to have been of sharpened wood,[45] and was set up singly or in groups[46] in the hope that a deer might become impaled. There is no evidence that deer were driven onto such spikes: the trapper rather used his knowledge of the deers' habits to set a spike where they would be likely to pass.

What is probably the introductory phrase of the lost law-text *Osbretha* 'deer-judgements' is followed by a long commentary which no doubt reflects some of the contents of the missing text.[47] It deals with the obligation of any trapper who sets a fixed spike or digs a deer-pit to give a public warning (*escaire*). According to a short commentary on warnings, the owners of the nine nearest landholdings must be informed.[48] The *Osbretha* commentary also refers to the trapper's duty to construct a fence (*ime*) around the trap to prevent accidents to people or livestock. But it is difficult to see how this could be done without reducing the effectiveness of the trap.

Another legal commentary states that if a person sets a spike on the edge of a path or track, and thereby causes the death or injury of a domestic animal belonging to somebody else, he must restore an equivalent animal if this is his first offence.[49] If it is his second offence, he must pay half the penalty-fine (*lethḟíach*) for the animal. For the third offence, he must pay the full penalty-fine (*lánḟíach*). It is implicit in this passage that if the spike has been set away from

[45] Adomnán refers to a wooden spike (*veru*) set overnight for this purpose (A. O. and M. O. Anderson, *Adomnan's Life of Columba*, 412 (new pagination: 148) = f. 84a). Triad 169 refers to a *sleg caille* 'spear of the wood', which may be a spear hidden to impale deer or other game – but the whole triad is difficult to understand (Meyer, *Triads*, 22 § 169).

[46] Early Norwegian law refers to a 'spear-hedge' (Larson, *The earliest Norwegian laws*, 396 § 9).

[47] *CIH* i 320.28–322.33 = *AL* iii 448.15–458.11. For a general discussion of the relationship between law-text and commentary, see p. 11 above.

[48] *CIH* i 285.14 = *AL* iii 272.8–9 *Eascaire in bera airndil dar nāe n-orba* 'notice of the fixed spike over nine holdings'.

[49] *CIH* iii 864.25–7.

a path or track the trapper is not held liable for any accidents: see discussion under 'deer-pits' below.

A quotation in *O'Davoren's Glossary* further stresses the trapper's responsibility to avoid endangering livestock through carelessness (*écell*).[50]

Deer-pits

The law-texts and commentaries contain many references to the deer-pit (*cuithech*),[51] and the person who digs deer-pits is known by the derivative term *cuthchaire*.[52] Trapping for meat and hides may be his permanent occupation: it is clear from two texts on royal households that a king was expected to have a number of professional trappers in his employment.[53]

According to the law-text on accidents, *Bretha Étgid*, a trapper who digs a deer-pit in a mountain or undivided land is not liable if a person or domestic animal falls into it. But the accompanying commentary specifies that he must give a warning to the king and the inhabitants of the territory.[54] In another version, the trapper is only required to warn the king or his steward (*rechtaire*).[55]

The mechanics of this form of deer-trap are unclear to me. It seems from the *Osbretha* commentary that a log (*cep*) formed part of the trap: perhaps it was balanced so as to strike the deer after it had fallen into the pit.[56] A quotation in *O'Davoren's Glossary* states that it is illegal to dig a deer-pit (*cuithech*) in another person's land without permission.[57] In the accompanying gloss, the *cuithech* is said to contain a pierced board (*clár toll*) or a log (*cep*).[58] The pit may be camouflaged: in *Bethu Phátraic* the term *cuithech* is used of a water-filled trap with some sort of covering (*brathlang*).[59] A

[50] *CIH* iv 1496.25–6 = O'Dav. 328 § 796.

[51] This word is formed from *cuithe* 'pit' + the suffix *-ech*.

[52] *Pace* L. Breatnach, 'On the agent suffix *-e* in Irish', it seems more likely that *cuthchaire* is from *cuithech* + agent suffix *-aire* than from *cuithe* + the root of *-cuirethar* 'puts' + agent sufix *-e*.

[53] *LL* i 116.3668, 119.3746 (*Tech Midchúarda*); O Daly, '*Lānellach tigi rīch*', 83 § 16.

[54] *CIH* i 285.13 = *AL* iii 272.8 *Eascaire na cuithighi do rīg ⁊ do thuaith.*

[55] *CIH* iii 972.32.

[56] *CIH* i 322.17.

[57] *CIH* iv 1483.16–17 = O'Dav. 271 § 472 *fer clandus cuithigh a tīr a cēili cen athcomarc.*

[58] The Egerton 88 version has *cepa*, but I follow Stokes in reading *cep* with Mac Fhirbhisigh. *Cep* is a masc. *o*-stem (from Latin *cippus*), so *cepa* would not be a possible nom. plur.

[59] Mulchrone, *Bethu Phátraic*, 114.2196. Cf. *TBC* St 148.4672–3.

Middle Irish topographical poem provides a legendary account of how a skilled trapper named Coba died in the pit (*cuithech*) which he had prepared to catch game (*fíad*).[60] He put his foot into the trap (*indell*) to test whether it was properly set. But his ankle, and then his leg and his arm became entrapped. Finally his neck was caught, and he died of strangulation.

Leg-traps

A ninth-century stone cross at Banagher, Co. Offaly, contains a fine representation of a stag with two of its legs caught in a trap.[61] The trap seems to be similar to a wooden device found in a bog at Drummacaladdery, Co. Donegal, which has been dated to the Late Bronze or Early Iron Age.[62]

Fig. 14. A deer in a trap from a ninth-century high cross at Banagher, Co. Offaly.

In the written sources of our period, I have come across only one reference to a leg-trap. This is in the *Canones Adomnani*, which lay down the Church's regulations on what food can and cannot be eaten. Here it is stated that a dead deer whose legs have been broken in a trap – but with only slight bleeding – may not be eaten.[63] Because the higher blood, which is described as 'the guardian and seat of life' has not flowed, the deer is regarded as carrion. However, its fat and hide may be used.

It is of course impossible to know to what extent such regulations were observed in practice. But it at least suggests that the intention of such leg-traps was to immobilise the deer without killing it,

[60]E. J. Gwynn, *Metrical Dindshenchas* iv 122–4.

[61]Harbison, *High Crosses* i 26 (discussion); i 377–9 (dating); ii fig. 66 (photograph).

[62]Mitchell, 'The relative ages of archaeological objects', 17. Another trap of this type has been found at Garvary, Co. Fermanagh (Edwards, *The archaeology of early medieval Ireland*, 64).

[63]*IP* 180 § 20.

so that when the trapper arrived he could slaughter and bleed it properly.

Hunting of wild pig

After deer-hunting, our sources devote most attention to the hunting of the wild pig (*Sus scrofa,* Old Irish *mucc allaid*).[64] There is less emphasis on the sporting aspect of this form of hunting, and the main motivation seems to have been to procure meat.[65] Wild pigs were commonly pursued with dogs,[66] but they might also be caught in a trap (*airches*),[67] or shot with an arrow.[68]

In literary texts the hunting of the wild pig, especially the boar (*torc allaid*), is associated particularly with tales of the *Fíana* and their leader Finn (Fionn). One of the best-known episodes is the death of Diarmaid Ó Duibhne, disembowelled by a wild boar at Beann Gulban (Benbulben, Co. Sligo).[69] Finn also features in a tradition concerning a famous boar which he had hunted on many occasions, but was finally killed by a peasant working at a drying-kiln. Finn felt the ignominy keenly, and uttered a verse, preserved in Triad 236:[70]

> It is not well that we fed our hounds,
> it is not well that we rode our horses,
> since a little peasant from a kiln
> has killed the boar of Druimm Leithe.[71]

The abundance of wild swine in Ireland is mentioned by Giraldus Cambrensis in the twelfth century,[72] but the archaeological evidence indicates that they did not survive long after the Norman

[64]According to Van Wijngaarden-Backer, 'Animal remains at Newgrange', 337, it is very difficult to tell from the evidence of the bones whether such pigs were feral forms of the domestic pig (*Sus domesticus*) or pure wild boar (*Sus scrofa*).

[65]E.g. Gwynn and Purton, 'The monastery of Tallaght', 146 § 51.

[66]Plummer, *Bethada* i 128 § xiii (26) trans. ii 124; i 165–6 § xxi (40) trans. ii 160–1.

[67]Calder, *Auraicept*, 10.113.

[68]Radner, *Fragmentary annals*, 50 s.a. 702.

[69]Ní Shéaghdha, *Tóruigheacht Dhiarmada agus Ghráinne*, 88.1557–9.

[70]Meyer, *Triads*, 30 § 236.

[71]There is also a reference to Druimm Leithe (Drumlea, Co. Tyrone) at Meyer, *Triads*, 6 § 51.

[72]O'Meara, *Giraldus: Topography*, 47 § 18 = Dimock, *Giraldi Topographia*, 57.12–13.

invasion.[73] Their extinction was presumably caused by a combination of over-hunting and the destruction of the oak-woods on which they largely depended for food.

Hunting of other land mammals

A ninth-century legal gloss includes badgers as well as deer and wild pigs among the three game-animals (*fíadmíla*).[74] The badger (*Meles meles*, Old Irish *brocc*)[75] was apparently hunted for food, as a poem in the *Bórama* contains a reference to *saill bruicc a Bérre* 'salted badger-meat (or badger-fat) from Bérre'.[76]

The Old Irish tale *Táin Bó Fraích* contains a description of a great hunt organised at Crúachu (Rathcroghan, Co. Roscommon) by the legendary king and queen of Connacht, Ailill and Medb.[77] As well as deer and wild boar, the hounds pursued hare (*Lepus timidus*, Old Irish *míl maige*[78]), fox (*Vulpes vulpes*, Old Irish *sinnach*), and otter (*Lutra lutra*, Old Irish *doburchú*[79]). Presumably, the last two were generally hunted for their pelts and because they were regarded as predators.[80]

As we have seen above (p. 186), legal commentary classes the hunting of the wolf as a public duty because of its depredations on livestock.

Hunting of marine mammals

It is clear from both written and archaeological evidence that marine mammals were valued as a source of food. Bones of various species of seal (Irish *rón*)[81] have been found in abundance

[73] Edwards, *The archaeology of early medieval Ireland*, 65.

[74] *CIH* iii 916.40–41 *Fiad cacha fedha .i. ut diximus, fiadhmīla .i. bruic, huis, muca aldti* 'the hunting of every wood i.e. as we have said, game-animals i.e. badgers, deer, wild pigs'.

[75] The Welsh cognate is *broch*. English *brock* 'badger' is probably a loan from Primitive Welsh * *brocc* (Jackson, *Language and history in early Britain*, 567 § 146).

[76] *LL* v 1277.37943 = Stokes, 'The *Bórama*', 46 § 29. Cf. Thurneysen, '*Tochmarc Ailbe*', 276 § 7.

[77] Meid, *Táin Bó Fraích* 3.53–8.

[78] Lit. 'the animal of the plain'. The other term for hare, *girrfhiadh* lit. 'short game', is not attested until the Mod.Ir. period.

[79] Lit. 'water-dog', cf. Welsh *dyfrgi* of the same meaning and etymology.

[80] Medieval Welsh law assigns values to the pelts of both the fox and the otter (Wiliam, *Llyfr Iorwerth*, 90 § 137.13 = *ALIW*, Venedotian Code, III xix §§ 8–9).

[81] The most frequent seals in Irish waters are the grey seal (*Halichoerus grypus*) and the common seal (*Phoca vitulina*).

in maritime sites from the early Christian period.[82] Adomnán's *Life of Columba* refers to the entitlement of the monastery of Iona to take seal-pups from a certain small island near Mull. They were no doubt valued for their flesh, as well as their hides and oil.[83] Adults were hunted with a seal-spear (*róngáe*).[84]

Irish *rosualt* (also *rossal*, *rosmáel*, *rasmóel*, etc.) is a borrowing from Old Norse *hrosshualr* 'walrus'.[85] In an additional item in the Yellow Book of Lecan version of *Cormac's Glossary*, it is described as a type of seal.[86] The walrus (*Odobenus rosmarus*) is not now classified as a seal, but is placed by zoologists in the same sub-order *Pinnipedia*. It is at present confined to Arctic waters, but was formerly also found in northern Scotland, especially in the Shetland Islands. It features in early Irish mythology – preserved in a gloss on a poem in praise of Columba – as a sea-beast whose vomiting affected the wellbeing of the surrounding land, sea, and air.[87]

A marine mammal which also attracted the attention of hunters was the porpoise (*Phocoena phocoena*), which is common around the Irish coast – particularly in summer – and may travel far up rivers. In Irish it is called *mucc mara* 'pig of the sea', presumably on account of its blubbery but edible flesh.[88] In the year 828 the *Annals of Ulster* record a great slaughter of porpoises on the coast of Ard Ciannachta (in the present Co. Louth), which was carried out by Norsemen.[89] A section of legal commentary on the duties and

[82] E.g. O'Kelly, 'Church Island', 133 (report by G. Roche).

[83] A. O. and M. O. Anderson, *Adomnan's Life of Columba*, 294 (new pagination: 74) = f. 42a. There is a reference to a sealskin container at *Thes.* ii 347.33.

[84] Stokes, *Lismore Lives*, 49.1641.

[85] Marstrander, *Bidrag*, 107.

[86] Meyer, *Sanas Cormaic*, 100 § 1125 *Rasmoel .i. rōn.*

[87] Stokes, 'The Bodleian Amra', 256 § 60 = *LU* 29.831–8; cf. E. J. Gwynn, *Metrical Dindshenchas* iii 428.25–36; Stokes, 'The prose tales in the Rennes Dindṡenchas (1)', 476 § 76.

[88] This term was also applied to the dolphin (*Delphinus delphis*). In the Old Irish glosses on Priscian, Latin *delphinus* is glossed *mucc mora* (*Thes.* ii 137.17–18). The dolphin is less likely to frequent rivers and estuaries than the porpoise, and so would have been more difficult to hunt. The term *socc sáil*, also meaning 'pig of the sea' is applied to the cuttlefish (*loligo*) at *Thes.* ii 101.34.

[89] *AU*[2] 284 s.a. 827 (recte 828) § 3 *Mucār már di muccaibh mora*, lit. 'a great pig-slaughter of the pigs of the sea'. There may also be a reference to the consumption of porpoise-flesh in *LL* v 1277.37940 = Stokes, 'The *Bórama*', 46 § 29 *stáec thuirc na tuinne* 'a steak of the boar of the sea'.

entitlements of a maritime king refers also to catches of salmon, seals, and porpoises.[90]

There is no evidence that the early Irish hunted whales. However, stranded whales were viewed with enthusiasm by those who lived near the coast. The *Annals of Connacht* record that in the year 1246 a whale (*míl mór*)[91] was stranded at Cúl Irra in Cairbre Dromma Cliabh (Co. Sligo). Its arrival 'brought great relief and joy to the whole territory'.[92] This entry reflects the principle expressed in Irish law that everybody in the community had a claim on a stranded whale. There were, however, certain procedures to be gone through, and it seems that there was no automatic general entitlement to a share. The head of the kin (*áige fine*) which owned the adjacent land was obliged to go the local king (*rí túaithe*) and fast against him so as to claim the whale on behalf of the community. The king could then formally assign the whale to the people, or else he could distrain it for his own use.[93] The king's entitlement to a stranded whale is perhaps linked with the idea that such windfalls are a direct result of his virtues as a ruler. This is expressed in one version of *Audacht Morainn* in the following terms: 'it is through the justice of the ruler that many creatures and many animals from the deep and great seas are cast up on lawful shores'.[94]

The consumption of the flesh of a stranded whale or other animal did not run counter to the Church's ban on carrion-eating, unless decomposition had set in. According to the *Canones Adomnani*, 'marine animals cast upon the shores, the nature of whose death we do not know, are to be taken for food in good faith, unless they are decomposed'.[95]

As well as whale-meat (*carnae máirmíl*),[96] other parts of the whale were used. The law-text on distraint refers to the sharing out of whalebone, which is called 'whale's eyelash' because of its similar

[90] *CIH* i 178.11–12.

[91] The whale is called *míl már* (*mármíl*) 'great animal' or *bled* (*bledmíl*) 'monstrous animal'. Different species are not distinguished.

[92] *AC* 86 s.a. 1246 § 2.

[93] *CIH* iii 888.10–14 = *AL* i 128.20–24 (mistranslated: *nō gaibaid a nathgabāil* means 'or he (the king) takes [it] in distraint'); cf. *CIH* v 1682.7–10.

[94] Kelly, *Audacht Morainn*, 61 § 22 (version in L^1 and N).

[95] *IP* 176 § 1.

[96] E.g. *CIH* vi 2292.28 = Binchy, '*Bretha Crólige*', 20 § 25.

appearance to human eyelashes.[97] Whalebone or baleen consists of horny plates which take the place of the teeth of the upper jaw in whales of the sub-order *Mysticeti.* These plates act as filters for the small planctonic organisms on which the whales live. In the mouth of a blue whale (*Balaenoptera musculus*) there are up to eight hundred such plates. Being both strong and flexible, whalebone has many uses. The glossator on this text refers to its use for making the backs of riddles and the breastwork of saddles.[98] In another gloss on the same text we are told that whalebone should be distributed for hoop-making in areas where there is no wood.[99] This implies that whalebone was used for some of the same purposes as hazel or willow rods.

The sub-order *Odontoceti* (toothed whales) includes whales ranging in size from eighteen metres in length to about five metres. Some of these whales have large teeth which were valued by the early Irish. For example, Heptad 37 includes a 'beautiful tooth' among the seven valuables which cannot be given in pledge by the head of a kin-group on behalf of its members.[100] It is clear from a ninth-century gloss on this heptad that the reference is to a whale's tooth (*détt míl máir*).[101] Further discussion on whale's teeth is found in *O'Davoren's Glossary.* Here it is claimed that the *fiacail airisi* is one of the three back teeth of the whale, but the context is unclear.[102]

Fishing

Many texts emphasize the importance of fish in the early Irish economy, and drawings of fish decorate manuscripts such as the Book of Kells (ff. 70^{v}, 71^{r}, 179^{v}, 250^{v}), the Book of Armagh (f. 33^{v}), and Rawlinson B 502 (f. 30^{v}).[103] The Old Irish wisdom-texts regard high

[97] *CIH* ii 371.30 = *AL* i 124.7 *fabra mīl mōir*; cf. *CIH* i 35.21 = *AL* v 250.4 *abrat bleidmīl.* The Old Irish form of the first element is *abrae* 'eyelash', which later acquired a prosthetic *f-*.

[98] *CIH* ii 371.31 = *AL* i 134.29–30 *re dēnam cūl criathar 7 clār sadall.* For the use of saddles among the early Irish, see p. 97.

[99] *CIH* ii 371.31–3 = *AL* i 134.30–31 *do dēnam circall de, in baile inā bī fid.*

[100] *CIH* i 35.21 = *AL* v 250.3–4 *fiacail gnaei.*

[101] *CIH* iii 906.2 *Fiacail gnoe .i. détt mīl māir*; cf. *CIH* i 35.28 = *AL* v 250.14 *.i. fiacail in mīl mōir.*

[102] *CIH* iv 1502.26–7 = O'Dav. 356 § 938. The lemma *Fiac*ail *airisi* is probably from some lost law-text, but the glossator's explanation casts little light on its meaning. I do not understand the form *airisi.*

[103] The fish was one of the symbols of Christ: see Meehan, *The Book of Kells*, 50.

fish-yields as a sign of a just king. *Audacht Morainn* states that 'it is through the justice of the ruler that abundance of fish swim in streams'.[104] Similarly, in *Tecosca Cormaic* we are told that the justice of the ruler brings fish into estuaries.[105]

No law-text on fishing rights has survived, but there are references in other law-texts which suggest that this topic was dealt with in considerable detail, possibly as part of the lost text *Muirbretha* 'sea-judgements'.[106] From the surviving material, it seems that the lawyers devoted special attention to the legal problems associated with estuarine weirs. The law-texts on distraint and on mill-races refer to *cáin inbir* 'the law of the estuary',[107] which is taken by an Old Irish glossator to regulate the use of public weirs.[108] Later legal commentary provides us with examples of the sort of problems which arose. One passage discusses the theft of fish from a weir, while another deals with the weir-owner who has built his weir too far out into the river: see under 'fishing methods' below.

Presumably, most fishing was carried out by local farmers who had rights to fish at particular places. The law-text on status, *Uraicecht Becc*, does however mention the professional fisherman (*íascaire*).[109] He is a person of low rank, with an honour-price of only one yearling heifer. A king or lord would be expected to employ one or more fishermen: they are included among the personnel of a king's household in *Suidigud Tige Midchuarda*[110] and in *Lánellach tige ríg 7 ruirech.*[111]

It is clear from the law-texts that there was not a universal public right to fish. According to *Di Astud Dligid 7 Chirt*, a law-abiding freeman (*recht*) is only permitted a single swift dip of a fishing-net in a stream.[112] The same text states that such a person is entitled to 'a salmon of the place' (*é áitte*),[113] i.e. a single salmon from near his

[104] Kelly, *Audacht Morainn*, 7 § 20.
[105] Meyer, *Tecosca Cormaic*, 2 § 1.23.
[106] See *GEIL* 276–7 and p. 296 below.
[107] *CIH* ii 369.9 = *AL* i 122.16; *CIH* ii 460.9–10 = Binchy, '*Coibnes Uisci Thairidne*', 68 § 9.
[108] *CIH* iii 888.16–17 *im cāin n-inbir .i. im coraid coitcinn.*
[109] *CIH* v 1616.26 = *AL* v 106.18.
[110] *LL* i 116.3663 (col. 2).
[111] O Daly, '*Lānellach tigi rīch*', 83 § 16.
[112] *CIH* i 241.20–2 = *AL* v 482.24; *CIH* iv 1504.19–21 = O'Dav. 365 § 975 *luathfobairt cacha srotha.*
[113] *CIH* i 241.20, 31–2 = *AL* v 484.8–9 *hæ āite .i. in bradān bīs isin āit, isinn inadh; slān a ænguin .i. cibed āit* 'a salmon of the place i.e. the salmon which is in the place, in

house.[114] These regulations may have been intended partly as conservation measures, but their main implication is that the fishing of some streams is privately owned. In other places, however, fishing is evidently free to all members of the community. An Old Irish gloss on *córus lín* 'regulation of nets' refers to a public fishing-place for the *túath.*[115]

Fishing methods

Fish have been caught by humans since the Stone Age, and many techniques have been employed: spearing, gaffing, strokehauling, tickling, netting, hooking with rod and line, poisoning, etc.[116] The Old Irish sources devote most attention to the catching of fish in weirs erected in rivers or estuaries.

A weir may be made entirely of stones, or entirely of stakes and wattling, or with a combination of these materials.[117] The two Old Irish terms for 'weir' reflect both methods of construction. The commonest term is *corae,* whose basic meaning is simply 'stone wall'.[118] The other term is *aire,* which means 'woven fence'.[119] Glossators take both terms to be synonymous.[120] In later sources, particularly in legal commentary, the word *sód* is also well-attested in the meaning 'weir'.[121] Not all weirs are designed simply to catch fish: legal commentary refers also to weirs constructed to direct the main current of a stream towards the bank to drive a water-mill.[122] Fishweirs and water-mills may be associated

---

the spot; killing one is free i.e. whatever place it may be'. For another interpretation of *áitt* here, see Mac Airt, 'Lexicographical notes', 265.

[114]See discussion of this passage by E. MacNeill, *Celtic Ireland,* 167.

[115]*CIH* iii 888.15 *Im cōrus līn .i. piscium .i. āit coitcend bīs don tuaith i n-uisce* 'about regulation of nets i.e. of fish i.e. a common place which is for the community in water'.

[116]For example, there is a reference at Carey, '*Scél Tuáin meic Chairill*', 102.65–6, to netting and spearing.

[117]See Went, 'Irish fishing weirs'.

[118]It is common in placenames: see Hogan, *Onomasticon,* s.v. *cora.*

[119]E.g. *CIH* ii 394.34–5 = *AL* i 202.1 *aire fri sruth* 'a weir against a stream'. *Aire* is vb.n. of *ar-fen* 'fences in front'; cf. *imbe* 'fence (of field)', vb.n. of *imm-fen* 'fences around' (*LEIA* A-42; *VKG* ii 517 § 719 *fe-n-*). For a discussion of recent discoveries of the remains of early fishweirs, see A. O'Sullivan, 'Intertidal survey on the Fergus estuary and the Shannon estuary'; 'Harvesting the waters: fishweirs in early Ireland'; 'Intertidal archaeological survey', 153–5.

[120]E.g. O'Mulc. 238 § 101 *aire .i. cora ēisc.*

[121]E.g. *CIH* ii 395.20 = *AL* i 206.1. See *LEIA* S-161 s.v. 2 *sod.* Ó Murchadha, '*Sódh* i logainmneacha', 130–1, notes an early instance of *sód* in the placename *Sódh Macáin.*

[122]*CIH* iii 972.24–5 = Appendix A, text 6 § 1.

structures: a Middle Irish saint's Life speaks of a mill on the river Dee (probably in the present Co. Westmeath) and its fishweir.[123]

Weirs are obviously most effective in catching migratory fish such as salmon and eels. The strategies of weir-builders vary: in some weirs, the intention is to force the fish into a channel so that they can be more readily caught with spears, hooks or nets. In a more complex arrangement, wickerwork baskets are set in gaps in the weir: fish which swim in are unlikely to get out and can easily be taken by the fishermen. The Irish term for such a basket is *ces*. The *Annals of Connacht* record a slaughter in the year 1225 of the people of Cúl Cearnadha by Aodh mac Cathail Chrobhdheirg and a band of Normans. Many of those fleeing were drowned at Dubh Conga (Ballycong, Co. Mayo), and the baskets of the weir were full of drowned children.[124]

Legal difficulties have long been associated with weirs. In his account of the Galway Fishery, Arthur Went details the various attempts to legislate for fishing at the weirs on the Corrib, and points out that the legal history of these weirs goes back at least to the year 1283.[125] Sources in Irish provide no information on specific weirs, but give a general account of the legal problems which were considered likely to arise.

Weirs may be owned by individuals: one Old Irish legal fragment refers to the fish-weir of a lord (*aire éisc flatha*).[126] A later gloss refers also to ownership by a kin-group (*fine*).[127] A particularly difficult problem is the legal relationship of a weir-owner with other weir-owners upstream or downstream. It is clear that attempts were made in Irish law to prevent any one weir-owner from getting an excessive share of the fish. The principle is quoted in *O'Davoren's Glossary*, presumably from an Old Irish law-text, that it is not proper to erect a weir over more than one third of the water.[128] Later commentary provides more detail, and states that a landowner who owns both sides of a river may construct a weir extending from one

[123]Meyer, *Betha Colmáin*, 62.13 § 59 *Muilenn Dée ⁊ a carad éisc laiss.*

[124]*AC* 12 s.a. 1225 *is amlaid dogebthea na cairr cona cessachaib ⁊ a llān intib do lenbaib arna mbāduth.* Cf. *AFM* iii 226 s.a. 1225.

[125]Went, 'The Galway fishery', 192.

[126]*CIH* iii 1018.6, vi 2163.6; *CIH* iv 1468.8 = O'Dav. 206 § 60.

[127]*CIH* ii 369.15 = *AL* i 130.4 *in cora coitcend na fine.*

[128]*CIH* iv 1468.9 = O'Dav. 206 § 60 *Nī tēchta nī bes* [*mó*] *nō trian inn uisce do aire .i. do ime.*

bank to a distance of one third of the breadth of the river. Alternatively, he may construct a weir from both banks, each extending to one sixth. A landowner who owns only one bank may construct a single weir extending to one sixth. If these limits are exceeded, the excess fish must be given to the weir-owners upstream or downstream, depending on which way the fish are migrating.[129] We are not told how the excess is to be calculated, but presumably the offending weir-owner's catch is compared with that of other weir-owners.

Another problem especially associated with weirs is that the trapped fish may easily be taken by persons other than the weir-owner. Legal commentary lays down the heavy fine of five *sét*s for stealing fish from a weir.[130] The theft of fish from a weir features in an obscure tale relating to Finn mac Cumaill which is preserved in a law-text. After beheading Lomna, Coirpre and his men went off with his severed head to a weir which was full of fish. They took some of the fish and cooked them, but did not give anything to the head. The head then complained: 'a speckled white-bellied salmon which issues from spawn under the seas . . . you have made a division which is not just, o Coirpre'.[131] Coirpre and his men were afterwards killed by Finn in vengeance for the beheading of Lomna.

Other texts refer to the catching of fish in a net (*lín*), by a netsman (*línaige*). Such nets were presumably made of linen thread, tied in meshes. For example, the Life of Saint Mo Ling refers to a miraculously heavy catch in which there was a salmon at every third mesh in the net.[132] We know that the activities of netsmen were in some circumstances controlled by law, as the law-text on distraint refers to 'the regulation of nets' (*córus lín*).[133] This is treated as a separate legal topic from 'the law of the estuary' (*cáin inbir*), but no

[129] *CIH* ii 395.18–21; v 1699.11–14 = *AL* i 204.27–206.2; *CIH* iii 1018.6–10; vi 2163.6–9.

[130] *CIH* i 261.13; iii 928.12–13 = *AL* iii 148.19–21; *CIH* iv 1189.36; *CIH* v 1874.20.

[131] *CIH* vi 2116.20–21 = Dillon, 'Stories from the law-tracts', 49.5–7 § ix *Orcc brecc brondfind brūchtas di magur fo muiribh . . . rorannais raind nāch cert a corbri. Orcc* seems here to have the meaning 'salmon'; however, salmon spawn in rivers, not in the sea: see p. 291 below. Another version of this episode is found in *Cormac's Glossary* (Meyer, *Sanas Cormaic*, 86–8 § 1018). E. J. Gwynn, ('Notes', 190) translates *brūchtas di magur* as 'which is bursting with spawn'.

[132] Stokes, 'The birth and life of St. Moling', 272 § 23 *co tarla bratān cach tres moccuil isin līn*; cf. Radner, *Fragmentary annals*, 4 s.a. 595.

[133] *CIH* ii 369.9 = *AL* i 122.15.

further details are given. According to *Berrad Airechta*, the earnings from net-casting (*tuillem línchuir*) are entirely immune from claim (*ruidles*) in Irish law.[134] This obviously applies only to net-casting which is not in breach of *córus lín* or other fishing regulations.

Some information on net-fishing is provided by the Old Irish canon law-text *Epistil Ísu* ('The Epistle of Jesus'). This text deals with the general regulations on Sunday observance in the western Church. However, as in the case of *Cáin Domnaig* ('The Law of Sunday'), the rules have been adapted to fit early Irish conditions. This is exemplified in the treatment of the offence of fishing between the two limits of Sunday, i.e. from vesper-time on Saturday to tierce on Monday. If nets are put into the water during this time, each man involved must pay a fine of one ounce of silver.[135] In addition, the nets must be burnt or surrendered to the stewards of the law (*do rechtairib na cána*). Likewise, the coracle (*clíab*), hide (*seiche*)[136] and clothing (*timthach*)[137] of the fishermen must also be burnt. It is clear from this passage that net-fishing was typically carried out by a number of men operating from a coracle made of wickerwork covered with a hide or hides.[138] The fisherman's hide is also mentioned in the context of fishing at a weir. Triad 230 refers to a *seiche corad* 'weir hide', apparently meaning the type of boat used to take in fish at a weir.[139]

As well as weir-fishing and net-fishing, our sources also refer to fishing by line and hook (*dubán*). This method is associated particularly with trout-fishing: see p. 295 below. In the ninth-century St. Gall Glosses, the Irish term for the three-pointed fish-spear or trident is given as *gáe gona éisc* 'a spear for killing fish'.[140] A well-preserved iron fish-spear of this type was found in a crannóg of the early Christian period at Strokestown, Co. Roscommon.[141]

[134] *CIH* ii 592.11 = Thurneysen, *Bürgschaft*, 8 § 13 = Stacey, '*Berrad Airechta*', 212 § 13.
[135] O'Keeffe, '*Cáin Domnaig*', 208 § 29. It is stated at § 33 that one third of such fines goes to God (i.e. for charity), one third to lords and churches, and one third to the enforcers of the law of Sunday and the guarantors of Sunday.
[136] The YBL version uses the word *codal*, with the same meaning as *seiche*.
[137] *Timthach* is normally used to mean 'clothing', but here it may include all the accoutrements, as well as clothing, used by the fishermen.
[138] For boat-construction, see p. 499.
[139] Meyer, *Triads*, 28 § 230.
[140] *Thes.* ii 91.17.
[141] It is described by Went, 'The pursuit of salmon in Ireland', 199, with an illustration on plate xxxv, opposite p. 244.

Salmon-fishing

There are many more references in our sources to the salmon (*Salmo salar*) than to any other fish.[142] Its prestige as a food is memorably expressed in a ninth-century triad which lists the three deaths which are better than life as 'the death of a salmon, the death of a fat pig and the death of a robber'.[143] The salmon, defined in one text as 'the crafty one of the water',[144] is also prominent in mythological material. For example, the hero Finn acquired his supernatural wisdom by licking his finger on which were traces of the cooked flesh of the 'salmon of knowledge'.[145] An entry in the *Annals of Clonmacnoise* suggests that the salmon had royal associations. Thus it is recorded that in the year 1061 Áed Ua Conchobair of Connacht destroyed the fort at Kincora (Cenn Corad) which had been the seat of Brian Bórama, and ate the two salmon which were in the king's fishpond there.[146] The implication is that this was a serious affront to Brian's descendants. A legal poem also associates salmon with kings: a king must give a salmon to his judge (*medam*) out of every abundant catch. In addition, the judge is entitled to the heads of all the king's salmon.[147]

Not surprisingly, our texts devote a good deal of attention to the extraordinary life-cycle and migratory patterns of the salmon, many details of which are still not completely understood by zoologists. The salmon begins life as an egg laid in gravel in a fast-flowing river. After about eighty or ninety days it develops into an alevin, still with yolk sac attached. When this sac has been absorbed – usually after about three weeks – it is called salmon fry, and on reaching finger length, parr. After a year or two, the parr turns silver and is called a smolt; it then starts its descent to the sea. At sea, it travels far out into the Atlantic where it feeds and makes rapid growth. After a year or more at sea, it returns for spawning, almost always to its natal river. The time of its return varies in different rivers, and may also depend on there being sufficient volume of water to allow it to leap up rapids or waterfalls. At the spawning beds the male ejects milt

[142]See Went, 'The pursuit of salmon in Ireland'.
[143]Meyer, *Triads*, 12 § 92 *Trí báis ata ferr bethaid: bás iach, bás muicce méithe, bás foglada.*
[144]McManus, 'Irish letter-names', 146 E: B^{24} *clesach uisce .i. éiccne.*
[145]Meyer, '*Macgnímartha Find*', 201 § 18. For other examples, see *DIL* E 144.38–46.
[146]*Ann. Clon.* 178 s.a. 1061.
[147]*CIH* iv 1268.20–22. The text has *ceand cach eo* where one would expect *ceand cach iach* 'the head of every salmon'.

which fertilizes the eggs released by the female. After spawning, salmon of both sexes are called kelts, and drop downstream in a weakened state. At this stage most die of disease or fall victim to predators. Some, however, reach the sea and there regain sufficient condition to return after one or more years for a second spawning. There are even records of salmon spawning three or four times.

It is uncertain how much was known about the salmon's life-cycle by the authors of our texts. Whoever composed the tale of the beheading of Lomna (quoted on p. 289 above) seems to have believed that salmon are spawned at sea rather than in rivers. This is perhaps not surprising as the young stages of a salmon in the river are very different in appearance from the returning fish after one or more years at sea. A similar idea that trout are spawned at sea may be reflected in the story of the mermaid Ambia (daughter of Cain son of Adam and Eve) which is preserved in a legal manuscript.[148] While she was asleep under the sea, a trout ejected its milt (*iuchair*) into her mouth, so that she became pregnant with twenty-two offspring.

A belief that salmon were spawned at sea may have given rise to the term *éo cétṡnáma* 'salmon of first swimming'. It is clear from two references that salmon were held to be most suitable for eating at this stage. For example, the late Old Irish tale *Scéla Cano meic Gartnáin* contains a fanciful account of fishing from King Gartnán's island of Inis moccu Chéin.[149] There were fifty fishing-nets, with ropes from them to the windows of the kitchen. There was a bell on the end of each rope in front of the steward, and four men for hauling in the 'salmon of first swimming'. Legal commentary on *Uraicecht Becc* likewise refers to 'half a salmon of first swimming' (*lethéo cétṡnáma*) among the possible foodstuffs which a client provides for his lord.[150]

Michael O'Brien must be correct in taking *éo cétṡnáma* to refer to what is now termed a 'first-run (or fresh-run) salmon', i.e. a salmon returning upriver from the sea.[151] It is at this stage that salmon are most prized for culinary purposes. The same *Uraicecht Becc* commentary refers also to *éo colgnama*, presumably another stage in the

[148] *CIH* ii 682.10–15.
[149] Binchy, *Scéla Cano*, 1.9–13.
[150] *CIH* ii 644.12; v 1599.39; vi 2266.38 = *AL* v 40.23–4 (mistranslated). The MSS have *letheo cetnama*(*dh*).
[151] M. O'Brien, '*Scéla Cano*'; M. O'Brien, '*Eo cétnama*'.

salmon's development. The *Dictionary of the Irish language* suggests that *colgnam* is for *colgṡnám*, a compound of *colg* 'sword, sting, bristle, awn, fury' and *snám* 'swimming'.[152] It is clear from the *Uraicecht Becc* citation that a salmon at the *éo cétṡnáma* stage is worth two at the *éo colgnama* stage. It is likely therefore that the latter is a spawning salmon when the flesh takes on an insipid or muddy flavour. The *colg* may refer to the 'kype', a vertical projection of cartilageneous tissue which projects from the tip of the lower jaw of the male salmon.[153]

Bethu Phátraic refers to the generally seasonal nature of salmon-fishing. It records a legend that Saint Patrick ordered some fishermen to cast their nets into the Sligo (Garavogue) River. They were doubtful, pointing out to him that salmon were never caught at that place in winter. Nonetheless, they threw out their nets and were rewarded with a catch of large salmon, which they gave to Patrick. He then blessed the river, so that fish are caught in it all year round.[154]

The importance of the salmon in early Irish society is reflected in the number of names which are applied to this fish. In Old Irish the usual word for salmon is *éo*, which is cognate with Welsh *eog*, Gallo-Latin *esox* of the same meaning. The variant *é* is also well attested in the early language.[155] From the late Old Irish period, the word *bratán* largely displaced *éo*, which was however retained into Classical Modern Irish as a poetical term, often used figuratively of a hero or king. The replacement of *éo* by *bratán* occurred throughout the Gaelic area, so that Modern Irish and Scottish Gaelic have *bradán* and Manx has *braddan*.

The etymology of the word *bratán* is uncertain. There is some evidence that it originally referred to a young salmon: thus a youthful warrior may be described contemptuously as a 'small salmon'

[152] *DIL* C 325.79.

[153] I owe this sugggestion to Patrick Timpson of the School of Science, Regional Technical College, Sligo, to whom I am grateful for many general comments on the section on fishing in this chapter.

[154] Mulchrone, *Bethu Phátraic* 87.1638–40 *benedixit fluminem conid hí Sliccech gamnach uisci nHéirend, ar gaibther íasc indi hi ccach ráithi* 'he blessed the river so that the Sligo (Garavogue) River is the *gamnach* ('stripper') of the waters of Ireland, for fish are caught in it at every season'. A *gamnach* is a cow still giving milk to a yearling calf (see p. 40), and is thus comparable to the Garavogue. This river is famous as one of the earliest in Ireland, with fresh-run salmon entering as early as November.

[155] Thurneysen, *Grammar*, 203 § 319 (c).

(*bratán bec*).[156] On the other hand, T. F. O'Rahilly argued that its basic meaning is 'the stabbing one', referring to the way in which the female prepares a bed in the gravel for spawning.[157] But this use of the root *brat-* is otherwise unattested in Irish: it normally means 'to rob, plunder'. Possibly, therefore, *bratán* was used originally of the salmon smolt, which is particularly voracious as it makes its way downstream to the sea.[158]

Many other terms are used for salmon in Old Irish, particularly in verse. Of these *maigre* is likely to have originally meant 'egg-bearing female', being a derivative from *magar* 'spawn, fry'.[159] In some instances, *maigre* retains an association with spawning. The poet Máelmuire Ó Lennáin compares human sexuality unfavourably with the breeding of the progeny of the salmon (*síl in maigri*) which grows from beautiful spawn (*ásas a hiuchra álaind*).[160] But in other attestations *maigre* is used simply as a synonym of *éo*.[161] Another common synonym is *éicne*, the origin of which has not been explained.

A rarer poetic term for salmon is *orcc*, which occurs in the story of the beheading of Lomna.[162] It is cognate with words for various types of fish in other Indo-European languages, including Latin *perca* 'perch'.[163] Another Old Irish term for salmon, *erc*, which is attested only in glossaries, is from the same root. The word *mugna*, of uncertain origin, is also explained as meaning 'salmon' in the legal phrase *ni blaisi mugna muna firfolta feisear* 'may you not taste salmon unless you know the true qualifications'.[164] This might be a reference to a qualified judge's entitlement to receive a salmon from the king.

Finally, it should be pointed out that because of its special prominence, words for salmon are sometimes substituted for the general

[156] *TBC* Rec. I 61.1992.
[157] T. F. O'Rahilly, '*Creach. Gad. Brad. Slad.*', 170.
[158] Alternatively, the *-án* suffix may have a passive force, so that *bratán* might mean 'that which is robbed or plundered'. It is suggested in *LEIA* B-80 that this could be a reference to the salmon as the prey most sought after by fishermen.
[159] *LEIA* M-9 s.v. *magar*; E. J. Gwynn, 'Notes', 190-1.
[160] *ACL* iii 238 § 3 = McKenna, *Aithdioghluim dána* i 302, poem 80 § 3. Cf. Thurneysen, 'Mittelirische Verslehren', 69 § 12 *a iuchra maigri*.
[161] Plummer, *Bethada* i 321 § x (22); trans. ii 312.
[162] *CIH* vi 2116.20–21 = Dillon, 'Stories from the law-tracts', 49.5–7 § ix.
[163] *LEIA* O-28 s.v. 2 *orc*.
[164] *CIH* iv 1245.24–5; *CIH* iv 1517.10 = O'Dav. 421 § 1266.

term *íasc* 'fish'. A clear example is found in the version of the miracle of the feeding of the five thousand in the tenth-century poem *Saltair na Rann.* The Latin Vulgate speaks only of 'two fish' (*pisces*) but the Irish poet assumes that they are salmon.[165] The *Annals of Ulster* contain an entry for the year 917 which I believe to show the same substitution. Due to snow and extreme cold and unwonted ice in that year, the chief lakes and rivers of Ireland were passable, i.e. frozen over. Because of these weather conditions, there was death of livestock and birds and fish (*ár di chethraib ⁊ énaib ⁊ éicnib*).[166] It seems to make better sense here to take *éicne* to refer to all types of fish which died under the ice rather than just the salmon.[167]

Trout-fishing

Two forms of the trout (*Salmo trutta*) are native to Ireland: the sea trout and the brown trout. The former migrates from the sea to spawn in gravel on a river-bed. In the case of the latter, however, migration is restricted to movements upstream or from a lake into a stream. The Irish name *brecc* 'speckled one' refers to the spotted sides characteristic of both forms of trout.

Judging from the written sources, it seems that weir-fishing was a regular method of catching salmon. The catching of trout, on the other hand, is associated with line and hook (*dubán*).[168] For example, a Life of Saint Máedóc of Ferns refers to *brecc dubáin* 'a trout caught by hook'.[169]

Eel-fishing

Unlike salmon and trout, the eel (*Anguilla anguilla,* Irish *escong*) spawns at sea. The eggs are laid in the Sargasso sea in the western Atlantic, and the larvae then drift eastwards for about three years. On reaching the continental shelf, the larvae metamorphose into elvers, which migrate to fresh water, maturing at eight to fifteen years. At this stage the mature eels (silver eels) return to the Sargasso sea for spawning.

[165]Stokes, *Saltair na Rann*, 112.7624 *dá oenbratān* 'two single salmon'. In the next verse they are simply described as *na dá iasc* 'the two fish'. Cf. Carney, *The poems of Blathmac, etc.* 12 § 33 *Bendachais dá híích* 'he blessed two salmon'.

[166]AU^2 366 s.a. 916 (recte 917) § 1.

[167]For other examples, see *DIL* E 74.3–7.

[168]This word looks like a derivative of *dub* 'black', but no etymology has been put forward.

[169]Plummer, *Bethada* i 244 § lxviii (194), trans. ii 237.

Silver eels may be caught in large numbers in weirs during their migration back to the sea. The main run occurs in late autumn during high floods and moonless nights. In his account of the Galway fishery, Arthur Went has shown that there were eel-weirs on the Corrib already in the year 1283, and probably earlier.[170] The eels were caught in a hood-shaped net (*cochall*, anglicized *coghill*).[171]

Other forms of freshwater fishing

It is clear from Old and Middle Irish sources that salmon, trout and eels were regarded as the most important freshwater fish. In his *Topography of Ireland*, Giraldus Cambrensis likewise stresses the great abundance of these fish in Irish rivers and lakes. He notes too the absence of fish such as pike (*Esox lucius*) and perch (*Perca fluviatilis*), which have subsequently been introduced.[172]

He also has an interesting passage on other freshwater fish, which has been discussed in detail by Arthur Went.[173] One of these fish is described as being larger and more round than the trout, with firm white flesh, pleasing to the taste.[174] Went suggests that this is the pollan (*Coregonus albula*), a bottom-living fish of larger lakes such as Loughs Neagh, Erne, Ree and Derg. The name is presumably Irish *pollán*, a derivative of *poll* 'hole, pool, depth'.[175]

Sea-fishing

There are a few references to sea-fishing from boats in Old Irish sources,[176] and bones from sea-fish such as cod (*Gadus morhua*) and ballan wrasse (*Labrus bergylta*) have been found in excavations at early Christian sites.[177] But in general, the written and archaeological evidence indicates a society which concentrated on the rich harvest of freshwater fish rather than on the fish of the sea.

The lost law-text *Muirbretha* 'sea-judgements' may have dealt with some aspects of sea-fishing. However, the three short quotations

[170] Went, 'The Galway fishery', 197–200.
[171] *Cochall* is a borrowing from Latin *cucullus* 'hood' (*LEIA* C-138). The term *cochall ēisg* 'hooded fishing-net' is used in *CIH* iv 1530.31–2 = O'Dav. 481 § 1592 (see v. ll.).
[172] O'Meara, *Giraldus: Topography* 37–8 § 5 = Dimock, *Giraldi Topographia*, 32.
[173] Went, 'Giraldus Cambrensis' notes on Irish fish'.
[174] O'Meara, *Giraldus: Topography*, 38 § 6 = Dimock, *Giraldi Topographia*, 33–4.
[175] *OED* s.v. *pollan*; cf. Beckett, *Fealsúnacht Aodha Mhic Dhomhnaill*, 188–9.
[176] E.g. O'Keeffe, '*Cáin Domnaig*', 204–6 § 24. Cf. Stokes, *Lismore Lives*, 49.1640 *murgha* 'sea-spear, harpoon'; in this story the prey is a seal.
[177] E.g. O'Kelly, 'Church Island', 133 (report by G. Roche). See McCormick, 'Stock-rearing in early Christian Ireland', 188 and table 6.5.

which have been preserved (with later commentary) from this text deal only with ownership of flotsam, jetsam and goods carried off by a stream or sea-current.[178] No mention is made of sea-fishing rights in this material. But we can speculate that the existence of a right to take flotsam which is 'beyond nine waves [from the shore]' suggests that there was an equal right to take fish at this distance.[179] On the other hand, the local king may have been entitled to a share of the catch (see p. 408).

It is well known that the Irish learnt a great deal about sea-faring from the Norsemen, and this is reflected in the many boating terms which have come into the Irish language from Old Norse.[180] This influence seems also to have extended to sea-fishing techniques, as we find Old Norse *dorg* 'fishing-line' borrowed into Irish as *doruba* (*dorgha, drugha*).[181] Similarly, the Irish words for cod (*trosc*) and ling (*langa*) are from Old Norse *þorskr* and *langa* respectively. The origin of the word *scatán* 'herring' is more problematical. It is likely to be a borrowing from Old English rather than Old Norse, as in meaning it is close to Old English *scead* 'shad' (*Alosa alosa, Alosa fallax*), larger relatives of the herring.[182] Herring-fishing was clearly practised by the inhabitants – probably mainly Norse-Irish in race – of the eastern and northern ports in the early thirteenth century. The *Annals of Loch Cé* record that in the year 1217 all the herring-fishermen (*scatánaig*)[183] of Ireland from Waterford to Derry went to the Isle of Man to fish. They committed acts of violence there, and were consequently put to death.

Collecting shell-fish

Shell-fish formed an important part of the diet of Mesolithic hunter-gatherers.[184] With the establishment of agriculture, the importance of such foodstuffs diminished, but they continued to supplement the diet of those living near the sea, particularly in

[178] *CIH* i 314.17–316.8 = *AL* iii 422.1–430.4. See *GEIL* 276–7.

[179] *CIH* i 314.17 = *AL* iii 422.1–2 *dar nāe tondaib*; cf. *CIH* i 178.20–1.

[180] Greene, 'The influence of Scandinavian on Irish', 79.

[181] Marstrander, *Bidrag* 93; *DIL* s.v. *doruba*.

[182] See *LEIA* S-36. Welsh has the same word for 'herring', *ysgadan*, which may likewise be a borrowing from Old English *scead*. The Modern English dialectal form *scad*, possibly of Norse origin (*OED* s.v.), refers mainly to the horse mackerel (*Trachurus trachurus*).

[183] *ALC* i 256 s.a. 1217. The manuscript has *scathánaigh*.

[184] Mitchell, *The Irish landscape*, 103–4; *Reading the Irish landscape*, 84.

times of famine or scarcity.[185] Among the activities forbidden on a Sunday in an Old Irish canon law-text was *cnúasach mara* 'sea-picking', which doubtless included hunting for shell-fish, crabs, shrimps, edible seaweed, etc.[186]

Because of the low status of such food, there is little mention of it in the literature. A quotation in the *Irish grammatical tracts* refers to hunting for periwinkles (*Littorina littorea*, Irish *fáechan*).[187] In an additional item in the Yellow Book of Lecan version of *Cormac's Glossary*, the word *giritán* is identified as having the same meaning as *fáechan*.[188]

Another shell-fish used for food is the common limpet (*Patella vulgata*). Its Irish name *bairnech* is a derivative of *bairenn* 'rock', and refers to the habitat of this species.[189]

Fowling

Bird-bones are quite common in excavations from the early Christian period in Ireland. Generally, bones from wild birds are more numerous than those of introduced species. In Lagore crannóg, for instance, about 19 per cent of the bird-bones (measured in bulk) is from domestic fowl, and the rest is from wild birds.[190] There is also sculptural evidence of the hunting of birds. The cracked panel reproduced below from the ninth-century South Cross at Castledermot, Co. Kildare, is likely to represent a hunting scene.[191]

Early Irish literature contains many references to game-birds, including woodcock (*Scolopax rusticola*, Irish *crebar*),[192] snipe (*Gallinago gallinago*, Irish *náescu*),[193] wild duck (*Anas platyrhynchos* etc., Irish *lachu*),[194] and red grouse (*Lagopus scoticus*, Irish *cerc*

[185] McCormick, 'Stock-rearing in early Christian Ireland', 189.

[186] O'Keeffe, '*Cáin Domnaig*', 206 § 24.

[187] Bergin, 'Irish grammatical tracts: II Declension', 50.270 § 6 *selg ar fáechnaib*.

[188] Meyer, *Sanas Cormaic*, 62 § 730; cf. Dinneen, *Foclóir* s.v. *gioradán*; Dwelly, *Faclair* s.v. *gioradan*.

[189] *LEIA* B-9 s.v. *bairenn*. *Bairnech* is cognate with Welsh *brenigen*, Breton *brenniken* of the same meaning.

[190] Hencken, 'Lagore crannóg', 229; McCormick, 'Stock-rearing in early Christian Ireland', 188 and table 6.4.

[191] Harbison, *High Crosses* i 39 (discussion); i 376–7 (dating); ii fig. 107 (photograph).

[192] E.g. O'Keeffe, *Buile Ṡuibne*, 40.1124; 55.1548.

[193] E.g. Meyer, *Sanas Cormaic*, 85 § 986.

[194] Many kinds of wild duck are found in Ireland, of which the commonest is the mallard (*Anas platyrhynchos*).

Fig. 15. What is probably a representation of a hunting scene on a ninth-century high cross at Castledermot, Co. Kildare. Two huntsmen pursue a boar (top left), an unidentified animal (middle left), a hare (middle right) and a deer (bottom right). The animal facing the spear-carrying hunter on the bottom left is probably his dog. On the top right is a representation of a large bird, possibly a swan or goose.

froích, male *cailech froích*). According to Giraldus Cambrensis, grouse (Latin *grutae*) were few and rather small in Ireland.[195]

In the Old Irish sagas, various warriors are portrayed as hunting the *géis*, a bird notable for its white plumage and generally taken to be a swan.[196] It is thought that the mute swan (*Cygnus olor*) is a post-Norman introduction, so it seems most likely that *géis* is to be identified with the whooper swan (*Cygnus cygnus*). Another Old Irish term for swan is *ela*,[197] which seems also to be used of the whooper swan, as a Life of Saint Colmán refers to these birds singing to his monks as they constructed a causeway.[198]

Various species of wild goose winter on Irish coasts, and must have been a useful source of meat to those living nearby. Old and Middle Irish texts distinguish the *gigrann* and the *cauth* (later *cadhan*).[199]

[195] O'Meara, *Giraldus: Topography*, 41 § 10 = Dimock, *Giraldi Topographia*, 47.5.

[196] E.g. *TBC* Rec. I 24.781. The phrase *gilithir géis* 'as white as a *géis*' is well attested: see *DIL* s.v. 1 *géis*.

[197] E.g. *Thes* ii 47.19 cygnis .i. *elaib*.

[198] Plummer, *Bethada* i 172 § iii (15); trans. ii 166 *dothiccdis eladha gacha trātha do chantain ciuil doibh* 'swans used to come every hour to sing to them'.

[199] For example, the two terms are distinguished in the YBL version of *Cormac's Glossary* (Meyer, *Sanas Cormaic*, 29 § 334; 60 § 709), and in the King and Hermit dialogue (Murphy, *Early Irish lyrics*, 16 (poem 8) § 25).

On the basis of modern usage,[200] the former is generally identified with the barnacle goose (*Branta leucopsis*),[201] and the latter with the brent goose (*Branta bernicla*). The shellfish *Lepas anatifera* 'goose barnacle' normally lives attached to boats and driftwood; it bears a certain resemblance to the barnacle goose, and in many languages goes by the same name.[202] It was widely held in medieval times that goose barnacles were the young of barnacle geese. In his *Topography of Ireland* Giraldus Cambrensis repeats this theory, and remarks that in some parts of Ireland bishops and religious men ate barnacle geese during times of fasting on the grounds that they were not classified as fleshmeat.[203] In Old and Middle Irish texts, the word *géd* is attested only in relation to the domestic goose (see p. 105), but it is likely that – as in the later language – *géd fíadain* 'wild goose' was applied to the grey-lag goose (*Anser anser*) and other grey geese.

Giraldus comments on the absence of partridges and pheasants.[204] He is doubtless correct about the pheasant (*Phasianus colchicus*), which is a post-Norman introduction, but the partridge (*Perdix perdix*) may be native.[205] He states also that the quail (*Coturnix coturnix*) is plentiful.[206] His account of an abundance of wild peacocks (Latin *pavones*) in the woods has been taken as a reference to the capercaillie (*Tetrao urogallus*).[207] The male of this large grouse – almost as big as a turkey – has a glossy blue-green breast somewhat similar to that of a peacock. The female is considerably smaller. This bird is known to have been common in the woods of Ireland and Scotland until hunting and loss of habitat brought about its extinction in both countries in

[200] Dictionaries and word-lists of Modern Irish and Scottish Gaelic sometimes give conflicting translations. For example, Dwelly (*Faclair*) translates both *giodhran* and *cathan* as 'barnacle goose'. The Irish surname Ó Cadhain is likewise sometimes anglicized Barnacle.

[201] E.g. Dinneen, *Foclóir* s.v. *giúghrann*. The Welsh cognate *gwyran* is also well attested in the same meaning (*GPC* s.v. *gwyran*; *VKG* i 101–2).

[202] For example, Scottish Gaelic *giodhran* means both 'barnacle' and 'barnacle goose' (Dwelly, *Faclair* s.v.).

[203] O'Meara, *Giraldus: Topography*, 41–2 § 11 = Dimock, *Giraldi Topographia*, 47–8.

[204] O'Meara, *Giraldus: Topography*, 47 § 17 = Dimock, *Giraldi Topographia*, 56.15–16.

[205] Van Wijngaarden-Bakker, 'The animal remains from Newgrange (1)', 359.

[206] O'Meara, *Giraldus: Topography*, 41 § 10 = Dimock, *Giraldi Topographia*, 47.11.

[207] O'Meara, *Giraldus: Topography*, 41 § 10 = Dimock, *Giraldi Topographia*, 47.4.

the eighteenth century.[208] In Scotland it bore the extraordinary name *capull coille* 'horse of the wood', possibly on account of the clopping noises made by the males at their display. This was borrowed into Scots as *capercaillie, capercailye,* etc., the earliest attestation being in 1530.[209]

Gerard Murphy must be correct in taking Irish *cailech feda* 'cock of the wood' to refer to the capercaillie.[210] The only other woodland grouse native to these islands is the black grouse (*Lyrurus tetrix*), but it is believed that it never extended its range to Ireland.[211] The identification of *cailech feda* with the capercaillie is also supported by a short Old Irish passage attributed to Flann Fína.[212] In answer to the question 'Who are the proudest you have met?', the reply is given: 'the men of Muskerry and cocks of the wood' (*Múscraige ⁊ coiligh fedha*). This suits the character of the cock capercaillie, who defends his territory with exceptional boldness, and will attack human beings as well as other birds. Groups of capercaillie cocks also engage in a complicated ritual prior to mating, in which each one displays formalized aggressive postures.[213] In a verse preserved in commentary to the *Amrae Choluimb Chille*, the celebrated lover Diarmait recommends to the beautiful Gráinne the flesh of the *cailech feda*, taken with a drop of mead.[214]

Trapping

The legal aspects of bird-trapping are dealt with in the law-text *Bretha Forma* 'judgements of trapping', which is also referred to as *Bretha Sén Forma* 'judgements relating to nets for trapping' and *Sénbretha* 'net-judgements'.[215] This text has not survived, but is probably the source of two quotations in *O'Davoren's Glossary*.[216] A short commentary on trapping birds on another person's

[208] Ussher and Warren, *Birds of Ireland*, 230. It was successfully reintroduced to Scotland from Sweden in 1837 (Fraser Darling, *Natural history in the Highlands and Islands*, 75–6).

[209] Craigie, et al., *A dictionary of the older Scottish tongue* s.v. *capercailȝe*.

[210] See G. Murphy, *Duanaire Finn* iii, glossary, 240 s.v. *coileach feadha*; Fox and Kelly, 'Capercaillie in Ireland'.

[211] Ussher and Warren, *Birds of Ireland*, 230–1.

[212] Meyer, *Tecosca Cormaic*, Preface vi–vii.

[213] Fraser Darling, *Natural history in the Highlands and Islands*, 132.

[214] Stokes, 'The Bodleian Amra', 264 § 73.

[215] For references, see *GEIL* 276 § 52.

[216] *CIH* iv 1502.15; 1526.1–2 = O'Dav. 355 § 932; 460 § 1480.

property is also based on this text.[217] The commentary lays down detailed regulations on the division of a bird's flesh and feathers between the trapper and the landowner. The proportion depends on whether the trapper has secured permission or not, and on whether the landowner is a king, lord, commoner or the Church. The higher the rank, the lower the proportion retained by the trapper. In the case of trapping on church land without permission, the trapper must do forty nights penance, as well as surrendering two thirds of the bird's flesh and all its feathers. But if he has secured permission, he only surrenders one third of the flesh and two thirds of the feathers.

It is clear from the second quotation preserved in *O'Davoren's Glossary* that there is no penalty for trapping a very small bird (*minnta*),[218] or two birds held to be nuisance, i.e. a heron (*corr*)[219] or a hawk (*séig*). See under 'hawking' below.

Hunting with missiles

There are some references in the sagas to the hunting of birds with missiles, usually stones released from a sling (*tailm*).[220] For example, Cú Chulainn is represented as having stunned twenty wild swans by this method.[221] In another saga, Cano mac Gartnáin unsuccessfully attempted to hit swans and wild duck;[222] in this case it is not clear what sort of missile was used – it may have been a spear.

In a series of glossed quotations on various legal topics, we find the phrase *aurchur la cuirr*, which may mean 'a shot after a heron (or crane)'.[223]

Hawking

The use of trained hawks and other raptors to catch game-birds was probably introduced to England from France during the ninth century. After the Norman invasion, hawking became increasingly popular among the aristocracies of England, Scotland, and Wales.

[217] *CIH* vi 2108.24–9.
[218] *Minnta* (also *minntán*) is used of any small bird, but particularly tits (*Paridae*).
[219] See p. 125.
[220] Knott, *Togail Bruidne Da Derga*, 4.132; Dillon, *Serglige Con Culainn*, 3.64. Another word for sling is *táball*: see Bauersfeld, 'Die Kriegsalterümer', 305–8, for a discussion of possible differences between a *tailm* and a *táball*.
[221] *TBC* Rec. I 24.785–7.
[222] Binchy, *Scéla Cano*, 6.150–7.171.
[223] *CIH* iii 786.15–16. See p. 125 above for a discussion of the meanings of *corr*.

This is reflected in the laws of these countries; for example, thirteenth-century Welsh law-texts contain provisions aimed at protecting hawks,[224] and devote much attention to the rights and duties of the royal falconer.[225]

There is no clear evidence of hawking in Old Irish texts, and it must be significant that the heroes of the eighth- and ninth-century sagas are generally portrayed as having hounds but not hawks. It is only later that the hunting hawk makes its definite appearance in Irish literature, as for example in the twelfth-century Life of Saint Colmán mac Lúacháin where there is a reference to two hunting hawks (*dá seabac selga*[226]) belonging to a king.[227] Early Irish law clearly regarded the hawk as vermin to be exterminated rather than an asset to be protected. Anybody is legally entitled to trap a hawk (*séig*) – even on another person's land without permission. A glossator explains that this is because of the hawk's depredations on young pigs and hens.[228] See p. 189 above.

Writing in the late twelfth century, Giraldus Cambrensis remarks on the abundance of hawks and falcons in Ireland,[229] and there are many references to the use of raptors for hawking in post-Norman documents. For example, the fourteenth-century Red Book of Ormond mentions the sparrowhawk (*speruarius*), the goshawk (*ostorus*), and the osprey (*nisus*).[230] A fifteenth-century medical manuscript in Irish contains a drawing of two men (wearing crowns), one of whom has a hawk on his arm. Nearby, there is a dog and a tree with three birds in it.[231]

[224] Williams and Powell, *Llyfr Blegywryd*, 114.10 = *ALIW*, Dimetian Code, II viii § 37.
[225] Williams and Powell, *Llyfr Blegywryd*, 13.21–14.31 = *ALIW*, Dimetian Code, I xiii §§ 1–15; Wiliam, *Llyfr Iorwerth*, 7–8 § 9 = *ALIW*, Venedotian Code, I x §§ 1–18. For the legal situation in medieval Scotland, see Gilbert, *Hunting*, 68–72.
[226] The word *seboc* seems to be an early borrowing from Old English *heafoc* 'hawk', perhaps via Welsh, *LEIA* S-59–60.
[227] Meyer, *Betha Colmáin*, 44.6–7 § 45. The translation of *fidchúach* as 'falcon' at Meid, *Táin Bó Fraích*, 13.329, is uncertain.
[228] *CIH* iv 1526.1–2 = O'Dav. 460 § 1480.
[229] O'Meara, *Giraldus: Topography*, 38 § 8 = Dimock, *Giraldi Topographia*, 34.16.
[230] N. B. White, *The Red Book of Ormond*, 183–4 (Appendix to Index).
[231] King's Inns MS 17 f. 1^v (De Brún, *Catalogue of Irish manuscripts in King's Inns Library*, 45–9).

Gathering

As in the case of hunting and fowling, there are legal restrictions on gathering. Normally, it is an offence to gather wild fruit or herbs on another person's land. It is stated in an Old Irish law-text that a fine – which later glossators take to be five *séts* – must be paid for taking wild garlic, seaweed or wild apples if they are privately owned.[232] According to a twelfth-century commentator on this text, a different fine (*éraic bairr* 'payment of tree-top') is due if the fruit is picked from the trees rather than from the ground.[233] It seems likely that this would be a heavier fine, though this is not made clear in the commentary. In certain circumstances, private ownership must give way to urgent need. For example, any law-abiding person (*recht*) is entitled to gather medicinal herbs required by an invalid, wherever they may grow.[234] Similarly, a hungry person may take a handful of hazel-nuts from a privately-owned wood.[235]

On common land a law-abiding person usually has unrestricted rights of gathering. There is a general entitlement to 'the pickings of any wood' (*cnúas cach ḟeda*) provided – as a glossator points out – it is not a private wood.[236] A similar entitlement to gather wild garlic on common land is also mentioned.[237] But even on common land, there is evidence that private ownership could be established – perhaps only temporarily – on a crop. The law-text on distraint refers to 'an appropriated tree which is in the wilderness'.[238] According to one of the glosses on this passage, the tree may have been appropriated because of its fine fruit.[239] There is a hint in a gloss on another text that the appropriation of wild apples may have been the prerogative of a king or lord.[240]

The evidence in the law-texts with regard to the ownership of seaweed is similarly ambiguous, and is made even more difficult to interpret because of the various uses to which seaweeds are put:

[232] *CIH* i 238.31 = *AL* v 474.1 *crim feam fiadubull dia rotēchtaidter*.
[233] *CIH* v 1874.9–10.
[234] *CIH* i 242.17–18 = *AL* v 486.12–13 *losa bro[th]chāin cacha muige* 'herbs for [an invalid's] broth from every plain'.
[235] *CIH* ii 571.2–3 = Binchy, 'An archaic legal poem', 157.45–8.
[236] *CIH* i 241.22 = *AL* v 482.26.
[237] *CIH* i 241.20 = *AL* v 482.23 *crim allda .i. arnā bī techtugud*.
[238] *CIH* ii 395.23 = *AL* i 202.2–3 *im crand ngabāla bīs i ndīthrib*.
[239] *CIH* iii 892.7 *is crand co mmes cain̄*. I find the other glosses difficult to understand.
[240] *CIH* i 238.34–239.1 = *AL* v 474.16–17.

human food, animal fodder, and soil fertilizer. In practice it seems likely that – as today – local custom governed the use of seaweed, with different conventions for seaweed cast up on the strand, seaweed growing on rocks exposed at low tide, and seaweed in deeper water.[241]

There is a general entitlement to the seaweed of every strand (*fem cacha trachta*) and the dulse of every rock (*duilix cacha cairrge*),[242] though a glossator adds the caveat in both cases that this does not apply if there is private ownership. A reference in a law-text on land-value also suggests that seaweed could be privately owned. Thus the presence of a 'productive rock' adds three cows to the value of the adjacent land-holding.[243] Later commentary places a lower value on access to seaweed, stating that it is a 'cow for a productive rock from which dulse or [other] seaweed is cut'.[244]

Our sources devote a good deal of attention to the gathering of nuts and wild fruit, generally referred to collectively as *mess.* According to the wisdom-texts, a good crop on the trees is one of the signs of a just ruler. Hence, the seventh-century *Audacht Morainn* states 'it is through the justice of the ruler that abundances of great tree-fruit of the great wood are tasted'.[245] Conversely, a ruler's injustice is believed to cause failure of tree-fruit.[246]

The annals regularly record years in which the tree-fruit was particularly abundant, sometimes distinguishing the nut-crop (*cnómess*) for human consumption from the acorn-crop (*daurmess*) for pigs.[247] Thus the *Annals of Ulster* record that in the year 835 there was such a heavy crop of nuts and acorns that the streams were dammed up and ceased to flow.[248] In an entry in the

[241] For law-cases involving the cutting or gathering of seaweed in nineteenth- and twentieth-century Ireland, see Osborough, 'Roman law in Ireland', 260–4.
[242] *CIH* i 241.25–6 = *AL* v 482.30–484.1.
[243] *CIH* ii 676.9 = Mac Niocaill, '*Tír cumaile*', 82.36.
[244] *CIH* iii 972.27 = Appendix A, text 6 § 3.
[245] Kelly, *Audacht Morainn*, 6 § 17 *Is tre fír flathemon ad- manna mármeso márfedo -mlasetar.* Cf. Meyer, *Tecosca Cormaic*, 2 § 1.22 *mess for crannaib* 'fruit on trees'.
[246] *CIH* i 219.18 = *AL* iv 52.8 *millead measa* 'destruction of tree-fruit'.
[247] See p. 83.
[248] *AU*[2] 294 s.a. 835 (recte 836) § 8 *Mes mōr eter cnōmes 7 daurmes 7 ro iadh glasa cor ansat di riuth.* Cf. *AU*[2] 504 s.a. 1066 § 3 *Cnōmhes mōr i nErinn uile ut rebellat fluminibus* 'a great nut-crop in all Ireland so that it hindered the rivers'.

Annals of Inisfallen, a three-fold distinction is made between the acorn-crop (*mess*), the nut-crop and the apple-crop.[249]

Nuts

A good deal of the countryside at our period was covered with scrub made up largely of hazel (*Corylus avellana*, Irish *coll*). The nutritious and palatable nut (*cnú, cnó*) of this tree can be kept for up to a year, and must therefore have been a particularly valuable winter-food. It is clear that there was a trade in hazel-nuts: the *Annals of Ulster* record that in the year 1097 nuts were so abundant that a *sesedach*[250] could be bought for a penny (*pinginn*).[251] This year was known as the 'year of the white nuts' (*blíadain na cnó finn*).

Another type of nut mentioned in the law-texts is the *cnó gnáe* 'beautiful nut', which is listed among the valuables which can be given as pledges.[252] This is possibly a reference to the nuts of various tropical plants (especially *Entada gigas*) which are brought to the coasts of Ireland from the West Indies by the North Atlantic drift. Such nuts are of course inedible, but have traditionally been prized as charms because of their attractive appearance. A legal commentator, probably writing in the twelfth century, puts forward the explanation that these are nuts which grow on a tree 'in the eastern world'.[253] The eastern world (*in domun toir*) is a common term in Irish literature for far-off parts of the world – viewed of course with a pre-Columbus perspective.

Wild fruit

The wild fruit most frequently mentioned in our sources is the wild apple (*fíaduball*). Because of its fruit, the wild apple-tree (*Malus sylvestris*, Irish (*fíad*) *aball*) is included among the 'seven nobles of the wood', i.e. the seven most valuable trees.[254] A legal commentator classifies common fruits into two categories: *cumra* 'sweet', and *fíadain* 'wild, sour, bitter'. He distinguishes the cultivated or sweet apple (*uball cumra*) from the wild apple (*uball fíadain*), and the cultivated plum (*áirne cumra*) from the sloe (*áirne fíadain*).[255]

[249] *AI* 402 s.a. 1310 § 3 *torad mor do meas 7 do cnoibh 7 d'ublaib.*
[250] See p. 583.
[251] AU^2 532 s.a. 1097 § 8. Part of this entry is a later addition.
[252] *CIH* ii 472.9 = *AL* v 406.22; *CIH* iv 1486.12 = O'Dav. 284 § 539.
[253] *CIH* iii 965.39.
[254] *CIH* i 78.18, 203.2 = *AL* iv 146.22.
[255] *CIH* v 1874.13–16. For a discussion on cultivated fruit-trees, see p. 259.

He applies this sweet/sour distinction to various other fruits, all of them wild. He classifies blackberries (*sméra*), bilberries (*fraochóga*), hazel-nuts (*cna*), and strawberries (*suba*) as 'sweet', and acorns (*dercain*), haws (*scechóra*), hips (*mucóra*) and rowanberries (*caora*)[256] as 'sour'. The implication seems to be that even the fruit which is classified as 'sour' might be used for food, though obviously less relished than sweet fruits.

The archaeological evidence confirms that many kinds of wild fruit, both sweet and sour, were consumed. Blackberry and elderberry seeds, dated by radio-carbon to the eighth or ninth centuries, have been excavated at Scotch Street, Armagh.[257] An eleventh-century pit in Winetavern Street, Dublin, contained a large variety of seeds of wild fruit, including rowanberries, blackberries, wild apples, sloes, hips and haws.[258] Some Old and Middle Irish literary references suggest that acorns – normally valued as food for pigs (see p. 83) – were consumed by humans when other more palatable food was in short supply.[259]

The importance of the bilberry (*Vaccinium myrtillus*) is clear both from archaeological and written evidence. Seeds from this fruit (Irish *fraochóg*, *fráechán*, anglicized *fraughan*, etc.) are fairly common in the excavations of Viking and Anglo-Norman Dublin.[260] Sources in the Irish language treat the bilberry as a valuable crop: a Middle Irish text on the entitlements of kings refers to *fráechmess Bríg Léith* 'the bilberry harvest of Brí Léith'.[261] Another member of the Heath family (*Ericaceae*) with edible berries is the cranberry (*Vaccinium oxycoccos*). It grows on bogs, and consequently the Irish names for its berry, *mónann* and *mónóg*, are derivatives of *móin* 'bog'. The ninth-century King and Hermit dialogue lists these two berries together: *mónainn derca, dercna froích* 'red cranberries, bilberries'.[262]

256 *Cáer* (later *caor*) has the general meaning 'berry', but here is likely to refer to edible berries of inferior flavour, particularly the rowanberry.

257 Hamlin, 'The archaeology of the Irish church in the eighth century', 297.

258 Mitchell, *Archaeology and environment in early Dublin*, 27.

259 E.g. Murphy, *Early Irish lyrics*, 14 (poem 8) § 22; O'Keeffe, *Buile Ṡuibne*, 12.334. For records outside Ireland, see Zohary and Hopf, *Domestication of plants*, 195–6.

260 Mitchell, *Archaeology and environment in early Dublin*, 25, 27.

261 Dillon, 'Taboos of the kings of Ireland', 8 § 1. For the name of this hill (near Ardagh, Co. Longford), see Hogan, *Onomasticon*, s.v. Brí léith; E. J. Gwynn, *Metrical Dindshenchas* v 183 (Index).

262 Murphy, *Early Irish lyrics*, 14 § 20. Cf. Dillon, *Stories from the Acallam*, 5.145 *mónainn máetha* 'tender cranberries'.

Stones of the wild cherry (*Prunus avium*) have been found in a number of excavations, most abundantly in the Late Bronze Age stratum at Ballinderry crannóg no. 2.[263] In my discussion of the Old Irish tree-list, I tentatively suggest that the tree named *idath*, which is included in the second class of trees, is to be identified with the wild cherry.[264] I also suggest that the *crann fir*, which is included among the third class, is the juniper (*Juniperus communis*).[265] A verse in the King and Hermit dialogue may contain references to the fruit of both these trees: *cáera ibair, idaith, fir* 'berries of yew, cherry(?), juniper(?)'.[266]

Wild herbs and roots

It is likely that a wide range of wild herbs and roots were eaten in our period, particularly in times of famine, and by landless people at the margins of society. Our sources concentrate on two wild plants of economic importance: wild garlic and watercress. In the saints' Lives, there are some references to other plants, such as nettle and sorrel, whose consumption is often associated with poverty.

There is also information on edible plants in other sources, particularly in nature poetry and Fenian tales. For example, the ninth-century King and Hermit dialogue idealizes the life of a hermit in the woods and pictures him subsisting on various berries, herbs and roots. But such material must be used with caution, as the poet's view of nature is a romantic one, and may not always be based on reality.[267]

Crim[268] (Wild garlic)

Wild garlic (*Allium ursinum*) is a frequent plant of woods and damp shady places; its usual habitat is reflected in the fairly common placename *Cremchaill* 'garlic wood'.[269] In our period

[263]Hencken, 'Ballinderry crannóg no. 2', 21; cf. Hencken, 'Ballinderry crannóg no. 1', 121.

[264]Kelly, 'The Old Irish tree-list', 115.

[265]Ibid. 119.

[266]Murphy, *Early Irish lyrics*, 12 § 15. The MS has *coera iob- caora fir* with *nó fidhuid* in the margin; I discuss Murphy's and my emendations in Kelly, 'The Old Irish tree-list', 115.

[267]For a general discussion of this poetical genre, see Ó Corráin, 'Early Irish nature poetry?'.

[268]*Crim* (*crem, craum*) is cognate with medieval Welsh *craf*, Old Breton *cram* of the same meaning, *LEIA* C-229 s.v. *crem*.

[269]Hogan, *Onomasticon*, s.v. *Cremchaill*; N. Williams, *Díolaim luibheanna*, 47.

it was evidently regarded as particularly important, as it is specifically mentioned in the law-text *Di Astud Chirt ⁊ Dligid.*[270] As in later times, wild garlic may have been prized for its medicinal properties.

A client is obliged to provide an annual *crimfeis*, 'garlic feast', for his lord.[271] If he fails to do so, he must pay his lord a fine of three *sét*s. Judging from later glosses, it seems that this feast consisted of garlic with cheese and milk, and that it took place before Easter (*ar cháiscc*).[272] It is implicit in another legal commentary that the garlic feast is a rather low-key affair, as the lord is accompanied by only three retainers.[273]

Biror[274] (Watercress)

Watercress (*Rorippa nasturtium-aquaticum*) features in ascetic diets – see p. 344 – and is mentioned with special frequency in the twelth-century poetry attributed to Suibne Geilt. He addresses this plant in the following terms:

a bhiorair, a barrghlasáin
do bhrú thobair luin

'o watercress, o green-topped one
from the edge of the blackbird's well'.[275]

In another poem he alludes to the fact that watercress continues to grow on in the winter:

sásadh biorair bairrghlais búain,
deogh uisge fhúair a glais ghlain

'a meal of green-topped longlasting watercress,
a drink of cold water from a pure stream'.[276]

[270] *CIH* i 238.31 = *AL* v 474.1; *CIH* i 241.20 = *AL* v 482.23.
[271] *CIH* v 1910.24–6; cf. *CIH* ii 498.25 = *AL* ii 326.7. See a brief discussion by Carey, 'Sequence and causation', 72.
[272] *CIH* iii 918.38.
[273] *CIH* iv 1269.1–2.
[274] *Biror* (later *bilar*) is cognate with Welsh *berwr*, Breton *beler* of the same meaning, *LEIA* B-52.
[275] O'Keeffe, *Buile Ṡuibne*, 35.982–3.
[276] Ibid. 64.1813–14.

Fochlocht (Brooklime)

Many texts refer to *fochlocht* (also *fothlucht*),[277] which is sometimes identified with the common edible water-plant brooklime (*Veronica beccabunga*).[278] This identification is supported by the reference in *Buile Ṡuibne* to *fothlocht fann foda* 'long prostrate fothlocht'.[279] A less satisfactory identification[280] is with lesser water-parsnip (*Berula erecta* = *Sium erectum*), which is frequent in east-central Ireland, and occasional elsewhere.[281] However, there appear to be no records of this plant being eaten, and it is reported as being a skin-irritant.[282] Furthermore, its mode of growth is not prostrate, as indicated in the *Buile Ṡuibne* reference quoted above. Great water-parsnip (*Sium latifolium*) is also an unlikely identification as it is said to be poisonous, and to have caused the death of cattle.[283]

Fochlocht was of sufficient importance as a foodstuff to feature in a proverbial expression quoted in the law-text *Bretha Nemed Toísech*.[284] The context refers to the relationship of a kin-member with a promiscuous woman, as a result of which a child of doubtful paternity may gain entry into the kin-group. Such a child is described metaphorically as *áth i fochlucht*. This metaphor seems to refer to a poisonous plant gathered along with edible plants. I believe that *áth* refers here to water-dropwort (*Oenanthe*), a genus which grows in similar situations to brooklime.[285] Water-dropworts are poisonous – especially hemlock water-dropwort (*Oenanthe*

[277]The etymology is unknown. The Scottish Gaelic form is *folachdan*, explained by Dwelly, *Faclair*, as 'water-salad, water-parsnip (*Sium augustifolium*)'.

[278]E.g. N. Williams, *Díolaim luibheanna*, 100. Its edibility is discussed – without much enthusiasm – by Mabey, *Food for free*, 133–4.

[279]O'Keeffe, *Buile Ṡuibne* 12.331. It is clear from this passage that *fothlocht* is a freshwater plant, and cannot be the seaweed laver (*Porphyra umbilicalis*), as suggested by Cecile O'Rahilly at *TBC* Rec. I 258 (note to ll. 1171–4).

[280]See *DIL* s.v. *fochlocht*.

[281]Webb, *An Irish flora*, 71.

[282]Mitchell and Rook, *Botanical dermatology*, 707.

[283]Forsyth, *British poisonous plants*, 43.

[284]*CIH* vi 2230.3. Cf. *CIH* vi 2116.4 = Dillon, 'Stories from the law-tracts', 48.

[285]Glossators explain *áth* (*áith*) as *áthaba* (*áithaba*), e.g. *CIH* vi 2116.8; iv 1139.12. Cf. Meyer, *Sanas Cormaic*, 7 § 63 (where there is a reference to the deadly nature of the plant); 87.21 § 1018. The second element of *áthaba* is probably the gen. sing. of *ab* 'stream'. In the later language, it may be preceded by *t-* or *d-*, e.g. Dinneen, *Foclóir*, s.v. *dáthabha, táthabha* (see Marstrander, 'Review of *Lexique: A* (ed. Vendryes)', 225 s.v. *áthaba*). The word has been equated with deadly nightshade (*Atropa bella-donna*), and hellebore (*Helleborus foetidus, Helleborus viridis*) as well as water-dropwort: see Beckett, *Fealsúnacht Aodha Mhic Dhomhnaill*, 269. N. Williams takes it to refer here to

crocata) – so if a plant became mixed up in a basket of brooklime, the result might be fatal. Hence the proverbial expression 'water-dropwort in brooklime'.

Other wild plants

The saints' Lives contain a number of references to the consumption of nettle (*Urtica dioica*, Irish *nenaid, nenntóc*) and sorrel (*Rumex acetosa*, Irish *samad*). For example, it is claimed in a Middle Irish Life of Saint Cóemgen that he lived for seven years with no food but nettle and sorrel.[286] In another Life of this saint, he is said to have had miraculous access to a supply of wild garlic, sorrel and other edible herbs throughout the year.[287] Here the word used for sorrel is *selgán*.[288]

The twelfth-century tale *Buile Ṡuibne*[289] gives a list of edible plants, which includes wood-sorrel (*Oxalis acetosella*, Irish *simsán, simsóg*). Because it has three leaves, the term *semar* (*semróc*) was sometimes applied to this plant as well as to trefoils such as *Trifolium repens* and *Trifolium dubium*. This gave rise to the tradition among later English writers that the Irish ate shamrock. The *Buile Ṡuibne* list includes other edible plants whose identity is obscure to me: *miodhbhun*,[290] *lus-bían*, and *biorragán*.

Another unidentified edible plant is *trechlam*, mentioned in *Táin Bó Cúailnge*.[291]

Wild roots

The roots of a number of wild plants were eaten; of these the most prized seems to have been *curar*.[292] This is probably to be identified with the common umbellifer *Conopodium majus* or pignut, which

hemlock (*Conium maculatum*), but points out that it is also used of water-dropwort (*Díolaim luibheanna*, 102, 120).

[286] Plummer, *Bethada* i 126 § viii (13); trans. ii 122.

[287] Plummer, *Bethada* i 165 § xx (39); trans. ii 160.

[288] Cf. Dwelly, *Faclair* s.v. *sealgag*; Cregeen, *A dictionary of the Manks language* s.v. *shughlaig*.

[289] O'Keeffe, *Buile Ṡuibne*, 12.331–34.

[290] This could refer to some plant with a sweet root or stem-base, i.e. *mid* + *bun*.

[291] *TBC* Rec. I 37.1173. The editor (p. 157) translates 'seaweed'.

[292] *Curar* is cognate with Welsh *cylor*, Breton *keler* of the same meaning (*LEIA* C-295). The Irish diminutive *cularán* is also used of various non-native plants with swollen tubers, and of cucumber. For a list of attestations, see Ní Chatháin, 'Swineherds, etc.', 207–9.

has an edible globular tuber.[293] In the ninth-century King and Hermit dialogue, *curar* is included among the sweet-tasting plants which are eaten at the coming of summer.[294] In *Tochmarc Ailbe* Finn asks the maiden Ailbe thirty questions, which include 'what is more brittle than *curar* ?'. To this she replies: 'the temper of a jealous woman'.[295] The tuber of the pignut is in fact particularly brittle and snaps easily.

Another edible root mentioned in the King and Hermit dialogue is *melle*, which is presumably a derivative of *mell* 'sphere, swelling'.[296] It has been identified with a plant called bitter vetch or heath-pease (*Lathyrus montanus*), which has small edible tubers.[297] This root is also mentioned in a poem in *Tochmarc Ailbe*.[298] In modern times it is called *corra meille* in Donegal[299] and *carra-meille* in Scotland.[300]

The early twelfth-century *Aislinge Meic Con Glinne* refers to *briscén* in a context which indicates an edible root.[301] I would take this to be the silverweed (*Potentilla anserina*), whose root was widely consumed in Ireland and Scotland until the early part of this century.[302] Various forms of the word survive in the modern dialects, such as *brioscán, blioscán, briosclán, brisgein*.[303]

Seaweed

As we have seen (p. 43), there is evidence that seaweed was valued as a feed for livestock. Some seaweeds also featured in the human diet, supplying iodine, iron, and other trace elements. Sometimes the texts do not specify a particular type of seaweed, and use a general word such as *femmar*. For example, in the saga *Táin Bó Cúailnge* Cú Chulainn offers Fergus a handful of

[293] Mabey, *Food for free*, 100; cf. Ní Chatháin, 'Swineherds, etc.', 209; Beckett, *Fealsúnacht Aodha Mhic Dhomhnaill*, 268 s.v. *cutharlán*; R. B. Breatnach, *Seana-chaint na nDéise II*, 99 s.v. *clutharacán*; Dinneen, *Foclóir* s.v. *cluthairicín*.

[294] Murphy, *Early Irish lyrics*, 14 (poem 8) § 23.

[295] Thurneysen, '*Tochmarc Ailbe*', 272 § 11.

[296] Thurneysen, 'MIr. *melle*'.

[297] Murphy, *Early Irish lyrics*, 14 (poem 8) § 20. See his glossary, s.v. *melle*.

[298] Thurneysen, '*Tochmarc Ailbe*', 278 § 8.

[299] Dinneen, *Foclóir* s.v. *carra mhilis*.

[300] Dwelly, *Faclair* s.v. *carra-meille*.

[301] Jackson, *Aislinge* 38.1194–5 = Meyer, *Aislinge*, 99.20. For a discussion of the whole passage, see p. 256 above under *cerrbacán* 'skirret'.

[302] Mabey, *Food for free*, 84; Carmichael, *Carmina gadelica* iv 118; Mitchell, *Archaeology and environment in early Dublin*, 35.

[303] Dinneen, *Foclóir* s.v. *briosclán*; Dwelly, *Faclair* s.v. *brisgean*; McDonald, *Gaelic words and expressions*, 50 s.v. *brisgein*.

watercress or seaweed (*dornd birair nó femair*), as well as various other foods.[304] More often, however, different types of seaweed are distinguished, though their identity is not always certain.

Duilesc (*Palmaria palmata*)

The most prized seaweed was undoubtedly *duilesc*, generally anglicized *dulse*.[305] It is mentioned quite frequently in legal material,[306] and is included in the eighth-century law-text *Críth Gablach* among the relishes which may be expected in a commoner's diet.[307] From a reference in *Aislinge Meic Con Glinne*, it seems that this seaweed was hung up in a sheaf (*scúap dulisc*) from which a handful was taken as required.[308]

Another word given in the glossaries for this seaweed is *fithrech* (*fithrach*).[309]

Medbán

Medbán[310] is probably to be identified with the reddish-brown seaweed laver *Porphyra umbilicalis*. In modern times it is normally cooked and then eaten with oatmeal, bread or potatoes. It has a high protein content and is rich in vitamins B and C. Edible seaweeds of the genus *Porphyra* have a worldwide distribution, and are especially popular in China and Japan.[311]

There is a reference to this seaweed in a poem attributed to Caílte, which is found in the Fenian tale *Cath Finntrága* from about the twelfth century. One verse praises the sea-produce off the islands of the southern coast of Ireland:

Iascach mara muiride
a críchaib Baí is Béire,
medbán Faíde fírglaine,
duilesc a cuanaib Cléire.

[304] *TBC* Rec. I 41.1314; Thurneysen, '*Tochmarc Ailbe*', 278 § 8.
[305] Mabey, *Food for free*, 159.
[306] E.g. *CIH* i 241.25 = *AL* v 484.1; *CIH* ii 384.31 = *AL* i 170.13; *CIH* iii 972.28 = Appendix A, text 6 § 3.
[307] *CIH* ii 563.4; iii 779.23 = *CG* 6.150.
[308] Jackson, *Aislinge*, 35.1077 = Meyer, *Aislinge*, 89.19.
[309] E.g. *CIH* iv 1506.21 = O'Dav. 374 § 1016 *Fithrach .i. duilesc.* The word is also given in Scottish Gaelic dictionaries, e.g. Dwelly, *Faclair* s.v. *fithreach.*
[310] The element *medb-* may mean 'sweet, honeyed, intoxicating' (*LEIA* M-48 s.v. *mid*).
[311] Mabey, *Food for free*, 160.

'Sea-fishing of the sea
from the regions of Dursey and Beare,
laver of truly clear Whiddy,
dulse from the bays of Cape Clear Island'.[312]

Medbán (*meadhbhán*) is still a common word for this seaweed in the south-west of Ireland.[313] But elsewhere the ordinary term is *slabhac*[314] (anglicized *sloke*) or *sl*(*e*)*abhacán*.[315] The same word is used in Scotland; for example, in his *Faclair*, Dwelly defines *slabhagan* as 'a kind of reddish seaweed, formerly cooked until it dissolved, and eaten with bread and butter'. This is obviously laver. In Donegal, *slabhac* is also applied to sea-lettuce (*Ulva lactuca*), an edible green seaweed similar in appearance to laver, but much inferior in flavour.[316]

Murraith

According to a Middle Irish commentator on *Félire Óengusso*, Saint Cíarán of Saiger was not only a vegetarian, but also avoided all milk-products. His meal every evening consisted of a piece of barley bread and a drink of spring water. As relish (*annlann*) he had two roots of sea-fern (*dā mecon do murrathaig*).[317] Possibly this refers to the edible seaweed known as dabberlocks (*Alaria esculenta*), which has a fern-like appearance and is attached to rocks by a branching holdfast. In Breton this seaweed is called simply *raden*, 'fern'.[318]

It must be admitted, however, that there is no other evidence of *raith* developing consonantal inflection. There is the further problem that in the metrical version of this text, *murrāthaig* with a long *ā* makes a better rime with *rom-gnāthaig*.[319] The identification is therefore uncertain.

[312] Dillon, *Stories from the Acallam*, 12.306–9.

[313] Dinneen, *Foclóir* s.v. *meadhbhán*.

[314] Dinneen, *Foclóir* s.v. *sleabhac*. It is the same word as Mid.Ir. *slibac* 'the pith inside a cow's horn' (*CIH* iv 1207.4 = *AL* iii 370.9).

[315] R. B. Breatnach, *Seana-chaint na nDéise II*, 361 s.v. *sleabhcán*.

[316] Cf. Ó hEochaidh, 'Seanchas iascaireachta agus farraige', 37–8.

[317] Stokes, *Félire*, 88.27.

[318] Quentel, 'Goémon et goémoniers', 204.

[319] Stokes, *Félire*, 88.38.

Cairrgín

Cairrgín or 'Irish moss' (*Chondrus crispus*) has useful medicinal and nutritional properties.[320] Dinneen claims that the name derives from Carragheen (Irish *Cairrgín* 'little rock'), in Co. Waterford.[321] But the word is widely used of this seaweed in both Irish and Scottish Gaelic, and seems more likely to be a straightforward derivative with *-ín* (Sc. G. *-ean*) from *carraig* 'rock',[322] with reference to its habitat on stones of the lower shore. The name *cosáinín na carraige* 'the small-footed one of the rock' is recorded for it in the Déise.[323]

I have found no references to the use of this seaweed in early Irish sources.

[320] Mabey, *Food for free*, 161–2; Mac Con Iomaire, *Cladaigh Chonamara*, 184–7.
[321] Dinneen, *Foclóir* s.v. *cairrgín*.
[322] Cf. McDonald, *Gaelic words and expressions*, 62 s.v. *carraigean*.
[323] Sheehan, *Sean-chaint na nDéise*, 204.

10
Diet and Cooking

The main purpose of farming is of course to provide food. In this chapter I deal briefly with the food eaten by the early Irish and the manner in which it was cooked. We are fortunate in the amount of information which Old Irish sources, particularly the law-texts and sagas, supply on this subject. From the Middle Irish period, the satirical tale *Aislinge Meic Con Glinne* is also invaluable, as much of this text consists of detailed descriptions of food.

Types of diet

It is clear from our sources that the staple diet of the early Irish consists of bread and milk. To provide flavour and essential nutrients absent from bread and milk, this diet is supplemented by a variety of other foodstuffs. The supplementary food is often called the *tarsunn*, a term which is generally translated 'relish, condiment'. According to the law-texts, it usually consists of vegetables, salted meat or honey.[1] A distinction is made in some texts between the *tarsunn* and the *annlann*;[2] the latter term is applied to a wider range of foods than the former, and includes butter, other milk-products, salted meat, suet, cabbage and seaweed.[3] A less common term for supplementary food is *sercol*, used in the law-text *Críth Gablach* of

[1]E.g. *CIH* ii 563.16; iii 779.36 = *CG* 7.168–9; *CIH* vi 2298.14–15 = Binchy, '*Bretha Crólige*', 36 § 45. Butter is not usually regarded as being part of the *tarsunn*, e.g. *CIH* ii 564.12; iii 780.19 = *CG* 8.205; *CIH* ii 565.14–15; iii 781.23 = *CG* 11.269. On the other hand, there is a reference to *tarsann imme 7 lomma* 'a relish of butter and milk' in Meyer, *Betha Colmáin*, 58.9 § 55. Commentary at *CIH* vi 2108.6 refers to a 'good *tarsunn*' (*dagtarsann*) which may include beer.

[2]E.g. *CIH* vi 2293.27–8; 2299.31–2 = Binchy, '*Bretha Crólige*', 22 § 27; 38 § 48; Jackson, *Aislinge* 38.1185–6 = Meyer, *Aislinge*, 99.6–8. The etymology of *tarsunn* is unknown (*LEIA* T-35), whereas *annlann* is thought to be a borrowing from Welsh *enllyn* 'anything eaten with bread' (*LEIA* A-76).

[3]E.g. Gwynn and Purton, 'The monastery of Tallaght', 144 § 45; *CIH* ii 483.12 = Thurneysen, 'Aus dem irischen Recht I', 353 § 10; E. J. Gwynn, *The rule of Tallaght*, 64 § 4; Stokes, *Félire* (commentary) 88.27; Meyer, *Betha Colmáin*, 60.17 § 58. Thurneysen, 'Aus dem irischen Recht I', 357, discusses the word *annlandtus*, a derivative of *annlann*, and sugggests that it here means 'Nierenfett' (suet). In a rigorous penitential diet (Binchy, 'The Old-Irish table of penitential commutations', 66 § 28) even milk may be classed as an *annlann*.

seaweed, onions and salt.[4] This word is a borrowing from Latin *ferculum* 'meal, titbit, delicacy'.[5]

Many references in sagas, saints' Lives, and other texts indicate that the main meal (*proind*) was normally eaten in the afternoon or evening.[6] The sagas make occasional mention of light meals and snacks at other times of day. For example, one tale recounts that early in the morning the warrior Óengus Gaíbúaifnech asked a woman who was preparing food for a drink (*deoch*) and a morsel (*mír*) – presumably of milk and bread.[7] In *O'Davoren's Glossary*, a distinction is made between a daytime meal (*díthat*) which may or may not include beer, and a nighttime meal (*feis*) which includes beer.[8] A glossator on the law-text *Bretha Crólige* likewise assumes that a man normally eats a loaf of bread in the day, and another at night.[9] Two versions of *Táin Bó Cúailnge* describe how Cú Chulainn arose late after a night of heavy snow, and then ate a *feis* and a *díthat*, and had a bath.[10] The implication seems to be that these two meals are normally separate, but that the hunger of the hero was so great that he ate them both at one sitting. Our sources also use the term *lón* for a light meal, especially as provisions for a journey on land or sea.[11]

The diet described in the texts is a balanced and healthy one. Cereals and milk provide the bulk of the carbohydrates and proteins required by the body, and animal-proteins are a useful supplement. Vegetables, fruit, seaweed, salt, and honey supply all essential minerals and vitamins. The main problem with this diet is the seasonal nature of many of the foods, and it must often

[4] *CIH* ii 563.3–4, iii 779.23–4 = *CG* 6.149–50 *serccol tarsuinn.*

[5] Eska, 'OIr. *sercol*'.

[6] E.g. *LU* 265.8760 = Windisch, 'Das Fest des Bricriu', 282.15 § 56; Stokes, *Félire*, 88.26; Gwynn and Purton, 'The monastery of Tallaght', 166 (note to 131.11). *Proind* is a borrowing from Latin *prandium*, which was used of the midday meal or luncheon.

[7] *LU* 137.4350.

[8] *CIH* iv 1499.1–2 = O'Dav. 339 § 854; cf. *CIH* ii 525.8–10 = *AL* iii 20.17–19; *CIH* iii 1080.29.

[9] *CIH* vi 2293.37 = Binchy, '*Bretha Crólige*', 22 § 27[6], *bairgen ferfuine i llō ⁊ bairgen i n-oidchi* 'a loaf baked for a man in the day and a loaf at night'. For loaf-size, see p. 330.

[10] *TBC* LL 15.528 *co tormalt feiss ⁊ díthait*; *TBC* St 19.543–4 *goro loing fes ⁊ dīthat an laoi.*

[11] E.g. *CIH* vi 2339.22 = L. Breatnach, *Uraicecht na Ríar*, 110 § 17. See *DIL* s.v. 1 *lón* (b). The examples given under 1 *lón* (a) seem to belong rather under 2 *lón* 'haunch, rump, etc.'.

have been difficult to obtain a balanced diet throughout the year. Food which can be kept for some time – such as cereals, salted meat, nuts, apples, hard cheese, onions – would therefore have been particularly appreciated. When there was failure of the cereal-crop or disease among cattle, there was a danger of famine (see p. 354 below).

The texts sometimes make a distinction between winter food (*gaimbiad*) and summer food (*saimbiad*), especially in the context of the food-rent (*bés*) which a client pays to his lord.[12] The former seems to have consisted mainly of meat and cereals. For example, the law-text on base clientship *Cáin Aicillne* refers to a winter food-rent consisting of a bullock, a flitch of bacon, three bushels of malt, and a half bushel of dried grain.[13] The term 'summer food', on the other hand, is associated particularly with dairy produce (sometimes called *bánbiad* 'white food').[14] A glossator on *Bretha Crólige* gives curds and butter as examples of summer food.[15] Similarly, the summer food-rent (*sambés*) in *Cáin Aicillne* contains a substantial element of dairy-produce: fresh milk, butter, and cream.[16] It also includes a bullock for roasting(?), another for stewing, vegetables, and a wether. Another law-text on clientship likewise refers to the 'wether of summer food' (*molt sambíd*), which a lord receives from his client in August.[17]

The general assumption in our sources is that a person of higher rank enjoys a greater variety of food than a person of lower rank. This is clearly illustrated in *Bretha Crólige,* which deals with the obligation of a person who injures another illegally to provide sick-maintenance (*othras*) for his victim. Such maintenance includes the provision of food appropriate to the injured person's status: the author quotes the principle *bíathad cáich fó míad* 'everyone is to be fed according to his rank'.[18] For example,

[12]E.g. *CIH* ii 563.10–11; iii 779.30 = *CG* 6.160.

[13]*CIH* ii 482.21–6 = Thurneysen, 'Aus dem irischen Recht I', 355–6 § 13.

[14]E.g. Jackson, *Aislinge,* 4.95–6 = Meyer, *Aislinge,* 9.14–15; Meyer, *Hibernica minora,* 67 § 10.

[15]*CIH* vi 2293.40–41 = Binchy, '*Bretha Crólige*', 22 § 27[12].

[16]*CIH* ii 483.30–31 = Thurneysen, 'Aus dem irischen Recht I', 356. A glossator to another text on clientship (*CIH* v 1910.26–7) states that the 'summer morsel' (*samfít*) which a client gives to his lord consists of curds (*gruth*), butter (*im*) and milk (*loim*).

[17]*CIH* iii 920.11.

[18]*CIH* vi 2300.5 = Binchy, '*Bretha Crólige*', 40 § 50.

a high lord (*aire ard*) is entitled to three condiments with his meals: honey, onions, and celery (unless these are forbidden by a physician).[19] The lower-ranking *bóaire febsa*, by contrast, gets only one condiment.[20] It is assumed that a lord eats more meat than a commoner: see p. 336.

FOOD-PRODUCTION AND CONSUMPTION

Food-trade

The evidence of our sources suggests that most food was consumed within the household which produced it, and there is little evidence of trade in foodstuffs in texts of the Old Irish period.

There must have been regular imports of the wine required for the celebration of the Eucharist, and it also featured as a luxury drink at feasts (see p. 358). The lost Old Irish law-text *Muirbretha* 'sea-judgements' evidently referred to the wine-trade, as Cormac states in his *Glossary* that the phrase *escop fína* 'wine-jar' occurs in this text.[21] He explains it as 'a vessel for measuring wine among Gaulish and Frankish traders' (*escra tomais fína la ceandaighaib Gall ⁊ Franc*). The importance of the wine-trade is also indicated by the prominence of Bordeaux (Latin *Burdigala*) – the centre of the wine-trade – in early Irish texts.[22] The name of this town was borrowed into Irish in the form *bordgal* and is used in the eighth-century *Félire Óengusso* in the meaning 'meeting-place, city'.[23]

The establishment of Norse towns on the eastern and southern coasts of Ireland in the ninth and tenth centuries must have stimulated trade, both local and international, in foodstuffs and other commodities. This burgeoning of trade is reflected in the borrowing of Old Norse *markaðr* 'market' into Irish in the form *margad*.[24] This word is used, for example, in the eleventh-century poem on

[19] *CIH* vi 2298.14–15 = Binchy, '*Bretha Crólige*', 36 § 45.

[20] *CIH* ii 563.16; iii 779.36 = *CG* 7.168–9.

[21] Meyer, *Sanas Cormaic*, 45 § 539 (emending *epscop* to *escop*, cf. *CIH* iv 1496.13 = O'Dav. 327 § 790 *eascop fina*). The word *escop* is a borrowing from Latin *scyphus* 'large jar'. Another loan-word connected with the wine-trade is *esarn* (*esern*) 'year-old wine', from Latin *exhibernum* (*vinum*). For a discussion of *escop* and *esarn*, see McManus, 'A chronology', 43–5.

[22] Ó Corráin, *Ireland before the Normans*, 71; Timothy O'Neill, *Merchants and mariners*, 44–8.

[23] Stokes, *Félire*, 20.72; 28.275; 254 § 27; 273.206; 275.254. See *DIL* s.v. *bordgal*; *LEIA* B-72 s.v. *bordgal*.

[24] Marstrander, *Bidrag* 154.

the Fair of Carmun (*Óenach Carmuin*), which describes three markets at this fair: one for food, one for livestock, and one for gold and fine raiment.[25] Evidence for trade in foodstuffs at this period is also provided by commentary on *Muirbretha*, which refers to the transport by ship of honey, wine, and salt.[26]

Food-rent

It seems from the law-texts that by far the most important transfer of foodstuffs in early Irish society was the food-rent which a client (*céile*) paid to a lord (*flaith*) in return for a fief.[27] This fief (*taurchrecc* or *rath*) usually consisted of livestock, but could also comprise land or farming equipment.[28] As far as we can judge from the texts, the system weighed heavily in favour of the lord, who could expect that a significant proportion of the food eaten in his household was produced on the farms of his clients. He received live animals, meat, milk-products, grain, malt, bread, and vegetables. The texts emphasize that all these foodstuffs must be of satisfactory size and quality.[29]

As well as food-rent and labour (see p. 446), the client must also provide winter-hospitality (*cóe*) for his lord. This involves the preparation of a feast between New Year's Day and Shrovetide for the lord and a retinue appropriate to his status.[30] In the fragmentary law-text on clientship, *Di Dligiud Raith 7 Somaíne la Flaith*, there are references to feasts provided by clients at other times of year.[31] For example, before Easter the client provides a 'garlic feast' (*crimḟes*),[32] and at the feast of Saint Martin or *Samain* he prepares a pig.[33]

[25] E. J. Gwynn, *Metrical Dindshenchas* iii 24.305–8.

[26] *CIH* i 315.27–8 = *AL* iii 426.23–5. For a brief account of *Muirbretha*, see *GEIL* 276–7.

[27] See Mac Niocaill, 'Investment in early Irish agriculture'; Patterson, 'Material and symbolic exchange in early Irish clientship'; Patterson, *Cattle-lords and clansmen*, 119–50. A man can be the base client of up to three lords at the same time. However, the fief received from a second lord (*flaith forgíallnae*) must not exceed two thirds of the fief received from the primary lord (*flaith cétgíallnae*). A fief received from a third lord (*flaith cuitrid*) cannot exceed one third of the primary lord's fief (*CIH* ii 488.1–3 = Thurneysen, 'Aus dem irischen Recht I', 367 § 28).

[28] *GEIL* 29–33.

[29] E.g. *CIH* ii 480.1–26 = Thurneysen, 'Aus dem irischen Recht I', 346–8 §§ 7–8.

[30] *CIH* ii 566.18; iii 782.24 = *CG* 13.334.

[31] *CIH* v 1910.22–1911.7.

[32] *CIH* iii 918.38; v 1910.24–5; cf. *CIH* ii 498.25 = *AL* ii 326.7. See p. 309 above.

[33] *CIH* iv 1513.41–2 = O'Dav. 407 § 1194; *CIH* iii 918.38; *CIH* v 1910.22–3. It is not always clear whether the text is referring to a food-rent brought to the lord's house

Feeding of guests, etc.

A person may consume another's food as a guest (*oígi*). As in many other societies, the early Irish householder was under obligation to provide food and drink to a visitor or passer-by. Failure to do so is perceived as the offence of *esáin* 'driving away', and entails payment of the offended person's honour-price.[34] The law-texts limit this obligation in various ways so that it is not abused. The low-ranking *fer midboth* and *ócaire* are exempt because of their limited resources,[35] and a criminal must not under any circumstances be fed or harboured.[36] In the case of the hospitaller (*briugu*), however, the obligation to provide hospitality is explicitly stated to be unlimited: he thereby gains additional status and privilege.[37]

The feeding of the poor has always been regarded as one of the Church's duties, and the saints' Lives recount many episodes in which a cleric provided food for the hungry. The law-text *Córus Bésgnai* calls on the laity to give alms to enable the Church to carry out this task, and also to cater for visitors to the monastery, described in the text as 'guests of God'.[38]

Legal commentary outlines the rules for feeding a craftsman (*sáer*) or other temporary employee.[39] A master-builder is entitled to salted meat (*saill*) as a relish (*tarsunn*) with his bread, whereas any assistant (*gillae*) who works with him only gets salt or a vegetable (*borrlus*).[40] He also has a larger ration of drink, which may be beer or fresh milk, than his assistants. Other craftsmen are allowed smaller rations than the master-builder. For example,

---

or a feast provided for the lord in the client's house. For example, did the lord visit his client's house merely to receive the *samfít* 'summer morsel' of curds, butter and milk mentioned at *CIH* v 1910.25–7? On the other hand, commentary at *CIH* v 1911.1–3 discusses the numbers which a lord may bring with him on four types of feast: the *lupait*, the *fuirec*, the *furnaide*, and the *forgab iter dā cairci*.

A commentator at *CIH* iv 1234.12–17 gives a chief poet (*ollam*) the option of having his food-entitlement brought to him in his own house during the autumn and winter. He explains that the roads may be dirty at this time of year.

[34] *GEIL* 139–40.

[35] *CIH* iii 778.14; 779.10 = *CG* 3.77; 5.130.

[36] *CIH* ii 451.24 = *BB* 74 § 39.

[37] *GEIL* 36–8.

[38] *CIH* ii 524.23–4 = *AL* iii 18.18 *fuirired n-aīged* [*n*]*Dē* 'feeding the guests of God'.

[39] *CIH* vi 2107.36–2108.23.

[40] *Borrlus* is perhaps to be identified with the leek, see p. 254 above.

the turner (*tornaire*) or the man who makes vessels from yew-wood (*ibrόracht*) gets three quarters of his ration.[41]

For feasting to mark special occasions, see p. 357 below.

Cooking methods and utensils

In ordinary households, the preparation of food was seen primarily as a woman's task. Indeed, a Middle Irish prophecy-poem includes 'a male cook in every house' among the signs of the coming of bad times.[42] However, in a large household, such as that of a monastery or king or other rich landowner, cooks seem usually to have been male. For example, a law-text includes the cook (*coic*) among the officers of the Church.[43] Another legal passage records a tale about a female landowner (*banchoairt*) who had thirty-five male cooks and thirty-five female cooks. But the main occupation of these female cooks is stated to be milking the herd of seven hundred cows; the implication is that the actual preparation of meals for the household is the task of the male cooks.[44] Similarly, the three chief cooks (*fulachtóiri*) in the household of King Conaire are male.[45]

The Old Irish sagas devote most attention to the meat-dishes characteristic of the feasts of kings and nobles – such food naturally fits in with the heroic tenor of these tales. In the law-texts, on the other hand, there is more emphasis on everyday food-preparation, particularly bread-making and milk-processing. There are many references to bread-making equipment such as the kneading-trough (*losat*), kneading-slab (*lecc*), griddle (*lann*) and griddle-turner (*lainnéne*).[46] The vocabulary used in the law-texts in relation to milk and milk-processing is also extensive: see below.

There is no evidence in our sources of elaborate dishes in the ancient Roman style with a large variety of ingredients. The ordinary person seems to have survived on a simple diet of bread and milk with a condiment. As we have seen above, those higher up the social scale enjoyed greater culinary variety, and may have used

[41] *CIH* vi 2108.12.
[42] *ACL* iii 241 § 7.
[43] *CIH* vi 2213.32.
[44] *CIH* vi 2117.36–2118.2 = Dillon, 'Stories from the law-tracts', 51 § xiv.
[45] Stokes, 'The destruction of Dá Derga's hostel', 295 § 123 = *LU* 234.7660.
[46] E.g. *CIH* i 285.36 = *AL* iii 274.18; *CIH* ii 472.3 = *AL* v 406.16.

exotic flavourings such as pepper (*scibar*).[47] But even in royal feasts the menu seems to have been a straightforward one of roast or boiled meat and bread, washed down with mead or beer. *Hisperica Famina* describes a simple feast of bread smeared with Irish oil (*oleum scottigenum*), joints of meat, milk, and beer.[48]

Food seems generally to have been served on some sort of a wooden board (*mías*)[49] – either a table or platter – and was probably eaten with the fingers after being cut up with a knife. Drink was served in a wooden mug (*ian*) or goblet (*escrae*). Those of high rank used a decorated drinking-horn (*corn*), perhaps imported from the Continent: see p. 56.

Milk and milk-products

In many parts of the world, the milk of domesticated mammals – particularly cows, sheep, and goats – is an important element in the human diet. Because of its perishable nature, milk which is not drunk within a day or so must be subjected to some preservative process. The commonest way of using excess milk is to make it into cheese, of which hundreds of types are known from various parts of the world. Milk may also be preserved by the introduction of certain bacterial cultures, as in the case of yoghurt and kefir. Cream may be preserved for a short time by clotting, or for longer periods by churning into butter.

Milk and milk-products are very commonly mentioned in our sources.[50] A tenth-century tale, *Tochmarc Ailbe*, counts milk (*mlicht*) as the best food, because it is 'good when fresh, good when old, good when thick, good when thin'.[51] The early twelfth-century comic tale *Aislinge Meic Con Glinne* is equally appreciative of this foodstuff and gives a vivid description of milk in its different forms, part of which reads: 'very thick milk, milk which is not very thick, milk which is thick but flowing, milk of medium thickness, yellow

[47]See p. 342.

[48]Herren, *Hisperica Famina*, 86.298–302. 'Irish oil' may possibly refer to butter in this context. For a brief discussion on the use of oil in early Irish cooking, see p. 359 (footnote).

[49]For examples from early texts, see *DIL* s.v. *mías*. This word is a borrowing from Latin *mensa* 'table' (*LEIA* M-47–8).

[50]For some reason – obscure to me – a marginal entry on p. 161 of the Trinity College Dublin MS H 4. 22 (no. 1363) attributes special knowledge about milk to the kings of Leinster (Abbott and Gwynn, *Catalogue*, 210).

[51]Thurneysen, '*Tochmarc Ailbe*', 272 § 22.

bubbling milk the swallowing of which requires chewing . . .'.[52] This text is an important repository of information on early Irish dairying, as it contains descriptions of many different types of milk-produce, particularly cheeses.

Milk

There are many references in our texts to the drinking of fresh milk (*lemlacht*),[53] and it was sometimes diluted with water to form a drink called *englas*. Another common treatment of milk was the addition of rennet (*binit*) from the stomach of a calf. As Cormac points out in his *Glossary*, this caused the milk to coagulate, giving a thickened milk (*as técht*, *as tiug*).[54] In later Irish this was known as *bainne clabair*, anglicized *bonnyclabber*.[55] Milk thickened with rennet can also be made into cheese: see p. 327–30

Fresh milk was used in the making of porridge (*lichtiu*)[56] and of a herbal broth (*brothchán*) heated for the nourishment of the sick.[57] The milk produced by a cow for the first few days after calving (*nús* 'colostrum, beestings') is strong-flavoured and yellow, and its consumption is not recommended by the author of the wisdom-text *Tecosca Cormaic*: he includes it among the foods which are worst for a person's body.[58]

Cream

Unlike milk, cream can remain in a palatable condition for a week or more without being treated in any way other than by being kept in a cool place. The gradual souring enhances the flavour, and many recipes in modern cookery books specifically call for soured cream.

Among the milk-products which a client must give to his lord, *Cáin Aicillne* includes the fill of a vessel with three *ól*-measures of cream (*úachtar*).[59] The cream is kept for three days and three nights before being handed over, and is tested after one and a

[52]Jackson, *Aislinge* 39.1208–10 = Meyer, *Aislinge*, 101.8–11.

[53]This word is a compound of *lem* 'lukewarm, tepid' and *lacht* 'milk'. The later form is *lemnacht*.

[54]Meyer, *Sanas Cormaic*, 12 § 125.

[55]For a discussion of the etymology of this term, see Malone, 'bonnyclabber'.

[56]E.g. *LU* 248.8120 = Windisch, 'Das Fest des Bricriu', 256.25 § 9 *littiu lemnachta*. For porridge, see p. 331 below.

[57]For a discussion of the nature of *brothchán*, see p. 349 below.

[58]Meyer, *Tecosca Cormaic*, 38 § 21.6.

[59]*CIH* ii 483.31–3 = Thurneysen, 'Aus dem irischen Recht I', 356; cf. *CIH* iii 778.6; 778.14 = *CG* 3.63; 3.75.

half days to check that it has not become rancid. The client must ensure that all thin milk has been removed and that the cream is not shaken or disturbed in any way.[60]

The term *croth* is applied in our texts to the film or layer on the surface of any liquid or viscous milk-product. It is often used of cream, as in the Old Irish table of penitential commutations, which refers to *ass cen chroith* 'milk without cream', i.e. skim milk.[61] In *Aislinge Meic Con Glinne* it is used of the yellow skin which has hardened on top of cream.[62]

Butter

The first stage in the production of butter (*imb*) is to keep a quantity of cream for a week or so in a cool place. The cream must then be churned until it separates into butter and buttermilk (*bláthach*). The lumps of butter are strained off, washed, and pressed into butter-pats, for which the Old Irish term is *mescán*.[63] Our sources indicate that the regular method of churning (*maistred*) involved the use of a churn-dash (*loinid*) and a wooden churn (*cuinneóc*). One law-text refers to the dairy-worker as *bé loinedo* 'the woman of the churn-dash',[64] and another refers to a person's right to cut wood on another's property to make a churn-dash.[65]

There is undoubtedly something miraculous about the sudden solidification of butter during churning, and it is not surprising

[60]The text has *cen fothana na ruba* 'without thin [milk] underneath or breaking'. The author's concern seems to be that no milk should be left in contact with the cream, as this causes the cream to become rancid rather than souring gradually. The glossator (*CIH* ii 484.3) explains *fothana* as *in medg* 'the whey', likewise stressing that the presence of sour milk can ruin the cream. The word *ruba* (*ro* 'great' + *be(n)-* 'strike') has a wide range of meanings, such as 'attack, breaking', and seems to refer to disturbance of the cream while souring. But see p. 752.

[61]Binchy, 'The Old-Irish table of penitential commutations', 66 § 28; cf. *CIH* iv 1484.21 = O'Dav. 276 § 494 *croich* (leg. *croith*) *.i. uachtar bainne* 'i.e. cream'.

[62]Jackson, *Aislinge*, 32.989–90 (and note) = Meyer, *Aislinge*, 83.3.

[63]In *Cormac's Glossary* (Meyer, *Sanas Cormaic*, 79 § 915) *mescán* is explained *.i. do mescad ind loma ássas* 'i.e. which develops from the mixing (=churning) of the milk'. It would thus contain the root *mesc-* 'mix' (*VKG* i 76; ii 577 § 779). Another possibility is that it is cognate with Latin *mergo* 'I plunge', Skt. *májjati* 'he plunges', etc. See *LEIA* M-42 s.v. *mesc-* and *IEW* 745 s.v. *mezg-*.

[64]*CIH* vi 2229.24 *bē loisde luineatha* 'woman of kneading-trough [and] churn-dash'; cf. *CIH* ii 508.21 = *AL* ii 366.1 (gloss on *Cáin Lánamna*).

[65]*CIH* i 241.24 = *AL* v 482.y. The glossator calls it *crosān in asa* 'the cross-stick of the milk'.

that many rituals and beliefs have become associated with butter-making. Butter is mentioned very frequently in Old and Middle Irish sources, and is often portrayed as a luxury or upper-class food. *Críth Gablach* specifies that a low-ranking person such as a *fer midboth* is not entitled to butter when visiting, but only to milk and cheese (*grus*) or cereals.[66] A visiting lord, on the other hand, must be given butter every day.[67] According to legal commentary on *Cáin Íarraith*, fresh butter was added to the porridge of the children of lords, but not to that of the offspring of commoners.[68] It seems also that the special monastic treat consumed on feast-days – called the *selann* – often consisted of butter.[69] Butter is generally included in the food-rent which a client must give to his lord. *Cáin Aicillne* specifies that a bullock given in rent must be accompanied by a pat of butter two fists in length and one fist in thickness.[70] Another text on rents speaks of a pat of butter the size of three fists (*imb tre-duirn*).[71] Butter seems to have often been kept in a bark-container (*rúsc*).[72]

Butter has the useful characteristic that it will keep in a cool place for a long time. For this reason, stored butter was an important insurance against hunger, and it is clear from the annals that the destruction of its butter-stores was a severe blow to a community. For instance, the *Annals of Connacht* record that in the year 1296 Aodh Ó Conchobhair and his allies ravaged the area around the monastery of Boyle, destroying corn (*arbanna*) and butter-stores (*imenna*).[73] The storing of butter in bogs may have originated as a method of keeping it both cool and safe. In later English accounts of Ireland, there are some references to the flavouring of bog-butter with garlic (presumably *crim* 'wild

[66] *CIH* iii 777.22–3 = *CG* 2.27. Binchy takes the word to be *grús*, but a short vowel is indicated by the imperfect rime *lis* : *grus* in a stanza quoted at Calder, *Auraicept*, 80.1064–7; 225.4020–3; Meroney, 'Studies in early Irish satire', 201 § 8.

[67] *CIH* ii 566.28; iii 782.34 = *CG* 14.350.

[68] *CIH* v 1759.37 = *AL* ii 148.21.

[69] E.g. Gwynn and Purton, 'The monastery of Tallaght', 146–7 § 52. See also *DIL* s.v. *selann*.

[70] *CIH* ii 483.34 = Thurneysen, 'Aus dem irischen Recht I', 356.7–8 § 13.

[71] *CIH* iii 920.12.

[72] The basic meaning of *rúsc* is 'bark (of tree)'. Commonly, the word is used of a container for butter (see *DIL* R 123.17–24).

[73] *AC* 194 s.a. 1296 § 3. For other examples, see *DIL* s.v. *imb*.

garlic'), but I have found no mention of this practice in early Irish sources.[74]

Buttermilk

The law-texts occasionally mention buttermilk (*bláthach*), the refreshingly sharp liquid left behind after the butter has been removed. The *fer midboth* must give his lord an *ól*-measure of cream, of fresh milk and of buttermilk or whey (*draumce*).[75] Mícheál Ó Sé makes the plausible suggestion that the word *bláthach* is a derivative of *bláth* 'flower', and refers to the sweet smell of fresh buttermilk.[76]

Cheeses

The versatility of milk as a food-source is demonstrated in particular by the astonishing range of cheeses which can be made by different treatments and additives. Consequently, among cheese-eating peoples there is generally a highly specialized vocabulary to distinguish various products. For example, one modern Kurdish community is reported to have four distinct terms for different types of sheep's milk cheese.[77] It is clear from the written sources that cheeses were of great importance in the early Irish diet, and there is a correspondingly rich variety of terms for cheese in the language. Often these terms are used in Old and Middle Irish texts without any hint as to the appearance or manufacture of the cheese in question: this means that the modern scholar must to some extent rely on guesswork when offering translations.

The simplest method of cheese-making is to stand a covered vessel of whole milk for a few days. After this time, the curds (*gruth*) will have risen to the top, with the whey (*medc*) below. The curds may be eaten at this stage: this is our 'cottage cheese'. Whey is a rather sour liquid, generally considered unpleasant to drink in its raw form. Diluted with water (*medcuisce*), it features prominently in monastic and penitential diets.[78] After the curds have been taken off, some fragments remain in the whey: this is called *gruiten* 'small curds', and is counted as a low-prestige food in a number of sources.[79] A more palatable whey drink is *tremanta*, which is mentioned three

[74]Lucas, 'Irish food before the potato', 30.
[75]*CIH* iii 778.13–14 = *CG* 3.75–6.
[76]Ó Sé, 'Old Irish buttermaking', 64. See also *LEIA* B-58 s.v. *bláth*.
[77]Ryder, *Sheep and Man*, 723.
[78]This is also called *englas midg* (Meyer, *Betha Colmáin*, 104.1 § 103).
[79]*CIH* v 1759.37 = *AL* ii 148.21; Meyer, *Sanas Cormaic*, 57 § 692.

times in *Aislinge Meic Con Glinne.* It is presumably to be identified with the later *treabhandar* (anglicized *troander*), which was made by boiling fresh milk and then adding sour buttermilk until the curds clotted. After straining, the liquid has a light acid taste, and will keep for a few days.[80] Another word which may have been applied to a type of whey is *drúchtán,* a derivative of *drúcht* 'dew', but its method of manufacture is uncertain.[81] *Draumce,* which is attested only in *Críth Gablach,* seems also to apply to some sort of whey drink.[82]

The commonest term in the law-texts which can be identified with 'cheese' is *grus,* which is clearly from the same root as *gruth* (curds). It is listed in *Uraicecht Becc* among the types of food which a householder provides for a visiting lord and his attendants.[83] In *Críth Gablach* it is included in the food-entitlements of various low-ranking persons.[84] The general word for cheese (*cáise* from Latin *caseus*) does not seem to occur in the legal material.

The law-text on base-clientship, *Cáin Aicillne,* describes the preparation of *milsén,* a soft cheese produced by heating a cauldron of fresh milk along with some butter.[85] The word *milsén* is a derivative of the adjective *milis* 'sweet', and obviously refers to the sweet taste which heating imparts to milk. Presumably, most *milsén* was made from cow's milk, but a law-text also refers to its production from goat's milk.[86]

A term for soft cheese widely used in *Aislinge Meic Con Glinne* is *máethal,* which is a derivative of the adjective *máeth* 'soft'.[87] However, such cheese could be of sufficiently solid consistency to be carried in a cloak. A Middle Irish Life of Saint Cóemgen describes

[80] Ó Sé, 'Old Irish cheeses, etc.', 85. I do not understand how Middle Irish *tremanta* acquired a final *-r* in the later language.

[81] See discussion by Ó Sé, 'Old Irish cheeses, etc.' 85.

[82] *CIH* iii 778.14; 779.1 = *CG* 3.75; 5.116.

[83] *CIH* ii 644.13; v 1599.41 = *AL* v 40.13.

[84] *CIH* iii 777.22–3; 777.34; 778.40 = *CG* 2.27; 2.44; 5.115; *CIH* ii 563.3; iii 779.22 = *CG* 6.148.

[85] *CIH* ii 483.30 = Thurneysen, 'Aus dem irischen Recht I', 356.1–2; *CIH* iv 1503.3 = O'Dav. 358 § 947. *Milsén* is also mentioned in the Rule of the Céli Dé (E. J. Gwynn, *The rule of Tallaght*, 64 § 5).

[86] *CIH* iii 920.28.

[87] *LEIA* M-7. In later Irish *máethal* (*maothal*) is applied to beestings, for which Old Irish has *nús.*

how some women carried cheeses of this type hidden in the corners of their cloaks.[88]

Various methods are used to preserve cheese, including salting, drying and pressing. Our sources – particularly the saints' Lives – contain many references to pressed cheeses. Thus *fáiscre grotha* 'curds which have been pressed' is frequently mentioned, and sometimes equated with Latin *caseus.*[89] A term showing a similar semantic development is *táth*: it literally means 'joining together, binding', and is used in the phrase *táth an grotha* 'the squeezing of the curds'.[90] *Táth* is also applied to the finished product, and is mentioned as a foodstuff in a number of texts.[91]

In general, hard cheeses keep longer than soft ones. Our sources contain many references to cheeses which are more or less hard. *Aislinge Meic Con Glinne* speaks of a dry cheese (*tírmcháise*),[92] and a cheese called *tanach* (*tanag*) was regarded as hard enough to be used as a deadly missile: the Old Irish tale *Aided Meidbe* attributes the death of Queen Medb to a blow from a piece of this cheese hurled from a sling.[93] A hard cheese which seems commonly to have had a rounded shape is *mulach,* as a hideous giant in *Togail Bruidne Da Derga* is said to have had buttocks which were the size of a *mulach.*[94] The diminutive *mulchán* is common in the later language, and variously anglicized as *mullagham, mulahaan,* and *mullahawn.*[95] It occurs in the nickname of a twelfth-century chieftain, *Mant na mulchán Ua Ruairc,* which could be translated 'Cheese-guzzler Ó Ruairc'.[96] In 1824 Ryland writes: 'cheese, made from skimmed milk, and called *mullahawn,* was formerly an article of commerce in Waterford, and was exported in large quantities, but it was of such a hard substance that it required a hatchet to

[88]Plummer, *Bethada* i 166 xxii (43); trans. ii 161.

[89]See *DIL* s.v. 2 *fáiscre.* Cf. Meyer, *Sanas Cormaic*, 27 § 312.

[90]Meyer, 'The death of Finn', 464.9.

[91]E.g. Van Hamel, *Immrama*, 38.430 = Stokes, 'The voyage of Mael Dúin, (1)', 490.13; Meyer, *Aislinge* 121.32 (later version).

[92]Jackson, *Aislinge*, 32.984–5 = Meyer, *Aislinge* 81.30–31.

[93]*LL* ii 462.14430–32 = Hull, '*Aided Meidbe*', 56.

[94]Knott, *Togail Bruidne Da Derga*, 11.351–2 = Stokes, 'The destruction of Dá Derga's hostel', 41 § 38. The text has *mét mulaig for got* 'the size of a cheese on a withe' – possibly the cheese is dried on a wickerwork tray; cf. Columella, *De re rustica* (ed. Forster and Heffner), 7.8.5.

[95]Cf. Dwelly, *Faclair* s.vv. *mulachag, mulachan* 'cheese'.

[96]*ALC* i 232 s.a. 1204. This entry records his death from a surfeit of sex, while convalescing after having been blinded.

cut it'.[97] The Scottish Gaelic form *mulchag* is likewise used of long-lasting cheeses: in South Uist it was customary to make cheese of this type on Mayday – *mulchag Bhealltainn* – and keep it for a full year.[98]

GRAIN-PRODUCTS

Bread

As we have seen (p. 219), the most highly regarded type of bread was made from wheatmeal. According to legal commentary, a bushel (*míach*) of wheat is worth a scruple (*screpal*) and gives twenty-eight loaves of bread, a bushel of barley is worth two pence and gives twenty-six loaves, and a bushel of oats is worth one penny and gives twenty-four loaves.[99]

Críth Gablach distinguishes two sizes of loaf: a standard loaf (*bairgen inraic*)[100] and a loaf baked for a woman (*bairgen banḟuini*), which is half this size.[101] According to the Old Irish Rule of Ailbe, the daily loaf of a monk weighs thirty ounces and is twelve inches – this probably refers to its diameter.[102]

A small loaf or bun is called a *tortíne*, a diminutive of *tort* (from Latin *torta* 'loaf').[103] A law-text on clientship refers to 'two small wheaten loaves' (*dá thoirtíne do thara*).[104] Another term for a bun is *srúbán*, which is stated in legal commentary to be one eighth of the size of a loaf.[105] It seems likely that it was so called because it was shaped like a snout (*srúb*).[106] Its small size is emphasized in

[97] Quoted by Lucas, 'Irish food before the potato', 28.

[98] McDonald, *Gaelic words and expressions*, 185.

[99] *CIH* iii 806.1–5. This commentary has further discussion on the quality of the different grains.

[100] Legal commentators regularly identify this as a loaf baked for a man (*bairgen ferḟuini*); e.g. *CIH* i 196.31 = *AL* iv 116.6 (Old Irish commentary); *CIH* v 1766.14 = *AL* ii 176.18–19.

[101] *CIH* iii 779.2 = *CG* 5.117–18; *CIH* v 1597.14–15 = *AL* v 30.24–5 (gloss on *Uraicecht Becc*). For the principle that a woman is normally entitled to half the food of her husband, see p. 351 below.

[102] O'Neill, 'The rule of Ailbe', 102 § 31a.

[103] Meyer, *Sanas Cormaic*, 104 § 1200; *LEIA* T-117 s.v. *tort*.

[104] *CIH* iii 921.13.

[105] *CIH* ii 370.23 = *AL* i 132.20. This is the *boim berrtha* i.e. the morsel (of bread and salted meat) which is paid for shaving. The term *srúb(h)án* is commonly used in post-Norman documents of a measure of oatmeal (anglicized *sroan*): see Dinneen, *Foclóir*, s.v. *srubhán*; *DIL* S 379.18–20.

[106] This view is rejected at *LEIA* S-188.

commentary on *Félire Óengusso* which says that the daily food of the ascetic Saint Cíarán consisted of a barley-bun (*srúbán éorna*) with fresh water and two roots of seaweed.[107]

Sowens

A useful food can be obtained by soaking the husks of bran from oats or barley for a few days so that a slight fermentation takes place. The resultant liquid can be drunk after straining, or boiled until it forms a jelly. There may be a reference to a dish of this type in a Latin Life of Saint Munnu, which describes the ascetic fare in the monastery of Cláeninis under the authority of abbot Silell. His monks were not permitted to riddle grain: it was mixed with water in a vessel along with its chaff (*cum sua palea*), and then cooked by adding stones which had been heated in a fire.[108]

Dishes derived from soaked husks are widely documented from the later period. For example, a decoction from barley or oats was called *cáthbruith*, which is presumably a compound of *cáith* 'husk, chaff' and *bruith* 'boiling'.[109] English documents from the sixteenth century onwards contain many references to *sowens*, a term which was applied both to a drink made from fermented husks, and to a jelly procured from this liquid by heating.[110] Generally *sowens* is associated with Ireland, Scotland, or northern England, and it is regularly implied that it was mainly consumed by the poor. In Wales the jelly from fermented husks was called *llymru*, a word which passed into the English language as *flummery*, which has acquired the additional meaning of 'empty, inflated talk'.

Porridge

Porridge (*lichtiu, littiu*)[111] is mentioned in relation to the diet of children in particular. For example, legal commentary on the feeding of fosterchildren refers to porridge made with oatmeal,

[107] Stokes, *Félire*, 88.y. For the identity of the seaweed, see p. 314 above.

[108] Plummer, *Vitae* ii 228 § vi. Alternatively, this passage may refer to a method of cooking porridge.

[109] See *DIL* s.v. *cáthbruith*; Beckett, *Fealsúnacht Aodha Mhic Dhomhnaill*, 262 s.v. *cáfraith*; Dwelly, *Faclair*, s.v. *càbhruich*. The word also occurs in Manx in the spelling *cowree* 'flummery, sowens' (Cregeen, *A dictionary of the Manks language* s.v.).

[110] *OED* s.v. *sowens*. The word is well attested in Modern Irish and Scottish Gaelic in forms such as *súghán* or *súbhán*, and is probably a derivative of Old Irish *súg* 'juice'. For the Irish evidence, see Mahon, 'Oatmeal as food'.

[111] For the development of *lichtiu* to *littiu* (later *leite*), see M. A. O'Brien, '*Uchtlach*, etc.'.

wheatmeal or barleymeal.[112] Generally, porridge is distinguished from a thinner meal-and-water gruel (*menadach*[113]), which features prominently in penitential diets (see p. 348 below). This sort of gruel is also called *brothchán uisci 7 mine* 'broth of water and meal', and is part of the ascetic diet associated with Saint Máel Rúain of Tallaght.[114]

Búaidrén

The identity of *búaidrén* is uncertain. Both editors of *Aislinge Meic Con Glinne* translate it as 'stirabout', as the word is clearly a derivative of *búaidrid* 'disturbs, stirs up'.[115] It is also mentioned in *Teagasg Maoil Ruain*, where it is said to be made from bread.[116] Possibly, *búaidrén* is some sort of bread-and-water gruel.

Beer

Beer (*cuirm*) is very commonly mentioned in early Irish sources, and was clearly a drink of great social importance. Many of our texts extol beer as an expression of hospitality which brings honour to the host who provides it. Thus a wisdom-text declares 'beer brings fame' (*adcota cuirm clotha*),[117] and the law-text on status, *Críth Gablach*, states that a king is expected to drink beer on Sundays – presumably with his household – and goes so far as to say that he who cannot ensure beer (*laith*) every Sunday is not a proper ruler.[118] In an Old Irish mythological text the reputation of a king is dismissed in disparaging terms because his guests did not leave his house 'with the smell of beer on their breath'.[119] The provision of beer is also expected of those of lower rank: *Críth Gablach* stipulates that a commoner of *mruigfer* rank should have a tub for fermentation (*bruth*) in his house,[120] and a mug of beer (*ian chorma*) should always be available.[121]

[112] *CIH* v 1759.41–1760.2 = *AL* ii 148.z-150.5.
[113] It is likely that this word is a derivative of *men* 'meal, flour' (*LEIA* M-36). It was borrowed into Old Norse in the form *minnþak* (Marstrander, *Bidrag*, 111^2).
[114] E. J. Gwynn, *The rule of Tallaght*, 2 § 1; 26 § 43.
[115] Jackson, *Aislinge*, 12.374 = Meyer, *Aislinge* 33.14.
[116] E. J. Gwynn, *The rule of Tallaght*, 4 § 3. He suggests that it is flummery, but this is not made from bread: see above.
[117] Smith, '*Senbriathra Fithail*', 11 § 11.
[118] *CIH* ii 569.29–30 = *CG* 21.542–4.
[119] Stokes, 'The second battle of Moytura', 68 § 36; Gray, *Cath Maige Tuired*, 32 § 36.
[120] *CIH* ii 563.20; iii 779.39 = *CG* 7.174–5.
[121] *CIH* ii 563.26–7; iii 780.5–6 = *CG* 7.184.

Because of a general scarcity of milk and vegetables in winter, beer must have been of considerable significance in the early Irish diet, providing vitamins and other nutrients not available elsewhere. It is noteworthy that in the law-text *Bretha Crólige* – which deals largely with the food of invalids – beer (*lind*) is treated as a substitute for vegetables and fruit. The invalid is entitled to beer only if there are no vegetables or fruit, or else at the specific instruction of the physician.[122] Beer features also in monastic diets, particularly at the celebration of Easter, though it was disapproved of by some ascetics (see p. 345).

The law-texts make it clear that the availablity of beer in a lord's house is to a large extent ensured by the fact that his clients must provide him with malt in their annual food-rent.[123] *Críth Gablach* stresses that only a fully fledged lord is entitled to malt from his clients. An aspiring lord (described as *fer fothlai*[124]) can only receive grain for food (*grán arbae biid*) on the grounds that he is not entitled to malt while still a commoner.[125] Some rents include beer rather than malt: the return (*somoíne*) on a rented knife consists in part of four mugs of beer with no dregs or froth.[126]

Large households – such as those of a king or monastic establishment – are assumed in our sources to have their own professional brewer (*scóaire*[127] or *ceirbsire*[128]). For example, an Old Irish text on the seating-arrangement of a royal household provides a place for the brewers,[129] and *Bethu Phátraic* records a tradition that Saint Patrick had his own personal brewer, a priest named Mescan.[130] Our sources provide us with detailed information on the technique of malting (see p. 246) but relatively little on brewing. It is clear from the ninth-century *Bethu Brigte* that the malt was first subjected to mashing (*imdell*) and then

[122] *CIH* vi 2299.38–9 = Binchy, '*Bretha Crólige*', 38 § 49.

[123] E.g. *CIH* ii 480.26 = Thurneysen, 'Aus dem irischen Recht I', 348; *CIH* iii 779.16–17 = *CG* 6.139.

[124] This term means 'man of withdrawal', i.e. he has acquired enough wealth to withdraw from the rank of *bóaire*, but cannot be considered as a lord: see *GEIL* 28.

[125] *CIH* ii 565.5; iii 781.13 = *CG* 10.254–5.

[126] *CIH* iii 921.13 *cethri cōid cormu cin huan cin descad.*

[127] An agent-noun from *scó* 'beer' (of unknown etymology, *LEIA* S-48 s.v. *scó*).

[128] In his *Glossary* (Meyer, *Sanas Cormaic*, 20 § 219) Cormac points out that *cerbseóir* (*ceirbsire*, etc.) is a derivative of Latin *cervisia* 'beer'.

[129] O Daly, '*Lānellach tigi rīch*', 83 § 11.

[130] Mulchrone, *Bethu Phátraic*, 155.3131; cf. Ó Riain, *Corpus genealogiarum sanctorum Hiberniae*, 118 § 671.9.

put to ferment (*for descthu*).[131] In the later medieval period the practice developed on the Continent of adding hops to preserve and provide flavour to beer. So far as I have been able to discover, there are no records in the Irish sources of any additives other than honey. Beer must consequently have been drunk quite soon after fermentation, though the wisdom-text *Tecosca Cormaic* warns against drinking immature beer (*nua corma*).[132] The brewer must be careful that he does not make nauseous beer (*scó sceithech*).[133] If a client knowingly provides bad beer (*míchuirm*) when his lord and attendants are feasting at his house, he must pay heavy fines and provide double the amount of good beer.[134] Before being drunk, beer may have to be strained: the Middle Irish voyage-tale *Immram Curaig Ua Curaig* includes among the three shouts of the house of a wealthy landowner 'the shout of the strainers straining the beer'.[135]

We have little information on different types of beer in our sources. Most beer was brewed from barley, but there is one reference in a law-text to the brewing of beer from wheat: *Bretha Nemed Déidenach* states that the reciter of poetry (*aí*) is entitled to 'a drink from the juicy drops of wheat (*tuirenn*)'.[136] As with modern beers, early Irish brews seem commonly to have had a reddish colour. Red beer (*derglaith*) is mentioned in an Old Irish hymn on Saint Brigit,[137] and also in legal commentary.[138] Some types of beer may have been brewed to foreign recipes. The tenth-century glossator Cormac mac Cuilennáin identifies *brocóit*

[131] Ó hAodha, *Bethu Brigte*, 7 § 21. See also the discussion in Binchy, 'Brewing in eighth-century Ireland' 4–5.

[132] Meyer, *Tecosca Cormaic*, 38 § 21.6.

[133] *CIH* v 1732.30 = *AL* ii 32.16; *CIH* iv 1526.20 = O'Dav. 462 § 1489.

[134] *CIH* v 1911.31–8; cf. *CIH* v 1732.30–1733.23 = *AL* ii 32.16–34.28.

[135] Stokes, 'The voyage of the Húi Corra', 26 § 1 = Van Hamel, *Immrama*, 96.5–6 *gáir na scacadóiri oc scacad lenna.*

[136] *CIH* iii 1129.20 = E. J. Gwynn, 'Privileges', 39.20–1 *dlighidh iar n-aisneis digh do thuirne sūighdhéraibh.* Athenaeus, *Deipnosophistae* (ed. Gulick), 4.152, records that the Gauls around Marseilles drank a honeyed wheat-beer which they called κόρμα (cognate with Irish *cuirm*). There is evidence of the brewing of beer from spelt wheat (p. 222) in Roman Britain: see Hillman, 'Evidence for malting spelt', 140.

[137] *Thes.* ii 337.3. The text has *derglaid*, which is glossed *ba lind derg* 'it was red beer'.

[138] *CIH* iv 1370.7 *deogh dergflatha*, for which Binchy suggests reading *deogh derglatha* 'a drink of red beer'. It is pointed out at *DIL* s.v. 2 *flaith* that this is probably the same word as *laith.*

as a special beer made from malt and honey.[139] He points out that the word *brocóit* is a borrowing from Welsh *bracaut* (*bragod*) 'bragget', a honey-beer which seems to have been intermediate in strength between mead and ordinary beer.[140] This drink is mentioned quite often in our texts. For example, in a comic section of the ninth-century Triads of Ireland one type of dispenser of drink is described as a *bolcṡrónach brocóite*, which might be translated as a 'swollen-nosed bragget-swigger'.[141] Another foreign word for beer is *beóir*, which is most probably a borrowing from Old Norse *bióṙr* – Marstrander comments on the frequent association between beer and Norsemen in Irish folk tradition.[142] Its earliest attestation is in the twelfth-century *Aislinge Meic Con Glinne*, which speaks of 'streams of beer and bragget' (*aibne beóri is brocóti*).[143]

Other liquors consumed were wine (p. 319) and mead (p. 113). There is no evidence that distillation in Ireland goes back earlier than the thirteenth century. Legends crediting Saint Patrick with the introduction of distilled liquor in the fifth century must therefore be discounted.[144] The Irish term *uisce beatha* (= Latin *aqua vitae* 'water of life') was anglicized *usquebaugh*, later abbreviated to *whiskey*. The fondness of the Irish for this drink is noted by various English commentators,[145] and there is also mention of its medicinal uses.[146] The earliest reference to whiskey in the annals contains mocking word-play on the name of this drink. The annalist records the death in 1405 of Risdeard Mag Raghnaill after drinking too much *uisce beatha*, and comments that it was *uisce marbtha* 'water of death' for him.[147]

[139]Meyer, *Sanas Cormaic*, 12 § 124; cf. *CIH* ii 607.37–8. See Russell, 'Brittonic words in Irish glossaries', 175.

[140]The medieval Welsh law-book *Llyfr Iorwerth* (ed. Wiliam, 63 § 94.28–30 = *ALIW*, Venedotian Code, II xxi § 1) states that a person of free *maenol* rank must give the king a vat of mead (*med*) or two vats of bragget (*bragaut*) or four vats of beer (*cwryw*).

[141]Meyer, *Triads*, 28 § 231. The implication seems to be that this type of dispenser (*dáilem*) is helping himself generously to his employer's liquor. Some MSS have *bolcsron*, but the meaning is unaffected.

[142]Marstrander, *Bidrag*, 117. See also Almqvist, *Viking ale*, 65–81.

[143]Jackson, *Aislinge*, 15.462 = Meyer, *Aislinge*, 37.25.

[144]Toussaint-Samat, *History of food*, 196.

[145]E.g. Morley, *Ireland under Elizabeth and James I*, 425, 428. See *OED* s.v. *usquebaugh*.

[146]Morley, *Ireland under Elizabeth and James I*, 420; cf. Wulff, *Rosa anglica*, 264.18.

[147]*AC* 392 § 7.

Meat

In our sources, meat features much more prominently in the diet of lords than in that of commoners.[148] Salted meat (*sall*[149] or *saillte*[150]) is given special prominence, and is generally from the pig. The law-texts contain many references to the flitch of bacon (*tinne*) in the food-rent which a lord receives from his clients.[151] The flitch of bacon on a hook (*tinne for croich*) is likewise given prominence among the signs of a farmer's prosperity.[152] Salted beef (*bóṡall*) is rarely mentioned. This is not surprising: Finbar McCormick has pointed out that the salting of an entire bullock requires a very large amount of salt; consequently it would make better economic sense to consume it fresh.[153]

Fresh meat (*carnae*) may be prepared in a number of ways. A simple method which is attested from both modern and ancient times in various parts of the world is to skin and cut up an animal and then boil the flesh in the skin over a fire.[154] An illustration in Derricke's *Image of Irelande*, published in 1581, shows meat being cooked in this way,[155] but I have found no account of this practice in Old or Middle Irish sources. Another method recorded from many parts of the world is to fill a hole in the ground with water and boil the meat by putting hot stones into the water. In Ireland, many such cooking-pits have been identified by archaeologists.[156]

[148]Contrast, for example, the meat-entitlements of the lord and commoner at *CIH* vi 2298.30–2299.21 = Binchy, '*Bretha Crólige*', 36–8 §§ 46–7.

[149]*Sall* (*saill*) doubtless contains the root *sal-* 'salt' (as in *salann*): see *LEIA* S-14 s.v. *saill.* Occasionally, *sall* seems to refer to fat-meat, e.g. *saill cen tsailliud* 'unsalted fat' (*CIH* iii 852.32; v 1773.20 = Thurneysen, 'Aus dem irischen Recht II', 245 = *AL* ii 202.19). In the later version of *Aislinge Meic Con Glinne*, a distinction is made between *senṡaill* 'matured bacon' and *úrṡaill*, which could mean 'freshly salted bacon' or simply 'fresh pork' (Meyer, *Aislinge*, 117.24–7).

[150]*Saillte* is past participle of *saillid* 'salts', and is sometimes used as a substantive 'salted meat', e.g. *CIH* ii 563.16; iii 779.36 = *CG* 7.168; *CIH* ii 566.29; iii 782.35 = *CG* 14.350.

[151]E.g. *CIH* iii 778.37; 779.16 = *CG* 5.110; 6.138; *CIH* ii 482.4 = Thurneysen, 'Aus dem irischen Recht I', 351 § 9.

[152]*CIH* ii 563.28; iii 780.7 = *CG* 7.186.

[153]McCormick, 'Stock-rearing in early Christian Ireland', 89.

[154]Toussaint-Samat, *History of food*, 10.

[155]Small, *The image of Irelande by John Derricke*, plate III.

[156]O'Kelly, 'Excavations and experiments in ancient Irish cooking-places'; O'Kelly, *Early Ireland: an introduction to Irish prehistory*, 223–7. See also Brindley, Lanting and Mook, 'Radiocarbon dates from Irish fulachta fiadh and other burnt mounds'; Ó Drisceoil, 'Burnt mounds: cooking or bathing?'.

The term *fulacht* is applied in our sources to a cooking-pit, particularly one which is out in the open.[157] Sometimes, it is stated to be in the vicinity of a dwelling-house,[158] but many references are to uninhabited areas. A law-text speaks of a *fulacht* in a wood,[159] and there is mention in other sources of a pit made by hunters (*fulacht fían* or *fulacht fíansa*) to cook the game which they have killed.[160] It is clear from the texts that the word *fulacht* is applied not only to a water-filled pit for boiling meat but also to an outdoor cooking-pit where meat is simply roasted on a spit or griddle over an open fire.[161] Mick Monk has identified such a pit in an early Christian site at Lisleagh, Co. Cork (pers. comm.). Chaff fragments at the edge of this pit suggest that the cooking-fire was lit by wisps of chaff: see p. 242 above.

Law-texts and sagas make frequent mention of the large cauldron (*coire*), clearly regarded as essential equipment in any prosperous household. *Críth Gablach* states that the highest grade of *bóaire* should have a bronze cauldron in which a boar can be cooked,[162] and a high-ranking lord should have a cauldron which can fit the flesh of a cow as well as a flitch of bacon.[163] A fleshfork (*aél*) is used to take the cooked pieces of meat out of the cauldron, and literary accounts of feasts record a tradition that the choicest portions go to those of highest rank and prestige.

Sometimes the meat was boiled into a broth (*enbruithe*), which presumably could be reheated a few times. In Saint Máel Rúain's ascetic rule it is stressed that a monk who is on a vegetarian diet must avoid the broth from meat, as well as the meat itself.[164] I have found no references to other ingredients associated with boiled

[157] The word *brothlach* has a similar meaning: see *DIL* B 205.79–80.

[158] E.g. Hull, '*Cáin Domnaig*', 160.10 § 1.

[159] *CIH* i 241.21–2 = *AL* v 482.26 *fulacht cacha caille* 'the cooking-pit of any wood'.

[160] E.g. Meyer, *Sanas Cormaic*, 46 § 562. See also *DIL* s.v. 1 *fulacht.* The holly-tree is described in legal commentary (*CIH* vi 2183.16) as *crann fulachta fiannsa* 'tree of the cooking-pit', presumably because its wood is especially suitable for making spits for outdoor cooking of game.

[161] For example, a glossator at *CIH* i 241.35 = *AL* v 484.13 refers to the use of a griddle (*indiūn*) at a *fulacht.* The apparatus used at *Fulacht na Morrígna* 'the cooking-pit of the [battle-goddess] Morrígain' is illustrated at col. 245 of YBL (facsimile p. 419); cf. *CIH* vi 2219.35 = Meyer, *Triads*, 16 § 120.

[162] *CIH* ii 564.7; iii 780.14 = *CG* 8.197. This category of *bóaire* is called a *mruigḟer.*

[163] *CIH* ii 567.24; iii 783.31–2 = *CG* 16.405–6; cf. Connolly and Picard, 'Cogitosus: Life of Saint Brigit', 14 § 3.2.

[164] Gwynn and Purton, 'The monastery of Tallaght', 145 § 49.

meat, but the use of the term *craíbechán* 'the branchy one' for a type of finely-chopped stew suggests that vegetables may sometimes have been added to provide flavour.[165]

Fig. 16. The cooking-spit of Deichen (bir nDeichin) reproduced from col. 245 of the fourteenth-century Yellow Book of Lecan (MS no. 1318 = H 2.16) in the Library of Trinity College, Dublin.

Another common method of cooking meat was to roast it on a spit (*bir*). The technique is shown in the Yellow Book of Lecan illustration (reproduced above) of *bir nDeichin*, the cooking-spit of Deichen, who was a legendary blacksmith at Tara.[166] The spit had a handle on the right by which the pieces of meat could be turned to ensure even roasting. This spit was of iron, but there are also references to wooden spits. A passage in *Aislinge Meic Con Glinne* describes the roasting of pieces of beef, mutton and ham on four straight spits of whitebeam.[167] The cook (Mac Con Glinne) donned a linen apron and hat, and rubbed honey and salt into the pieces of meat before roasting them over a fire of ash-wood. He was so deft at turning the spit that not one drop of juice fell into the fire, and all the flavour remained within the pieces of meat.

Fish may also be cooked by this method. The Life of Saint Brendan of Clonfert describes how a man named Doburchú caught a trout and grilled it on a spit over a fire of dry bracken.[168] An entry in *Cormac's Glossary* likewise refers to the cooking of a fish on a griddle (*indeóin*).[169]

[165]E.g. Jackson, *Aislinge*, 32.991–3 = Meyer, *Aislinge* 83.5–7; *ACL* i 247 § 285 = *CIH* ii 629.30–31. Or it might have something to do with the type of cow described as a *craíbech*: see p. 526.

[166]YBL col. 245; cf. *CIH* vi 2219.35 = Meyer, *Triads*, 16 § 120. A similar cooking-spit (*bir bruinnes* 'the spit which springs forth(?)') is reproduced from the Book of Leinster on p. 356.

[167]Jackson, *Aislinge*, 24.735 = *Meyer, Aislinge*, 63.1 *la cethri bera fírdírge findchuill.* For the identity of *findcholl* see p. 380. Spits of rowan (*caírthend*) are mentioned at *LL* ii 445.13882; Stokes, 'Find and the phantoms', 300.158.

[168]Plummer, *Bethada* i 81 § liii (163), trans. ii 79. The name *Doburchú* means 'otter': see p. 282.

[169]Meyer, *Sanas Cormaic*, 88.7 § 1018.

Our sources naturally concentrate on the choicer portions of meat, particularly in the context of feasting (see p. 358). However, there is a certain amount of information on the uses of offal, i.e. intestines, head, heart, liver, kidneys, tail, etc. *Aislinge Meic Con Glinne* makes frequent mention of the consumption of tripe (*inbe*), and it was obviously quite highly regarded as a food. The same text also gives prominence to the *maróc*[170] and the *indrechtán*,[171] both of which seem to be sausages of some kind. As Lucas points out,[172] the descriptions in the text suggest long sausages made from an animal's intestine, but no hints are given as to the composition of the stuffing. In a list of items in a food-rent, the law-text *Cáin Aicillne* includes *muccrecht saillte*, which the glossator identifies as *cáelán* 'intestine'.[173] It seems to be some sort of salted pork sausage.[174] Other types of offal are mentioned in a legal passage on the entitlements of a judge (*medam*) from his king.[175] These include the head of a sheep, and the liver and chest-cartilage (*longbronn*) of a pig.[176]

For the consumption of the eggs of hens and geese: see p. 104f.

VEGETABLES AND FRUIT

As we have seen in Chapter 8, our texts show little variety of vegetables and fruit in cultivation. The most frequently mentioned vegetable is *cainnenn*, which probably refers to some type of bunching onion, suitable for eating raw (p. 252). Watercress (*birar*), sorrel (*samad*) and wild garlic (*crim*) were regularly collected from the wild, and were no doubt also eaten raw. Another method of preparing vegetables was to cook them in a soup or broth. A dish of this type is described in commentary

[170]Marstrander, *Bidrag*, 74, 132, takes *mar*(*óc*) to be a borrowing from Old Norse *mörr* 'suet, sausage'.

[171]The etymology of *indrechtán* is unknown.

[172]Lucas, 'Irish food before the potato', 17–18.

[173]*CIH* ii 479.25 = Thurneysen, 'Aus dem irischen Recht II', 371. *Caolána* 'intestines' are also included among the types of meat which a client gives to his lord between Epiphany and Christmas (*CIH* v 1910.33).

[174]In *O'Davoren's Glossary* (*CIH* iv 1514.25–6 = O'Dav. 410 § 1210), the *mucrecht* is defined as *an caolán tóna* 'the lower intestine'. The first element of this compound is obviously *mucc* 'pig', but what is *recht*?

[175]*CIH* iv 1268.20–34. The first stanza is edited by Watkins, 'Indo-European metrics and archaic Irish verse', 230–1.

[176]For a discussion of the meaning of *longbronn*, see *DIL* L 200.71–8, and Jackson, *Aislinge*, 55, note to l. 388.

on *Félire Óengusso*, dating from about the eleventh century.[177] According to this tradition, Saint Columb Cille met an old woman one day collecting nettles in the churchyard at Iona. She had only one cow, which had not yet borne a calf, so she had to survive on nettle broth (*praissech nennta*). The saint ordered his cook to feed him thereafter solely on nettle broth without milk or butter. However, the cook bored a hole in the stirring-stick (*crann súaite*) which he used to mix the broth, and secretly added melted butter through the hole each day. Because of the butter, the saint throve on this dish, and the monks commented on his healthy appearance. Eventually the trick was revealed to the saint, but because of the cook's good intentions he blessed his successors for ever.

This anecdote provides us with some clues as to the evolution of vegetable dishes in Irish cooking. *Braisech* (*praisech*) is a borrowing from Latin *brassica* 'cabbage', and originally referred simply to the vegetable. In *Aislinge Meic Con Glinne*, however, *braissech* is also used of cabbage eaten in the form of a soup, with a layer of animal fat on top.[178] This is a similar dish to the nettle broth with melted butter mentioned in the commentary on *Félire Óengusso*. A further development seems to have been the thickening of the soup with meal.[179]

Apples – both cultivated and wild – are often mentioned in our texts, and there is occasional mention of the cultivated plum. Hazel-nuts and wild fruit of various types were also important supplementary food.[180]

For medicinal herbs see p. 257, and for a discussion of the dietary importance of honey see p. 113.

SALT

Sodium chloride or common salt (*salann*) is vital to the human body, so it is not surprising that it is frequently mentioned in our sources. Legal and other texts include a sack of salt among the three essential sacks in a prosperous household. For example, a

[177] Stokes, *Félire*, 146.

[178] Jackson, *Aislinge*, 15.467 = Meyer, *Aislinge*, 37.30. In Welsh, *bresych* (also a borrowing from Latin *brassica*), means (a) 'cabbage', (b) 'vegetable soup', e.g. *bresych ffa* 'bean soup', *bresych pys* 'pea soup' (*GPC* s.v. *bresych*).

[179] See *DIL* P 199.35.

[180] See p. 306.

rich landowner in the voyage-tale *Immram Ua Corra* was famed for his hospitality because he was never without three sacks: a sack of malt for preparing beer, a sack of wheat for preparing bread, a sack of salt for making every food taste well.[181] Writing in the late seventh century, Adomnán refers to a piece of rock-salt (*petra salis*) which had been blessed by Saint Columb Cille.[182]

Salt is most commonly obtained from salt-mines or by evaporating sea-water in salt-pans. In maritime areas where neither method is practicable, salt may be obtained in small quantities by burning seaweed or shore-vegetation, and boiling up the resultant ash with water. There are no references to Irish salt-mines or salt-pans in pre-Norman texts,[183] so it is possible that most salt was obtained from ash. Support for this theory comes from the use in the law-texts of the term *murlúaith* 'sea-ash' for salt. *Críth Gablach* refers to the use of 'sea-ash' for salting joints of meat after they have been cut up.[184] Because of the labour involved in extracting salt from seaweed, it is easy to understand the economic sense of importing salt from countries where it can be obtained with greater ease. The twelfth-century *Aislinge Meic Con Glinne* refers to 'English salt' (*salann Saxanach*) in a context which implies that it was regarded as being of particularly high quality.[185] From approximately the same period, legal commentary speaks of ships with cargoes of salt and other goods, but provides no further details.[186] Salt was transported in cakes rather than in granular form: a saint's Life refers to 'two lumps of salt' (*dá chloich salainn*).[187]

Being a food which is essential but not universally available, salt is especially vulnerable to the attentions of tax-gatherers, and much

[181]Van Hamel, *Immrama*, 96.11–12 = Stokes, 'The voyage of the Húi Corra', 26 § 1. Cf. *CIH* ii 564.1–3; iii 780.8–10 = *CG* 8.189–91; *CIH* iii 1058.3–4.

[182]A. O. and M. O. Anderson, *Adomnan's Life of Columba*, 340 (new pagination: 104) = ff. 58b–59a.

[183]See Scott, 'Some conflicts and correspondences', 115–16.

[184]*CIH* ii 564.2; iii 780.9 = *CG* 8.190; cf. *CIH* i 242.21 = *AL* v 486.17. Early Norwegian law refers to black salt (*svartasalt*) procured by burning seaweed, and used for cattle-licks (Larson, *The earliest Norwegian laws*, 427). There is also the possibility that *murlúaith* 'sea-ash' is a metaphorical expression for any form of sea-salt. See Lucas, 'Washing and bathing in ancient Ireland', 99–100.

[185]Jackson, *Aislinge* 23.734 = Meyer, *Aislinge*, 61.29. A half-quatrain quoted in *Cormac's Glossary* (Meyer, *Sanas Cormaic*, 93.16–17 § 1059) refers to salt-production on the coast of the Isle of Man. See also Timothy O'Neill, *Merchants and mariners*, 84.

[186]*CIH* i 315.27 = *AL* iii 426.23.

[187]Stokes, *Lismore Lives*, 72.2408.

research has been carried out on the history of salt-taxation.[188] So far as I have been able to discover, the only reference to such a tax in early Irish sources is in the eighth-century text about Conall Corc and the Corco Luigde, which speaks of a salt-tax (*salanngabál*) on the Aran Islands.[189]

Pepper

Pepper (*Piper nigrum*) is a forest vine native to tropical India.[190] The powerful flavour of its dried berries was discovered in prehistoric times, and already by the fifth century BC pepper was being exported from India to the Mediterranean area.[191] Pepper was introduced to Ireland in the early Christian period, and was called *scibar*, a borrowing from Latin *piper*.[192]

It does not seem to have been of great significance in the early Irish diet as it is only occasionally mentioned in the texts. Nonetheless, it was sufficiently well known to be included in a list of proverbial phrases in *Aislinge Meic Con Glinne*, all of which refer to some useless or absurd undertaking. One of these is given as *longad i scellaib scibair* 'dining on pepper-seeds'.[193] This phrase raises some questions about the form in which pepper was consumed in early Irish cooking. Pepper is exported in two forms: dried immature berries ('black pepper') and the ripe seed from which the outer skin has been removed ('white pepper'). In his interpretation of this passage, Meyer apparently took the *scella scibair* to be the husks of the ripe pepper, which would obviously be dry and inedible. Jackson, on the other hand, took the phrase to refer to the actual seeds of the pepper, which would be much too strong-flavoured to be eaten on their own. As the word *scell*

[188]See Adshead, *Salt and Civilization*; Toussaint-Samat, *History of food*, 457–79.

[189]Meyer, 'Conall Corc and the Corco Luigde', 60.23–6; Hull, 'Conall Corc and the Corco Luigde', 900. See *DIL* G 1.65–6.

[190]Sweet pepper (*Capsicum annuum*) is native to America, and was unknown in Europe until about the sixteenth century.

[191]Toussaint-Samat, *History of food*, 491.

[192]E.g. *Thes.* ii 136.28 *piperi .i. scibar.* The phonetic details of this borrowing have been discussed by Pedersen (*VGK* i 236), Vendryes (*LEIA* S-44 s.v. *scibar*) and McManus, 'A chronology', 41[47]. The Latin word was later borrowed a second time in the form *pipur*, Mod.Ir. *piobar.*

[193]Jackson, *Aislinge*, 28.858–9 = Meyer, *Aislinge* 71.30. Another absurdity in this list is *salann for lúachair* '[sprinkling] salt on rushes', cf. *CIH* vi 2223.16 *scris for aīn* 'sprinkling [salt] on rushes'.

can mean both 'shield, outer layer, husk' as well as 'seed', it is not certain which explanation is correct.

Restricted diets

The early Irish Church inherited the practice of fasting from Jewish tradition, and it played an important part in early Irish regimes of mortification. Monastic poetry testifies to the existence of a strong ascetic tradition in the early Irish Church, often tinged with melancholia and an obsessive fear of damnation.[194] In this spiritual atmosphere, it is easy to understand how the mortification of the body was seen as a means of escape from the torments of Hell.

We are fortunate that a considerable number of religious texts have survived from between the seventh and ninth centuries in both Old Irish and Latin which describe various restricted diets. These texts provide a great deal of information on early Irish food, and hence on farming practice. Those in Latin have been edited by Ludwig Bieler in *The Irish Penitentials* (Dublin 1963). This book also contains a translation of *The Old Irish penitential*[195] and of *The Old Irish table of penitential commutations.*[196] Another important source is a ninth-century untitled text known as *The monastery of Tallaght* which contains traditions relating to Máel Rúain, the founder of Tallaght and a prominent member of the Céle Dé reform movement, and to his disciple Máel Díthruib.[197]

Monastic diets

Our sources are unanimous in stressing that the ration (*pít*)[198] of monks and nuns should be meagre, and there are references to the

[194]E.g. Murphy, *Early Irish lyrics*, 58 (poem 25); 62–4 (poem 27); Stokes, *Félire*, 264–88 (Epilogue).

[195]*IP* 258–77. This is a revision by D. A. Binchy of E. J. Gwynn's edition of the text in *Ériu* 7 (1914) 121–95.

[196]*IP* 277–83. This translation is by D. A. Binchy, who edited the text in *Ériu* 19 (1962) 47–72.

[197]Ed. Gwynn and Purton, 'The monastery of Tallaght'. Gwynn edited two other texts from the same tradition in *The rule of Tallaght.* The first, *Teagasg Maoil Ruain* 'the teaching of Máel Rúain', is a seventeenth-century paraphrase of the *Monastery of Tallaght* with additional material. The second is *Ríagail na Céle nDé* 'the rule of the Céli Dé', which Gwynn suggests goes back to a ninth-century original (Introduction p. vii).

[198]The term *pít* (also *fít, fiit*) refers mainly to food which has been rationed in a religious context. However, the compound *samfít* 'summer ration' is used of the milk-products which a client must give to his lord (*CIH* v 1910.26; cf. *CIH* ii 498.25; v 1801.37 = *AL* ii 326.7).

importance of diet in subjugating their sexual desires.[199] According to *Teagasg Maoil Ruain*, typical monastic fare at Tallaght consists of bread with a condiment (*annlann*) of fish, butter, cheese or a hard-boiled egg. If available, vegetables such as cabbages or leeks may also be taken. However, the text specifies that the cabbage must be dressed with milk rather than with butter. If there are apples, each monk is allowed three or four large ones or five or six small ones.[200] It is held to be safer for the soul to eat a small ration of food which is of good quality (*mín*) than a large ration of food which is coarse (*anmín*).[201]

Attitudes towards the consumption of meat varied considerably in the early Irish Church. Some texts record traditions of monastic vegetarianism. For example, Tigernach of Clúain Eois (Clones) is said to have lived on barley-bread, water-cress and hot water,[202] and the *Litany of Irish saints* states that Fintan's monks consumed only water and the herbs of the land.[203] However, even under Máel Rúain's stern régime at Tallaght, vegetarianism seems to have been optional, though monks who were not vegetarian were only allowed to consume the flesh of wild pig or deer.[204] Ordinarily, those who had vowed not to eat meat were expected to avoid even a meat broth (*enbruithe*),[205] but at Easter the whole community of Tallaght was advised to eat a small piece of meat. This provision was to allow for the ban on meat to be relaxed in the monastery in the event of scarcity of bread or other vegetable food during the year. It was felt that if a monk had avoided meat at Easter, he would not be entitled to break his vow at any other time of year and would then be expected to submit to the 'white martyrdom' of death by starvation.[206]

[199] E.g. Gwynn and Purton, 'The monastery of Tallaght', 149 §§ 59–60; 151–2 § 63.
[200] E. J. Gwynn, *The rule of Tallaght*, 2–4 § 2.
[201] Gwynn and Purton, 'The monastery of Tallaght', 152 § 63 (see note to 152.16 on p. 171); E. J. Gwynn, *The rule of Tallaght*, 74 § 39.
[202] Stokes, *Félire*, 110 § 4.
[203] *LL* vi 1700.52230–31.
[204] Gwynn and Purton, 'The monastery of Tallaght', 129 § 6. The text at lines 24–5 seems to imply that the flesh of the domestic animals of the monastery was fed to guests. In later tradition (E. J. Gwynn, *The rule of Tallaght*, 6 § 8) it is stated that Máel Rúain did not allow venison, liver or lard to be eaten at Easter, but that he had to relax this prohibition on account of a famine.
[205] Gwynn and Purton, 'The monastery of Tallaght', 145 § 49.
[206] Gwynn and Purton, 'The monastery of Tallaght', 132 § 12; 146 § 51. For a discussion of the rationale behind this provision, see their note on pp. 166–7. The same

There was also ambiguity in the Church's attitude towards beer. In some traditions, beer features prominently in the celebration of Easter. For example, one of the miracles in *Bethu Brigte* involves the rapid fermenting of eighteen vatfuls of beer for feasting in the churches from Easter Sunday to Low Sunday.[207] Máel Rúain, on the other hand, forbade the drinking of 'the liquor which causes forgetfulness of God' in his monastery. His monks were not allowed to drink beer anywhere in the district of Cúalu, but this rule was relaxed in the case of long journeys outside this area. On one occasion Dublitir, the abbot of Finglas, asked Máel Rúain to allow the monks of Tallaght to drink beer at the three chief festivals of the Church (Christmas, Easter and Pentecost). Máel Rúain refused, whereupon Dublitir said that his own monks would drink beer on these occasions and be in Heaven along with Máel Rúain's monks. Máel Rúain retorted that his monks would not need to be cleansed by the fire of Doomsday, whereas Dublitir's monks would perchance have something for the fire to cleanse![208]

In most monastic diets there is some relaxation on special occasions. Thus *Ríagail na Céle nDé* states that some delicacy should be provided on Sundays, on high festivals and on the feasts of the apostles.[209] The extra delicacy is often called the *selann*, and seems usually to have consisted of butter, but honey and salted meat are also mentioned.[210]

Máel Rúain was not in favour of excessive fasting, and taught that a regular moderate ration was preferable to a total fast.[211] He also warned against the sudden reduction in a monk's ration, for fear of causing illness and piles.[212] On the other hand, the author of *The monastery of Tallaght* records with admiration the extraordinary privations of the recluses of Iona who survived on a water-based gruel (*tiuglagin*). On one occasion the abbot noticed that their

idea is expressed in E. J. Gwynn, 'An Irish penitential', 148 § 14. The possibility of eating meat in Lent when there is no other food is discussed in *Teagasg Maoil Ruain* (E. J. Gwynn, *The rule of Tallaght*, 14 § 23).

[207] Ó hAodha, *Bethu Brigte*, 7 § 21.

[208] Gwynn and Purton, 'The monastery of Tallaght', 129–30 § 6.

[209] E. J. Gwynn, *The rule of Tallaght*, 74 § 39. See also ibid. 2 § 1.

[210] Gwynn and Purton, 'The monastery of Tallaght', 146–7 § 52; Meyer, *Sanas Cormaic*, 90 § 1051; *CIH* iv 1526.8–12 = O'Dav. 461 § 1484; E. J. Gwynn, *The rule of Tallaght*, 89 (Notes).

[211] Gwynn and Purton, 'The monastery of Tallaght', 155 § 68.

[212] Ibid. 152 § 63.

colour was bad, so he added a lump of butter to the daily gruel of each recluse until his proper colour was restored. Thereafter the recluses were alternately brought near to death and then back to life again by the addition or withholding of butter in their diet.[213]

Lenten fasting

The early Irish Church observed three Lents: forty days before Christmas, forty days before Easter, and forty days after Pentecost.[214] These were occasions of severe fasting in the monasteries. For example, it is recorded that Máel Díthruib of Tallaght lived on bread and water during these periods for the sake of his father's soul.[215] Lay people were also instructed to fast during the three Lents; in particular, they were to abstain from fresh and salted meat.[216]

The spring Lent (*corgas erraig*) commemorates Christ's fast in the wilderness, and is therefore obviously of greater significance in the Church's year than the winter Lent (*gamchorgas*), which was associated with Elias, and the summer Lent (*samchorgas*), which was associated with Moses. Our texts therefore emphasize the dietary restrictions of the spring Lent. For example, *Bretha Crólige* states that the condiment to which a freeman of middle or high rank is entitled consists only of vegetables during the spring Lent – no mention is made of such restrictions during the other Lents.[217] Similarly, Máel Rúain instructed that the Lenten penance of bread and water be relaxed by the addition of a sip of whey-water during the summer Lent.[218]

The special severity of the diet during the spring Lent fits in with the annual farming cycle. At this time of year, the store of grain and salted meat may already be getting low, and milk-production is not yet properly under way. To restrict consumption as far as possible at this period therefore makes a virtue of necessity.

Twice-weekly fasting

The Irish names for Wednesday and Friday testify to the observance of twice-weekly fasts throughout the year in the early Irish Church. Wednesday is *Cétaín* 'the first fast', and Friday is *Aín* or

[213]Gwynn and Purton, 'The monastery of Tallaght', 147 § 52.
[214]Plummer, *Vitae* i Introduction p. cxx, footnote 1; J. Ryan, *Irish monasticism*, 391–2.
[215]Gwynn and Purton, 'The monastery of Tallaght', 163 § 86.
[216]E. J. Gwynn, 'An Irish penitential', 150 § 36 = *IP* 265 § 36.
[217]*CIH* vi 2298.33 = Binchy, '*Bretha Crólige*', 36 § 46.
[218]Gwynn and Purton, 'The monastery of Tallaght', 143 § 42.

Aín dídine 'the last fast'. The consumption of meat was forbidden on these days.[219]

Penitential diets

The texts supply much detail on the diets imposed on lay or clerical penitents as part of the expiation for their sins. In general, the most rigorous diets are for high-ranking ecclesiastics who have committed grave sins. According to an eighth-century Old Irish penitential,[220] a bishop who is guilty of fornication must do seven years on bread and water, or twelve years on a slightly less severe diet called *dobrit* (= *dobrith*), in which the bread and water must have been accompanied by a small quantity of some other food.[221]

For the same sin, a priest or deacon spends three and a half years on bread and water, with extra rations during Church festivals and on Sundays. The penance is further reduced in the case of one who is subject to monastic discipline but is neither a priest nor a deacon. He does penance for only three years, and gets a larger measure of bread and whey-water, as well as a prime *bochtán* of good milk.[222]

Severe dieting can damage the health. In particular, the early Irish practise of speeding up a penance by increasing its severity has obvious physical dangers. For example, the Old Irish table of penitential commutations allows for the commutation (*arrae*) of a year's penance to a three-day black fast (*dubthredan*) during which the sinner neither ate, drank, nor slept, and was subjected to other torments.[223] An even more ferocious commutation was to spend seven months bound in fetters day and night on a diet of water and water-cress (*birar*): this was regarded as the equivalent of seven years strict penance.[224]

[219]E.g. *CIH* ii 523.12 = *AL* iii 14.20.

[220]E. J. Gwynn, 'An Irish penitential', 140 §§ 2–6. Much the same regulations are to be found in the *Penitential of Cummean* (*IP* 112–14).

[221]The plausible suggestion is made in *O'Davoren's Glossary* (*CIH* iv 1494.13 = O'Dav. 319 § 745) that *dobrith* is a compound of *dobur* 'water' + *ith* 'corn'. It is to be distinguished from *doborfiit* (*dobur* 'water' + *pít* 'ration, morsel') which occurs in a passage on penance at *Thes.* ii 38.29 (with correction on p. 493, supplement).

[222]For *bochtán* as a liquid measure, see Appendix B, p. 577. In the corresponding passage in the *Penitential of Cummean* (*IP* 114 § 3), he gets a Roman pint (*sextarius Romanus*) of milk.

[223]Binchy, 'The Old-Irish table of penitential commutations', 66 § 27 = *IP* 280 § 15 (translation only).

[224]Binchy, 'The Old-Irish table of penitential commutations', 66 § 35 = *IP* 280 § 19a (translation only).

The danger to life inherent in penitential diets is discussed in a remarkable passage in *The monastery of Tallaght*.[225] The author claims that a meeting of the saints of Ireland took place in Mag Léna to discuss the problem of penitents dying while on bread and water. In the traditional manner, the saints fasted against God to force Him to respond to their concerns.[226] He sent an angel who told the saints that the fruit and plants of the earth had lost their strength and force on account of the falsehood and sin and injustice of men. The angel explained that in former times, when men were obedient to God's will, water was as nourishing as milk was in their own day. He therefore instructed the saints to mix some meal with butter to make a gruel to sustain penitents. After that, three types of gruel were used in penitential diets. Those whose sins were lightest got gruel upon water (*menadach úas usciu*), whereas those whose sins were most serious got gruel under water (*menadach ís usciu*), i.e. a very weak mixture. Those whose sins were of intermediate seriousness got gruel between two waters (*menadach eter dá usce*) – this is explained as a gruel in which the meal neither floats nor sinks to the bottom of the vessel.

According to the law-text *Di Chetharślicht Athgabálae*, a form of penitential diet is imposed on a slave or low-ranking servant who has committed a serious offence. Until his crime is paid for by his master or employer, he is held in fetters or chains and fed on a *bochtán* of milk and an *aircháelán* or half-loaf of bread. However, on Christmas or Easter or Sunday he is entitled to 'the loaf of a noble day' (*bairgen úasallaithi*) with its condiment (*annlann*).[227]

Food for invalids

Bretha Crólige is mainly concerned with the food-entitlements of persons who have been illegally injured, but whose capacity to eat has not been affected. But it also discusses the ill-effects of certain foods on invalids, particularly if there are associated stomach problems.[228] Honey is therefore forbidden if the patient is suffering from diarrhoea (*lír*). Likewise, the invalid should not be supplied with meat preserved in sea-salt or with whale-flesh (*carna*

[225] Gwynn and Purton, 'The monastery of Tallaght', 157–8 §§ 73–4.

[226] For the use of fasting in early Irish law to exert pressure on a person of high rank, see *GEIL* 182–3.

[227] *CIH* ii 363.26–7 = *AL* i 106.1–3.

[228] *CIH* vi 2292.27–32 = Binchy, '*Bretha Crólige*', 20 § 25.

máirmíl), because 'sea-produce impels one to drink'. Horse-flesh is also forbidden because it stirs up sickness in the stomach of one who has been wounded. Beer may be excluded at the discretion of the physician.[229]

The importance of vegetables in the diet of invalids is repeatedly stressed. *Imus* 'celery(?)' is held to be especially valuable as it 'prevents sickness and does not stir it up, prevents thirst and does not affect wounds'.[230] Many texts refer to the health-giving qualities of *brothchán*, which seems to have been made by heating milk with oatmeal and herbs.[231] According to the law-text *Di Astud Chirt 7 Dligid*, any law-abiding freeman is entitled to collect herbs for this dish (*losa brothcháin*), even on private property.[232] The cow whose milk goes to make this dish (*bó brothcháin*) also has special legal protection. Even if the owner of the cow owes a debt or liability to another person, the animal cannot be distrained to discharge this debt as long as it is producing milk for an invalid's *brothchán*.[233]

Invalid's craving

It is recognised in the law-texts that an invalid may develop a craving for a particular food or drink – this is called *mían ngalair* 'the desire of sickness'. If the desired foodstuff is not available in the invalid's home, there is a legal entitlement for it to be obtained from a neighbour. This seems to be the point of an anecdote in *Bethu Brigte*. Brigit's fostermother was gravely ill, so Brigit and another girl were sent to get beer for her at the house of a man named Báethchú. He refused them, so when he returned to his ale-vat, he found that it had miraculously been emptied, and a spider's web covered it.[234] The girls then filled three vessels from a nearby well, and found that the liquid was tasty and intoxicating beer which immediately cured the fostermother.

[229] *CIH* vi 2299.38–9 = Binchy, '*Bretha Crólige*', 38 § 49.
[230] *CIH* vi 2298.14–18 = Binchy, '*Bretha Crólige*', 36 § 45. Binchy translates 'does not infect wounds', but the author seems merely to be saying that *imus* does not interfere with the healing process.
[231] Cf. Dinneen, *Foclóir* s.v. *brachán*; P. L. Henry, 'A linguistic survey', 163.
[232] *CIH* i 242.17 = *AL* v 486.12–13.
[233] *CIH* i 38.21 = *AL* v 260.4; cf. *CIH* iii 906.6.
[234] Ó hAodha, *Bethu Brigte*, 2 § 8. The text is here incomplete: see the editor's note on p. 40.

As we have seen above, honey is not recommended in the case of an invalid with a stomach infection. However, in other circumstances it may help to restore a sick person to health.[235] Honey can therefore be requested from a beekeeper for a neighbouring invalid even during the 'three years of immunity' when a hive is normally free of all obligations to neighbours.[236]

Pregnant woman's craving

The principle of an invalid's entitlement to food is extended in the law-texts to include the sudden craving of a pregnant woman, known medically as pica.[237] A glossator to the main law-text on distraint explains the phrase *mír méin* 'desired morsel' as referring to the desire of a pregnant woman,[238] and commentary on *Bretha Étgid* discusses some of the legal aspects of a pregnant woman's craving for food.[239] One version of this commentary makes particular mention of her sudden desire for beer triggered off by the smell of malt (*túth mbracha*).[240] The law-text *Bretha Nemed Déidenach* provides an example of this phenomenon in the story of a pregnant woman who entered a house where a feast was being prepared for a king.[241] At the smell of the beer her child leapt in her womb, and she felt a sudden craving for a drink. Her request was refused by the brewer (*scóaire*) because the casks had been sealed. However, the child in her womb – the future poet Aithirne – uttered a short satirical verse which caused the hoops of the casks to burst and the beer to run ankle-deep throughout the house. The woman was then able to satisfy her desire by drinking three draughts of beer. The passage concludes with the observation that if any poet is refused beer he can obtain the same effect by reciting Aithirne's verse.

For a pregnant woman's craving for chives (*foltchép*), see p. 257.

[235] See discussion at *BB* 99–100 (Notes to § 6).

[236] *CIH* ii 444.28 = *BB* 52 § 6.

[237] I am grateful for some of these references to Donnchadh Ó Corráin who delivered a paper on pica at the International Celtic Congress at Swansea in 1987.

[238] *CIH* ii 387.34; iii 890.38; v 1692.15 = *AL* i 180.5; cf. *CIH* i 242.17 = *AL* v 486.12.

[239] *CIH* i 270.25–271.9; iii 940.28–941.24 = *AL* iii 204.8–206.y.

[240] *CIH* iii 1068.3; iv 1256.25; cf. *CIH* iv 1517.8–9 = O'Dav. 420 § 1265.

[241] *CIH* iii 1118.34–1119.16 = E. J. Gwynn, 'Athirne's mother'. There is a similar story in *Betha Finnchua* (Stokes, *Lismore Lives*, 85.2818–30).

THE DIET OF WOMEN AND CHILDREN

In Irish legal tradition, a wife is normally felt to enjoy half the status of her husband, and hence is entitled to half his honour-price (*díre*).[242] Following this logic, the author of *Bretha Crólige* states that a wife on sick-maintenance is entitled to half the food (*lethbíathad*) which would be due to her husband. However, a concubine on sick-maintenance can only claim one third or one quarter of his food.[243]

From the age of seven to ten, a child on sick-maintenance is entitled to what is termed the 'soft food of fosterage' (*máethbiad altramma*), which the glossator explains as 'egg-yolk, butter, curds and porridge'.[244] One can compare the twelfth-century commentary on the fragmentary text on fosterage, *Cáin Iarraith*, which details the type of food which fosterparents are expected to provide for the children in their care. As in the case of adults on sick-maintenance (see p. 318 above), fosterchildren are to be fed according to rank.[245] All children are entitled to porridge (*lite*), with the addition of dairy-products. The child of a commoner is to be fed oaten porridge made with buttermilk or water.[246] A blob (*tumad*) of small curds (*gruiten*) is allowed with this dish, though the author specifies that the child must only be allowed his bare sufficiency. The child of a lord, on the other hand, gets his full sufficiency of porridge made from new milk and barleymeal with a blob of fresh butter. The child of a king gets his full sufficiency of porridge made from new milk and wheatmeal with a blob of honey. Obviously, we cannot assume that these regulations were always rigidly enforced. The value of the passage lies in the general idea which it gives us of typical diets for children of different social classes.

[242] *CIH* iii 779.7 = *CG* 5.125.

[243] *CIH* vi 2301.21–4 = Binchy, '*Bretha Crólige*', 44 § 56.

[244] *CIH* vi 2300.38–9 = Binchy, '*Bretha Crólige*', 42 § 52 *maotbiad altruma .i. in buidecān 7 im 7 maotla 7 lictiu.*

[245] *CIH* v 1759.36–1760.2 = *AL* ii 148.20–150.5; cf. *CIH* i 82.22–5. It seems likely that in this passage *mac* (usually 'son') refers to a child of either sex.

[246] The text does not actually say that the porridge is of oats, but this can be deduced from the fact that children of higher rank get porridge of barley or wheat.

TABOO FOODS

Certain foods – especially of animal origin – are held to be taboo among particular racial or religious groups. Thus dog-flesh is treated with abhorrence in European countries, but much relished in China. Pork is eaten in the Christian and post-Christian communities of Europe and America, but shunned by Jews and Muslims. Various types of food-taboo are documented in early Irish sources, both secular and religious.

Judaeo-Christian traditions regarding clean and unclean food are transmitted in a number of texts of our period. For example, the *Canones Adomnani* deal almost exclusively with diet, and are on some points explicitly based on the Old Testament. For example, the author quotes from the Book of Exodus to support his assertion that it is wrong to eat the flesh of a pig which has tasted human flesh or blood.[247] But other provisions in this text doubtless reflect eating-habits which have no special association with Christianity. A distaste for carrion, for example, is natural in the human: his senses of smell and taste discourage him from treating it as food. On this topic, the *Canones Adomnani* are merely codifying and refining instinctive behaviour. There is also a less powerful but nonetheless widely felt revulsion at the consumption of food which has been interfered with by other animals. This is exemplified in the same text's condemnation of eating marrow from deer-bones which have been gnawed by wolves.[248] Other texts forbid the consumption of food or drink contaminated by a dog, cat, fox, raven, eagle, crow, cock or hen.[249]

Christianity probably also introduced eating-habits which conflicted with earlier practice. Thus an Old Irish penitential prescribes penance for three and a half years on those who eat horseflesh.[250] As Gwynn demonstrated in his note on this passage, this stern prohibition does not represent the unanimous view of the early Church: in an Irish context it may have been aimed at the consumption of horseflesh during rites which had incurred

[247] *IP* 176 § 7, which is based on Exodus 21:28. See *IP* 9 § 9 for a brief discussion on the date and authorship of this text.

[248] *IP* 180 § 19.

[249] *IP* 160–62 §§ 16–18; E. J. Gwynn, 'An Irish penitential', 146 i § 3.

[250] E. J. Gwynn, 'An Irish penitential', 146 i § 2. Irish Canon I (*IP* 160 § 13) prescribes four years on bread and water for this offence. Cf. A. O. and M. O. Anderson, *Adomnan's Life of Columba*, 252 (new pagination: 48) = f. 26b.

ecclesiastical disapproval. Writing in the late twelfth century, Giraldus Cambrensis describes a 'horrible usage' of a people in Tír Conaill, who – as part of a king's inauguration ritual – ate the flesh of a white mare with whom the new king had just had sexual intercourse.[251] The archaeological evidence shows that the Church's ban on the eating of horseflesh was sometimes ignored.[252] The bones of horses – presumably butchered for human consumption – have even been found in the middens of monastic sites at Moyne, Co. Mayo,[253] and at Church Island, Co. Kerry.[254] On the other hand, it is noteworthy that legal commentary states that the flesh of a horse, like that of a dog, has no value.[255]

Early Irish literature also provides information on a type of food-taboo which anthropologists have identified in many societies throughout the world.[256] This taboo is linked with the idea of a totem-animal which has intimate associations with an individual or a family. Often, a person may take the name of his totem-animal: the best-known example from Irish tradition is *Cú Chulainn* 'the hound of Culann', whose totem-animal was the dog. There was a solemn prohibition (*gess*) against his eating the flesh of a dog,[257] and when he was tricked by his enemies into doing so, his death followed soon afterwards. In another tale the totem-animal whose flesh is taboo is probably to be identified with the badger.[258]

[251] O'Meara, *Giraldus: Topography*, 109–10 § 102 = Dimock *Giraldi Topographia*, 169. See Schröder, 'Ein altirischer Krönungsritus'.
[252] E.g. Hencken, 'Ballinderry crannóg no. 1', 233–4; Hencken, 'Ballinderry crannóg no. 2', 72; cf. *CIH* vi 2292.28 = Binchy, '*Bretha Crólige*', 20 § 25 (reference to the consumption of *carna eich* 'horseflesh').
[253] McCormick, 'Stock-rearing in early Christian Ireland', 244.
[254] O'Kelly, 'Church Island', 133 (report by G. Roche). Margaret McCarthy of the Department of Archaeology, University College Cork, refers me to the recent find of butchered horse bone at an early medieval church site at Dunshaughlin, Co. Meath (pers. comm.).
[255] *CIH* i 306.41; iv 1208.36.
[256] Toussaint-Samat, *History of food*, 76.
[257] *LL* ii 445.13883–4 *Geiss dó dano cárna a chomanma do ithi* 'it was taboo for him to eat the flesh of that which had the same name [i.e. a dog]'.
[258] Mac an Bhaird, 'Tadhg mac Céin and the badgers', suggests that the flesh of the badger was taboo to Tadhg mac Céin because the name *tadg* originally meant 'badger'. It seems from other references (p. 282) that there was no general taboo on the consumption of badger-flesh. For the taboo on crane-flesh among the Irish, see p. 127.

Hunger and famine

As we have seen, the staple diet of the early Irish consisted of bread and milk-products. If the cereal-crop failed or disease struck the cattle, hunger prevailed. The annals contain many records of scarcity of food, often associated with unusual weather-conditions in spring or summer, or with the disruptive effects of warfare. Famines often persisted over a number of years, as it took time for the survivors to re-establish agricultural production.[259]

We know from the study of more recent famines – such as the Great Hunger of the 1840s – that death usually comes from the diseases which take hold on a weakened population, rather than from simple starvation. The same pattern can be discerned in the terse records of the early annals. The *Annals of Ulster* describe the year 764 as being one of excessive snowfall, drought, famine and dysentery (*riuth fola*) throughout Ireland – it is clear that the disease flourished because the resistence of the hungry people was low.[260] Another entry in the same annals records a common feature of all famines: the highest mortality is among the young, the old and the weak.[261]

The effects of hunger and starvation in a community are well documented in our sources. A Middle Irish tale entitled *Erchoitmed Ingine Gulidi* 'the excuse of the daughter of Gulide' gives a vivid description of a hungry household in cold spring weather. Their old food (i.e. from the previous year) is used up, and there is no new food. There is scarcely pasture for one cow or one goose or one bee. The milk in their stripper cows (*gamnacha*) has dried up, and there is only watery milk (*englasa*) in their milch cows. Their dogs are famished and their cats are ravenous.[262] Other texts describe the attempts of hungry people to feed themselves on wild plants, such as nettles, water-cress and sorrel (p. 311).[263] In some cases the only way to stave off immediate starvation was by killing and eating the

[259]For example, the great scarcity and famine (*ascalt mōr 7 fames*) of AD 764 was followed by a shortage of bread (*defectio panis*) in 765. See AU^2 216 s.a. 763 (recte 764) § 4; 218 s.a. 764 (recte 765) § 9.
[260]AU^2 216 s.a. 763 (recte 764) §§ 4, 7, 10. For a general discussion on the effects of famine in Ireland, see Crawford, *Famine: the Irish experience 900–1900.*
[261]AU^2 282 s.a. 824 (recte 825) § 7.
[262]Meyer, *Hibernica minora*, 66 § 6.
[263]Cf. Lucas, 'Nettles and charlock as famine food'. For the consumption of bark, see Lucas, 'The sacred trees of Ireland', 45–6.

milch cows, thereby creating further problems for the future.[264] A less destructive measure was the bleeding of cattle and horses: this must have been widely practised in hard times (see p. 53).

Ordinary taboos and patterns of conduct tend to break down during famines, and the annals record such abnormal behaviour as cannibalism, the eating of dog-flesh, and the selling of children in exchange for food.[265] Ties of friendship loosen as famine takes hold: the *Annals of Ulster* describe the hungry summer of 1433 as *samhradh na mearaithne* 'the summer of non-recognition'.[266] Other annals highlight a problem which often increases mortality during a famine. This is the natural tendency of starving people to migrate in search of food, thereby spreading disease and putting extra strain on the food-resources of other areas.[267]

Our texts provide little information on famine relief. If the crops of an individual early Irish farmer failed in a particular year, he could hope for help from his kin-group (*fine*), as the principle of kin-solidarity was deeply embedded in early Irish tradition. Destitute people could also expect assistance from their local monastery in accordance with the constantly reiterated Christian precept of feeding the hungry.[268] Indeed, famine victims might become permanently attached to the Church as base tenants (*dáermanaig ecalsa*).[269] But when famine appeared on a wide scale, these options were closed off. In some cases even the monasteries – often places of wealth and abundance – felt the rigours of involuntary hunger. Monastic writings contain a number of references to the effects of famine on dietary regulations.[270]

[264]Cf. *TBC* Rec. I 40.1275–6.
[265]E.g. AU^2 158 s.a. 699 (recte 700) § 6; *CS* 318 s.a. 1112.
[266]*AU* iii 130 s.a. 1433. See *DIL* M 106.15–21.
[267]E.g. *AI* 176 s.a. 1005 § 2; AU^2 558 s.a. 1116 § 5.
[268]E.g. *CIH* ii 524.24–5 = *AL* iii 18.19–20 *puipir do biathad* 'feeding paupers'.
[269]*CIH* ii 522.6-7 = *AL* iii 10.23. Cf. *CIH* iii 746.15–16.
[270]E.g. E. J. Gwynn, *The rule of Tallaght*, 6 § 8; Gwynn and Purton, 'The monastery of Tallaght', 146 § 51.

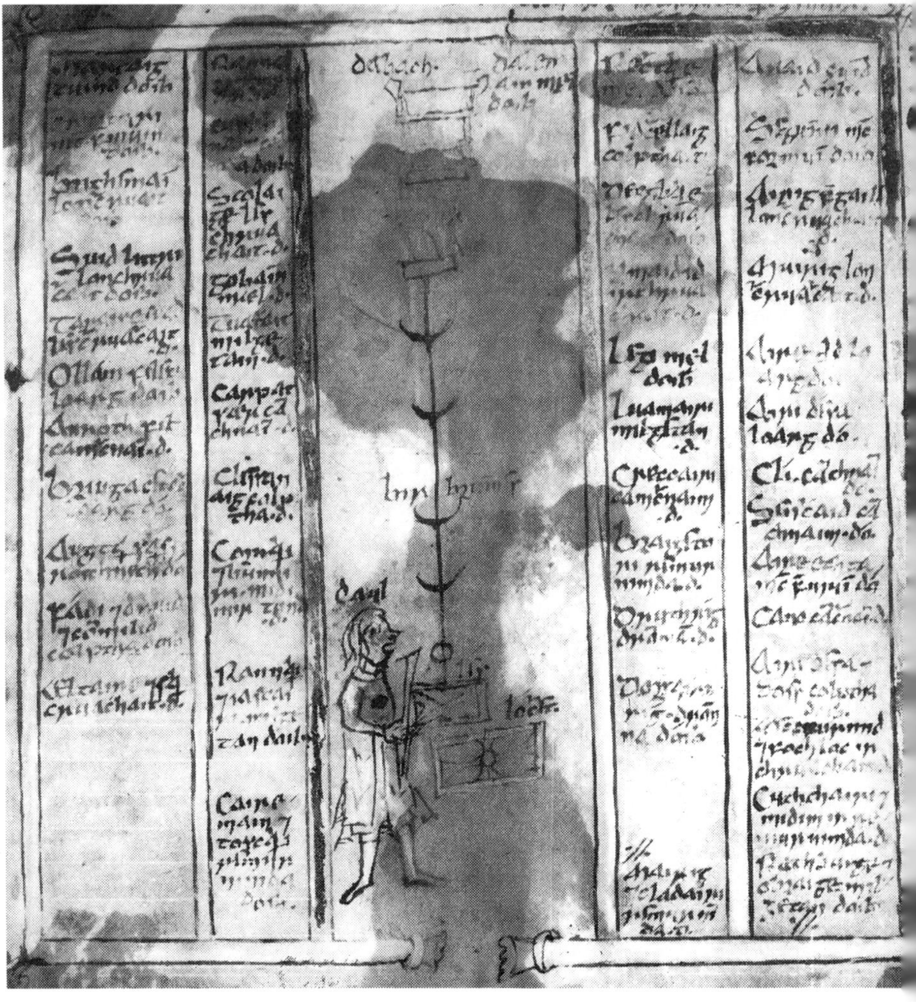

Plate 5. An illustration of *Suidigud Tigi Midchúarda* 'the seating of the house of the mead-circuit' from f. 29^{a} of the twelfth-century Book of Leinster (MS no. 1339 = H 2.18) in the Library of Trinity College, Dublin. It has been reproduced by kind permission of the Board of the College.

FEASTING

The feast (*fled*) was a very important institution in early Irish society, and is frequently mentioned in legal and other sources. One of the duties of a client was to feast his lord along with a retinue appropriate to his status (see p. 320 above). A client was also obliged to attend a commemorative feast in the event of his lord's death.[271] The food-rents which a lord received from his clients would enable him to provide regular feasts, and thereby enhance his prestige in the community.

The sagas pay particular attention to the feasts provided by kings, or to which kings were invited by lords or other dignitaries.[272] Sometimes such feasts were held to celebrate a seasonal festival, such as *feis na Samna* 'the feast of Samain', held by the king at the beginning of November.[273] It was also customary to celebrate the consolidation of a king's reign with an inauguration feast. This was viewed as a sort of marriage-feast symbolizing the new king's mating with his territory, thereby bringing fertility to its soil and livestock.[274]

Some texts give descriptions of the feasts provided by kings for their own households and subjects. *Suidigud Tige Midchúarda* 'the seating of the house of the mead-circuit' specifies where the different ranks and professions are to sit, and also what cut of meat is appropriate to each person's rank.[275] Versions of this text survive in the twelfth-century Book of Leinster[276] and the fourteenth-century Yellow Book of Lecan.[277] In both, the text is accompanied by a rough diagram of the seating-arrangement. I reproduce above part of the diagram in the Book of Leinster – unfortunately blotched in the manuscript – which shows a low-grade bard (*daul*) holding

[271] *CIH* ii 434.29; cf. gloss at 435.1 and commentary at *CIH* iii 984.27; vi 2162.39.

[272] E.g. *TBC* Rec. I 17.545.

[273] E.g. Watson, *Mesca Ulad*, 2.30.

[274] O'Rahilly, 'On the origin of the names *Érainn* and *Ériu*', 14–21; Binchy, 'The fair of Tailtiu and the feast of Tara', 134–8.

[275] Cf. E. J. Gwynn, *Metrical Dindshenchas* i 24.153–26.180. The seating-arrangement in King Conchobar's household is described in O Daly, '*Lānellach tigi rīch*'.

[276] *LL* i 116–20.3637–3789 (f. 29[ab]). Text and translation are given in Gilbert, *Facsimiles of National Manuscripts of Ireland* ii plate LIII.

[277] The text is reproduced in the facsimile of YBL, pp. 418–20 (cols. 243–7).

what may be a poetic staff.[278] Above him is the cooking-spit, *bir bruinnes.* As is common in the feasts of early Irish tradition, the main dish at this feast is evidently pork.[279] The most important guests are entitled to the best cuts, such as the the *loarg* 'haunch' or the *lónchrúachait* 'tenderloin steak'. Those of somewhat lower status get the *colpthae* 'shank' or the *leschrúachait*, which is probably to be identified with the centre-cut loin steak.[280] Low-ranking persons get such inferior cuts as the *milgetan* 'belly', the *dronn* 'chine', or *remur n-imda* 'shoulder fat'. Obviously, this is a literary text and not to be taken as the record of an actual feast. Nonetheless, it is important in the general light which it sheds on how the early Irish regarded the various cuts of meat.

Other texts provide similar but less detailed accounts of the cuts of meat to which persons of different rank are entitled. A Middle Irish tale describes a magical cauldron called the *coire ansicc*, in which pigs and beeves were cooked.[281] Each guest who plunged a fleshfork into the cauldron would automatically get the cut of meat appropriate to his or her status: a king would get a haunch (*loarg*), a queen would get a loin (*les*), the lowest rank of lord would get a shank (*colpthae*), etc.

The drink served in the mead-circuit was clearly mead (*mid*), which is discussed on p. 113. The diagrams in both the Book of Leinster and the Yellow Book of Lecan include a representation of a two-handled vat (*dabach*) from which the mead is dispensed by the butlers (*dáilemain*) who are assigned a position beside it. In general, mead seems to have enjoyed greater prestige than the more commonplace beer (*cuirm*). Wine (*fín*) is occasionally mentioned as a festive drink in both monastic and secular contexts.[282] For example, a Latin Life of Saint Cíarán of Saiger describes how he miraculously provided a meal of cheese, vegetable, fish, honey and oil, accompanied by a sufficiency of

[278]For the *daul* (*dul*) 'low-grade bard, satirist', see L. Breatnach, *Uraicecht na Ríar*, 34–5, 39–40; McCone, 'A tale of two ditties', 140 (note 5).

[279]Cf. Thurneysen, *Scéla Mucce Meic Dathó*, 6.11–13.

[280]For a discussion of the identity of the various cuts of pork in this text, see Sayers, 'A cut above: ration and station in an Irish king's hall'.

[281]Stokes, '*Scél na Fír Flatha*, etc.', 187–8. A similar passage is found at *CIH* ii 349.36–350.5 = *AL* i 48.6–17. A fuller version, with additional information on the cuts appropriate to those of low status, is at *CIH* iii 880.17–26.

[282]For references to the wine-trade, see p. 319.

wine.[283] In the comic tale, *Fled Dúin na nGéd* 'the feast of the fort of the geese', the guests are offered wine, as well as mead and beer.[284]

Feastdays were of great importance in the Church year, and it was customary to celebrate them with special food and drink. Even a slave who has committed an offence and is held in chains on short rations is legally entitled to 'the loaf of the noble day' (*bairgen úasallaithi*) with an appropriate condiment.[285] The major feast in the Church year is Easter, and many texts stress the importance of beer on this occasion. For example, *Bethu Phátraic* refers to the consumption at Easter of a haunch of meat with its accompanying dishes (*loracc cona timthuch*) as well as a pitcher of beer (*cilornn cormma*).[286]

[283]Plummer, *Vitae* i 230 § xxx *galmula et holus et pisces et mel et oleum et . . . vinum.* There are very few references to the use of oil (Irish *ola(e)*, Latin *oleum*) in Irish cooking. In general, societies which use butter in cooking do not use oil, and vice versa: see Toussaint-Samat, *History of food*, 120–3. For 'Irish oil', see p. 323 above.
[284]Lehmann, *Fled Dúin na nGéd*, 5.136.
[285]*CIH* ii 363.27 = *AL* i 106.1–2. The glossator takes the noble days to be Easter, Christmas and Sunday.
[286]Mulchrone, *Bethu Phátraic*, 75.1385–6; cf. Ó hAodha, *Bethu Brigte*, 7 § 21.

11
Farm Lay-out

There are no contemporary maps or plans of seventh- or eighth-century Irish farms, so our knowledge of their lay-out must be pieced together from the written word and from archaeological excavation. Both types of evidence indicate a pattern of scattered farmsteads in the more fertile parts of the country, with considerable uncultivated areas (*díthreb*) of mountain, woodland, and marsh. The law-texts regularly represent the typical land-holding as being a rectangular block surrounded by four neighbouring farms.[1] This is obviously a simplification, but it at least demonstrates how contemporary observers perceived the general pattern of land-settlement.

There seem to have been few concentrations of population. There is documentary evidence which indicates the presence of ports,[2] which would naturally have attracted a population dependent on trading, boat-building, and fishing. However, it is significant that the political centre of each *túath* or kingdom at this period was the *óenach*, an assembly of freemen which took place annually at a fixed site, often a low hill.[3] Trading seems also to have been carried out on such occasions, as well as horse-racing, music, and story-telling.[4] These assemblies thus fulfilled many of the functions associated with towns in other societies, i.e. government, political discussion, trade, social intercourse, and entertainment.

It is probable that the largest concentrations of population were to be found in and around the major monasteries. Great areas of land were owned by the Church, and managed by abbots in the same way as secular land-owners, with monastic clients and tenants providing food-rent and labour (see p. 446). The population at the monastic settlement of Armagh was large enough to be described in

[1] E.g. *CIH* ii 412.1–3; v 1712.4–6 = *AL* i 260.9–11. See discussion at *BB* 92 (note to § 2).

[2] E.g. Meyer, *Tecosca Cormaic*, 2 § 1.25 *bárca do thochor i port* 'ships coming to port' (included among the things which are best for a king).

[3] *GEIL* 4. One text refers to a flower-covered hill of assembly on which no heather or furze or tree ever grew (*CIH* v 1561.17–19; vi 2117.4–6 = Dillon, 'Stories from the law-tracts', 50 § xii).

[4] E. J. Gwynn, *Metrical Dindshenchas* iii 18.233–6; 20.237–268; 24.305–8.

the eighth-century *Liber Angeli* as a town (Latin *urbs*).[5] The growth of population associated with the monasteries is also reflected in the development of the Irish word *cathair*. Originally it referred to a fortified residence of a king or lord, but already in the law-texts it is used of any built-up place.[6] In some instances it is used specifically of a monastic settlement.[7] Evidence of the monasteries as commercial centres is to be found from the tenth century onwards, and Charles Doherty has argued that some monasteries had taken over the *óenach* already by AD 800.[8]

FARMHOUSE

Our most detailed written source of information on early Irish farmhouses is the law-text *Críth Gablach* of about 700 AD. The author of this text gives us a remarkably comprehensive survey of the possessions of persons of different rank, and includes some generalizations about the size of their houses.

He provides most detail on the house of a *mruigfer* (the most prosperous grade of *bóaire*). A man of this rank is expected to live in a house with a diameter of twenty-seven feet.[9] His outhouse (*airchae*) has a diameter of seventeen feet. Every household item (*inchruth*) is in its proper place, including a cauldron with a spit, a vat for brewing beer, mugs, kneading troughs, a tub, washing-vessels, and various tools and pieces of farming equipment (see p. 463). There should be a candle in the candlestick, and a fire should always be burning.[10] The house has a thatched roof – we are not told what

[5]Bieler, *Patrician texts*, 184–6.

[6]E.g. *CIH* i 42.15 = *AL* v 272.6.

[7]E.g. *Thes.* ii 328.4; Mulchrone, *Bethu Phátraic*, 70.1270.

[8]Doherty, 'Exchange and trade in early medieval Ireland', 81; Doherty, 'Some aspects of hagiography', 302–3.

[9]*CIH* ii 564.4; iii 780.11 = *CG* 8.193 *tech .uii. traiged fichet* 'a house of twenty-seven feet'. I take this figure to refer to the diameter of the house: the archaeological evidence indicates that the houses of the early Christian period were generally circular. However, from about AD 800, there is also evidence of rectangular houses (Edwards, *The archaeology of early medieval Ireland*, 22–7). Note that a Middle Irish description of houses, etc. at *CIH* iii 1058.15-16 states that the *ócaire* has a main house of sixteen feet in length, and a larder (*airecal*) of twelve feet in length (*.6. troighthi .x. fod a thighe mhōir; .2. troigh .x. fod a airecail .i. tech a bhīdh*). These figures may refer to the long side of rectangular structures. For a discussion of the length of the Irish foot (*traig*), see Appendix B, p. 565.

[10]In excavations, the fireplace is shown to be positioned centrally, e.g. Monk, 'Excavations at Lisleagh Ringfort', 58.

material is used, but it is probably of reeds[11] – and the floor is strewn with rushes (*aín*).[12]

Críth Gablach also provides information on the houses of lower-ranking members of society. The *ócaire*, the lowest grade of adult freeman, has a house with a diameter of nineteen feet.[13] His outhouse has a diameter of thirteen feet, and it is here that his food-rent is divided up for his lord.[14] The text stresses that the *ócaire*'s house is larger than another category of house, the *tech nincis*, which is provided for a fosterson who undertakes the maintenance of an elderly landowner when his own relatives are unable or unwilling to do so.[15] The *tech nincis* is seventeen feet in diameter.[16] It is constructed of wattling (*fíthe*) from ground level up to the roof-pinnacle (*cléithe*).[17] From the level of the lintel (*fordorus*) upwards there are feathers between every second band of wattling, i.e. a layer of feathers is set under the thatch.[18] There are two doorways, one with a hurdle (*cleth*), and the other with a wooden door (*comlae*). According to one version of this passage, half of the interior of the house is taken up with the bed-cubicle (*imdae*), and the other half is paved (*plait*).[19] The larder (*cuile*) seems to be a separate structure.[20]

[11] See p. 385. Other texts refer also to a roof made with wooden shingles (*slinn*), e.g. *CIH* vi 2099.25, 28; Meid, *Táin Bó Fraích*, 3.70. There seems to be no early evidence of thatching with corn-straw: see p. 240.

[12] *CIH* ii 564.31; iii 780.37 = *CG* 9.232.

[13] *CIH* iii 778.33 = *CG* 4.103. Chris Lynn (pers. comm.) observes that the archaeological evidence indicates an average diameter for a dwelling-house of about 6 metres, i.e. nearly 18 feet. For a detailed discussion of the dimensions of houses, see Lynn, 'Houses and related outbuildings in early Christian Ireland'.

[14] *CIH* iii 778.33–4 = *CG* 4.104.

[15] *CIH* iii 778.29–32 = *CG* 4.98–102; *CIH* iii 894.11–13. Binchy suggests that the second word is *ind-chís* (*CG* 27–8 (note to l. 98f.)), but its formation and meaning are uncertain.

[16] The text at *CIH* iii 894.11 has *.uii. traighe* 'seven feet', but this must be a scribal error, as the corresponding passage in *Críth Gablach* has *.uii. traighe .x.* 'seventeen feet' (*CIH* iii 778.29 = *CG* 4.98).

[17] Compare the house of rods (*tech slat*) mentioned in commentary on *Uraicecht Becc* (*CIH* v 1613.5 = *AL* v 94.15–16).

[18] There are also some literary references to the use of feathers in roofing, e.g. Stokes, '*Scél na Fír Flatha*, etc.', 195 § 32.

[19] *CIH* iii 894.12–13. Presumably *plait* here means 'paved area' (*DIL* P 190.77). At *CIH* iii 894.17–18, the *ploit nincis* is said to be seven fists in width at the bottom, six fists at the middle and four fists at the top. Perhaps this is to be taken as the dimensions of the paved area within the house.

[20] *CIH* iii 894.13.

In his recent excavations of a seventh-century house-site at Glenarm, Co. Antrim, Chris Lynn has discovered a remarkable level of agreement between the archaeological findings and the Old Irish descriptions of the *tech nincis*.[21] He has shown that the house was roughly circular and constructed with a double layer of wattling. The rough side of each layer was turned inwards so that both the outside and inside walls had a smooth surface. Insulating material was packed inside the cavity between the two layers of wattling: there is no evidence of daub. He makes the ingenious suggestion that the wooden door was for the doorway leading outside, whereas the hurdle was for the doorway into the larder or shed.

Críth Gablach provides us with little additional information on the houses of lords. The implication seems to be that a lord's house (*dún*) was much the same as that of a prosperous commoner. The diameter of the house of the lowest grade of lord, the *aire déso*, is twenty-seven feet – the same as that of the *mruigfer*.[22] Even the highest grade of lord, the *aire forgill*, has a house of only thirty feet in diameter.[23] The most significant difference between the house of a lord and that of a commoner is the presence of defensive earthworks, the digging of which is listed among the duties owed by a client to his lord.[24]

In *Críth Gablach*'s scheme, the king's house (*dún*) is considerably larger. It has a diameter of thirty-seven feet, and contains twelve bed-cubicles,[25] whereas a lord's house has only eight bed-cubicles.[26] A double system of earthworks protects it on all sides.[27] Literary sources contain many accounts of the grandeur of kings' houses, and it is obviously difficult to distinguish fact from fantasy.

FARMYARD

Literary and legal texts are in agreement that a house is normally surrounded by an enclosed area know as a *les*, conventionally

[21] Lynn, 'Deer Park Farms'.

[22] *CIH* ii 566.19–20; iii 782.26 = *CG* 14.336–7. Cf. *CIH* iii 1108.5–6.

[23] For more detailed discussion, see Charles-Edwards, 'Críth Gablach and the law of status', 66–7.

[24] *CIH* iv 1493.12–13 = O'Dav. 314 § 721; *CIH* v 1906.15–16.

[25] *CIH* ii 570.12 = *CG* 23.575.

[26] *CIH* ii 566.20; iii 782.26 = *CG* 14.337; *CIH* ii 567.22; iii 783.29 = *CG* 16.401.

[27] *CIH* ii 570.6–9 = *CG* 22.566–72.

translated as 'farmyard' or 'courtyard'.[28] Traces of the roughly circular earthen bank surrounding the *les* are common throughout the country; tens of thousands have been identified, either singly or in clusters. They are generally between 100 and 150 feet in diameter.[29] In archaeological terminology, they are often called 'ring-forts', which is a rather misleading appellation, as they were clearly not designed to provide protection against human aggressors. Archaeologists also use the more suitable term 'rath', from Irish *ráth* 'earthen rampart'.

There is no archaeological evidence of the housing of livestock within the *les*. In the texts, on the other hand, there are some references which imply that livestock were regularly housed in the *les*, particularly at night. For example, the eighth-century tale *Fled Bricrenn* refers to a dunghill (*otrach*) in the middle of a *les*.[30] Such a dunghill would no doubt consist largely of animal-dung cleared from night-shelters, and ultimately spread on the land as fertilizer. According to *Críth Gablach*, a farmer of *bóaire* rank is expected to have a sheep-pen (*lías caírech*), a calf-pen (*lías lóeg*), and a pig-sty (*muccḟoil*).[31] I would suggest that these structures were generally to be found inside the *les*.[32] According to Irish Canon VI, a 'dog of the four doors' (*canis .iiii. hostiorum*) was expected to guard the dwelling-house, and the pens (*caulae*) of the sheep, the calves, and the cows.[33] This presupposes that these four structures were within a single enclosure, which I would identify with the *les*.

It is difficult to glean much precise information in the written sources on the housing of animals. The *lías* 'pen' for calves and sheep seems to have been a wattled structure secured by stakes driven into the ground.[34] As Lucas points out in his *Cattle in ancient Ireland*, there is no evidence that such pens were roofed. He suggests that they were normally round, as in the

[28]The Modern Irish spelling is *lios*. The Welsh cognate *llys* means 'court, royal residence, courtyard, enclosed space, etc.' (*GPC* s.v. *llys*).

[29]Norman and St Joseph, *The early development of Irish society*, 41.

[30]*LU* 255.8380 = Windisch, 'Das Fest des Bricriu', 265.1 § 25.

[31]*CIH* ii 563.7–8; iii 779.27–8 = *CG* 6.156.

[32]Obviously, in a *crannóg* or lake-dwelling there would not have been enough space to keep livestock at night in the same enclosure as the dwelling.

[33]*IP* 174 § 4 = *GEIL* 354 (5) § 4; cf. *CIH* vi 2216.37 *cú chethardoruis* 'dog of four doors'.

[34]E.g. *TBC* LL 37.1376; Stokes, 'The prose tales in the Rennes Dindṡenchas, (1)', 468 § 67.

case of dwelling-houses.[35] He quotes, however, a passage from the seventh-century *Hisperica Famina* which speaks of square sheep-folds.[36]

The evidence with regard to cow-houses is uncertain. It is noteworthy that *Críth Gablach* refers to calf-pens, sheep-pens and pig-sties, but not to cow-houses. Writing in the twelfth century, Giraldus Cambrensis comments that on account of the mildness of the climate the Irish never build stalls for their cattle.[37] But the sources in the Irish language do not altogether support his statement. In a number of cases, the word *tech* 'house' is used in contexts which imply a roofed structure for cattle. For example, in a ninth-century text on the reciprocal services of the kings of Cashel, we find a description of the domestic buldings appropriate to a high-ranking person. These consist of a house of forty feet (in diameter), an outhouse (*airchae*) of twenty-three feet, an ox-house (*tech ndam*) of twenty-four feet, and a drying-kiln (*áith*) of thirty feet.[38] Legal commentary from approximately the same period as Giraldus refers to the offence of removing thatch from the cattle-house (*tech na n-innile*).[39] Another commentary lists the five farm buildings as the dwelling-house (*teg mór*), the cow-house (*bótheg*), the pig-sty (*foil muc*), the calf-pen (*lías láeg*), and the sheep-pen (*lías cáerach*).[40] The commentator's use of *teg* (*tech*) 'house' for the cows' building implies that it is a more substantial structure than the *lías* 'pen' of the calves and sheep. The *Annals of Inisfallen* record that during Lent in the year 1028, there was such a heavy fall of snow that neither people nor cattle left their houses for three days and three nights.[41] Again, the context implies that some form of roofing provided the cattle with shelter from the snow.

[35] Lucas, *Cattle in ancient Ireland*, 27–8.

[36] Herren, *Hisperica Famina*, 88.313 *quadrigonas . . . aulas.*

[37] O'Meara, *Giraldus: Topography*, 53 § 26 = Dimock, *Giraldi Topographia*, 67.1–2 *nec armentis unquam stabula parari solent.*

[38] O'Keeffe, '*Dál Caladbuig*', 19 § 2.

[39] *CIH* vi 2191.11.

[40] *CIH* iii 760.21 = *AL* v 514.y-z. Thurneysen discusses this list at *Irisches Recht*, 70.

[41] *AI* 194 s.a. 1028 § 3.

The domestic *bóthech* 'cow-house' is to be distinguished from the *bódún* (*badún*) 'cow-fortress', where the cows of the whole neighbourhood could be brought for protection from cattle-raiders.[42] Sometimes, they were not safe even within such enclosures. The *Annals of Connacht* record a raid in 1236 on the stronghold of Rind Dúin on Lough Ree. The *badún* was overrun and many cattle from the surrounding territory were driven off.

There is little information on the early Irish pig-sty. It was presumably circular, as the basic meaning of *foil* – the commonest term for this structure in Old Irish – is 'ring'.[43] The word *cráo* (*cró*) may be used of a calf-pen, sheep-pen or kennel, but is most commonly applied to a pig-sty.[44] For example, the *Annals of Connacht* report an episode in 1342 in which Ó Birn struck Ó Conchobhair on the shoulder with a stick in the doorway of the church of Elphin. Ó Birn shouted 'Into your sty, boar' (*at chrú, a chullaigh*), to which Ó Conchobhair retorted 'May that bring no good for my swineherd' (*narab fearr dom mhucaidhe*).[45]

The term *airleng* is used specifically of the hurdle which closes the pig-sty.[46] Possibly, the pig-sty itself was sometimes of stone, but there seems to be no evidence on the materials used in its construction. According to *Críth Gablach*, the penalty-fine (*díre*) for the destruction of a pig-sty is five *séts*, which must be paid to the owner in pigs.[47] In addition the culprit must restore the sty. Heptad 10 makes it clear that if a farmer houses pigs in a dwelling-house (*tech*), no penalty-fine is payable in the event of its destruction.[48]

Críth Gablach tells us nothing about the structures provided for horses on the early Irish farm. Heptad 4, however, refers to the driving of a horse into its *inne*, which may mean 'stable'.[49] A typical farmyard would also have contained a hen-coop (see p. 103).

[42]*DIL* s.v. *badún*. This word was anglicized *bawn*, and is well attested in English documents from the sixteenth century onwards. In Modern Irish the meaning of *bán* has developed to refer to the green in front of any farmhouse.

[43]See Binchy, 'A text on the forms of distraint', 82 § 1.

[44]Its cognates in other Celtic languages refer mainly to the pig-sty, e.g. Welsh *crau* (singulative *crowyn*), Old Breton *crou*, Cornish *crow*: see *LEIA* C-240 s.v. *cró*; Greene, '*Cró, crú* and similar words', 3–4.

[45]*AC* 288 s.a. 1342 § 3.

[46]*CIH* iv 1470.39 = O'Dav. 217 § 125; cf. *CIH* iii 825.14.

[47]*CIH* ii 564.41–2; iii 781.6–7 = *CG* 10.245–6.

[48]*CIH* i 13.9 = *AL* v 162.20.

[49]*CIH* i 6.1 = *AL* v 136.9–10. See p. 93 above.

Bechbretha refers to the possible presence of bee-hives within the *les*.[50]

There are also references to structures other than buildings and enclosures within the *les*. The *corróc* 'pit' mentioned in legal commentary on Heptad 64 may refer to a souterrain for keeping dairy-products and other perishable foodstuffs.[51] The paved area at the entrance of the *les* was called the *airdrochat*, and it was regarded as important that it be kept clean.[52] At the beginning of an episode in the *Genealogies of the saints*, Diarmait of Lecc na Sinnach is described as clearing his *airdrochat* with a shovel.[53] Likewise in a description of miserable conditions in the tale *Erchoitmed Ingine Gulidi* it is said that the pavements were dirty.[54]

There are few references to cultivation within the *les*, and it seems likely that crops were almost always grown in the garden or fields. The main exception is in the case of a fosterson who undertakes the maintenance of an elderly landowner (see p. 362 above). He lives in a house defined as a *tech nincis*, which is surrounded by a yard (*les nincis*). In the yard there is room for four raised beds.[55] Each raised bed (*indrad*) is twelve feet in length and eight feet in width. No information is given on the crops grown, but such small dimensions suggest vegetables rather than cereals. According to legal commentary on *Bechbretha*, a fruit-tree might also be grown within a *les*.[56]

In the case of persons of higher rank, cultivation within the *les* was regarded as shameful. A Middle Irish poet includes 'ploughing the lawn of his *les*' among the three failures of a minor lord.[57]

[50] *CIH* ii 455.31 = *BB* 84 § 50.

[51] *CIH* i 56.32 = *AL* v 326.14.

[52] For an excavated *airdrochat* from the early Christian period at Glenarm, Co. Antrim, see Hamlin and Lynn, *Pieces of the past*, 46.

[53] Ó Riain, *Corpus genealogiarum sanctorum Hiberniae*, 111 § 668.

[54] Meyer, *Hibernica Minora*, 66 § 6 *It salcha na herdrochait.*

[55] *CIH* iii 894.13–15.

[56] *CIH* i 105.6.

[57] O'Donoghue, 'Advice to a prince', 46 § 12 *Tri meth ócthigirn . . . urland a liss d'ar.*

Garden

A reference in the seventh-century law-text *Bechbretha* suggests that the vegetable garden (*lubgort*)[58] was often situated outside the *les*.[59] This would make sense if I am correct in believing that a typical *les* contained a dwelling-house, outhouse, and various structures for housing animals. There would be little room for vegetable-growing, and constant danger of interference from livestock, even if the garden were fenced in.

Apple-trees and plum-trees may also have been grown in the garden. *Bethu Brigte* describes how Saint Brigit miraculously caused an alder-tree growing in a garden (*lubgort*) to bear heavy crops of apples and plums.[60] For a discussion of the vegetables and fruit of the early Irish, see Chapter 8.

Fields and greens

The law-texts distinguish an area outside the *les*, called the *airlise*, which means literally 'that which is in front of the *les*'. According to *Críth Gablach*, this area extends the length of a spear-cast on every side.[61] It is obviously difficult to identify such an area archaeologically. However, Norman and St Joseph note that some raths are surrounded by small roughly rectangular fields or enclosures 'radiating outwards from the rath like the petals of a flower'.[62] Perhaps, the area within is to be identified with the *airlise*.[63] According to Old Irish commentary on *Bretha Comaithchesa*, a land-holding is divided evenly between the heirs. However, in addition to his share

[58]This compound goes back to the Common Celtic period; cf. the cognates Welsh *lluarth*, Breton *liorz*, Cornish *lowarth* (VKG ii 3). In later Irish, *lubgort* is generally replaced by *garrda* (*garrga*), a borrowing from Old Norse *garðr*. For example, in commentary based on *Bechbretha* at *CIH* i 105.6, a distinction is made between *garrgha* and *lis*, just as the text of *Bechbretha* distinguishes between *lubgort* and *les*.

[59]*CIH* ii 455.31 = *BB* 84 § 50 *Beich bíte i llugburt no i llius* 'bees which are in a garden or in a courtyard'. Compare Bieler, *Patrician texts*, 174.6–7 = *Thes.* ii 239.16 *Ochter nAchid . . . cona llius ⁊ a llubgort* 'Óchter Achid . . . with its yard and its garden'.

[60]Ó hAodha, *Bethu Brigte*, 12 § 33.

[61]*CIH* ii 564.22–3; iii 780.28–9 = *CG* 9.219–20. For the use of *airchor* 'spear-cast', as a measure of length, see Appendix B, p. 567.

[62]Norman and St Joseph, *The early development of Irish society* 64–5.

[63]In an eleventh-century record of a grant of land in the Book of Durrow, the term *airlise* (*erlese*) is used of the area within a monastic enclosure (Best, 'An early monastic grant', 137–9). Church buildings were often sited at approximately the centre of a round or oval outer enclosure, e.g. Swan, 'The early Christian ecclesiastical sites of County Westmeath', 23, 25, 26.

of land, the oldest heir gets the farm-houses (*tige*), the yard (*lis*) and the surrounding enclosed area (*airlise*).[64]

In general, it is likely that the *airlise* of a typical farm contained various enclosures for grazing and cultivation (including a garden). There may also have been a kiln (*áith*) for drying corn, and a barn (*saball*)[65] for storing it. A prosperous farmer of *bóaire* rank is expected to own both a drying-kiln and a barn.[66] As archaeologists have found no clear evidence of such structures within the farmyard, the likelihood is that they were normally situated outside.[67] This arrangement would certainly reduce the danger of fire spreading from the kiln to the farmhouse and other farmyard buildings. According to *Críth Gablach*, the penalty-fine (*díre*) for the destruction of a kiln is a cow and a yearling heifer (i.e. 2½ *séts*).[68] In addition, the culprit must restore the kiln. The penalty-fine for a barn is five *séts*, likewise with obligation to restore.[69] A prosperous farmer might also have his own mill at a convenient site near his house: see p. 482 below.

In the case of a king's dwelling, the *airlise* seems also to have included an area of assembly. Thus a law-text lists seven things which reveal the injustice of a king, including 'the expulsion of a synod from his *airlise*'.[70] This implies that a synod of clerics might be expected to meet in a king's *airlise*.

Another difficult term is *faithche*, conventionally translated 'green'. This translation suggests a large expanse of grass, which suits some contexts. For example, there are references in literary sources to the *faithche* of the royal residences at Temair, Emain and Crúachain.[71] Horse-racing takes place on such greens,[72] and

[64] *CIH* ii 575.14.

[65] This word is a borrowing from Latin *stabulum* 'stable', which also gives Welsh *ystabl*, Breton *staol*, both meaning 'stable'. The change of meaning from 'stable' to 'barn' in the Irish form could be taken as evidence of the lack of substantial roofed structures for sheltering livestock on early Irish farms. But see the discussion on *bóthech* above.

[66] *CIH* ii 563.7; iii 779.27 = *CG* 6.156. He also owns a share in a water-mill (*muilenn*): see p. 482.

[67] There is however a possible case of a drying-kiln within a farmyard at Letterkeen, Co. Mayo (Edwards, *The archaeology of early medieval Ireland*, 32).

[68] *CIH* ii 564.39; iii 781.4 = *CG* 10.241–2.

[69] *CIH* ii 564.41; iii 781.6 = *CG* 10.244–5.

[70] *CIH* i 219.16–17 = *AL* iv 52.6.

[71] *DIL* s.v. *faithche*.

[72] E.g. *LL* iv 1011.29583.

there may be a sacred tree (*bile*) or trees.[73] A law-text refers to the playing of ball-games on the lawn (*airscar*) of a *faithche.*[74] Another law-text speaks of the *faithche* of privileged persons, such as a king, hospitaller (*briugu*) or poet.[75] A monastic settlement also normally has a *faithche.*[76]

In the context of a typical farm, the law-texts sometimes use the term *faithche* of the whole area which surrounds the dwelling. Here it is best translated 'infield'. Medium and long distances are often imprecisely defined in the early Irish system of measures (see Appendix B, p. 560), and in *Bechbretha* the extent of the *faithche* is stated to be 'as far as the sound of a bell or the crowing of a cock reaches'.[77] The parts of the farm outside this range are described as *sechtar faithchi* 'beyond the *faithche*', i.e. outfield.[78]

In this usage, the *faithche* includes tilled fields. Thus *Bretha Comaithchesa* refers to trespass by a pig in a *gort faithche,* i.e. a tilled field within the *faithche.*[79] According to a commentator on Heptad 64, a *faithche* may contain hills (*tulcháin*), places of assembly (*inada airechtais*), roads (*slighti*), margins (*imli*), recesses (*cúla*), hollows (*cabána*), and dark places (*inada diamra*).[80]

The written sources tell us a great deal about field-boundaries (see next section), but rather little about the fields themselves. The term *gort* is generally applied to a tilled field or garden, e.g. *ithgort* 'cornfield', *glaisengort* 'woad-garden', *lubgort* 'vegetable garden', *aballgort* 'orchard'. Sometimes, however, it is applied to enclosed pasture. For example, the tale *Fled Bricrenn* refers to

[73]E.g. *TBC* LL 43.1581 *óen na prímbili bís for faidchi prímdúni* 'one of the chief trees which is on the green of a great fort'.

[74] *CIH* i 280.33 = *AL* iii 252.17 *blā liathrōite urscar faiche prīmcathrach* 'the immunity from liability of a ball is the lawn of the green of a principal *cathair*'. It is probable that *prímchathair* refers here to the residence of a king.

[75] *CIH* i 50.28–30 = *AL* v 302.x-z.

[76]In a heptad at *CIH* i 207.2 = *AL* iv 6.9 it is implied that a church without a green (*ceall gen faithche*) is exceptional.

[77] *CIH* ii 454.13–14 = *BB* 82 § 46. A glossator on Heptad 64 (*CIH* i 57.15 = *AL* v 328.8) suggests a much smaller area: *na ceithri guirt is nesa don baili* 'the four tilled fields nearest to the farm'.

[78] *CIH* ii 454.26–7 = *BB* 82 § 48. Later commentators introduce a further parallel distinction between *maigen* 'place around the residence' and *sechtarmaigen* 'the area outside the *maigen*', e.g. *CIH* ii 456.1–7 = *BB* 171, Appendix 1 (h). But this seems pure casuistry.

[79] *CIH* i 72.18 = *AL* iv 108.10.

[80] *CIH* i 57.10–13 = *AL* v 328.3–6.

horses being put to graze in a *férgort* lit. 'grass-field'.[81] The terms *achad* and *clúain* seem to be used exclusively of fields for grazing: the latter is applied particularly to low-lying pasture near water.[82]

Some archaeological reports relating to early Irish fields have already been published, and further work is in progress. The most dramatic discovery has undoubtedly been the vast system of Neolithic fields (*c.*3000 BC) on the Mayo coast, which has been mapped out and partially excavated by Séamas Caulfield.[83] For the early Christian period, an important source of information is the aerial photography of J. K. St Joseph, which shows up field-systems as lines in growing crops or as slight ridges in the ground which can still be discerned in certain lights.[84] Such fields are generally small and squarish or irregular in shape; they are of the type which is sometimes called the 'Celtic field'.[85]

On the other hand, farming procedures described in the law-texts suggest that the open-field system may also have been in operation at this period. The eighth-century author of *Críth Gablach* clearly regarded it as normal practice for four low-ranking farmers – probably relatives – to plough in co-operation. In a typical arrangement, each contributed four oxen for co-operative ploughing (*comar*). One might expect that this practice would lead to co-ownership of a large open-field in which each farmer had separate strips. This would facilitate ploughing, fencing, crop-protection, and harvesting. In the case of a person who had the resources to farm on his own – such as the *mruigḟer* – one would similarly expect him to sow his sixteen bushels of seed-corn in one or two large fields rather than in many small fields.[86] Small fields, on the other hand, would suit minor crops such as the woad mentioned in both legal and literary sources (see p. 264).

[81] E.g. *LU* 258.8514 = Windisch, 'Das Fest des Bricriu', 271.7 § 36; cf. Stokes, *Lismore Lives*, 85.2846 *férghort cluana*; Thurneysen, *Scéla Mucce Meic Dathó*, 23 § 23.10 *a ngurt ḟeoir.*

[82] Joyce, *Irish names of places* i 233–6.

[83] Caulfield, 'The neolithic settlement of North Connaught'.

[84] Norman and St Joseph, *The early development of Irish society*, 62, 63, 64, 67, 69.

[85] See Rackham, *The history of the countryside*, 158–60, 250. As Rackham shows, the problem about the term 'Celtic field' is that many fields of this type date from pre-Celtic times in both Ireland and Britain.

[86] According to *Críth Gablach* (*CIH* ii 564.6–7; iii 780.13–14 = *CG* 8.196–7), a *mruigḟer* normally has 'sixteen bushels in the ground' (*sé méich deec i talmain*).

The law-texts regularly describe the typical cultivated field as having two headlands (*airchinn*) and two sides (*tóeb*). According to glossators, the headlands are 'short' and the sides are 'long'.[87] In the most detailed reference to field-size in our sources, *Bretha Comaithchesa* gives the length of a headland as being 'three *forrach*s and the spear-cast of a youth'.[88] The precise length which the author had in mind is uncertain,[89] but it is at least clear that he is thinking of a fairly large rectangular field. Field-crops seem to have been grown on a raised bed (*indrad* or *immaire*).[90] As we have seen on p. 367 above, a raised bed in one particular case was expected to be eight feet wide. This seems to agree approximately with a fragment of legal commentary which equates the width of an *immaire* with seven sods of the plough.[91] In a typical field there was a drainage trench (*etarche*) between each raised bed.[92]

In their book *The early development of Irish society*, Norman and St Joseph refer to unusual field-patterns near Lady's Island (Co. Wexford) which show up in aerial photography.[93] They are rectilinear and present the appearance of a chess-board slightly squashed to one side. The authors draw attention to their close similarity to field-patterns in Scandinavia and suggest that they were laid out by Norse colonists. For a discussion of Norse influence on Irish farming, particularly as reflected in linguistic borrowing, see p. 249.

WALLS AND FENCES

The main law-text on farming, *Bretha Comaithchesa*, gives precise descriptions of what constitutes a proper field-boundary (*imbe n-inric*) in law. Four types are distinguished: the stone wall (*corae*), the trench-and-bank (*clas*), the bare fence (*nochtaile*), and the oak fence (*dairimbe*).[94]

[87]E.g. *CIH* i 197.27 = *AL* iv 124.6–7; *CIH* ii 412.6 = *AL* i 274.26–7. MacNeill, 'Law of status or franchise', 287 (footnote), claims that the length of the side was twice that of the headland, but this is not explicitly stated in the texts.

[88]*CIH* i 78.9–10; 200.12–13 = *AL* iv 138.15–16; cf. *CIH* iii 907.30.

[89]For discussion, see Appendix B, p. 566.

[90]For a discussion of these terms, see Mac Mathúna, 'Some words for "(man-made) ridge" in Irish'.

[91]*CIH* i 137.30 *lethet cach imairi dībh no .uii. fōit arathair.*

[92]Meyer, *Sanas Cormaic*, 43 § 516.

[93]Norman and St Joseph, *The early development of Irish society*, 67–70.

[94]*CIH* i 73.7–18; 195.20–196.2 = *AL* iv 112.1–18. Cf. *CIH* vi 2133.7–29 (commentary). See also Ó Corráin, 'Some legal references to fences and fencing in Early Historic Ireland'.

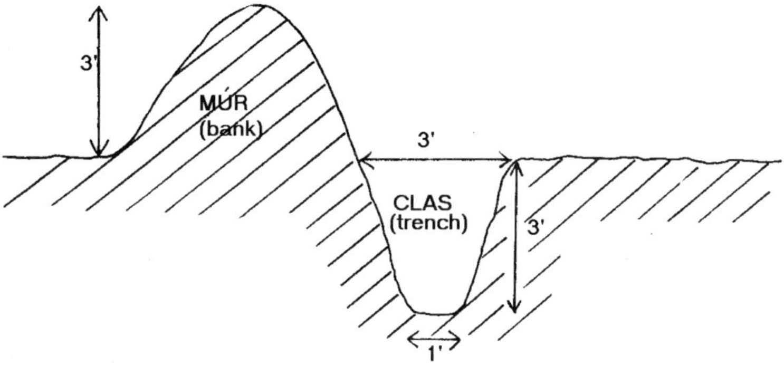

Fig. 17. A suggested cross-section of a trench-and-bank

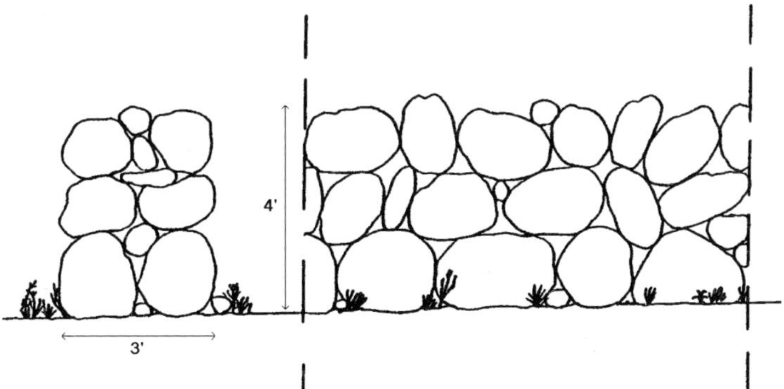

Fig. 18. A suggested front-view and cross-section of a 'wall of three stones' (corae trí liac).

The author specifies that the stone wall should be three feet in thickness and twelve fists (i.e. four feet) in height. The tool used in its construction is an iron lever (*socc*). It must be a 'wall of three stones' (*corae trí liac*). This probably means that it should be made of three layers or 'courses', as in my illustration.[95]

Field-boundaries of the trench-and-bank (*clas*) type would no doubt have been used in less stony terrain than the *corae*.[96] The bank (*múr*) should be three feet in height. The trench should be three feet deep, three feet wide at the top and one foot wide at the bottom. It is constructed with a spade (*rámae*).[97]

Trees or bushes may sometimes have been planted in the bank to strengthen it; even if this were not done, furze, blackthorn, whitethorn, bramble, etc. would grow up naturally.[98] In time, larger trees would be likely to establish themselves. In Old Irish legal material, there are some references to the presence of trees in field-boundaries. For example, an Old Irish gloss on *Bechbretha* speaks of the value of trees growing in boundaries, which is taken to be greater than that of trees growing in woods.[99] Commentary on *Bretha Comaithchesa* refers to the way in which trees may grow up on a stone wall (*corae*), and ultimately destroy it.[100]

A less permanent type of field-boundary is the *nochtaile* lit. 'bare fence', also called the *felmae* or *felm*.[101] It should be a defence against animals of all sizes: 'a small pig should not be able to penetrate because of its closeness, and an ox should not be able

[95] Commentary at *CIH* vi 2133.11 gives a totally different interpretation of the *corae trí liac*, taking it to consist of three stones at the bottom, two stones in the middle and one stone on top. But this contradicts the Old Irish text which gives the wall a uniform thickness of three feet. The interpretation at *CIH* iii 907.27 also seems impractical.

[96] A glossator (*CIH* i 199.12 = *AL* iv 130.18) notes that the *clas* is used as a field-boundary in soft ground (*a tīr buic*).

[97] *CIH* i 65.12; vi 2193.15 = *AL* iv 72.2.

[98] Cf. Groenman-van Waateringe, 'Field boundaries in Ireland', 288

[99] *CIH* iii 923.36–7 = *BB* 58 § 15^{c} (B) *is mō dīri in craind bīs i netarbib a thīre oldās in crann bīs i fidbaid, cid aonfidh.* Cf., also, *CIH* i 202.31; ii 582.19–20 = *AL* iv 150.10–11.

[100] *CIH* i 65.30 *brisid craind an coradh.*

[101] It is clear from *CIH* i 65.13, 20 = *AL* iv 72.3, 5; *CIH* i 73.11; 195.27 = *AL* iv 112.6 that the author of *Bretha Comaithchesa* regarded the terms *felmae* (*felmad* MS) and *nochtaile* as interchangeable. The form *felm* is attested in the proverbial expression *cúaille feda i feilm n-argait* 'a wooden stake in a fence of silver', e.g. *CIH* vi 2115.38; 2116.4 = Dillon, 'Stories from the law-tracts', 48 § ix; Meyer, *Sanas Cormaic*, 87 § 1018; *CIH* vi 2230.3. Another word for a wattle fence is *udnocht* (e.g. *Thes.* ii 294.16): see discussion of its etymology by Russell, 'The Celtic preverb **uss*', 98.

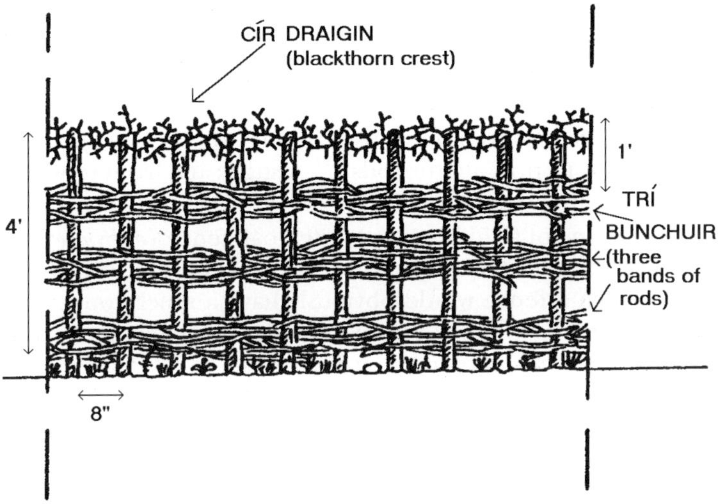

Fig. 19. A suggested front view of a bare fence (nochtaile).

to penetrate because of its firmness and height'.[102] The standard *nochtaile* consists of four-foot[103] stakes set about eight inches apart ('the length of a foot as far as the joint of the big toe').[104] A billhook (*fidbae*) is used to cut the stakes,[105] and they are driven into the ground by three blows of a mallet (*forchae*).

Three bands of pliable rods (*trí bunchuir*) are woven between the stakes: one at the bottom, one in the middle, and one at the top. The stakes project three fists (i.e. one foot) above the wattling,[106] and a crest of blackthorn – the early Irish equivalent of barbed wire – tops the fence. The author stresses that this blackthorn crest (*cír draigin*) should make the fence impregnable to livestock.[107] The

[102] *CIH* i 73.11–12; 195.27–8 = *AL* iv 112.7–8.
[103] I take 12 fists (*dá dorn .x.*) to be equivalent to 4 feet: see Appendix B, p. 564.
[104] *CIH* i 73.15; 195.32; vi 2133.15–16 = *AL* iv 112.13–14; *CIH* iv 1493.1–2 = O'Dav. 313 § 715.
[105] *CIH* i 65.13; vi 2193.16 = *AL* iv 72.3.
[106] *CIH* i 73.15–16; 195.32–3 = *AL* iv 112.14–15 *trī duirn fot an cuaille uas fenamain.*
[107] *CIH* i 73.16–17; 195.33–196.1 = *AL* iv 112.15–16. This crest (*corona spinarum*) is also mentioned in a passage on fences for keeping out hens in Wasserschleben, *Die irische Kanonensammlung*, 215, bk. 53 ch. 9.

seventh-century *Hisperica Famina* likewise advises that blackthorn (*prunus*) be used in the construction of fences.[108]

The material used in the construction of a *nochtaile* is not specified in this account, and must have varied in different localities. It is likely that hazel was commonly used, as this tree is elsewhere said to be valued for its nuts and its rods.[109] A suggested front-view of a typical *nochtaile* is given on p. 375.

The fourth type of field-boundary – the oak fence (*dairimbe*) – is said to have the same dimensions as the *nochtaile*, but with posts of oak.[110] Such a fence would obviously have a much longer life than the *nochtaile*, as posts of split oak can last for many years. It is constructed with the aid of an axe (*biáil*).[111]

It seems probable that the four types of field-boundary distinguished above correspond fairly closely to the reality of early Irish farming, though naturally the variations would have been greater than the text indicates. It is worth noting that the basic design of the stone wall and trench-and-bank types survives in many parts of the country today. The invention of barbed wire has ousted fences of the *nochtaile* type, but the four-foot height above the ground is still normal.

As well as the description in *Bretha Comaithchesa* of the ordinary fences of the early Irish countryside, we also have a legal reference to fences which are unusual, either because of their situation or their construction. Heptad 46 lists seven fences in the law of neighbourhood (*comaithches*) for which it is difficult to arrange sureties.[112] This is presumably because these fences raise

[108] Herren, *Hisperica Famina*, 76.180 *spinosis densant septa prunis* 'they make fences thick with prickly thorns'. In this context, it seems likely that *prunus* refers specifically to blackthorn (OIr. *draigen*).

[109] *CIH* i 202.25; ii 582.12 = *AL* iv 150.1 *a mes 7 a cháel.*

[110] *CIH* i 73.17–18; 196.1–2 = *AL* iv 112.17–18. Middle Irish commentary at *CIH* vi 2133.22–6 (cf. *CIH* i 66.22–3 = *AL* i 76.17–18) has a totally different explanation of the construction of the *dairime*, apparently taking it to consist of cut oak-trees laid horizontally so as to form a barrier: the top of each tree is placed on the base of the next one in the line (*ind in craind ar bun in crainn* [*e*]*ili*). The commentator claims that this type of fence is peculiar to woodland.

[111] *CIH* i 65.13; vi 2193.15 = *AL* iv 72.2.

[112] *CIH* i 41.24–6 = *AL* v 270.13–15 [*A*]*tāit .uii. naile la* Féniu *a comaithches nād usa adrodmaith*er *a rātha* 'there are seven fences among the Irish in the law of neighbourhood for which it is not easy to admit their sureties'. In Old Irish *aile* generally refers to a fence of stakes and wattle, but in this heptad it may also include stone or sod constructions.

particular problems for the persons involved in their upkeep. Most of these fences are in or near water and are therefore more difficult to construct than ordinary fences and especially liable to damage in storms and floods. The first in the list is the wave-fence (*tondaile*), which presumably must stretch far enough out to sea to prevent cattle getting past at low tide. A fence of this type would obviously need to be sturdily built and would require frequent repair. There are similar difficulties of upkeep with the shore-fence (*trachtaile*), the stream-fence (*sruthaile*), and the lake-fence (*lochaile*). There is a reference to the vulnerability of such fences in an Old Irish elegy on an unnamed ruler. His death is described in the metaphor *rofóenad felm thuile*, which could be translated 'the fence has been flattened in the flood' or 'the fence at the high-tide mark has been flattened'.[113]

The nature of the fifth fence, the *toraile*, is unclear, as the meaning of the element *tor* is doubtful.[114] One glossator takes it to be constructed with sods on stones,[115] but another glossator equates it simply with a stone wall.[116] The sixth fence is the boundary fence (*tórandaile*). Special legal significance obviously attaches to a fence or wall which marks the boundary between two holdings.[117] The last fence in the heptad is the drought-fence (*tartaile*). The glossator explains that 'it is raised in dry weather and falls in the wet'.[118] This suggests a bank constructed largely of mud, which would collapse in wet weather.

If a fence (or other field-boundary) is improperly constructed it is described as an *animbe* 'defective fence'.[119] The landowner must

[113] Meyer, *Über die älteste irische Dichtung* ii 25 (b).

[114] *DIL* distinguishes 1 *tor* 'sorrow', 2 *tor* 'tower', 3 *tor* 'host', and 4 *tor* 'bush' (only in late sources).

[115] *CIH* i 41.33 = *AL* v 270.26 *fōtt for cloich.*

[116] *CIH* iii 906.12 *Toraile .i. cora cloch.* Possibly he is thinking of a fishing-weir, often constructed with both stones and branches: see p. 287.

[117] The glosses *.i. cloch for fōtt nō fōtt for carruig* 'i.e. stone on sod, or sod on rock' at *CIH* i 41.33–4 = *AL* v 270.26–7 may refer to the fact that a boundary is likely to be marked by a substantial wall of stones and sods. The fragmentary version of the heptad at *CIH* iii 906.12 has *tornaile* which may be a scribal error for *tórandaile.* Commentary at *CIH* i 95.20 has *dronaili* 'firm fence'.

[118] *CIH* i 41.34 = *AL* v 270.27–8 *.i. donīther isin tart ⁊ tuitid isin fliuch.* In other commentary at *CIH* i 95.30 and *CIH* vi 2138.29 it is equated with the word *garad*, and classed as one of the three temporary fences (*imeda urcradacha*). The other two are the *galam* (= *lochaile*) and the *lod* (= *sruthaile*).

[119] *CIH* ii 462.1 = Binchy, '*Coibnes Uisci Thairidne*', 72 § 14; *CIH* i 198.8; ii 579.10 = *AL* iv 124.z.

pay for any injury caused to a person or animal as a result of the defect.[120] Naturally, it is a serious offence to damage another person's fence.[121] Even if there are no further consequences – such as damage to crops by livestock – the culprit must pay a fine. According to *Bretha Comaithchesa,* this amounts to a yearling bullock for three stakes, a yearling heifer for five stakes, a two-year-old heifer for eight stakes, and five *séts* for twelve stakes.[122] He must also restore the broken fence and is responsible for its upkeep until the end of the year.

Gates and stiles

Wooden or iron field-gates were not a feature of the early Irish farm, and the word *geata* 'gate' is a late borrowing from English. It is likely that the most usual way of closing livestock into a field or enclosure was by a movable hurdle secured by some sort of tie. In his *Glossary,* Cormac cites the word *udim* (*aidemm*),[123] and takes it to refer to the knotted withe (*roid*) through which a stick (*crand*) passes to secure a hurdle (*clíath*).[124] A passage on distraint in the law-text *Bretha Nemed Toísech* refers to the use of a splinter of wood (*oirdne*) to secure the *udim* of a cattle-pound.[125] Our sources provide no information on how livestock gained access to a field bounded by stone walls; probably a section of the wall was removed and rebuilt as required.

Access for people into fields was sometimes provided by a stile (*céim*). For example, the author of the Old Irish text known as *The monastery of Tallaght* describes the abbot Dublitir and his confessor coming out of a field (*gort*) into the monastic green (*faithche*) by way of a stile.[126] Full or partial responsibility for injuries caused by a defective stile devolves on the owner.[127]

[120] *CIH* i 201.3–4; ii 580.29–30 = *AL* iv 140.26–7.
[121] *CIH* i 239.16 = *AL* v 474.12–13 *coscrad aile* 'destruction of a fence'.
[122] *CIH* i 79.10–12; 203.29–204.7 = *AL* iv 152.7–10. The text refers also to the wattling (*intech*) interwoven between the stakes. A slightly different list of fines for damage to fences is given at *CIH* ii 580.34–8 = *AL* iv 144.29–34.
[123] See Thurneysen, *Grammar,* 163 § 252; *DIL* A 105.23–38.
[124] Meyer, *Sanas Cormaic,* 110 § 1269 (reading the variant *roid,* i.e. *ro* 'great' + *id* 'withe', as suggested by Plummer, 'Some passages in the Brehon laws. IV', 118).
[125] *CIH* vi 2226.40 (*tathrad* is a misprint – read *tabhrad*); *CIH* iv 1531.18–21 = O'Dav. 484–5 § 1615.
[126] Gwynn and Purton, 'The monastery of Tallaght', 130.7–8 § 7.
[127] This seems to be the implication of *CIH* i 201.4–5; ii 580.31–2 = *AL* iv 140.28–9. Meyer, *Cáin Adamnáin* 28 § 41, takes *céim* to mean 'door-step' (in a list of places where

Trees and woodland

It is clear from our sources that trees were of enormous significance in the early Irish economy. They supplied apples and nuts for human consumption, and acorns for pigs. Their foliage was an important foodstuff for livestock, and there is evidence that branches of holly and ivy were cut for this purpose during the winter (see p. 46).

Legal commentary distinguishes three categories of wood: firewood (*connad*), rods (*cáelach*), and timber suitable for sawing into planks (*clárach*).[128] There are many references to the use of wood in the manufacture of furniture, handles of tools and weapons, carts, plough-frames, boats and domestic vessels.[129] As we have seen above, rods and wooden posts were used to make farmhouses, sheds, and enclosures for livestock. Oakwood was regularly used in the construction of church-buildings: a law-text on status, *Uraicecht Becc*, makes special mention of the wright who is capable of building a *daurthach*, literally 'oak house'.[130] According to another law-text, a standard beam (*clár n-inraic*) is twelve feet in length, two feet across, and one foot in thickness.[131]

Wood was by far the most important form of fuel. Our sources contain a number of references to the use of sticks or split logs for burning. By contrast, there is little mention of turf as a fuel (p. 397). In an Old Irish passage which details the provisions which must be made for a fosterson who looks after an elderly landowner (p. 362), it is stated that he gets seventeen trees as firewood. Each tree is to be split into four.[132] Wood was also made into charcoal (*gúal*) for smelting. According to *Críth Gablach*, a farmer of *mruigḟer* rank is expected always to have a sack of charcoal in his house for use in the manufacture or repair of iron implements.[133]

---

fatal accidents may occur), but it may also refer to other types of steps, including stiles.

128 *CIH* i 261.8; iii 928.8 = *AL* iii 148.12.

129 The use of pottery seems to have almost entirely ceased in Ireland during the Iron Age: see Raftery, 'The conundrum of Irish Iron Age pottery'.

130 *CIH* v 1615.22 = *AL* v 102.24.

131 *CIH* iii 921.27–8.

132 *CIH* iii 894.16. This wood-allowance is called *cúal nincis*.

133 *CIH* ii 564.3; iii 780.10 = *CG* 8.191 *míach gúaile fri herna*.

The Old Irish tree-list

The main law-text on farming, *Bretha Comaithchesa*, contains a list of twenty-eight trees and shrubs. They are arranged in four classes in accordance with their economic value.

Complete versions of the tree-list are found in four manuscripts of *Bretha Comaithchesa*.[134] There are some slight variations in the species listed and in their classification.[135] I reproduce here the list in the form in which it is found in the oldest manuscript.[136]

Class 1: **Airig Fedo** 'nobles of the wood'

daur 'oak' (*Quercus robur, Quercus petraea*)
coll 'hazel' (*Corylus avellana*)
cuilenn 'holly' (*Ilex aquifolium*)
ibar 'yew' (*Taxus baccata*)
uinnius 'ash' (*Fraxinus excelsior*)
ochtach 'Scots pine' (*Pinus sylvestris*)
aball 'wild apple' (*Malus sylvestris*)

Class 2: **Aithig Ḟedo** 'commoners of the wood'

fern 'alder' (*Alnus glutinosa*)
sail 'willow, sally' (*Salix caprea, Salix cinerea*, etc.)
scé 'whitethorn' (*Crataegus monogyna*)
cáerthann 'rowan' (*Sorbus aucuparia*)
beithe 'birch' (*Betula pubescens, Betula pendula*)
lem 'elm' (*Ulmus glabra*)
idath 'wild cherry (?)' (*Prunus avium*)

Class 3: **Fodla Fedo** 'lower divisions of the wood'

draigen 'blackthorn' (*Prunus spinosa*)
trom 'elder' (*Sambucus nigra*)
féorus 'spindle-tree' (*Euonymus europaeus*)
findcholl 'whitebeam' (*Sorbus aria*)
caithne 'arbutus' (*Arbutus unedo*)
crithach 'aspen' (*Populus tremula*)
crann fir 'juniper (?)' (*Juniperus communis*)

[134] *CIH* i 78.18–79.9; 202.19–203.14 (two versions); ii 582.6–22; v 1857.26–42 = *AL* iv 146.18–150.14. Parts of the list are included in *CIH* ii 571.3–8 = Binchy, 'An archaic legal poem', 158.51–70. A complete version is also found in Calder, *Auraicept*, 88–90.1152–7; 232.4248–52.

[135] For example, the positions of whitethorn and aspen are reversed in the versions at *CIH* i 202.28–9; ii 582.16–18.

[136] Rawlinson B 487 (Bodleian Library, Oxford) f. 67a = *CIH* i 78.18–79.9. The tree-names are given in their Old Irish spelling when this is known.

Class 4: **Losa Fedo** 'bushes of the wood'

raith 'bracken' (*Pteridium aquilinum*)
rait 'bog-myrtle' (*Myrica gale*)
aitenn 'furze, gorse, whin' (*Ulex europaeus, Ulex gallii*)
dris 'bramble' (*Rubus fruticosus* aggregate)
fróech 'heather' (*Calluna vulgaris, Erica cinerea,* etc.)
gilcach 'broom' (*Sarothamnus scoparius*)
spín 'wild rose (?)' (*Rosa canina* etc.)

There are no problems about the identification of most of the trees and bushes in the above list, but my translations of *idath* and *crann fir* are uncertain.[137] There is also some uncertainty about my identification of *spín* as wild rose, as this word (or a derivative) is attested in the names of a number of thorn-bearing trees and shrubs.[138]

This tree-list displays considerable botanical knowledge, and few native trees are omitted. It is interesting that the author includes the arbutus (*caithne*), as this tree is now very rare outside the Killarney area.[139] However, place-name evidence suggests that it was formerly much more common in the west of Ireland. Thus the Irish name for Smerwick in the Dingle peninsula is *Ard na caithne* 'the hill of the arbutus',[140] and there is a *Doire na caithne* 'oak-wood of the arbutus' (Derrynacaheny) near Inchicronan in Co. Clare.[141] In both these areas the arbutus is now extinct.

Economic importance of the 'nobles of the wood'

An Old Irish commentary on *Bretha Comaithchesa* summarizes the economic importance of the seven 'nobles of the wood'.[142] The value of the oak is said to derive from 'its acorns and its [use for] woodwork' (*a mes ⁊ a ṡaíre*).[143] There are many references to the fattening of pigs on acorns (p. 83), and a later commentator claims that a single oak can provide enough to fatten one pig in a good

[137] I discuss the evidence in Kelly, 'The Old Irish tree-list' 115, 119.
[138] Ibid. 122–23.
[139] Ibid. 118–19.
[140] Hogan, *Onomasticon*, s.v. *Árd na caithne*; An Seabhac, '*Triocha-Céad Chorca Dhuibhne*, II', 99.
[141] Westropp, 'The forests of the counties of the Lower Shannon Valley', 278.
[142] *CIH* i 202.24–7; ii 582.12–15 = *AL* iv 148.z-150.4.
[143] It seems preferable to read *saíre* 'woodwork' rather than *soíre* 'nobility, dignity', though the latter translation would also make sense in view of the high prestige of the oak.

year. This commentator describes the oak as *in Temair feda* 'the Tara of the wood', on account of its being the tallest woodland tree.[144] Another use of the oak which is mentioned quite often in legal material is the provision of bark for tanning leather.[145] For example, one legal commentary refers to the tanning of shoes and a horse's bridle.[146]

The hazel is esteemed because of 'its nuts and its rods' (*a mes 7 a cháel*). Hazel-nuts were a prized element in the human diet, and the hazel is consequently described in legal commentary as *in briugu feda* 'the hospitaller (food-provider) of the wood'.[147] Hazel-rods are strong, pliable and quick-growing, and were widely used in the construction of fences, enclosures and house-walls. For example, the wattling in the walls of the seventh-century farm-house excavated by Chris Lynn at Glenarm, Co. Antrim, has been identified as hazel.[148] A cartload of rods (*carr caíl*) is included in the annual rent which a client must give to his lord.[149] The type is not specified in the texts, but it is likely that the rods would mainly be of hazel, perhaps with some bundles of sally-rods for basket-making.

The Old Irish commentator provides two reasons why the holly-tree is included among the nobles of the wood. The first is obscure to me, and possibly refers to the cutting of holly-branches as winter fodder instead of grass.[150] The second use is for making chariot shafts (*feirtsi carpait*). Later legal commentary refers to *in crann fulachta fíannsa* 'the tree of the open-air cooking-pit'.[151] This is undoubtedly the holly, as its hard wood is especially suitable for making cooking-spits.

144 *CIH* vi 2183.19–21. An undated scrap of vellum between cols. 336 and 337 in YBL (Trinity College Dublin MS H 2. 16 = no. 1318) is omitted in the photo-lithographic reproduction. It contains the following record: *Tuile feidlecan do thecht in iarthur connacht isin bliadain sea conár fhágsat duille for daraig a crich í fhlaithbertaig uile* 'an abundance of moths came into west Connacht this year so that they did not leave a leaf on an oak in the whole territory of O'Flaherty'. This is a reference to the moth *Tortrix viridana*, whose caterpillars sometimes defoliate large areas of oakwood.

145 E.g. *CIH* i 202.20; ii 582.7 = *AL* iv 148.19.

146 *CIH* iii 972.7 *coirtegh brōg ł coirtegh adastair.* Cf. Stokes, *Lismore Lives*, 28.943 *do coirted chuaran* '[oak-bark] for tanning shoes'.

147 *CIH* vi 2183.16; cf. *CIH* ii 570.41 = Binchy, 'An archaic legal poem', 157.34–5.

148 Lynn, 'Deer Park farms', 14.

149 *CIH* v 1906.16; cf. *CIH* ii 570.10 = *CG* 22.573.

150 The text has *fer for araili* (*araile* v.l.) which might mean 'grass substitute'. See discussion at Kelly, 'The Old Irish tree-list', 109–10. For the hook used to cut holly, see p. 491 below.

151 *CIH* vi 2183.22–33. For this method of cooking, see p. 337 above.

According to this commentary, the yew is valued for one reason alone: 'its noble artefacts' (*a aicdi sáera*). The law-texts contain a number of references suggesting that yew-wood was the preferred material for domestic vessels,[152] and a text on status, *Uraicecht Becc*, distinguishes the expert in yew-work (*saí ibrórachta*) as a special category of craftsman.[153]

The inclusion of the ash among the nobles of the wood is attributed to 'support of a royal thigh and half-material(?) of a weapon' (*folach rígslíasta ⁊ lethárad airm*). This suggests that ash-wood was valued because it was used to manufacture furniture and spear-shafts. Other texts refer also to oars and yokes of ash.[154]

The pine is regarded as a noble tree because of 'its resin in a bowl' (*a bí i tulchubu*).[155] Pine-resin was used to make pitch for caulking boats, preserving wood, etc. This reference suggests that the pine-tree – which later died out as a native species – was still fairly common in Ireland of the eighth or ninth centuries. There are a number of literary references to the use of pine-beams in house-building,[156] and a gloss identifies the mast of a ship as being of pine.[157]

The apple-tree is prized for 'its fruit and its bark' (*a mes ⁊ a rúsc*). The importance of the apple in the early Irish diet is obvious (see p. 260), but the mention of the bark is mysterious. Perhaps the reference is to the use of the inner bark of the apple-tree to dye cloth yellow.[158]

[152] E.g. *CIH* iii 1130.40 = E. J. Gwynn, 'Privileges', 41.23 (see his short note on the passage on p. 230); *CIH* iv 1235.14; Best, 'The settling of the manor of Tara', 134.14 § 10. Cf. *CIH* vi 2183.17 (commentary) *int ēochrann aicdide* 'the yew-tree of artefacts'.

[153] *CIH* v 1615.35–6 = *AL* v 104.11.

[154] See *DIL* s.v. *uinnius*.

[155] The MSS have *a bī a tulca* (*CIH* i 202.27 = *AL* iv 150.4) and *a bí a tulcuma* (*CIH* ii 582.14–15).

[156] *DIL* s.v. *ochtach*. A later hagiographical source (Plummer, *Vitae* ii 257 § xvi) describes how monastic builders spent four days in the woods of Connacht searching vainly for pines to construct a refectory. Pines were eventually provided through the miraculous intervention of Saint Samthann. This story doubtless reflects the actual decline in abundance of this tree.

[157] *CIH* iv 1516.24–5 = O'Dav. 418 § 1253. The glossator uses the later word for pine, *crann giúis*.

[158] I owe this suggestion to John Tierney of the Department of Archaeology, University College Cork.

Economic importance of other trees, bushes, etc.

Our sources provide much less information on the uses of trees and bushes in the three lower classes. There are references to the use of alder-wood for the manufacture of shields, masts and tent-poles.[159] No doubt because of its lightness, willow is often mentioned in the context of house-building,[160] and there are a number of references to livestock being tied or restrained by a twisted willow withe (*gat*).[161] A Middle Irish gloss suggests that elm-branches were fed to cattle (p. 42), and there is a possibility that ropes or thongs were made from elm-bark.[162]

In some versions of the tree-list, other plants of economic importance are added to Class 4. Thus *eidenn* 'ivy' (*Hedera helix*) is given as an eighth member of this class.[163] As we have seen above (p. 46), there is evidence of ivy-branches being cut for cattle-fodder in the winter. Another category of plant which is sometimes included in Class 4 is *lecla*[164] or *aín*.[165] This is probably to be identified primarily with the soft rush (*Juncus effusus*), widely used as a floor-covering. For example, *Críth Gablach* refers to 'fresh rushes for strewing' (*aín nue dia esair*)[166] and includes 'knives for cutting rushes' (*scena búana aíne*) among the tools which a prosperous farmer should own.[167] According to *Cáin Aicillne*, the inner pith of the rush is used in candle-making (see p. 55).

The term *aín* seems also to be used of other economically important wetland plants of rush-like appearance. For example, in legal commentary on church buildings, it makes better sense

[159] See *DIL* s.v. *fern*.
[160] See *DIL* s.v. *sail*.
[161] See p. 494. Cf. *TBC* Rec. I 31.988; 32.1025. Withes were also used for plough traction: see Lucas, 'Irish ploughing practices (4)', 196.
[162] *LL* iv 834.24902–3 *co súanmib rũisc* 'with ropes of bark'; see *DIL* s.v. *súainem*. Cf. Wiliam, *Llyfr Iorwerth*, 93 § 140.48 = *ALIW*, Venedotian Code, III xxii § 110 *raf luyf* 'a rope of elm[-bark]'.
[163] *CIH* i 203.13; v 1857.40 = *AL* iv 148.2.
[164] *CIH* i 202.32; ii 582.21 = *AL* iv 150.12 (wrong sentence-division). *Lecla* is glossed by the word *lúachair* 'rush' in the quotation of this passage at Pearson, 'A medieval glossary', 73 § 152. For a discussion on the etymology of *lúachair*, see N. Williams, 'Some Irish plant names', 454–6.
[165] *CIH* ii 571.7 = Binchy, 'An archaic legal poem', 158.69.
[166] *CIH* ii 564.31; iii 780.37 = *CG* 9.232.
[167] *CIH* ii 563.22; iii 780.1 = *CG* 7.177.

to translate *aín* as 'reeds' rather than 'rushes'.[168] The value of a timber church (*daurthach*) which is fifteen feet long and ten feet wide is given as ten dry heifers if it has a roof of *aín* (*tuige aíne*).[169] Its value is doubled to ten milch cows if it has a roof of shingles (*tuige slinned*). As rushes are poor thatching material, it is probable that *aín* refers here to the reed *Phragmites communis.* In Adomnán's seventh-century *Life of Columba,* there is an account of the career of a man called 'Librán of the reed-plot' (*Libranus harundineti*), who was a native of the district of the Connachta. Adomnán explains that he was so named because for many years he worked in a reed-plot, gathering reeds.[170]

The law-text on distraint refers to the illegal cutting of plants on privately owned *murmag* 'sea-plain'.[171] A glossator identifies these plants as growing 'on the edge of the dunes' (*ar ur na dumaige*), and names them as *muraín* and *magra. Muraín* is probably to be regarded as a compound of *muir* 'sea' + *aín* 'rush' and identifed with sea-rush (*Juncus maritimus*) which is very common in salt-marshes and sandy shores; this identification is supported by another gloss which equates it with *murlúachair* (lit. 'sea-rush').[172] *Magra* may possibly be bent or marram (*Ammophila arenaria*), a common seaside grass.[173]

Damage to trees

According to *Bretha Comaithchesa,* if a person damages a tree belonging to another, he must pay the owner a penalty-fine

[168] *Gilcach* in Class 4 of the tree-list is sometimes taken to mean 'reed' (*DIL* G 82.11), but the meaning 'broom' is more likely in this context: see discussion at Kelly, 'The Old Irish tree-list', 122.

[169] *CIH* vi 2099.23–8 = Petrie, 'Ecclesiastical architecture', 364–5.

[170] A. O. and M. O. Anderson, *Adomnan's Life of Columba,* 434 (new pagination: 162) = f. 92b.

[171] *CIH* ii 384.20; 620.7; iii 890.22 = *AL* i 166.27 *im rasas do murmaige* 'for stripping(?) your sea-plain'. The word *rasas* has not been satisfactorily explained.

[172] *CIH* ii 385.25–6 = *AL* i 172.14–16. This glossator is careful to distinguish *fír-lúachair* 'true rush' (i.e. *Juncus effusus*) from *murlúachair* (= *muraín*) mentioned in a previous gloss (*CIH* ii 384.38 = *AL* i 170.24).

[173] The glossator at *CIH* ii 384.39; v 1690.16 = *AL* i 170.24–5 (partially mistranscribed) evidently equates *magra* (*magru*) with *muirnech* 'bent'. This grass is also called *muirín* (e.g. *CIH* ii 385.26 = *AL* i 172.16; cf. *CIH* ii 709.36). In his *Foclóir* Dinneen gives both *muirín* and *muiríneach* in this meaning.

(*díre*).[174] This is fixed at five *séts* if the tree is in Class 1,[175] a milch cow if it is in Class 2, a yearling heifer if it is in Class 3, and a sheep if it is in Class 4. In addition, the culprit must pay compensation (*aithgein*), which varies in accordance with the severity of the damage. In the case of a Class 1 tree, he must pay a milch cow for cutting it at the base, a two-year-old heifer for cutting it at a fork, or a yearling heifer for cutting a branch.[176] One would expect the compensation for damage to trees of Classes 2–4 to be proportionately lower, but the text does not give full figures.[177]

A more complete set of fines for damage to trees is set out in Old Irish commentary on *Bretha Comaithchesa*,[178] though it disagrees with the main text in minor points of detail regarding Classes 2–4. For example, the commentary gives a sheep as the fine for cutting a branch of a Class 2 tree, whereas the text has a yearling heifer. This commentary also refers to the offence of *aurbe*, which seems to mean the complete destruction of a tree – this could be a more serious offence than base-cutting in the case of regularly coppiced trees such as hazels. It entails a fine of five *séts* for trees of Classes 2 and 3, and a yearling heifer for Class 4. There is no information on the *aurbe* of a tree of Class 1.

This commentator recognises the absurdity of fining somebody for cutting a single rush or bracken-frond. He therefore states that there is no penalty for cutting one stem (*óengas*) of any plant in Class 4. He also deals with the illegal removal of the bark from an oak for tanning (*coirtged*).[179] If the culprit has stripped off enough to tan a pair of woman's sandals he must pay a cow-hide, if enough for a pair of man's sandals he must pay an ox-hide. In addition, he must cover the wound with a mixture of smooth clay, cow-dung and fresh milk until there has been the width of two fingers' new

[174] *CIH* i 203.2–14 = *AL* iv 146.21–148.2.

[175] Cf. *CIH* i 239.11 = *AL* v 474.6–7. According to the glossators, the general fine for offences in this list is 5 *séts*.

[176] *CIH* i 203.2–3 = *AL* iv 146.22–3. Medieval Welsh law applies a similar three-fold scale of fines: 120 pence for a whole oak, 60 pence for a main branch, and 30 pence for a side branch (Wiliam, *Llyfr Iorwerth*, 90 § 138.1–5 = *ALIW*, Venedotian Code, III xx §§ 1–3).

[177] For more detailed discussion on payments for damage to trees, see Kelly, 'The Old Irish tree-list', 108–9, 114, 116, 120.

[178] *CIH* i 202.16–33; ii 582.1–22 = *AL* iv 148.13–150.14.

[179] *CIH* i 202.19–22; ii 582.6–9 = *AL* iv 148.19–22. In legal glosses, this offence is called *snomad* 'barking', e.g. *CIH* ii 383.8 = *AL* i 164.2.

growth on all sides. This technique is similar to the modern one of painting an air-excluding preparation over the wound where a bough has been sawn off.

In assessing the fines for tree-damage, *Bretha Comaithchesa* takes into account the species but not the size of the tree. So a person who cuts down a ten-foot oak must apparently pay the same fines as the person who cuts down an eighty-foot oak. Later commentary – which may derive from a lost law-text *Fidbretha* 'tree-judgements'[180] – gives a very brief outline of a fairer but more complex system, in which there is not only a distinction between the different classes of tree or bush, but also between a large tree (*crann mór*) and a slender tree (*crann cáel*).[181]

There is no mention in *Bretha Comaithchesa* or its commentary of any seasonal variation in the penalties for illegal cutting of trees. In *Bechbretha*, however, it is stated that at some times of year, the lopping (*tairdbe*) of a bough from another person's tree is regarded as an equally serious offence to base-cutting (*bunepe*).[182] This suggests that the author of the text knew that most types of tree are more likely to develop rot or disease if a cut is made during the growing season.[183]

In legal material a distinction is regularly made between trees which are classed as *nemed* 'sacred, privileged', and those which are not.[184] The penalty for damage to a sacred tree (*fidnemed*) is much higher than for an ordinary tree (*fid comaithchesa*). For example, the penalty-fine (*díre*) for an apple-tree classed as *nemed* is given as twenty *séts*, four times greater than for an ordinary apple-tree.[185] According to a glossator on *Bechbretha*, a tree growing on the land of a person of high status (i.e. a *nemed*-person) is thereby counted as a *fidnemed*.[186] The use of the term *fidnemed* is not confined to legal material. For example, the *Annals of Ulster* record the destruction

[180]For references to *Fidbretha*, see *GEIL* 274 § 46.
[181]*CIH* vi 2106.3–14.
[182]*CIH* ii 446.19–20 = *BB* 60 § 16.
[183]This seems to be the interpretation of the A glossator, who distinguishes the fine for cutting during the growing season (*beodatu*) from that of the period of dormancy (*marbdatu*). See discussion at *BB* 109–10 (note to § 16).
[184]E.g. *CIH* ii 446.14–15 = *BB* 58 § 15; *CIH* iv 1286.20–30 = *BB* 176–7 (Appendix 2 (c)); *CIH* i 239.11 = *AL* v 474.7; *CIH* ii 383.6–9 (glosses) = *AL* i 162.28–9.
[185]*CIH* v 1537.21. Cf. *CIH* ii 570.41–571.1 = Binchy, 'An archaic legal poem', 157.37–41 (and note on p. 162).
[186]*CIH* ii 446.17–18 = *BB* 58 § 15^{c} (A).

by lightning in AD 996 of the monastery of Armagh, including its timber building, stone church, porch and *fidnemed*.[187] This may be – as Binchy suggests – a reference to a single especially venerated tree which grew near the church buildings.[188]

Venerated trees may also grow on secular sites, and be a strong focus of local pride. Traditions concerning such trees are recorded in the Middle Irish metrical and prose *Dindṡenchas*. For example, one passage refers to five outstanding trees: *Éo Rossa* (a yew), *Bile Dathi* (an ash), *Éo Mugna*[189] (an oak), *Cráeb Uisnig* (an ash), and *Bile Tortan* (an ash).[190] A tree's status may make it a target for enemy attack: the *Annals of Ulster* record how the Ulstermen chopped down sacred trees (*biledha*) at the royal inauguration site of the Cenél nÉogain at Telach Óc in AD 1111.[191] The veneration of trees described in Irish sources can be compared with the cult of sacred groves well attested among the Celts of Britain and the Continent.[192] There is the difference, however, that early Irish tradition lays much more stress on single trees rather than on trees in a group.

In certain limited circumstances, the law allows a person to cut wood on another's property without liability. Thus he may cut barrel-hoops, rods for spancelling animals, material for a churn-dash, shafts for weapons, sticks for driving animals, branches to construct a bier to carry off a dead body, and wood for an ox-yoke or plough.[193] He may collect enough brushwood (*crínach*) for one night's cooking.[194] He may also cut rods to make

[187] *AU*² 426 s.a. 995 (recte 996) § 1. According to other annals, the destruction was caused by the Airgíalla rather than by lightning (e.g. *CS* 234 s.a. 994; Stokes, 'The annals of Tigernach (2)', 350).

[188] Binchy, 'An archaic legal poem', 162.

[189] The basic meaning of word *éo* is 'yew' (cf. the cognates Welsh *ywen*, Breton *ivin*, English *yew*, etc., *IEW* 297), but it may also be used of particularly fine or venerated trees of other species. In this sense, *éo* is of similar meaning to *bile*. For an account of the *Éo Mugna*, see Vendryes, *Airne Fíngein*, 4.45–5.66.

[190] Stokes, 'The prose tales in the Rennes Dindṡenchas, (2)', 277–9 § 160. See also E. J. Gwynn, *Metrical Dindshenchas* iii 146–49; iv 240–7; Bieler, *Patrician texts*, 162 § 51.

[191] *AU*² 552 s.a. 1111 § 6; cf. *AU*² 534 s.a. 1099 § 8 (describing the cutting down of another sacred tree, the *Cráeb Telcha*).

[192] See discussion at *BB* 109 (note to § 15).

[193] *CIH* i 241.22–4 = *AL* v 482.26–9. I follow the glossator's interpretation of *arad cach fedna* as *.i. cuing ⁊ cechta* 'i.e. yoke and ploughbeams'.

[194] *CIH* i 241.21 = *AL* v 482.25; *CIH* ii 571.2 = Binchy, 'An archaic legal poem', 157.46.

a hurdle (*cleth*) or a basket (*clíab*); but if he cuts too many he must pay a fine of five *séts*.[195]

Woodland

Our sources give an impression of a fairly well wooded countryside, and place-names such as *fid* 'wood', *caill* 'wood', and *daire* 'oak-wood' are common.[196] In the law-texts of the seventh and eighth centuries it is taken for granted that a farmer has access to considerable amounts of wood for many purposes – the annual requirements of hazel-rods alone must have been very considerable. Consequently, woodland was much valued: the wisdom-text *Tecosca Cormaic* states that 'a wood is good at every season' (*maith fidbad cecha ráithe*).[197]

In general, it seems that the main woods were on poorer marginal land, often hilly.[198] The author of the seventh-century law-text *Bechbretha* pictures a farmer's outfield being bounded by a great wood (*ruud már*) or difficult country (*écmacht*) or undivided land (*dírann*).[199] On the other hand, woods on good land were not unknown. The law-text on land-values distinguishes three types of cultivable land (*etham*), all of which are good for grazing and the cultivation of corn, flax, woad and other crops (see p. 394 below). The third type is defined as 'land cultivable by labour; that is, land clearable by axe'.[200] Middle Irish commentary on Heptad 64 refers also to the existence of damp pastures (*clúainti*) and tilled fields (*guirtíni*) within a wood.[201]

The texts do not suggest that large areas of the country were under woodland. On the contrary, the general picture we get is of woods and copses, very often privately owned, whose resources are limited and need careful protection by the law. The author of a ninth-century series of geographical triads clearly regarded large woods as unusual in the Ireland of his day.[202] He lists the three wildernesses of Ireland as *Fid Mór hi Cúailngi* 'the great wood in

[195] *CIH* i 238.33 = *AL* v 474.4. See gloss and commentary on this passage at *CIH* i 239.4–9 = *AL* v 474.20–7.
[196] See Hogan, *Onomasticon* s.vv.
[197] Meyer, *Tecosca Cormaic*, 34.104 § 16.
[198] *CIH* iii 901.12 *aimrēd cach rūd* 'every great wood is uneven'.
[199] *CIH* ii 454.26–8 = *BB* 82 § 48. Cf. *CIH* ii 573.27–8; iv 1437.41.
[200] *CIH* ii 675.26–7 = Mac Niocaill, '*Tír cumaile*', 82.10 *Etham frichnama: tīr inbēla sōn.*
[201] *CIH* iv 1210.28; v 1852.42 = *AL* v 324.17.
[202] Meyer, *Triads*, 4 § 43.

Cooley (Co. Louth)',[203] *Fid Déicsen hi Tuirtri* 'the wood of Déicsiu in Tuirtre',[204] and *Fid Moithre hi Connachtaib* 'the wood of Moithre in Connacht'.[205] The implication is that such extensive woods were rare.[206]

A simile in the twelfth-century tale *Cath Ruis na Ríg* gives us some idea of how woodland was managed at this period.[207] An army in which all the lesser warriors had been slain, leaving only the great champions, is compared with an oak-wood in the middle of a plain in which all the underwood (*cáel*) had been removed, leaving only the great oaks (*railge romóra*). This suggests that – as in medieval England – it was a common practice to coppice the underwood to yield a crop of rods every decade or so, and to keep a few large trees to provide timber when they had grown to maturity.[208]

For the inhabitants of woods, see p. 425.

Roads

Modern advertisements for the sale of agricultural land stress where possible its convenient road-access. The importance of access was not lost on the authors of the Old Irish law-texts. According to the text on land-values, if a piece of land is beside a highway leading to a monastery or to the house of a lord, three cows are added to its value.[209] If there is a road leading to a wood or sea or mountain, a cow is added to its value. If there is a cow-track leading to a

[203]This is the same wood as *Fid Conailli*: see Hogan, *Onomasticon* s.vv. *Fid Conailli, Fid Mór*.

[204]Lit. 'the wood of gazing': Meyer suggests 'Spy-wood'. Tuirtre (the territory of the Uí Tuirtri) lay at this period to the west of Lough Neagh and comprised the present districts of Cookstown, Stewartstown (Co. Tyrone), Moneymore and Magherafelt (Co. Derry). The location of the wood is unknown, but may have been on the slopes of Slíab Calland (Slieve Gallion). The Uí Tuirtri seem to have moved to the east of the Bann about AD 1000: see Eoin MacNeill, 'The Vita Tripartita of St. Patrick', 29. There was also a *Fid Mór* 'great wood' to the west of the Sperrin Mountains, probably near Strabane (Co. Tyrone): see Mulchrone, *Bethu Phátraic*, 92.1761; *PTBA* 180.14.

[205]*Moithre* is presumably a derivative of the *o*-stem *mothar* 'thicket, dense woodland'. Connacht is mentioned as a source of timber in saints' Lives, e.g. Plummer, *Vitae* i 136 § lxxi; ii 257 § xvi.

[206]For other woods mentioned in the early sources, see Hogan, *Onomasticon* s.vv. *caill, caille, fid*.

[207]LL iv 773.23090–4 = E. Hogan, *Cath Ruis na Ríg*, 42 § 34; cf. Todd, *Cogadh Gaedhel re Gallaib*, 198.

[208]Rackham, *The history of the countryside*, 65–7.

[209]*CIH* ii 676.8 = Mac Niocaill, '*Tír cumaile*', 82.35.

cattle-pond (*gelestar*) or to another piece of land or to a highway, a two-year-old heifer is added to the value.

Another law-text distinguishes five categories of road.[210] They are – in descending order of importance – highway (*slige*), road (*rout*),[211] byroad (*lámraite*), curved road (*tógraite*), and cow-track (*bóthar*). The highway is the widest of these roads: it must be constructed so that two chariots (each drawn by two horses) can pass. At the other end of the scale, the cow-track must fit two cows, one sideways and one lengthways. The author of the text points out that if the track is narrower than this, a cow's calf and yearling cannot fit beside her, and so may be crushed by the cow which is following.

Our sources stress the political as well as the economic significance of the road-system. Major roads are regularly associated with kings. A law-text lays down the general principle 'to every king his road',[212] and legal commentary gives the three 'lands' of a king as 'a road, unshared land, and the sea'.[213] The construction and maintenance of roads is often mentioned as a public duty. The seventh-century *Life of Saint Brigit* by Cogitosus describes how a king issued an edict to the various peoples (*túatha*) under his control that they should construct a firm wide road – partly over boggy terrain – with a foundation of branches, rocks and earth. It was to be capable of bearing the weight of chariots, mounted horses, wagon-wheels, and rushing people.[214] A text on the law of neighbourhood specifies that local client farmers are responsible for clearing the roads at the time of winter-visiting (*cóe*) by lords, and at the time of an assembly (*óenach*).[215] According to the text on roads, this clearing consists of digging out the ditches on either side, filling in potholes, and cutting away bushes.[216] Roadside fences must also be maintained: presumably each farmer is responsible for the fence

[210] *CIH* iii 893.22–30 = Appendix A, text 4 § 1. *Cormac's Glossary* of *c.*900 gives seven categories (Meyer, *Sanas Cormaic*, 96 § 1082): *sét, rout, rámat, slige, lámrotae, túagrotae,* and *bóthar.* In this list *rámat* is classed as a type of road or way, but elsewhere it is used of the clearing in front of a king's fort. For discussion of the meanings of *rámat,* see Appendix A, p. 543.

[211] *Rout* may be used as a general term for any sort of road or track, or – as here – for a secondary road.

[212] *CIH* i 293.8 = *AL* iii 304.25 *cach rīg a rāmut.*

[213] *CIH* ii 455.27–30 = *BB* 171, Appendix 1 (g); cf. *CIH* v 1549.36–7 *mac for rōt, rígh a díre* 'a child on the road, the fine [for his death] belongs to the king'.

[214] Connolly and Picard, 'Cogitosus: Life of Saint Brigit', 23 § 30.

[215] *CIH* i 201.40–202.1; ii 580.7–9 = *AL* iv 144.23–5.

[216] *CIH* iii 893.32–3 = Appendix A, text 4 § 2.

abutting on his own property. Failure to carry out these duties incurs a fine, stated in the legal glosses to be five *séts*.[217]

Reference is also made to private roads not under the control of the king. Thus the text on roads accounts for the curved road (*tógraite*) in the following terms: 'a well-off man hires a way (*conar*) towards a wood(?) or mountain'. Those who use this route must pay a toll of one animal every second year.[218] In such cases, the duty of maintenance may perhaps fall on the clients of the owner of the road, and the fine for failure to carry out this work would be paid to him.

Tradition relating to various important roads has been preserved in many texts. For example, the *Triads of Ireland* lists three major highways[219] and the *Metrical Dindshenchas* lists five.[220] The great highway (*Slige Már*) is reputed to have followed the line of eskers (*Eiscir Ríata*) across the central plain from Dublin to Galway.[221]

Causeways

The extensive bogs of central and western Ireland were a serious impediment to transport and communication. It was therefore often necessary to construct a causeway (*tóchar*) across such terrain.[222] A number of these have been excavated by archaeologists. For example, a causeway of about one kilometre in length was found during turf-cutting at Corlea, Co. Longford, and has been dated to about 150 BC. It was made of logs of oak laid on top of brushwood, and would have been capable of supporting wheeled vehicles.[223]

The Old Irish mythological tale *Tochmarc Étaíne* contains an account of the construction of a causeway over the Bog of Lamraige, which had reputedly never been crossed before. In

[217] *CIH* i 239.15 (and gloss) = *AL* v 474.12; *CIH* ii 368.31 = *AL* i 122.14; *CIH* ii 381.30 = *AL* i 156.27–8; *CIH* ii 401.13–14 (and gloss[11]) = *AL* i 230.23.

[218] *CIH* iii 893.27–8 = Appendix A, text 4 § 1 (4).

[219] Meyer, *Triads*, 6 § 49.

[220] E. J. Gwynn, *Metrical Dindshenchas* iii 278.41; cf. Vendryes, *Airne Fíngein*, 9.121–11.137.

[221] E. J. Gwynn, *Metrical Dindshenchas* iii 282.89–92. For a general discussion, see O Lochlainn, 'Roadways in ancient Ireland'. The geological term esker (from Irish *eiscir*) refers to the winding gravel ridges left by the retreat of the ice cap: see Quin and Freeman, 'Some Irish topographical terms', 85–7.

[222] The element *tóchar* is common in placenames: see Hogan, *Onomasticon*, s.v. *tóchar*. See also Lucas, 'Toghers or causeways'; Raftery, *Trackways through time*.

[223] Mitchell, *Reading the Irish landscape*, 148.

this case the foundation of the causeway consisted of tree-trunks and branches, over which layers of clay, gravel and stones were placed.[224] According to commentary on the law-text on status, *Uraicecht Becc*, the ability to construct a causeway adds two cows to the honour-price of a master wright.[225]

Bridges

In normal conditions it is possible for vehicles and pedestrians to cross many rivers at a ford (*áth*) where the riverbed is firm and the water shallow. The crossing can be made easier by building up the riverbed with stones, planks or hurdles. Thus the ford at Dublin was called *Áth Cliath* 'the ford of the hurdles' from early times. This was doubtless because hurdles were placed horizontally on the riverbed to provide a convenient path for those crossing.[226]

A bridge (*droichet*) requires fairly sophisticated technology if it is to span a river of any width. Recently, oak timbers from a massive bridge spanning the Shannon have been discovered at Clonmacnois, and have been dated by dendrochronology to *c.*800.[227] The seventh-century law-text on watermills speaks of 'the regulations relating to a bridge' (*cáin droichit*)[228] and of 'the ditch which is around a bridge' (*clad bís im droichet*).[229] From approximately the same period, the law-text on distraint refers to 'the construction (lit. raising) of a bridge' (*ocbáil droichit*).[230]

According to commentary on *Uraicecht Becc*, ability to build a bridge adds two cows to the honour-price of a master wright.[231]

[224] Bergin and Best, '*Tochmarc Étaíne*', 178 §§ 7–8.
[225] *CIH* v 1613.2 = *AL* v 94.11–12.
[226] Mitchell, *Archaeology and environment in early Dublin*, 15.
[227] Moore, 'Ireland's oldest bridge'.
[228] *CIH* ii 460.10 = Binchy, '*Coibnes Uisci Thairidne*', 68 § 9.
[229] *CIH* ii 461.30 = Binchy, '*Coibnes Uisci Thairidne*', 70 § 13. In the notes to §§ 9 and 13 Binchy (quoting Thurneysen, *Bürgschaft*, 84) points out that in such contexts *droichet* could equally refer to a trackway of planks over marshy ground. At *CIH* iii 920.42, the term *droichet* is used of the causeway leading to a house; cf. Jackson, *Aislinge*, 35.1097 and note.
[230] *CIH* ii 371.23 = *AL* i 124.7.
[231] *CIH* v 1613.6 = *AL* v 94.16.

LAND-VALUES

The worth of a given area of land obviously varies in accordance with the quality of the soil, road-access, water-supply, and other factors. The assessment of its value becomes of legal significance in cases where it is to be sold or rented, or divided among heirs. A law-text, probably written in the eighth century, distinguishes six categories of land.[232] Three of these are classed as *etham* 'arable', and three as *ainetham* 'non-arable'.[233] The text gives the value in cows of a *cumal* area of each type of land. Unfortunately, the extent of a *cumal* is unknown: see discussion in Appendix B, p. 574.

In descending order of value, the six categories of land are:

1. Best arable land (*etham remibí ethamnaib*). This is defined as level land which is good for corn, milk, flax, woad, honey, madder, and fruit.[234] It should not require manuring (*tuar*) or clearance (*slige*), and should be free of *gláma*, a term which seems to include various thorny or prickly weeds. The weed-free state of the land may be tested by loosing a horse into it. If brambles (*dris*), thorns (*draigen*), burdock (*gleslige*) or thistles (*omthann*)[235] stick to the horse's mane or coat, the land cannot be classed as best arable.

A *cumal* of this type of land is worth twenty-four milch cows.

2. Hilly arable (*etham taulchach*). The second grade of land is defined as being 'good for every plant and for every crop'. There is access to water, and there are ash-trees on every second piece of ground.[236] This description would suit fertile hilly farmland such as that of south Tipperary or east Wicklow. The ash-tree is

[232] *CIH* ii 675.18–676.16. Mac Niocaill has edited this text along with a later commentary on the same topic (*CIH* iii 843.3–21) in *Ériu* 22 (1971) 81–6. Another short section of commentary (*CIH* iii 972.24–9 = Appendix A, text 6) also belongs with these texts.

[233] *Etham* is from *ith* 'corn' + agent-suffix *-am*, i.e. 'corn-producer'.

[234] The text has *cumrad*, which is probably to be taken as a scribal variant of *cumrae* 'sweet, non-bitter, pleasant-flavoured, fragrant', used substantively. This adjective often describes fruit such as apples and plums, and seems to be used of any sweet fruit at *CIH* vi 2299.39 = Binchy, '*Bretha Crólige*', 40 § 49. *Cumrae* can also refer to vegetables such as onions and celery, so it is possible that the author of our text intended to convey that land classified as best arable should be capable of growing fruit-trees and vegetables, as well as the crops he has already named.

[235] The MS has *oman*: for the emendation, see Mac Niocaill, '*Tír cumaile*', 84[6].

[236] *CIH* ii 675.25 = Mac Niocaill, '*Tír cumaile*', 82.8–9 *is fuinnside cach la maigin and* 'every second place is ashy'.

well known as an indicator of a limey soil and hence a reasonable level of fertility.

A *cumal* of this type of land is worth twenty milch cows.

3. Arable land requiring labour (*etham frichnama*). This is defined as 'land clearable by axe (*tír inbéla*) in which every plant flourishes'. The fact that the author distinguishes this type of land as a separate category suggests that there was still a considerable amount of wooded land with good agricultural potential in eighth-century Ireland.

   A *cumal* of this type of land is worth sixteen milch cows.

4. Rough land (*ainmín*). The most valuable non-arable land is described as 'hard land, ferny plain and untilled plain' (*cotattír, rathenmag 7 ógmag*).[237] This definition suggests flattish land, which is either stony or covered with bracken or rushes, but nonetheless provides some grazing. It is likely that damp pasture (*lénae* or *clúain*) belongs in this category.

   A *cumal* of this type of land is worth sixteen dry cows, i.e. half the value of arable land requiring labour.

5. Very rough land (*antrenn*). Lower on the scale of value is *antrenn*, which is given the description 'heathery mountain and furze there' (*slíab fraích 7 aitenn i suidiu*). This clearly refers to mountain pasture, suitable particularly for sheep. A *cumal* of this land is worth twelve dry cows.

   Our text does not include wooded uncultivable land in its classification. This may be because such land would generally be regarded as common or waste, and would therefore rarely come up for sale or rent.[238] See the discussion on the ownership of mountain, moorland and woodland on p. 406.

6. Shallow land (*andomain*). The least valuable of the six types is said to be land which is *andomain* 'shallow', i.e. with no depth of soil. The author of our text describes it as 'black land and bog' (*duibthír 7 móin*). A similarly low value is assigned to such land in Heptad 37, which includes shallow land (*tír n-andomain*)

[237] The adjective *óg* 'whole, untouched' may be used in the sense of 'untilled': see *DIL* O 110.17–18 and note by Mac Niocaill, '*Tír cumaile*', 84[9].

[238] On the other hand, the eighth-century *Additamenta* to Tírechán's material on Patrick in the Book of Armagh records that Cummen and Brethán bought the property of Óchter Achid 'including wood, plain and riverside pasture' (*iter fid 7 mag 7 lēnu*) (Bieler, *Patrician texts*, 174.7). For further details of this transaction, see p. 420 below.

among the seven least profitable pledges which the head of a kin-group can give on behalf of his kin.[239] In glossaries, this term is applied to other types of wetland as well as bog. For example, O'Davoren identifies it with marsh (*cuirrech*), mire (*seiscenn*), or bog (*móin*).[240]

A *cumal* of this type of land is worth eight dry cows. As such land is clearly unsuitable for grazing, its value must derive from other factors. Probably the most significant of these was the production of turf (peat) for fuel.

Turf eclipsed wood as a fuel in later times, but in the seventh- and eighth-century law-texts, references to turf-cutting are few and ambiguous. The clearest of these is in the text on mill-races: here the ditch of a turf-bog (*clad fótbaig móna*) is included among the seven ditches which are exempt from liability in the case of accidental drowning.[241] It seems safe to assume that the author is thinking of a ditch made during the cutting of turf for use as fuel. Other references in Old Irish law-texts are more doubtful. For example, one text includes *tochar puirt* in a list of offences on another person's property. The glossator takes it to be a reference to illegal turf-cutting, for which a fine of five *séts* is due.[242] But he may be wrong; in another law-text the glossator takes a similar phrase to refer to jetsam cast up on the seashore.[243] This appears to be a quotation from the lost law-text *Muirbretha* 'sea-judgements', so the meaning 'jetsam' would suit the context well.[244]

[239] *CIH* i 35.22–3 = *AL* v 250.6 (glossed *curraigh ⁊ seisccend*).

[240] *CIH* iv 1471.28–9 = O'Dav. 220 § 140. Cf. *CIH* ii 621.15 *Andomuin .i. . . . móin ⁊ lothrach* 'shallow [land] i.e. . . . bog and marsh'.

[241] *CIH* ii 461.29–30 = Binchy, '*Coibnes Uisci Thairidne*', 70 § 13. He reads *fót[b]aig*, following the glossator.

[242] *CIH* i 239.10–11 = *AL* v 474.6. One can compare *CIH* ii 370.4 = *AL* i 122.20 *im chōrus puirt i n-aimseraib tochuir*, where the glossator makes a similar connection with turf-cutting. In both cases, however, the text may actually refer to the ownership of jetsam: see next note.

[243] *CIH* i 315.4 = *AL* iii 424.13 *Dīles tochur do fir puirt* 'what is cast up belongs to the owner of the shore'. The word *port* (Latin *portus*) can mean 'place, locality, abode, bank, mound, shore, harbour'. *Port mónad* is used for 'turf-bank' in a Middle Irish gloss at *CIH* ii 370.6 = *AL* i 132.4. In later Irish the derivative *portach* is used with the same meaning, and also in the more general sense of 'bog'. In Scottish Gaelic, 'turf-bank' is *bac mòna*, where *bac* is from Old Norse *bakki*.

[244] For discussion on *Muirbretha*, see *GEIL* 276–7 § 53.

The first explicit mention in Irish of turf as fuel is in the early twelfth-century tale *Aislinge Meic Con Glinne*.[245] Here the miserable hospitality of the monastery of Cork is said to include two sparks of fire in the middle of a wisp of oat-chaff and two sods of wet turf (*dá ḟót do úr-mónaid*). Post-Norman sources – both in Irish and English – contain many references to the burning of turf. For a discussion of turf-cutting implements, see p. 479.

Wetland which could not be drained for turf-cutting might be of use as a source of reeds for thatching, basket-making, etc. (see p. 385).

After distinguishing the above six land-types, the author of the law-text on land-values goes on to list other factors which increase the value of a piece of land.[246] Five *séts* are added if there is a mill-site, a mine for copper or iron ore, a permanent cattle-pond (*gelestar*) or access to an estuary. Twice this sum is added if there is a great wood (*ruud*) with a ditch or stone wall.[247] Presumably, this means that the wood is adjacent to the property, and separated by an effective barrier. The owner of the property would have convenient access for wood-cutting, food-gathering, grazing, etc.

The value of a coastal property is increased by three cows if there is access to a 'productive rock'. The later commentator is doubtless correct in taking this to refer to a rock from which dulse or other seaweed may be collected.[248] Three cows are also added if there is a stream at the side or headland of the property, with undivided land (*dírann*) beyond. The same applies if the property is bounded by mountain (*roilbe*).

As we have seen on p. 390 above, a road or cattle-track may also add to the value of the property in certain circumstances.

[245] Jackson, *Aislinge*, 6.159–60 = Meyer, *Aislinge*, 15.3.

[246] *CIH* ii 676.1–15 = Mac Niocaill, '*Tír cumaile*', 82.27–83.4. Commentary based on this passage, with some variations, is found at *CIH* iii 843.11–15 = Mac Niocaill, '*Tír cumaile*', 83.16–20, and at *CIH* iii 972.24–9 = Appendix A, text 6.

[247] For a discussion of the boundary earthworks of English woods, see O. Rackham, *Trees and woodland*, 114–16.

[248] *CIH* iii 972.27–8 = Appendix A, text 6 § 3. For edible seaweeds, see p. 313 above.

12
Land-tenure

Land comes under the control of individuals, families or larger groups in a variety of ways. It may be bought or inherited or acquired as a dowry, or it may be rented for a greater or lesser period. In some circumstances, such as in the aftermath of war or famine or in underpopulated areas, it may be possible to take over unoccupied land without incurring opposition. Land is more often taken by force of arms; this is sometimes known in Irish as *tír claidib* 'sword-land'[1] or *orbbae claidib* 'sword-inheritance'.[2] World history abounds with records of the violent appropriation of land: often the invaders are themselves fleeing from other aggressors or from some natural disaster.

Early Irish literature preserves various traditions of large-scale prehistoric population movement (*tochumlud*) and expulsion (*indarbe*). For example, a tenth-century tale-list includes titles such as *Tochomlod Múscraige de Maig Bregoin* 'the migration of the Múscraige from Mag mBregoin', *Tochomlod na nDése ó Themraig* 'the migration of the Dési from Tara',[3] and *Tochomlod Dáil Ríatai i nAlbain* 'the migration of the Dál Ríatai to Scotland'.[4] The migration to Scotland is an undoubted historical fact. However, as Francis John Byrne makes clear in his article 'Tribes and tribalism in early Ireland', the other migrations are probably mere fictions, composed to explain the presence of peoples of the same name in different parts of the country.[5] And even where there are genuine historical records of the displacement of one dynasty by another, Byrne points out that 'such movements hardly involved wholesale migration but merely the unsettling of a few royal families. They belong to dynastic politics rather than to pure tribalism. The basic population remained undisturbed, merely exchanging one set of overlords for another, and even the noble families of the displaced dynasty usually left some branches clinging to remnants

[1] E.g. *LL* vi 1455.44008 = *CGH* 151; Stokes, '*Cóir Anmann*', 324 § 76.
[2] E.g. *LL* iii 601.18471.
[3] See Meyer, 'The expulsion of the Déssi'; Meyer, '*Tucait indarba na nDéssi*'.
[4] Mac Cana, *Learned tales*, 49, 60.
[5] Byrne, 'Tribes and tribalism', 142–3.

of property in the scenes of their earlier supremacy . . .'.[6] In the annals, records of population movement are few and uninformative. One that is likely to be historically accurate is the flight of the famine-stricken people of Ulster (*Ulaid*) in 1047. According to the *Annals of Tigernach* and the *Chronicum Scotorum*, these refugees migrated to Leinster (*Laigin*),[7] apparently because of their king's alliance with the king of Leinster, Diarmait mac Maíl-na-mBó.[8] There is, however, no evidence that they were given land in Leinster, and most of them may have ultimately returned to their own territory.[9] The same annals record movement of population in the opposite direction – also caused by famine – in the year 1116. Hunger was so severe in the province of Leinster that it was almost emptied, and the people scattered throughout Ireland.[10]

Most of our information on early Irish land-tenure comes from the law-texts of the seventh and eighth centuries. Thus the law-text on status, *Críth Gablach* – dated to about AD 700 – gives details of the amount of land which is typically owned by commoners of the rank of *bóaire, ócaire,* etc.[11] There is, however, no means of assessing approximately what proportion of the population were landowning commoners, what proportion were servile tenants on other people's land (see p. 441), and what proportion were lords (see p. 448). Legal material is also unhelpful with regard to variations in the practice of land-tenure in different parts of the country: the authors of the law-texts present a picture of uniformity which can hardly have existed in reality.

Land-tenure and its limitations

The law-texts refer to different types of land-tenure, and provide us with a good deal of information on factors which limit an individual's control over his or her land. A distinction is made between a person's inherited share of kin-land (*fintiu*), and land which he

[6] Ibid. 143. For suggestions on the movement of certain families in Leinster AD 300–550, see Smyth, *Celtic Leinster,* 145 (plate V).

[7] Stokes, 'Annals of Tigernach (2)', 387; *CS* 276 s.a. 1045. Cf. *AI* 176 s.a. 1005 § 2 for a similar exodus by the Ulaid.

[8] Byrne, 'Tribes and tribalism', 141.

[9] For population movement in post-Norman Ireland, see p. 427 below.

[10] Stokes, 'Annals of Tigernach (3)', 37; *CS* 318 s.a. 1112. Cf. AU^2 558 s.a. 1116 § 5.

[11] See discussion on farm size on p. 421.

has acquired personally. As one would expect, a landowner has greater freedom to dispose of personal land than kin-land, but the kin-group is nonetheless entitled to a proportion of personal land if it is sold or bequeathed (see p. 402 below).

If a landowner is a client (*céile*), his control over his land is further diminished, as he cannot sell it against the wishes of his primary lord (*flaith cétgíallnae*).[12] The same restriction applies if he is a monastic client.[13] Early Irish law also recognizes the need to provide adult dependants with protection against irresponsible decisions on the part of the landowner, and the law-texts provide information on the entitlement of wives or sons to veto or defer contracts which might affect their economic wellbeing. For example, an adult son can legally prevent his father from disposing of land on which his survival depends.[14] Likewise, a principal wife who has brought the same amount of property into a marriage as her husband can annul a contract by him (apart from certain arrangements essential for the wellbeing of the household).[15] There is therefore no question of his selling off land against her wishes. A principal wife who has brought no property into the marriage is entitled to 'disturb' (i.e. defer) her husband's foolish contracts.[16] A wife of lower status who has brought in no property is entitled only to defer foolish contracts relating to food, clothing, cattle and sheep, i.e. matters concerning the day-to-day running of the house and farm.[17] She has evidently no legal say in the sale or purchase of land.

A landowner's control of his land may be subject to certain rights enjoyed by his neighbours. A neighbour may have a right of way across his land if there is no other access. The law-text *Bretha Comaithchesa* describes how the passage (*imirche*) of cattle by right of way should be organized. The cattle-owner must provide three men to ensure that no damage is done by the cattle, and the landowner must likewise provide three men to help with the

[12] *CIH* iii 750.32–3. For primary, secondary and tertiary lordship, see p. 320 (footnote).

[13] *CIH* ii 591.3.

[14] *CIH* i 227.7–10 = *AL* v 436.22–7.

[15] *CIH* ii 506.1 = *SEIL* 19 § 5.

[16] *CIH* ii 512.29–31 = *SEIL* 46 § 22. See Thurneysen's discussion at *SEIL* 46–9.

[17] *CIH* ii 512.22–4 = *SEIL* 45 § 21.

operation.[18] *Cormac's Glossary* deals specifically with a neighbour's access to a cattle-pond (*gelestar*) across another person's land. A cow-track (*bóthar*) may be constructed for this purpose, and a fence (*imbe*) erected around the pond so that the cattle do not graze the land.[19] Similar interference with a neighbour's land is permitted if it is necessary to dig a mill-race for a mill, though appropriate compensation must be paid.[20]

Landownership by a woman is subject to special limitations. Such a woman is generally a 'female heir' (*banchomarbae*) who has inherited a share of land because there were no male heirs. By virtue of her possession of land, she enjoys the usual landowner's entitlements to buy, sell, hire labour, etc. However, she has less freedom to bequeath land to her children than a male landowner: see under p. 415 below.

Royal land and church-land differ from other types of land in that they come under the control of individuals by virtue of office. A king's control over royal land is absolute while he is in office. Church-land, on the other hand, is controlled by an abbot or monastic superior (*airchinnech*) whose decisions must be approved by the monks.[21] For example, one law-text states that an abbot's contract can be annulled by his monks within ten days. The abbot, likewise, can annul a monk's contract within five days.[22] Another law-text describes the sale of land against the wishes of the monks (*tír do reic sech manchu*) as 'a fist around a sun-beam' – i.e. an absurdity.[23]

The kin-group which originally gave a monastic site to the Church retains certain privileges: see under 'church-land' on p. 404 below.

[18] *CIH* i 205.1–16 = *AL* iv 156.17–21; cf. *CIH* iii 1125.18–19 = E. J. Gwynn, 'Privileges', 33.21–2 (with discussion in the notes on p. 225).

[19] Meyer, *Sanas Cormaic*, 54–5 § 674.

[20] *CIH* ii 460.32–5 = Binchy, '*Coibnes Uisci Thairidne*', 68–70 § 10.

[21] See Hughes, *The Church in early Irish society*, 126, for the administration of church property by abbots, and ibid. 124, for its administration by bishops.

[22] *CIH* iv 1348.27–30 = McLeod, *Early Irish contract law*, 124 § 1.

[23] *CIH* iii 750.32–3; iv 1375.31–2 *is dorn im gae ngrēine.*

Kin-land

It is clear from the law-texts of the seventh and eighth centuries that much of the farmland – perhaps most of it – was classified as kin-land (*fintiu*) at this period.[24] Placename evidence confirms the later importance of this type of landownership. Many thousands of townlands bear the name of the family which at one time held the land, e.g. *Baile Uí Dhubhda* (Ballydowd, Co. Mayo), *Baile Mic Aodhagáin* (Ballymacegan, Co. Tipperary).

Kin-land is divided among kin-members, who in general farm as individuals. However, each kin-member is legally obliged to maintain the integrity of his share of this land. One law-text stresses that he may not sell it or otherwise alienate it, or dispose of it secretly or encumber it with liabilities or unfavourable contracts.[25] Another law-text affirms that a kin-member is not free to sell his share of the kin-land without the permission of other members of the kin.[26] There are thus severe limitations on the major decisions which a kin-member can make with regard to the use or disposal of his holding. There are even restrictions in the case of extra land which he has acquired through his own successful farming or the profits of a profession. If such land is sold or bequeathed, a proportion – varying between one third and two thirds – must go to the kin.[27]

The other side of the coin is the legal obligation on the kin to take over responsibility for a kin-member who has got into trouble. Kinsmen are expected to help and support one another,[28] and may be required to pay a fine or debt incurred by a relative.[29] If a kin-member becomes an absentee (*esert*) and neglects to fence his holding, a near kinsman (*fine chomocus*) may be distrained to do the job instead.[30] But if a kin-member exploits or cheats his kin, he runs the risk of being expelled from the group and thereby loses his legal rights in society.[31]

[24] For a general discussion on kin-land, see Charles-Edwards, *Early Irish and Welsh kinship*, 61–73, 415–30.

[25] *CIH* ii 489.16–18 = Thurneysen, 'Aus dem irischen Recht I', 370 § 33. I follow him in interpreting *nadi 'mfuich cintaib nā coraib* as 'who does not damage it (the kin-land) with liabilities or [unfavourable] contracts'.

[26] *CIH* i 247.24 = *AL* v 510.7–9. Cf. *CIH* ii 591.3–4

[27] *CIH* ii 533.17–20 = *AL* iii 48.15–19. For details, see *GEIL* 100.

[28] *CIH* ii 489.8–9 = Thurneysen, 'Aus dem irischen Recht I', 370 § 32.

[29] *CIH* ii 411.22–3 = *AL* i 260.1–3.

[30] *CIH* i 75.24–7 = *AL* iv 128.3–7.

[31] *GEIL* 13.

For the division of kin-land among heirs, see under 'land-acquisition' on pp. 411–15

Royal land

The law-texts normally recognise three grades of king in early Irish society: the petty king (*rí túaithe*) who ruled a single kingdom (*túath*), the overking who ruled three or four kingdoms, and the provincial king who was overlord of a whole province.[32] It can be assumed that the vast majority of those who attained kingship in our period came from royal or noble kin-groups with extensive land-holdings. A typical king would thus own considerable lands as a consequence of his membership of his own kin-group.

In addition, certain lands in each kingdom were set aside specifically for the use of the king during his reign. Such land is described in the law-texts as *mruig ríg* 'king's land'.[33] As an example, a legal glossator refers to *Tír Mugain i n-Élib*, which is 'king's land' for the king of Cashel.[34] The accompanying commentary records a tradition that any cattle found on 'king's land' during the day of his inauguration are forfeit to the king (see p. 138). Legal glossators emphasize the king's duty of holding the regular assembly (*óenach*) on 'king's land'.[35] Non-legal sources refer also to land allotted to the king's heir (*ferann rígdamna*).[36]

Other lands in the kingdom were apparently distributed among members of the king's household. I have been unable to find any references to this practice in the Old Irish law-texts, but it is clear from later sources that certain professionals obtained land in return for their services to a king. For example, in an address to his patron Aodh Mág Uidhir, the sixteenth-century poet Eochaidh Ó hEóghusa asks to be given a farm in an area safe from the raids

[32] *GEIL* 17–18. For more detailed discussion, see L. Breatnach, 'The first third of *Bretha Nemed Toísech*', 36–7.

[33] E.g. *CIH* i 4.4 = *AL* v 128.15; *CIH* i 39.32 = *AL* v 264.14; *CIH* i 54.12 = *AL* v 314.14. Cf. *CIH* ii 691.33 *ferann taīsigechta*. For such land in the later medieval period, see Simms, *From kings to warlords*, 129–30.

[34] *CIH* i 40.2–6 = *AL* v 264.25–266.3.

[35] *CIH* i 4.10–11 = *AL* v 128.26–7; *CIH* i 54.18 = *AL* v 316.20–21.

[36] Stokes, 'The adventure of St. Columba's clerics', 132 § 1 *ferann rīgdamnachta*; *LU* 101.3138 *ferann rīgdomna*; *AI* 308 s.a. 1176 § 7 *fearann tānisteachta*.

of the O'Neills and the O'Donnells.[37] In his 'Lawes of Irelande',[38] Sir John Davies writes *c.*1610 that 'the chief lord [equivalent to Old Irish *rí*] had certen landes in demesne which were called his *loughty* or mensall landes wherein hee placed his principall officers, namely his brehons, his marshall, his cupbearer, his phisicion, his surgeon, his chronicler, his rimer, and others, which offices and possessions were hereditary and peculiar to certen septs and families'.[39]

Church-land

The Church owned or controlled a great deal of land and it is likely that a number of agricultural innovations spread from Church farms. It has been shown in Chapter 7 (p. 222) that many terms relating to crop-cultivation are borrowings from Latin, and so are likely to have originally had Church associations. The impact of the Church on the early Irish landscape can be demonstrated by the number of place-names which record the existence of church-buildings.[40] In many cases such buildings formed the nucleus of a Church land-holding, described in legal and annalistic sources as a *termonn.*

The details of how the Church came to acquire so much land are not recorded. However, the law-texts and the semi-legendary Lives of the saints provide us with some idea of the processes which gave rise to the Church's extensive holdings. The seventh-century Lives of Patrick in the Book of Armagh record various traditions of gifts of land made to the Church in the fifth century. For example, a king is said to have granted in perpetuity to the monks of bishop Assicus enough grazing for a hundred cows with calves as well as for twenty oxen.[41] This would clearly have been a substantial acreage.

Another entry records the gift by the Uí Ḟíachrach of part of the plain between the rivers Gleór and Ferne to Patrick for ever,

[37]Bergin, *Irish bardic poetry*, 136–8, poem 33. See also P. A. Breatnach, 'The chief's poet', 59–66.

[38]Morgan, 'Lawes of Irelande', 311. *Loughty* is Irish *lucht tighe* 'household': see Simms, *From kings to warlords*, 176.

[39]See also McErlean, 'The Irish townland system of landscape organisation', 329.

[40]The elements *mainister, cell, eclas, tempall, domnach, daurthach,* and *airecal* are common in place-names: see Hogan, *Onomasticon.*

[41]Bieler, *Patrician texts*, 140.23–5.

together with the serfs working on that land.[42] It is also recorded that Binén, a priest and anchorite, was given an inheritance from his mother's kin. On this land he founded a church, in which Patrick himself is reputed to have celebrated the first Mass.[43]

The giving of land to the Church was obviously felt to be a meritorious act which benefited the soul of the donor. The law-text *Di Astud Chirt 7 Dligid* refers to 'land which is given to the Church for the sake of one's soul'.[44] But another law-text, *Córus Bésgnai*, is careful to stress that a person cannot bequeath land to the Church if he should thereby cause undue economic loss to his relatives. Hence, he is not entitled to bequeath land without the permission of his kin-group, unless it is land which he has himself purchased.[45]

If an old or sickly landowner cannot find a relative or other person to look after him, he may make an arrangement to bequeath part of his land to the Church in return for maintenance.[46] It is clear from the law-text *Bretha Nemed Toísech* that this maintenance is normally carried out by a monastic client (*manach*). The Church does not take possession of its share of the land until the owner dies.[47]

Even when kin-land has legally been transferred to the Church, the kin-group retains certain rights in relation to Church appointments. For example, commentary on *Córus Bésgnai* deals with appointments to the post of abbot.[48] Priority is given to the most suitable member of the kin-group of the patron saint of the church (*fine érluma*). Failing this, the post goes to the most suitable member of the kin-group which originally owned the land (*fine grín*).[49] In cases where the kin-group of the patron saint has

[42]Bieler, *Patrician texts*, 170.26–8. It is not certain that Ferne is a river, but this seems to suit the context best.

[43]Bieler, *Patrician texts*, 172.4–10.

[44]*CIH* i 224.19 = *AL* v 428.4. For a discussion on donations to the Church for the sake of the soul in the writings of Hieronymus and Augustine, and their influence on early Irish law, see Bruck, *Kirchenväter und soziales Erbrecht*, 76–104, 163–95.

[45]*CIH* ii 534.20–21 = *AL* iii 52.8–10; cf. *CIH* ii 532.28–30 = *AL* iii 44.14–17. See discussion by Mac Niocaill, 'Christian influences on early Irish law', 153–4.

[46]*CIH* ii 535.1–2 = *AL* iii 52.29–31.

[47]*CIH* vi 2213.3–7 = L. Breatnach, 'The first third of *Bretha Nemed Toísech*', 16 § 21.

[48]*CIH* v 1820.8–30; cf. *CIH* vi 1929.6–14. See discussion by Etchingham, 'The implications of *paruchia*', 153–60.

[49]Cf. *CGH* 24 *Cenēl Nath-ī i mMaigib Ailbe ō Chill Ēogain, is leo grian na cille* 'the race of Nath-í in the plains of Ailbe from the church of Éogan, theirs is the land of the church'.

also supplied the land (*fine érluma 7 grín imalle*) there is obviously no conflict of interests.

Common land

Much of Ireland is mountainous or boggy, and consequently of limited use for farming. Such land is not worth cultivating, but may provide rough grazing, wood, turf, herbs, berries, etc. In the present day, most of it is classed as commonage; however, this does not mean that everyone is entitled to graze livestock, hunt, shoot, fish, mine, extract gravel, or cut turf on it. There are normally restrictions on all these activities. As it is particularly difficult to legislate for the use of common land, disputes can easily arise.

It is obvious that the authors of the Old Irish law-texts found similar difficulties in defining people's rights on common land, and there are different emphases in different texts. The general assumption seems to be that ownership of mountain or other rough land is vested in the *túath*. Consequently, each freeman of the *túath* is entitled to a share in its use. One law-text includes 'the undivided land above all' (*dírann úas cách*) among the equal entitlements of every law-abiding freeman (*recht*).[50] The accompanying gloss defines it as 'the common mountain' (*in slíab coitchenn*).

Other texts similarly imply that there are common rights on such lands. Old Irish commentary to *Bretha Comaithchesa* points out that certain offences (*caithchi*) against land do not apply in a great wood (*ruud*), mountain (*roilbe*) or commonage (*foach túaithe*) because such land is undivided (*dírann*).[51] Similarly, the law-text *Bretha Étgid* gives it as a general rule that cattle are free to graze on mountain (*dind*).[52] Another law-text, *Bretha Nemed Déidenach*, declares the principle that every landowner is entitled to a share in undivided plain, mountain or water and must have access thereto. I take this to mean that the landowner is entitled to a share in the use of such undivided land.[53]

[50] *CIH* i 241.28 = *AL* v 484.4–5.

[51] *CIH* i 198.4–6 (*a n*[*d*]*īrind*); ii 579.6–8 (*a ndīrainn*) = *AL* iv 124.26–7.

[52] *CIH* i 291.1–5 = *AL* iii 296.11–18. The glossator allows for the possibility that mountain may sometimes be privately owned.

[53] *CIH* iii 1125.16–17 = E. J. Gwynn, 'Privileges', 33.19–21. Gwynn, on the other hand, suggests in his note on this passage (*Ériu* 13, 225) that a landowner with an inadequate holding may actually appropriate undivided land. But there seems to be no evidence of such a principle in other texts.

In a tenth-century source, *Immacallam in dá thúarad*, we also find evidence of the existence of common land. The sage Ferchertne prophesies various disasters which will befall the country: these include overpopulation and the extension of privately owned land into mountain (*roilbe*).[54] The implication is clearly that it is an offence against the community for individuals to encroach upon such land. In another prophecy-text, the prohibition of access to turf and wood (*crosad móna is feda*) is included among the signs of bad times.[55]

But the law-texts also recognise private rights on common land. For example, the text on distraint refers to 'an appropriated tree which is in the wilderness'.[56] This is clearly a tree on unowned land which has through some agreed procedure been designated as private property. In the gloss it is described as a marked tree (*crann crosta*), so it seems that a member of the community was entitled to lay claim to a tree for felling or fruit-picking by marking it with a cross (*cros*). No doubt there were limitations to this entitlement, particularly in less wooded parts of the country, but no details have survived in the texts.

The complexity of the law with regard to common land is also illustrated by a passage in *Bechbretha* which deals with the finding of a stray swarm of bees in great wood (*ruud*), undivided land (*dírann*) or inaccessible country (*écmacht*).[57] According to this text, the finder becomes the sole owner of the bees. However, even though the author evidently regards such land as ownerless, parties other than the finder also have a stake in the honey. Thus both the finder's church and the head of his kin-group are entitled to one ninth of the honey. Old Irish commentary adds the further stipulation that the finder's lord (*flaith*) is likewise entitled to a share.[58]

The extent of an Irish king's rights on common land is difficult to judge. The law-texts of the seventh and eighth centuries do not assign any specific privileges on common land to either local kings

[54] Stokes, 'Colloquy', 44 § 227 *brogfaiter crīcha i rroilbi* 'boundaries will be extended into the great mountain'.

[55] *ACL* iii 240 § 6. A legal gloss at *CIH* ii 370.5 = *AL* i 132.2 refers to 'common bog' (*mōin coitcend*).

[56] *CIH* ii 395.23 = *AL* i 202.2–3 *im crand ngabāla bīs i ndīthrib.*

[57] *CIH* ii 455.1–4 = *BB* 84 § 49.

[58] *CIH* iii 924.23 = *BB* 84 § 49[i] (B). See discussion in Notes.

or overkings. In later Irish legal material, however, we find evidence of some degree of royal control over such areas. In legal commentary – probably at least as early as the eleventh century – the three 'lands' *(bruigi)* of a king are given as a road (*rót*), undivided land (*dírann*), and the sea (*muir*). By virtue of his office, the king is entitled to a share of any valuables found in these places.[59]

WATER

Legal disputes may arise in connection with the ownership of water, whether in a well, a stream, a lake, an estuary or the sea. The problem may relate to the actual water itself – for drinking by livestock or people – or it may be about the ownership of fish caught in the water. As we have seen above (p. 287), there is evidence in early Irish legal material of public fishing places, as well as private rights over streams and estuaries. There are also references to the right of a farmer to construct a path across a neighbour's land so that his cattle can reach water (p. 401). Legal commentary deals further with the situation of a farmer who owns land on either side of a commonly owned stretch of water (*usce coitchenn*).[60] He is entitled to access across the water – presumably by bridge, ford or causeway.

As we have seen in the previous section, the sea is classified in legal commentary as one of the three 'lands' of a king. This gives him certain privileges in relation to shipwrecks off his territory. For example, commentary on *Muirbretha* 'sea-judgements' states that a local king (*rí túaithe*) is entitled to one third of the cargo of a ship bound for another territory which is driven by the wind onto his shore. Fractions of this third go to the provincial king (*rí cúicid*) and to the king of Ireland (*rí Éirenn*).[61] It is implicit in another legal commentary that a king receives a substantial revenue from the ships and boats which use the harbours (*caillte*) of his territory. Of this income, seven *cumal*s must be paid to his church.[62] There seems also an assumption that the king is entitled to a share of salmon, seals and porpoises caught in the sea.[63]

[59] *CIH* ii 455.27 = *BB* 171 (g); *CIH* iii 959.22–4 = *BB* 184 (b).

[60] *CIH* i 98.41.

[61] *CIH* i 315.5–11 = *AL* iii 424.16–25. Cf. O'Keeffe, '*Dál Caladbuig*', 20 § 9 *dlegait Mūscraige trian cach thurchairthe dia cuirend tond hi tīr nĒirenn acht rand ríg Temrach* 'the Múscraige are entitled to a third of everything which is cast up by a wave on the land of Ireland, except for the share of the king of Tara'.

[62] *CIH* i 177.34–9.

[63] *CIH* i 178.11–12.

BOUNDARY-MARKS

So far as we know, maps or diagrams showing the extent of land-holdings did not exist in pre-Norman Ireland. Boundaries had therefore to be publicized in the community and related to easily identifiable landmarks. An Old Irish commentary to *Bretha Comaithchesa* lists twelve types of landmark (*blá*) which may serve to mark a boundary.[64] They include a rock (*ailblá*), a ditch (*cladblá*), a tree (*fidblá*), water (*linnblá*), and a road or lane (*blá réime*). Trees are particularly useful for this purpose:[65] the commentary refers to a venerated tree (*bile feda*) or a great oak (*rail*) or a mound with a tree on it (*dumae crainn*) or even the stump of a large tree (*bun n-omnai*). Rivers and streams are also obviously convenient as permanent indicators of a boundary, and it is noteworthy that in an eighth-century record of a land-transaction (see next section), the boundary of the holding is mainly delineated by streams.[66]

A boundary may be marked by a naturally occurring rock or boulder, but this commentary also refers to stones placed by human agency, such as a gravestone (*ail lechta*).[67] In other legal sources there are many references to standing stones, sometimes incised with ogham, which serve both to mark the holding's boundary and the grave-mound (*fert*) of the landowner.[68] For example, an early legal poem refers to standing stones (*gaill*) which fasten title to land,[69] and a legal commentary states that the ogham in a standing stone (*gallán*) gives the force of possession to the rightful heir.[70]

64 *CIH* ii 581.11–42 = *AL* iv 142.8–144.17. The etymology of the term *blá* (also *blaí*) is uncertain: see *LEIA* B-55 s.v. 4 *blá, blaí.*

65 As in medieval England: see Rackham, *The history of the countryside*, 230.

66 Bieler, *Patrician texts*, 172.11–22; cf. *CIH* ii 581.36 = *AL* iv 144.10.

67 *CIH* ii 581.17 = *AL* iv 142.17.

68 For discussion see Charles-Edwards, *Early Irish and Welsh kinship*, 261–2; Charles-Edwards, 'Boundaries in Irish law'; McManus, *A guide to Ogam*, 163–5.

69 *CIH* ii 570.37 = Binchy, 'An archaic legal poem', 157.17–18.

70 *CIH* vi 2143.39–40. Cf. *CIH* v 1566.6–7 *int oghom isin gollán . . . amal fiadain hé* 'the ogham in the standing stone . . . is like a witness [of ownership]'.

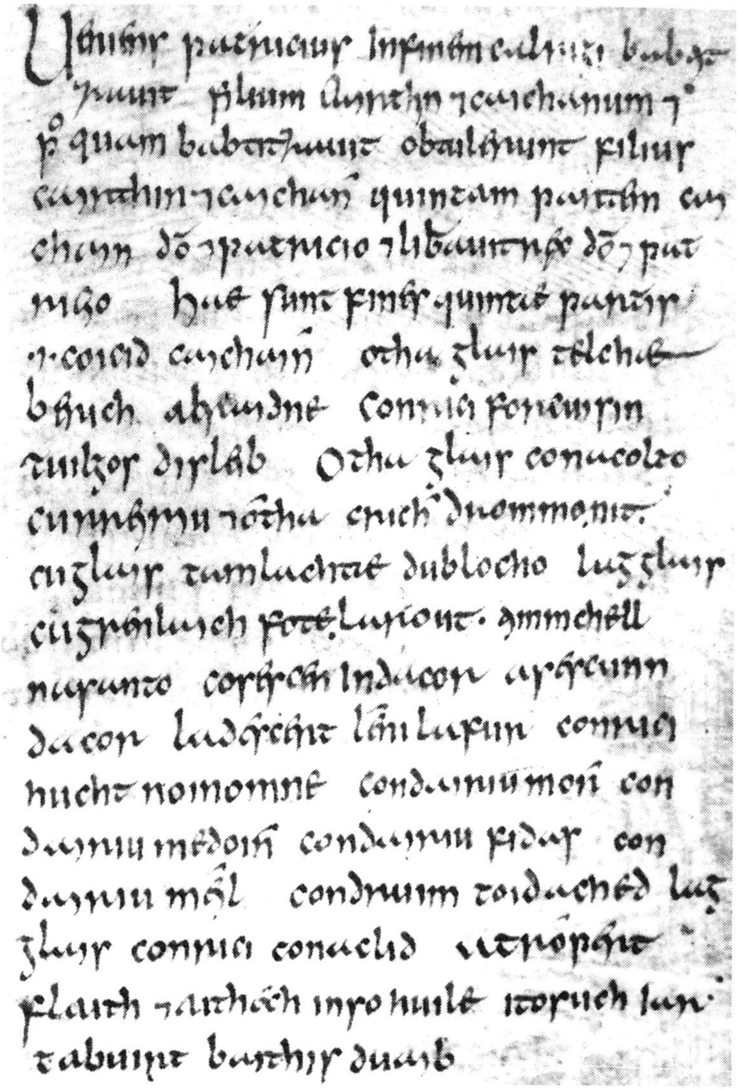

Plate 6. This shows part of f. 17b of the ninth-century Book of Armagh in the Library of Trinity College, Dublin, and has been reproduced by kind permission of the Board of the College. It is the earliest record of a land-transfer which survives in our sources.

LAND-ACQUISITION

When ownership of land changes, it is obviously of particular importance that this fact should be recorded. Our sources refer to the rôle of the *senchae* 'historian, custodian of local tradition' in memorizing information on landownership.[71] Written evidence is also mentioned in this connection, particularly with regard to church-land. One text includes a written document (*scríbend*) among the three things which secure title to church-land (*tír n-ecalsa*),[72] and another text lists written evidence among the ten 'immovable rocks' which confirm ownership of land.[73]

Charters recording the details of land-transactions have survived from early medieval Wales and Brittany, and have proved a rich source of information on the social history of these countries.[74] By contrast, charters do not occur among the surviving documents of the Old Irish period. The nearest which we possess are records of land-transfers to the Church in the eighth-century *Additamenta* in the Book of Armagh.[75] I reproduce on page 410 the most detailed of these. The text first describes how Patrick came to the territory of the Calrige (Calry, Co. Sligo) and baptized Macc Cáirthin and Caíchán. It then states that they gave to Patrick and to God the fifth part of Caíchán's estate, with the permission of the local king.[76] The boundaries of this fifth are then carefully recorded, and include such landmarks as the Great Oakwood (*Daire Mór*), the Slope of the Nine Trees (*Ucht Noí nOmne*) and the Stream of Conaclid (*Glais Conacolto*).[77] As the *Additamenta* were written about three centuries after Patrick's time, this passage is obviously not to be taken as a genuine record of a fifth-century land-transfer. On the other hand, the amount of detail suggests that the original

[71] *CIH* iv 1376.2 = *AL* v 368.2–3; *CIH* iii 747.3–4; 748.13, 37; 749.8. See also *DIL* s.vv. *senchae, senchaid.*

[72] *CIH* iii 917.8.

[73] *CIH* iii 747.5; cf. *CIH* vi 2199.9 = Thurneysen, 'Aus dem irischen Recht III', 361 § 46.

[74] See W. Davies, *An Early Welsh microcosm: studies in the Llandaff charters*; W. Davies, *The Llandaff charters*; W. Davies, 'The Latin charter tradition in western Britain, Brittany and Ireland in the early medieval period'.

[75] Bieler, *Patrician texts*, 172.11–26.

[76] Note the implication that the king could have blocked the transaction. See discussion by Doherty, 'Some aspects of hagiography', 309–13.

[77] None of the place-names in this passage has been identified. This list of boundary landmarks can be compared with the 'perambulations' of Anglo-Saxon charters (Rackham, *The history of the countryside*, 9).

author was describing a place of which he had firsthand knowledge. The connection with Patrick may thus be fictitious, but the actual boundaries of this ecclesiastical holding are likely to be genuine. It is possible that this and other records in the Book of Armagh were copied from earlier documents with the intention of emphasizing the Patrician connection of various monasteries, thus placing them within the orbit of Armagh.[78]

Land-transactions of the eleventh and twelfth centuries were copied into blank pages of the eighth-century Book of Kells.[79] For example, one short text – Kells Charter III – records the purchase of land in about the year 1092 by the priest of Cenannas (Kells) and his kinsmen. The boundaries are given in detail and a list of sureties appended. The seventh-century Book of Durrow also contains a late eleventh-century record of a grant of land given by the monastic community of Glenn Uissen (= Killeshin) to that of Dairmaig (Durrow).[80] This practice of recording land-transactions in manuscripts is mentioned in legal commentary on proofs of land-ownership, written in about the twelfth century.[81] The commentator distinguishes the *senlebar* 'old manuscript', *maclebar* 'copy', and *fíarduille* 'curved leaf', i.e. scroll.

Male inheritance

The assumption in the law-texts and other sources is that the typical landowner is in possession by inheritance. There are however many gaps in our knowledge of how land-inheritance worked in early Irish society. This is partly because two crucial law-texts on the subject, *Córus Fine* 'the regulation of the kin-group'[82] and *Maccṡlechtae* 'son-sections',[83] survive only in fragments.

The normal procedure seems to have been for an inheritance of land (*orbae*) to be divided among the owner's sons on his death. However, genealogical tradition provides evidence of the possible division of land by a father during his own lifetime. For

[78] Doherty, 'Some aspects of hagiography', 304.

[79] Mac Niocaill, *Notitiae as Leabhar Cheanannais*; Mac Niocaill, 'The Irish "charters"'; Herbert, 'Charter material from Kells'.

[80] Best, 'An early monastic grant in the Book of Durrow'.

[81] *CIH* vi 2143.32–7. For post-Norman land-charters in Irish, see Hardiman, *Ancient Irish deeds and writings*.

[82] *CIH* ii 728.17–746.16.

[83] *CIH* i 107.9–110.13; iv 1296.17–1301.16; v 1546.26–1550.14.

example, Catháer of the Laigin is said to have given land to each of his sons, except for Fíachu who was too young. Catháer asked his other sons that they should each keep Fíachu for a month until he was old enough to receive an inheritance. The division was clearly made during Catháer's lifetime, as the story goes on to tell how Fíachu wept in his father's presence on account of the arrangement.[84] Land may also be divided up in the owner's lifetime if he is an imbecile, senile or physically incapable.[85]

Not all sons are eligible for a share in their father's land. *Maccṡlechtae* excludes sons who fall into the category of 'sons of darkness'.[86] A son may be excluded because of doubt as to his paternity, such as a son conceived in the bushes (*macc muini*) or the son of a prostitute. A proclaimed or outlawed son (*macc fócrai*) is also excluded, as is an impious son (*macc ingor*) who fails to look after an aged father.[87] One who has slain a kinsman likewise forfeits his share of an inheritance.[88]

The division of an inheritance among the heirs (*comarbai*) is made by the youngest. The eldest gets the first choice, the second eldest the second choice, and so on.[89] The youngest gets the last choice: it is therefore in his interests to divide the inheritance as fairly as possible. According to Old Irish commentary on *Bretha Comaithchesa*, the eldest son gets the farm-buildings (*tige*), the yard (*lis*), and the area around the yard (*airlise*) in addition to his share of land.[90] If a man predeceases his father before the division of the inheritance, his sons get the share which would have been his.[91] Where there are no eligible sons, the inheritance must go to some other relative or relatives. If there are daughters, it may go to one or more of them: see p. 415 below.

[84] *CGH* 70–1 = *LL* vi 1353.40440–1354.40453. Cf. Dillon, *Lebor na Cert*, 164.

[85] *CIH* iv 1276.19–23 = Smith, 'Advice to Doidin', 68. Cf. Meyer, *Triads*, 26 § 205.

[86] *CIH* iv 1296.36–9; cf. *CIH* i 232.6–7 = *AL* v 452.15–16. For further discussion on excluded sons, see *GEIL* 102–3.

[87] *CIH* ii 534.26 = *AL* iii 52.17.

[88] *CIH* iv 1301.17–20.

[89] *CIH* iv 1289.11 *rannaid ósar ⁊ dogoa sinser* 'the youngest divides and the eldest chooses'. However, in an eighth-century poem (O Daly, 'A poem on the Airgialla', 180 § 6), it is the eldest who makes the division 'in accordance with Irish law' (*íar Fénechas*). Thomas Charles-Edwards suggests that the eldest might be responsible for the division when he is also a political leader. See p. 430 below for references to the division of land by the *ceann fine* in post-Norman Ireland.

[90] *CIH* ii 575.14.

[91] *CIH* ii 739.17–18; v 1547.13–17.

If a whole family dies out, the legal situation is complex: how is the property to be distributed among the wider circle of relatives who might have a claim on it? No detailed account of the procedures has survived, and it is sometimes impossible to reconcile the evidence in various texts – there may have been variation in practice. Nonetheless, the underlying principle is clear: a kinsman's entitlement to a share in the land is based on the nearness of his relationship with the extinct family. For example, according to the late recension of *Cóic Conara Fugill*, if there is extinction of a *gelḟine* (the descendants on the male line of a grandfather), three quarters of the inheritance (*díbad*) are distributed among the next circle of kinship, i.e. the *derbḟine* (the descendants on the male line of a great-grandfather). The remaining quarter is divided in the proportion of three to one among the *íarḟine* (the descendants on the male line of a great-great-grandfather) and the *indḟine* (the descendants on the male line of a great-great-great-grandfather).[92] It is obvious that in such cases, minute portions of land might be distributed among very large numbers of relatives.

The same text describes a five-year procedure for dividing inherited land.[93] This seems to be designed for cases where the heirs have inherited equal shares of a reasonably large holding, and do not have a detailed knowledge of the quality of the land.[94] In the first year, the share of each heir is temporarily marked out by a plough: this is described as the 'ox-division' (*damrann*). In the second year, the shares are exchanged. In the third year, the shares are measured. In the fourth year, house-posts are established in each share.[95] Finally, in the fifth year, lots are cast to decide the portion of land which goes to each heir.[96]

A notable feature of the early Irish system of land-inheritance is a kinsman's entitlement to a re-sharing of the kin-land in certain

[92] *CIH* iii 1034.37–1035.5 = Thurneysen, *Cóic Conara Fugill*, 44 § 70. See also discussion by Baumgarten, 'Kindred metaphors'.

[93] *CIH* iii 1034.3–11 = Thurneysen, *Cóic Conara Fugill*, 42 §§ 64–5. Cf. *CIH* ii 740.12–18.

[94] *CIH* ii 740.12–13 *Ferann dībaid . . . nochan aithnidh doib é* 'inherited land . . . which is not known to them (the heirs)'.

[95] This hardly means that each heir is required to erect a permanent dwelling-house. The intention seems rather to be symbolic: the erection of any sort of structure demonstrates a claim to ownership.

[96] Old Irish commentary on *Bretha Comaithchesa* describes the same procedure in relation to the straightforward division of an inheritance among sons, *CIH* ii 575.4–12.

circumstances. The details are not altogether clear from the texts, but it seems that if one branch of a kin is particularly prolific, its members may demand a re-sharing at the expense of a less prolific branch with excess land. The intention is to secure if possible a viable land-holding for all kinsmen.[97]

Our sources also refer to an extra share of land, called the *cumal ṡenorbai* 'the *cumal* of the old inheritance'.[98] According to an Old Irish legal gloss, this is land valued at one *cumal* which goes to the head of a kin-group. In return, he takes on liability for the crimes or debts of the kin-group, and provides hospitality for a visiting king or bishop.[99] A later gloss gives his share as one seventh of the inheritable land (*tír ndíbaid*).[100]

Female inheritance

Some law-texts are very dogmatic in relation to a woman's lack of legal rights, and stress the extent of her contractual incapacity.[101] Other texts give prominence to circumstances in which a woman's rights are similar to those of a man.[102] Because of these different emphases, it is particularly difficult to generalize about female inheritance in our period.

Normally, a daughter can only inherit movable property.[103] However, if she has no brothers, she may take over part or all of the land as a female heir (*banchomarbae*),[104] but she cannot inherit more land than the worth of fourteen *cumals*.[105] Some legal restrictions on a woman are of necessity relaxed in the case of a female heir: she obviously could not manage her farm without the right to make purchases, sales and other essential contracts. If she marries a landless man, he does not acquire control over her property. In such

[97] The evidence is discussed by Charles-Edwards, *Early Irish and Welsh kinship*, 64–70.
[98] *CIH* i 215.34 = *SEIL* 141.
[99] *CIH* iii 912.2–3 = *SEIL* 141 (v) H[1]. For more detailed discussion of the *cumal ṡenorbai*, see Plummer, 'Notes on some passages in the Brehon laws IV', 113–14; Thurneysen, *Irisches Recht*, 80; Dillon, *SEIL* 141–2; Charles-Edwards, *Early Irish and Welsh kinship*, 513.
[100] *CIH* i 216.9–10 = *SEIL* 141 (v)[a].
[101] E.g. *CIH* ii 443.29–444.6 = Thurneysen, *Irisches Recht*, 35 § 38.
[102] E.g. *CIH* vi 2295.2–4, 2295.39–2296.2 = Binchy, '*Bretha Crólige*', 26 § 33; 28 § 35.
[103] *CIH* ii 736.28–9; iv 1154.30–31; vi 2039.39–40 = *SEIL* 133 § 6; 168 § 23.
[104] For a fuller discussion of the rights of the female heir, see *SEIL* 129–79 and Charles-Edwards, *Early Irish and Welsh kinship*, 82–7. If there is a number of daughters and no sons, it is likely that the land is divided between them.
[105] *CIH* i 217.21 = *SEIL* 155 (xv). A *bóaire* is expected to own land worth fourteen *cumals*: see p. 421 below.

a case, according to the law-text on marriage, *Cáin Lánamna*, 'the man goes into the position of the woman and the woman goes into the position of the man'.[106]

There are special restrictions on a female heir's right to bequeath land to her husband or sons. Normally, it reverts on her death to her own kin-group. But if her husband is an alien – such as a Briton, who would have no land in the *túath* – she is entitled to pass on an inheritance to her son.[107] According to an Old Irish glossator, this consists of land worth seven *cumal*s, i.e. the property-qualification of an *ócaire*, the lowest grade of adult freeman.[108] There is also evidence that if a female heir has no son, she may bequeath land to her daughter or daughters.[109] There was obviously strong pressure on a female heir to marry a cousin on her father's side: this would ensure that all the land stayed within the kin-group.[110]

It is likely that the opportunities for an early Irish woman to acquire personal land were limited. In legal theory, however, there seems to be no reason why an exceptional woman should not acquire personal land from professional earnings or from the profits of successfully farming her share of kin-land.[111] In such cases, she can bequeath her personal land in the same way as a male landowner. One text refers to the division of land – apparently personal land – between a woman's son(s) and her kin-group after her death.[112] Legal commentary also refers to possession by a woman of 'inheritance of hand and thigh' (*orbae cruib 7 slíasta*): this is apparently land which a parent has acquired through his or her own exertions and which may be given to a

[106] *CIH* ii 515.23–4 = *SEIL* 57 § 29.
[107] *CIH* ii 431.30–31 = *AL* iv 284.19–21. See also the section on 'adoption' below.
[108] *CIH* iii 917.30–31.
[109] *CIH* i 215.15–32 = *SEIL* 135–6 § 9 (and glosses, cf. *CIH* iii 911.21–3). Cf., also, *CIH* ii 378.18 = *AL* i 146.y *Athgabāil aile do ingin im comorbus a māthar* 'distraint of two days for a daughter about the inheritance of her mother'.
[110] *CIH* i 216.35–217.10 = *SEIL* 150 (xii) = Charles-Edwards, *Early Irish and Welsh kinship*, 518.25–38. Cf. *CIH* iii 912.26–8. See Ó Corráin, 'Marriage in early Ireland', 11–12; Ó Corráin, 'Women and law in early Ireland', 52–6.
[111] For a brief discussion of the legal position of exceptionally gifted or qualified women – such as a female physician, a female wright, or a holy woman – see *GEIL* 77.
[112] *CIH* i 217.11–20 = *SEIL* 151–4 = Charles-Edwards, *Early Irish and Welsh kinship*, 518.29–34 (and discussion on p. 83). The Old Irish gloss (*CIH* iii 913.3–4) has *is iarna ēcaib randtair iter macu 7 a brāithri* 'after her death, it is divided between sons and her kinsmen'.

son or daughter.[113] A passage in the eighth-century *Additamenta* in the Book of Armagh describes a complicated land-transaction whereby a nun used the valuables which she would have brought into her marriage to buy half an estate. This half is described as being fully her property (*ógdíles*). The passage thus recognizes a woman's entitlement to acquire personal land as a paternal gift.[114]

The law-texts preserve some traditions which suggest that it was possible for a woman to be a large landowner. For instance, the legendary Dúanach of the Ulaid is described as a *banchoairt* 'female landowner', and is said to have possessed seven hundred cows.[115] Another legendary female landowner mentioned in legal material is Medb of the Ulaid. She is represented as being a female hospitaller (*banbriugu*), i.e. a person who acquires special status by providing hospitality to allcomers.[116] Legal problems connected with the division of her property among her three sons after her death are discussed in two fragments of commentary.[117] There is no mention of any of her property reverting to her father's kin-group.

Adoption

The entitlement of an aging or infirm landowner to adopt an heir is hedged in with restrictions. The adoption (*fóesam*) must be a legal contract bound by sureties from the adopter's kin, and authorized by the head of the kin.[118] The adopted son does not gain automatic rights to a full share of the kin-land, but only to what is stipulated in the contract.[119] Indeed, an Old Irish legal commentary stresses that he is not entitled to any land at all unless the kin-group has actually failed in its duty of caring for the kin-member

[113] *SEIL* 152.
[114] Bieler, *Patrician texts*, 174.3–14. This passage seems to be a contemporary account of an actual transaction: see discussion on p. 420.
[115] *CIH* vi 2117.36 = Dillon, 'Stories from the law-tracts', 51 § xiv.
[116] Cf. Meyer, *Sanas Cormaic*, 87 § 1018 *batir banbrugaid són* 'they were female hospitallers'. For a brief discussion of the rôle of the *briugu* 'hospitaller', see p. 422 below.
[117] *CIH* i 106.39–107.3; v 1546.11–24.
[118] *CIH* ii 459.12–14 = Binchy, '*Coibnes Uisci Thairidne*', 66 § 6.
[119] *CIH* ii 431.14 = *AL* iv 284.17–18. Cf. Wasserschleben, *Die irische Kanonensammlung*, 116–17 bk. 32, chs. 21-2.

who is in need of maintenance (*goire*). In a case of unjustified adoption, an adopted son is only entitled to payment for his work (*lóg frichnama*).[120]

It seems to have been a recognized procedure for a sister's son to be adopted so as to undertake the duty of maintenance. In return, he is entitled to the 'inheritance of a sister's son' (*orbae niad*).[121] As a result of this arrangement, the sister's son is often referred to as the *gormac* (Modern Irish *garmhac*), which literally means the 'dutiful son'. According to an Old Irish gloss, an adopted sister's son can only inherit land worth seven *cumal*s, i.e. the minimum property-qualification of an *ócaire*.[122]

Marriage

Property – including land – may change hands as part of a marriage agreement.[123] The bulk of the bride-price (*coibche*) seems to have gone to the bride's father;[124] however, there is also evidence that a proportion is kept by the bride herself.[125] If her father is not alive at the time of her marriage, the head of her kin gets half the bride-price if it is her first marriage, a third if it is her second marriage and a quarter if it is her third marriage.[126] If a marriage breaks up through the fault of the wife, the bride-price must be given back. But if the husband is to blame, it is retained.[127]

Our sources contain many references to the acquisition of a bride in return for livestock and precious metal. In the eighth-century tale *Táin Bó Fraích*, King Ailill demands sixty horses with golden bridle-bits and twelve cows with white red-eared calves as bride-price (*tindscrae*) for his daughter Findabair.[128] There is also evidence in both legal and literary texts that land might be given for this purpose. For example, a law-text refers to land given as a bride-price (*coibche mná*),[129] and a genealogical text records a prophecy

[120] *CIH* v 1546.5–9.
[121] *CIH* ii 431.30–31 = *AL* iv 284.y *orba niad .i. fearand gormheic.*
[122] *CIH* iii 917.30–31.
[123] For a general discussion, see Ó Corráin, 'Marriage in early Ireland'; Ó Corráin, 'Women and law in early Ireland', 46–50.
[124] *CIH* i 294.40 = *AL* iii 314.5 *cach aithair a cētcoibche* 'to every father his [daughter's] first bride-price.
[125] *CIH* i 222.7–8 = *AL* iv 60.10–12.
[126] *CIH* i 222.28–223.2 = *AL* iv 62.9–11.
[127] *GEIL* 73–5.
[128] Meid, *Táin Bó Fraích*, 7.162–5.
[129] *CIH* i 247.21. = *AL* v 510.5–6.

concerning a bride-price (*tindscrae*) of land which would be given to the Déssi for Eithne Úathach, daughter of Crimthann.[130] A saint's Life describes how King Domnall of Tara gave land worth eighty cows as bride-price (*tochrae*) for the daughter of the king of the Uí Ḟailgi. She evidently acquired control over this land, as she was able to pass it on to Saint Colmán mac Lúacháin and his successors for ever.[131]

The main law-text on marriage, *Cáin Lánamna*, distinguishes three types of property arrangement in a marriage. In a 'marriage of a woman on man-contribution' (*lánamnus mná for ferthinchur*), the woman has brought in little or nothing.[132] In a 'marriage of a man on woman-contribution' (*lánamnus fir for bantinchur*),[133] a landless man has married a propertied woman: see p. 416 above. In the case of a 'marriage of joint contribution' (*lánamnus comthinchuir*), both parties have contributed land, livestock and equipment.[134] Presumably, in this type of marriage, the wife's father has received a bride-price (some of it going to the bride), and may also have given property to the couple from his own personal resources. It seems unlikely that his kin-group would in normal circumstances permit him to transfer part of his share of kin-land to his daughter and son-in-law.

Purchase

In the law-text *Críth Gablach* it is assumed that if a successful farmer of *bóaire* rank accumulates more livestock than his holding can support, he will try to buy further land by selling the excess animals.[135] Another law-text, *Córus Bésgnai*, refers to the purchase of land through the profits of a profession (*dán*).[136] Legal commentary lists the professions which are likely to bring in sufficient profit for land-purchase as poetry (*filidecht*), learning (*léigenn*), craftsmanship in wood or stone (*saírsecht*), smithcraft (*goibnecht*), metalwork (*cerdacht*), and law (*brethemnas*).[137] A

[130] *CGH* 345; Stokes, *Cóir Anmann*, 362 §§ 169–70.
[131] Meyer, *Betha Colmáin*, 90–2 § 89.
[132] *CIH* ii 512.22 = *SEIL* 45 § 21.
[133] *CIH* ii 515.23 = *SEIL* 57 § 29.
[134] *CIH* ii 505.35 = *SEIL* 18 § 5. Note that a legal passage on the property-qualifications of various ranks differentiates the cattle (*crodh*) owned by an *ócaire* from those of his wife (*CIH* iii 1058.7–8).
[135] *CIH* ii 565.2–3; iii 781.10–11 = *CG* 10.250–1.
[136] *CIH* ii 533.19–20 = *AL* iii 48.18–19.
[137] *CIH* ii 533.33 = *AL* iii 50.6–7; *CIH* ii 738.6.

Middle Irish poem describes how the metalworker Goscen was given as a reward for his labours a holding of land well supplied with wood, water and bog.[138]

The purchaser of a holding of land may also acquire the serfs who work there. Thus the eighth-century *Additamenta* in the Book of Armagh refers to the purchase of the estate of Óchter Achid by Brethán and Cummen.[139] The account of the transaction mentions Colmán of the Britons, the abbot of Slane who died in AD 751, so it is likely to be a contemporary and accurate record.[140] The site was evidently for religious use, as Cummen is described as a nun. No information is given about Brethán: he may have been a cleric or a relative of Cummen. The text records that half the estate belongs to Cummen 'in building and in man' (*i ndoim i nduiniu*). It seems that she purchased her half with the goods which she would have brought into her marriage, i.e. her *tinól*.[141] This was apparently not quite enough, so she made a mantle which was sold to Éládach son of Máelodor for a brown horse, which was in turn sold to Colmán of the Britons for a *cumal* of silver. This *cumal* went as an additional payment (*forlóg*) for the purchase of Óchter Achid.

Exchange

For reasons of convenience, a farmer might decide to give some of his holding in exchange for a piece of land belonging to another. Obviously, he would need to be quite sure that he was getting a good bargain. Triad 72 includes exchanging for bad land (*cóemchlód fri drochferann*) among the three unfortunate things of husbandry, while Triad 73 includes exchanging for good land (*cóemchlód fri dagferann*) among the three fortunate things.[142] As we have seen in the section on male inheritance (p. 414), the exchange of portions of land was part of the procedure in the division of an inheritance among a wider circle of the kin.

[138] E. J. Gwynn, *Metrical Dindshenchas* iv 318.
[139] Bieler, *Patrician texts*, 174.3–14.
[140] AU^2 206 s.a. 750 (recte 751) § 6 *Mors Colmān na mBretan m. Faelāin abbatis Slaine.*
[141] For this term, see *SEIL* 127.
[142] Meyer, *Triads*, 8 §§ 72–3.

Farm-size

There are no contemporary records which give the acreage of – for example – an individual farmer of *bóaire* rank. Archaeological evidence can provide detailed information about the dimensions of buildings and enclosures on a particular site, but is of less assistance when it comes to the question of how much land was owned by the people who lived there.

The law-text on status, *Críth Gablach*, provides general information on the amount of land owned by different categories of person, using the *cumal* as the unit of value.[143] According to *Críth Gablach*, the lowest grade of adult freeman, the *ócaire*, is expected to own land worth seven *cumal*s (*tír .uii. cumal*).[144] Above him, a *bóaire* is expected to have fourteen *cumal*s of land.[145] This amount seems to have been regarded as a standard comfortable farming unit in both legal and non-legal sources.[146] As we have seen above, a female heir (*banchomarbae*) cannot inherit more than this no matter how much land her father possessed.[147] The most prosperous grade of *bóaire* (the *mruigḟer*) has twenty-one *cumal*s of land.[148]

Críth Gablach does not discuss the amount of land owned by the various grades of lord, ranging from the *aire déso* 'lord of vassalry' to the *aire forgill* 'lord of superior testimony'. This is no doubt because their status in society depends primarily on their possession of clients (*céili*), to whom they give fiefs of livestock, land or equipment in return for food-rent, labour and other services. The focus of interest is therefore on the number of their clients rather than the extent of their lands. Nonetheless, it can be assumed that early Irish lords generally had larger farms than commoners of *ócaire* or *bóaire* rank. According to commentary on *Uraicecht Becc*, a lord is expected to have twenty-eight *cumal*s of land, i.e.

[143] Presumably, a person's share of kin-land and any personal land which he may own are counted together. Perhaps rented land may also be included.

[144] *CIH* iii 778.25 = *CG* 4.91. Binchy's emendation of the text here is confirmed by *CIH* iii 784.22. For a discussion of the *cumal* as a unit of value, see Appendix B, p. 574.

[145] *CIH* ii 563.6; iii 779.26 = *CG* 6.153.

[146] E.g. *CGH* 207.40 *cach ra n-orbba bóhairech* 'every other *bóaire*'s inheritance'.

[147] *CIH* i 217.21 = *SEIL* 155 (xv).

[148] *CIH* ii 563.18; iii 779.37 = *CG* 7.172. Commentary on *Uraicecht Becc* allots twenty-eight *cumals* to the highest category of *bóaire tuíseo* '*bóaire* of precedence' (*CIH* v 1611.38; vi 2276.24–5 = *AL* v 90.10). He is presumably a commoner who has obtained the property-qualifications of a lord: see *GEIL* 28.

double the amount farmed by an ordinary *bóaire*.[149] The same commentary states that even the highest grade of lord, the *aire forgill*, has only one ploughteam.[150] He has thus no greater ploughing capacity than the most prosperous commoner of *bóaire* rank.[151] This suggests that a lord's superior socio-economic position is not dependent on a significantly larger cereal-production than that of a commoner. The extra land at his disposal seems therefore to be devoted to stock-rearing: the surplus stock is distributed to his clients, who in return supply him with a regular food-rent (p. 320).

According to *Uraicecht Becc*, a hospitaller (*briugu*) has twice the land of a lord (*flaith*),[152] and the accompanying commentary to the text enumerates this as fifty-six *cumal*s of land.[153] Some glossators on this text seem reluctant to accept that the *briugu*, a non-noble, could have so much land.[154] However, this figure fits in with what we know of the rôle of the *briugu* in early Irish society.[155] He is a rich landowner of non-noble birth who acquires extra status by providing hospitality. Unlike a lord, he does not receive food-rent from clients. He therefore needs a great deal of land to produce enough food to satisfy large numbers of guests.

A section of legal commentary contains an interesting observation on the proportion of a low-ranking farmer's land used for different purposes.[156] Of the seven *cumal*s owned by the average *ócaire*, one *cumal* is ploughed for cereals,[157] and another provides aftergrass (*athlompaire*) for winter. Three *cumal*s are grazed by various types of livestock, which are specified in the commentary as seven cows, three dry cattle, ten sheep, five pigs,

[149] *CIH* ii 645.10, 20, 34; 646.24 = *AL* v 44.14, 29; 46.16–17; 48.26; cf. *CIH* iii 1057.19, 27; *CIH* v 1562.31, 36; 1563.5, 11–12. The figure of forty-nine *cumals* at *CIH* ii 685.27 is a mistake for twenty-eight *cumals*; cf. *CIH* v 1563.11–12.

[150] *CIH* ii 646.27 = *AL* v 48.27. *Críth Gablach* likewise allots only one ploughteam to the *aire tuíseo* (*CIH* ii 567.27; iii 783.34 = *CG* 16.409).

[151] According to *Críth Gablach*, the most prosperous *bóaire* – the *mruigfer* – has a full ploughteam (*CIH* ii 563.25; iii 780.4 = *CG* 7.181).

[152] *CIH* ii 653.35 = *AL* v 76.1–2.

[153] *CIH* ii 654.15 = *AL* v 78.1.

[154] Compare the glosses at *CIH* ii 653.35; v 1608.11; vi 2273.36; 2324.19, where scribal insertions and alterations suggest uncertainty. Note also that in another commentary (*CIH* v 1564.7), the *briugu* has only twenty-eight *cumal*s of land.

[155] See *GEIL* 36–8.

[156] *CIH* iii 1058.10–12. This commentary introduces a further distinction, irrelevant in the present discussion, between *ócaire*s with six, seven, and eight *cumal*s of land.

[157] The commentary mentions only barley (*éornae*). Cf. *CIH* v 1543.22, 26.

two horses, three hens and a cock.[158] No information is given about the use of the remaining two *cumal*s of land; some of it may have been woodland (p. 389).

Rent of land

If a farmer does not have enough land for his needs he may make an agreement to rent land (*fochraic thíre*) from another farmer. In a marriage into which both husband and wife have contributed property (*lánamnus comthinchuir*), this is one of the agreements which either partner can make without the consent of the other.[159] The same provision is made in the case of a grown-up son whose father still controls the farm. Even against his father's wishes, he may rent land if there is not enough room for him to farm along with his father.[160] According to legal commentary, if a farmer rents out land to another with permission to plough, graze, and use water, he is entitled to one third of all the produce. If there is only permission to graze, he gets every seventh cow at the end of the year.[161]

The rent of land may form part of the contract between a client and his lord. The commonest form of clientship seems to have consisted of the advance of a fief (*taurchrecc* or *rath*) of cattle in return for an annual food-rent and other services.[162] However, it is clear from a law-text on the profit (*somoíne*) due for various forms of renting that a lord could also give a fief of land or farming equipment to his client.[163] A client may rent land from a lay lord (*flaith*) or from the Church, and pays his rent in cattle.[164]

Landless people

The Old Irish law-texts concentrate on landowners and their dependants, and it is generally assumed that wealth is to be equated with ownership of extensive lands grazed by numerous cattle. Loss of land normally brings about loss of rank: *Uraicecht Becc* ordains that

[158] *CIH* iii 1058.12 *tīr trī gchumal aige a n-irchomhair ḟeóir fri hearnaile* 'he has land of three *cumal*s for grass for [the various] classes [of livestock]'.
[159] *CIH* ii 506.16 = *SEIL* 19 § 5.
[160] *CIH* i 45.34 = *AL* v 284.27–8; cf. *CIH* vi 2193.5–6 = Thurneysen, 'Aus dem irischen Recht III', 311 § 7.
[161] *CIH* i 257.30–34 = *AL* iii 126.20–128.3; cf. *CIH* iii 778.26–7 = *CG* 4.93–4.
[162] For further discussion of clientship, see *GEIL* 29–33 and Charles-Edwards, *Early Irish and Welsh kinship*, 337–63.
[163] *CIH* iii 921.9, 20, 21, 25, 32, 35.
[164] *CIH* iii 921.32–4.

a freeman who sells off his land loses his free status.[165] Conversely, if an outsider (*deorad*) buys a land-holding he is reclassified as a person of legal standing (*aurrad*).[166]

Bretha Étgid recognizes the special circumstances of a partnership between a landless man and a landowner.[167] According to the accompanying commentary, the former is called a *carpat ar imram* 'chariot in motion', and has cattle but no land.[168] The latter is called a *foltach fuithirbe* 'one with substance in fields', and has land but no cattle.[169] The commentator gives the example of one partner with twenty-eight *cumal*s of land, and the other with twenty-four cows, but presumably less affluent people could also enter such partnerships. Neither partner is entitled to an honour-price until a formal agreement (*coindelg*) has been made between them. They then each have half the honour-price which is due to a person of their combined wealth. It is possible that the term *mruigfer* 'landman', which is used in *Críth Gablach* of the most affluent grade of *bóaire* (see p. 447 below), may also be used in a quite different sense of a farmer who has land but no stock, i.e. the *foltach fuithirbe* of legal glosses and commentary. In *Bretha Déin Chécht*, the *bruider* (presumably a later spelling of *mruigfer*)[170] is ranked below the *fer midboth* 'man of middle hits'.[171] In the law-text on status *Uraicecht Becc*, the *mruigfer* is placed above the *fer midboth*, but below the *ócaire* or *bóaire*.[172] In both texts, this category of person is identified as a *foltach fuithirbe* in the accompanying gloss.

It is probable that certain professionals could also attain wealth and status without owning land, though as we have seen above (p. 419) the assumption in legal commentary is clearly that a successful poet, smith, builder, or lawyer is likely to put his

[165] *CIH* ii 638.10 = *AL* v 20.1–2; cf. *CIH* ii 585.5–10 = *AL* iv 352.11–17.

[166] *CIH* v 1640.18.

[167] *CIH* i 260.13; iv 1188.24; 1261.4–6 = *AL* iii 140.y–142.2.

[168] Commentary on *Cóic Conara Fugill* likewise refers to the arrangement whereby a *sáerbothach* 'free cottier' grazes and waters his cattle on another's land in return for a supply of milk (*CIH* iii 1038.28–31 = Thurneysen, *Cóic Conara Fugill*, 54 § 121).

[169] The adjective *foltach* is a derivative of *folud* 'wealth, property, etc.', and *fuithirbe* (*fuithribe*) means 'ridge, field-boundary, field' (see Mac Mathúna, 'Some words for "(man-made) ridge" in Irish', 445–6).

[170] Cf. *CIH* vi 2308.26 = Binchy, '*Bretha Déin Chécht*', 28 § 12 (commentary) *bruigfeoir*.

[171] *CIH* vi 2308.35 = Binchy, '*Bretha Déin Chécht*', 30 § 13. For the *fer midboth*, see p. 445 below.

[172] *CIH* ii 655.11; v 1609.12; vi 2274.17; 2325.21 = *AL* v 78.25 *mbruidfear, mbruigfer*.

profits into land-purchase. One could imagine that there were nonetheless many landless craftsmen in monastic settlements who made an adequate living by manufacturing articles for their clerical patrons. The merchant (*cennaige*) could also become rich without owning land, though it must be admitted that this profession does not feature prominently in the written sources.[173] The law-texts make special mention of the position of the hermit, sometimes called a *díthir Dé* 'landless one of God', or simply a *díthir* 'landless one'.[174] His prestige is so great that his evidence can overturn any contract,[175] and his oath cannot be annulled by a counter-oath (*frithnoíll*).[176]

But, in general, landlessness entails legal restrictions. For example, *Berrad Airechta* points out that a contract made with a landless person (*díthir*) is invalid unless authorized by a propertied superior.[177] Landlessness is often associated with displaced or destitute people at the margins of society. The law-texts refer to the type of wandering down-and-out known as the *sinnach brothlaig* 'fox of a cooking-pit',[178] and the *ríascaire* 'marsh-dweller', who travels from marsh to marsh or from mountain to mountain.[179] Such persons survive through scavenging or charity.[180] A law-text on lordship warns of the folly of granting a 'fief of the wood' (*rath caille*),[181] i.e. a fief of cattle to a wood-dwelling wanderer (*loingsech*).[182] A lord who does so cannot expect any profit (*somoíne*) on his investment. There are a number of legal references to the *raitech* 'man of the road, vagrant', who is

[173] *GEIL* 7. Note, however, that a story preserved in *Cormac's Glossary* and in the *Metrical Dindshenchas* describes the drowning of Brecán, a renowned merchant (*cennaige*) of the Uí Néill, who was said to be engaged in mutual trading (*comchennach*) between Ireland and Scotland (Meyer, *Sanas Cormaic*, 28 § 323; E. J. Gwynn, *Metrical Dindshenchas* iv 82.37–44; MacCana, *Learned tales*, 146–7). Cf. also the mention of the *cennaige* (Latin *negotiator*) in a ninth-century text, *The monastery of Tallaght* (ed. Gwynn and Purton, 149–50 § 61).

[174] E.g. *CIH* v 1818.16.

[175] *CIH* v 1570,5. The reading *fiadnaisi fir dīthre* here seems better than *.f. fir dīthreb* of *CIH* iv 1150.37.

[176] Meyer, *Triads*, 22 § 165; cf. 40 § 165.

[177] *CIH* ii 593.35–8 = Stacey, '*Berrad Airechta*', 215 § 37.

[178] *CIH* vi 2193.26–7 = Thurneysen, 'Aus dem irischen Recht III', 317 § 11; *CIH* ii 585.30–31 = *AL* iv 354.11–12.

[179] *CIH* ii 585.27–9 = *AL* iv 354.7–10.

[180] E.g. *CIH* ii 585.9 = *AL* iv 352.15–16.

[181] *CIH* ii 433.16–17.

[182] *CIH* iii 918.11.

described in glosses as 'an unattached person who travels from place to place',[183] or 'one who is exiled from his kin'.[184] The law-text on legal entry, *Din Techtugud*, deals with the case of a land-holding being taken over by a *raitech*.[185] He is allowed the use of one third of the land for three days until a decision is made about his credentials. As a glossator points out, the *raitech* may be able to prove that he is a genuine heir who was abducted by foreigners when young, or that he has been away for many years at study.[186] If he is found to have no genuine claim, he must leave and pay a fine for the use of the land. Commentary on *Din Techtugud* distinguishes the *raitech ascnama methasa* 'vagrant who seeks a land-holding' and the *raitech déirge a methas* 'vagrant who leaves a land-holding'.[187] The latter is no doubt to be identified with the *esert* 'absentee' (see p. 402).

Landlessness is sometimes associated in our sources with predatory war-bands of young men. Kim McCone has demonstrated that membership of a war-band (*fían*) was commonly regarded as a prelude to the settled life of a landowner.[188] An observation in the wisdom-text *Tecosca Cormaic* contrasts the lifestyle of the warrior and the farmer: *fénnid cách co trebad* 'everyone is a warrior till farming' i.e. a farmer is as unlike a warrior as it is possible to be.[189] A general low esteem for landlessness is suggested by the maxim *ferr mruig mlichtaib* 'land is better than milk-yields', i.e. it is better to own the permanent asset of land rather than the temporary asset of cattle.[190]

[183] *CIH* ii 363.30–1 = *AL* i 106.7–8 *duine dīlmain bīs for sibal a hinadh d'inadh.*

[184] *CIH* i 33.7; v 1843.6 = *AL* v 244.17 *loingsech fine*; cf. *CIH* iii 905.39.

[185] *CIH* i 212.32–213.17 = *AL* iv 28.1–5. See Charles-Edwards, *Early Irish and Welsh kinship*, 269–70.

[186] *CIH* i 33.7–10 = *AL* v 244.16–20.

[187] *CIH* i 213.6–7 = *AL* iv 30.7–8; cf. *CIH* i 212.35 = *AL* iv 28.6; *CIH* i 33.7 = *AL* v 244.17. *Methas* 'land-holding, home territory' is attested with both *o*- and *u*-declension. There is no justification for the suggestion at *AL* vi 565 that it refers to 'marches, unappropriated ground between two territories'.

[188] McCone, 'Werewolves, cyclopes, *díberga*, and *fíanna*'; McCone, *Pagan past and Christian present*, 203–32.

[189] Meyer, *Tecosca Cormaic*, 46 § 31.10. For the idiom with *cách co*, see *DIL* C 273.19–21.

[190] Smith, '*Senbriathra Fithail*', 87 § 90 (YBL version).

Nomadry

Nomadry is characteristic of people living in difficult terrain such as the Arctic or the Sahara, where grazing may be sparse or seasonal, and climatic conditions unsuitable for agriculture. To secure adequate grazing, the entire population must move at intervals with all their livestock and possessions. Nomadry is to be distinguished from transhumance, the practice of driving livestock in summer to upland grazing. As we have seen above (p. 43), the accounts of transhumance in early Irish sources indicate that those who accompanied the cattle were mainly young people, and it can be assumed that enough able-bodied adults stayed behind to look after the crops and homestead. It is clear, too, that all livestock did not necessarily leave the farm: *Críth Gablach* refers to the constant presence of sheep 'without migration' (*cen immirgi*) in the green of a prosperous farmer.[191] The general picture is thus of transhumance as a means of reducing grazing-pressure on the home farm during the summer months. Nomadry, by contrast, does not at all fit in with the types of farming described in the Old Irish law-texts, with their constant emphasis on crop-cultivation, manuring of the soil, fencing of land, and other indicators of a settled lifestyle.

In a recent article, 'Nomadry in medieval Ireland', Katharine Simms has described the very different conditions which obtained in parts of the country in the turbulent fifteenth and sixteenth centuries.[192] In this period, there is evidence of the widespread harassment and expulsion of settled agriculturalists by wandering pastoralists, who were organized in groups called *creaghts*.[193] Kenneth Nicholls has linked the growth of nomadry with a declining population and under-utilization of land: he sees this development as stemming from the great plagues of the fourteenth century and the collapse of the colonial economy in Ireland.[194]

[191] *CIH* ii 564.7–8; iii 780.14–15 = *CG* 8.198.

[192] Simms, 'Nomadry in medieval Ireland'.

[193] She makes the plausible suggestion that the term *creaght* (also *creat*, *keryaghte*) is an anglicization of Irish *caoraigheacht* (*cáeraigecht*) 'mass, mustering'.

[194] Nicholls, *Land, law and society in sixteenth-century Ireland*, 9–12.

LAND-TENURE IN LATER GAELIC IRELAND

Legal glosses and commentary date mainly from between the eleventh and fourteenth centuries, and are therefore approximately half a millennium later than than the law-texts on which they are based. One might therefore expect that they would be a rich source of information on the development of Irish laws and institutions during the intervening period. However, as we have seen in the Introduction (p. 11), the glossators and commentators are mainly preoccupied with the correct interpretation of the Old Irish law-texts, and in elucidating the principles which are laid down there. The merits of alternative interpetations of key passages are discussed, but direct contradiction is rare.

Nonetheless, these later sources do give occasional evidence of change. In relation to the history of land-tenure in Ireland, the use of certain terms for landowning commoners is significant. Throughout the Old Irish law-texts, a distinction is made between two main grades of landowning commoner: the *bóaire* and the *ócaire*. The legal capacity of both is constantly stressed, and the law protects them from exploitation by their lord or lords.[195] But by the time of the main *Uraicecht Becc* glosses – probably written in about the twelfth century – the distinction between *bóaire* and *ócaire* has clearly been lost.[196] This may reflect a general squeezing out of the class of landowning commoner in Irish society. As we have seen above (p. 408), legal glosses and commentaries contain hints of the expansion of the power of lords and kings at the expense of commoners. The use of the term *bíattach* points in the same direction.[197]

Bíattach means 'one who supplies food (*biad*)', and has two distinct meanings in Middle and Early Modern Irish. Firstly, *bíattach* refers to the rich hospitaller who gains status by supplying food and

[195]See *GEIL* 31–2.

[196]Thus the glossator at *CIH* v 1610.40 = *AL* v 86.23 explains the second *bóaire* as 'the best *ócaire*'.

[197]For discussion on this term, see Price, 'The origin of the word *betagius*'; Mac Niocaill, 'Vocabulaire social irlandais', 514–5; Mac Niocaill, 'The origins of the betagh'; Curtis, 'Rental of the manor of Lisronagh, 1333', 47, 51–3; Otway-Ruthven, 'The organization of Anglo-Irish agriculture in the Middle Ages', 3, 12; Nicholls, 'Anglo-French Ireland and after', 373–80.

hospitality – he is the equivalent of the Old Irish *briugu*.[198] Secondly, *bíattach* refers to the person who pays a food-rent to a lord. In the latter usage, *bíattach* seems to be equivalent to the *céile* 'client' of the Old Irish law-texts, and it is noteworthy that a Middle Irish glossator explains the *dóerchéile* 'base client' as a *dóerbíatach*.[199] The fourteenth-century lawyer Giolla na Naomh mac Duinn Shléibhe Mhic Aodhagáin clearly takes the *bíattach* to be the ordinary category of commoner, as it is the *bíattach* who swears in murder cases 'on behalf of the grades of commoner'.[200] One can compare the similar position of the *bóaire*, who is viewed as the typical commoner in the Old Irish law-texts.

Both Irish and Anglo-Norman sources suggest that the *bíattach* is in a less economically favourable position than the *bóaire* or even the *ócaire* of the earlier period. The thirteenth- or fourteenth-century *Cath Maighe Léna* refers to *biataigh* 'under base tribute of hard service',[201] and a sixteenth-century record of a legal dispute describes them as 'humbly paying food and tribute'.[202] In Anglo-Norman sources, the word is latinized *betagius* or anglicized *betagh*, and regularly applied to a servile tenant. For example, a bull of Pope Urban IV dated 1261 defines the *betagii* as *laici ascripti glebe* 'laymen bound to the land',[203] and in the rental of the Manor of Lisronagh, it is stated that they 'shall do services at the will of the lord'.[204] The situation of these *betagii* is thus more to be compared with the *dóerfuidir* or *senchléithe* of the Old Irish law-texts than with the *bóaire* or *ócaire*.[205]

Another aspect of land-tenure in which change may be detected is that of inheritance. According to the Old Irish law-texts, an inheritance is to be divided in equal shares by the youngest heir. But already in the ninth-century, a legal glossator refers to a significant departure from this general principle in that the eldest son is entitled to the farmhouse and farmyard in addition

[198] See p. 422 above.
[199] *LL* iv 829.24747–8 = Stokes, 'Colloquy', 39 § 200^5 L.
[200] *CIH* ii 691.11–12 *.4. biadthaig dēc ar son na ngrādh fēne.* The *fer buailtedh* 'man of cattle-enclosures' swears on behalf of the noble grades: *.4. fir dēg aga mbia buailti ar son na ngrādh flata*; cf. *CIH* ii 691.28.
[201] Jackson, *Cath Maighe Léna*, 60.1562–3.
[202] *CIH* v 1620.29 *do beth an[a] mbiadtachaib dæra ac hīc bīdh 7 cīsa co umal.*
[203] Price, 'The origin of the word *betagius*', 186.
[204] Curtis, 'Rental of the manor of Lisronagh, 1333', 47 (trans. p. 53).
[205] See *GEIL* 34–6.

to his own share (see p. 413 above). Later sources preserve the theory – and sometimes also the practice – of equal division by the youngest. In his *Land, law and society in sixteeenth-century Ireland*, Kenneth Nicholls points out that 'while in some areas the division was still in the sixteenth century being made by the youngest of those entitled to share . . . in others it was by this date being made by the chief or *ceann fine* himself, who could assign to himself the best or largest portion'.[206] In his 'Lawes of Ireland', written *c.*1610, Sir John Davies gives an account of the methods of land-division in Gaelic Ireland, which includes a reference to the rôle of the head of the kin-group (*ceann fine*, anglicized *canfinny*):[207]

> '. . . The rest of the landes being distributed among severall septes every sept had a cheefe or *canfinny* [Irish *ceann fine*] as they called him with a *tanist* [Irish *tánaiste*] of that sept both which were chosen by the chiefe lord or captaine of the countr-eye. All the rest of the landes except the porcions of the cheefes & tanistes discended in course of gavelkinde & were partible among the males onlye in which division the bastardes had their porcions as well as the legitimate. The particion was ever made by the *canfinny* or cheefe of the sept and was made to continewe sometimes for 3 yeares, sometymes for 7 yeares & for shorter or longer tearmes which being ended they made a newe division & seldom or never did make a particion in perpetuity. This custom of gavelkinde . . . doth argue that the iland was inhabited by the Old Britaines . . . which custome the Walshemen have reteyned untill this daye'.

This passage is also interesting in that the author stresses the frequency with which kin-land is re-divided among the heirs. In the earlier period, by contrast, the texts suggest that kin-land was normally divided only once in each generation.[208] The author notes the similarity between the rôle of the kin-group in land-inheritance among the Irish and the Welsh. He takes this as evidence that Ireland was originally inhabited by the ancestors

[206] Nicholls, *Land, law and society in sixteenth-century Ireland*, 18–19.
[207] Morgan, 'Lawes of Irelande' 311.
[208] See discussion by Charles-Edwards, *Early Irish and Welsh kinship*, 63.

of the Welsh.[209] However, it is more likely that both peoples inherited their kin-structure from Common Celtic times.[210]

OFFENCES AGAINST LAND

I have discussed damage to crops by livestock (p. 140) and by human interference (p. 236). I deal here with various offences against land, ranging from a person's illegal presence thereon to physical damage by mining, sod-cutting, etc.

Trespass

There are references in the law-texts to the offences of hunting, fishing or food-gathering on private land (see Chapter 9) but no suggestion that a person's mere presence is counted as an offence. On the contrary, it looks as if a person crossing another's land may even acquire legal entitlements. For example, if a person finds a stray swarm of bees on the infield (*faithche*) of another's farm, he is entitled to a quarter of the honey for a year. If he finds a swarm on the outfield (*sechtar ḟaithchi*) he is entitled to a third of the honey.[211]

The law is, however, very strict with intruders in a private yard or house: the assumption is that such persons have criminal intentions. According to *Críth Gablach*'s account of offences against a prosperous farmer of *mruigḟer* rank, a visitor is permitted to open the yard (*les*) from outside – presumably to see if anyone is at home and to announce his presence. However, if he enters the yard without permission he is liable for a fine of five *sét*s. The same fine is due if he opens the door of the house without permission, and he must pay one cow if he peers into the house.[212] The law-text *Di Astud Chirt 7 Dligid* refers briefly to trespassing in a lord's enclosure (*dul tar dún*) and in a church enclosure (*dul tar chill*).[213] According to a glossator on this text, an intruder must pay ten *sét*s for crossing the wall (*múr*) or ditch (*clad*) of a monastery, and five *sét*s for crossing the rampart (*lúth*) of a lord's fort.

[209] Compare O'Rahilly's theory that Ireland was inhabited by P-Celtic-speaking peoples before the Goidelic invasion (T. F. O'Rahilly, *Early Irish history and mythology*, 205–8).

[210] See the detailed comparison undertaken by Charles-Edwards, *Early Irish and Welsh kinship*.

[211] *CIH* ii 454.13–28 = *BB* 82 §§ 46–48.

[212] *CIH* ii 564.15–16; iii 780.21–3 = *CG* 8.209–11.

[213] *CIH* i 239.14–15 = *AL* v 474.11.

In modern times, one of the most powerful arguments against public access to farmland is the landowner's possible liability if there is an accident. This issue is covered in some detail in Middle Irish commentary on the main law-text on accidents, *Bretha Étgid*. The general conclusion is that the onus is on the casual passer-by to avoid dangerous situations on farmland. In relation to such accidents, commentary distinguishes between a person who is engaged in some specific business (*torbach*) and one who is not (*esbach*). For example, if a cow which has recently calved (*nuídlech*) attacks a casual passer-by on her own ground (*láithrinn*),[214] the farmer is free from responsibility for the injuries, even if the victim has not provoked the cow in any way.[215] On the other hand, if the victim is on business – as a farm employee, for instance – the owner of the cow must pay half the appropriate fine for any injuries. The rules change when the cow ceases to be classed as a *nuídlech*, and the 'madness of her newly-calved state' (*meracht a nuídlechais*) has left her. In this situation, a casual passer-by can reasonably expect that he should not be attacked; if he is, the owner of the cow must pay half the fine for his injuries. The owner must likewise pay the full fine if the cow injures somebody who is on business.

There is little detailed information in the Old Irish law-texts themselves on the liability of landowners in the case of accidental death or injury on their land. *Bechbretha* specifically excludes liability if a passer-by is stung while looking over bee-hives at swarming times or otherwise interfering with the bees. On the other hand, the beekeeper **is** liable if a bee stings somebody without provocation as he passes by the hives.[216]

Illegal entry

If a person wishes to claim land which is occupied by somebody else, he must initiate the procedure of *tellach*, normally translated 'legal entry'.[217] He must first make formal entry with two horses, accompanied by a witness and sureties. If the occupant makes no

[214]The term *láithrinn* is not attested outside legal commentary. It refers to the proper place for a domestic animal, tool, etc.

[215]*CIH* i 275.21–6 = *AL* iii 228.21–8. Similar principles apply to injuries caused by a bull when the cows are in heat: *CIH* i 276.3–8 = *AL* iii 232.1–9.

[216]*CIH* ii 449.4–11 = *BB* 66 § 27–8. The glossator limits the beekeeper's liability to the *torbach*, the passer-by on business. The text, however, does not make this distinction.

[217]This procedure is described in the law-text *Din Techtugud* (*CIH* i 205.22–213.37 = *AL* iv 2.1–32.17, partially retranslated at Watkins, 'Indo-European metrics and

move to vacate or submit to arbitration, the claimant enters again after ten days with four horses and two witnesses. After a further ten days, he makes his final entry with eight horses and three witnesses. If the occupant still fails to react, the claimant acquires legal ownership. To demonstrate his ownership, he must spend a night on the property, kindle a fire, and tend his animals. If the claimant is a woman, the procedure is similar, though ewes are substituted for horses and in her last entry she brings a kneading-trough and a sieve for baking to symbolize her right to ownership.

The availability of the procedure of *tellach* gives great power to an individual claimant. In order that this power may not be abused, the seriousness of the offence of 'illegal entry' (*tellach n-indligthech*) is stressed in the texts. It is a public scandal (*gó thúaithe* 'a falsehood of the kingdom') if a claimant who has made an illegal entry is not fined.[218] Entry is illegal if it is carried out in spite of an offer to submit the case to arbitration, or in defiance of a contract (*cor bél*) or an agreement among heirs.[219] It is likewise illegal if the claimant brings the wrong number of livestock, or brings cows instead of horses.[220]

Illegal distraint

The Old Irish law-texts devote a great deal of attention to the legal procedure of *athgabál* 'distraint'.[221] It allows an individual to enforce a claim against another by means of the formal seizure of property belonging to him. As in the case of legal entry, distraint must be carried out in accordance with a strict set of rules. For example, distraint cannot be carried out on a holy day, or in defiance of a legitimate postponement or across a territorial boundary.[222] A person who carries out an illegal distraint (*athgabál étechtae*) is fined.[223]

---

archaic Irish verse', 221, 227–8, 234–5). For further discussion and references see *GEIL* 186–9, and Charles-Edwards, *Early Irish and Welsh kinship*, 259–73.

218 *CIH* i 213.27–9 = *AL* iv 32.3–6. Legal commentary at *CIH* iv 1189.38; 1262.36 = *AL* iii 146.z-148.1 also refers to the fine for illegal entry into land (*fiach techtaigthi tīre*).

219 *CIH* i 23.26–24.3 = *AL* v 210 (Heptad 24).

220 *CIH* i 211.36 = *AL* iv 20.6–7.

221 See *GEIL* 177–86 for discussion and references. For further detail, see Binchy, 'Distraint in Irish law'.

222 Cases where distraint is illegal are listed in Heptads 39–41 (*CIH* i 36.23–38.28 = *AL* v 254.1–260.14).

223 *CIH* i 213.27 = *AL* iv 32.4.

A claimant normally distrains the defendant's cattle, but he can also take other movable property or land.[224] The texts provide no information on the point, but it seems likely that this only occurs when a defendant has no cattle, or if all his cattle belong to categories excluded in the law of distraint. A low-ranking defendant such as a slave, tenant-at-will (*fuidir*) or herdsman, is unlikely to own any property worth taking; the claimant may therefore distrain him personally by keeping him in fetters and on meagre rations until his owner or master pays what is due.[225]

Illegal habitation

It is obviously against the law to set up house on somebody else's land, and this offence is mentioned in *Bretha Comaithchesa*.[226] However, if the rightful owner of the land tolerates this illegality without taking action, the squatter may acquire ownership after a prescriptive period (*rudrad*) whose length varies according to the circumstances of the case.[227] Another text lists various methods by which the landowner can indicate in the presence of witnesses his opposition to a squatter's presence.[228] He may drive out his cattle or distrain them, or put in his own cattle.[229] He may knock down the fences. He may give voice to his objections by complaint, shouting or groaning. He may even carry out a physical assault (*fogal*) on the squatter. According to a glossator, he is entitled to injure or kill the squatter if his verbal objections are ignored.[230]

Similar regulations apply to the erection without permission of a mill or drying-kiln on another's land.[231]

Illegal sod-cutting, etc.

The law-texts contain a number of references to illegal digging on another person's property. As with many other farming offences, the fine is fixed at five *séts* in *Di Astud Chirt 7*

[224] *CIH* ii 409.1–2 = *AL* i 258.4–5.
[225] *CIH* ii 363.23–8 = *AL* i 104.35–106.4.
[226] *CIH* i 75.5 = *AL* iv 124.11. This offence is called *attrab* 'squatting' at *CIH* ii 571.8 = Binchy, 'An archaic legal poem', 158.74.
[227] *CIH* i 53.28–54.17 = *AL* v 314.8–21 (Heptad 61). See *GEIL* 109 for a brief discussion on *rudrad*, with further references.
[228] *CIH* iii 749.27–38; 756.8–11; iv 1376.21–4.
[229] This quotation is given twice in *O'Davoren's Glossary* (*CIH* iv 1469.18–19 = O'Dav. 211 § 96; *CIH* iv 1481.17 = O'Dav. 263 § 419). Read *cartad, ecor, airchellad* 'expelling [cattle], putting in, distraining'.
[230] *CIH* iii 749.29–30.
[231] *CIH* iii 759.9–11.

Dligid.[232] Legal commentary distinguishes the offence of *fótbach* 'sod-cutting' from *forscris* 'scraping'; the latter term probably refers to the removal of a thin layer of sod (*scraith*) from the ground.[233] The same commentary also includes the offence of *forraite*, which may refer to the illegal construction of a road (*raite*) on somebody else's land. Alternatively, it may be the offence of 'purpresture' or encroachment on somebody else's road.[234]

Illegal ploughing

The law-text *Bretha Comaithchesa* warns the farmer not to plough on his neighbour's land.[235] According to commentary, he must pay a fine of five *sét*s for this offence, with forfeiture of whatever has been sown.[236]

Illegal mining and quarrying

Old Irish literary and legal texts contain numerous references to metals, and it is clear that they were of great importance in the early Irish economy. Gold and silver were particularly significant as a mark of high rank. For example, the law-text on status, *Críth Gablach*, states that a 'lord of precedence' (*aire tuíseo*) is expected to own two bridles, one of gold and the other of silver.[237]

The Old Irish law-texts contain a number of references to the private ownership of mines. The text on distraint lists the offence of digging in somebody else's silver-mine (*fothlae t'airgetlaig*).[238] A ninth-century gloss expands this reference to include the illegal digging of the ore of iron (*iarn*), copper (*umae*), tin (*créd*) and gold (*ór*).[239] The same text also refers to the offence of digging into a cliff (*claide alla*) in search of copper or of iron-ore.[240] According to glossators, the fine for illegal mining is five *sét*s.[241] In addition,

[232] *CIH* i 239.10 = *AL* v 474.5; *CIH* iv 1527.27 = O'Dav. 467 § 1513 *tubae tíre* 'cutting the land'. The glossator takes five *sét*s to be the regular fine for offences in this list.
[233] *CIH* iv 1189.28–30; 1262.26–9. Legal commentary also mentions the offence of *forscris* (once *forscrath*) in the context of mining, *CIH* i 270.17; iii 940.13; iv 1256.7 = *AL* iii 202.19.
[234] Rackham, *The history of the countryside*, 265-7.
[235] *CIH* i 75.4 = *AL* iv 124.11.
[236] *CIH* iii 866.23–6; cf. *CIH* i 257.38 = *AL* iii 128.8–10.
[237] *CIH* ii 567.25; iii 783.32–3 = *CG* 16.407.
[238] *CIH* ii 384.19 = *AL* i 166.26.
[239] *CIH* iii 890.20–1. A list of metals in Recension A of *Audacht Morainn* (Kelly, *Audacht Morainn*, 67 § 39b) refers also to lead-ore (*méin lúadi*).
[240] *CIH* ii 390.8 = *AL* i 184.12–13.
[241] *CIH* i 239.20 = *AL* v 474.32; *CIH* ii 384.35 = *AL* i 170.18.

the culprit must restore the metal which he has taken, even if it has been worked into a finished article.[242]

A difficult passage in the law-text *Bretha Étgid* deals with entitlement to open a mine (*méin*).[243] According to the accompanying commentary, a person is free to dig for ore if he has permission from the landowner or if it is a mine which is not owned (*méin arna fuil techtugud*).[244] The reference in the text to the 'middle trench' (*midchlas*) seems to mean that a miner must dig a single trench or shaft, and not damage the sward or cause a hazard over a wide area. If he does so, his offence is compared in the commentary with illegal trench-digging (*clad indligthech*), discussed in the next section.[245]

It seems from legal commentary on *Bretha Étgid* that the fine for removing stones from another person's land is five *séts*.[246]

Illegal trench-digging

It is normally illegal for a person to dig a trench across somebody else's land. However, if a person is constructing a mill he may find that there is no route for the mill-race other than through the property of a neighbour. In such a case he may insist that the neighbour allow him to carry out the work, though naturally he must pay a fee (*fochraic*). This is fixed at a *sét* worth ten scruples in the case of arable land (*etham*), and at half that rate in the case of rough land (*ainmín*).[247] The entitlement to dig a mill-race on a neighbour's land does not apply if it belongs to the Church or to a lord or king or if there is a burial-mound (*fert*).[248]

[242] *CIH* i 270.21 = *AL* iii 204.1–2 *cid ina tindib, cid ina tanailgib, cid 'na aic*[*d*]*e urrlaim bes* 'whether it be in ingots, in sheets(?), or as a finished article'.

[243] *CIH* i 270.15; iii 940.11; iv 1256.3 = *AL* iii 202.16.

[244] A glossator on *Di Chetharšlicht Athgabálae* (*CIH* ii 390.35 = *AL* i 190.2) refers to a mine which is 'common to the people of the territory' (*coitcend do tuaith*).

[245] *CIH* i 270.17 = *AL* iii 202.20.

[246] *CIH* i 261.9–10 = *AL* iii 148.16–17; cf. *CIH* iv 1189.29–30; 1262.28–9.

[247] *CIH* ii 460.32–461.9 = Binchy, '*Coibnes Uisci Thairidne*' 68–70 § 10. A *sét* is usually worth twelve scruples: see Appendix B, p. 589.

[248] *CIH* ii 461.12–13 = Binchy, '*Coibnes Uisci Thairidne*', 70 § 11 *nemed cille no dūin nō* (*a*) *maigen f*[*e*]*irt* lit. 'privileged land of a church or of a fortress or the precinct of a mound'. I believe that *nemed* in this context refers to any land owned by the Church or by a lay dignitary: see discussion at *BB* 107. One of the duties of a client was to dig the burial-mound of his lord (*fertad do flatha, CIH* i 51.34 = *AL* v 306.4; *CIH* iv 1503.35 = O'Dav. 362 § 961; cf. *CIH* ii 434.29). As we have seen above (p. 409), a landowner may be buried on the boundary of his property.

Illegal burning

If a farmer burns vegetation carelessly, fire may spread outside his own land. The law-text *Di Astud Chirt 7 Dligid* lists the offence of *loscad luise,*[249] which an Old Irish glossator defines as 'fire into the land of neighbours' (*teine fo thír comaithech*).[250] According to a later glossator, if the burnt area extends to the length of twelve stakes the culprit must pay a fine of five *sét*s and restore any damage: the implication seems to be that if a smaller area is burnt, the offence is regarded as too trivial for legal action.

There is also a reference to the offence of burning land in commentary to *Bretha Étgid.*[251]

Pollution of waterways

Legal commentary refers to the offence of throwing the carcase of an animal into a stream without having burned it.[252] Commentary on animal diseases likewise specifies that the corpse of a rabid dog must be burned to ashes before being thrown into a stream: see p. 215.

249 *CIH* i 239.10 = *AL* v 474.5; cf. *CIH* iv 1513.22–3 = O'Dav. 405 § 1185 *Luisi .i. lasadh.*
250 *CIH* iii 916.34.
251 *CIH* i 261.5 = *AL* iii 148.9; cf. *CIH* iv 1189.26; 1262.24.
252 *CIH* i 107.4. I am indebted to Liam Breatnach for this reference.

13
Farm labour

In previous chapters, there has been discussion of a wide range of activities on the early Irish farm: ploughing, harrowing, sowing, weeding, harvesting, tending livestock, cutting wood, fencing, etc. In this chapter I examine what the texts tell us about who actually carried out these tasks.

Slaves

There are many references to slavery in law-texts, wisdom-texts, saints' Lives, annals, and sagas. We can conclude, therefore, that the institution of slavery was of considerable importance in early Irish society, and that much of the hard work in well-off households was carried out by slaves. A task for young slaves of either sex was the herding of domestic animals. In our only firsthand account of life as a slave in Ireland, Patrick refers briefly in his *Confession* to his task of herding livestock, sometimes in woods and on the mountain, where he endured snow, frost and rain.[1] The ninth-century *Bethu Brigte* is not an historically authentic document like Patrick's *Confession*, but it does nonetheless give us a good deal of insight into the social organisation of the period in which it was written. It is significant, therefore, that Brigit, the daughter of a slave and therefore a slave herself, is represented as having been required to herd pigs by her master Dubthach (who was also her father).[2] Legal commentary on the text on joint-herding (*comingaire*) also refers to the use of a slave (*dóer*) for the task of herding livestock.[3]

Adult slaves were put to heavy work on the farm and in the kitchen. The male slave (*mug*) is associated particularly with the cutting and carrying of wood.[4] The Old Irish law-text *Bretha Étgid* states that there is an exemption (*blaí*) with regard to slave-work (*mugsaine*): this means that a slaveowner need not pay the fine

[1] Newport White, *St. Patrick: his writings and life*, 35–6 § 16; Bieler, *The works of St. Patrick*, 25 § 16; Howlett, *The book of letters of Saint Patrick the bishop*, 60.

[2] Ó hAodha, *Bethu Brigte*, 2 § 9 *ingaire mucc*.

[3] *CIH* vi 2139.4.

[4] Cf. Stokes, 'The second battle of Moytura', 68–70 § 37; Gray, *Cath Maige Tuired*, 32.161–4 § 37.

for any damage which his slave may do during the course of his proper work.[5] According to the accompanying commentary, the slave cuts wood, gathers it into a bundle (*cúal*), ties it with a withe (*gat*), and brings it home on his back.

In the texts, the female slave (*cumal*)[6] is sometimes represented as being engaged in milking (*mlegon*) or churning (*maistred*).[7] She is often associated with the laborious processes by which grain is prepared for bread-making. In one version of the heroic tale *Táin Bó Cúailnge*, a female slave is said to work at the quern (*bró*), the kneading-trough (*losat*) and at other servile work (*dáeropair*).[8] The law-text *Bretha Étgid* refers to her work at the kneading-slab (*lecc*) and kneading-trough.[9] The accompanying commentary speaks also of her use of the sieve (*críathar*), presumably for sieving the meal which she has ground.

The horizontal water-mill was introduced to Ireland in the early Christian period (see p. 482 below), and must have to some extent reduced the need for female slave-labour. The tenth-century scholar Cormac mac Cuilennáin provides a pseudo-etymological explanation for the word *cumal*, taking it to be from Latin *cum mola* 'with the quern'.[10] He goes on to observe that the *cumal* is 'the woman who grinds at the quern, for this is the work which female slaves used to do before water-mills were made'.[11] He thus implies that quern-grinding by female slaves was a practice of bygone times, outmoded by technological development. But it is noteworthy that he does not say that slavery itself was outmoded. In fact, the evidence of the written sources indicates that the enslavement of both males and females was practised in Ireland from the fifth century (and no doubt before) until about the twelfth century.[12] The growth of the slave-trade in the Irish Sea

[5] *CIH* i 265.1 = *AL* iii 174.8.

[6] For the use of the term *cumal* as a unit of value, see Appendix B, p. 591.

[7] E.g. Ó hAodha, *Bethu Brigte*, 1 § 1; 3 § 12.

[8] *TBC* LL 42.1529–31. The corresponding passage in *TBC* Rec. I 40.1269–71 speaks only of work at the quern.

[9] *CIH* i 285.36 = *AL* iii 274.18.

[10] Meyer, *Sanas Cormaic*, 28 § 324; Ní Dhonnchadha, 'An edition of *Cáin Adomnán*' (thesis), 151. In another pseudo-etymological gloss (Meyer, *Sanas Cormaic*, 109 § 1249), the Latin word *trulla* 'kneading-trough' is linked with Irish *tráill* 'slave', a borrowing from Old Norse *þræll* (Marstrander, *Bidrag*, 76; *LEIA* T-124 s.v. *traill*).

[11] Meyer, *Sanas Cormaic*, 28 § 324; cf. E. J. Gwynn, *Metrical Dindshenchas* i 22.109–20.

[12] For slavery in the later period, see Holm 'The slave trade of Dublin, ninth to twelfth centuries'.

area in the eleventh century seems to have brought many foreign slaves into the country. The twelfth-century *Lebor na Cert* contains a number of references to foreign slaves in the tributes or stipends of various kings. For example, the stipend given by the king of Ailech to one of his subsidiary kings includes 'seven women from over the great sea'.[13] Legal commentary from about the same period also refers to foreign slaves.[14]

SEMI-FREEMEN

The Old Irish law-texts regularly make a distinction between the slave and the *fuidir*, a term generally translated 'semi-freeman' or 'tenant-at-will'.[15] For example, one law-text strongly advises a lord (*flaith*) not to free his *fuidir*s or release his slaves.[16] If he does so, there will be 'failure of corn and milk and fruit'. This suggests that the author regarded the labour of both slave and *fuidir* as essential for the prosperity of high-ranking farmers. There may also have been a fear that the abolition of slavery or servile tenancy could lead to anarchy and other social ills.

According to legal commentary, there is no fixed limit to the labour which the *fuidir* must perform for his lord. He must do whatever the lord requires, provided that he is not subjected to physical hurt.[17] In this respect his situation is totally different from that of a client (*céile*). The client is a freeman who receives a fief from his lord, and in return provides food-rent and carries out certain fixed tasks (see p. 446).

It is clear from the short law-text on the *fuidir* that this term is used to describe persons in a variety of servile situations.[18] In general, the lord provides maintenance for the *fuidir* in return for labour, and is responsible for any fines or liabilities incurred by him or his

[13]Dillon, *Lebor na Cert*, 68.1013. There are also references to foreign slaves at 40.591; 68.997; 88.1313; 98.1444.

[14]E.g. *CIH* vi 2083.12 *gaill daora.*

[15]Apart from one occurrence in E. J. Gwynn, *Metrical Dindshenchas* iv 318.17 (spelled *fuither*) this word is not attested outside legal sources. For a detailed discussion of the position of the *fuidir*, see Charles-Edwards, *Early Irish and Welsh kinship*, 307–36.

[16]*CIH* i 231.15–17 = *AL* v 450.12–15 *saerad fuidre . . . fuaslucad do mogaib.* The *fuidir* and *mug* are also listed separately at *CIH* ii 359.29 = *AL* i 84.28–9; *CIH* ii 363.23 = *AL* i 104.35.

[17]*CIH* v 1880.3–4 = Thurneysen, *Irisches Recht*, 68 Kommentar IV.

[18]*CIH* ii 426.1–429.12 = Thurneysen, *Irisches Recht*, 62–7. The text distinguishes ten types of *fuidir.*

family.[19] The ordinary *fuidir* is so dependent that he cannot make a contract without the permission of his lord.[20] In addition, he is not entitled to receive payment for any offence against himself, his son or his daughter, or to inherit property (*díbad*). These assets go to his lord. If he is the victim of theft, the culprit must pay to the lord the penalty-fine (*díre*) for the object stolen; the *fuidir* himself only receives restitution (*aithgein*).

Some *fuidir*s may be given a fief (*rath*) of livestock or land in return for which they provide food-rent to their lord as well as labour.[21] A *fuidir* may attain a less dependent relationship with his lord, in which case he pays for his own offences.[22] This type of *fuidir* is also entitled to receive a penalty-fine for an object belonging to him which is stolen; only one third of this sum goes to the lord.

A *fuidir* is permitted to sever his relationship with his lord, provided that he leaves no outstanding debts or liabilities. He must also surrender to the lord two thirds of the produce of his husbandry. A person who is classed as a base *fuidir* (*dóerḟuidir*) does not have this entitlement.[23] In fact, it is difficult to argue that there would have been much practical difference between such a *fuidir* and a slave. It is clear from the text that some categories of *dóerḟuidir* were condemned criminals who had been ransomed from death by their lord.[24]

The law-texts regularly treat the *bothach* 'cottier' as a person on the same social level as the *fuidir*,[25] and it seems likely that he is expected to work for his lord under similar conditions. If a *bothach* or *fuidir* family occupy the same land for three generations, its members are reduced to the status of *senchléithe*, a term meaning

[19] *CIH* ii 426.1–6 = Thurneysen, *Irisches Recht*, 63 § 1.

[20] *CIH* ii 491.24 = Thurneysen, 'Aus dem irischen Recht I', 375 § 38.

[21] E.g. *CIH* v 1879.39 = Thurneysen, *Irisches Recht*, 68 Kommentar IV *cūic samaisci doberar a rath fuidre*; *CIH* v 1879.26 *ferunn tucad i rath fuidre.*

[22] *CIH* ii 426.21 = Thurneysen, *Irisches Recht*, 63 § 2. This is the *fuidir laiss mbiat .u. treba* 'the *fuidir* who has five households (?)'. For a discussion of this difficult term, see Charles-Edwards, *Early Irish and Welsh kinship*, 319–24.

[23] *CIH* ii 429.9–10 = Thurneysen, *Irisches Recht*, 67 § 11.

[24] For further discussion, see *GEIL* 35. Legal commentary (*CIH* vi 2009.15 = Thurneysen, *Irisches Recht*, 68 Kommentar III) refers to the *daorfuidhir tar muir* 'base *fuidir* from overseas' who must be a slave.

[25] *GEIL* 35.

'ancient dwelling'. Such persons are bound to their lord and cannot renounce their tenancy. If the land is sold, they transfer to the new owner.[26]

SERVANTS

Our sources pay little attention to low-ranking servants, and it is often unclear whether a servant in a particular context is to be regarded as a slave or as an employee. However, there are enough references to show that the notion of doing servile work for pay was well established in early Irish society. For example, the ninth-century text on Sunday observance, *Cáin Domnaig*, refers to a workman who does unnecessary work of his own volition on Sunday. He must pay various fines for this offence (see below p. 455), and also forfeits his own payment (*fochraic*) for a year.[27] A servant's freedom to change his or her employer seems to be recognized in *Cormac's Glossary*, compiled about AD 900. He explains the word *amus* 'servant' as meaning 'one who goes from one employer to another'.[28]

The law-text on distraint emphasizes the legal disabilities of the farm-worker. The shepherd (*áugaire*) and the cowherd (*búachaill*) are included in a list of men of low status along with the slave (*mug*) and the semi-freeman (*fuidir*).[29] If such a person incurs a debt or liability, the plaintiff carries out distraint by putting a fetter on his foot and keeping him on short rations. This procedure is different to the normal method of distraint (see p. 170) in which the plaintiff takes livestock belonging to the defendant. The assumption is that a person at the economic level of a shepherd or cowherd would not own enough livestock or other property to be distrained. Consequently, he must rely on his master to pay off his debts or liabilities.

Legal commentary provides some information on the cowherd and other farm-workers. One passage (probably based on the lost law-text *Cáin Bóslechtae* 'the law of cow-sections') gives the penalty

[26] *GEIL* 35–6. There is an intricate discussion on the distinctions between the *fuidir*, *bothach* and *senchléithe* in commentary at *CIH* iii 745.18–35.

[27] Hull, '*Cáin Domnaig*', 168.74–6 § 8.

[28] Meyer, *Sanas Cormaic*, 1 § 12. This is a pseudo-etymological gloss based on the interpretation of *amus* (*amos*) as *am-ḟos* 'unstable, flighty'.

[29] *CIH* ii 363.23–4 = *AL* i 104.35–6.

for killing a cowherd as seven *cumal*s.[30] As we have seen (p. 28), cattle-raiding was widespread in early Irish society. It is obvious, therefore, that the cowherd would be in particular danger from cattle-raiders. It is clear from another legal commentary that herdsmen are expected to protect the animals under their charge from wolves.[31] Legal commentary sometimes makes a distinction between the cowherd (*búachaill*) and the milker (*bligre*),[32] and one gloss even separates the work of the cowherd (*búachaill bó*) and the calfherd (*búachaill láeg*).[33] A ninth-century text on the reciprocal services of the kings of Cashel refers to a *rechtaire for blicht* 'a steward for milk', who seems to have been a dairy-manager for the royal herds.[34]

Old and Middle Irish texts of many types contain references to the swineherd (*muccid*). For example, the tale *Togail Bruidne Da Derga* gives a vivid description of the swarthy and forbidding appearance of the three swineherds of King Conaire, appropriately named Dub ('black'), Donn ('brown') and Dorchae ('dark').[35] Each wears a greenish tunic and a black cloak and has a *gabulgice*, which probably refers to some sort of forked pole used in herding swine. Our texts also contain occasional references to the groom (*echaire*)[36] and the bridle-boy (*gillae glomair*).[37]

In addition to those who work with stock, our sources distinguish a number of other servants who specialize in particular tasks. In general, such specialized servants are represented as working for a king or some other dignitary who can afford to maintain a large labour-force. For example, the household of King Conchobar is said in an Old Irish poem to include men with the professions of hunter (*selcid*), fisherman (*íascaire*),[38] trapper (*cuthchaire*) and fencer(?) (*etarpuige*), as well as various entertainers, manufacturers

[30] *CIH* iv 1367.4 (trans. Thurneysen, 'Aus dem irischen Recht V', 388). The same fine is exacted for the killing of a ploughman (*airem*).

[31] For the details, see p. 186 above.

[32] E.g. *CIH* i 10.3 = *AL* v 152.20–21; *CIH* iv 1276.3. The same distinction is made in O'Keeffe, 'Colman mac Duach and Guaire', 46.29.

[33] *CIH* ii 508.26 = *AL* ii 366.7.

[34] O'Keeffe, '*Dál Caladbuig*', 20 § 10.

[35] Stokes, 'The destruction of Dá Derga's hostel', 288 § 112 = *LU* 232.7583–9.

[36] Meid, *Táin Bó Fraích* 11.283.

[37] *CIH* v 1607.26 = *AL* v 72.y; *CIH* v 1730.29–30 = *AL* ii 26.5.

[38] Cf. *CS* 56 s.a. 565, where there is a reference to the fisherman of Comgall, abbot of Bangor.

and attendants.[39] The diagram of the royal banqueting hall at Tara (*Tech Midchúarda*) – probably dating from about the tenth century – shows an even greater variety of specialized servants.[40] Among the workers who are present in the Tara hall but not in Conchobar's household are the ditch-digger (*cladaire*),[41] the rampart-builder (*ráthbuige*),[42] and the labourer (*obraige*). It must be stressed that these lists of workers are found in literary texts and therefore have no pretensions to strict historical accuracy. Nonetheless, they give us some idea of the sort of work-force which was felt to be appropriate for a powerful monarch.

We can assume that in a large household such as that of a king or lord, the work of the indoor and outdoor servants would be organized by the steward (*rechtaire*). In the law-texts he is represented as a person of considerable importance, with an honour-price half that of his employer.[43] Other important functionaries in large households are the charioteer (*arae*)[44] and the messenger (*techt*).[45]

We are told in a Middle Irish tale that Odba, the royal swineherd (*rígmuccaid*) of King Conn Cétchathach, also acted as a deer-hunter (*fer selga oss n-alta*).[46] He never lived in a house, but spent his life 'in woods and steep valleys hunting and herding swine'. In practice it seems certain that – like Odba – servants in the household of a king or lord often carried out a number of different tasks for their master. For example, it would be highly unlikely that a ploughman (*airem*) would be allowed to take his ease during the parts of the year when his services were not required for ploughing: his master's steward would no doubt put him to other work. According to *Críth Gablach*, it is acceptable in the month of sowing for a king to

[39] O Daly, '*Lānellach tigi rīch*', 83 § 16. The word *etarpuige* (*etarbaige*) is otherwise unattested. It may be an agent noun from *etarbae* 'fence'.

[40] *LL* i 116. The attendance at the *Tech Midchúarda* is also described in a poem at *LL* i 117.3703–120.3789 = Petrie, 'Tara Hill', 199–204.

[41] In later Irish, this word acquired the meanings 'rascal, cheat, coward'.

[42] Cf. *CIH* ii 585.28 = *AL* iv 354.9. L. Breatnach, 'On the agent suffix *-e* in Irish', takes *ráthbuige* (*ráthmaighe*, etc.) to be a compound of *ráth* 'rampart, fort' and *-buige*, an agent noun from the root of the verb *bongid* 'breaks, cuts'. For references in saints' Lives to the rampart-builder, see Plummer, *Vitae* i pp. xcviii–xcix.

[43] *GEIL* 65.

[44] See p. 497 below.

[45] See p. 96 above.

[46] Stokes, 'The prose tales in the Rennes Dindshenchas, (2)', 55 = *LL* iii 759.22601–6.

go about with a reduced retinue of three persons, i.e. his judge and two others.[47] The implication is that all other able-bodied members of his household are at work in the fields.

FREEMEN

Our sources provide a good deal of information on the work of freemen of commoner rank. Three main categories are distinguished in the law-texts: the *fer midboth*, the *ócaire*, and the *bóaire*.

Fer midboth

The term *fer midboth* 'man of middle huts' is normally applied to a male teenager who has a small honour-price in his own right and limited legal capacity.[48] He generally works along with his father. If, however, his father is dead, he is allowed to inherit his share of land, though he does not acquire the full legal status of a landowner until he reaches the age of twenty.[49] If, on the other hand, his father lives on and he does not inherit until he himself is old, his honour-price remains at the level of *fer midboth* until such time as he inherits.[50]

Ócaire

According to the law-text on status, *Críth Gablach*, an *ócaire* has a small farm worth seven *cumals*.[51] It is likely that he does all the regular farm-work himself, with the assistance of his wife, children, and possibly other dependent relatives. As a typical *ócaire* has only one ox, he cannot conveniently carry out the vital operation of ploughing from his own resources.[52] He therefore usually makes a co-ploughing agreement (*comar*) with other farmers of similar rank.[53] He is also likely to make an agreement of joint herding

[47] *CIH* ii 569.25–7 = *CG* 21.535–8. Normally, the lowest grade of king (*rí túaithe*) has twelve persons in attendance in public, and nine in private (*CIH* ii 568.14 = *CG* 18.454–5). Higher grades of king have larger numbers in attendance.

[48] The 'middle huts' may refer to temporary habitations on his father's land. For discussion on the *fer midboth*, see Binchy, *CG* Legal Glossary 89–90. In some MSS, the spelling is often *fer midbotha*, e.g. *CIH* iii 777.16, 20.

[49] *CIH* iii 778.8–9 = *CG* 3.67–9.

[50] *CIH* iii 778.10–11 = *CG* 3.69–70.

[51] For this measure of land-value, see p. 574.

[52] *CIH* iii 778.27 = *CG* 4.95.

[53] For a discussion of *cyfar* 'co-ploughing' in medieval Welsh law, see Charles-Edwards, *Early Irish and Welsh kinship*, 446–56. Langdon, *Horses, oxen and technological innovation*, 238–40, notes that there is little evidence of cooperative ploughing in medieval England.

(*comingaire*) whereby the livestock of a number of less well-off farmers are looked after in a single herd. This arrangement has the obvious labour-saving advantage that herding duties can be shared among a few households. On the other hand, there may be difficult legal problems to resolve if an animal belonging to one owner kills or injures an animal belonging to another (see p. 176).

It is assumed in *Críth Gablach* that the *ócaire* is in a client relationship with a lord, from whom he receives a fief (*taurchrecc*).[54] A client (*céile*) pays off his fief with both goods and services. He must provide his lord with food-rent (*bés tige*), consisting of live animals, meat, grain, malt, bread, milk, milk-products, and vegetables. He must also carry out certain fixed tasks. Thus he – or possibly someone working on his behalf – must join the reaping-party (*meithel*) in his lord's cornfields,[55] and help to excavate the rampart around the lord's dwelling.[56] If he fails in his duties, he may be fined, or sink to a condition of greater dependency, as a *fuidir* or a slave. According to the law-texts, a man may have contracts of clientship with up to three lords.[57]

[54] *CIH* iii 778.34 = *CG* 4.105. A client's fief seems normally to be of livestock, but may also be of land or farming equipment. Early Irish law distinguishes two categories of client: base and free. The rights and duties of the base client (*céile gíallnae, dóerchéile*) are described in *Cáin Aicillne* (*CIH* ii 479.23–502.6; v 1778.34–1804.11 = Thurneysen, 'Aus dem irischen Recht I'), and those of the free client (*sóerchéile*) in the fragmentary *Cáin Ṡóerraith* (*CIH* v 1770.15–1778.33 = Thurneysen, 'Aus dem irischen Recht II', 238–60). As well as the fief, the base client receives property known as the *séoit taurchluideo* 'chattels of subjection' from his lord. These are of the same value as the client's honour-price. If the client pays his food-rent to his lord and carries out his other duties of labour and service for at least seven years, the fief becomes his own property on his lord's death. The contract of base clientship cannot be terminated by either party except through the payment of stiff fines. In free clientship, no 'chattels of subjection' are received by the client, and either party can terminate the arrangement without penalty. The free client must restore the fief to his lord's heirs on his death. For further discussion on base and free (= noble) clientship, see *GEIL* 29–33; Charles-Edwards, *Early Irish and Welsh kinship*, 337–63.

[55] The law-text *Di Dligiud Raith 7 Somoíne* refers to the *meithel gíallnae* 'reaping-party of base clientship' (*CIH* ii 434.30 = Crigger, 'A man is better than his birth', 326 § 7; cf. *CIH* v 1910.14 = Crigger, 'A man is better than his birth', 339 § 11; *CIH* ii 382.20–1 = *AL* i 156.33). According to *Críth Gablach*, a *fer midboth* must carry out half a day's work (*lethdrécht*) in his lord's cornfields at three days' notice (*CIH* iii 778.20 = *CG* 4.84–5).

[56] *CIH* iv 1493.12–13 = O'Dav. 314 § 721; *CIH* v 1906.15–16 = Crigger, 'A man is better than his birth', 326; cf. *CIH* ii 570.9 = *CG* 22.570.

[57] *CIH* ii 488.1–3 = Thurneysen, 'Aus dem irischen Recht I', 367 § 28. See *GEIL* 32.

The client must carry out certain public works, such as road-maintenance (p. 391), patrolling for wolves (p. 186), and clearing the site for an *óenach* or assembly (p. 458). It seems also that he could be summoned by royal decree to surrender part of his harvest. *Críth Gablach* refers to the king's entitlement to issue an ordinance for the provision of crops (*rechtge fri tuar toraid*), presumably in times of general shortage.[58] A short legal passage describes the rights of a judge (*brithem*) attached to a king or chieftain.[59] These include the entitlement to a day's ploughing (*lá n-air*), a day's reaping (*lá mbúana*) and a day's fencing (*lá n-imbi*).[60] The text does not specify who carries out these tasks, but it is likely that they fall on the client. The Church also benefits from his labour in the form of first fruits (*prímiti*) and tithes (*dechmad*).[61] At his death, he is expected to bequeath the amount of his honour-price to the Church. In the case of an *ócaire*, this comes to the sum of three *séts*.[62]

Bóaire

Further up the social scale is the *bóaire*. *Críth Gablach* provides most detail on the highest category of *bóaire*, who is known as a *mruigḟer*, lit. 'landman', and owns land worth twenty-one *cumals*.[63] He possesses full ploughing equipment, and therefore does not need to become involved in a co-operative ploughing arrangement, with its potential for friction and litigation. He is nonetheless the client of a lord or lords, and must carry out the various duties of clientship which have been outlined above. It is likely that the *mruigḟer* can depute a servant to fulfil some or all of his obligations to labour for his lord. According to a gloss on *Cáin Ṡóerraith*, the free client provides 'a man for every heifer every third year' (*fer cacha samaisce cach tres blíadain*).[64] It is unclear how much labour is involved in this arrangement. One could take it to refer to a third

[58] *CIH* ii 569.16 = *CG* 21.523. For another view on this difficult passage, see Ní Dhonnchadha, 'The *Lex Innocentium*', 63.

[59] *CIH* iv 1268.20–1269.20. The first stanza is edited by Watkins, 'Indo-European metrics and archaic Irish verse', 230–1.

[60] *CIH* iv 1269.10–11.

[61] *CIH* ii 531.3–24 = *AL* iii 38.7–40.28. Cf. *IP* (Irish Canon III) 168 §§ 4–5.

[62] *CIH* ii 532.8 = *AL* iii 42.19–20.

[63] *CIH* ii 563.18; iii 779.37–8 = *CG* 7.172. The text also distinguishes a less affluent *bóaire febsa*, with fourteen *cumals* of land (*CIH* ii 563.5–6; iii 779.25–6 = *CG* 6.152–3).

[64] *CIH* v 1770.24–5; cf. *CIH* ii 435.25; 436.13 = Crigger, 'A man is better than his birth', 329 (12); 331 (7).

of a day's work annually for each heifer given in the fief. As the heifer has half the value of a milch cow (p. 63), this would mean that a fief of twelve milch cows entitles the lord to eight man-days of labour annually.[65]

Flaith

The law-texts present a picture of a very class-conscious society with rigid ideas about the distinctions between a commoner and a lord (*flaith*). A lord is clearly not expected to carry out physical work,[66] and the *Annals of the Four Masters* record with evident horror that there was such a shortage of labour in the year 1085 as a result of famine that some rich men had to till their own land.[67] Nonetheless, a lord would certainly have kept a close eye on the vital activities of his farm.

Our sources are emphatic as to the impropriety of a king engaging in physical labour. According to *Críth Gablach*, if a king is found working with a mallet, axe or spade, his honour-price is reduced to that of a commoner (*aithech*).[68] The inference seems to be that any physical work is demeaning for a king. One can compare the legal commentary on loss of rank which refers contemptuously to the 'kingship of the three handles' (*ríge na trí larg*), which are listed as the handle of a flail, the handle of an axe and the handle of a billhook.[69] The commentator is likewise emphasizing the incompatibility of kingship and manual labour.

Women

The lines of demarcation between male and female work in the house and farm are not always clearly drawn in our sources, and it seems that many tasks could be performed by either sex. Much information on this subject is provided in the law-text *Cáin Lánamna*, which deals with the division of the assets of a marriage in the event of a divorce. The aim of the author is to establish a fair division on the basis of the amount of land and livestock

[65]The *bóaire febsa* receives a fief of twelve cows (*CIH* ii 563.9–10; iii 779.29 = *CG* 6.159). The value of the *mruigfer*'s fief is given as two *cumals* (*CIH* ii 564.11; iii 780.17 = *CG* 8.202); it is uncertain how many cows this represents: see p. 592.

[66]The sons of lords are trained for a life of leisure, hunting, and warfare: see p. 452.

[67]*AFM* ii 924 s.a. 1085 *co ndernaitt aittreabhaigh diaroilibh daoinibh saidhbhribh innte.*

[68]*CIH* ii 569.21–3 = *CG* 21.530–33. Cf. *ó rígh go rámhuinn* 'from king to spade', i.e. from the highest to the lowest (T. F. O'Rahilly, '*Rámha*, etc.', 363[5]).

[69]*CIH* i 254.4–5 = *AL* iii 106.14–15.

contributed by each party, and the amount of work (*aurgnam*) for which each has been responsible. Even if a wife has brought no property into the marriage, she is entitled to a proportion of the milk-products, young animals (*indud*), corn and salted meat.[70] However, this only applies if she has shown herself to be a hard worker (*márdéntaid*) during the marriage.

It is clear from *Cáin Lánamna* that it is regarded as normal for a farmer's wife to be involved in the major tasks of the farm, such as ploughing (*ar*), reaping (*búain*), looking after livestock in enclosures (*croud*),[71] and fattening pigs (*méthad*).[72] Literary sources likewise assume that it is regular for a husband and his wife (perhaps with older children) to be working together in the fields.[73] In many operations the husband is associated particularly with the earlier stages, whereas the wife is responsible for the finished product. For example, *Cáin Lánamna* states that if there is a divorce in a marriage of joint property, the wife is only entitled to one sixth of any wool still in fleeces, but she gets one third of any wool which has been combed, and half of any cloth.[74] The implication is that the husband has borne the brunt of the labour of rearing and shearing the sheep, whereas the wife has been responsible for combing the wool, spinning it into thread, and weaving it into cloth.[75] Similar ratios apply in relation to flax and woad: again the implication is that the husband has had most responsibility for these crops in the field, whereas the wife has been concerned with the preparation of linen and the processing of the woad-leaves for use in dyeing.[76] Indeed, one saint's Life

[70] *CIH* ii 515.6–8 = *SEIL* 56 § 27. For *indud*, see Ó Corráin, 'Irish law and canon law', 162.

[71] Greene, '*Cró, crú* and similar words', 5, takes *croud* (later *crúd*) to be a derivative of *cráo* (*cró*) 'enclosure'. He demonstrates that its basic meaning is 'to bring livestock into an enclosure and to care for them therein'. He compares Scottish Gaelic *crothadh uan* 'to bring lambs into an enclosure' and Icelandic *at króa lömb* of the same meaning. In Munster Irish *crúdh* developed the meaning 'milking'.

[72] *CIH* ii 509.1–3 = *SEIL* 33 § 14. When used of animals, the adj. *méth* is most commonly applied to pigs: see *DIL* s.vv. *méth, méthaid.*

[73] E.g. Ó hAodha, *Bethu Brigte*, 14.458–61 § 39.

[74] *CIH* ii 510.15–17 = *SEIL* 36 § 15.

[75] The list of women's possessions at *CIH* ii 379.4–12 = *AL* i 150.3–13 includes various implements connected with spinning, weaving and sewing. In early Ireland, as in medieval Europe generally, the distaff (*cuicél*) is regarded as the woman's implement *par excellence*, e.g. *CIH* ii 587.27 *cucæl . . . do cach mnaī* 'a distaff to every woman'.

[76] The preparation of woad for dyeing is treated as a woman's task in *Vita Sancti Boecii* (Plummer, *Vitae* i 95 § xxvii).

records a tradition that it was not proper for men to be present in a house when cloth was being dyed.[77] In legal commentary on the rearing of livestock in *Cáin Lánamna*, a distinction is regularly made between the husband's task of tending the pens in which the animals are kept (*líasrad*), and the wife's task of strewing rushes beneath them (*esrad*).[78]

The man has the major rôle in the sowing of cereals, harvesting, threshing, and drying the grain in a kiln. The grinding of the grain in a quern is usually the job of a woman – whether wife, daughter or slave – though an Old Irish text on Sunday-observance, *Epistil Ísu*, refers to the quern being used by a male servant (*feramus*) or a female servant (*banamus*).[79] The miller who grinds in a water-mill is normally male,[80] but it is a woman who kneads the dough in the kneading-trough (*losat*).[81] Both sexes may be involved in brewing. The professional brewer (*scóaire* or *ceirbsire*) is male,[82] but there is also a detailed description in *Bethu Brigte* of brewing carried out by nuns to provide beer to celebrate Easter.[83]

Livestock are herded by people of either sex, often by the young. However, literary and legal references suggest that the professional herd is usually male (p. 442). The law-text on status *Míadṡlechtae* expresses a strong prejudice against the freeman who devotes his time to herding cattle. Thus the *bógeltach* 'cow-grazer' is not entitled to honour-price (*díre*) or free status (*soíre*) because 'it is the work of a child or a woman that he does'.[84] Many texts emphasize the strong association of women with milking and dairying (*togairṭ*).[85] Hence, it is the wife rather than the husband who fattens up the

[77] Stokes, *Lismore Lives*, 121.4063–5.

[78] *CIH* ii 508.11–12 = *AL* ii 364.18–19 (discussed briefly by Thurneysen at *SEIL* 29); *CIH* v 1806.16 = *AL* ii 372.3–4; *CIH* i 176.1 = *AL* ii 414.29–30; *CIH* i 176.12–13 = *AL* ii 416.10. *Líasrad* is clearly a derivative of *lías* 'pen, fold', and *esrad* is from *esair* 'strewing, litter'.

[79] O'Keeffe, '*Cáin Domnaig*', 204 § 23.

[80] However, there is a Middle Irish reference to a female miller (*banmuilleóir*) at Stokes, *Acallamh na Senórach*, 132.4855.

[81] Hull, '*Noínden Ulad*', 28.10 § 1.

[82] See p. 333 above.

[83] Ó hAodha, *Bethu Brigte*, 7 § 21.

[84] *CIH* ii 585.16 = *AL* iv 352.23 *is gnīm meic no mnā dognī.*

[85] Ó hAodha, *Bethu Brigte*, 3 § 12; Connolly and Picard, 'Cogitosus: Life of Saint Brigit', 13 § 1.3; Mulchrone, *Bethu Phátraic*, 7.153; *CIH* iii 895.12–13. See also Lucas, *Cattle in ancient Ireland*, 42–3; Bitel, *Land of women*, 124–5.

pet pig on milk.[86] However, it seems that the professional milker (*bligre*) may be male or female.[87]

In general, the work of a woman is concentrated in or near the home to a much greater degree than that of a man. As in most societies, she has the major rôle in food-preparation: *Cáin Lánamna* refers to her job of feeding (*bíathad*) the household.[88] She has charge of the hearth, whereas the collection and splitting of firewood is a male task.[89] In view of her central rôle in child-rearing and housework, it is not surprising that the person who goes out to labour for a lord is assumed to be a man.[90]

Our sources tell us little about the daily routine of high-ranking women, but the emphasis seems to be on needlework and embroidery.[91] Physical toil is viewed as being beneath their dignity: it is clear from a reference in *Táin Bó Cúailnge* that the idea of noblewomen (*mná sáera*) working at querns is regarded with horror.[92]

Children

In our sources, the farming chore most commonly associated with children is that of herding livestock.[93] The Lives of the saints contain references to boys herding pigs,[94] cows,[95] sheep,[96] and calves.[97] Often, the boy is represented as taking his orders from his mother.

[86] *CIH* ii 509.1–3 = *SEIL* 33 § 14.

[87] A topographical tale at *LL* iii 747.22312; 748.22323 refers to a female milker (*bligrióir*) named Odras. On the other hand, Saint Brigit's milker is reputed to be a man named Colmán (Ó Riain, *Corpus genealogiarum sanctorum Hiberniae*, 144.300 = *LL* vi 1663.50985). Legal commentary at *CIH* iv 1276.5 refers to an oath by a *bligre*. As the oath of a woman is normally invalid (*GEIL* 202) this suggests that the *bligre* is regarded as male in this context.

[88] *CIH* ii 509.2 = *SEIL* 33 § 14.

[89] *CIH* i 82.6.

[90] E.g. *CIH* ii 435.25 *in fear gacha samaisce* 'a man for every dry heifer' i.e. a day's work (*manchaine*) by a man in return for each dry heifer (or equivalent) in the lord's fief.

[91] Many examples are given in *DIL* s.vv. 2 *druine, druinech, druinechas.*

[92] *TBC* Rec. I 40.1269–70.

[93] O'Keeffe, '*Cáin Domnaig*', 206 § 26. Cf. *CIH* ii 585.16 = *AL* iv 352.23; *CIH* vi 2139.4.

[94] Plummer, *Vitae* i 172 §§ viii–ix; Mulchrone, *Bethu Phátraic*, 25.421–2.

[95] E.g. Plummer, *Vitae* i 201-2 §§ iv–v; ii 77 § vii. *Búachaill* 'one who tends cows' (*LEIA* B-107) later acquired the meaning 'boy'.

[96] Plummer, *Vitae* i 235 § iii; Mulchrone, *Bethu Phátraic*, 7.134.

[97] Plummer, *Vitae* ii 178 § xxvii.

Vita Sancti Fechini recounts how the young Féchín is entrusted by his mother with the care of seven cows and a calf.[98]

Middle Irish commentary on the law-text on fosterage, *Cáin Íarraith*, provides some general information on the training which boys and girls of different social classes should receive.[99] Children of noble or royal birth get no training in the work of the kitchen or farm. The boys are prepared for a life of leisure, hunting and warfare. They learn the board-games *fidchell* and *brannuigecht*, and how to swim, ride a horse and throw a spear. The girls learn how to sew, cut cloth and embroider.[100]

The training of children of commoner rank (i.e. the offspring of a *bóaire* or *ócaire*) is quite different. According to one version of the commentary, the girls are taught the use of the quern and the kneading-trough, and how to look after lambs and kids. The boys are taught how to dry grain (*tírad*) and to split firewood.[101] The other version also includes wool-combing (*círad*) and the care of lambs, calves, kids and young pigs among the skills which boys must learn.[102] In a section of legal commentary on miscellaneous topics, it is stated that an adopted son (*mac fáesma*) is put to work at fencing (*ime*) and weeding crops (*gortglanad*).[103]

Clergy and monastic clients

Just as a secular lord has clients from whom he receives rent and services in return for a fief, so the Church has monastic clients who fulfil the same rôle.[104] The monastic client is of such vital importance in the economic organisation of the Church that the word *manach* – a borrowing from Latin *monachus* 'monk' – is often used of a monastic client, a layman dependent on the Church. Out of

[98] Plummer, *Vitae* ii 77 § vii; cf. i 202 § v; ii 206 § 2.

[99] Fragments of this Old Irish text survive, interspersed with commentary dating from about the twelfth century. It is likely that the commentary reflects the content of the missing portions of text. For a discussion of the reliability of commentary as historical evidence, see p. 11 above.

[100] *CIH* v 1760.33–4 = *AL* ii 154.5–7; *CIH* v 1761.24–5 = *AL* ii 156.28–30. There is a shorter version at *CIH* i 82.13–14.

[101] *CIH* i 82.5–6.

[102] *CIH* v 1760.21–2 = *AL* ii 152.10–12. It is difficult to reconcile the evidence of the two versions on the training of young commoners, and there is probably a scribal error in the MS tradition. It is certainly strange that there is no mention in this version of a girl being trained in the care of domestic animals.

[103] *CIH* ii 660.40. *Gortglanad* is literally 'field-cleaning'.

[104] See discussion by Charles-Edwards, 'The Church and settlement', 171–5.

context, therefore, we cannot tell whether *manach* is being used in the sense of 'monk' or 'monastic client'. The institution of monastic clientship was clearly so widespread that the abstract *manchaine* is applied not only to the services which a monastic client carries out for the Church but also to the services which a secular client carries out for his lord. In the law-texts, the monastic client is also called a *bachlach*.[105] The opening paragraph of *Bretha Comaithchesa* compares the *bachlach* who is under the control of a monastic superior (*airchinnech*) with the secular client (*aithech*) who is under the control of a lord (*aire*).[106] Similarly, in an Old Irish passage on advocates, the term *bachlach* is applied to a commoner of *bóaire* rank who is a client of the Church – the secular equivalent being an *aithech*.[107] *Bretha Nemed Toísech* stresses that the contract which a monastic client (*manach*) makes with his superior (*airchinnech*) lasts until the end of his life.[108]

It is also clear from legal sources that some of the Church's dependants were at a social level corresponding to a *fuidir* rather than to a freeman of *ócaire* or *bóaire* rank.[109] *Córus Bésgnai* refers to the lack of legal capacity of a *dáermanach ecalsa*, i.e. a base *manach* of the Church.[110] According to the accompanying gloss, such a person may have entered into dependency on the Church by being ransomed from execution or saved from famine. An entry in *O'Davoren's Glossary* suggests that the heaviest work on church land – such as the construction of a stone pavement (*clochán*), a stone wall (*caisel*) or a causeway (*tóchar*) – would be carried out by dependants of this type.[111]

As we have seen above (p. 448), persons of noble or royal rank are not expected to do physical work. In religious texts, on the other

[105] *Bachlach* is a derivative of *bachall* (from Latin *baculum*) 'crozier, pastoral staff'; cf. the Breton cognate *baelec* (modern *beleg*) 'priest' (*VKG* i 225–6 § 140 (1)). In later sources, *bachlach* is applied to low-ranking servants or slaves, e.g. *CIH* vi 2307.9 = Binchy, '*Bretha Déin Chécht*', 26 § 6²; Dillon, *Lebor na Cert*, 98.1451.

[106] *CIH* i 64.8–9 = *AL* iv 68.3–5; *CIH* iv 1485.28–9 = O'Dav. 281 § 524.

[107] *CIH* iii 896.24 = L. Breatnach, 'Lawyers in early Ireland', 11–12 *ó bóairechuib aithech 7 bachlach*.

[108] *CIH* vi 2231.3–7.

[109] In later sources, a low-ranking tenant on Church land is called a *scolóc*: see *DIL* S 101.76-86. The basic meaning of the term is 'student, pupil', but it was commonly applied to the lowest grade of monastic tenant: see Doherty, 'Some aspects of hagiography', 314.

[110] *CIH* ii 522.1 = *AL* iii 10.16.

[111] *CIH* iv 1515.3 = O'Dav. 412 § 1221 s.v. *manach*.

hand, there is strong emphasis on the spiritual benefits of labour, even for high-ranking clerics.[112] For example, saints' Lives refer on many occasions to abbots such as Colmán, Comgall, and Mo Chóemóc at work in the fields,[113] and even in his old age, Bishop Mo Ling is reputed to have joined the monks in digging the soil.[114] A legal gloss refers to a cleric (*clérech*) filling his shoes with straw in preparation for work in the cornfield in autumn.[115]

Our sources also provide evidence of debate within the Church as to the purpose of physical toil in the religious life. The ascetic abbot Máel Rúain of Tallaght is quoted as saying – in response to a query from the anchorite Máel Díthruib – that his monks must work at reaping, flailing, hammering (stakes) or ditch-digging, and that their daily recitation of fifty psalms is to be regarded as 'additional work' (*foropair*).[116] Saint Mo Chuta's excessive demands on the labour of his monks is implicitly criticized in Lives of this saint. He allowed no oxen or other domestic animals into his monastery, so the monks had to prepare the fields with spades rather than ploughs. Another saint, Fínán, complained to Mo Chuta that he had turned his monks into brute beasts, and insisted that they be released from their shameful toil.[117]

The purpose of the early Irish monks' labour was to achieve self-sufficiency for their community as well as to gain spiritual benefits. A Life of Mo Lua represents the saint as addressing his monks in the following terms shortly before his death: 'My dearest brothers, cultivate the land well and labour well, so that you may have enough food and drink and clothing. For where there will be sufficiency among the servants of God, there will be stability; and where there will be stability in service, there will be religion. For the end of religion will be eternal life . . .'.[118] A passage in eleventh-century commentary on *Félire Óengusso* also deals with the idea of monastic self-sufficiency.[119] It recounts how fifty bishops from the Britons of

[112] J. Ryan, *Irish monasticism*, 360–64.
[113] Plummer, *Vitae* i 264 § xv; ii 11 § xxii; 177 § xxvi.
[114] Plummer, *Vitae* ii 204 § xxix.
[115] *CIH* ii 421.14–16; v 1721.14–16 = *AL* i 300.21–4.
[116] Gwynn and Purton, *The monastery of Tallaght*, 133 § 16. I take *forchae* here to refer to the mallet used to drive in fencing-stakes, etc: see p. 492.
[117] Plummer, *Bethada* i 296–7 § xxi (32) (trans. ii 287–8); Plummer, *Vitae* i 188 § xlvi.
[118] Plummer, *Vitae* ii 223 § li.
[119] Stokes, 'On the Calendar of Oengus', p. xl; Stokes, *Félire*, 54. A fuller version of the story is to be found in *LL* v 1243.36874–903.

Cell Muine came on a pilgrimage during Lent to Saint Máedóc of Ferns.[120] The steward brought fifty wheaten loaves with leeks and whey-water. However, this meal was refused by the chief bishop,[121] who demanded a pig and an ox instead. Máedóc agreed to this, but on the following morning remonstrated with his visitors for eating meat during Lent. The chief bishop explained to him that the pig had been fed on its mother's milk, and the ox had eaten nothing but the grass of the earth. The only work needed for their rearing had been protection from wolves and robbers. It was therefore not sinful to eat them because of the small amount of labour involved. In the case of the bread, on the other hand, there were three hundred and sixty-five additional labours (*foropre*)[122] for each loaf, i.e. there was expenditure of sweat towards each loaf on every day of the year.[123] The bishop claimed that this amount of labour was greater than Máedóc's dignity. Máedóc therefore vowed to eat no bread unless it was produced by the work of his own hands. He and his monks[124] then made a clearing (*ceppach*) in woodland to the south of Ferns, and bread was baked from grain grown on this land. The author concludes the story by underlining the moral: no one should eat something without having contributed work or prayer.[125]

REST, HOLY DAYS, FESTIVALS, ETC.

Sunday

It seems likely that in pre-Christian times normal work stopped only for pagan festivals or other communal gatherings. One of the effects of the introduction of Christianity was the establishment of the Judaic idea of a weekly day of rest during which only certain essential or emergency tasks were allowed. It is clear from the texts that the early Irish Church treated the proper observance of Sunday

[120] Cell Muine is St. Davids in Wales: see Hogan, *Onomasticon* s.v. *Cell Muni*.

[121] The *Félire Óengusso* version has simply *int eps*cop, but the *LL* version has *int eps*cop *uasal* 'the high bishop' (*LL* v 1243.36883).

[122] Note Stokes, *Félire*, 470 (Corrigenda to p. 55, l. 41) where the translation of *foropre* as 'weevils(?)' is corrected to 'extra labours'.

[123] The rest of the story is found in the *LL* version only.

[124] The text simply has *siat* 'they' (*LL* v 1243.36900). Presumably this refers purely to Máedóc and his monks – there was hardly assistance from the fifty bishops.

[125] The text (*LL* v 1243.36902) has *lubra ł ernaigthi* 'work or prayer', but it is possible that *ł* should be emended to 7 in view of the emphasis on physical work throughout the story.

as a very serious matter indeed. Ninth-century ecclesiastical legislation entitled *Cáin Domnaig* 'the law of Sunday' adapts the Church's teaching on this matter to Irish conditions, and places severe penalties on the person who fails to observe it.[126] An offender must pay a fine of four dry heifers, and forfeits the clothing which he is wearing as well as any equipment or valuables which he may have with him. If he is riding a horse, it too is forfeit. The fine is divided between the local lords and the person who prosecutes the offender. If somebody witnesses another person working on a Sunday, and fails to prosecute, he is held to be as guilty as the actual offender and must pay the same fine.[127]

The text concludes with a strongly worded general denunciation of those who violate the sanctity of Sunday, and describes them as 'worse than demons'. On account of their activities, God brings plagues on the fields. There is failure of corn, milk, fruit, fish and every produce of sea and earth. There are famines, hungers, unseasonable weather, and raids for slaves by foreigners with avenging swords.[128] Similar abhorrence of the violation of Sunday is expressed in many other texts.[129]

Cáin Domnaig devotes much attention to defining the exact circumstances in which work is permitted on a Sunday. In general, travel (*imthecht*) is forbidden except for essential journeys, such as going to church, seeking a physician or midwife, chasing a robber, going in the direction of a shout or scream, etc. Most farmwork is banned, and there must be no loading of an ox or horse or man, and no splitting of firewood. It is, however, permitted to tend cattle, rescue a cow from wolves or a marsh, bring a bull to a cow which is in heat, drive cattle to water, follow a swarm of bees, or deal with broken fences or trespassing cattle.[130] In the house, there must be

[126] For an edition of this text, see Hull, '*Cáin Domnaig*'. It is similar in content to the second part (i.e. §§ 17–33) of *Epistil Ísu* 'the epistle of Jesus', ed. O'Keeffe, '*Cáin Domnaig*', 200–11.

[127] Hull, '*Cáin Domnaig*', 162 § 2.

[128] Hull, '*Cáin Domnaig*', 171 §§ 10–11.

[129] E.g. Plummer, *Vitae* i 43 § xxx; 263 § xii; Van Hamel, *Immrama*, 106.361–71 § 14 = Stokes, 'The voyage of the Húi Corra', 50 § 61.

[130] Hull, '*Cáin Domnaig*', 160–2 § 1. I follow him in taking *aurbaidi mruigrechta* 'breaches of land-law' to refer mainly to problems of broken fences and trespassing cattle which must be dealt with immediately. This is supported by the statement in *Epistil Ísu* (O'Keeffe, '*Cáin Domnaig*', 210) that it is permitted to go to protect a field of corn or pasture which is being damaged, presumably by cattle which have broken in.

no cleaning, grinding, or churning. It is, however, permissible to get water or fire. The regulations on cooking are complex, and – as in modern sabbatarian practice – varying degrees of severity are applied. According to *Cáin Domnaig*, baking (*fuine*) is banned, but a person is permitted to go to a cooking-pit (*fulacht*), presumably to cook food in water. On the other hand, *Epistil Ísu* forbids the boiling of food (*berbad bíd*).[131] The rule of the monastery of Tallaght is similarly strict: the monks are not allowed to eat cabbage which has been boiled on a Sunday, or a vegetable which has been cut on a Sunday, or bread which has been baked on a Sunday, or nuts or blackberries which have been picked on a Sunday.[132]

Some texts deal with the correct behaviour when the duties of Sunday-observance and hospitality come into conflict. *Cáin Domnaig* states that a person is permitted to get food for guests on a Sunday, and to minister to them for the sake of Christ.[133] The prohibition on Sunday grinding may also be relaxed in the interests of hospitality. In his writings of the late eleventh century, bishop Patrick of Dublin records among the 'wonders of Ireland', the tradition of the miraculous mill which will grind on a Sunday if a guest has arrived, but will otherwise refuse to move on that day.[134] *Epistil Ísu*, however, allows no such exceptions, and lays down severe penalties for Sunday grinding, whether in a mill or hand-quern. For example, if a water-mill belonging to a church is used on a Sunday, the culprit must pay the fine of one *cumal*.[135]

Other holy days

Similar restrictions apply in the case of certain other holy days. Thus *Epistil Ísu* forbids unnecessary activity on Christmas Day or on Little Christmas (Epiphany).[136] However, it is clear that there is no ban on cooking at Easter or Christmas, as both are times of feasting and the relaxation of penitential diets (see p. 359).

[131] O'Keeffe, '*Cáin Domnaig*', 202 § 17.

[132] Gwynn and Purton, *The monastery of Tallaght*, 132 § 13.

[133] Hull, '*Cáin Domnaig*', 160 § 1 (lines 8 and 13).

[134] Aubrey Gwynn, *The writings of bishop Patrick 1074–84*, 66 § xxii.

[135] O'Keeffe, '*Cáin Domnaig*', 204 § 23. Middle Irish commentary on the law-text on water-mills restricts milling to six days. In the seventh-century text itself, however, milling goes on all week. For discussion, see Binchy, '*Coibnes Uisci Thairidne*', 59.

[136] O'Keeffe, '*Cáin Domnaig*', 208 § 28 *Ciped laa didiu forsa mbē notlaic mór nó notlaic stéille, is amal domnach insin ⁊ ní himthíagar and* 'On whatever day Great Christmas or Christmas of the Star falls, it is like Sunday and one should not go about thereon'.

Assemblies and festivals

Our sources provide some information about other times in which normal farmwork is curtailed or suspended. Freemen would be expected to assemble for the *airecht,* at which legal business is transacted.[137] A larger assembly is the *óenach* (usually translated 'fair'), to which numerous references are made in legal, annalistic, and literary sources. It is clear from *Críth Gablach* that every king is expected to hold an *óenach* for the people of his *túath,*[138] and other law-texts stress the subjects' duty of clearing and preparing the site.[139] Though not explicitly stated in the legal sources, it seems likely that an *óenach* is normally held each year in early August.

The most prominent fair in our texts is *Óenach Tailten* 'the fair of Tailtiu', convened by the king of Tara.[140] In pre-Viking times, it seems to have been an annual event, unless some exceptional circumstance prevented its celebration. For example, the *Annals of Ulster* record that in the year 811, the monastic community of Tallaght fasted against the king of Tara in protest against the violation of the sanctuary of Tallaght.[141] This sanction was sufficient to ensure that not a single horse or chariot came to the fair. The same annals complain in 873 'that the Fair of Tailtiu was not held, although there was no just and worthy cause for this – something which we have not heard of from ancient times'.[142] Thereafter, the celebration of the fair seems to have been very sporadic; in 1006 Máel Sechnaill II held the fair after a break of at least seventy-nine years.[143] This fair was described in glowing and no doubt exaggerated terms by the contemporary poet Cúán Ua Lothcháin. Nonetheless, his verses provide us with some general information on the history and function of *Óenach Tailten,* which was held in the kalends of August at Tailtiu (now Teltown, Co. Meath).[144] Its primary purpose was to celebrate the festival of *Lugnasad,* which marked the beginning of the harvest. Cúán refers to the festival's

[137] *CG* Legal glossary 73.
[138] *CIH* ii 569.2–4 = *CG* 20.502–5.
[139] E.g. *CIH* ii 381.30 = *AL* i 156.28.
[140] See Binchy, 'The fair of Tailtiu and the feast of Tara'. Cf. Meyer, *Triads*, 4 § 35.
[141] AU^2 266 s.a. 810 (recte 811) § 2.
[142] AU^2 330 s.a. 872 (recte 873) § 6.
[143] E. J. Gwynn, *Metrical Dindshenchas* iv 160.189–90, and discussion in his notes, p. 419.
[144] E. J. Gwynn, *Metrical Dindshenchas* iv 146–62.

association with Lug, a prominent deity in the Celtic pantheon.[145] In *Cormac's Glossary*, compiled about AD 900, the word *Lugnasad* is interpreted as *Lug-násad* 'Lug's festival', an etymology which is generally accepted by modern scholars. In Cormac's explanation, the festival is explicitly identified with 'the fair which used to be held by him (Lug) at the beginning of autumn every year at the coming of *Lugnasad*'.[146]

Another topographical poem provides an account of the week-long triennial fair of Carmun, which may have been held in the Curragh of Kildare.[147] Like the fair of Tailtiu, this was an early August celebration of the beginning of the harvest, with horse-racing, music, and story-telling. There may be some historical basis in the poet's claim that this festival was attended by women as well as by men, though he is at pains to stress that they stayed in separate assemblies.[148] He also states that legal matters were discussed,[149] and that there was a market of food, a market of livestock, and a 'great market of foreigners' where gold and fine raiment could be bought.[150]

In her pioneering book, *The festival of Lughnasa*, Máire MacNeill provides a large amount of information from early Irish sources and from later folklore on traditional gatherings at the beginning of August or thereabouts. It is clear that this is the major agricultural festival of the year in Irish tradition. In modern times, it survives under various guises, such as the annual pilgrimage to Croaghpatrick, Co. Mayo, or the Fraughan Festival in Glencullen, Co. Dublin.[151] *Lugnasad* is essentially a festival to celebrate the ripening of the fruits of the earth: in a note by a sixteenth-century scribe it is described with some exaggeration as 'the day of ripening

[145] E. J. Gwynn, *Metrical Dindshenchas* iv 150.46.

[146] Meyer, *Sanas Cormaic*, 66 § 796 *Lugnasad .i. nāsad Loga maic Ethlend .i. aonach nofertha lais im taite foghmair in gach bliadhain im thoidecht Lugnasad.*

[147] E. J. Gwynn, *Metrical Dindshenchas* iii 2–24. See his notes on pp. 470–1 for a discussion on the location of Carmun and the authorship of the poem.

[148] E. J. Gwynn, *Metrical Dindshenchas* iii 18.225–8.

[149] E. J. Gwynn, *Metrical Dindshenchas* iii 18.213–16.

[150] E. J. Gwynn, *Metrical Dindshenchas* iii 24.306–8 *marggad bíd, marggad beóchraid,/ marggad mór na nGall ngrécach/ i mbíd ór is arddétach.*

[151] As Máire MacNeill points out (*The festival of Lughnasa*, 38–9) the fruit of the bilberry (Irish *fraochán*, anglicized *fraughan*) features prominently in customs associated with the festival of Lughnasa, presumably because it is one of the earliest fruits of the season.

of all fruits'.[152] Because a good harvest of fruit and corn is believed in Irish tradition to depend on the justice of the king,[153] it is obvious that he must figure prominently in such a celebration. This point is particularly stressed in Cúán Ua Lothcháin's poem on the *Óenach Tailten*.[154]

Our sources provide less information about other festivals. The spring festival of *Imbolc* was celebrated early in February.[155] It is likely that this festival was originally associated with the earth goddess *Brigit* 'the exalted one',[156] but was taken over by the Church early in the Christian period. The goddess was transformed into a saint, whose mythical career is described in *Bethu Brigte* and other texts, and whose feast is celebrated on 1st February.[157] However, many pagan elements are found in traditional lore on Brigit.[158] The tenth-century scholar Cormac mac Cuilennáin provides a gloss *Oímelc .i. oi-melg .i. isi aimser andsin tic as caírach* '*Oímelc* i.e. sheep-milk, i.e. that is the time that sheep's milk comes'.[159] *Oímelc* here seems to be a glossator's deformation of the word *Imbolc* to provide an explanatory etymology.[160]

The festival of *Beltaine* at the beginning of May celebrates the coming of summer. It is the time when cattle are driven to the hills for summer-grazing (p. 43). Máire MacNeill points out that the customs associated with this festival contain strong undertones of anxiety, and are mainly directed towards averting the dangers which

[152]Flower, *Catalogue* ii 261; Meyer, *Hibernica minora*, 49[10] *Lā aipchi na n-uili thorad*.

[153]This is the central theme of the wisdom-text *Audacht Morainn* (Thurneysen, 'Morands Fürstenspiegel'; Kelly, *Audacht Morainn*).

[154]E. J. Gwynn, *Metrical Dindshenchas* iv 160.209–16; cf. iii 18.217–18.

[155]See Vendryes, '*Imbolc*'; Hamp, '*Imbolc, Óimelc*'.

[156]*LEIA* B-87 s.v. 1 *brí*.

[157]Stokes, *Félire*, 58. By the sun (i.e. length of day) and because the Julian year was slightly too long, 1st February in the seventh century corresponds approximately to 30th January in the ninth century and, following the Gregorian calendar-change of 1582, to about 4th February thereafter. We therefore celebrate the Feast of Saint Brigit three days earlier by the sun than was the case in the seventh century. For this information, I am grateful to Dan McCarthy of the Department of Computer Science, Trinity College, Dublin.

[158]Mac Cana, *Celtic mythology*, 34–5.

[159]Meyer, *Sanas Cormaic*, 86 § 1000.

[160]E.g. *LEIA* O-17 s.v. *oimelc*. Cf. Van Hamel, *Compert Chon Culainn*, 43 § 55. Hamp, '*Imbolc, Óimelc*', 109, suggests that one might read *oimelc* with a short *o-*, possibly reflecting an original **ommolg* (**uss-molgo-*) 'milking'; cf. Russell, 'The Celtic preverb **uss*', 109.

threaten cattle and dairy-processes.[161] Many *Beltaine* rituals for the protection of cattle and other livestock have been recorded by folklorists in the nineteenth and twentieth centuries, and some are probably of great antiquity.[162] Cormac mac Cuilennáin refers to a tradition that chanting druids used to drive cattle between two fires at *Beltaine* to protect them from diseases.[163] Admittedly, Cormac compiled his glossary at least five centuries after the introduction of Christianity to Ireland; but he may nonetheless have preserved here a tradition going back to pagan times.

The festival of *Samain* at the beginning of November marks the start of winter.[164] It features prominently in early Irish mythological literature as a time when the barriers between the natural world and the Otherworld are removed, and many tales deal with encounters at *Samain* between mortals and the fairy people.[165] *Samain* is a time for public assembly (*óenach*) and feasting (*feis*).[166] The main dish at such feasts seems to have been the *banb samna* 'the piglet of *Samain*': see p. 85.

From the early Irish farmer's point of view, the festivals of *Beltaine*, *Lugnasad* and *Samain* are of particular economic significance in that the value of young animals goes up at the time of a festival. For example, a female calf is worth two scruples from its birth until *Samain* when it acquires the value of three scruples. At the following *Beltaine* its value rises again to four scruples, and then to six scruples at the next *Samain*. It attains its maximum value of twenty-four scruples at its sixth *Beltaine*.[167] Likewise, the value of sheep and pigs may go up at the festivals of *Beltaine*, *Lugnasad*, or *Samain*.[168]

Finally, a ninth-century monastic rule preserves a tradition of a celebration called *Féil na n-airemon* 'the festival of the ploughmen'.

[161] Máire MacNeill, *The festival of Lughnasa*, 69.
[162] Danaher, *The year in Ireland*, 109–27.
[163] Meyer, *Sanas Cormaic*, 12 § 122; cf. *CIH* ii 606.17–20 and Van Hamel, *Compert Chon Culainn*, 43 § 55. One pseudo-etymological explanation given for *Beltaine* is *beil-tine .i. tene bil* 'i.e. fortunate fire'. For further discussion on the fire ritual at *Beltaine*, see Binchy, 'The fair of Tailtiu and the feast of Tara', 129.
[164] *Félire Óengusso* (Stokes, *Félire*, 232) places *Samain* on 1st November. This date corresponds by the sun to about 4th November in the twentieth century. See p. 460 above.
[165] MacCana, *Celtic mythology*, 127.
[166] *DIL* S 49.11–17.
[167] *CIH* iii 845.22–38 = Appendix A, text 3 §§ 1–6.
[168] *CIH* iii 845.39–846.14 = Appendix A, text 3 §§ 7–12.

It is said to have been instituted by the sixth-century founder of the monastery of Iona, Columb Cille, for his monks. This festival is celebrated when the crops have reached their full growth (*is and forcenntae a n-ás*) after three months of tending and watering.[169] Máire MacNeill has suggested that it is to be equated with *Lugnasad*, at the beginning of August.[170] However, as we have seen, *Lugnasad* is a celebration of the ripening of crops rather than their attainment of full growth. One would expect the festival of the ploughmen to be held in June, i.e. three months after the normal time of ploughing and sowing. Possibly, it is to be equated with Saint John's Eve at midsummer, which has traditionally been celebrated with bonfires, music and dancing.

[169] Gwynn and Purton, *The monastery of Tallaght*, 156 § 68.
[170] Máire MacNeill, *The festival of Lughnasa*, 2–3, 40.

14
Tools and Technology

Our sources contain many references to simple agricultural tools such as the spade, shovel, mattock, sickle, threshing-stick, etc., and also frequently mention more complex machinery such as the plough and the water-mill. In the law-texts the concern of the authors is often with the legal consequences of injuries caused accidentally by agricultural equipment. There is also emphasis on the necessity for farmers of various ranks to own tools and other equipment appropriate to their status in society. For example, the early eighth-century law-text *Críth Gablach* gives much information on what a farmer should possess, and gives an especially valuable account of the equipment of the *mruigfer*, a prosperous commoner.[1]

In this chapter I refer frequently in the footnotes to S. Rees, *Agricultural implements in prehistoric and Roman Britain* i–ii, and to K. D. White, *Agricultural implements of the Roman world.* These references are included for the purpose of comparison, and because of the linguistic evidence of influence from the Roman world on the tools used in early Christian Ireland, e.g. Irish *rastal* 'rake' (Latin *rastellus*), Irish *coltar* 'coulter' (Latin *culter*), Irish *pistal* 'yoke-peg(?)' (Latin *pistillum*), Irish *súst* 'threshing-stick' (Latin *fustis*), Irish *srathar* 'pack-saddle' (Latin *stratura*). But it must be stressed that the use of agricultural tools in Ireland goes back to Neolithic times (see p. 3 above), so Romano-British influence is only one element in the evolution of Irish tools.

[1] *CIH* ii 563.20–5; iii 779.39–780.4 = *CG* 7.174–82. Compare the lists of agricultural tools and other equipment in Cato, *De agri cultura* (ed. Hooper and Ash), 10.1–14.5.

Plate 7. The left half of the bottom margin of f. 5^{v} of a fifteenth-century Irish medical manuscript (MS 17) in the Library of King's Inns, Dublin, showing two men wearing English-style headgear. The man on the right is using a spade. It is not possible to identify the tool used by the other man; it may be a two-handled instrument for grubbing up bushes. The original of this illustration was made available through the courtesy of the Council of King's Inns, and is published by their permission.

CULTIVATION

Spade

Tools of cultivation are mentioned in many texts, but little precise information is given. The commonest term for 'spade' is *rámae* (later also *rámann*). The etymology of *rámae* provides evidence about the development of this tool: it contains the same root as *ráid* 'rows', and has the basic meaning of 'oar'.[2] It seems, therefore, that the early Irish spade originated as an oar-shaped digging implement of wood.

Spade-dug cultivation ridges at Carrownaglogh, Co. Mayo, go back at least to the early Bronze Age.[3] Wooden spades have not turned up in excavations in Ireland, but a Bronze Age wooden shovel has been found in Co. Cork (see below). Wooden spades have been found at Roman sites in Britain; on some of them, the marks of wear on the blade are still clearly discernible.[4] The commonest spade-type found at Roman sites in Britain is a wooden spade fitted with an iron shoe at the base of the blade. The iron shoe of such a spade has been found in the excavation of a crannóg at Ballinderry, Co. Westmeath.[5] Spades with all-iron blades are less common in Roman Britain,[6] and have not been identified from the early Christian period in Ireland.

Another word which is generally taken to refer to a spade in Old Irish is *fecc*.[7] For example, in the Otherworld wanderings which brought them to Purgatory, the Uí Chorra brothers came across a man with a fiery *fecc* in his hand, ceaselessly digging (*oc rómur*). This

[2] *LEIA* R-1 s.v. *rá-*. Welsh *rhaw* 'spade' shows the same semantic development. Another Irish word for spade is *rúam*, of which *rúamar* 'digging' is presumably a derivative (*LEIA* R-48 s.v. *ruam*; T. F. O'Rahilly, '*Rámha*, etc.', 365). O'Rahilly takes *rúam* to be 'the Ivernic counterpart of Welsh *rhaw*' but supplies no conclusive evidence.

[3] Illustrated at Mitchell, *The Irish landscape*, plate 27 (opposite p. 192); *Reading the Irish landscape*, 197.

[4] Rees, *Agricultural implements* i 320–2. There are also uncertain records of wooden spades or shovels from Neolithic and Bronze Age sites in Britain, ibid. 320–1.

[5] Hencken, 'Ballinderry crannóg no. 2', 47. This spade is also illustrated by Edwards, *The archaeology of early medieval Ireland*, 61.

[6] Rees, *Agricultural implements* i 326–9.

[7] It is presumably from the same root as *feccaid* 'bends', so we can guess that it was originally applied to a spade with the blade set at an angle to the handle. In later Irish *feac* was applied both to 'spade' and 'spade-handle' (*DIL* F 51.60-3).

was his penance for having dug on a Sunday during his lifetime.[8] This tool is also mentioned in a legal passage which lists the standard belongings (utensils, clothing, etc.) which are expected to be in the possession of a monk on his death: they are called his *íargrinne* 'after-bundle'.[9] Three tools are included: his billhook (*fidbae*), his spade (*fecc*) and his shovel (*slúasat*).

For spades used in excavation, mining and turf-cutting, see p. 479.

Shovel

The blade of the shovel is more concave than that of the spade, and is designed to remove loose material during agricultural operations, mining, etc. A wooden pick and shovel have been found in a Bronze Age copper mine (*c.*1700–1600 BC) at Mount Gabriel, Co. Cork.[10] A wooden shovel from the early Christian period has been found at Moynagh Lough, Co. Meath.[11] In an episode in the *Genealogies of the saints* (see p. 148), there is a reference to the use of a shovel (*slúasat*) to clear the paved entrance-path (*aurdrochat*) in front of a house,[12] and an entry in the *Annals of Loch Cé* records the removal of snow with shovels.[13] The early twelfth-century tale *Aislinge Meic Con Glinne* refers to the use of the shovel to clear away manure.[14]

The word *slúasat* is also attested in the meaning 'oar, paddle'. T. F. O'Rahilly suggests that this meaning is the original one, and compares the semantic development of *rámae* from 'oar' to 'spade'.[15]

Fork

Heavy three-pronged agricultural forks are known from Roman sites in northern Europe, but this type has not been found in

[8]Van Hamel, *Immrama*, 106.361–2 = Stokes, 'The voyage of the Húi Corra', 50 § 61. Stokes takes *oc rómur* to mean 'rowing', but 'digging' – as in Van Hamel – makes better sense.

[9]*CIH* ii 574.36–575.3.

[10]William O'Brien, *Mount Gabriel: Bronze Age mining in Ireland*, 144–9.

[11]Bradley, 'A separate-bladed shovel from Moynagh Lough'.

[12]Ó Riain, *Corpus genealogiarum sanctorum Hiberniae*, 111 § 668 = *LL* vi 1579.48483–1580.48491.

[13]*ALC* ii 484.7 s.a. 1588.

[14]Jackson, *Aislinge*, 35.1096 = Meyer, *Aislinge*, 91.9. Cf. Plummer, *Bethada* i 325 § xiii (42); trans. ii 316.

[15]T. F. O'Rahilly, '*Rámha*, etc.', 363–4. He takes *slúasat* to contain the Indo-European root *pleu-* 'to sail, float'.

Roman Britain.[16] So far as I have been able to discover in the published records, no agricultural fork has been identified in any pre-Norman site in Ireland.

Latin *furca* 'pitchfork' has been borrowed into Irish as *forc*, but seems only to be used of a type of three-pronged weapon.[17]

Mattock/Hoe

In modern archaeological writings, the mattock is generally distinguished from the hoe by certain differences of construction and function. The term 'mattock' is used of an implement with a right-angled digging blade on one side of the head and an axe-type blade on the other side. The term 'hoe' is used of an implement with a right-angled digging blade on one side of the head, but no axe-type blade on the other side. There may, however, be another type of supplementary blade (or blades) on this side. For example, one Romano-British type has a digging blade on one side and two fork-like prongs on the other.[18] In archaeological terms this is nonetheless classed as a hoe.

It does not seem to be possible to make any such distinction between a mattock and a hoe in our texts. I therefore use the term 'mattock' to translate Old Irish *bacc*, the basic meaning of which is 'bent, crooked'.[19] Some texts make an explicit distinction between a *bacc* and a *rámae* (spade). For example, a legal gloss gives the two 'valuables of the field' as the *bacc* and the *rámae*.[20] A topographical poem on the origin of the plain of Mag Raigne attributes its clearance to a legendary Roman named Raigne. His tools were an axe (*túag*), a mattock (*bacc*) and a spade (*rámann*).[21] The clearance of the plain of Mag Femin was similarly effected by the brothers Femen and Fera with an iron shovel (*slúasat*) and

[16] Rees, *Agricultural implements* i 304.

[17] See *DIL* s.v. *forc*.

[18] Rees, *Agricultural implements* i 309.

[19] *Bacc* and its derivative *baccán* are used of a variety of crooked and curved objects: see *DIL* s.vv. and *LEIA* B-2 s.v. *bacc*. The word *fecc* seems also to have the meaning 'bent', but all the evidence points to it being a spade rather than a mattock: see p. 465 above.

[20] *CIH* ii 473.7 = *AL* v 410.17 *do sētaib faithchi .i. bac ⁊ rāma*.

[21] E. J. Gwynn, *Metrical Dindshenchas* iii 194–6. In the prose version at *LL* iii 702.21120–4 he used a spade (*rúam* or *rámann*) and a mattock (*bacc*); cf. Stokes, 'The prose tales in the Rennes Dindṡenchas, (1)', 434–5 § 43.

a mattock (*bacc*).[22] They took turns shovelling (*ic fuilged*) and working with the mattock (*ic baccad*).

Drag-hoe/Rake

Classical Latin *rastrum*, usually translated 'drag-hoe', is applied to a toothed implement for preparing a seedbed, breaking up clods, removing weeds or covering over seeds.[23] Its Late Latin diminutive *rastellus* (*rastellum*) is used of a wooden rake,[24] and has been borrowed into Irish as *rastal.*[25] In post-Norman times, the influence of English agricultural practice brought in the word *ráca*, a borrowing from English *rake.*[26]

Plough

There are many references to ploughs and ploughing in the Old Irish law-texts, and it is clear that ploughing was regarded as the normal means of preparing land for the sowing of cereals.[27] According to the law-text on status, *Críth Gablach*, a farmer of *mruigfer* rank is expected to own a complete set of ploughing equipment, as well as six oxen.[28] A farmer of lower rank who does not have enough oxen to pull a plough is expected to make a co-ploughing arrangement (*comar*) with other farmers of similar rank.[29]

Both the archaeological and the linguistic evidence attest to the antiquity and wide use of the plough in Europe and the Middle East. Two-ox wooden ploughs are illustrated in a tomb of about 1400 BC at Thebes in Egypt,[30] and stone share-points from the Shetland Islands have been dated to the period 2000–500 BC.[31] Rock-engravings in the Italian Alps from the late Bronze Age or early Iron Age show ploughing scenes with from two to six oxen at

[22] *LL* iii 749.22355–8.

[23] Rees, *Agricultural implements* i 318; K. D. White, *Agricultural implements of the Roman world*, 52–6.

[24] K. D. White, *Agricultural implements of the Roman world*, 56–7.

[25] *LEIA* R-7 s.v. *rastal.* In an additional item in the Yellow Book of Lecan version of *Cormac's Glossary*, a pseudo-etymological explanation of the word is given: *rastal .i. ris talmuin benas* 'i.e. which strikes against the earth' (Meyer, *Sanas Cormaic*, 99 § 1121).

[26] *DIL* s.v. *ráca.*

[27] The saint who insisted that his monks tilled by spade rather than by plough was clearly regarded as exceptional: see p. 454.

[28] *CIH* ii 563.25; iii 780.4 = *CG* 7.181–2; *CIH* ii 564.5; iii 780.12 = *CG* 8.194–5.

[29] *CIH* iii 778.27–8 = *CG* 4.95–6; *CIH* iii 779.13 = *CG* 5.133–6.134; *CIH* ii 563.9; iii 779.29 = *CG* 6.158. For co-ploughing, see p. 445.

[30] Singer et al., *A history of technology* i 540.

[31] Rees, *Agricultural implements* i 27–9.

work.[32] In one of these illustrations five oxen are yoked abreast, and in another there is a crude but unmistakeable representation of six oxen yoked in tandem in three pairs.

The root *ar-* 'to plough' is found in most Indo-European languages, and gives Old Irish *arathar*, Welsh *aradr*, Breton *arar* 'a plough'.[33] The date of the introduction of ploughing to Ireland is uncertain: G. F. Mitchell makes the tentative suggestion of *c.*700 BC.[34] At this period the type of plough in use was a coulterless scratch-plough, known to archaeologists as an 'ard'. Its share is drawn through the ground at a shallow level, loosening the soil but generally not turning the sod.[35] Cross-ploughing is necessary, so the fields associated with this machine tend to be squarish in shape, and are sometimes described by archaeologists as 'Celtic fields' (see p. 371). Because of the scratch-plough's limited capabilities, it is likely that much supplementary work had to be done with spades or mattocks before the seedbed was ready for sowing.

The earliest ploughs of this type had shares of wood, stone or bone.[36] It is possible that iron ploughshares were introduced to Ireland along with other iron tools at the beginning of the Iron Age, i.e. *c.*400 BC. However, the earliest iron shares which have so far been discovered in Ireland date from the seventh century AD. The action of the ploughshare in the ground has obvious affinities with that of the snout of a rooting pig, and this is reflected in the terminology.[37] Hence, the Irish word for ploughshare, *socc*, originally meant 'pig'.[38] The same development of meaning seems to

[32]Payne, *Yr aradr gymreig*, 14 (illustrated in frontispiece); O'Loan, 'A history of early Irish farming (3)' (illustrated opposite p. 136).

[33]Cognates in other Indo-European languages include Latin *aratrum*, Greek ἄρατρον, Armenian *araur*, Lithuanian *árklas*, Old Norse *arðr*, all meaning 'a plough' (*IEW* 62). However, no derivatives of the root *ar-* 'to plough' are found in Sanskrit, Avestan or other Indo-Iranian languages.

[34]Mitchell, *The Irish landscape*, 158, 163; *Reading the Irish landscape*, 143, 150.

[35]L. White, *Medieval Technology*, 41–2.

[36]Rees, *Agricultural implements* i 6–48.

[37]Cf. *CIH* ii 563.27–8; iii 780.6–7 = *CG* 7.185–6 *srúb tuirc . . . srúb n-arathair* 'the snout of a boar . . . the snout of a plough'.

[38]It survives in this meaning in *socc sáil* 'cuttlefish' (lit. 'pig of the sea') which glosses *loligo* at *Thes.* ii 101.34: see p. 283 above. Irish *socc* and Welsh *hwch* 'pig, sow' go back to **sukko-*, from the same root as Latin *sus* 'pig', English *sow*, Greek ὗς, etc. (*LEIA* S-158).

have taken place among the Celtic-speaking peoples of the Continent, as Modern French *soc* 'ploughshare' likewise derives from **sukko-* 'pig'.[39]

There is archaeological evidence that the coulter (Latin *culter*) has been in use in Britain at least as early as the Roman period.[40] The coulter is a blade fitted to the plough-beam so as to cut the soil vertically ahead of the share. The effect is to increase the penetrating power of the plough, and to facilitate the making of a furrow. The need for cross-ploughing is reduced or eliminated, so fields associated with the covltered plough tend to be longer (see p. 372). Mitchell has suggested that the coulter (Irish *coltar*)[41] was introduced to Ireland about the time of the first Christian missions.[42] However, there is no mention of the coulter in Old Irish texts, and it is particularly noteworthy that it is not included in the lists of agricultural tools and machinery in *Críth Gablach*.[43]

In a recent paper, Niall Brady has argued on the basis of both archaeological and written evidence that the coulter did not appear in Ireland until about the tenth century.[44] This would certainly fit in with the fact that the coulter, while ignored in the seventh- and eighth-century law-texts, is mentioned in later legal glosses and commentary. For example, commentary on *Uraicecht Becc*, dating from about the twelfth century, states that one of the three 'snouts' which a person of *bóaire* rank should have is 'the snout of a coulter under the ground' (*srúb cuiltir fo thuinn*).[45] Glosses from about the same period on another law-text define the plough-irons (*erna arathair*) as the ploughshare (*socc*), the coulter (*coltor*) and the goad (*brot*).[46] The coulter is also mentioned in non-legal texts: an eleventh-century Life of Máedóc of Ferns describes how the saint miraculously saw from a great

[39] *IEW* 1038; Lambert, *La langue gauloise*, 198.

[40] Rees, *Agricultural implements* i 60.

[41] Ir. *coltar* 'coulter' is a borrowing from Latin *culter* 'coulter (one attestation in Pliny), knife, razor'. In Irish the word seems always to be used in an agricultural sense, apart from *LL* vi 1435.43309, where it may simply mean 'knife', as in the majority of Latin attestations. Welsh *cwlltwr* 'coulter' is likewise from Latin *culter*: see *LEIA* C-160 s.v. *coltar*.

[42] Mitchell, *The Irish Landscape*, 166; *Reading the Irish landscape*, 153.

[43] Lucas, 'Irish ploughing practices (1)', 55, suggests that there is an implicit reference to the coulter at *CIH* iii 778.27 = *CG* 4.95, but this seems unlikely.

[44] Brady, 'Reconstructing a medieval Irish plough', 37.

[45] *CIH* v 1612.2 = *AL* v 90.17.

[46] *CIH* ii 467.18 = *AL* v 392.10–11.

distance that a ploughman at his monastery had slipped between the ploughshare and the coulter while the team of oxen was in full career. He lifted his hand, and blessed the ploughman and the oxen, thereby instantly stopping them in their tracks, and saving the man from serious injury.[47]

The plough was further improved by the addition of a mouldboard, which turned the sod either to the left or right, depending on how it was attached. The archaeological evidence suggests that it was in use in Roman Britain,[48] and Mitchell is of the opinion that furrows at a site which could be as early as the seventh century at Goodland, Co. Antrim, are characteristic of those produced by mouldboard-ploughs.[49] So far as I can discover, the written sources in Irish contain no evidence to confirm the use of the mouldboard at this period. Our sources are also silent on the attachment of wheels to the plough-frame, which obviously improved its manoeuvrability. Such a plough may have originally been developed among the Continental Celts, as Latin *carruca* 'wheeled plough' is a Celtic word, which ultimately gave French *charrue* 'plough'.[50] The wheeled plough was well-known in medieval Wales, and a thirteenth-century law-text lists the wheels (*oluyneu*) among the parts of a plough.[51]

There are few references in Irish written sources to the wooden parts of a plough, and it is difficult to interpret this information. For example, a legal glossator lists the wooden parts as *cécht*[52] (beam), *cuing* (yoke) and *piscatal*.[53] The word *piscatal* is otherwise

[47] Plummer, *Bethada* i 222 § xli (115); trans. ii 215. Cf. Macalister, *Lebor Gabála* iii 60.1052 § 16.

[48] Rees, *Agricultural implements* i 2, 86; Rees, *Ancient agricultural implements*, 13.

[49] Mitchell, *The Irish landscape*, 174; *Reading the Irish landscape*, 164.

[50] *Carruca* is a derivative of *carrus* 'cart' (Dottin, *La langue gauloise*, 243; Lambert, *La langue gauloise*, 192). Writing in the first century AD, Pliny (*Natural History* (ed. H. Rackham), 18.48.172) speaks of the recent invention in the Raetian area of Gaul of a plough with wheels (*rotulae*). For discussion, see K. D. White, *Agricultural implements of the Roman World* 141–2; L. White, *Medieval technology*, 50.

[51] Wiliam, *Llyfr Iorwerth*, 95 § 145.2 = *ALIW*, Venedotian Code, III xxii § 239.

[52] In the Saint Gall Glosses (*Thes.* ii 149.24), *cécht* is used to explain Latin *buris* 'ploughbeam'. But in general, it seems that *cécht* is applied to the whole wooden basis of the plough, i.e. sharebeam, ploughbeam, and yokebeam. For example, a Life of Saint Máedóc speaks of a ploughteam (*seisrech*) with plough (*go gcécht*) and with plough-irons (*go n-iarann n-arathair*) (Plummer, *Bethada* i 186 (23); trans. ii 180).

[53] *CIH* v 1874.34. In the other version of this text (*CIH* i 241.35–6 = *AL* v 484.14) the glossator does not include the *piscatal*.

unattested, but it may be a variant of *pistal*, which generally means 'peg'.[54] Another reference in legal commentary is of little help, as it merely states that there is a fine of two ounces of silver for stealing a *cuing* (yoke), an *eiredach* (goblet), a *scorán* (?), and a *pistal*.[55] *Pistal* is presumably the same word as medieval Welsh *pystyl*,[56] which is mentioned among the parts of a plough in many law-texts. Unfortunately, the identity of the *pystyl* is uncertain,[57] but it is clear that it is something which is attached to the yoke. One can compare the reference in a Middle Irish poem to the *pistal cuinge*, which most probably refers to the peg which secures the yoke to the yoke-beam of a plough or the pole of a cart.[58] *Scorán* is possibly to be compared with medieval Welsh *yscur*, used of a part of a plough in the law-text *Llyfr Iorwerth*. Its editor, Aled Wiliam, suggests that it is a piece of wood projecting from the long yoke and used as a handle.[59]

The problem of how to utilise the power of oxen (or other draught animals) to maximum effect exercised the ingenuity of the makers of ploughs and carts for thousands of years. The most straightforward way of attaching the yoke to the oxen is to strap it in front of their horns, and this method is known to have been used in the Alpine region in Pliny's time.[60] However, Columella advocates the use of the withers-yoke, and condemns the head-yoke as both cruel and inefficient.[61] It is noteworthy that the Old Irish tale *Tochmarc Étaíne* 'the wooing of Étaín' also contrasts these two methods of yoking oxen, clearly regarding the head-yoke as the more primitive.[62]

[54]In an Old Irish tale, a *pistul* seems to be a long iron bolt (Knott, *Togail Bruidne Da Derga*, 38.1280 = *LU* 236.7744).

[55]*CIH* iii 1107.5–6.

[56]Both words are taken to be from Late Latin *pistillum*: see *LEIA* P-9 s.v. *pistol*.

[57]For detailed discussion in Welsh of the problem of the identity of the *pystyl* (*pestl*), see Payne, *Yr aradr gymreig*, 147–8. See also Wiliam, *Llyfr Iorwerth*, 95 § 145.3 (and notes p. 131) = *ALIW*, Venedotian Code, III xxii § 242; Williams and Powell, *Llyfr Blegywryd*, 97.11–12 (and notes p. 217) = *ALIW*, Dimetian Code, II xxxv § 59; Emanuel, *The Latin texts of the Welsh Laws*, 490.23; *GMWL* 255–6 s.v. *pystyl*; Jenkins, *The law of Hywel Dda*, 305.

[58]Thurneysen, 'Mittelirische Verslehren', 101 § 181.

[59]Wiliam, *Llyfr Iorwerth*, 97 § 148.9 (and notes on p. 132). For further references, see *GMWL* 302 s.v. *yscur*.

[60]Pliny, *Natural history* (ed. H. Rackham), 8.70.179.

[61]Columella, *De re rustica* (ed. Ash), 2.2.22–4.

[62]Both head-yokes and withers-yokes have been found in excavations from the early Christian period in Ireland (Rees, *Agricultural implements* i 72–3).

The text describes how the fairy people under the command of Midir made a causeway (*tóchar*) in a single night across the bog of Lámraige.[63] The clay, gravel, and stones for this causeway were hauled by oxen, and it was observed that the fairy people placed the yokes on the withers of the oxen. The author claims that until that time the men of Ireland used to put the strain on the foreheads of their oxen. However, following the practice of the fairy people, the king of Tara was the first of the men of Ireland to put the yoke on their withers. He was thereafter known as *Eochaid airem*, i.e. Eochaid the ploughman.

In the law-text *Críth Gablach* the halter (*cennos*)[64] is included as part of the standard ploughing-equipment owned by every farmer from the rank of *ócaire* upwards.[65] Another law-text refers to injury inflicted by an ox which has broken out of its halter.[66] In *Cormac's Glossary*, the word *esem* is used of the strap which binds the yoke to the oxen.[67] The word *amlach*, attested only once, may refer to the halter and straps together.[68]

It is regular for two oxen – or other draught animals – to be yoked together when working. Early illustrations from the Mediterranean area show many scenes of two yoked animals ploughing or drawing vehicles,[69] and in India today ploughing is often carried out by a yoked pair of buffaloes. The archaeological and linguistic evidence from early Christian Ireland likewise emphasizes the importance of the yoked pair. The design and dimensions of many yokes from this period indicate that they were used by two animals.[70] Similarly, the Old Irish term for a yoked pair of oxen or horses is *córait*, which is generally taken to be a derivative of the adjective *cóir* 'proper, fitting'.[71] The word for yoke, *cuing*, reflects the practice of yoking

[63] Bergin and Best, '*Tochmarc Étaíne*', 178 §§ 7–8; cf. Stokes, *Cóir Anmann*, 330 § 103.

[64] This word is also attested in the spellings *cendas*, *cennas*. The first element is obviously *cenn* 'head'. It is suggested in *DIL* C 130.51 and in *LEIA* C-68 that it is a special use of *cennas* 'headship, leadership'.

[65] *CIH* iii 778.27; 779.13 = *CG* 4.95; 6.134.

[66] *CIH* i 242.27 = *AL* v 486.24; cf. *CIH* i 284.14 = *AL* iii 268.5.

[67] Meyer, *Sanas Cormaic*, 43 § 514.

[68] *CIH* iii 898.17 = Binchy, 'A text on the forms of distraint', 80 § 10.

[69] E.g. Singer et al., *A history of technology* i 540; K. D. White, *Agricultural implements of the Roman world*, 128.

[70] Rees, *Agricultural implements* i 72–3.

[71] *LEIA* C-207.

in pairs, as it has the basic meaning of 'co-yoke, shared yoke'.[72] There is also literary evidence: a Latin Life of Saint Ailbe refers to ploughing by two oxen.[73]

From the accounts of farming by Greek and Latin authors, it seems that in the light, dry soils of the Mediterranean area it was generally possible for two oxen to plough without excessive strain.[74] However, the expansion of agriculture in the heavier damp soils prevalent in central and northern Europe lead to a need for ploughing with larger teams. It is significant that the author of *Críth Gablach* assumes that ploughing is normally carried out with four oxen. He states that a farmer of *ócaire* rank owns a quarter of a ploughing outfit, which he specifies as an ox, a ploughshare, a goad and a halter.[75] It is implicit that he expects him to plough in co-operation with three other farmers of similar rank: there is no suggestion that he might merely make an arrangement for two-ox ploughing with one other farmer. A Middle Irish law-text on status takes the same view, and fixes the proper number for a ploughteam at four oxen (*arathar cona fedain téchta .i. .iiii. doim*).[76] There are also references to ploughing with four oxen in saints' Lives,[77] and in the *Metrical Dindshenchas.*[78]

The problem of simultaneously harnessing four (or more) oxen is obviously much more complex than merely coping with two oxen. In later medieval times it seems that the harnessing of two pairs of oxen working in tandem was general in western Europe: there is a particularly clear illustration in the Luttrell Psalter from about 1340.[79] This method has the technical difficulty that power must

[72]It is from **ko(m)-iung-*, i.e. *ko(m)* 'with, co-' + the stem *iu(n)g-*, which appears in most Indo-European languages with the meaning 'yoke', e.g. Welsh *iau*, Latin *iugum*, Greek ζυγόν, Sanskrit *jugám*, Gothic *juk*, etc. (*LEIA* C-273 s.v. *cuing*; *IEW* 509). *Cuingir* 'yoked pair' in later Irish is a derivative of *cuing* (*DIL* C 596.2).

[73]Plummer, *Vitae* i 59 § xxxiv . . . *duos boues ad arandum.*

[74]However, Pliny – writing in the first century AD – refers to the pulling of the plough by eight oxen in Italy (*Natural History* (ed. H. Rackham), 18.47.170). Lynn White suggests that he is referring primarily to the heavy soils of the Po Valley (*Medieval technology*, 42).

[75]*CIH* iii 778.27–8 = *CG* 4.95–6.

[76]CIH ii 707.33.

[77]Plummer, *Vitae* i 36 § iv; 203 § x.

[78]E. J. Gwynn, *Metrical Dindshenchas* iv 300.18 *in ceathardham* 'the team of four oxen'. See also Lucas, 'Irish ploughing practices (1)', 53–4.

[79]Reproduced in Payne, *Yr aradr gymreig*, plate 6 (opposite p. 33); Langdon, *Horses, oxen and technological innovation*, 17.

be transferred from the leading pair of oxen to the plough without interfering with the work of the pair behind them. Its main advantage lies in the greater manoeuvrability of a ploughteam working in tandem.

The Irish archaeological evidence indicates the practice of yoking four oxen abreast rather than in tandem. Undated finds, probably from the early Christian period, include short yokes obviously intended for a pair of oxen, and longer yokes which would require at least four oxen. For example, a yoke from Co. Mayo of 213 cm. (= 6 ft. 2 ins.) in length and 9 cm. (= 3½ ins.) in width can hardly have been used with less than four oxen.[80] Following a suggestion by G. Fenton, both A. T. Lucas and Niall Brady have argued that this long yoke would have been used to link two outer oxen, while the inner oxen would have had shorter yokes linking them together just behind the outer pair.[81] This interpretation would explain some difficult passages in the eighth-century tale *Togail Bruidne Da Derga.* One passage refers to a giant whose shins are as thick and long as an outer yoke (*cuing imechtair*).[82] Another passage refers to a warrior whose shins are as long as an inner(?) yoke (*cuing úarmedóin*).[83] In a third passage the author makes a distinction between a *cuing imechtair* and a *cuing úarmedóin.*[84]

The Welsh law-texts from about the thirteenth century also provide evidence of the manufacture of yokes designed for more than two oxen. Thus there are references to a *meiau* (*meyyeu*) 'mid-yoke' of eight feet and a *hiriau* (*hyryeu*) 'long yoke' of sixteen feet.[85] It is clear, therefore, that ploughing with four (or more) oxen abreast was practised in Wales at this period. Writing in the late twelfth century, Giraldus Cambrensis states that in Wales it is more usual for ploughs or carts to be drawn by four rather than by two oxen.

[80]Rees, *Agricultural implements* i 73.

[81]Lucas, 'Irish ploughing practices (1)', 57; Brady, 'Reconstructing a medieval Irish plough', 32; Jenkins, *The law of Hywel Dda*, 303.

[82]Knott, *Togail Bruidne Da Derga*, 11.350–51 = *LU* 210.6824. The same phrase is used at Stokes, *Acallamh na Senórach*, 37.1300; 164.5943 and O'Keeffe, *Buile Ṡuibne*, 51.1448 § 45.

[83]Knott, *Togail Bruidne Da Derga*, 16.520 = *LU* 213.6954. The version in D iv 2 has *iarmedoin*: see discussion in Knott's vocabulary s.v. *cuing*.

[84]Knott, *Togail Bruidne Da Derga*, 38.1279–81 = *LU* 236.7742–4.

[85]Wiliam, *Llyfr Iorwerth*, 96–7 § 148.6–7 (and note on p. 132 s.v. *eytheuyc*). Cf. Williams and Powell, *Llyfr Blegywryd*, 71.13–15 = *ALIW*, Dimetian Code, II xx § 8; Wiliam, *Llyfr Iorwerth* 60 § 90.27–8 = *ALIW*, Venedotian Code, II xvii § 6. Rowlands, 'Mesur tir', 273, takes the Welsh foot to be 10.125 in.

It is clear that he is speaking of oxen yoked abreast, because he describes how the driver (Latin *stimulator*)[86] walks backwards in front of the oxen, and encourages them by voice, while the ploughman steers the plough.[87] This description would hardly suit two pairs of oxen in tandem.

Irish written sources lack detailed accounts of ploughing methods. The twelfth-century *Lebor Gabála* refers to a *cethardam* 'team of four oxen', managed by a *cennairem* 'front ploughman' and a *tónairem* 'rear ploughman'.[88] It is probable that the front ploughman walked backwards ahead of the team like the Welsh *geilwad*, while the rear ploughman guided the plough from behind like the Welsh *amaeth*. A commentary on *Félire Óengusso*, from about the same period as *Lebor Gabála*, describes a miracle performed on the ploughing-field (*fairche in arathair*) by Bishop Étchén.[89] The bishop was ploughing when Columb Cille arrived to have episcopal orders conferred on him. Columb was unimpressed by the ploughman bishop, and sought to test his powers. He first asked him for the ploughshare. Étchén gave it to him, and miraculously continued to plough just as well as before. He then asked him for the outer ox (*imechtraid*). Étchén gave it to him also, and immediately called on a stag (*dam allaid*) from the wood to take the ox's place. These miracles convinced Columb of the bishop's credentials, and he was subsequently content to receive sacerdotal rather than episcopal orders. It is probable that the author of this story envisaged four oxen yoked abreast. If there were only two, it would hardly make sense to refer to an outer ox. The usual Irish term for a team of oxen is *seisrech*, which is ultimately a compound of the numeral *sé* 'six' and a root meaning 'to bind'.[90] Some literary references likewise

[86]In Welsh of this period, he is known as the *geilwad* 'caller', whereas the man who holds the plough is the *amaeth* 'ploughman'.

[87]Dimock, *Descriptio Kambriae*, 201.9–13; *GMWL* 153 s.v. *geylwat*. See Payne, 'The Welsh plough team to 1600', 241–2.

[88]Macalister, *Lebor Gabála* iii 60.1049–50 § 16. A triad in a text from about the fourteenth century gives the three ruins of ploughing as *tōineiream anand ēgubhaidh 7 cēacht dealam dībhālach 7 capall ēagcruaidh aimrēigh* 'a feeble incompetent rear ploughman, and a wretched damaged plough, and a weak troublesome horse' (Mac Niocaill, 'Vocabulaire social irlandais', 541 § 19). I read *anand* as *anfand*; cf. §§ 2, 11.

[89]Stokes, *Félire*, 72.18–28.

[90]Paul Russell points out to me that *seisrech* is from * *swes-* 'six' (*LEIA* S-59 s.v. 3 *sé*) + *reg-* 'to bind' (*LEIA* R-15 s.v. 2 *reg-*). The term may sometimes be used of more than

suggest that six oxen were yoked together for certain tasks. The eighth-century tale *Fled Bricrenn* gives an account of a phantom so mighty that he holds a fistful of stripped oak-trees in each hand. Each oak is described as 'the burden of a yoke of six oxen' (*eire cuinge sesrige*), i.e. so big that it would need six oxen to drag it along the ground.[91] The author seems to envisage a single yoke (*cuing*) with six oxen abreast, though he does not say so explicitly. A later reference in a saint's Life gives no indication as to the method of yoking, but undoubtedly assumes that a *seisrech* of oxen are simultaneously yoked to a plough.[92] According to *Críth Gablach*, a farmer of *mruigfer* rank is expected to own six oxen.[93] It has been suggested that he would use only four of these at any one time, allowing two to rest in turn.[94] But the literary references raise the possibility of the simultaneous use of all six.

From the thirteenth century, new breeds of strong horses started to replace oxen for ploughing and other heavy tasks (see p. 95). In later Irish, therefore, the term *seisrech* came to be applied to a team of six horses. The frequent references by English commentators to an Irish practice of attaching the plough to a horse's tail have never been fully explained. Such a method of ploughing seems impractical as well as cruel, yet it was apparently so widespread that it was legislated against on a number of occasions in the first half of the seventeenth century.[95] There is, however, no mention of it in sources in the Irish language. Sian Rees raises the possibility that this form of ploughing may have had some ritual significance, and compares a Bronze Age carving at Tegneby in Denmark in which the beam of a plough appears to be tied directly to a horse's tail.[96] But the evidence from Ireland indicates a regular farming practice rather than an occasional ritual act.

six oxen, e.g. *LL* v 1294.38501–2 = Stokes, 'The *Bórama*', 68 § 72 *dā dam déc in cach ṡesserig* 'twelve oxen in every team'.

[91] Windisch, 'Das Fest des Bricriu', 295.22 § 81 = *LU* 273.9061–3. The same phrase is used in Knott, *Togail Bruidne Da Derga*, 25.833 = *LU* 221.7213–14.

[92] Plummer, *Bethada* i 38–9 § xxiv (74); trans. ii 38. Cf. Plummer, *Vitae* ii 152 § xxx *sex boues in aratro.*

[93] *CIH* ii 564.5; iii 780.12 = *CG* 8.194–5.

[94] Charles-Edwards, 'A comparison of Old Irish with medieval Welsh land-law', 296.

[95] G. O'Brien, *The economic history of Ireland in the seventeenth century*, 38–9; Lucas, 'Irish ploughing practices (2)', 72–9; E. E. Evans 'Some problems of Irish ethnology: the example of ploughing by the tail'.

[96] Rees, *Agricultural implements* i 74.

Harrow

The function of the harrow is to level and compact the soil after it has been ploughed. In medieval illustrations and writings, the two processes are often treated together; for example, the late eleventh-century Bayeux tapestry shows what is probably a mule[97] pulling a wheeled plough and below it a horse pulling a square harrow with four rows of iron teeth.

No parts of a harrow have been identified from early Christian Ireland, but there are written references which show that harrows were in regular use. For example, a proverbial phrase for a useless undertaking is given in legal commentary as *ar cen foirse* 'ploughing without harrowing'.[98] Similarly, the law-text *Críth Gablach* states that a high-ranking lord should have full ploughing equipment, and then adds that he should also have *dá chapall do foirtsiud*, which Binchy translates 'two horses for harrowing'.[99] Ploughing and harrowing are also mentioned together in a strange passage in the *Annals of Ulster* for the year 1013 which describes the humiliating torture inflicted on a group of Norse captives by Gilla Mo-Chonna, king of southern Brega.[100] The king yoked most of his captives to a plough, and forced two others to follow, harrowing (*ic foirsed*) from their scrotums (*asa tíagaib*).[101]

In later Irish sources, the word *clíath* 'hurdle' is used in the sense of harrow, sometimes in the phrase *clíath fuirsidh*.[102] *Clwyd* (*cluyd*), the Welsh cognate of *clíath*, can likewise have the same meaning. In the list of tools and utensils in the thirteenth-century law-text *Llyfr Iorwerth* a distinction is made between two types of harrow.[103] The more valuable is the *og*, which presumably refers to a harrow

[97] L. White, *Medieval technology*, 63. Presumably the designers of the tapestry included only one animal pulling each machine because of the difficulties of representing two or more animals in side view.

[98] *CIH* iii 750.34.

[99] *CIH* ii 567.27 *dā chapal do fortsed*; iii 783.34 *dā capal do foirtsid*, = *CG* 16.410. Binchy (*CG* 35, note to l. 410) argues for a connection with the root *u̯ert-* 'to turn' (*IEW* 1156–7), but this does not seem to suit the levelling action of the harrow.

[100] AU^2 444 s.a. 1012 (recte 1013) § 4.

[101] I follow the interpretation of *tíag* 'bag' suggested by Ó Súilleabháin, 'Nótaí ar thrí fhocal ó na hAnnála', 23.

[102] See *DIL* C 240.49–52. In *AFM* vi 2186.20 s.a. 1600, *clíath* seems to be equated with *práca*, a borrowing from English *brake* 'heavy harrow for crushing clods' (*OED*).

[103] Wiliam, *Llyfr Iorwerth*, 95 § 145.4 = *ALIW*, Venedotian Code, III xxii §§ 249–50.

with iron teeth of the type illustrated in the Bayeux tapestry.[104] Of slightly less value is the *draenglwyd* (*draen* + *clwyd*) 'thorn-harrow', which obviously consisted of a weighted frame with thorn branches underneath. Dafydd Jenkins describes it as 'a medieval counterpart of the chain-harrow'.[105]

MINING, EXCAVATION, TURF-CUTTING, ETC.

Certain tools may be used both for tilling the soil and for other purposes. Thus the law-text *Bretha Comaithchesa* refers to the use of the spade (*rámae*) in the digging of a boundary ditch (*clas*).[106] Legal commentary likewise implies that the spade and the shovel are used in mining for metal ores.[107]

Spades for sod-cutting have been identified from Roman Britain, but none has so far been found in Ireland from the early Christian period. Rees suggests that they were standard military tools introduced to Britain by the Romans for cutting the turf for rampart construction.[108]

As we have seen in Chapter 11 (p. 397), the cutting of turf (peat) for fuel is rarely mentioned in early Irish sources. Most references to this fuel are to be found in post-Norman writings. It is not surprising, therefore, that there is no information on turf-spades in Old or Middle Irish texts. In later Irish, the usual term is *sleaghán* (anglicized *slane*), which seems to be a derivative of *sleg* 'spear'.[109] Another term which is attested once in the context of turf-cutting is *rúam*, elsewhere used in the general meaning 'spade'. It occurs in a passage which describes the death of King Raigellach of Connacht at the hands of a party of turf-cutters who set about him with their spades (*dona rūamaib*).[110]

[104] The word *og* (also *oged*) is cognate with words for 'harrow' in many Indo-European languages, e.g. Latin *occa*, Greek ὀξίνη, Old High German *egida*, etc. (*IEW* 22).

[105] Jenkins, *The law of Hywel Dda*, 301.

[106] *CIH* i 65.12; vi 2193.15 = *AL* iv 72.2. According to commentary on this text (*CIH* i 199.12; ii 579.30 = *AL* iv 130.18–19) the spade is worth a scruple, and need only have been tempered once (*a leagadh ænfecht*). By contrast, the axe and billhook must have been tempered twice and the *socc* (lever?) three times: see p. 492 below.

[107] *CIH* iv 1257.16 = *AL* iii 204.5.

[108] Rees, *Agricultural implements* i 331.

[109] T. F. O'Rahilly, 'Mid.Ir. *lága*, etc.', 152, compares Mod.Ir. *láighe, lágha* (angl. *loy*) 'spade' which seems originally to have had the meaning 'spear'. See *DIL* s.v. *láige*.

[110] O'Grady, *Silva Gadelica* i 396.17; trans. ii 431. See discussion by T. F. O'Rahilly, '*Rámha*, etc.', 364.

In Scottish Gaelic, the term for turf-spade is *tréisgir*, a borrowing from Old Norse *torfsceri*, lit. 'turf-cutter'.[111]

Harvesting and processing of cereals

Sickle[112]

Our sources provide us with little information on the sickle (*serr*, *corrán*).[113] Writing in the first century BC, the Roman writer Varro distinguishes three methods of cutting corn in the Italy of his day: near ground level, half-way up the stalk, and close to the ear.[114] A study of Irish sickles from the early Christian period convinced Michael Duignan that the third method was that which was most widely practised in this country: see p. 238.[115]

A scythe is essentially a long-bladed sickle. It seems originally to have been developed as a grass-cutting implement: Roman authors distinguish between a *falx faenaria* for cutting hay and a *falx messoria* for cutting corn.[116] Over fifty scythe-blades have been found in excavations from Roman Britain, and the consensus among archaeologists is that their main function was to cut grass, but that they could also have been used for cutting corn.[117] The absence of any archaeological or documentary evidence of the scythe in early Christian Ireland supports the view that little or no hay was cut at this period (see p. 47). When the scythe made its appearance in Ireland in post-Norman times, it was called a *speal*, a word which is likely to be of Middle English origin.[118]

[111] McDonald, *Gaelic words and expressions*, 248. Forms in other dialects include *taraisgeir*, *tairisgil*, *tairisgein*, and *toirsgian* (modelled on *sgian* 'knife').

[112] Some archaeologists make a distinction between a sickle and a reaping hook, but for the purposes of this book the terms are taken to be synonymous: see discussion at Rees, *Agricultural implements* ii 438.

[113] *Serr* is cognate with Welsh *serr* 'sickle, billhook' (*LEIA* S-95). *Corrán* is a derivative of *corr* 'point' (*LEIA* C-213); it is used of a variety of hooks, hinges and projections (*DIL* s.vv. *corrán*, *corránach*). There seem to be no clear attestations of *corrán* in the meaning 'sickle' in Old or Middle Irish. However, the fact that it is the regular word for 'sickle' in Modern Irish, Scottish Gaelic and Manx indicates that this meaning is early.

[114] Varro, *Rerum rusticarum* (ed. Hooper and Ash), 1.50.1–2.

[115] Duignan, 'Irish agriculture in early historic times', 140.

[116] K. D. White, *Agricultural implements of the Roman world*, 72–4.

[117] See discussion at Rees, *Agricultural implements* ii 473–80.

[118] T. F. O'Rahilly ('*Sp-* in Gaelic', 26, and 'Sc. *fàl*, etc.') maintains that *speal* goes back to Middle English *pilen*, *pelen*, 'to shave, cut off'. However, M. A. O'Brien, 'Ir. *spel*, etc.', raises the possibility of a relationship between *speal* and the Old Irish personal name *Spelán*.

Threshing implements

The earliest type of implement for removing the grains from cornstalks was the threshing-stick. In Old Irish, the commonest term for this implement is *súst*, and in the later language the derivative *súistgal* is occasionally employed to describe violent movement. *Súst* (also *súist*) is a borrowing from Latin *fustis* 'rod',[119] and is one of the many Latin loanwords in Irish which are associated with the cereal-crop (see p. 222). There are also a few attestations of the native word *flesc* 'rod' used in the same sense. For example, the law-text *Bretha Nemed Toísech* states that the minimum penalty payable to a low-ranking poet is 'a bushel of barley grain after a rod has stripped (i.e. dehusked) it' (*miach do éorna ith oro fema flesc*).[120] Another attestation of *flesc* in this meaning occurs in a gloss on *Bretha Comaithchesa.*[121]

The flail, made of two sticks fastened together, is thought to have been developed for the purpose of threshing corn in about the fourth century AD, possibly in Gaul.[122] It is not clear from either written or archaeological evidence when the flail was introduced to Ireland.[123] But by the time of the eleventh- or twelfth-century legal commentators, it seems that the standard threshing implement in this country was the two-piece flail rather than the simple threshing-stick. Commentary on *Bretha Étgid* discusses accidental injury inflicted by one thresher on another thresher who is working by his side or opposite him. Injury may be caused if the head flies off (*má dá cind foceird in tśúist*), which strongly suggests a two-piece flail.[124] I reproduce on page 483 from a fifteenth-century

[119] *Fustis* was likewise borrowed into Welsh in the form *ffust* 'stick, flail' (*LEIA* S-203 s.v. *súist*).

[120] *CIH* vi 2215.31–2, translated by L. Breatnach, *Uraicecht na Ríar*, 31. This passage is also quoted at *CIH* iv 1498.7–8 = O'Dav. 336 § 833 and (partially) at *CIH* iv 1500.33 = O'Dav. 348 § 898. In the latter quotation *flesc* is glossed by the word *traigh* 'foot'. Could this be a reference to the threshing of corn by treading (usually under the hoofs of oxen)?

[121] *CIH* i 71.11 = *AL* iv 94.16 (wrong word-division).

[122] For a discussion of the flail in Ireland, see Ó Danachair, 'The flail and other threshing methods'. See also Duignan, 'Irish agriculture in early historic times', 141.

[123] The Latin word normally applied to the flail is *flagellum*, which is also used of various type of whip or scourge. *Flagellum* is borrowed into Old Irish as *sroigell*, but the meaning 'flail' is not attested (see *DIL* s.v. *sroigell*).

[124] *CIH* i 273.31 = *AL* iii 220.19. Another commentary refers to the *lorg* 'handle' of a *súist* (*CIH* i 254.4 = *AL* iii 106.15): this is probably the handstaff – *colapa* or *lámhchrann* in later Irish – of a two-piece flail.

manuscript a drawing of a sheaf of corn being threshed by men wielding two-piece flails.[125]

Another beating instrument which should be mentioned here is the flax scutching rod (*flesc lín*). It is included in a long list of implements connected with spinning, weaving and embroidery in *Di Chetharślicht Athgabálae.*[126] It is implicit in the glosses that all the items in this list belong to women.

Corn-grinding machines

The quern is a simple machine of great antiquity for grinding corn into flour.[127] The term *bráo* (later *bró*) is commonly used of the quern, but may also be applied to a millstone or whetstone.[128] Consequently, the quern is sometimes called a *lámbró* 'hand-quern' to distinguish it from other grinding stones.[129]

The quern is not often mentioned in our sources, and is regarded as the typical implement of the female slave (see p. 439). By contrast, great attention is devoted to the water-mill, undoubtedly the most complex piece of technology which an early Irish farmer would regularly encounter.[130] Archaeological excavation and the study of the written sources have provided us with a good deal of information on water-mills in early Christian Ireland. Such mills were sufficiently important and widespread for a special law-text, entitled *Coibnes Uisci Thairidne* 'kinship of conducted water', to be written in the seventh century on some legal aspects of milling and mill-construction.[131]

[125] De Brún, *Catalogue of Irish manuscripts in King's Inns Library*, 46.
[126] *CIH* ii 379.7 = *AL* i 150.6.
[127] See Caulfield, 'The beehive quern in Ireland'.
[128] *DIL* s.v. 1 *bró*; *LEIA* B-92 s.v. 1 *bró*.
[129] For example, a law-text (*CIH* i 238.32–3 = *AL* v 474.3–4) distinguishes three grindings which are illegal because the owner's permission has not been obtained: *bleith a muileand* 'grinding in a mill', *bleith a lāmbrōin* 'grinding in a hand-quern', and *bleith for lī[m] brōin* 'grinding (i.e. sharpening a tool) on a whetstone'.
[130] The law-text *Uraicecht Becc* gives the millwright (*sáer muilinn*) the same status as the lowest grade of lord, the *aire déso* (*CIH* v 1615.35–7 = *AL* v 104.11–12). This puts him in the same bracket as the builder of wooden churches, boat-builder, etc.
[131] *CIH* ii 457.11–462.18 = Binchy, '*Coibnes Uisci Thairidne*', 64–72.

Plate 8. The right half of the bottom margin of f. 5^{v} of a fifteenth-century Irish medical manuscript (MS 17) in the Library of King's Inns, Dublin, showing two men wielding flails. The original of this illustration was made available through the courtesy of the Council of King's Inns, and is published by their permission.

Mills were often owned by several people: the eighth-century author of *Críth Gablach* assumes that even the lowest grade of adult freeman, the *ócaire*, has a share in a mill.[132] The higher-ranking *mruigḟer* may have a share in a mill,[133] or have his own mill.[134] Legal commentary refers also to a *muilenn túaithe* 'mill of the *túath* or kingdom', and states that anyone who finds something on the ground must report the fact at this location, as well as at six other specified places in the *túath*.[135] Presumably, the *muilenn túaithe* is a large mill owned by a professional miller (*muilleóir*), and likely to be frequented by many people. There are a number of references to the occupation of miller in our sources.[136] Rarely, a miller might be female: see p. 450 above.

If a mill is destroyed, the culprit must pay the honour-price of the owner, and restore it.[137] Even to grind corn in another person's mill without permission is regarded as a serious offence: the culprit must pay five *sét*s, and forfeit the meal which he has ground.[138] The law-text on distraint *Di Chetharṡlicht Athgabálae* lists 'the eight parts which serve a mill'[139] in a passage which has been discussed in detail by A. T. Lucas and Gearóid Mac Eoin[140] One of these parts is the *oircél*, which Lucas and Mac Eoin identify with the chute which directs the water towards the vanes of the mill-shaft. What must be a particularly large mill is described in an eighth-century tale as *muilenn deich n-oircél* 'a mill of ten chutes'.[141]

There have been excavations of water-mills at a number of sites in Ireland. These include the mill at Little Island, Cork, which is

[132] *CIH* iii 778.28 = *CG* 4.96.
[133] *CIH* ii 564.4; iii 780.11 = *CG* 8.192.
[134] *CIH* ii 564.36; iii 781.2 = *CG* 9.238.
[135] *CIH* iv 1211.27; vi 2171.38.
[136] E.g. Ó hAodha, *Bethu Brigte*, 19.67; Meyer, *Triads*, 30 § 234; Van Hamel, *Immrama*, 36.358 § 14 = Stokes, 'The voyage of Máel Dúin (1)', 482 § xiv.
[137] *CIH* ii 564.37–8; iii 781.3–4 = *CG* 10.240–1. If the mill has a number of owners, each doubtless gets a proportion of his honour-price from the culprit.
[138] *CIH* ii 564.36; iii 781.2 = *CG* 9.238–10.239. The reference is to a *mruigḟer*'s mill, but it is to be assumed that the same fine is payable in the case of anybody's mill.
[139] *CIH* ii 374.19–20; iii 889.1–2; v 1684.33–5; 1901.28–30 = *AL* i 124.17–19. The names of eight parts of the actual mechanism of the mill are preceded by three other essentials for milling: the water-source (*topur*), the mill-race (*tuidin*(?)) and the land of the mill-pond (*tír linde*).
[140] Lucas, 'The horizontal mill in Ireland', 29–34; Mac Eoin, 'The early Irish vocabulary of mills and milling'. Cf. Mac Eoin, 'The death of the boys in the mill'.
[141] *LU* 263.8699 = Windisch, 'Das Fest des Bricriu', 279.22 § 52. Read *noirccēl* with Egerton 93 (variant given by Windisch) rather than *forcél* with *LU*.

the earliest tidal mill so far discovered in Europe: it has been dated by dendrochronology to about AD 630.[142] The remains of a mill at Drumard, Co. Derry, have been dated by the same method to about AD 782.[143] The archaeological evidence thus broadly indicates that the watermill was introduced to Ireland from the Roman world in the sixth or early seventh century, and the evidence of milling terminology in Irish supports such a date. Of particular significance is the fact that the only Old Irish word for 'mill', *muilenn*, is a borrowing from Latin *molina* of the same meaning. Among the words for parts of the mill listed in the law-text on distraint are *milaire* 'pivot-stone(?)', which is a borrowing from Latin *miliarium* 'mile-stone, central column of wine-press or oil-mill',[144] and *cup* 'hopper(?)', which may be a borrowing from Latin *cupa* 'vat, cask, barrel'.[145]

The term *marcmuilenn* 'horse-mill' is found only in glossaries, and is defined as ' a mill which a horse turns'.[146] Such mills are often mentioned in Latin sources,[147] but there is no other evidence of their use in early Christian Ireland.

Wood-cutting

The cutting and processing of wood was clearly of very great importance in the early Irish economy. A prosperous farmer is therefore expected to possess various tools for working with wood: the small axe (*eipit*), large axe (*biáil*), billhook (*fidbae*), saw (*tuiresc*), adze (*tál*), and auger (*tarathar*).[148] To sharpen these tools he has a whetstone (*liae forcaid*).[149]

[142]Rynne, 'Milling in the 7th century: Europe's earliest tide mills'; Rynne, 'The introduction of the vertical watermill into Ireland'; Rynne, 'Early medieval horizontal-wheeled mill penstocks'; Baillie, *Tree-ring dating and archaeology*, 192.

[143]Baillie, 'A horizontal mill of the eighth century AD at Drumard, Co. Derry'. He illustrates a suggested reconstruction of this mill on p. 28 of this article. The mechanism is also illustrated at Edwards, *The archaeology of early medieval Ireland*, 63; Mytum, *The origins of early Christian Ireland*, 198.

[144]Mac Eoin, 'The early Irish vocabulary of mills and milling', 16.

[145]Mac Eoin, 'The early Irish vocabulary of mills and milling', 16. On the other hand, it is suggested in *LEIA* C-294 that *cup* is from another Latin word, *cupa* 'the bar of an oil-mill'.

[146]*CIH* ii 617.12; iii 1075(b).6; O'Mulc. 271 § 817 *Marcmuilenn .i. muilenn imsuí ech.*

[147]Moritz, *Grain-mills and flour in Classical antiquity*, 74–102; cf. Langdon, *Horses, oxen and technological innovation*, 117–18.

[148]*CIH* ii 563.22–4; iii 780.1–3 = *CG* 7.178–80.

[149]*CIH* ii 563.24; iii 780.3 = *CG* 7.179–80 (and note on p. 29); cf. *CIH* i 290.27–8 = *AL* iii 294.20 *blā liac līmad, no ruitech* 'grinding is the immunity of whetstones or of

Axe

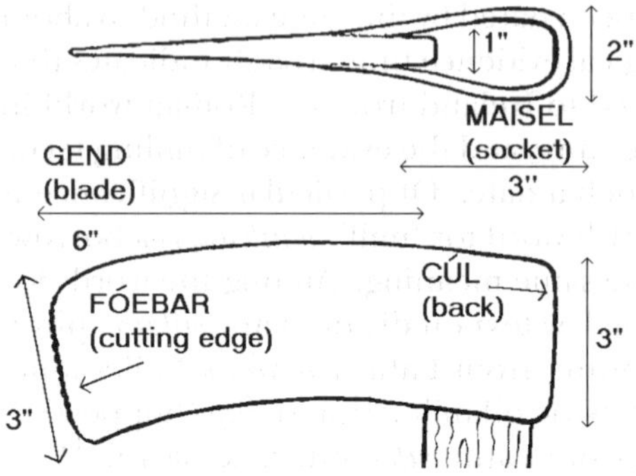

Fig. 20. A suggested side and top view of the axe described at CIH iii 921.25–32.

The large axe (*biáil*) is the most commonly mentioned tool in our texts.[150] Its crucial importance is indicated in a ninth-century triad which lists the three which are best in a house as 'oxen, men, axes'.[151] According to *Críth Gablach*, the penalty-fine (*díre*) for stealing or destroying somebody else's axe is a two-year-old heifer (*colpthach*), double the normal penalty-fine for a billhook.[152] Another law-text provides a detailed description of the size and manufacture of a proper axe (*biáil innraic*),[153] which I have used as the basis of a speculative illustration above.[154] The text states that the *gend* should be an *airtem* in length, which is probably six inches.[155] I take *gend* here to refer to the blade of the axe.[156] The cutting-edge (*fóebar*) of the blade is three fingers in width, i.e.

cranks', i.e. if a person is injured while sharpening a tool on a whetstone, or if he is struck by the crank (called by the glossator the *crand cam* 'crooked stick'), the owner of the machine is not liable. Another term for whetstone is *límbró*, e.g. *CIH* i 238.33 = *AL* v 474.4 *for lī[m] brōin*.

[150]Old Irish *biáil* is cognate with Welsh *bwyell*, Breton *bouhal*, etc. (*LEIA* B-48).

[151]Meyer, *Triads*, 28 § 227 *Trí ata ferr i tig: daim, fir, béla.* It seems that *i tig* here means 'in a household, among the resources of a household'.

[152]*CIH* ii 564.42; iii 781.7 = *CG* 10.246–7.

[153]*CIH* iii 921.28–32 *Is ē in biāil innraic: airtem fod a genned, .iii. mēr lethed a foebuir, trī mēr ina maisil, trī mēr lethad a cūil, ordlach a leth[ad] isa maisil frit, dā mér a llethud anall. A dēiccsi fo tri gortaib .i. donnghorad 7 bānghorad 7 a gorad foberta; is hē nī cluī ibar na daur ratus fora iarn* . . . Note that *mér* is omitted before *lethad* in the transcription at 921.29.

[154]For illustrations of Irish domestic axes from excavations of the Iron Age and early Christian period, see Raftery, *La Tène in Ireland*, 239; Hencken, 'Ballinderry crannóg no. 1', 153; Hencken, 'Lagore crannóg', 106.

[155]For a discussion of this unit of length, see p. 564.

[156]Elsewhere, *gend* (*genn*) is generally used of a wedge: see *DIL* G 69.72–83. However, it does not seem likely that a wedge as long as six inches would be required to secure the axehead to the handle. For further discussion of the meanings of *gend*, see p. 502.

approximately three inches.[157] The back (*cúl*) of the axehead is also three fingers.

The length of the *maisel* of the axe is three fingers. The meaning of this term is uncertain, but I suggest that it refers to the loop of iron which forms the socket for the axe-handle. It is one inch (*ordlach*) across where it is attached to the blade, and widens to two inches at the other side. Although *maisel* is otherwise unattested, a derivative *maisled* occurs in legal commentary on accidental injury caused by axes.[158] The commentator deals with four situations where such an injury may be immune from liability. The first is when the axe flies out of the hand of the user and hits another person. The second is when the head flies off the handle. The third is when somebody is hit by a chip of wood. I suggest that the fourth type of accident, *maisled,* is when somebody is hit by the blunt side of the axehead. The implication seems to be that if a person holding an axe injures another with the blade, he cannot be immune from liability and must be guilty of either culpable negligence or malicious intent.

In the description of a proper axe, it is further specified that the axehead should be manufactured with three heatings:[159] red heat (*donngorad*),[160] white heat (*bángorad*), and the heat of tempering (*gorad foberta*).[161] Its iron should not be dented by oak or yew, i.e. the toughest woods which would be encountered.

Another word which may have the same meaning as *biáil* is *rodb,* used in heroic literature of a warrior who defeats his opponent 'like a axe(?) cuts a tree-trunk' (*amail treagdas rodhb omna*).[162] The word *biáil* is also used of large axes designed for military purposes. Thus the law-text *Bretha Étgid* refers to the slave in a king's house who wields a double-bladed axe (*biáil imfáebur*) to protect the royal

[157]For the finger (*mér*) as a measure close to the inch, see p. 561.

[158]*CIH* i 285.29–32 = *AL* iii 274.7–11. It is suggested in the translation at *AL* iii 275.15 (and in the glossary at *AL* vi 550) that *maisled* refers to injury caused by the block on which wood is cut, but this seems an unlikely accident.

[159]*CIH* iii 921.30–31. On the other hand, legal commentary on *Bretha Comaithchesa* refers to an axe worth one scruple which has only been tempered twice (*CIH* i 199.13; ii 579.30 = *AL* iv 130.20 *a leagadh fa dī*).

[160]The adj. *donn* is usually translated 'brown', but here it obviously refers to the dull red of heated iron.

[161]*Fobairt,* lit. 'bringing under', is used of the tempering of iron by immersion in water. For other examples, see *DIL* F 182.73–82.

[162]*TBC* St 99.3124; 156.4926. See Thurneysen, '*Rodb*-'.

household.[163] The *gallbiáil* 'foreign axe' may be primarily a weapon of war. An obscure quotation in *O'Davoren's Glossary* speaks of a foreign axe with two black spikes or 'ears' which is worth sixteen scruples.[164] These spikes are presumably set in the opposite direction to the blade so that an attacker may inflict damage with a back-stroke as well as with his first blow. In another legal reference, however, *gallbiáil* seems to be used in a non-military context.[165]

The written sources also provide evidence of a smaller domestic axe. As we have seen above, *Críth Gablach* distinguishes the *biáil* from the *eipit*, a word which also occurs in a list of domestic tools in *Auraicept na nÉces.*[166] It is clearly a participial formation from the compound *as-ben* 'cuts', which is commonly attested in the verbal noun *eipe* 'cutting (especially of trees)'. It is probable, therefore, that the *eipit* is a light axe or hatchet suitable for chopping firewood. I would equate it with the *túag connaid* 'firewood axe' of later sources, whose small size is apparent from an episode recorded in the *Annals of Connacht* for the year 1243.[167] The annalist recounts how Giolla gan Ionathar Ua Miadhaigh tricked Hugo de Lacy into inspecting the castle-moat at Durrow. As Hugo was stooping down to measure the ground Giolla gan Ionathar killed him with a *túag connaid* which he had hidden under his armpit. Another murder with this type of axe is recorded in the same annals for the year 1424.[168] Seán mag Raghallaigh was using a *túag connaid* to chop up firewood, and with it initiated a fatal attack on Tadhg Ó Fallamhain. Yet another axe-killing is recorded for the year 1232: Conchobhar mac Aodha mheic Ruaidhri was killed in Túatha Síl Muireadhaigh by a man with a

[163] *CIH* i 285.23 = *AL* iii 272.23.

[164] *CIH* iv 1485.40–41 = O'Dav. 282 § 532 *Cailech .i. cluas,* ut est *gallbiāil innraic mitir a fíu .ui. screpail .x. cona dīb dubhchailchib .i. cona dīb cluasaib ⁊ cona srūibh .i. na cluasa īchtaracha* 'a projection/flange i.e. an ear, that is a proper foreign axe with its two black cups is worth sixteen scruples, i.e. with its two ears and with its snout i.e. the lower ears'. This use of the word *cailech* (presumably Latin *calix* 'chalice, cup, projection') does not seem to be otherwise attested. Perhaps *dub-* here means 'black, death-dealing'. For an addendum on *clúasa*, see p. 752. In legal commentary at *CIH* ii 669.28 the *gaillbiáil*, which is defined as a *crostúag lethan* 'broad cross-axe', is valued at fourteen scruples.

[165] *CIH* iii 895.22.

[166] Calder, *Auraicept*, 292.5725–8.

[167] *AC* 79 s.a. 1243 § 4. The editor points out that this story properly belongs under the year 1186, and relates to this earl's father.

[168] *AC* 466 s.a. 1424 § 4.

white axe-handle (*samthach*).[169] To prevent the identity of the perpetrator becoming known, the people of the Túatha whitened the handles of all their axes.

Legal glosses and commentary provide conflicting information on the commercial value of a domestic axe. According to one commentary on *Bretha Comaithchesa*, an axe used in cutting wood for an oak-fence is worth one scruple.[170] On the other hand, a gloss on the same passage assigns a value of three or four scruples to the carpentry axe (*túag ṡaírsi*) or firewood axe (*túag connaid*).[171]

Billhook

In the view of a legal glossator, the two principal 'irons of husbandry' (*erna trebtha*) are the large axe and the billhook.[172] The Old Irish term for the latter is *fidbae*, which has the basic meaning 'wood-cutter'. However, like its cognates in other Celtic languages,[173] the term has become specialized in the sense of a medium-sized hooked tool for cutting rods, etc.

In excavations from the early Christian period, billhooks of different shapes and sizes have been discovered.[174] Likewise, the archaeological evidence from Roman Britain indicates a tool with much variation in design.[175] Nonetheless, the author of an Old Irish legal passage on tools felt capable of providing precise measurements for a standard billhook (*fidbae n-inraic*).[176] Its socket (*cráu*) is a fist in diameter: this means that the user can take a comfortable hold of the wooden handle which fits into the socket.[177] The curved part of the blade (*croch*) measures three

[169] *AC* 42 s.a. 1232 § 6. *Samthach* is used of the handle of an axe or of a billhook, e.g. *CIH* iii 894.17 *samtach fidba nō bēlai.*

[170] *CIH* i 199.13; ii 579.30–1 = *AL* iv 130.19–20.

[171] *CIH* ii 669.26–9.

[172] *CIH* ii 467.18 = *AL* v 392.11. According to a commentator on *Bretha Comaithchesa* (*CIH* i 199.14; ii 579.32 = *AL* iv 130.21–2), a billhook for fencing is worth one scruple and must have been tempered twice (*a leagad fa dī*) or subjected to long heating(?) (*sithaile*) or intense(?) tempering (*beoleagadh*).

[173] Gallo-Latin *vidubium* (whence Modern French *vouge*), Welsh *gwyddif*, Old Breton *gwedom* (*VKG* i 389; Lambert, *La langue gauloise*, 200).

[174] E.g. Hencken, 'Ballinderry crannóg no. 1', 128; Hencken, 'Lagore crannóg', 105.

[175] Rees, *Agricultural implements* ii 467–73.

[176] *CIH* iii 921.22–5.

[177] *Dorn dia chrāu* lit. 'a fist for its socket'. Compare the use of the term *dornchlae* (*dorn* + *clí*(?)) 'sword-hilt'. A rivet (*seim*) keeps the handle securely in the socket (Meyer, *Triads*, 22 § 172).

fingers: this presumably means that this part of the billhook is the length of an average man's three fingers.[178]

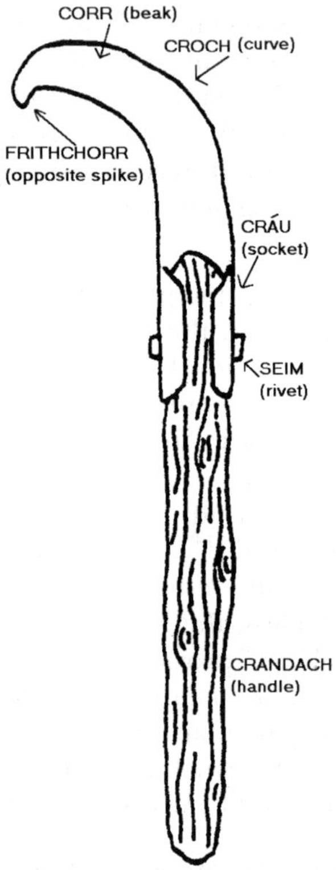

Fig. 21. A suggested view of the billhook described at CIH iii 921.22–5.

There are also three fingers in the beak (*corr*) of the tool, and one finger in its 'opposite spike' (*frithchorr*). Finally, the author gives the thickness of the metal at three points on the back of the blade: 'an inch is the width of its back at its handle (*crannach*), half an inch at its middle (*medón*), one third of an inch at its beak (*corr*)'.[179] It is clear, therefore, that the author is describing a billhook with a cutting edge (*fóebar*) on one side of the blade only.

It is difficult to ascertain what is meant by *frithchorr*. I suggest that it refers to the bottom of the tip of the blade: this may have a slight inward curve so that the user can draw branches towards him. In some Romano-British examples this beak is curved sideways, giving the tool a cutting edge in two planes. Another possibility is that *frithchorr* refers to a short spike or talon on the back of the blade. This is an occasional feature in both early and modern billhooks,[180] but I cannot see how this arrangement can be reconciled with the dimensions set out in our text.

[178]The text reads *Trī mēoir ina croch rēisiu fod a foebuir*; there seems to be an omission after *reisiu* which would have told us the length of the cutting edge (*fóebar*). *Croch* (read dat. sing. *croich* (*cruich*)) is a borrowing from Latin *crux* 'cross', and usually refers to a cross or a cross-like object. However, it can also be used of a curved or hooked object, cf. *CIH* ii 563.28; iii 780.7 = *CG* 7.186 *tinne for croich* (*cruich, croith* MSS) 'a flitch of bacon on a hook'. For *mér* 'finger' as a unit of measurement, see p. 561.

[179]*Ordl- lethit a cūil oca crand- lethordl- ar medōn trian oca chuirr.*

[180]Rees, *Agricultural implements* ii 467, 469; K. D. White, *Agricultural implements of the Roman world*, 87, 93, 94, 96.

Saw

Small saw-blades have been found occasionally in excavations from the early Christian period in Ireland,[181] and it seems likely that *tuiresc* in the list of the *mruigfer*'s equipment in *Críth Gablach* has been correctly identified in this meaning.[182]

Other tools for cutting wood, etc.

The list of tools in *Críth Gablach* also includes the adze (*tál*), which consists of a small blade set at right angles to the handle. This tool is used particularly for domestic carpentry – a law-text on distraint refers to 'those who work with adze and axe' (*áes táil 7 béla*).[183] Another wood-working tool in the list is the auger (*tarathar*), used for boring holes.[184]

Our sources also mention other cutting tools used on the farm. The law-text *Di Chetharšlicht Athgabálae* includes the 'hook of a woman householder's house' (*cromán tige bantrebthaige*) along with other farm tools and appliances.[185] According to the accompanying glosses this is an iron hook (*bac*) or sickle (*corrán*) for cutting ivy or holly, presumably for winter fodder (see p. 46). The context is not altogether clear, but the sense seems to be that this tool is of particular importance to a woman in poor economic circumstances so that she can feed her cattle in winter. Elsewhere, a legal glossator refers to this type of hook in a general farming context with no specific association with women.[186]

The list in *Críth Gablach* includes knives for cutting rushes (*scena búana aíne*).[187] The sharpening of a four-inch knife for rush-cutting is described in a difficult passage in another law-text:[188] 'this is the

[181] Hencken, 'Lagore crannóg', 108–9; O'Kelly, 'Two ring-forts at Garryduff', 46–7; Mytum *The origins of early Christian Ireland*, 248.

[182] *CIH* ii 563.22; iii 780.1 = *CG* 7.178 (see vocabulary to *CG* s.v. *tuiresc*).

[183] *CIH* iii 898.1 = Binchy, 'A text on the forms of distraint', 78 § 5.

[184] *CIH* ii 563.22; iii 780.1 = *CG* 7.178. It is cognate with Welsh *taradr* of the same meaning (*LEIA* T-30).

[185] *CIH* ii 374.10; v 1684.27 = *AL* i 124.15. *Crom(m)án* is a derivative of *cromm* 'bent, crooked', and is cognate with Welsh *cryman* 'billhook, pruning hook, sickle'. In Modern Irish, *baintreabhach* means 'widow', but in the earlier language it refers to any woman who is the head of a household, whether widowed or married to a landless man. In a gloss (*CIH* v 1684.28) she is identified as a *banchomarbe* 'female heir'.

[186] *CIH* i 243.5 = *AL* v 488.22.

[187] *CIH* ii 563.22; iii 780.1 = *CG* 7.177.

[188] *CIH* iii 921.14–16. The rest of the passage deals with the rent (*somoíne*) which the recipient of a knife must give to its owner.

proper quality of a knife: after it is ground so that it is a sharp blade, it is drawn three times across a leather strop (*fraig*)[189] from level with your face down to your two knees, and sharpened then by striking against your palm; you then cut with it three carts of rushes from one grinding'. In a legal gloss a piece of meat is said to be the length of the haft of a knife (*em na sceine*).[190]

Fencing and wall-construction

The author of *Bretha Comaithchesa* lists the four tools which must be given as a fine (*smacht*) for failure to maintain the four types of field-boundary.[191] In three cases, the tool is obviously that used in the construction of the field-boundary: a spade (*rámae*) for the trench-and-bank, a billhook (*fidbae*) for the bare fence, and a heavy axe (*biáil*) for the oak fence. The tool which is to be given for the stone wall is a *socc*, which normally means 'ploughshare'. I follow Thurneysen's suggestion that *socc* refers here to an iron lever.[192] This would be equivalent to the modern pin-spar used in the building of stone walls. According to a commentator on this passage, the *socc* is worth one scruple – as are the other three tools – and must have been tempered three times (*a leaghadh fa trí*).[193]

Another fencing tool is the mallet (*forchae*), which is used to drive the stakes into the ground.[194] It is probable that the head of this tool was normally of stone, but hard wood may also have been used. For a discussion on the methods of erecting the fences and other field-boundaries described in the law-texts, see p. 372.

[189] This is the only attestation of *fraig* with the meaning 'leather strop (for sharpening a knife or razor)': see *DIL* s.v. 5 *fraig*. The first part of the passage is quoted at *CIH* v 1566.9–10, where *tar fraig*id is glossed *.i. tar criss* 'i.e. across a belt'.

[190] *CIH* ii 370.24 = *AL* i 132.20.

[191] *CIH* i 65.13 = *AL* iv 72.2; *CIH* vi 2193.15 = Thurneysen, 'Aus dem irischen Recht III', 314 § 8; *CIH* i 199.11–14 = *AL* iv 130.17–22.

[192] Thurneysen, 'Aus dem irischen Recht III', 315.

[193] *CIH* i 199.12–13; ii 579.31 = *AL* iv 130.19. By contrast, the axe and the billhook need only be tempered twice, and the spade once. A glossator at *CIH* ii 669.26–9 assigns a value of three scruples to each of these four tools.

[194] *CIH* i 73.15 = *AL* iv 112.13. Gwynn and Purton (*The monastery of Tallaght*, 133.14 § 16 and glossary s.v. *forrach*) take *fer na forchae* to mean 'a man for the measuring-rod', but it seems more likely to mean 'a man for the mallet'. Mallets would have been widely used in house-building and carpentry as well as fencing. The heavy iron blacksmith's hammer is called an *ord*.

ANIMAL RESTRAINTS, GOADS, ETC.

Restraints

Our sources mention many devices for reducing the mobility of domestic animals. The *búarach* 'cow-spancel' is used to immobilize the hind legs of a cow during milking, and is therefore particularly associated with milkmaids in the literature. A story preserved in a legal context tells of a mythical female landowner (*banchoairt*) who had seven hundred cows.[195] These cows were milked by thirty-five women-cooks (*banchoici*), each of whom had a spancel. The spancels were kept hanging from a cross-beam in seven groups of five, and each cook was able to recognise her own spancel. It is likely that the early Irish spancel commonly consisted of a stick, a rounded piece of wood and a rope or withe.[196] The Middle Irish text entitled *Bórama* refers to a bronze spancel, but this was probably exceptional.[197] At all events, a typical spancel was of solid enough construction to deliver a mortal blow. This is clear from a passage on place-lore which describes how a professional buffoon (*drúth*) tried to rape a girl named Sampait who was driving her cattle to be milked. She struck her assailant a blow with her spancel, making pulp of his head.[198]

Another form of restraint is the *airchomal*, which is explained in *Cormac's Glossary* as a method of tying the front legs of an animal together.[199] In our sources it is normally associated with horses. For example, the Old Irish tale *Scéla Cano meic Gartnáin* contains a reference to the weaving of an *airchomal* for a horse named Grip.[200] This text refers to part of this device as the *id*, which Binchy takes to be a ring or band. It seems therefore that a rope or thong is passed through one or more rings to reduce the horse's mobility. An *airchomal* has three parts, which – according to a legal glossator – comprise two rings (*dá id*) and a thong (*léithre*).[201] The term *airchomal* may also be used in relation to other animals. Legal

[195] *CIH* vi 2117.36–2118.2 = Dillon, 'Stories from the law-tracts', 51 § xiv.

[196] Lucas, *Cattle in ancient Ireland*, 44.

[197] *LL* v 1270.37729 = Stokes, 'The *Bórama*', 40 § 12. Note, however, the references to iron hobbles (*ferrei sigilli*) in Herren, *Hisperica Famina*, 76.169; 88.315.

[198] *LL* iii 745.22259. In another version (E. J. Gwynn, *Metrical Dindshenchas* iv 24.20) she strangled the buffoon with her spancel; cf. Stokes, 'The prose tales in the Rennes Dindṡenchas, (2)', 36–7 § 86.

[199] Meyer, *Sanas Cormaic*, 68 § 812.

[200] Binchy, *Scéla Cano*, 9.238.

[201] *CIH* i 241.38 = *AL* v 484.18. See *DIL* s.v. *leithriu* (*léithriu*?).

commentary on the restraint of domestic animals speaks of the use of the *airchomal* on calves.[202] Another law-text refers to its use to reduce the mobility of trespassing hens.[203]

Writing in the tenth century, Cormac distinguishes the *airchomal* from another type of restraint called the *langpheitir*,[204] which he defines as a tie between hind-leg and fore-leg.[205] He takes the word to be a borrowing from the English language, and etymologizes it as *lang* 'long' + *pheitir* 'fetter'.[206] In a detailed discussion, Marstrander argues on the contrary that the word is ultimately of Norse origin, and cites a hypothetical Old Norse **lang-fjöturr* 'long fetter'.[207] Although there is other evidence of farming terminology of Norse origin in Middle Irish (see p. 249), Marstrander's case remains doubtful.[208]

Other methods of restraint mentioned in legal glosses and commentary are a running noose (*sás co rith*) around an animal's neck, a horse-hair tie (*rith rúainne*) around its teeth, or a willow withe (*gat salach*) in its mouth.[209] Another commentary prescribes methods of restraining smaller animals and fowl.[210] A hen is restrained by having a hood (*cochall*) placed over its head, a goat has some sort of leather 'shoe' (*bróc*) put on its legs, and a pig has a form of saddle (*srathar*) strapped to its back. A fierce dog is held by at least one chain (*slabrad*): see p. 115.

[202] *CIH* vi 2134.4 *urcomail imna gamn*aib. Other versions have *urcholl fona gamnaib* (*CIH* i 68.30 = *AL* iv 86.2), and *urchall-fona gamhnaib* (*CIH* ii 670.6). On the basis of these readings, *DIL* A 177.52 postulates a form *aircholl* 'spancel, fetter', but further attestations are needed.

[203] *CIH* iii 897.35 = Binchy, 'A text on the forms of distraint', 78 § 3.

[204] Meyer, *Sanas Cormaic*, 68 § 812. An older term for this type of spancel seems to have been the *id ḟata* 'long spancel' (e.g. *TBC* Rec. I 67.2201 *id* [*ḟ*] *ata*). *Langpheitir* survives in Modern Irish in forms such as *lonncaidir, lonncairt*, etc.

[205] Legal glossators, on the other hand, take it to be a tie between legs and head, e.g. *CIH* i 240.10 = *AL* v 478.17 (*laingfitir*); *CIH* ii 386.5 = *AL* i 174.7 (*langḟithil*); *CIH* iii 844.35; vi 2136.40 (*laingfiter, -ir*).

[206] Meyer, *Sanas Cormaic*, 68 § 812 *Langp*[*h*]*eitir .i. anglis insein; lang .i. fata, pheitir im- glas.* (*Anglis* is presumably an Irish form of *englisc* 'English'.)

[207] Marstrander, *Bidrag*, 23–7.

[208] In a brief discussion on Norse loan-words in Irish, Greene argues against Marstrander's explanation of *langpheitir* (Greene, 'The influence of Scandinavian on Irish', 75–6).

[209] *CIH* i 240.10 = *AL* v 478.17; *CIH* ii 386.5–7 = *AL* i 174.8–10; *CIH* iii 844.35–6; vi 2137.1.

[210] *CIH* i 68.29–30 = *AL* iv 86.1–2. Cf. *CIH* ii 670.6–7; vi 2134.3–4.

A general term for a tie is *nasc*, which occurs in a number of farming contexts. For instance, a *nasc* is included in the list of farming tools and utensils in *Auraicept na nÉces*,[211] and a Middle Irish Life of Saint Senán refers to ties, pens and milking enclosures as necessary farming equipment.[212] A law-text on status refers to a low-ranking craftsman called a *nascaire*, 'maker of ties or rings', but it is not clear for what purpose they were intended.[213] Possibly, the tie was for the necks of livestock.

Goad

Every farmer from the rank of *ócaire* upwards is expected to own a goad (*brot*),[214] which is included along with the share and the coulter among the 'irons of ploughing' in a legal gloss.[215] A goad is an iron spike which may be attached to a stick for prodding oxen or draught horses into action.

A satirical verse on the ploughman of Mag Muccín accuses him of finding the goad light, but the plough too heavy (*étrom lais in bruitín, rothrom lais in cécht*).[216] He evidently had the reputation for overuse of the goad on his oxen, and laziness in his manipulation of the plough.

Rods

Rods of various types have doubtless been used to drive and control animals since the beginning of domestication. Early Irish sources refer particularly to the *echlasc* (*echḟlesc*),[217] a rod for driving horses.[218] It may have had a metal hook (*baccán*) at the top.[219] Some literary references suggest that it was possible to stick the *echlasc* into the ground for use as a temporary tether.[220] In the sagas, a rod used by charioteers is called a *deil*.[221]

[211] Calder, *Auraicept*, 292.5725.
[212] Stokes, *Lismore Lives*, 57.1907–58.1908.
[213] *CIH* v 1616.24 = *AL* v 106.17.
[214] *CIH* iii 778.27 = *CG* 4.95.
[215] *CIH* ii 467.18 = *AL* v 392.11.
[216] Thurneysen, 'Mittelirische Verslehren', 83 § 68.
[217] Both forms are from *ech* 'horse' + *flesc* 'rod'.
[218] E.g. *CIH* iii 1080.24.
[219] Meid, *Táin Bó Fraích*, 2.30–31.
[220] E.g. *TBC* Rec. I 47.1534.
[221] *DIL* D ii, 5.53–61.

BUTCHERING, ETC.

The *mruigfer*'s tools also include a spear for killing livestock (*gaí gona cethrae*).[222] This seems to have been an implement specifically designed for the slaughter of domestic animals. The texts give no information as to how such spears differed from those used in hunting or warfare.

Other tools may also be used to kill and cut up domestic animals. A legal glossator states that a bullock which a client gives as food-rent to his lord should have been slaughtered by an axe (*túag*), and not have died through disease.[223] There is no suggestion that a special type of axe is used. Similarly, it is implicit in a legal passage on tools that a knife (*scian*) may be used for such diverse tasks as cutting up meat and harvesting rushes.[224] Likewise, the Old Irish tale *Táin Bó Fraích* speaks of the flesh of a heifer being cut up by adze (*tál*) and axe (*biáil*) in preparation for a stew.[225] It is implicit in the law-text *Bretha Étgid* that an iron implement (*iarn*) is normally used for the slaughter (*airlech*) of domestic animals.[226] According to the glossator, if a slaughterer uses another person's implement for this purpose, he is responsible for any damage to it caused by a stone or tooth.

The skulls of cattle found in excavations from the early Christian period often show shattering of the frontal bone which would be consistent with a stunning blow of a hammer (*ord*) or heavy stone.[227]

After curing, a pig's carcase hangs up in the farmer's house. Thus *Críth Gablach* refers to a flitch of bacon on a hook (*tinne for croich*).[228]

TRANSPORT TECHNOLOGY

Chariot, cart

The Old Irish law-text on roads edited on p. 537 of this book illustrates the importance of the chariot (*carpat*) as a means of

[222] *CIH* ii 563.24; iii 780.3 = *CG* 7.180. Both MSS have *gai* which could be taken as nom. sing. or nom. plur. In his glossary to *CG* (s.v. *gáe*) Binchy takes it to be nom. plur.

[223] *CIH* ii 481.3 = *AL* ii 238.z.

[224] *CIH* iii 921.9–18.

[225] Meid, *Táin Bó Fraích* 9.227.

[226] *CIH* i 288.15–17 = *AL* iii 286.5–9.

[227] McCormick, 'Stock-rearing in early Christian Ireland', 96.

[228] *CIH* ii 563.28; iii 780.7 = *CG* 7.186.

transport.[229] In this text – as in early Irish literature generally – the chariot is associated particularly with persons of high rank, the king and the bishop. In another law-text, the term *cairptech* 'one who has a chariot', appears to be used as a synonym for a lord.[230] The *arae* 'charioteer' is mentioned in *Uraicecht Becc* as a servant who enjoys prestige and privilege in accordance with the rank of his employer.[231] The chariot-builder (*carpatṡáer*) has the status of the second grade of *bóaire*, and is thus a freeman with an honour-price of three *séts*.[232]

Much research has been carried out on the archaeological and literary evidence relating to the chariot among the Celtic peoples,[233] and consequently the Irish material has been investigated in some detail.[234] It seems that the early Irish chariot was a light two-wheeled vehicle drawn by two horses. Each wheel (*droch*) had a metal rim (*fonnad*), and a pole (*sithbe*) connected the body of the chariot (*cret*) to the yoke (*cuing*). Old Irish commentary on the law-text *Bretha Comaithchesa* refers also to the shafts of the chariot (*feirtsi carpait*), which are said to be made from holly.[235] It is unclear where these shafts were situated: Greene suggests that they projected from the back of the chariot,[236] whereas Sayers argues for an identification with the swingletree between the horses and the chariot.[237]

[229]For discussion of the etymology of *carpat*, see *LEIA* C-41. It is cognate with Latin *carpentum* 'chariot', a borrowing from Gaulish: see below.

[230] *CIH* vi 2302.10 = Binchy, '*Bretha Crólige*', 46 § 58.

[231] *CIH* v 1617.12 = *AL* v 108.20.

[232] *CIH* v 1616.17 = *AL* v 106.7–8.

[233]The Continental Celts are known to have been especially skilled in the construction of horse-drawn vehicles. In Latin sources, there are many words of Celtic origin which refer to vehicles of various types, e.g. *benna, carpentum, carrus, carruca, essedum, petorritum, reda, covinnus* (Dottin, *La langue gauloise*, 223–302; Lambert, *La langue gauloise*, 201).

[234]See Greene, 'The chariot as described in Irish literature'; Harbison, 'The Old Irish "Chariot"'; Sayers, 'Old Irish *fert*, etc.'; Sayers, 'Textual notes on descriptions of the Old Irish chariot and team'.

[235] *CIH* i 202.26; ii 582.13–14 = *AL* iv 150.2–3.

[236]Greene, 'The chariot as described in Irish literature', 67–8.

[237]Sayers, 'Old Irish *fert*, etc.'. However, Langdon (*Horses, oxen and technological innovation*, 12) holds that the swingletree was unknown in Europe until the twelfth century.

The reproduction below from the badly weathered ninth-century Cross of the Scriptures at Clonmacnois gives some idea of a contemporary view of a two-horse chariot.[238]

Fig. 22. A two-horse chariot from a ninth-century cross at Clonmacnois, Co. Offaly.

The Old Irish law-texts often distinguish the chariot from the *carr* or *fén* 'cart',[239] which seems to have been a much heavier vehicle, probably with four wheels. It was used to carry loads such as rods, rushes, manure, and corn.[240] Typically, a cart was drawn by two oxen.[241] However, one law-text refers to a *capall cairr* 'cart-horse',[242] and a passage in a Latin saint's Life speaks of a light *carrus* drawn by a single horse, which was used to transport vessels for milk and butter to and from a monastery.[243] When a family moved house, their belongings were carried in an ox-drawn *carr*.[244]

[238]Harbison, *High Crosses* i 48 (discussion); i 372–3 (dating); ii fig. 135 (photograph).

[239]Old Irish *carr* is cognate with Welsh and Breton *carr* of the same meaning (*LEIA* C-41–2 s.v. 1 *carr*). The Gaulish cognate was borrowed into Latin as *carrus* 'cart'. Old Irish *fén* is cognate with Welsh *gwain*, etc. (*VKG* i 59; *IEW* 1119).

[240]*CIH* ii 570.10 = *CG* 22.573 *carr coíl, carr oíne* 'a cart-load of rods, a cartload of rushes'; *CIH* ii 384.15; v 1690.9 = *AL* i 170.1–2 *do chairr .i. ailich no arba* 'of your cart, i.e. for manure or corn'.

[241]E.g. Bieler, *Patrician texts*, 112.7; 120.2–3.

[242]*CIH* ii 471.18 = *AL* v 406.1.

[243]Plummer, *Vitae* ii 26 § xvii. A *currus* drawn by a single horse is also mentioned at *Vitae* ii 52 § xxix.

[244]*CIH* iii 897.14 = Appendix A, text 2 § 1 (10).

Another Old-Irish word for the farm-cart is *á*, which appears in an amusing tongue-twister quoted in legal commentary. According to the preamble it is the record of a conversation overheard in Connacht.[245]

'In esser dom to á?'	'Will you lend me your cart?'
'Tó, mani má mo á'.	'I will, if my cart doesn't break.
'Ara tairi mo á mó?'	'Will my cart come back soon?'
'Mani má to á, tó'.	'If your cart doesn't break, it will'.

Pack-saddle

Smaller loads can be secured onto a pack-saddle (*srathar*),[246] strapped to the back of a horse or ox.[247] Loads may of course also be carried on the human back: legal commentary refers to the carrying by a slave of a load of firewood held together by a withe (*gat*).[248]

Boat

By lakes, rivers or the sea, a boat may be a useful or essential adjunct to farming. The law-text *Di Chetharṡlicht Athgabálae* distinguishes the small wickerwork coracle (*clíab*) from the boat (*náu*).[249] According to one glossator, if a person destroys the former he must pay a penalty-fine of five *sét*s to the owner, whereas if he destroys the latter he must pay ten *sét*s.[250]

Ropes

There are many references in our sources to ropes, and the author of *Críth Gablach* assumes that a prosperous farmer of *mruigḟer* rank possesses a number of ropes (*lomna*).[251] Another law-text states that a workman (*gniae*) is expected to have a rope: an accompanying gloss defines it as a 'rope for carts and burdens' (*loman carr*

[245] *CIH* vi 2112.37–9 = Dillon, 'Stories from the law-tracts', 43 § i. A closely similar version is quoted in *Cormac's Glossary* (Meyer, *Sanas Cormaic*, 8 § 70). The verse has been edited by Watkins, '*In essar dam do á*'. I quote his translation here.

[246] Like Welsh *ystrodur* of the same meaning, *srathar* is a borrowing from Latin *stratura* 'pack-saddle'. See *LEIA* S-182 s.v. *srathar.*

[247] E.g. *LU* 95.2938 = O'Grady, *Silva Gadelica* i 234.8 (trans. ii 265.z); *CIH* vi 2137.5.

[248] *CIH* i 265.12–13 = *AL* iii 176.12–15.

[249] *CIH* ii 384.15; v 1690.8 = *AL* i 166.21 (where *clíab* is inaccurately translated 'basket').

[250] *CIH* v 1690.8. For a brief reference to the boat-builder, see *CIH* v 1615.27 = *AL* v 104.1. For the use of bovine hides in boat-building, see p. 55 above.

[251] *CIH* ii 563.22; iii 780.1 = *CG* 7.178.

7 beart).[252] A legal heptad refers to the use of a rope (*súainem*) to allow a person to descend a cliff.[253]

There is no precise information in our sources on rope-making, but twisted willow withes are likely to have been a common material. There is also evidence of ropes made from bark. For a brief discussion, see p. 384 above.

Other farm tools

In the *Críth Gablach* list of tools and utensils which a *mruigfer* would be expected to own, we find the words *dias fidchrann* occurring after *tál, tarathar, tuiresc* 'adze, auger, saw' and before *eipit* 'hatchet'.[254] In his translation, Eoin MacNeill separates the two words, taking *dias* to mean 'a pair of shears', and *fidchrann* to mean 'a trestle(?)', which he explains as 'a strong wooden frame to hold large timber for sawing, etc.'[255] D. A. Binchy, on the other hand, takes the two words together, and tentatively translates 'a pair of wooden shafts, i.e. a wooden shears'.[256]

Neither explanation is satisfactory. *Dias*, usually meaning 'two persons', is only rarely used to mean 'two things'.[257] There is therefore little probability that *dias* by itself was used for 'shears', as MacNeill suggests. Furthermore, the word *deimes* occurs in this meaning in both Old and Modern Irish, as well as in Scottish Gaelic and Manx. For example, the ninth-century glossator on Priscian uses it to explain Latin *biceps* 'shears (for cutting hair, fleece)'.[258]

Binchy's suggested translation 'wooden shears' is also unlikely. Wooden spades (p. 465) and wooden ploughshares (p. 469) can be used to break open the soil, but a cutting implement can hardly have been made of this material. In early Irish tool-making, the main use of wood seems to have been to provide handles. An exception is the wooden tongs (*tenchor crainn*), which is included in a short list of tools and utensils in legal commentary. The commentator states that a penalty-fine (*díre*) of five *séts* must

[252] *CIH* ii 374.8 = *AL* i 140.3.
[253] *CIH* i 32.1 = *AL* v 236.15.
[254] *CIH* ii 563.23; iii 780.1–2 = *CG* 7.178.
[255] E. MacNeill, 'Ancient Irish law', 291 § 89 (and footnote).
[256] *CG* 29, note to l. 178.
[257] *DIL* D ii, 66.13–15.
[258] *Thes.* ii 145.26.

be paid by a person who steals or destroys this implement. It is distinguished from the iron tongs (*tenchor iairn*) which has twice this value.[259] Wood is also the material used for the beam, foreshare, stilt, and handle of the plough, but it is unlikely that the words *dias fidchrann* refer to any part of this machine, as the plough is dealt with separately at the end of the *Críth Gablach* list.[260]

It is possible that the word *dias* in our list has the meaning 'point' – it is well attested as the point of a sword – in which case it may refer to a chisel or gouge.[261] In the Middle Irish text *Cóir Anmann*, there is another attestation of *fidchrann* in a context where it seems to refer to some sort of wooden bar, but unfortunately no information is given as to its function.[262] A remote possibility is that *días fidchrann* is a wooden reaping tool. Latin sources refer to the use in the Mediterranean area of *mergae* 'reaping-boards', a pair of wooden plates which are forcibly brought together to remove the ears of corn from the stalks. Another Roman reaping tool is the wooden *pecten* 'comb'.[263]

Another Irish word which has been taken to refer to a farm tool or implement is *usca*. The *Dictionary of the Irish language* gives it a separate head-word and quotes 'heather-brush' as a possible meaning.[264] It occurs among the items listed under the heading *ogam tírda* 'rustic ogham' in the text on the Ogham alphabet in *Auraicept na nÉces*.[265] The author has collected words beginning with each letter of the ogham alphabet from various contexts, such as birds, colours, churches, forts, saints, arts, etc. The words collected in the rustic ogham section are mostly the names for tools, such as *biáil* (axe), *eipit* (hatchet), *ord* (sledge-hammer), or *tál* (adze); but a few are miscellaneous words connected with the farm such as *machae* (milking enclosure), *cúal* (bundle of firewood) or *dabach* (tub). I suggest that the word intended by

[259] *CIH* v 1537.23 *Dīre theanchuirnn croind .u.s. Dīre teancuir irnd .x.s.*
[260] *CIH* ii 563.25; iii 780.4 = *CG* 7.181–2 *óg n-arathair cona uili chomopair* 'full ploughing outfit with all its appurtenances'.
[261] *DIL* D ii, 65.20–34.
[262] Stokes, *Cóir Anmann*, 400 § 268.
[263] For a discussion on both these tools, see K. D. White, *Agricultural implements of the Roman world*, 110–15; Columella, *De re rustica* (ed. Ash), 2.20.3.
[264] *DIL* s.v. *usca*.
[265] Calder, *Auraicept*, 292.5728. For a discussion of the ogham alphabets in *Auraicept na nÉces*, see McManus, *A guide to Ogam*, 137–40.

the author is not *usca* but *úsca* 'lard, grease, tallow',[266] a very important animal-product in any early Irish household: see p. 55.

Finally, the rustic ogham list includes *ngend* (= *gend*).[267] This word is well attested in Irish of a wedge (of wood or metal) used to tighten a door, fetter, etc.[268] Its cognates in other Celtic languages are applied to a variety of tools used for splitting or tightening, e.g. Breton *genn* 'wedge for splitting wood'; Welsh *gaing* 'chisel, wedge (of wood or iron)'; Cornish *gen* 'iron wedge, miner's chisel'.[269] I suggest above (p. 486) that *gend* may also be applied to the blade of an axe, i.e. the part which does the splitting.

There is another uncertain attestation of *gend* in a section of legal commentary dealing with the punishment of setting adrift, which was regarded as the appropriate penalty for some serious crimes.[270] The culprit is put far out to sea in a boat with a single oar and given enough gruel for a day and a night. To ward off the sea-birds he is supplied with a *geann orda trí ndornd*.[271] This seems to refer to a wedge or similar pointed tool of about one foot (i.e. three fists) long, which could be used to keep the birds away from the food. Mary Byrne suggests translating 'a hammer-shaped(?) (hammer-headed?) wedge three hands long (or broad?)', taking *orda* to be an adjective from *ord* 'sledge-hammer'.[272] It is difficult to see how a wedge or other blade could be hammer-shaped or hammer-headed, but I have no better explanation to offer.

[266]Note that the *dathogam* 'colour-ogham' list includes *ūsgdha* 'tallow-coloured' (Calder, *Auraicept*, 290.5701).
[267]For the non-radical initial *ng-* in ogham, see McManus, *A guide to Ogam*, 37–9.
[268]*DIL* G 69.72–83; 70.1–4.
[269]*IEW* 437–8; *GPC* s.v. *gaing*; Hemon, *Geriadur*, s.v. *genn*.
[270]See *GEIL* 219–21 for a discussion of this punishment.
[271]*CIH* ii 744.34. The other versions have *ordu trī ndornd* (*CIH* i 109.12) and *airtim trī ndorn* (*CIH* iv 1301.40).
[272]M. Byrne, 'On the punishment of sending adrift', 99.

Appendix A
The texts

Practically all Old Irish law-texts provide some information on early Irish farming; in Appendix 1 of my *Guide to early Irish law* I provide a list of law-texts, with information on editions and translations.[1] A few of these law-texts are particularly important sources of information for the present book, and I give a brief account of them here.

The main law-text on farming is *Bretha Comaithchesa* 'judgements of neighbourhood', which can be dated on linguistic grounds to about the eighth century. It deals with the division of land among heirs, fencing, trespass by livestock and pets, damage to trees and bushes, and other problems which may arise between neighbouring farmers. The Irish text is printed in *CIH* i 64.6–79.12; 191.1–205.21. There is a translation in *AL* iv 69–159, but unfortunately it is often inaccurate; a new edition and translation by Thomas Charles-Edwards is expected to appear within a few years in the Early Irish Law Series. It will include an edition and translation of the important *Comingaire*-text on joint herding of livestock (*CIH* i 192.1–33; ii 576.24–577.24 = *AL* iv 100.4–102.19). A legal poem which deals with some of the topics discussed in *Bretha Comaithchesa* has been edited and translated by D. A. Binchy in the journal *Celtica* 9 (1971) 152–68.

The law-text *Críth Gablach*, from about the same period as *Bretha Comaithchesa*, provides a very valuable account of early Irish society, and gives us much insight into farming practice at the time. Binchy edited this text without translation, but with a legal glossary and textual notes, in the Medieval and Modern Irish Series (Dublin 1941, reprinted 1970). The most accurate translation is that by Eoin MacNeill in *PRIA* 36 C (1923) 281–306. The text is printed in *CIH* ii 563.1–570.32; iii 777.6–783.38; 952.1–953.9.

A great deal of general information on farming is provided by three law-texts on the relationship between lords and clients. *Cáin Aicillne* 'the law of base clientship' (*CIH* ii 479.23–502.6; v 1778.34–1804.11) is edited and translated into German by Rudolf Thurneysen in *ZCP* 14 (1923) 338–94. In *ZCP* 15 (1925) 239–53, the same scholar edited and translated the fragmentary *Cáin Ṡóerraith* 'the law of free fief' (*CIH* v 1770.15–1778.33). *Di dligiud raith ⁊ somaíne la flaith* 'on the law relating to the fief and profit of a lord' (*CIH* ii 432.21–436.32) is edited and translated by Bette-Jane Crigger in her doctoral thesis 'A man is better than his birth', 312-58.

[1] *GEIL* 264–83.

Particular aspects of farming are dealt with in a number of law-texts. A text beginning *Cis lir fodla tíre?* discusses the value of various types of land (*CIH* ii 675.18–676.16). Under the title '*Tír cumaile*' it has been edited and translated by Gearóid Mac Niocaill in *Ériu* 22 (1971) 81–6, along with later commentary on the same topic (*CIH* iii 843.3–21). In Text 6 below, I provide an edition and translation of another related commentary (*CIH* iii 972.24–9). A section of commentary on diseases and defects which may appear after purchase in cattle, sheep, horses, pigs, goats, dogs, and poultry (*CIH* iii 1000.26–32) has been edited and translated by Neil McLeod, *Early Irish contract law*, 299 [f]. The law-text *Bretha Crólige* (*CIH* vi 2286.24–2305.3) deals with illegal injury, and with the obligations of the culprit to arrange for the care of his victim. It is a particularly useful source of information on the early Irish diet (see p. 316–19), and has been edited and translated by Binchy in *Ériu* 12 (1938) 1–77. Another law-text on illegal injury, *Bretha Déin Chécht* (*CIH* vi 2305.4–2316.39), starts with an important passage on early Irish cereals. This text has been edited and translated by Binchy in *Ériu* 20 (1966) 1–66.

The Old Irish law-texts on dogs (*Conṡlechtae*) and on cats (*Catṡlechtae*) have only survived in fragmentary quotations. However, these are accompanied by glosses and quite extensive commentaries, so we can get some idea of the contents of the original texts.[2] There was also a law-text entitled *Cáin Bóṡlechtae* 'the law of cow-sections', which is equated in some sources with *Cáin Dar Í* 'the law of Dar Í (a female saint)'.[3] Commentary on the distraint of cattle also associates Saint Mo Ling with this law, and calls it *Cáin Dar Í ⁊ Mo Ling* 'the law of Dar Í and Mo Ling' (*CIH* ii 715.14–15).[4] *Cáin Bóṡlechtae* (*Cáin Dar Í*) is known to have dealt with offences against cattle and cowherds,[5] and it is possible that it also treated other legal matters relating to cattle. There is no evidence that *Cáin Bóṡlechtae* was in the *Senchas Már* collection of texts,[6] and it is not mentioned in the list of legal topics at *CIH* vi 2102.31–2103.32.

In this Appendix, I edit with translation seven legal passages which have particular relevance to early Irish farming. Texts 1, 2, 5, 6 and 7 have not to my knowledge been edited or translated previously. In general, I print the texts as they stand in the manuscripts, though occasionally with a different word-division. So as to present the reader with as coherent a text as possible, I sometimes restore between square brackets letters or words

[2]See *GEIL* 275 §§ 48–49.

[3]See *GEIL* 275–6 § 50; Ó Riain, 'A misunderstood annal: a hitherto unnoticed *cáin*'.

[4]The tradition of Mo Ling's care for wild and domestic animals is mentioned at Plummer, *Vitae* ii 201 § xxiii.

[5]*CIH* iv 1367.4 (trans. Thurneysen, 'Aus dem irischen Recht V', 388); *Thes.* ii 306^5.

[6]*GEIL* 244.

which I believe to have been in the original. I also on a few occasions use round brackets to indicate letters or words which should be omitted.

Texts 1, 2, 4, 5, and 7 are in Old Irish. On linguistic grounds, I would date Text 4 to the early eighth century (see p. 537 below), whereas texts 1, 2, 5, and 7 seem later, probably from the late eighth or ninth centuries. Texts 3 and 6 are Middle Irish commentary, from about the twelfth century.

I would like to record here my deep appreciation of those who read Texts 1, 2, 4, and 5 with me in a seminar at the Dublin Institute for Advanced Studies, and made many valuable comments and suggestions.

Text 1

Defects of cows and sheep

This particularly difficult Old Irish law-text is found at pp. 145^{b}-146^{a} of H 3. 18 (no. 1337), in the library of Trinity College, Dublin. This is a composite manuscript consisting mainly of legal material copied in the fifteenth and sixteenth centuries.

Our text is printed at *CIH* ii 674.37–675.17. It details various cow-defects (*bóainmea*) which a buyer should watch for (§§ 1–2), and also describes some of the good features of the ideal cow and calf (§§ 3–4). There is then a discussion about milk-yield and the value of cows in relation to the number of calves which they have borne (§ 5), and about the three categories of cow given in payment (§ 6). Finally, there is a treatment of the good qualities of sheep (§ 7) and their bad qualities (§ 8).

My paragraphing is only partly justified by the lay-out in the manuscript; the further subdivision is intended to facilitate cross-reference to the translation and discussion.

Text

§ **1.** (1) Dian cri[e] boin, bo boid, bo ineach cin boainmead. (2) It é ainmead bo *ar*an athchuir: (3) .i. maine be comblicht a sini, (4) mat gairde, (5) ma t*r*ifine, (6) mat goinig, (7) mad luach, (8) mana cara a læg, (9) mat finchad, (10) ma tarbchad, (11) mat lond, (12) mat sen*ach*, (13) mat lethcolpach, (14) mat atloscc, (15) mat cæch, (16) mat comgith a féra, (17) mana biatha a læg co maith ⁊ a mbligre, (18) ma raed*r*a, (19) mana dart*har*: as e fe*r*lond[1] bo an*n* sin.

§ **2.** (1) Ma údgilti, (2) mana beth a da noga fuirr-i, (3) nis fuirr a læg, mana epell- as, (4) nibi buirech, nadbid buirech, (5) nabid becech, (6) nabi lon*n*, (7) nabid lugach, (8) nibi modísach.

§ **3.** (1) Bo og allainn asi diriud diainim, (2) asi coir cuin*n*riug alaloig, (3) nis tercorpai sealb na mbrug na bligri na medam na fiada na feichme.

§ **4.** (1) Bo occ al*uinn* co læog a cuirp folei, (2) nibi læg fothla na tothla[2] (na tothla)[3] na tudomain; (3) ni teind fearbolc bil belltaine; (4) ni teit as o fiach seice it*er* geart ⁊ as ⁊ induth.

§ **5.** (1) Linatside a t*r*i sine mesair da ordlach dec, a ceathramad dia læg; (2) cach æ l*eth* leithe thige; (3) asi bo fiach feine aire .xx.s hi sin; (4) bo ceathramad loig air .xxx.s.; (5) bo coic*id* læg aire .xl.s.; (6) [bó ṡessid] lo[í]g aire .l. (aire) .s.; (7) bo (aire) .uii.mad loig aire sescat .s.

[1] *Read* forlann.

[2] *There is a gap for about three letters after the first* na tothla.

[3] *Omit as dittography.*

§ 6. (1) Teora bai fil- la *Féniu*: (2) bo fiach feini, bo eirce do thir insin; (3) bo dire; (4) bo feig flatha; (5) ni cach bo fiach feine bo ineirce do thir; (6) ni cach bo (bo)[4] eirce do thir bo in*n*isi.

§ 7. (1) Aic m*ac*cu a cairaich slan noicc nin*n*raic sochraid solomrad alahuan no treasuan, (2) cærad find no dub no l*acht*na.

§ 8. (1) Nip b*r*isc a c*r*oicean*n*, (2) nip f*or*find no f*or*ofin*n*,[5] ma dub no lachtna; (3) nip toich, (4) nip gungabl*ac*,[6] (5) nip congalfinnach, (6) nip daintach, (7) nip anc*r*ad, (8) nip ladrach, (9) nip l*eth*eirlach, (10) nip romeidleach; (11) ropai uaitsiu a cairaich indraicso æs ⁊ coland ⁊ oland.

Translation

§ 1. (1) If you buy a cow, [let her be] a docile cow, a perfect cow without cow-defects. (2) These are the defects of a cow so that she is to be returned: (3) i.e. if her teats do not produce a full milk-yield, (4) if they are short, (5) if she is three-teated, (6) if she is a gorer, (7) if she is a kicker, (8) if she does not love her calf, (9) if she is shaggy(?), (10) if she is hostile to the bull, (11) if she is fierce, (12) if she refuses [to suckle her calf], (13) if she is half-shanked, (14) if she is lame, (15) if she is blind in one eye, (16) if she is a breaker into grasslands(?), (17) if she does not feed her calf well and her milker, (18) if she is not in-calf(?), (19) if she is not bulled: that is a cow which is too strong [for the bull].

§ 2. (1) If she is udder-damaged(?), (2) if she does not have her two ears(?), (3) her calf should not run onto her(?), unless there may be death from it(?), (4) she is not a bellower, (5) she is not a bawler, (6) she is not fierce, (7) she is not a kicker(?), (8) she is not —(?).

§ 3. (1) A young beautiful cow, which is upstanding and faultless, (2) which is proper [and] in control of a second calf, (3) neither land-holding nor land nor milker nor judge nor witness(?) nor litigant(?) corrupts her.

§ 4. (1) A young beautiful cow with the calf of her body at her side, (2) it is not a weaned calf or a calf which has been put onto a cow, or a —(?), (3) the calf is not castrated(?) before Mayday; (4) it does not go out as payment separately from her, along with dung and milk and increase.

§ 5. (1) Her three teats fill a measure of twelve inches, her fourth teat is for her calf, (2) each of them —(?); (3) she is a cow for debts of the Féni: 20 scruples for her in that case; (4) a cow of a fourth calf: 30 scruples for her; (5) a cow of a fifth calf: 40 scruples for her; (6) a cow of a sixth calf: 50 scruples for her; (7) a cow of a seventh calf: 60 scruples for her.

§ 6. (1) There are three cows in Irish law: (2) a cow for debts of the Féni, that is a cow of payment for land; (3) a cow for honour-price; (4) a

[4] *Omit as dittography.*
[5] *Read* fofind; *see discussion to § 8 (2) on p. 518 below.*
[6] *The* b *is subscript.*

cow for payment to a lord; (5) not every cow for debts of the Féni is a cow of payment for land; (6) not every cow of payment for land is a cow of the milking-place [i.e. a milch cow(?)].

§ 7. (1) Invoke sureties for a healthy young proper well-shaped easily sheared ewe with her second or third lamb, (2) a white or black or dun-coloured sheep.

§ 8. (1) May her skin not be brittle, (2) may she not be white-backed or white-bellied, if she is black or dun-coloured; (3) may she not be —(?), (4) may she not have a misshapen posterior, (5) may she not be cluster-haired, (6) may she not be full-mouthed, (7) may she not be ill-shapen, (8) may she not have defective hoofs, (9) may she not have uneven tufts, (10) may she not be greatly bleating, (11) may it (the payment) be given by you for this sheep, proper with regard to age and body and wool.

Discussion

§ 1. (1) dian cri boin. I read *día crie,* i.e. *día* 'if' + 2 sing. pres. subj. of *crenaid* 'buys'. Rolf Baumgarten suggests that one might keep *dian* of the MS, with the *-n* indicating the nasalisation of the initial *c-* /g/; cf. *Thes.* i 105.26 *ancondammucbaitis* 'when they used to exalt me' for regular *a condammucbaitis.*

bo boid. I read *bó boíd* 'a docile cow'. One could also take *bo* here and in the next phrase to be for *ba,* 3 sing. pres. subj. of the copula, i.e. 'let her be docile, let her be perfect'.

bo ineach. The adjective *inich* (also *inech, iních, infich*) 'perfect, complete, flawless' is from *in-* + *fíach,* i.e. 'fit for payment' (*DIL* I 267.61–268.4, 268.18–26); cf. *inraicc* 'proper, of standard quality' from *in-* + *reicc,* i.e. 'fit for sale'. There are no other attestations of *inich* in the Old Irish law-texts, but it is common in legal commentary, usually in connection with domestic animals. For example, legal commentary at *CIH* v 1627.33 = *AL* iii 376.9–10 states of a sheep with a defective udder, 'it is to be returned [to the former owner] and a perfect sheep of the same nature given in its place' (*is a hatchur ar cūlu 7 cura iních aicenta dara hēisi*). The spelling *inech* is occasional, e.g. *CIH* vi 1980.14 *aithgin inech* 'complete restitution'. In many of its attestations, *inich* qualifies the noun *bó,* e.g. *CIH* iii 839.18 *bō infich*; 977.1 *bō inich*; 999.3 *bō inich*; 999.17 *bō iních*; v 1773.2–3 *lōgh tēora mbō ninich,* etc.

Another possibility would be to take *boinech* to be an adjective formed from *bó* 'cow', meaning 'cow-like, with the attributes of a cow'. The *-n-* could be explained as coming in by analogy from the acc./dat. sing. *boin,* cf. *boinín* 'calf'. However, there are no other attestations of this form.

(2) it é ainmead bo. Redundant final *-d* is common in this MS, e.g. § 7 (2) *cærad.* I therefore read *ainmea,* which I take to be nom. plur. of *ainim* 'defect'. I likewise read acc. plur. *cen bó-ainmea* 'without cow-defects' in (1) above. The noun *ainim* does not occur in the Old Irish Glosses. *DIL* A 150.70 suggests that it was originally a fem. *ā*-stem – presumably *ainem* – and these examples of nom./acc. plur. *ainmea* support this view.

aran athchuir. One possibility would be to read *ara n-athchur* 'for their return', i.e. if the cows suffer from any of the defects described below, the buyer can insist on returning them to the seller, and getting suitable replacements. But the rest of the paragraph refers to a single cow, so I follow Rolf Baumgarten's suggestion of taking *aran* to be for *ara* 'so that' with 3 sing. pres. indic. of the copula (Thurneysen, *Grammar*, 486 § 798 *arndid, arnid, arin*), and *athchuir* to be predicative genitive of *athchur*, vb.n. of *ad-cuirethar* 'returns'; cf. *CIH* vi 2230.34 *conid nathcuir* (read *conid athchuir*).

Another solution would be to emend to *aran athchuirthi* 'so that she is to be returned', with verbal of necessity of the same compound; cf. *CIH* iv 1235.27 = Text 7 § 2 (16) *as athchuirthe* (= *-thi*) '[a defective horse] is to be returned'.

(3) maine be comblicht a sini. Lit. 'if there is not a full milking of her teats' (reading gen. plur. *sine*).

(4) mat gairde. Good sense is obtained by reading *mat gairti* 'if they (the teats) are short'. Shortness of teats may indicate a poor milk-yield.

(5) ma trifine. Binchy reads *ina tri fine* here (*CIH* ii 674.38), but the *in* of the MS could equally be read as *m*. I would emend to *mad trifne* 'if she is three-teated'. The original may have had *matrifne*, with assimilation of *-d* and *t-*.

Legal sources contain many references to the defect of 'three-teatedness', i.e. with one teat absent or non-functional (see p. 202). A cow with this defect is described as *triphne* (*trefne*) from *trí* + *sine* (Thurneysen, *Grammar*, 139 § 226(b)), e.g. *TBC* Rec. I 62.2041 *oc blegon bó triphne*; *TBC* St 69.2142 *ic bleogan bo trefne*; *TBC* LL 57.2104 *ic blegu[n] bó trí sine.* The abstract in *-echt* is attested at *CIH* iii 839.35 *trif.s.*; 977.5 *treifnecht*; 999.10 *trefneacht.* The reformation *tresinechas* (= *treṡinechas*) occurs at *CIH* iii 839.12.

(6) mat goinig. Read *mad guinid* 'if she is gorer'. For a discussion of goring by cattle, see p. 179.

(7) mad luach. *Luach* is from *lua* (earlier *lue*) 'heel, hoof'. A law-text on horse-defects likewise states that if a buyer finds that a horse is a kicker (*luach*) he can return it to the seller or obtain compensation (*CIH* iv 1235.25 = Text 7 § 2 (12)).

(9) mat finchad. The word *finchad* is otherwise unattested. One possibility would be to emend to *finnach* (*findach*) 'hairy, shaggy'. The idea may be that a cow with a long shaggy coat is to be avoided (see p. 35). Another possibility would be to read *fincha(i)de*, from *finnach* + the adjectival suffix *-de* (Thurneysen, *Grammar*, 220–2 § 347–8). This form is attested only in the Middle Irish History of Philip and Alexander of Macedonia (Book of Ballymote version, p. 496a29): *Adconncadar . . . mnā ⁊ firu finchaide amal cethri* 'They saw . . . men and women, hairy like animals'.[7] Liam Breatnach suggests, on the other hand, that *finchad* may be a compound of *fine* 'kin' + *cath* 'fight, fighter', cf. (10) *tarbchad.* She would be a cow which was by temperament hostile to other cattle (her 'kin').

(10) ma tarbchad. The *u*-stem *cath* 'fight, battle; battalion' is common in compounds, particularly as the first or second element of personal names, e.g. *Cathgal* 'battle-valour', *Fáelchad* 'wolf-fighter' (M. A. O'Brien (ed. R. Baumgarten), 'Old Irish personal names', 224–5; *LEIA* C-47 s.v. 1 *cath*). *Tarbchad* here means 'hostile to the bull', obviously a serious defect in a cow. This is presumably the same condition as *ingenas búaile* 'virginity of the cattle-enclosure', included among the defects of cattle in legal commentary: see p. 201.

As in the case of (5) *ma trifine*, the final -*d* of *mad* is assimilated with the following *t*-.

(12) mat sen*ach*. Aspects of a cow's behaviour towards her calf are treated in (8) *mana cara a lœg* 'if she does not love her calf', and in (17) *mana biatha a lœg co maith* 'if she does not feed her calf well'. Here, the author deals with the cow which is not merely indifferent to her calf, but actively hostile.

The verb *sénaid* 'denies, rejects' is well attested of a cow refusing to suckle her calf. For example, legal commentary at *CIH* v 1655.10 = *AL* i 24.20–1 has: *madh a n-aimsir lachta no tsēndais na ba a laoghu* 'if in the time of milk, the cows used to refuse their calves'. The MS has *sen*-, which could be expanded as *sénach* 'rejecting, refusing', though this adjective is otherwise unknown. Another possibility would be to read *séntach*, attested in later Irish in the meaning 'negative'.

(13) mat lethcolpach. The word *colpthae* (later *colpa*) means 'shank, lower leg', so *lethcholpthach* 'half-shanked' must refer to a cow with a leg injury or deformity which causes her to limp. The ability to travel long distances to summer pasture would have been necessary on many farms, so a cow with any walking difficulty would have been avoided. Legal commentary at *CIH* vi 2072.10 also refers to a cow which is *lethcolp*- (= *lethcholpthach*). In two other instances, a scribe has omitted the final stroke indicating -(*th*)*ach*: *CIH* ii 715.39 *lethcolb* and *CIH* vi 2163.31 *leccolb.* At *CIH* vi 2163.29 *bacach* 'lame' has been substituted for *lethcholpthach.*

(14) mat atlosc. The dilemma is whether to read *mad athlosc* 'if she is very (or chronically) lame', or to follow Binchy's suggestion (*CIH* ii 674[t]) that the second -*at* is an erroneous repetition by a scribe, and simply

[7] Meyer, 'Die Geschichte von Philipp und Alexander von Macedonien', 101.16-18. The -*de* of *finchaide* is subscript, but seems to be in the hand of the main scribe.

read *mad losc* 'if she is lame'. The latter option seems more likely, as any degree of lameness would be unacceptable in a cow (cf. (13) above). Comparison can be made with a law-text on the proper characteristics of a horse, which states that if it turns out to be lame (*losc*) it can be returned to the seller (*CIH* iv 1235.25 = Text 7 § 2 (11)).

(15) mat cæch. Blindness in one eye could easily be overlooked by a buyer. Legal commentary contains a number of references to the cow or other animal which suffers from the defect of being blind in one eye (*cáech*), e.g. *CIH* ii 715.38; vi 2072.10; 2163.29.

(16) mat comgith a féra. Possibly for *mad comgith i féra* 'if she is a breaker into grasslands', i.e. a cow which is especially liable to break through fences. A bovine with this temperament is described in *Bretha Comaithchesa* as a *foirrgid* 'reacher, stretcher' (*CIH* i 73.1; 195.10 = *AL* iv 108.13). According to accompanying commentary, a *foirrgid* is defined as one which goes from its own good grass into the good grass of another, or from its own bad grass into the good grass of another across a fence (*CIH* i 195.14–18 = *AL* iv 110.29–34).

I take *comgith* to be from the compound *con-boing* 'breaks, smashes' with the agent suffix *-ith* (*-id*). This compound is used in the context of fence-breaking in a quotation in *O'Davoren's Glossary* from an unidentified source: *conbongat etarba* 'they destroy a fence' (*CIH* iv 1476.40–1 = O'Dav. 241 § 281). *Fér* 'grass' is occasionally used in the plural to mean 'pastures, grasslands' (*DIL* F 84.12–14).

(17) mana biatha a læg co maith 7 a mbligre. According to § 5 (1), the milk from a cow's three teats is for human consumption, while that of the fourth teat is reserved for the calf.

Mbligre is for Old Irish *mligre* 'milker' from the root *mlig-* (later *blig-*) 'to milk'. Initial *mbl-* is a common intermediate spelling between OIr. *ml-* and later *bl-*. This example and § 3 (3) *bligri* are the only occurrences of the word in an Old Irish law-text, but it is well attested in commentary, e.g. *CIH* i 10.5 = *AL* v 152.19 *bligre*; *CIH* vi 2168.25 *bligri*. An acc. plur. *bligre* occurs in *Ríagail na Céle nDé* (E. J. Gwynn, *The rule of Tallaght*, 66 § 7). Gwynn suggests emending to *bligreóiri*, acc. plur. of *bligreóir* 'milker', but this is unnecessary.

The second element seems to be the agent suffix -(*a*)*ire*, but I know of no other case where it is syncopated. The regular form *bligire* (*bliugere* MS) is attested at Ó Riain, *Corpus genealogiarum sanctorum Hiberniae*, 144.300 = *LL* vi 1663.50985.

(18) ma raedra. It is unclear what defect is meant by *ma raedra*, which seems to be a verbal form. It is tempting to link it with the condition of *radrachus* (see p. 201), which is defined in legal commentary at *CIH* iii 998.37 as *dairthir í 7 ni anand innti* 'she is bulled and it does not stay in her', i.e. she fails to conceive or aborts. But I have no suggestions as to the relationship between *raedra* and *radrachus*. No etymology has been proposed for *radrachus*.

(19) mana dart*har*. I read *mani darthar*, 3 sing. pres. subj. passive of *dairid* 'bulls'. If the cow has not been bulled (and the seller claims that she has been), the buyer obviously has a claim against him. Legal commentary at *CIH* iii 1001.36–1002.5 (= McLeod, *Early Irish contract law*, 301 [r-t]) deals with various aspects of such claims.

as e fe*r*lond bo an*n* sin. It seems impossible to make sense of *ferlond* of the MS. The word could also be read as *frelond*, but this form is equally incomprehensible to me. Following Binchy's suggestion at *CIH* ii 674^{u}, Liam Breatnach would read *is é forlann bó in sin* 'that is a cow of superior strength'. Such a cow is not bulled because she is powerful enough to drive off the bull, cf. (10) *ma tarbchad*. *Forlann* is well attested in the meaning 'superiority in strength or numbers': see *DIL* F 345.36–58.

§ 2. (1) ma údgilti. One could take *úd* to be for *úth* 'udder', and *-gilti* (= *-gilte*) to be a past participle of *geilid* 'grazes'. The expression 'udder-grazed' may refer to a cow with some defect or disease of the udder, such as mastitis spreading from a teat (see p. 201).

(2) mana beth a da noga fuirr-i. One possibility would be to take *oga* to be a scribal error for *ó* (OIr. *áu*) 'ear'; for non-historical *-g(h)-* cf. *dogogha* for *do-goa* (*CIH* iv 1289.11), and perhaps *lugach* for *luach* (*CIH* ii 675.2 = § 2 (7) below). There are some attestations of the dual of *ó*, e.g. *mo dā n-ō* 'my two ears' (Meyer, *Sanas Cormaic*, 93.19 § 1059); *dā nóó eich* 'the two ears of a horse' (Meyer, 'The oldest version of *Tochmarc Emire*', 436); *dā auo fuiri* 'two ears on it (a vessel)' (Meyer, *Sanas Cormaic*, 34 § 414). One could read *mani bet a dá n-áu fuiri* lit. 'if there are not her two ears on her'. Cattle were sometimes marked on their ears (*ina n-oaib*) to denote ownership (*CIH* iv 1253.2).

A difficulty about this interpretation is that one would expect any serious ear-damage to be immediately obvious to the buyer, who would then avoid this particular cow. Other possibilities are that *oga* is from *og* 'point, horn(?)', *og* 'egg', or *óg(ae)* 'wholeness'; but I cannot make sense of the passage with any of these words.

The MS has *fuirri* with a stroke over the second *r*. This seems more likely to be a mistake for *fuirri* 'on her' (OIr. *fuiri*) than the abbreviation of some other word.

(3) nis fuirr a læg. I take *nis* to contain the negative particle *ní* + 3 sing. fem. infixed pronoun Class A, referring back to *bó* 'cow'. The form *-fuirr* looks like a 3 sing. pres. subj., possibly from *for-reith* 'runs onto'. This compound is difficult to distinguish from *fo-reith* 'helps'; however, there are some clear attestations, e.g. *CIH* iii 1128.22, 25 = E. J. Gwynn, 'Privileges', 38.10–11, 14 *forreith forreither*. In the present context, it seems to refer to some aspect of a calf's behaviour towards its mother, perhaps an overaggressive sucking at her teats.

mana epell- as. One could read *-epell*tar, 3 sing. pres. subj. passive of *at-bail* 'dies', but the context does not seem to suit. The future stem *ebl-* is attested for both *aigid* 'drives' and *ailid* 'rears' (Thurneysen, *Grammar*, 403–4 § 649), but a future is impossible after *mani.*

(4) nibi buirech, nadbid buirech. I have no explanation for the repetition of *búirech* here, and can only put it down to scribal dittography. Yet the use here of the formula *nībi . . . nādbi* is difficult to explain away. *Níbi* is neg. 3 sing. consuetudinal present of the copula, and *nádbi* is the corresponding relative form. One can compare the *Uraicecht Becc* phrase *nībi briugu nādbi cēdach* 'he is not a hospitaller who is not of hundredfold wealth' (*CIH* v 1608.14 = *AL* v 76.10). We find a reversal of this formula in the same text in the phrase *nādbi cainfoltach nībi cainfuillmech* 'whoever does not fulfill his contracts well, he does not make good profits' (*CIH* v 1618.9 = *AL* v 112.7); cf. *beas ændānach bid ændīreach* 'he who has one art has one honour-price' (*CIH* v 1617.5 = *AL* v 108.11). Because of these examples, it is tempting to impose a *níbi . . . nádbi* pattern on § 2 (4)-(8). However, although there is variation between *nībi* and *nādbi* in this passage, it seems impossible to establish a series of linked pairs. This leaves us with the very unsatisfactory option of omitting *nadbid buirech* as a dittography, and emending *nabi(d)* of (5)-(7) to *níbi.*

There is no problem with the meaning of *nībi búirech* 'she is not in the habit of bellowing', i.e. she is not a bellower. If a cow is particularly prone to bellowing, it is probable that she is ill, and that further symptoms will appear. One can compare § 8 (10), which advises against the purchase of a sheep which is *romeidleach* 'greatly bleating'.

(5) nabid becech. The verb *béiccid* 'bawls, shrieks, howls' is used of noises made by various animals, including goats, sheep, and cattle. In general, it refers to a higher-pitched cry than *búirid* 'bellows'. An additional item in the Yellow Book of Lecan version of *Cormac's Glossary* (Meyer, *Sanas Cormaic*, 98 § 1104) has *raibceth cethra .i. ro-bēiced* 'bellowing of cattle i.e. great bellowing'. See *DIL* R 3.49–53.

(6) nabi lon*n*. This is a repetition of § 1 (11) *mat lond* 'if she is fierce'.

(7) nabid lugach. *DIL* L 235.78–9 suggests taking *lugach* here to be for *luach* 'a kicker', as in § 1 (7). There is a word *lugach* which is generally taken to be a derivative of *lugae* 'oath', and to mean 'in the habit of swearing, contentious', e.g. Meyer, *Tecosca Cormaic*, 44 § 27.16 . . . *osé gáeth gúach líath lond lugach labar* '. . . and he clever, lying, grey-haired, violent, swearing, garrulous'. Obviously, this use of *lugach* could not be applied to a cow, so the *DIL* explanation is the more plausible. For scribal insertion of *-g-* see discussion on § 2 (2) *noga* above.

(8) nibi modísach. I have no suggestions for *modísach.*

§ **3. (1) bo og allainn.** I take *og* to be for OIr. *oac* (*óc*) 'young' rather than *og* 'whole, perfect'; cf. § 4 (1) *bo occ al*uinn.

asi diriud diainim. Read *osí díriug diainim* 'which is upstanding and faultless'. *DIL* D ii, 138.17–18 takes *diriud* to mean 'unblemished', but the scribal *-g/-d* variation is so common that we are justified in taking it to be for *díriug* (*díriuch*) 'straight, upstanding'. Cf. Meid, *Táin Bó Fraích*, 8.201–2 *is é díriuch dianim* (of the warrior Fráech).

(2) cuin*n*riug alaloig. I read *cuindriug ala-loíg* 'in control of a second calf', taking *cuindriug* to be dat. sing. of *cuindrech* 'correction, control' (*VGK* ii 596 § 795; *LEIA* C-273). A cow which is successfully rearing a second calf has obviously proved her fertility and her competence as a mother, and is therefore worth buying. Compare § 7 (1) which recommends a ewe on her second or third lamb.

(3) nis tercorpai. *Nis tercorpai* is obviously a verbal form preceded by the negative particle *ní* with the 3 sing. fem. infixed pronoun Class A referring back to *bó* 'cow', as in the case of § 2 (3) *nis fuirr a lœg*. I take it to be from **do-arcorpai*, an otherwise unattested compound of *corpaid* 'corrupts, defiles, debases', with the preverbs *to-* and *ar-* (giving *ter-* in prototonic position). The compound *ar-corpai* is occasionally attested, e.g. *arcorbi talmain* 'it (the love of God) debases the world' (Hull, '*Apgitir Chrábaid*', 60.24 § 6).

The meaning of *-tercorpai* in this context is uncertain. The simplex *corpaid* (*corbbaid*) is often used of corruption in the physical sense, by poison, dirt, excrement, or illicit sexual intercourse. In a Middle Irish tale it is applied to laceration by thorns: *droigni is drisi romchoirb* 'thorns and brambles have defiled me' (O'Keeffe, *Buile Ṡuibne*, 58.1633). The metaphorical use of *corpaid* (and *ar-corpai*) is also well attested, e.g. *mani choirbet anfolta* 'unless unfulfilled duties disqualify' (*CIH* vi 1929.8); *fo bīthin arrocorpait géssi Con*aire 'because the taboos of Conaire have been broken' (*LU* 207.6750 = Stokes, 'The destruction of Dá Derga's hostel', 33 § 26).

The present example is particularly difficult because *-tercorpai* has a list of six nouns as its subject. The first two – *selb* 'holding' and *mbrug* 'land' – seem to refer to the land on which the cow has grazed. Possibly the land has caused some defect or injury to the cow in question. The other four nouns are various categories of person: *bligri* 'milker', *medam* 'judge', *fiada* 'witness' (or 'lord, owner'), and *féichem*(?) 'litigant'. The implication seems to be that an act by one of these people could render the cow unsuitable for sale. For example, the judge might have previously awarded the cow to a third party, thereby making the sale invalid.

sealb. *Selb* 'ownership, property' is used in a variety of meanings in legal material. I tentatively take it to mean 'land-holding' in this instance.

mbrug. OIr. *mruig*, later *brug* 'land'. As in § 1 (17) *mbligre*, the MS has the intermediate spelling with *mbr-*.

bligri. OIr. *mligre* 'milker', cf. § 1 (17) *mbligre.* It is unclear to me why the milker should be mentioned at this point. Possibly, it means that a rough or inept milker could frighten or physically damage a cow, thereby making it difficult for a subsequent owner to milk her.
medam. *Medam* is occasionally attested in the sense of 'judge, arbiter'. It is regularly glossed by the more common word *brithem* 'judge, jurist', e.g. O'Donoghue, 'Advice to a prince', 49 § 32; *CIH* iv 1268.20 (see footnote and gloss) = Watkins, 'Indo-European metrics and archaic Irish verse', 230 (91).
fiada. As *fíada* is listed here between *medam* 'judge' and *féichem* 'litigant', it is possible that it means 'witness', and refers to some law-suit involving the cow: a witness could have given evidence which disqualifies her from being sold. Another possibility is that *fíada* here means 'lord', i.e. the cow is due to be given as rent to a lord, and therefore cannot be sold. One can compare Text 2 § 1 (3) where it is stated that a cow due for rent (*bó chís*) cannot be distrained. Finally, *fiada* could simply mean 'owner': this usage is common in legal commentary, e.g. *CIH* i 289.33 = *AL* iii 292 (footnote) *ō fiadad na ndam* 'from the owner of the oxen'; *CIH* iii 937.31 = *AL* iii 192.11 *a dā fiadad* 'of their two owners (of dogs)'.
feichme. The MS could be read as *feichme* or *feichine.* The original probably had *féichem* 'litigant, creditor, debtor, contractor'. The cow may be due to be paid to a creditor of the person who is trying to sell her: the sale cannot therefore proceed.

§ 4. **(1) co læog a cuirp folei.** Liam Breatnach suggests reading *co lóeg a cuirp fó leith* 'with the calf of her body at her side', taking *fó* to be for *fo* + *a.*

(2) nibi læg fothla na tothla. In the MS there is a gap for about three letters after *na tothla,* and then *na tothla* is repeated. I omit the second *na tothla* as dittography.

Both *fothla(e)* and *tothla(e)* are verbal nouns of compounds of *tlenaid* 'takes away, steals'. Old Irish commentary on *Bretha Comaithchesa* refers to *smachta fothla ⁊ tothla* 'fines for *fothla* and *tothla*' (*CIH* i 194.7 = *AL* iv 106.20), which are explained as two forms of animal-trespass: see discussion on p. 139 above. One might argue, therefore, that the author is advising against the purchase of a cow whose calf is liable to break into other farmers' land. However, a more likely explanation is provided by the use of the phrase *lóeg fothlai* in Hull, *Longes mac nUislenn*, 45.90 § 7. Here Hull suggests the translation 'weaned calf', which suits our context well: the calf should be at its mother's side, i.e. still dependent on her for milk.

I have found no other occurrences of the phrase *lóeg tothla,* but one might guess that it refers to a calf which is not the cow's own, but has been put on her for rearing.

na tudomain. I have no expanation for *tudomain.*

(3) **ni teind fearbolc bil belltaine.** Following a suggestion by Thomas Charles-Edwards, I translate: 'the edge of May does not cut [its] scrotum', i.e. the calf is not castrated before Mayday. I take *teind* to be from the verb *teinnid* 'cuts, breaks, etc.', and *bil* to be the noun meaning 'rim, edge'. *Ferbolc* is well attested in the meaning 'bag, satchel, sheath', though not in the meaning 'scrotum'. However, as we have seen above (p. 478), *tíag* 'bag' may be used in this sense.

(4) **ni teit as o fiach seice.** I take this to mean that the calf should not be separated from its mother, and given as payment (*o fiach*) for some fine or debt. *Seice* is the 3 sing. fem. conjugated preposition from *sech* 'past, in addition to, without'.

it*er* geart 7 as 7 induth. Included in the value of a cow are her products, particularly dung (*gert*), milk (*as*) and calves (*induth*). *Gert* may be applied to both milk and dung (see p. 230), but in this context seems to refer specifically to dung.

§ 5. (1) **Linatside.** The anaphoric pronoun -*side* seems unnecessary here. Liam Breatnach suggests that one might read *sidi,* nom. plur. of *side* 'blast (of wind)', with the extended meaning 'gush, flow'; cf. the diminutive *sidén* 'blast, gush' (*DIL* S 221.77–9).

mesair da ordlach dec. *Mesar* 'measure, vessel used for measuring' is a borrowing from Latin *mensura* 'measure' (*LEIA* M-41). A twelve-inch measure seems to have been standard for liquids: see p. 571.

(2) **cach æ l*eth* leithe thige.** It is difficult to know how to interpret this phrase. *Cach æ* 'each of them' seems to refer back to the teats; but how then are we to take *leth leithe thige*? One could take *leithe* and *tige* to be abstract nouns from *lethan* 'broad' and *tiug* 'wide', perhaps providing further information on the dimensions of the milk-vessel mentioned in (1). But *leithe* and *tige* are late forms: in an OIr. text one would expect *leithet* and *tiget.* Furthermore, the distinction between breadth and width could only apply to a rectangular vessel, whereas the archaeological evidence is of rounded vessels made of wooden staves.

(3) **asi bo fiach feine aire .xx.s. hi sin.** The syntax is a problem here. I would place a colon after *feine* and translate 'she is a cow for debts of the Féni: 20 scruples for her in that case'. The word *aire* is written very clearly in the MS here (and in (5)–(7)), so there is no justification for reading *arre* 'payment'. The *bó ḟiach Féine* is also mentioned in § 6 (2), and the sense seems to be that such a cow is of high enough quality to be given for a debt (*fíach*) among the *Féni* 'freemen' (for a discussion of the meanings of the term *Féni*, see *BB* 133–4, note to § 33).

It is implicit that this *bó ḟiach Féine* has produced three calves, as the rest of the paragraph assigns regularly increasing values to cows which have produced four, five, six, and seven calves. The author's view that a cow increases its value with each calf – up to seven calves – is difficult to reconcile with our other sources. In the Old Irish law-texts the ordinary value of a fully grown milch cow is one ounce of silver, i.e. 24 scruples (see p. 58). In Middle Irish legal commentary (*CIH* iii 845.33–6 = Text 3 §§ 5–6), she is said to be worth 20 scruples when she bears her first calf at the age of four, 22 scruples when she bears her second calf at the age of five, and 24 scruples when she bears her third calf at the age of six. There is no mention of any further increase in value; on the contrary, other legal commentary takes a cow's potential (*saíltinche*) into account when assessing her value (*CIH* i 306.26), and this will naturally decrease as the cow ages. The explanation may be that the author of our text is thinking of the cumulative value of **all** the calves of a cow.

I can see no alternative but to take *.xx.s.* to be an abbreviation for *fiche screpul* 'twenty scruples'. The abbreviation *.s.* regularly signifies *sét*, a unit of value equal to half a milch cow (see p. 589), but this is obviously inapplicable in our text. *Screpul* is generally abbreviated *sc-p* in the MSS, but there are instances of *.s.* at *CIH* v 1636.11, 12, 13, 15, etc.

(4) bo ceathramad loig air .xxx.s. As ordinals are usually inflected in Old Irish, it is likely that the original text read *bó chethramaid loíg airi trícha screpul* 'a cow of a fourth calf: 30 scruples for her'.

(6). The MS has *log aire .l. aire .s.* which makes no sense. Following Binchy's suggestion at *CIH* ii 675.9, I omit the second *aire* as dittography. On the basis of the two preceding sentences, I suggest that the original had *bó ṡessid loíg airi coíca screpul* 'a cow of a sixth calf: 50 scruples for her'.

§ 6. (2) bo fiach feini, bo eirce do thir insin. The first of the three categories of cow distinguished in § 6 is the *bó ḟiach Féine* 'a cow for debts of the Féni', already mentioned in § 5. It is defined as 'a cow of payment for land'. This presumably means that such a cow is of high enough quality to be given in a payment relating to land, e.g. rent for land, a fine for a breach of land-law, etc. *Éirce* is gen. sing. of *éraic*, vb.n. of *as-ren* 'pays'. In § 6 (5) the same phrase is repeated with the substitution of *in(d)éirce*, gen. sing. of *in(d)éraic*, vb.n. of *in-éren* of the same meaning.

(3) bo dire. Such a cow must be of higher quality than the *bó ḟiach Féine*, as she is paid for serious offences involving the honour-price (*díre*) of the injured party.

(4) bo feig flatha. I read *bó ḟéich flatha* 'a cow for payment to a lord'. It is likely that the best cow on a client's farm would be reserved for his lord's rent. One can compare Heptad 41, which includes *bo ciss flatha no eclasa* 'a cow intended as rent for lord or church' among the seven categories of cow which cannot be distrained: see discussion below to Text 2 § 1 (3) *bo chis*.

(5) ni cach bo fiach feine bo ineirce do thir. The rest of this paragraph reads like an exercise in logic, but the point of it is lost on me. Having stated in (2) that a *bó fíach Féine* is a *bó éirce do thír*, he qualifies this here by stating that not every *bó fíach Féine* is a *bo inéirce do thír* (= *bó éirce do thír*). He seems to be saying that a cow might in some circumstances be eligible to be given in a payment or fine, but not in a case involving land.

(6) ni cach bo (bo) eirce do thir bo in*n*isi. I omit the second *bo* as a dittography. Here we are told that not every *bó éirce do thír* is a *bó innisi* 'a cow of the milking place'. This may mean that a cow need not necessarily be in milk when given in payment in a case involving land. *Innisi* (= *indise*) is gen. sing. of *indes* 'milking place, milking enclosure'. One can compare *CIH* iii 897.1–3 which refers to three *nemidh indisi* 'immune ones of the milking place', i.e. cows which cannot be distrained. Cf. 897.9–10 *at e neimidh innisi*(*n*) *insin* 'they are immune ones of the milking place'.

Liam Breatnach suggests that one could also – assuming a greater degree of dittography – take (5) and (6) together and read *ní cach bó fíach Féine, ní cach bó éirce do thír bó indise* 'not every cow for debts of the Féni, not every cow of payment for land is a cow of the milking place [i.e. a milch cow]'. This would make excellent sense.

§ 7. (1) aic m*accu*. *Aic* is 2 sing. imperative of *ad-guid* 'invokes': see Thurneysen, *Grammar*, 375 § 588. The phrase *ad-guid maccu i* 'invokes sureties for' is well attested in the law-texts in relation to contracts (*DIL* A 59.20–36). In the present example, the buyer is urged to invoke sureties (i.e. call on the seller to appoint sureties) to guarantee the quality of the sheep.

a cairaich. Read *i caírich*, acc. sing. of *cáera* 'sheep'; cf. § 8 (11).

alahuan no treasuan. I take *alahuan* and *tresuan* to be independent datives meaning 'with her second lamb' and 'with her third lamb' respectively. For the use of *ala-* and *tres-* in compounds, see Thurneysen, *Grammar*, 248 §§ 394–5.

(2) cærad find no dub no l*achtn*a. For the redundant *-d* of *cærad*, cf. § 1 (2) *ainmead.* For a discussion on sheep-colours, see p. 70.

§ 8. In § 2 the author uses the negative 3 sing. consuetudinal pres. of the copula in his list of undesirable characteristics of cows. In this paragraph the subjunctive form *nip* is used. There seems to be no distinction in meaning.

(1) nip b*r*isc a c*r*oiceann*n*. The adj. *brisc* means 'brittle, easily broken'. In the present context it probably refers to an unhealthy tightness in the sheep's skin. An animal in good health has loose skin.

(2) nip f*or*find no f*oro*finn*n*, ma dub no lachtna. I suggest omitting the stroke over the first *f* of *f*or*ofinn*, as I can make no sense of this form. One would then read *fofind* 'with white below' in contrast to *forfind* 'with white above'. Compare the juxtaposition of the adjectives *focháel* 'slender below' and *forlethan* 'broad above' in literary descriptions of a beautiful woman, e.g. *TBC* Rec. I 2.32 *agad fochóel forlethan* 'a face which is broad above and slender below'.

I believe that the author is advising the buyer to avoid sheep which have the mixture of colours characteristic of the mouflon, the ancestor of the domestic sheep (see p. 70).

(3) nip toich. *Toich* is a common adjective meaning 'proper, right', but the context demands some undesirable characteristic of a sheep.

(4) nip gungabl*ac*. In the MS, the *g* is written below the first *a*. The identity of this adjective is uncertain.

In later Irish, the noun *gúnga* (also *gúng*) 'posterior, loins' is well attested. A derivative adj. *gúngach* is commonly used of a misshapen posterior, either too large (*DIL* G 176.8–14), too small (Ó Cuív, *Cnósach*, 132) or in a crouching or contorted position (Dinneen, *Foclóir* s.v.). In the present context, *gungablach* must refer to some physical defect in the sheep's hindquarters. Liam Breatnach suggests that it might be a compound of *gúng* + *gablach* 'splay-forked', i.e. with the hind legs awkwardly spaced.

(5) nip congalfinnach. *DIL* C 440.73–441.7 distinguishes two words *congal*: (1) a fem. *ā*-stem 'conflict, fight', and (2) a masc. *o*-stem 'group, cluster'. Both *DIL* and *LEIA* (C-193) suggest that the second meaning of *congal* may be an extension of the first, with a semantic development from 'fight' to 'hostile group' to 'cluster'.

The compound *congalfinnach* 'cluster-haired' seems here to refer to a sheep whose fleece is short and curly, and therefore of little use for spinning. The adj. *finnach* is from *finna* 'hair, fur, fleece'. Such sheep may also have been regarded as less hardy: Aristotle (*Historia animalium* (ed. Thompson), 8.10) states that sheep with curly fleeces survive the winter very poorly.

(6) nip daintach. When used in connection with a horse, the adj. *daintech*, lit. 'toothed', has the meaning 'liable to bite, vicious' (see p. 153). However, a sheep poses no threat as a biter, so some other explanation is needed here. One possibility is that *daintech* refers to a sheep which is 'full-mouthed', i.e. whose teeth are fully grown. Such a sheep is too old to be worth buying. In his *De re rustica* (ed. Forster and Heffner), 7.3.7, Columella warns against buying a ewe of over three years with projecting teeth, as she is likely to be infertile.

Seán Ua Súilleabháin makes an alternative suggestion that the sheep which is *daintech* is prone to nibble at its skin, perhaps indicating skin-disease.

(7) nip anc*rad*. Binchy reads *ancrud*, but the MS clearly has the suprascript abbreviation for *ra* rather than *ru*. It is uncertain whether the original had *ancrud* (*an-* 'bad' + *cruth* 'shape, form') with *u*-flexion or *ancraid* with *i*-flexion; cf. Thurneysen, *Grammar*, 219 § 345.

(8) nip ladrach. I read *ladrach* rather than Binchy's *ladarach* (*CIH* ii 675.16). The MS has *ladach* with what seems to be the suprascript abbreviation for *ra* over the *d*. Strictly, this should be expanded as *ladraach*, but one of the *a*s is clearly redundant. The spelling *lad(h)rach* is well attested in the early language (see *DIL* L 18.1–8), and in both Modern Irish and Scottish Gaelic.

Ladrach is an adjective formed from *ladar* 'the space between the fingers or toes (of humans or animals)'. In the present context it must refer to some defect or disease affecting the sheep's hoof. It could, for example, refer to a sheep with foot-rot. Mícheál Ó Dochartaigh MRCVS of the Faculty of Veterinary Medicine, Dublin, refers me to the term *ladhairíneach* (a derivative of *ladhar*) which is used of bovine 'foul-in-the foot' in Ring, Co. Waterford. R. B. Breatnach, *Seana-chaint na nDéise II*, 250, notes that *ladhairíneach* is also used of a blister between a person's fingers or between the claws of a hen. In his *Faclair*, Dwelly quotes the phrase *tarbh ladhrach* 'a bull with large hoofs', so another possibility is that the author of our text is warning against buying a sheep with large or misshapen hoofs.

(9) nip l*eth*eirlach. *Airlae* (also *aurla, urla, orla, erla, irla*) means 'lock of hair, tress, tuft'. A sheep which is *letheirlach* 'half-tufted' must therefore have a fleece characterized by uneven tufts. The health of a sheep – or other mammal – can generally be judged from the state of its coat. A sheep whose fleece hangs in uneven tufts is likely to be old or unhealthy. Aristotle maintains (*Historia animalium* (ed. Thompson), 8.10) that short-fleeced sheep withstand the rigours of winter better than those with shaggy fleeces.

(10) nip romeidlech. The adj. *meiglech* (*meidlech*) is a derivative from *meigel* 'bleating': see *DIL* M 81.36–54. A sheep which bleats excessively is likely to be unhealthy; cf. § 2 (4) *nibi buirech* 'she (a cow) is not a bellower'.

(11) ropai uaitsiu. I would read 3 sing. pres. subj. *robé uait-siu* lit. 'may it be from you', i.e. let payment be made by you for a sheep which is proper with regard to age, body and wool. Binchy suggests the emendation *ron-bia* 'we will have (lit. to us will be)', but I do not see how this can fit the context.

indraicso. I doubtfully divide *indraic-so*, taking *indraic* 'proper, fit for sale' to be an adj. agreeing with acc. sing. *cairaich*. Liam Breatnach suggests that one might take *indraicso* to be gen. sing. of the abstract *indracus* (*inraccus*).

Text 2
Livestock not to be distrained

As we have seen in the section on illegal distraint (p. 170), there are many restrictions on the circumstances in which a person can remove livestock from somebody else's farm in the pursuit of a legal liability. An animal which cannot legally be distrained is described as *nemed* 'privileged' or *deorad* 'immune from legal process'. The passage edited here lists twenty-five reasons why livestock might be *nemid athgabálae* 'immune from distraint'

This passage is found at p. 376[a] of the Trinity College Dublin manuscript H 3. 18 (*CIH* iii 897.10–29). It is part of an Old Irish text on various aspects of distraint which is printed at *CIH* iii 896.9–901.13. Most of this text remains untranslated, but Liam Breatnach has edited and translated 896.19–41 in *Brehons, serjeants and attorneys* (ed. D. Hogan and W. N. Osborough, Dublin 1990) 10–13 and D. A. Binchy has edited and translated 897.29–898.33 in *Celtica* 10 (1973) 72–86.

This passage must be studied in conjunction with the Old Irish Heptad no. 41 (*CIH* i 38.19–21 = *AL* v 260.1–4) which lists *.uii. neimid cethre ata urcuillte i nathgabāil* 'seven immune cattle which are forbidden in distraint'. Also of relevance is a section of commentary at *CIH* ii 715.5–15. Both these sources include cases of immunity which are not among the twenty-five *nemid athgabálae.* For example, the commentary states that a mare should not be distrained for nine days after the birth of her foal, but our passage does not refer to this category. Heptad 41 includes a sick cow (*bó co ngalur*) and a cow which supports an invalid (*bó brothcháin*)[1] among the immune cattle, but both are absent from our list.

Early Irish authors often arranged their material in numbered lists, generally in threes, fours, sevens or tens. The arrangement of material in sevens is particularly common in the law-texts, and one of the longest texts in the *Senchas Már* collection – the Heptads (*inna sechtae*) – consists largely of groups of seven. Sometimes, extra items have been added to a heptad, either by the author or a later redactor. For example, Heptad 53 purports to list seven departures in the law of marriage (*.uii. nindsguithi . . . a cāin lānamna*) but in fact the single surviving version contains eleven items (*CIH* i 48.27–32 = *AL* v 296.13–20).

Arrangement in numbers larger than ten is unusual in legal material. However, one text lists the sixteen signs of bad pleading (*CIH* vi 2342.1–6), and another gives seventeen indirect signs of a criminal's guilt (*CIH* iv 1359.26–1364.8). Legal commentary at *CIH* v 1905.38–1906.17 catalogues the fifty living services which a client must perform for his lord. In editing such long lists it is not always easy to arrive at the correct number

[1]For the *bó brothcháin*, see p. 349.

of items, as the scribes do not number them individually. In my edition of this passage on the twenty-five *nemid athgabálae*, I have suggested one method of allotting numbers to the different categories of livestock, but other arrangements are also possible. No matter which way the categories are arranged, there always seems to be some overlapping. For example, I am uncertain whether I am correct in taking the twenty-fifth item to comprise boars, stallions, bulls and rams during their mating seasons. There is overlapping here with items 14 and 22.

Text

§ 1. Cis l*ir* nemid ath*gabala*? A .u. xx.it; cadead-side? Nī. (1) nuidlich co ndichet a nus, (2) bo fodhadha cum*aile*, 7 (3) bo chis, (4) bo focrecha, 7 (5) uairtnech, 7 (6) craibech, 7 (7) bo mblegar fri pit, 7 (8) gamnach, 7 (9) laulgach .i. nis ninathg*abala* ar 'musfuachat fria f*or*us; orggai[d] in gaimen in loeg inge m*ana* be a cethrai lais olchenai.

(10) Neimid da*no* cen mbes carr f*or* loarcaib ic imirgib, 7 (11) bai occ omulc, (12) daim cen mbiti fo mamu,[2] (13) capaill senchilli no tigerni na nathco*m*airc nech, ar is deorad.

(14) Tairb d*ano*, cene *m*bis dair f*or* buaib is deor*ad* cadhesin. (15) Ech buada, cein *m*bis im aige ni b*eir* cid a cinaid fadeisin. (16) Ech robaid co neser la scis. (17) Ech fir consni enech tuaithi, 7 (18) allmair a c*um*at.

Neimid cær*ach*: (19) molt midi, (20) curu darla a broth, (21) reithi iarna slaini 7 (22) cein bes f*or* cairib. Neim*id* muc: (23) muca denma, 7 (24) muca cene *m*biti f*or* mesruth; (25) cull*aig*h muc 7 ech, 7 tairb 7 rethe;[3] is deorath ca*ch* æ a aimsire.

§ 2. Cia mét do cethraib folo*n*gat cinaid duine? .i. boin[4] nama. Cid ara nab-[5] eich 7 cairich? Fo bith a llobra f*r*ia nimain 7 a ndocoi*m*eta ar conaib 7 ar thaitib; mil nad focair a aidid 7 fo[f]ich a f*or*us.

§ 3. Muca dano, ni b*er*at cinaid fo bith confechad a forus; acht ata re do mucaib i mb*er*at duinichin .i. man beit bai no eich lais: re duith, ar isi re insin nad focladat. Atat rei aili nad mb*er*at cid a cin[t]a fesin .i. re bidi f*or* mes, ar id muc[a] denma and; ni fil acht aidbriuth foraib co tabeir comrach iar*um* f*or*a saill no f*r*ia mart.

[2] *Read* fo mámaib.

[3] *Read* rethi.

[4] *Read* boí.

[5] *Perhaps read* cid arná berat. *See discussion on p. 531.*

TRANSLATION

§ 1. How many are immune from distraint? Twenty-five. What are they? Not hard: (1) a newly calved cow until her beestings have gone, (2) a cow which supports a champion(?), and (3) a cow for rent, (4) a cow for payment(?), and (5) an *úairtnech*, and (6) a *craíbech*, and (7) a cow which is milked for a meagre ration, and (8) a cow with a yearling calf, and (9) a milch cow; i.e. they are not to be distrained because they injure themselves at their pound; the yearling injures the young calf if the rest of its cattle are not with it.

(10) They are immune while there is a cart on their haunches at migrations, and (11) cows being milked, (12) oxen while they are under yokes, (13) horses of an old church or of a lord which no one demands, for it is an immune [animal].

(14) Bulls are themselves immune while the cows are in heat. (15) A prize horse, while it is racing it does not even bear liability for its own offence. (16) A horse with a warning that it may be requisitioned, with a respite. (17) The horse of a man who strives for the honour of the territory, and (18) the same for [a horse] from overseas.

Immune sheep: (19) a fattened(?) wether, (20) a ewe which has come into heat, (21) a ram after recovery [from castration], and (22) [a ram] while on the ewes. Immune pigs: (23) pigs of production, and (24) pigs while they are feeding on acorns; (25) boars and stallions and bulls and rams; each of them is immune in his time [of mating].

§ 2. How many animals bear liability for the offence of a person? i.e. cows only. Why do horses and sheep not bear [the offence]? Because of their weakness when being driven and the difficulty of keeping them from wolves and from thieves; an animal which does not proclaim its slaughter and which damages its pound.

§ 3. Pigs, moreover, do not bear liability for an offence because they damage their pound; but there is a time for pigs when they bear liability for a human offence, i.e. when he (the defendant) does not have cows or horses: the time of farrowing, for that is a time when they do not root up [the ground]. There are other times when they do not bear liability for even their own offences, i.e. the time when they are feeding on acorns, for they are then pigs of production; there is only a claim on them and he (the plaintiff) books for later their salted meat or their carcase.

DISCUSSION

§ 1. (1) nuidlich co ndichet a nus. Our text places the newly-calved cow (*nuídlech*) at the head of the list of twenty-five categories of immune animal. Heptad 41, which only lists immune cows, likewise puts the *nuīthlech iar toud* 'cow after calving' in first place (*CIH* i 38.19 = *AL* v

260.2).

The form *-dichet* is 3 sing. perfective pres. indic. of *téit* 'goes'. For *nús* 'beestings, colostrum', see p. 38 above.

(2) bo fodhadha cum*aile*. As suggested at *DIL* F 209.14, it makes best sense to emend to *fothuda*, gen. sing. of *fothud*, vb.n. of *fo-suidethar* 'supports, maintains'. One can compare the reference to *im boin fosuidethar carru* 'about a cow which supports champions' in the main law-text on distraint, *Di Chetharślicht Athgabálae* (*CIH* ii 372.1; iii 888.31; 892.10; v 1683.38 = *AL* i 124.8). Thurneysen, 'Ir. *fotha*(*e*) und *foth*', compares *athgabāil fothuda cairr* 'distraint for the maintenance of a warrior' from the same text (*CIH* ii 399.31 = *AL* i 226.32). A glossator explains this as *in bo meith* 'the fat cow'.

The identity of the third word in the phrase is uncertain. The manuscript has *cum-e* which Binchy expands as *cumaile* (*CIH* iii 897.11). The most straightforward interpretation would be to take it as gen. sing. of *cumal* 'female slave', and to translate 'a cow for supporting a female slave'. But it is very difficult to envisage a defendant preventing the distraint of one of his cows on the legal grounds that the milk was necessary for the maintenance of a female slave. In Old Irish commentary on *Di Chetharślicht Athgabálae*, we find the same phrase in the form *bō fothuda cumaile*, with the last word spelt out fully in the manuscript (*CIH* iii 914.26). No doubt because he could make no sense with *cumal* 'female slave' here, Binchy suggests in a footnote that one might emend to *bo fothuda caurad* 'a cow which sustains a champion (*caur*)'. This emendation seems too drastic, but raises the possibility that the word is the *o*-stem *cumal*(*l*) 'champion'. This is poorly attested in Irish (*DIL* s.v. *cumall*), but has been compared with the common Gaulish element *camulo-* (Evans, *Gaulish personal names*, 160–1; *LEIA* C-287 s.vv. *cumal, cumall*). In our text one might therefore read *bó ḟothuda cumail* 'a cow which supports a champion'. It is even possible that there was an *ā*-stem *cumal* 'champion', in which case no emendation would be necessary at *CIH* iii 897.11 or 914.26.

It would obviously be of general advantage to the community that a cow which was providing milk for a valued warrior should not be distrained. Compare also item (17) below.

(3) bo chis. The meaning of this phrase is clarified by the corresponding category in Heptad 41, which is defined as *bo cīss flatha nō eclasa* 'a cow intended as rent (*cís*) for lord or church' (*CIH* i 38.21 = *AL* v 260.3). It would clearly not be in the interests of either lord or church if a cow which had been designated as rent was distrained by some other person. For *cís flatha* 'rent due to a lord', cf. *CIH* iv 1502.30 = O'Dav. 354 § 928.

(4) bo focrecha. The main law-text on distraint, *Di Chetharślicht Athgabálae*, states that immune livestock should not be distrained, and includes *bō focreaca iter dā sealbh* among three examples (*CIH* v 1735.22 = *AL* ii

42.15). In this case it seems best to emend *focreaca* to *fochreca*, and take it to be gen. sing. of *fochraic* 'payment, fee, hire, rent'. In the Old Irish Glosses, this word has the *ā*-stem gen. sing. *fochricce*, but the *i*-stem gen. sing. *fochreca* is well attested in legal material (e.g. *CIH* i 70.26 = *AL* iv 92.4). A literal translation would thus be 'a cow of payment between two holdings'. The glossator suggests various circumstances in which two farmers might have an economic interest in a single cow (*CIH* v 1735.22–4 = *AL* ii 42.16–18). One farmer might own the body, and the other might own the milk. Alternatively, the cow might be one which is the subject of a disagreement (*bō cosnuma*) or which is being sued for (*bō aidbriudha*). In other law-texts the term *fochraic* is used of the cow which is given as rent for land (*CIH* iii 778.26–7 = *CG* 4.94), or as payment for trespass (*CIH* i 70.19–26 = *AL* iv 92.3–4).

In conclusion, I would read *bó fochreca* in our passage, and take it to mean that the defendant's cow cannot be distrained by the plaintiff if it is due in payment to a third party. This prohibition may also apply – as the glossator on *Di Chetharṡlicht Athgabálae* suggests – to a cow held in co-ownership.

(5) uairtnech. This category of cow is also included in Heptad 41 (*CIH* i 38.20 = *AL* v 260.2). The glosses on this heptad read *.i. in bō blegar risin uartan. risin salann* nō *risin mbalgum* 'i.e. the cow which is milked with (or for?) the *úartan*, with the salt or with the sup'.[6] The main problem is the identity of *úartan*, from which *úairtnech* appears to be a derivative. The word *úartan* is included in a list of domestic and farming items in the *ogam tírda* 'rustic ogham', but no clue is given as to its identity (Calder, *Auraicept*, 292.5726). No convincing etymology has been put forward: see *LEIA* U-10 s.v. *úartan*.

As in the case of Heptad 41, a glossator on *Cáin Lánamna* associates *úairtnech* with milking. § 12 of *Cáin Lánamna* discusses the division of a household's milk and milk-products in the event of a divorce (*CIH* ii 508.15–18 = *SEIL* 31). A small proportion – one thirty-sixth – goes to the farm-workers (*do urgnamtaib fotrebtha*). According to the glossator this is divided between the cowherds and calfherds, except for the extra *úairtnech* share, which goes to the cowherds (*cenmothā cuitigh uairtneach d'imarcraid do brith do buachaillib bō*). In spite of Thurneysen's discussion at *SEIL* 31–2, the meaning of *úartan* and *úairtnech* remains obscure. The gloss on Heptad 41 suggests some connection with the salting of milk in the cheese-making process. Perhaps the *úairtnech* cow is immune from distraint because of its special rôle in cheese-production. One can compare the passage in Virgil's *Georgics* (ed. Fairclough), 3.394–403, which describes the feeding of salted herbage (*salsas . . . herbas*) to milch cows. They consequently drink more, and produce greater quantities

[6] Máirín Ní Dhonnchadha suggests reading *selann* 'relish (with food)': see p. 345.

of salt-flavoured milk, which is made into cheese. Virgil also mentions the addition of salt to the milk itself. Aristotle (*Historia animalium* (ed. Thompson), 8.10) refers to the practice of mixing salt in the food of sheep so that they drink more and increase their milk-yield.

In *Cnósach*, 114, Ó Cuív suggests that *úairtnech* may be the same word as Modern Irish *fuairthneach* 'third-year stripper'.

(6) craibech. In Heptad 41, the term *craibech* is explained *.i. in bō bleg*ar *frisin craoib do croth*ad *ria .i. iar mbreith a laoig fo .c.oir* 'i.e. a cow which is milked with a branch being shaken at her, i.e. immediately after the birth of her calf' (*CIH* i 38.23–4 = *AL* v 260.7–8). The glossator may be correct in taking *craibech* here to be from *cráeb* 'branch, bough, bush'. Derivatives of *cráeb* include *cráebach* 'branchy, branching', *craíbech* 'branches, bushy place' (also 'fence' at *CIH* i 99.26–7), and *craíbechán* 'vegetable stew'. But I can see no reason why a cow should be described as branchy or branching, and why she should be immune from distraint.

A connection with *crob* 'hoof, etc.' seems unlikely.

(7) bo mblegar fri pit. The word *pít* (*fít*) is commonly used of a small quantity of food, particularly a monastic ration (see p. 343). The Heptad 41 glossator provides the explanation *.i. fri feit na mnā* nō *risin roinn mbicc* 'i.e. for the woman's ration or for the small portion' (*CIH* i 38.25 = *AL* v 260.10). This suggests that it was illegal to distrain a cow if it could be demonstrated that her milk was required to feed someone on a meagre ration. The gloss *fri feit na mnā* might refer to milk sustaining a nursing mother.

Alternatively, this might be a cow which is drying off, and therefore only producing a small quantity of milk. To subject a cow in this condition to the stress of distraint might cause sickness, especially mastitis.

(8) gamnach. The *gamnach* 'stripper' is a cow which has a yearling calf (*gamain*) at heel. Such a calf is notoriously difficult to control: see the Old Irish tetrad quoted on p. 40 above. It is no doubt because of the troublesome calf that it was considered inadvisable to distrain a stripper.

(9) laulgach. Our text has previously excluded certain milch cows, such as (1) a cow which has just calved, and (7) a cow which is milked for a meagre ration. Having made these exceptions, it is not likely that the author should then proceed to exclude the entire category of milch cow. It seems, therefore, that he is thinking of the distraint of milch cows and strippers together. He envisages a situation in which milch cows suffer injury in the pound, but the precise circumstances are unclear: see discussion under *ar 'musfuachat* below.

nis ninath*gabalae*. *Nís* (*n*-) is here 3 plur. pres. indic. negative of the copula (Thurneysen, 'Air. *nis*'; Thurneysen, *Grammar*, 485 § 796). The form *inathgabálae* means 'capable of distraint' (*in-* + *athgabál*).

ar 'musfuachat fria f*oru*s. The pound (*forus*) is the structure – presumably of wickerwork or stones – in which distrained animals are kept by the plaintiff. According to *Di Chetharšlicht Athgabálae*, different types of animals should not be kept in the same pound (*CIH* ii 422.16 = *AL* i 268.20). In § 3 of our text, the author refers to the damage which distrained pigs may do to a pound by digging.

Here, on the other hand, it is the impounded animals which suffer injury. One possibility is that the milch cows, perhaps deprived of their calves, are in a distressed condition and injure themselves by crashing against the walls of the pound. Or the author may be thinking of fighting between milch cows and strippers. I take *'musfuachat* to be 3 plur. pres. indic. of *fo-fich* 'injures, attacks' with the reflexive *imm-* 'each other, mutually, themselves' + 3 plur. infixed pronoun (Thurneysen, *Grammar*, 517 § 841). Formally, it could also be from *imm-fuich* 'annuls, impugns, abrogates', but the meaning does not suit.

orggai[d]. The author is concerned that if a yearling and a calf are impounded together without adult cattle from the herd – particularly the mother of the calf – it is likely that the yearling will injure the calf.

(10) neimid da*no*. Binchy suggests that *neimid da-* of the MS should be emended to *neimid dam* 'privileged oxen' (*CIH* iii 897^{e}). There is no doubt that the ox was the main cart-pulling beast of the early Irish farm. However, there are also some references to the cart-horse (*capall cairr*): see p. 96. It is possible, therefore, that the intention is to ban the distraint of either oxen or horses while engaged in moving household goods. I expand *da-* as *dano*, the enclitic particle 'indeed, too'.

The word *immirge* 'migration' is used of the moving of a household in a variety of circumstances, such as to a new home or in the process of seasonal transhumance. Heptad 42 states that such migration is usually a protection against distraint (*nochis gnāth is turrtugud do athgabāil imirgi*), but lists eight exceptions to this rule (*CIH* i 38.29–39.19 = *AL* v 260.15–262.4).

(11) bai occ omulc. The form *omulc* is probably dat. sing. of **omalc* 'milking', verbal noun of an unattested compound *uss-mlig-*. The further compound *ind-uss-mlig-* is well attested in the meaning 'mulcts, levies', e.g. *inombligh* 'he levies' (*CIH* iv 1510.29 = O'Dav. 393 § 1119). It is possible that **omalc* is to be identified with *oimelc*, the only attestation of which is in an entry in *Cormac's Glossary* (Meyer, *Sanas Cormaic*, 86 § 1000 s.v. *óimelc*).[7] See Russell, 'The Celtic preverb **uss*', 109; Hamp, '*Imbolc, Óimelc*'.

I am uncertain what exactly is meant by *bai occ omulc* 'cows milking'. The author has previously listed many situations in which a milking cow

[7]See p. 460.

may not be distrained. Perhaps he means here that the plaintiff may not drive off the cows while they are actually in the process of being milked. Earlier in the text (*CIH* iii 897.1–2), it is stated that it is proper for cattle to be distrained early in the morning. There are three circumstances in which the cattle are classed as *nemidh indisi* 'immune ones of the milking-enclosure'. One of these is given as *dia ragbaither indis riut* 'if there is a reaching of the milking-enclosure before you'. This implies that the plaintiff must arrive before milking commences.

(13) capaill senchilli no tigerni. The word *capall* is applied to a work-pony, as distinct from the more valuable *ech* of (15)–(18). The compound *senchell* 'old church' is common in placenames (Hogan, *Onomasticon*, s.v. *senchell*), and is no doubt to be identified with *andóit* 'old church, church with subordinate foundations'. *Andóit* is from late Latin *antitatem* (Classical Latin *antiquitatem*): see McManus, 'A chronology', 63; McManus, 'On final syllables', 147[25].

na nathco*mairc* nech. If a plaintiff wishes to pursue a case against a high-ranking *nemed,* such as a cleric or lord, he does not use the normal method of distraining his livestock. Instead, he must fast against him, and can only distrain his livestock if there is no response to his fast (*GEIL* 182–3). The author may be stressing that even the humble work-pony of such a high-ranking defendant is immune from normal distraint. I take the verb *ad-comairc* 'asks, demands' to refer here to the plaintiff's notice of intention to distrain livestock, though I have found no other examples of its use in this context. The usual term is *apad.*

deorad. Here and in (14) and (25), the adj. *deorad* is applied to an animal which is immune from distraint, with exactly the same meaning as *nemed.* This usage is not attested in *DIL* s.v. *deorad.*

(14) tairb d*ano*, cene *m*bis dair f*or* buaib. The distraint of a bull in the time of bulling (*tarb i n-aimsir dára*) is also banned in the main text on distraint, *Di Chetharṡlicht Athgabálae* (*CIH* iv 1458.7; v 1735.33 = *AL* ii 44.6).

(15) ech buada. Horse-racing is often mentioned in the law-texts and other sources (see p. 99). A prize horse cannot be distrained while it is racing (*im aige*), even for an offence which it has committed itself (cf. § 3). Presumably this refers to the day – or period of days – when the horse is due to race.

(16) ech robaid co neser la scis. I am uncertain of the interpretation of this phrase. One possibility would be to take *ech robaid* to mean 'a horse of warning', i.e. a horse which is used to bring a warning of attackers or some other danger. But if so, it is difficult to make a connection with *co neser la scis.*

I therefore doubtfully take it to mean that the horse cannot be distrained because a warning has been given to the owner that it is to be requisitioned (*co neser*) after a period of grace (*la scís*). Presumably this requisitioning is ordered by a king or lord, cf. *CIH* ii 570.3 = *CG* 22.561–2. I take the form *-eser* to be 3 sing. pres. subj. pass. of a lost verb related to *íasacht* 'lending' (cf. Watkins, '*In essar dam do á*', 162).

(17) ech fir consni enech tuaithi. 'The man who strives for the honour of the territory' must be the *aire échta* (see p. 116). Cf. *aire échta . . . do dígail enechruccai túaithe* 'the *aire échta* for avenging the honour of the territory' (*CIH* ii 566.34–5 = *CG* 14.358–60).

(18) allmair. As we have seen in Chapter 2 (p. 90), a high value is assigned in legal and other sources to the horse from overseas (*ech allmuir*). For example, an Old Irish gloss at *CIH* iii 889.6 has *allmairi .i. eich ⁊ sēoit maithi cheanae* 'foreign (goods) .i. horses and other valuables'. Cf. *CIH* iv 1484.30 = O'Dav. 277 § 501 *cullach .i. ech bretnach* ut est *nō cullach allmuire* 'a stallion, i.e. a British horse, that is, (or) a stallion from overseas'.

(19) molt midi. One would expect this to refer to a wether being fattened up for slaughter, just as (23)–(24) refer to pigs in the process of being fattened. I have no plausible explanation for *midi.* It could be gen. sing. of *mide,* rarely attested in the meaning 'middle, centre', and probably the same word as *Mide* 'Meath' (*LEIA* M-50). One possibility would be to emend to *mithisi* or *mithisen,* taking it to be gen. sing. of *mithis* (also *mithise*); cf. *mithig* 'timely, seasonable'. A *molt mithise* would thus be a fat wether ready for killing. One can compare the reference in *Togail Bruidne Da Derga* to four *mucca midisi* being cooked in a cauldron (Knott, ed. 24.815 (see v.ll. in glossary) = Stokes, 'The destruction of Dá Derga's hostel', 186 § 87). These are presumably fat pigs. One can compare Meyer, *Sanas Cormaic,* 33 § 397 *la dā muic midisen,* but cf. ibid. 90 § 1052 *cota mmuic midligen.*

Máirín Ní Dhonnchadha draws my attention to the third stanza of a legal poem on the rights of judges (*CIH* iv 1268.20–6): ⁊ *giall cach fīrferba/ in cach midisi aimsirda/ i mortaidth*er *morcarrnda/ dleagait ara cēill* 'and a pledge for every true(?) milch cow at every profitable season, in which great meat is slaughtered, they are due for their (i.e. the judges') intelligence'.[8]

(20) curu darla a broth. Read *cáera i tarla a bruth,* lit. 'a ewe in which may happen her heat'. Spellings with *cau-* and *cu-* are common in the manuscripts, e.g. *CIH* ii 368.6 *cauru;* v 1627.25 *cura;* 1900.17 *curu.* The form *-tarla* is best taken as 3 sing. perfective pres. subj. of *do-cuirethar* 'happens'. It could also be for the 3 sing. perf. *-tarlae.*

Broth of the manuscript is no doubt a variant spelling of *bruth* 'blaze, glow; fury, ardour; boiling; brewing (of ale); glowing mass (of metal), etc.' (*DIL* B 216.78–217.40; *LEIA* B-106 s.v. *bruth*). For the *o/u* variation, cf. *bruthach, brothach* 'fiery'; *bruthgal, brothgal* 'fury'. There are no

[8] For a discussion of the phrase *ferba fíra* 'milch cows', see Lindeman, 'Archaic Irish *ferba fíra*'. See also P. Kelly, 'Dialekte im Altirische?'.

other attestations of *bruth* (*broth*) in the sense of 'sexual heat in animals', but cf. *oíbel*(*l*) 'spark, flame; heat in animals'.

(21) reithi iarna slaini. I take this to mean 'a ram after recovery', i.e. a wether which has recently been castrated, and is therefore not in a fit condition to be distrained. It could be argued that a male sheep which has been castrated is no longer a ram (*reithe*) but a wether (*molt*). However, the commentary at *CIH* ii 715.11–12 unambiguously explains *rethe iar slān* with the gloss *.i. iar slān*ugud *iar mbuai*n *as* 'i.e. after recovery after being castrated'. The phrase *buain as* 'cutting out' is used of the castration of a bull-calf at *CIH* ii 482.12 = *AL* ii 246.12.

At *CIH* ii 715.7–8 the same commentary makes a similar reference to the immunity from distraint of a stallion (or boar) which has been castrated: *cullach iar trogh co tēora nōmaithi .i. iar mbein aneich dā ndēnann a trogh, a uirge* 'a stallion/boar after castration (?) for twenty-seven days, i.e. after the removal of that which causes parturition, his testicles'. The difficulty here is the fact that *trog* normally means 'parturition, offspring' rather than 'castration'; cf. *CIH* ii 715.5–7 *at urcoillte don lāir iar trogh .i. is urcoillte don lāir do gab*āil *i nathgab*āil *iar trogh a serraig uaithi co cenn nōmaithe* 'they are forbidden [to distrain] the mare after parturition i.e. it is forbidden for the mare to be taken in distraint for nine days after giving birth to her foal'. In the previous citation, the commentator tries to make the connection from *trog* 'parturition' to 'that which causes parturition', i.e. 'his testicles' (*a uirge*). The problem may lie in a corruptly transmitted text, from which the commentator is trying to extract sense. Perhaps the original read *cullach iar slán* (or *iar sláini*) 'a stallion/boar after recovery'.

(22) for cairib. The guttural inflection of *cáera* 'sheep' is well attested, e.g. nom. plur. *caīrich* (*CIH* iii 897.23 = § 2 of this text), gen. plur. *caīreg* (*CIH* vi 2292.2 = Binchy, '*Bretha Crólige*', 18 § 23), dat. plur. *ċaīrchaib* (*Thes.* i 339.20). However, there is also evidence of non-guttural inflection. This is particularly clear in the dat. dual and plur.; e.g. *di chaírib* (*Thes.* ii 239.19; *LU* 60.1729), *for caīrib* (*CIH* iii 914.28), *dona cáoiribh* (Bergin, 'Irish grammatical tracts: II Declension', 165 § 202). *DIL* C 20.57 quotes a non-guttural nom. plur. form *caīre* from *AL* iv 96.26 (= *CIH* i 72.13). But the other version of this text has *caīrigh* (*CIH* i 194.15). This suggests that *caīre* here is merely a scribal variant of *caīrigh* with the common falling-together of *-e, -i, -idh, -igh* in final syllables.

(23) muca denma. I take *denma* to be gen. sing. of *dénum* 'making, production'. The author here distinguishes between *muca dēnma* and (24) *muca . . . for mesruth* 'pigs on acorns'. In § 3, however, he explicitly identifies pigs on acorns as *muca dēnma.* Commentary at *CIH* ii 715.9–11 suggests that the distinction lies in the type of food on which the pigs are being fattened. Thus the *mucca for mes* are defined *.i. a cēin beiti ac ithe measa* 'i.e. while they are eating acorns', whereas the *dagmuca denma* 'fine pigs of production' are explained as *na demuca a cēin bitht*er *aca mēthad gen mes* 'the fine pigs while they are being fattened without acorns'. For the diet of pigs, see p. 82.

(24) muca cene *m*biti f*or* mesruth. See preceding note.

(25) cull*aigh* muc ⁊ ech ⁊ tairb ⁊ rethe. I take it that the author here distinguishes as a single final category the four common male domestic animals – boars, stallions, bulls and rams – which cannot be distrained at the mating season. He has already dealt with the bull in (14) and with the ram in (22).

Binchy restores *cull-* of the MS as the singular *cullach*, but I feel that it makes better sense to read *cullaig muc* ⁊ *ech* 'males of pigs and of horses'. This fits in with the following plural *tairb* 'bulls'. I consequently emend *rethe* to the plural *rethi* 'rams'.

§ 2. It is usual to distrain cows for a person's offences. In this paragraph, the author explains why horses and sheep are not normally distrained.

boin nama. Read *boí namá*. The spelling *boin* of the manuscript is likely to have originated from the misreading of the length-mark over the *i* as an *n*-suspension.

cid ar(a)na b*erat* eich ⁊ cairich? The reading *cid ara nab-* of the manuscript suggests the 3 sing. prototonic pres. indic. pass. of *as-beir* 'says'. This phrase is very common in definitions in the law-texts, e.g. *CIH* iii 778.22 = *CG* 4.87–8 *cidh ara neper ócaire?* 'why is an *ócaire* so called?'. However, it does not suit the present example. One possibility would be to follow Binchy's suggestion at *CIH* iii 897[g], and emend to *cid arná bertar?* 'why are they (horses and sheep) not brought [on distraint]?'. Because of the use in § 1 (15) and in § 3 of the phrase *beirid cinaid* 'it bears the offence' – i.e. the animal is distrained for the offence – I read *cid arna berat?* 'why do they (horses and sheep) not bear [the offence]?'.

fo bith a llobra fria nimain. I translate 'because of their weakness when being driven'. The author clearly feels that cattle are physically better able to withstand being rounded up and driven to a pound than are horses or sheep.

ar conaib. I take *cú* here to refer to the *cú allaid* 'wolf', the main predator on sheep: see p. 186.

ar thaitib. I take this to be a reference to the *echtháid* 'horse-thief', who is condemned in law-texts and wisdom-texts: see p. 166.

mil nad focair a aidid. This phrase does not seem to fit in with what precedes, and it is likely that some material has been omitted. Liam Breatnach suggests that we should read *nach míl*, as at *CIH* iii 897.29 (= Binchy, 'A text on the forms of distraint', 78 § 1).

Aided means 'violent death', and in the context of animals is well attested in the sense of 'slaughter' (*DIL* A 104.12–15). The meanings of *fúacair* (*fo-úacair*) includes 'proclaims, challenges, announces, reveals, warns', but it is difficult to see which of them is appropriate here. Possibly the phrase refers back to the sheep of the previous sentence; the author may be suggesting that sheep are unsuitable for distraint because they do not give enough warning when attacked by wolves.

7 foich a forus. In § 1 (9), there is a reference to cows injuring themselves against their pound (*'musfuachat fria forus*), and in § 3 pigs are said to damage their pound (*confechad a forus*) by rooting it up.

I take *fo*[*ḟ*]*ich* here to be 3 sing. relative of *fo-fich* 'injures, damages'. It seems to be a reference to the damage which a distrained horse may do to a pound.

§ 3. In § 2, the author discusses why it is normal to distrain cows rather than sheep or horses. In § 3, he explains why pigs are generally unsuitable for distraint, but gives one exception to this rule.

confechad a forus. The compound * *con-fich* is not attested, but it obviously has the same meaning as *fo-fich* in this context, cf. § 2.

re duith. The form *duith* is gen. sing. of *doth* 'birth, parturition, hatching (of birds and animals)'. The author points out that there is no danger of a sow rooting up the ground before farrowing, and therefore can be distrained without damage to the pound. On the other hand, it would seem unwise to move a sow which is about to farrow from her own sty to the unfamiliar surroundings of another farm. This is the approach of the commentator at *CIH* ii 715.8–9: he includes among livestock which are immune from distraint *muca andaigthe .i. na muca in tan bīt ag annad, ag breth orc* 'farrowing sows, i.e. sows when they are farrowing, bearing piglets'. Admittedly, the word *andud* (*annad*) is not otherwise attested in the meaning 'farrowing' (*DIL* s.v. *andud* (d)), but its meaning is clear from the gloss *ag breth orc*.

muc[a] denma. In § 1 (23)–(24) a distinction is made between *muca dēnma* and *muca cēne mbīti for mesruth*, and in commentary at *CIH* ii 715.9–11 the former are taken to have been fattened without acorns, and the latter with acorns: see notes to § 1 (23). In this passage, on the other hand, the author identifies acorn-fattened pigs as *muca dēnma*.

aidbriuth. Binchy reads *aidbriuch* here (*CIH* iii 897.28), but the *-th* is clear in the manuscript. *Aidbriuth* (*aidbriud*) is verbal noun of *ad-fíri* 'sues, draws attention to an offence, claims against'.

comrach. I take this to be for Old Irish *cuimrech* 'bond', verbal noun of *con-rig* 'binds'. The spelling *comrach* is found later in the same text: *dobeir comrach forræ* 'he puts a bond on it' (*CIH* iii 900.23 = *AL* ii 130.1).

The idea seems to be that the plaintiff cannot distrain the pigs because they are being fattened. He can however lay claim to their meat (salted or unsalted) after they have been slaughtered.

Text 3

Value of cattle, sheep and pigs

This commentary on the values assigned in Irish law to cattle, sheep and pigs at various ages is found in the Trinity College Dublin manuscript H 3. 18 (no. 1337), p. 328B. It is printed in *CIH* iii 845.22–846.14.[1] The first six words are in large script, so it is clear that this is a quotation, probably the opening phrase, from a lost Old Irish legal passage on the values of these domestic animals. The remainder is written in the smaller script normally used for Middle Irish glosses and commentary. In general, commentary explains and expands upon the content of the Old Irish law-texts (see p. 11). Where they can be confirmed, the values given in this commentary agree with those of the extant Old Irish law-texts. For example, according to § 4 of this commentary the three-year-old dry heifer (*samaisc*) has a value of twelve scruples, i.e. half that of a mature milch cow. This agrees with the value assigned to this heifer in an Old Irish law-text at *CIH* vi 1975.4: see p. 63. In this commentary, the values of livestock are given in terms of the scruple (*screpul*), and the penny (*peinginn*). There are three pence in a scruple.

A condensed version of §§ 1–10 of this commentary is to be found in the Trinity College Dublin manuscript H 3. 17 (no. 1336) p. 658[c]. It is printed in *CIH* vi 2106.15–33. There is also a fragment on f. 57[d] of Rawlinson B 506 at the Bodleian Library, Oxford. It comprises §§ 1–3 and part of § 4; it is printed in *CIH* i 180.10–19.

The words *Bel(l)taine*, *Lugnasad*, and *Samain* refer to the festivals marking the beginning of May, August and November respectively.

Text

§ 1. BID LAEG O ERRUCH TRIA SAMRUD 7rl-. Da screp*ul* ar*in* læg on ló b*er*ar isi*nn* errach co samai*n* iar cind; da scrip*ul* air risi*n* re-si*n*, cidh firen*n* cid boi*n*en*n* he. T*r*i scrip*uil* arin mboi*n*enn o samain cosi*n* mbelltai*ne* is nesa, da scr*e*p*ul* beos ari*n* firen*n*; 7 bliad*ain* i[s] slan doib and-side.

§ 2. .iiii. scrip*uil* ari*n* mboi*n*enn on bell*taine* co samai*n* 7 a da trian ari*n* firen*n*, 7 bliad*ain* co leith i[s] slan doib and-sidhe. .ui. scrip*uil* ari*n* mboi*n*en*n* ond samai*n*-si*n* co bellt*aine* 7 .iiii. scrip*uil* ari*n* fire*nn*, 7 da bliad*ain* i[s] slan doib and-sidhe.

[1] It is translated by McLeod, 'Interpreting early Irish law (2)', 91–3.

§ **3.** .uiii. scrip*uil* ari*n* mboi*n*end on belltai*ne*-si*n* cosi*n* mbell*taine* aile i cind bliad*na* 7 a [da][2] tria*n*-sidhe ari*n* firen*n*, 7 t*r*i bliad*na* i[s] slan doib [ann]-side.

§ **4.** Samaisc o si*n* coro dairt*er*, 7 da screp*ul* .x. ui*r*re; 7 o dairfidht*er* is se scrip*uil* .x. uirri co ruca lægh, 7 o b*er*as lægh; .xx. scrip*ul* uirri, 7 .iiii. bliad*na* i[s] slan di and-sidh.

§ **5.** .xx. scrip*ul* ui*r*ri beos o si*n* amach co rucca in lægh tanaisi; 7 o beras, da scr*e*p*ul* ar .xx.it ui*r*ri, 7 .u. bliad*na* i[s] slan dí and-sidhe.

§ **6.** Da scrip*ul* ar .xx.it ui*r*ri beos co ruca i*n* t*r*es læg; 7 o b*er*as, .iiii. scr*e*p*uil* ar .xx.it ui*r*ri, 7 se bliad*na* i[s] slan dí and-sid; 7 noch[a] roig comlogh risi*n* mboi*n*enn ari*n* firen*n* *no*coro caithea i*n* feren*n* t*r*ian a aisi d'i*m*arcraid sech i*n* re rosoig a lanlogh ari*n* mboi*n*end.

§ **7.** Peingi*n*d ari*n* uan on lo(gh) b*er*ar he isi*n*n errach [co belltaine]; cid firend cidh boi*n*en*n* int uan, nochon uil a ndethfir risi*n* ré-si*n*.

§ **8.** Pei*ngi*n*n* co leith ari*n* mboi*n*en*n* o si*n* amach cosi*n* lugnassad is nesa, 7 pei*ngi*n*n* beos ari*n* firen*n* reisi*n* re-si*n*. Screp*al* ari*n* mboi*n*end o si*n* cosi*n*n lugnassad aile i ci*n*d bliad*na*, 7 da pi*n*gin*n* ari*n* firen*n* resi*n* re-si*n*; *no* comadh p*eingi*n*n* co leith, 7 bliad*ain* 7 raithe i[s] slan doib and-sidhe.

§ **9.** Da scr*e*p*al* ari*n* mboinen*n* co cend da bliad*na*, 7 a da t*r*ian ari*n* firend, 7 da bliad*ain* i[s] slan doib and-sidhe.

§ **10.** T*r*i scrip*uil* ari*n* mboi*n*enn o si*n* amach, 7 a da t*r*ian ari*n* firen*n* *no* cor thocaithea i*n* firen*n* t*r*ian a aisi d'i*m*arc*r*aidh sech i*n* mboi*n*en*n*, 7 o caithfes is comlogh orru; is amlaidh dono i*m* æs muc.

§ **11.**[3] Comlogh in firen*n* 7 in boi*n*enn in cein bit i coimit*echt* lachta, 7 trian et*ar*u o sin amach co hind a f*or*barta, 7 comlog iat o si*n* amach; 7 lethlog boi*n*in*n* a cho*m*aisi f*or*i*n* tarb do g*r*es.

§ **12.**[4] Pingi*n*d ari*n* lúrcc co belltai*ne*, di pingi*nn* arin mbanb co lugnasadh, 7 scr*e*p*al* aire im samai*n*, 7 ce(o)theora pei*n*gin*ne* co leth aire i*m* belltai*ne*, 7 t*r*i sc*r*ip*uil* log cacha muici a cin*n* da bliad*na*.

TRANSLATION

§ **1.** IT IS A CALF FROM THE SPRING THROUGH SUMMER, ETC. Two scruples for a calf from the day it is born in the spring until the following November; two scruples for it during that time, whether it is male or whether it is female. Three scruples for a female from November until the next May, two scruples still for a male; and they are then a year old.

§ **2.** Four scruples for a female from May until November and two thirds [of that amount] for a male, and they are then a year and a half old. Six

[2] *The* da *can be restored from the version at* CIH i 180.18.

[3] *In the margin opposite this paragraph is written* .g.ii. (= gné aile *'another version'*).

[4] *In the margin opposite this paragraph is written* .g.ii. (= gné aile *'another version'*).

scruples for a female from that November until May and four scruples for a male; and they are then two years old.

§ **3.** Eight scruples for a female from that May until the following May and two thirds [of that amount] for a male; they are then three years old.

§ **4.** She is a dry heifer from then until she is bulled, and she is worth twelve scruples; and when she is bulled she is worth sixteen scruples until she bears a calf, and when she bears a calf she is worth twenty scruples; and she is then four years old.

§ **5.** She is worth twenty scruples from then on until she bears a second calf; and when she bears it she is worth twenty-two scruples; and she is then five years old.

§ **6.** She is worth twenty-two scruples from then until she bears a third calf; and when she bears it she is worth twenty-four scruples; and she is then six years old; and a male does not reach equal value with a female until the male has lived for a third of his age beyond the time at which the female has reached her full value.

§ **7.** One penny for a lamb from the day in which it is born in the spring [until May]; whether the lamb is male or whether it is female, there is no dictinction between them during that time.

§ **8.** One and a half pence for a female from then on until the following August [i.e. of the same year], and a male is still worth one penny during that time. A scruple for a female from then until the August of the following year, and a male is worth two pence during that time; or it may be one and a half pence; and they are then a year and a quarter old.

§ **9.** Two scruples for a female to the end of two years, and two thirds [of that amount] for a male, and they are then two years old.

§ **10.** Three scruples for a female from then on, and two thirds [of that amount] for a male until the male has lived a third of his age beyond that of the female; and when he has passed this, they are of equal value; the same applies in respect of the age of pigs.

§ **11.** A male and a female are of equal value while they are sucking (lit. in the accompanying of milk), and there is a third between them until the completion of their development, and they are of equal value from then on; and a bull is always half the value of a female of the same age.

§ **12.** A penny for a piglet until May; two pence for a young pig until August, and a scruple for it in November, and four and a half pence for it in May, and the value of every pig after two years is three scruples.

Discussion

§ **7. [co belltaine].** In a footnote to *CIH* iii 845.39, Binchy suggests that *co belltaine* be supplied here. The other version is of no assistance as it only gives part of this paragraph (*CIH* vi 2106.28). In support of Binchy's suggestion is the statement in § 12 that a piglet (*lúrcc*) is worth a penny from birth until May (*co belltaine*).

§ **8. no comadh *peinginn* co leith.** Commentators use the form *comad* (*gomadh, gumadh*) to introduce an uncertainty or an alternative version. Here the commentator raises the possibility that a male sheep has the value of one and a half pence – rather than two pence – from the first to the second August of its life.

The other version amalgamates §§ 8 and 9 (*CIH* vi 2106.28–31), and does not give an alternative value for the male sheep.

§ **11.** In the margin this paragraph is marked *gné aile* 'another version', so it is likely that it was added to §§ 1–10 from another source. It is noteworthy that the version in *CIH* vi 2106.15–33 stops at § 10.

§ 11 sums up the general principles in relation to the differences in value between male and female livestock of the same age. A calf (or lamb) of either sex has the same value until it is weaned. Thereafter, the female is worth an extra third. This applies 'until the completion of their development' (*co hind a forbarta*). As has been noted in § 6, a male bovine must be an extra third older than a female before it attains its maximum value of twenty-four scruples. In § 10 the same principle is applied to sheep and (by implication) to pigs.

The final sentence of § 11 gives an exception to the above rule. A bull – unlike an ox – has only half the value of a cow of the same age. See discussion on the value of older male bovines in Chapter 1, p. 66 above.

§ **12.** As in the case of § 11, this paragraph is marked *gné aile* 'another version' in the margin. According to § 10, the value of a pig is assessed by the same criteria of age and sex as a sheep. However, this paragraph provides a method of assessing the value of pigs irrespective of their sex.

Text 4
Roads and tracks

This text is found in the Trinity College Dublin manuscript H 3. 18 (no. 1337) at p. 373[a], and is printed at *CIH* iii 893.22–35. It provides a valuable account of the early Irish road-system: the author has arranged five categories of road or track in order of importance (§ 1), starting with the highway (*slige*) and working down to the cow-track (*bóthar*). There is also discussion of road-maintenance (§§ 2–3) and a short definition of the *rámut* 'open space' in front of a king's fortress (§ 4). This text has some general similarities with the section entitled *De itineribus* 'concerning ways' in Isidore's *Etymologies.*[1]

Another version of this text is found in *Cormac's Glossary*, which was compiled by the scholar Cormac mac Cuilennáin about AD 900 (Meyer, *Sanas Cormaic*, 96 § 1082). In most sections, it agrees quite closely with our text. However, it introduces the further category of the *sét*, which is defined as a 'path for one animal', and is therefore narrower than a cow-track.

There are to my knowledge six extant versions of § 1082 of *Cormac's Glossary.* The version in the Leabhar Breac (LB p. 271[a]) was printed by Stokes, *Three Irish Glossaries* (London 1862) 38. The version in the Yellow Book of Lecan (YBL pp. 279–80) was translated by John O'Donovan (Stokes, (ed.) *Sanas Chormaic: Cormac's Glossary* (Calcutta 1868) 141–2) and printed with some variants from LB by Meyer, *Sanas Cormaic*, 96. The version in the Bodleian manuscript Laud 610 (f. 83[b]) was edited and translated with variants from both YBL and LB by Stokes, 'On the Bodleian fragment of Cormac's Glossary', *Transactions of the Philological Society* (1891–94) 190–3 (separate pagination: 36–9). The version in f. 183[a] of the Book of Uí Maine (UM) was transcribed from a photograph (with many gaps) by Meyer in 'Cormacs Glossar nach der Handschrift des Buches der Ui Maine', 318. There are also versions in the Trinity College manuscript H 2. 15B (no. 1317) at p. 35 (H^1) and at p. 100 (H^2).

This text contains a number of early linguistic features, which indicate a date of composition around AD 700. The early form of the preverb *di-* (generally *do-* in Classical Old Irish) is preserved in § 1 (1) *di-s-cuet.* The *-et* ending of this form is also early, becoming *-at* in Classical Old Irish. Another indication of an early date is the 3 plur. suffixed pronoun in § 1 (5) *iurrt[h]us* (read *iurrus* with Laud).

In the discussion below, I have given readings from the *Cormac's Glossary* version when they cast some light on the interpretation of our text.

[1] Isidore, *Etymologiarum* (ed. Lindsay), 15.16.1–13.

Text

§ 1. Róda, cis l*ir*-side? Nī, a .u. .i. slighi ⁊ ród ⁊ lamraite ⁊ tograide ⁊ bothar. (1) Caide int slige? Nī, discuet da carput sech in aile; doronad fri imairecc da carp*at* .i. carp*at* rig ⁊ carr*pat* esp*uic* ara ndichet cechtar nai sech araile. (2) Ród: docuet carp*at* ⁊ da oeneochde imbi; doronad fri echraite mendoto a medo*n*. (3) Lamraiti .i. it*er* di slig*id* .i. slig*e* tar deisc*er*t in men*n*oto ⁊ araili t*ara* tuaisc*er*t; f*r*i lesa f*r*i caii doronad. (4) Tograiti: focren fear trebur conair do ascnam rait no sleibi. Caidi a fochraicc? anaim cach duine[2] i*m*ateit ca*ch* la mbliadhain. (5) Bothar: doallad di ba f*or*, araile f*or* tarsna, araili f*or* fot, ara tall*a* a llaogu no i[3] nga*m*nai ana narrad; ar ma[d] i*n*a ndiaidh be[i]t, iurrt[h]us in bo bias i*n*na diaidh.

§ 2. Ataat tri glanta do ca*ch* æ, teora aimsir[4] i nglant*ar*, teora aiscsiu[5] ara nglant*ar*: aimsir echruathair, aimsir cue, aimsir catha. It e a t*r*i glanta: glanadh a feda, gl*an*ad a uisc*i* ⁊ gl*an*ad a cocclaid.

§ 3. Cadeat inna huisc-[6] ara nglant*ar*? Arna uaillnid[7] eochu oc dul do aonuch; a coclaid, arna h*esar*laider[8] nech f*or* oc dul for tres.

§ 4. Ramhut .i. urscur bis ar dun rig; cach duine dua taicmaic tir dleg*air* do a glanad.

Translation

§ 1. Roads, how many are there? Not hard, five, i.e. highway, and road, and by-road, and curved road, and cow-track. (1) What is the highway? Not hard, two chariots can go on it past each other; it has been constructed for the meeting of two chariots, i.e. the chariot of a king and the chariot of a bishop so that each of them can go past the other. (2) Road: a chariot and two horsemen can pass on it; it has been constructed as a horse-road of the locality internally. (3) By-road, i.e. between two highways, i.e. the highway past the south of the locality and the other past the north; it has been constructed for errands and winter-visiting. (4) Curved road: a well-off man hires a way towards a wood(?) or mountain. What is its rent? an animal every second year from every herd which passes. (5) Cow-track: Two cows fit on it, one sideways, the other lengthways, so that their calves and their yearlings fit along with them; for if they are behind them, the cow which is following them will injure them.

§ 2. There are three clearings for each of them, three times in which it is cleared, three reasons for which it is cleared: the time of a rush of horses,

[2] *Read* díni.
[3] *Read* a.
[4] *Read* aimsera.
[5] *Read* aicsin.
[6] *Read* aicsin.
[7] *Read* éilnea?
[8] arna h-laider MS.

the time of winter-visiting, the time of battle. These are its three clearings: clearing its wood, clearing its water, and clearing its ditch.

§ 3. What are the reasons for which it is cleared? So that it may not soil horses going to an assembly; digging its ditch, so that no one may be thrown out(?) on it going to battle.

§ 4. Clearing i.e. the open space which is in front of the fortress of a king; every person to whose land it extends is obliged to clear it.

Discussion

§ 1. In *Cormac's Glossary*, the roads and tracks are placed in a different order. In YBL the introductory sentence reads: *Atāt trā ilanmand forna conairib .i. sēt* ⁊ *rout* ⁊ *rāmat* ⁊ *slighe* ⁊ *lāmrotæ* ⁊ *tuagrotæ* ⁊ *bōthur* 'there are indeed many names for ways, i.e. path and road and clearing and highway and by-road and curved road and cow-track'.

(1) discuet. This is 3 sing. perfective pres. indic. of *téit* 'goes' with 3 sing. fem. infixed pronoun *-s-* (Thurneysen, *Grammar*, 473 § 769). The pronoun refers back to the fem. noun *slige*. *Cormac's Glossary* has *doscuet* (YBL, H^1, H^2), *doscuchad* (LB), *doscuat* (Laud), *doscuaid* (UM).

da carput. The manuscripts of *Cormac's Glossary* have readings without *da*, e.g. *doscuet carp-* (YBL), *doscuaid carpat* (UM). It is probable, therefore, that *da* in our text should be omitted. One would translate 'a chariot can go on it past another' with no change in the basic meaning of the passage. The emendation avoids the use of a dual subject with a singular verb. This usage is exceptional, though some instances are provided by Thurneysen, 'Singularisches Verb bei dualischem Subjekt im Irischen'. See Thurneysen, *Grammar*, 349 § 539.

(2) ród. *Cormac's Glossary* has the older disyllabic form *rout*.

da oeneochde. The manuscript of our text treats *de* as a separate word, and Binchy prints this word-division at *CIH* iii 893.25. However, the only way I can make sense of this passage is to take *oenechdae* to be one word, as in *Cormac's Glossary*. One would therefore distinguish a substantivized adjective *óenechd(a)e* 'horseman'. For the use of the suffix *-dae*, *-de*, see Thurneysen, *Grammar*, 220–2 §§ 347–8. Compare *óenechaid* 'horseman' (*Thes.* ii 104.27; 109.26), formed with the agent suffix *-id*.

fri echraite mendoto. It seems best to take *echraite* to be a compound of *ech* 'horse' + *raite* 'road'. A less likely alternative would be to emend to *echrada*, acc. plur. of *echrad* 'horses', an *ā*-stem collective formed from *ech* 'horse' (Thurneysen, *Grammar*, 169 § 264). *Mendoto* (*mennota*) is gen. sing. of *mennot* 'place of abode, locality, neighbourhood' (*DIL* s.v. *mennat*).

(3) lamraiti. I read *lámraite* following our text, rather than the *Cormac's Glossary* readings *lamrotæ* (YBL, H^1), *lamróta* (LB), *lamroitea* (Laud), *lamroda* (UM), *lamrotea* (H^2). The noun *raite* (fem. *iā*-stem or masc. *io*-stem) is well attested in the meaning 'road' in the law-texts, e.g. *CIH* i 239.15 = *AL* v 474.12 *glanad raite* 'clearing a road (or roads)', where the meaning is confirmed by the gloss *.i. smacht nemglanta na rōt* 'i.e. the fine for failure to clear roads'. *Raite* is also attested in the meaning 'land which has been cleared but not cultivated, rough grazing' e.g. *CIH* i 55.19; 57.19 = *AL* v 320.8; 320.19 *frīthe raiti* 'lost property found on rough land'. Here a glossator defines *raite* as land which lies between cultivated land (*faithche*) and wilderness (*dírann*). There is no doubt that only the meaning 'road' can apply in the case of *lamraite*.

This compound of *lám* 'hand' and *raite* 'road' is attested in legal commentary at *CIH* v 1549.37–8 in a discussion on the legal position with regard to a child found dead on a road (*mac for rōt*). Here a distinction is made between children found on a highway (*for rōt mōr*) or on small by-roads (*for lāimraitib beacaib*).

fri caii. Winter-hospitality (*cóe, cáe*) is briefly discussed on p. 320.

(4) tograiti. *Cormac's Glossary* has *tuagrottæ* (YBL), *tuagróta* (LB), *tuagroitea* (Laud), *tuaghroda* (UM), *tuagrotæ* (H^1), and *tuagrotea* (H^2). *Tóg-raite* is a compound of *tóg* (later *túag*) 'curve, bow, hoop' and *raite* 'road'. Siobhán Ní Laoire points out to me that a road must be constructed in curves up a steep hill, and this fits in with *do ascnam . . . sleibi* 'towards a mountain' at the end of the sentence.

do ascnam rait no sleibi. The only problem in this phrase is the identity of *rait,* which corresponds to *raotæ* (YBL), *roitoi* (LB), *raitea* (Laud), *raithea* (UM), *rotæ* (H^1), and *raodæ* (H^2) in the manuscripts of *Cormac's Glossary.* With a question mark, *DIL* R 124.34–6 takes it to be from *ruud* (later *rúd*) 'wood', which fits the context very well. I would read *roídeo,* unattested gen. sing. with palatal *-d-* of *ruud:* see discussion at *BB* 156–7. In all the other attested instances of the gen. sing. of this word, the *-d-* has been depalatalized, e.g. *Thes.* ii 290.11 *roída; CIH* i 57.28 *raoda*; 59.14 *raeda,* etc.

anaim ca*ch* d(u)ine i*m*ateit ca*ch* 'la mbliadhain. The manuscripts of the *Cormac's Glossary* version of this passage give a variety of readings, but both YBL and Laud support the emendation of *duine* in our text to *díni,* gen. sing. of *díne* 'herd (especially of cattle)'. Thus YBL has *anam cach dīne nō cach mīl immidrēt cach ala bliad*ain 'an animal from every herd or from every [kind of] beast which goes over it (the road) every second year'. No other examples are given in *DIL* of this use of *ainim(m)* 'soul, life' in the meaning 'animal'. However, the derivative *anmandae* is well attested in this meaning (see *DIL* s.vv. *anmandae,* 1 *anmann*).

For the driving of cattle and other livestock to summer pasture in hills or other rough land, see p. 43 above.

(5) doallad di ba for. I read *do-alla dí boí for* 'two cows fit on it'. For the non-historic final *-d* of *do-alla,* see L. Breatnach, 'The first third of *Bretha Nemed Toísech*', 3. I emend *ba* to *boí*; cf. *dī baī* (*Thes.* i 8.28). I take *for* to be the dat. 3 sing. masc.-neut. conjugated preposition of *for* (Thurneysen, *Grammar*, 275 § 437; cf. Kelly, 'A poem in praise of Columb Cille', 15 § 12c and note).

Cormac's Glossary has *talla dā boin fair* (YBL, Laud, UM, H[1], H[2]), *talla dī boi*n (LB). Thurneysen (*Grammar*, 217 § 340) says that *dá boin* is probably a neologism. The reading of our text supports his view.

iurrt[h]us in bo. The MSS of *Cormac's Glossary* have *iurthass* YBL, *iurtais* LB, H[1], *iurrus* Laud, *írus* UM, *iurthais* H[2]. As Thurneysen makes clear (*Grammar*, 270 § 429) the superior reading is *iurrus* of Laud; cf. *írus* UM. It is 3 sing. *s*-fut. of *oirgid* 'kills, injures' with the 3 plur. suffixed pronoun. The -*t*(*h*)- of the other MSS is analogical.

The author is pointing out that if the cow-track is too narrow, the young calves or yearlings cannot fit beside their mothers. In such circumstances, the cow behind may trample them.

§ 2. **teora aimsir i nglant*ar*.** Reading *aimsera* with *Cormac's Glossary*, I translate 'three times in which it (i.e. each of the roads) is cleared'. I take -*glantar* to be 3 sing. pres. indic. pass. of *glanaid*. However, some of the *Cormac's Glossary* readings indicate the 3 plur. *i nglanatar* 'in which they are cleared' and this is the translation given by Binchy, '*Aimser chue*', 19. Thus YBL has *a nglanait-* and Laud has *i nglanaiter*. On the other hand, LB and UM have *i nglantar*.

teora aiscsiu. It is necessary to emend *aiscsiu* of our text to *aicsin*, nom. plur. of *accais* 'cause, reason'. Likewise, § 3 *inna huisc-* must be emended to *inna* (*h*)*aicsin*. No spellings with metathesis of -*cs*- are given in *DIL* under *accais*, *aicsenach*, or *aicsendaid*.

Cormac's Glossary has *teora tucaite* here, with *tucait* 'cause, reason' rather than *accais*.

aimsir echruathair. Lit. 'time of horse-rush'. The roads must be cleared for the occasions when many horses are expected to use them. This would apply in particular when an assembly is to be held. O'Donovan's translation 'time of horse-racing' (*Sanas Chormaic: Cormac's Glossary*, 142) does not seem to fit in with the reference in § 3 to the soiling of horses going to an assembly.

aimsir cue. See discussion of this passage, at Binchy, '*Aimser chue*', 19–20.

aimsir catha. *Cormac's Glossary* has *aimser coctha* 'time of war'.

glanadh a feda. The LB version of *Cormac's Glossary* shows the expected lenition *glanad a ḟeda*.

O'Donovan's translation 'cleansing of its brushwood' makes good sense. One can compare the reference in *Di Chetharṡlicht Athgabálae* to the duty of *cartad raite* 'clearing out a road' (*CIH* ii 368.31 = *AL* i 122.14). According to glossators, this must be done *i n-aimsir cua* ⁊ *aenaig* 'in time of winter-hospitality or of an assembly' or *i n-aimsir coctha* 'in time of war'. One glossator describes the clearing of roads as 'cutting away their brambles and thorn-bushes' (*a ndrisi 7 a ndraigne do bēim dib*). Cf. *CIH* iii 1092.8–9.

A less likely explanation would be take *glanad a feda* to refer here to the surface-cleaning of wooden planks used to make roads. Such planks were used particularly in the construction of a causeway (*tóchar*) across boggy terrain (see p. 392).

gl*a*nad a uisc*i*. Lit. 'clearing [away] its water'. I take this to mean that water-filled potholes in the road should be repaired with an even surface of stones and gravel.

gl*a*nad a cocclaid. I take *coclaid* to be gen. sing. of *coclad*, a compound of *com-* 'with, double' + *clad* 'ditch, bank'.[9] One can compare *CIH* ii 401.26 = *AL* i 232.18 where a road (*rōt*) is glossed *.i. ima mbi clad* 'i.e. about which there is a ditch'. It is obviously important that the ditch on either side of a road should be regularly dug out so that water can drain off.

No clear attestations of the compound *coclad* in the sense of 'double ditch' are given in *DIL* C 283.46–53. There are some attestations of the expression *tine i coclad* (or *-aid*), e.g. *CIH* iv 1530.27 = O'Dav. 480 § 1590; *LL* iii 586.18129. O'Davoren takes there to be a connection with *coicilt* (earlier *coicill*) vb.n. of *con-ceil* 'spares, preserves'. Possibly, however, *tine i coclad* refers to a fire kept alive by being banked up.

§ 3. cadeat inna huisc- ara nglant*ar*. I read *haicsin*, following the UM and Laud versions of *Cormac's Glossary*. The other versions have *achuis* (YBL), *acuis* (H^1, H^2), *acsi* (LB).

At this point *Cormac's Glossary* continues: *arnā hēilnet a cairpthiu oc dul for caī* 'so that they (the roads) may not defile their chariots going on winter-visiting'. It is likely that a similar wording was present in an earlier version of our text, but was omitted by a copyist at some stage in the transmission.

arna uaillnid eochu. The form *uaillnid* is particularly difficult. *Cormac's Glossary* has *arnā huilled echradæ* (YBL), *arnélln- a echraide* (LB), *arnā huallnet echraide* (Laud), *arna uallnet eachraighe* (UM), *arnā hēillet echrudæ* (H^1), *arnā huillet echradæ* (H^2). I tentatively read *arná éilnea echu* 'so that it (the road) may not soil horses', taking *-éilnea* to be 3 sing. pres. subj. of *as-lena.*

a coclaid. In § 2, it seems best to take *coclaid* as gen. sing. of *coclad* 'double ditch'. I follow *DIL* C 283.54–8 in taking *coclaid* in § 3 to be verbal noun of *con-claid* 'digs up, excavates'.

The same form is attested in an Old Irish law-text which lists the 'offences of roads', i.e. the duties connected with road-maintenance for which fines must be paid if they are not performed. The text reads:

[9] *Clad* refers both to the excavated trench and the bank which has been built up with this material: see p. 374.

caithghe raite .i. ime fris a ænur 7 a slaige 7 a nurscartadh 7 a coclaidh 7 glanadh a ngreallach a n-aimsir cua 7 ænaigh 'offences of roads, i.e. a fence towards him alone (i.e. maintainance of a road-fence on his land?), and tree-removal, and clearing them, and digging out their ditches, and clearing their mires in the time of winter-visiting and of an assembly (*CIH* i 201.40–1 = *AL* iv 144.23–5). The first item in this list presumably stresses that the landowner is the sole person responsible for the fence which lies between his land and a road. In a footnote to *CIH* i 201.41, Binchy suggests emending *slaige* to *slaide* 'clearing'; I would take this to be a reference to the duty of removing trees and bushes from the road, cf. § 2 *glanadh a feda. Coclaid* must refer to the digging out of the roadside ditches. *Glanadh a ngreallach* refers to the removal of muddy or boggy material from the road, and its replacement by stones and gravel; cf. § 2 *glanad a uisci.*

arna h-laider nech f*or*. The YBL version of *Cormac's Glossary* has *arnā esarlaither nech fair* and the other versions are in close agreement. Binchy suggests reading *arna huallaider* in our text (*CIH* iii 893.35). In the light of the *Cormac's Glossary* readings, however, it seems better to read *arná hesarlaider nech for* 'so that no one may be thrown out on it'. But what is this verb? The connection with *as-cuir-* suggested at *DIL* s.v. *-esarlaither* seems doubtful.

As in § 1 (5), *for* is the dat. 3 sing. masc.-neut. conjugated preposition of *for.*

§ **4. ramhut.** In the law-texts *rámat* has two distinct meanings: (1) cleared area, open space, (2) road. One can compare the use of the English word *street,* which is normally applied to a major city thoroughfare, but may also be used of the paved area outside a farmhouse.

In our text, *rámat* is not included among the five types of road, and is given a separate treatment after the topic of roads has been concluded. It is defined as a cleared area in front of the king's *dún* or fortress; presumably it was a place where people gathered for public business or for audience with the king. In *Cormac's Glossary,* exactly the same definition is given: *aurscor bís ann ar dun rig* (YBL) 'a clearing which is in front of the fortress of a king'. However, *rámat* is there included among the seven types of way (*conar*).

In the law-text *Uraicecht Becc,* one of the qualifications of the higher grade of *briugu* 'hospitaller' is stated to be *trī rāmata lais* 'he has three roads' (*CIH* ii 654.24; v 1608.33–4; vi 2325.4 = *AL* v 78.8). The glossator explains this as *trī rōid ar amus a thige ara rēideann*[10] *cāch chuici* 'three roads to his house so that all can drive to him', i.e. his house should be accessible to wheeled vehicles. Liam Breatnach compares a passage in *Bretha Nemed Toísech* where a judge is advised to be *fer trī ruite, fer trī raite,*

[10]The MSS have *rætiund, reeann, redeand.*

fer trī rāmada 'a man of three paths(?), a man of three roads, a man of three highways' (*CIH* vi 2221.26; cf. *TBC* Rec. I 115.3801–2). The variety of ways mentioned may imply that his house should be accessible to various classes of person.

The law-text *Di Chetharślicht Athgabálae* makes a distinction between a *rout* (*rót*) and a *rámat* (*CIH* ii 401.13–14 = *AL* i 230.23). According to the glossator, the *rout* is a small road with a ditch, whereas the *rámat* is a large road without a ditch.

In *Bretha Étgid*, the word *rámat* is used in a context which would suit either the meaning 'road' or 'open space'. This is in the phrase *cach rig a ramut* 'to every king his *rámat*' (*CIH* i 293.8 = *AL* iii 304.25). The glossator takes it to refer to the king's entitlements with regard to royal roadways, and distinguishes a main road (*prímrót*) and a minor road (*forrót*). This would certainly fit in with other legal references to royal associations with roads (p. 391). On the other hand, the definition of *rámat* in our text raises the possibility that the author is thinking of the king's control of the open space in front of his fortress.

cach duine dua taicmaic tir. The compound *do-acmaing, do-acmaic* [* *to-ad-com-icc*] is well attested in the meaning 'reaches, extends to'. In the manuscript *taicmaictir* is written as one word (thus, *CIH* iii 893.36). However, it is clear from *Cormac's Glossary* that the division should be *taicmaic* 'reaches' and *tír* 'land'. Thus the YBL version has *cach comaithech asa tīr dotroich dleg*ar *de a glanad* 'every neighbouring farmer whose land reaches it is obliged to clear it'.

Text 5

Calves with two mothers

It is clear from sources such as the text on *comingaire* 'joint-herding' that it was regular for livestock belonging to two or more farmers to be looked after in the same herd. This arrangement has an obvious potential for disagreement if an animal belonging to one of the farmers is killed by another animal, particularly if the culprit cannot be identified (see p. 176). The text edited here treats of another legal quandary: how to deal with the situation where a calf belonging to farmer A is nourished by a cow other than its mother which belongs to farmer B.

Two versions of this Old Irish text survive. A glossed version is found at p. 430[a] of H 3. 18 (now no. 1337) in Trinity College Dublin,[1] and is printed at *CIH* iii 967.35–968.15. I provide a translation below. A version without glosses is found at ff. 1[d]–2[a] of the British Library [formerly Museum] manuscript Egerton 88, which was compiled by Domhnall Ua Duibhdhabhoireann (O'Davoren) and his pupils between AD 1564 and 1569. This version is printed at *CIH* iv 1267.17–31. Apart from the absence of glosses, it agrees fairly closely with the H 3. 18 version. However, it sometimes has better readings, so I reproduce it below. In Egerton 88, §§ 1–5 are written in the large letters which are normally used in legal manuscripts to distinguish Old Irish law-text from later glosses and commentary. § 5a is written in small letters: see note on § 5a below.

In the discussion I refer to the H 3. 18 version as **H**, and the Egerton 88 version as **E**.

Text (H 3. 18)

Teora breatha ma[i]c da bo ⟨.i. laegh da bo⟩

§ 1. Bó f*or*li*ng* a[r][2] laegh a seitche ⟨.i. carus laegh bo eile a naenbuailidh ł a naentig⟩; is *m*athair-side laigh ⟨.i. is leisin math*air* b*eir*is i*n* laegh-sin⟩. Ni la boi*n* ailes ⟨.i. nocho leisin *m*boi*n* roailest*ar*, ge rochar⟩.

§ 2. Bó b*eir*is marbhlaegh dob*er*ar[3] ar laegh naile ⟨.i. is leisin *m*boin b*er*es a haenur eisidhe⟩; is math*air*-side[4] namá, ar ni ar les laigh dognith*er acht* ar les la*chta*[5] na bó ⟨.i. rug i*n* marblaegh⟩.

[1] *I have indicated the glosses by angle-brackets ⟨ ⟩. In the manuscript, glosses and text are undifferentiated.*

[2] ar E.

[3] dober E.

[4] maithir-side E. *Read* máthar-side.

[5] blechta E.

§ 3. Bó do*beir* nech f*or* laegh arhaile ⟨.i. a cheile⟩, i*na* égnairc ⟨.i. i*na* egmais i*n*tí isa laegh hé⟩, mana nardama i[c] coin*n*dhíul[6] ⟨.i. mana roib i*n* fer isa laegh he i*na* aititi*n* ag díul fo boi*n* i*n* fir eile⟩, is leis a aenur in laegh; ma doroghma[7] im*orro* ⟨.i. dia roib i *n*aididi*n* a laigh fuithe⟩, is roind i *n*de etarru ⟨.i. i*n* laegh, cen*mo*tha a logh i*n* tan t*u*cadh fo boi*n* hé do br*eith* d'i*m*arcraid d'fir bunaid⟩.

§ 4. Os masgaraidh co*m*amsa,[8] focerdad crandchur etarru dus cía da lína dob*er*ar[9] lulghaidh dara eisi na bó a[c] coi*n*diul[10] ⟨.i. is a*n*d dognith*er* crandchur etarru i*n* inbaidh b*er*ar l*et*hlogh i*n* laigh o*n* fir isa bo dob*er*ar fair in tan dob*er*ar fae⟩.

§ 5. Ocus manasgara comamsa,[11] [is][12] na bó dasaighed[13] i *n*gamnachus i*n* laigh; is fer[14] na bo-sin: i*n*erenar[15] ní ass diaraile. Mad i*n*dlaegha[16] a n[d]ís, is crandchur dodaran*na*[17] doib ⟨.i. focul tig arin corp⟩.

§ 5a. Mad i*n*dlaegh[a a ndis is cranchor],[18] ł ma gamnacha a ndhís iad is la fer mbunaid in laegh mana t*u*ca i*n* fer eile a l*et*hlogh dó i*n* tan dob*er*ar fo boi*n* he; 7 da t*u*ca, is crandchur focerdad f*or*sin laegh; gidbe dib aga mbía i*n* gamnach, is e dob*eir* ineraic diaraile asi*n* lægh.

[6] a cuindiul E.
[7] ma rodma E.
[8] os ma caruit comsa E.
[9] dobera E (*sic leg.*).
[10] ocundiul E.
[11] osma nes carait comsa E.
[12] is E, *om.* H.
[13] doeisit E.
[14] 7 is ferr E. *Read* os fer.
[15] ni erenur E.
[16] indlaeg E.
[17] nodronda E.
[18] *The material between square brackets here has been supplied from* E.

TRANSLATION

Three judgements of the 'child' of two cows ⟨i.e. the calf of two cows⟩

§ 1. A cow which takes over the calf of her companion ⟨i.e. which loves the calf of another cow in the same enclosure or cow-house⟩; the latter is the mother of the calf, ⟨i.e. that calf belongs to the mother which bears [it]⟩. It does not belong to the cow which rears [it], ⟨i.e. it does not belong to the cow which reared [it], though she loved [it]⟩.

§ 2. A cow which bears a dead calf, which is put onto another calf, ⟨i.e. it belongs to the cow which bears [it] only⟩; it belongs to the mother alone, for that is not done for the sake of the calf but for the sake of the milk of the cow ⟨i.e. which bore the dead calf⟩.

§ 3. A cow which somebody places on the calf of another ⟨i.e. of his neighbour⟩ in his absence ⟨i.e. in the absence of the person whose calf it is⟩, if he has not acknowledged it at sucking ⟨i.e. if the man whose calf it is has not allowed it to be sucking under the cow of the other man⟩, the calf belongs to him alone; if he has acknowledged, however, ⟨i.e. if he has allowed his calf to be under her⟩, there is division in half between them, ⟨i.e. of the calf, except that its value when it was brought under the cow is to be given in addition to the original owner⟩.

§ 4. If they part from joint farming(?), they cast lots between them to find out which of them should give a milch cow instead of the cow which is suckling ⟨i.e. when lots are cast between them is when half the value of the calf is taken by the man who owns the cow which is put on top of it (the calf) at the time that it is put under her⟩.

§ 5. If they do not part from joint farming(?), the calf belongs to the cow which remained without a new calf; as for the owner of that cow: something is paid from him to the other. If they are both in calf, lots divide it for them, ⟨i.e. a verdict which applies to the body⟩.

§ 5a. If they are both in-calf cows there is lot-casting, or if they are both strippers, the calf belongs to the original owner, unless the other man gives its half value when it is put under the cow; and if he gives it, they cast lots on the calf; whichever of them owns the stripper gives payment to the other for the calf.

TEXT (EGERTON 88)

Teora br*eth*a m-c da bo

§ 1. Bo forling ar laog a seitce; is maithuir-side laog. Ni la boin ailes.

§ 2. Bo beris marblaogh dober ar laog naile; is maithir-side nama, ar ni ar leas laog dognithar *acht* is ar les blechta na bo.

§ 3. Bo dob*eir* nech for laog araile ana ecnairc, man nardama a cuindiul; ma rodma im*orro* is roind i nde aturru.

§ 4. Os ma caruit comsa focerdat cran*n*cor eturr*u* dus cia dia lina dobera lulgaid dar eisi na bo ocundiul.

§ 5. Os manes carait comsa is na bo doeisit i ngamnachus in laog ⁊ is ferr na bo-sin ni erenur ni as diaraile. Mad indlaog a ndis is cran*n*cur nodronda doib.

§ 5a. (*in smaller script*) Mad indlaogha a ndis, is cranchor; no madh gamnach*a* a ndis, is la fer mbun*aid* in laogh mana thabhr*a* in fer .ii.e a lethlogh dó in tan dobeir fo boin é; ⁊ da tuga, is cran*n*chor focerdat forsan laogh; cidbe dib aga mbé in gamnach is é dobeir ineiric diar.ii.e asan laogh.

Discussion

Teora breatha ma[i]c da bo. The manuscripts have *mac* (H) and *m-c* (E). I read gen. sing. *maic* rather gen. plur. *mac*, as there is no nasalization of the following *dá*. *Mac* 'son, child' is occasionally applied to the offspring of cattle. For example, a Life of Saint Cíarán of Saighir refers to an ox (*dam*) which was the 'child' of a cow (*mac bó*) belonging to Saint Brénainn (Plummer, *Bethada* i 122 § xxxi (61); trans. ii 118). The phrase *mac dá bó* 'child of two cows' also occurs in legal commentary at *CIH* vi 2113.20 = Dillon, 'Stories from the law-tracts', 44 § iii.

In a Latin Life of Saint Cíarán of Clúain mic Nois such a calf is described as *vitulus inter duas vaccas* 'a calf between two cows' (Plummer, *Vitae* i 203 § viii). While being milked, both of them used to lick this calf (*ambe diligebant eum*). A calf of two cows (*vitulus vaccarum duarum*) is also mentioned in a Latin Life of Saint Cainnech (Heist, *Vitae*, 197–8 § 59). In *CIH* vi 2217.30 the *mac dá bó* is described in the poetical phrase *dalta dá deil* 'fosterling of two teats'.

Tēora bretha maic dā bō 'three judgements of the "child" of two cows' is clearly the title of this short text, but it is difficult to work out which exactly are the three judgements. I suggest that the first judgement deals with the case of a cow taking over another's calf of her own volition (§ 1), the second with the deliberate placing of a calf on another's cow in various circumstances (§§ 2–3), and the third with the farming agreement, and how ownership of the calf is apportioned (§§ 4–5a).

§ 1. **bo forling a[r] laegh.** Some cows have a particularly strong maternal instinct, and are liable to take over the calves of other cows, even when they already have their own calves. This paragraph emphasizes that the ownership of the calf stays with its actual mother even if it has been nourished by another cow. The literal meaning of the compound *for-ling* is 'leaps on', and this is the only example of its use in connection with the taking over of calves by cows: see *DIL* s.v. *for-ling*.

It is probable that the original text had *for lóeg*, as one would expect a compound with the preverb *for-* to be followed by the preposition *for* rather than *ar*; cf. *CIH* ii 593.15 *foriada-side forsa ndliged*; *CIH* vi 2199.10

= Thurneysen, 'Aus dem irischen Recht III', 361 § 46 *fortongat for bīu.* For scribal variation between *ar* and *for* elsewhere in the text, cf. § 2 *ar laegh* H, *ar laog* E; § 3 *for laegh* H, *for laog* E.

a naenbuailidh ł a naentig. It is noteworthy that the H glossator envisages that cows belonging to different owners may be present in the same enclosure or cowhouse.

§ **2. bó b*eir*is marbhlaegh.** It is obviously important for the farmer to ensure that a cow whose calf is stillborn should take on another calf. This could be a twin or a calf whose mother has died or is sick, or a calf whose mother has rejected it. If a substitute calf is not provided, the cow will give no milk (see p. 38). As our text points out, 'that is not done for the sake of the calf but for the sake of the milk of the cow'.

is math*air*-side nama. It seems best to read *is máthar-side namá* 'it (the calf) is of the mother only', emending *math*air H, *maithir* E to gen. sing. *máthar.* The H gloss has *is leisin mboin beres a haenur ēisidhe* 'it belongs to the cow which bears [it] alone'.

§ **3. mana nardama i[c] coin*n*dhíul.** I would restore Old Irish *mani n-árdama oc cuindíul,* taking *-árdama* to be perfective 3 sing. pres. subj. (prototonic) of *ad-daim* 'acknowledges, permits, endures', with 3 sing. masc. infixed pronoun, Class A (referring to *lóeg*). The form *cuindíul* is dat. sing. of *cuindéol,* a compound of *com-* + *déol,* verbal noun of *dinid* 'sucks'; there may be an attestation of *cuindéol* (*cuineol* MS) in relation to pigs at *CIH* iv 1479.28 = O'Dav. 254 § 360. The preverb *com-* often has the sense 'co-, joint' (Thurneysen, *Grammar*, 502–4 § 830), so one might expect *cuindéol* here to mean 'co-sucking', i.e. the calf is taking milk from two cows. However, this meaning does not suit the context. From the legal point of view, it is irrelevant if the calf is continuing to suck from its own mother; what matters is the fact that it is sucking from a cow belonging to another farmer. It seems best therefore to take *cuindéol* here to be merely a more emphatic form of the simplex *déol*; cf. *boingid* 'breaks', *con-boing* 'smashes, breaks completely'. It is noteworthy that the glossator simply has *ag díul* 'sucking'.

is leis a aenur in laegh. This phrase is absent in E, but must have been in the original, as it is essential for the sense of the passage.

madoroghma. I read *ma 'd-rodma* (cf. *ma rodma* E) for *ma ad-rodma,* perfective 3 sing. pres. subj. (deuterotonic) of *ad-daim* 'acknowledges'.

§ **4. os masgaraidh co*m*amsa.** The problem here is the divergence between the readings of the two MSS: E's *os ma caruit comsa* is difficult to reconcile with H's *os masgaraidh com*amsa. On the basis of both MSS, I suggest reading *os ma scarait comsai* 'and if they part from joint farming', referring to the two farmers whose cattle have been looked after together. *Coms*(*a*)*e* (later *coimse*) is well attested in the meaning 'farming partnership, joint husbandry', e.g. *CIH* iii 778.28 = *CG* 4.96; *CIH* ii 490.1 = Thurneysen, 'Aus dem irischen Recht I', 372 § 34; see also *DIL* s.v. *coimse.*

I take *comamsa* of H to be a scribal error for *comsa*, but it is also possible to view it as the scribal substitution of a synonym, cf. § 2 *lachta* H, *blechta* E. *Com(m)ámus* (gen. sing. *comámsa*) is from *com-* 'co-' + *mám(m)us* 'yoke, duty', and is used of a partnership or union, especially marriage.

In his discussion of this passage at *SEIL* 14, Thurneysen indicates a preference for E's reading *carait*, nom. plur. of *carae* 'friend, relative'. As farming partnership must often have involved kinsmen, this interpretation is also possible.

focerdad crandchur etarru. Many issues relating to farming law are resolved by the casting of lots (*crandchur*), e.g. *CIH* ii 577.21 = *BB* 134 (note to § 34). In this case, the herd is being broken up because the two farmers have parted from their joint husbandry (*comsa*). In the normal course of events, this would result in the separation of the *mac dá bó* from the cow which has nourished it. As this cow's milk-yield may depend on the presence of this calf (§ 2 *ar les lachta na bó*), separation is obviously not in the interest of the owner. Lots are therefore cast to see who is to get this cow (and calf). He in turn reciprocates by restoring a milch cow (*lulgach*) – with another calf – to the other farmer. Consequently, neither farmer loses out.

§ 5. ocus manasgara comamsa. E has *os manes carait comsa.* As in § 4, I suggest combining elements from both H and E, and read *os mani scarat comsai* 'and if they do not part from joint farming'.

dasaighed. I can make no sense of *dasaighed* H, which could be from *do-saig* 'approaches, seeks'. The E reading *doeisit* suggests Old Irish *do-essid*, 3 sing. perf. of *saidid* 'sits, remains'.

gamnachus. *Gamnachas* is an otherwise unattested abstract noun from *gamnach* 'stripper', i.e. a cow which has an older calf (*gamain*), but has not come into calf again in a particular year (see p. 40). It is logical that the calf under discussion in this text – the *mac dá bó* – should be assigned to the cow which is not due to calve.

inerenar. *In-érenar* is 3 sing. pres. indic. passive of *in-éren* 'compensates'; cf. *CIH* ii 510.32 = *SEIL* 38 § 17 *ineranar.* E has *ni erenur ni as diaraile* 'nothing is paid from him to the other', with exactly the opposite meaning to the H version. However, this seems unfair on the owner of the cow which is not a *gamnach*; so I take *ni* E to a scribal miscopying of *in.* This interpretation is confirmed by the 5a commentary in both manuscripts, which states that the owner of the *gamnach* gives a payment (*inéraic*) for the calf to the other farmer.

is fer na bo-sin. E has ⁊ *is ferr na bo-sin.* I read *os fer inna bó-sin* 'as for the owner of that cow'.

is crandchur dodaran*n*a doib. If both cows are in-calf (*indlaeg*), ownership of the *mac dá bó* is determined by lot-casting. The H gloss describes this as *focul tig arin corp* 'a verdict which applies to the body'. I take *focul* 'word, award, test, etc.' to refer to the verdict provided by the lot-casting;

corp presumably refers to the (living) body of the calf. For this usage of *do-icc*, see *DIL* D ii, 301.25–32.

I would emend *dodaranna* H, *nodronda* E to *nod-ranna* 3 sing. pres. indic. of *rannaid* 'divides', preceded by *no-* + *-d-*, 3 sing. masc. infixed pronoun Class C (agreeing with *lóeg*).

§ 5a. In E, §§ 1–5 are unglossed and written in the large script normally reserved in legal manuscripts for Old Irish text, whereas § 5a is written in the smaller script used for glosses and commentary. The scribe of the H version, on the other hand, makes no distinction between text, glosses and commentary. I take § 5a to be commentary based on §§ 4 and 5. The use of the independent pronouns *iad* (H) and (*h*)*é* (H, E) indicates that the language is Middle Irish.

madh i*n*dlaegh[a a ndis is cranchor]. As it stands, the H commentary treats in-calf cows and strippers as having the same entitlement, which contradicts § 5. The full version is to be found in E, and I supply the missing words from this manuscript.

Text 6
Land-values

Under the heading '*Tír cumaile*' in *Ériu* 22 (1971) 81–6, Gearóid Mac Niocaill provided an edition and translation of an Old Irish text on land-values from the Trinity College Dublin manuscript H 3. 18 (*CIH* ii 675.18–676.16) together with a later commentary on the same topic, also from H 3. 18 (*CIH* iii 843.3–21). The short section of commentary edited here belongs with these two texts. It is found at p. 432[b] of H 3. 18 (*CIH* iii 972.24–9) and lists various factors which increase the value of a *cumal* area of land. Values are given in cows.

For a general discussion of the legal material on land-values, see p. 394 above.

Text

§ **1.** Ceithre ba thor*m*aiges ar thír cum*ail*e dai*n*gen cluid *no* coradh. .iiii. ba ar lansodh muili*n*d, dí ba ar lethsod muili*n*d.

§ **2.** Bó ar thibraid na traighen*n*; 7 dí ba ar mhéin íarai*n*d; .iiii. ba ar mei*n* umha. Di ba ar inber eisg nanta. Bó ar slighi nairchi*n*dta go rú ł a roilbe ł ar romuir.

§ **3.** Bo ar charraigh thoraigh dia mbent*ar* duileasg *no* femnach; 7 na tri ba adrubartais ar p*r*imród dochum chille *no* dui*n*e. Conadh .iiii. ba .xxit. uile sin ar thír chum*ail*e.

Translation

§ **1.** A securely constructed bank or wall adds the value of four cows to a *cumal* area of land. Four cows for a full mill-weir; two cows for a half mill-weir.

§ **2.** A cow for a well which does not dry up; and two cows for an iron mine; four cows for a copper mine. Two cows for an estuary of permanent(?) fish. A cow for a fixed road to a wood or onto a mountain or as far as the sea.

§ **3.** A cow for a productive rock from which dulse or [other] seaweed is cut; and the three cows which you mentioned for a principal road to a church or fort. That is twenty-four cows in all for a *cumal* of land.

Discussion

§ **1. daingen cluid no coradh.** The literal meaning is 'a secure structure of a bank or wall'. *Daingen* 'secure, firm' may be used substantively of a boundary, e.g. *CIH* ii 422.15 = *AL* i 268.19. See *DIL* D i, 28.33–6 and R. B. Breatnach, *Seana-chaint na nDéise II*, 135 s.v. *daingean.*

The other commentary has *.iiii. bæ ar lānimbe claide ł corad* (*CIH* iii 843.11) which I would translate 'four cows for a full boundary of a bank or wall', rather than 'four cows for a full fence, ditch or stone wall' (Mac Niocaill, '*Tír cumaile*', 86). One can compare *ime di clud no coraid* 'a boundary of bank or wall' in the Old Irish text (*CIH* ii 676.4 = Mac Niocaill, '*Tír cumaile*', 82.31).

lethsod muilind. This is the only instance in legal material where a distinction is made between a full mill-weir (*lánṡód muilinn*) and a half mill-weir (*lethṡód muilinn*). The other commentary merely states *dā baī ar sōdh muilinn* 'two cows for a mill-weir' (*CIH* iii 843.12–13 = Mac Niocaill, '*Tír cumaile*', 83.19). Presumably our commentator is contrasting a weir which spans the whole river and one which projects only half way across the river. For a discussion of fish-weirs and mill-weirs (sometimes dual-purpose structures) see p. 287.

§ **2. dí ba ar mhéin íaraind; .iiii. ba ar mein umha.** The Old Irish text does not make our commentary's distinction between the value of an iron-mine and a copper-mine, and gives 5 *séts* (= 2½ cows) as the additional value provided by the presence of either type of mine (*CIH* ii 676.6 = Mac Niocaill, '*Tír cumaile*', 82.33).

ar inber eisg nanta. Estuaries are usually good places for fishing, and two law-texts refer to *cáin inbir* 'the law of the estuary' which regulated fishing-rights (see p. 289). The Old Irish text on land-values states that 5 *séts* are added to the value of a piece of land if it is next to an estuary (*CIH* ii 676.11–12 = Mac Niocaill, '*Tír cumaile*', 82.y). This is slightly more than the two cows of our commentary.

I am uncertain about the meaning of *nanta*, and tentatively suggest that *anta* is gen. sing. of *anad*, vb.n. of *anaid* 'stays, remains'. The commentator may be stressing that the estuary should be a regular rather than an occasional source of fish.

ar slighi nairchindta. According to the Old Irish text (*CIH* ii 676.13 = Mac Niocaill, '*Tír cumaile*', 83.1), a cow is added to the value of the land if there is a fixed road which reaches a wood or sea or mountain (*rōt n-airceand rosaid rūd nō romuir no roillbe*). A 'fixed road' presumably refers to a properly constructed regular access-route. Our commentator calls it a *slighe airchindta* with the same meaning. See *DIL* s.vv. *airchenda, airchenn.*

The value of roads is treated rather differently in the other commentary (*CIH* iii 843.11–12 = Mac Niocaill, '*Tír cumaile*', 83.16–17). Three cows are added to the value of the land if there is a road which leads to both wood and mountain (*rosoich rūdh* [7] *roilbe*). However, if it only leads to one of these, the added value is a cow and a dry heifer (*samaisc*), i.e. 1½ cows.

§ **3. ar charraigh thoraigh.** Read *ar charraig thoraid* 'for a productive rock' (see *DIL* T 254.46–51). Our commentary gives one cow as the additional value of a seaweed-producing rock, whereas the Old Irish text gives three cows as the value of direct access to the sea where there is such a rock (*CIH* ii 676.9 = Mac Niocaill, '*Tír cumaile*', 82.36).

primród dochum chille no duine. This corresponds to the Old Irish text's *sligi adcumaid o thīr co flaith nō mainistir* 'a road which extends from the land to a lord or monastery' (*CIH* ii 676.8 (and footnote[d–d]) = Mac Niocaill, '*Tír cumaile*', 82.35). In both text and commentary the added value is three cows.

adrubartais. The form is a Middle Irish 2 sing. perf. of *as-beir* 'says, speaks' (*DIL* A 426.7–17). The commentator is presumably addressing a colleague in a law-school who has previously worked on this material.

.iiii. ba .xx.it. The number of cows listed in this commentary totals twenty-four. The commentator may be thinking of the twenty-four cows which represent the value of a *cumal* of best arable land in the Old Irish text (*CIH* ii 675.34 = Mac Niocaill, '*Tír cumaile*', 82.22). However, in the Old Irish text features such as roads, mines, etc. add extra value to a piece of land, and are not counted as part of the land's basic worth.

Text 7

Horse-qualities

This Old Irish legal passage on the qualities which a buyer should look for in a horse is to be found on f. 20[a] of the fifteenth-century manuscript no. 1242 (23 P 3) in the Library of the Royal Irish Academy, Dublin (*CIH* iv 1235.19–27). It is written in a script intermediate in size between that used for ordinary legal text, and that used for glosses and commentary. Some letters on the left-hand margin are illegible or indistinct due to rubbing. A transcript was published by Smith, 'Further light on the *Finnsruth Fithail*', 331.

Text

§ 1. (1) Co ber breith um techta eich? (2) Each mor sla*n* og urasa ⟨.i.[1] im leigsin a marcaidh fuirri⟩, (3) an ardceandach airreachtach, (4) beoc*r*aide brui*n*dleathan bairneach, (5) beac a tiathug*ud*, (6) suiligh slea*m*ain sei*m*chosach socineoil, (7) slan gaiti as, slan goisti, slan [t]aib,[2] (8) soraig a thucht,[3] (9) socomail i lai*m*.

§ 2. (1) Ni bi cnoc ana leac uma druim, (2) nibi mamdrui*m*neach, (3) nibu calaceimneach, (4) nib roisil, (5) nibi roard, (6) nib ocheall, (7) nib i*m*lei*m*neach, (8) nib robeil, (9) nib docearrdach, (10) nip leasc, (11) nip losc,[4] (12) nib luach, (13) nip luatchairceach, (14) ni bi cu cua na fothach na hetrocht na crithach, (15) forlim[5] slan soi*m*ri*m*e somul; (16) dia mbe nach ae, as athchuirthe no is fuillithe.

Translation

§ 1. (1) How should I give judgement about the proper qualities of a horse? (2) A large healthy young docile horse ⟨i.e. with regard to allowing its rider to mount it⟩, (3) swift, high-headed, mettlesome, (4) lively-hearted, broad-chested, fiery, (5) small its ——(?), (6) gentle, calm, narrow-legged, of good stock, (7) healthy with regard to castration, healthy with regard to halter, healthy with regard to side, (8) of pleasant appearance, (9) easy in the hand.

§ 2. (1) There is not a lump or stone about its back, (2) it is not yoke-backed(?), (3) it does not walk stiffly, (4) it should not be too low, (5) it should not be too tall, (6) it should not be ——(?), (7) it should not be in the habit of leaping about, (8) it should not have an excessively large mouth,

[1] *In his transcription Binchy has an* ⁊ *before the* .i. *but I cannot see it in the MS.*

[2] *There is a line-break after* slan. *A letter is illegible due to rubbing before* aib. *Read* taíb?

[3] *About three letters have been erased after* thu, *probably as a correction of a dittography.*

[4] *The initial* l *is indistinct.*

[5] *There may be one or two letters obscured before* forlim.

(9) it should not be clumsy, (10) it should not be lazy, (11) it should not be lame, (12) it should not be a kicker, (13) it should not be tail-swishing(?), (14) it is without a hollow or glanders or weakness or shivering, (15) sound leaping, a fleet [horse] of good riding; (16) if there be any of these, it is to be returned or compensation is to be paid.

Discussion

§ 1 (1) **Co ber breith.** The phrase *Co ber br(e)ith?* 'how should I give judgement?' is particularly common in the fragmentary law-text *Findśruth Fíthail*, e.g. *CIH* iii 786.25; 787.13, 20, 24; 788.12, 30; 789.1, 4, 6; vi 2131.1; 2132.29; 2135.4, 37; 2136.40ff. It also occurs in *Bretha Nemed Déidenach* (*CIH* iii 1113.19, 33 = E. J. Gwynn, 'Privileges', 16.11, 16.27). See Smith, 'Fithal and Fland Fína', 34–8; Smith, 'Further light on the *Finnsruth Fithail*'.

One could also read *co bér brith* 'how shall I give judgement?'.

(2) **og.** I take *og* here to be for OIr. *oac* (*óc*) 'young' rather than *og* 'pure, whole'. Cf. Text 1 § 3 (1), § 4 (1).

.i. im leigsin a marcaidh fuirri. This Modern Irish gloss was added to explain the word *urasa* (*airassa*) 'easy (to mount)'. *Marcaidh* is for *marcaigh*, gen. sing. of *marcach* 'rider'. *Léigsin* is a late form of the verbal noun of *léicid* 'leaves, lets, allows': see *DIL* L 81.56.

(3) **ardceandach.** The adjectival compound *ardchenn* 'high-headed' is well attested in early Irish texts as a complimentary epithet of horses, e.g. *ech líath . . . ardchend* 'a grey . . . high-headed horse' (*LU* 262.8659–61 = Windisch, 'Das Fest des Bricriu', 278.12–14 § 49). For other instances see *DIL* A 387.29–32. The form *ardceandach* (= *ardchennach*) is a derivative in *-ach* with the same meaning. Virgil, *Georgics* (ed. Fairclough), 3.79, likewise advises that a horse should have a high neck (*ardua cervix*). The abstract *cennairde* 'high-headedness' is attested in legal commentary at *CIH* iv 1275.39 (= vi 2096.27), apparently in the context of identifying a stolen horse.

In a medieval treatise on horses, *gobairde* 'highness of the muzzle, "star-gazing"', is classed as a defect.[6]

(4) **beocraide.** This is a compound of *béo* 'living, lively' + *cr(a)ide* 'heart'. When used of a horse, it must have some such meaning as 'plucky, high-spirited, valiant'. It is also used in a description of horses in *TBC* LL 80.2918; Cecile O'Rahilly translates 'with lively heart'.

bruindleathan. This is a compound of *brú* 'belly, chest' + *lethan* 'wide'. Another instance of the use of this compound in a description of a horse is to be found in *LU* 329.10812 = Bergin and Best, '*Tochmarc Étaíne*', 174 § 3. Varro, *Rerum rusticarum* (ed. Hooper and Ash), 2.7.5, advises that a

[6] Ó Cuív, 'Fragments of two mediæval treatises', 48 § 11.

horse should have a broad chest (*pectus latum*), as does Xenephon, 'On the art of horsemanship' (ed. Marchant), 1.7.

bairneach. *Bairnech* is an adj. from *barae* 'anger, passion, vehemence'. It is also used as a complimentary epithet of a horse in *LU* 262.8667 = Windisch, 'Das Fest des Bricriu', 278.20 § 50. It means 'fiery, spririted, mettlesome'.

(5) beac a tiathug*ud*. I have no suggestions for *tiathugud*.

(6) sei*m*chosach. The compound *séimchosach* 'narrow-legged' is likely to refer here to a horse whose slim legs indicate a capacity for speed.

socineoil. A horse should be *sochineóil* 'of good stock'. Early Irish texts contain many references to good breeding in horses, and particularly emphasize the importance of foreign stock: see p. 90.

(7) slan gaiti as. I am indebted to Liam Breatnach for suggesting that good sense can be obtained by taking *gaiti* (*gaite*) as gen. sing. of *gat*, vb.n. of *gataid* 'steals, removes', i.e. castrates. One can compare the reference in Text 2, § 1 (21) to *reithi iarna slāini* 'a ram after recovery [from castration]'.

slan goisti. This probably means that the horse has not been chafed or injured by its halter (*goiste*).

slan [t]aib. One could take the illegible letter to be a *t*, and read *slán taíb* 'healthy with regard to side', i.e. the horse has no injuries on its side from beatings, etc.

(8) soraig a thucht. Read *soraid a thucht* 'pleasant its appearance'.

(9) socomail i lai*m*. *Socomail i láim* 'easy in the hand' presumably means that the horse responds well to pressure from the rider's hand on the reins. For this adj., see *DIL* s.v. *socamail*.

§ 2. I start a new paragraph at this point, though the scribe does not provide *Ní* with a capital *n*. However, the text falls naturally into two halves: § 1 (1)–(9) dealing with horses' good qualities, and § 2 (1)–(16) dealing with horses' faults.

In the text on the defects of cattle and sheep edited above (Text 1 §§ 2, 4, 8), the negative 3 sing. consuetudinal pres. *níbi* and the negative pres. subj. *níp* are used with no apparent distinction in meaning. Likewise in § 2 of this text, the author appears to have switched between *níbi* and *níb* (*níp*). However, in some cases, a later scribe may be responsible for inaccurate transcriptions. For instance, it seems unlikely that the original text had *níb roísel, níbi roard*. One would expect either *níbi roísel, níbi roard* or *níb roísel, níb roard*.

(1) Ni bi cnoc ana leac uma druim. There are other references in our sources to animal-diseases characterized by lumps or pustules (*cnuicc*): see p. 201 (black-quarter?) and p. 203 (cow-pox?). *Lecc* may refer to some form of bony growth on the back; one can compare the horse-defect called *lec ós crú* 'stone above the hoof', which I suggest may be ringbone: see p. 211. For *anā* (= *iná*) 'than', see *DIL* I 213.19–28.

(2) nibi mamdrui*m*neach. I read *mamdruimneach* rather than Binchy's *maindruimneach.* Perhaps *mámdruimnech* 'yoke-backed' means that the horse's back has suffered from the effects of the yoke (*mám*). Alternatively, it may be a use of *mám* 'hill, lump, gap'.

(3) nibu calaceimneach. It seems best to read *níbi caladchéimnech* 'it is not stiff-walking'. However, it could also be the pret. *níbu* used jussively.

(4)–(5). Varro, *Rerum rusticarum* (ed. Hooper and Ash), 2.7.4, likewise advises that a horse should be neither too big nor too small.

(6) nib ocheall. I have no explanation for the otherwise unattested adjective *oche(a)ll*, obviously referring to some horse-defect. Possibly emend to *dochéill* 'senseless'.

(7) nib i*m*lei*m*neach. The adj. *léimnech* 'leaping' is well attested as a good quality in horses, e.g. *TBC* LL 80.2917 *ar dá n-echaib lúatha lémnecha* 'behind two speedy leaping horses'. Here, however, *imléimnech* 'leaping about' is clearly regarded as a defect. Presumably, the author is thinking of a horse which is of a highly nervous temperament, and constantly bucks and rears when mounted.

(8) nib robeil. Perhaps read *robéoil* 'with too large a mouth'. One can compare the adj. *sobéoil* (*sobéil*) 'with a good mouth': see *DIL* S 310.64–5. Having a slender mouth (*gopchóel*) is regarded as a good feature in a horse (*LU* 260.8593 = Windisch, 'Das Fest des Bricriu', 276.4 § 45).

(9) nib docearrdach. *Cerd(d)ach* means 'skilful, accomplished', so it is likely that *docherdach* refers here to an awkward clumsy horse.

(13) nip luatchairceach. The meaning of this compound is uncertain. The second element seems to be *cairchech*, an adj. from *cairche* 'hair (esp. of animals), tail'. One could take it to be *lúath-chairchech* 'swift-tailed', i.e. a horse whose constantly swishing tail indicates nervousness or bad temper. *DIL* s.v. *lúathchairchech* quotes Plummer's suggestion that this compound means 'foot-hairy', in which case it would be necessary to emend to *lúa-chairchech.*

In descriptions of the horses of champions, *caschairchech* 'curly-tailed' is given as a characteristic to be admired (*LU* 260.8594 = Windisch, 'Das Fest des Bricriu', 276.5 § 45; *LL* iv 833.24871). The abstract *caschairchige* 'curly-tailedness' ia attested in legal commentary at *CIH* iv 1275.40 (= *CIH* vi 2096.28).

(14) ni bi cu cua. Liam Breatnach proposes reading *ní bí cu cúa* 'it is not with a hollow'. It is suggested at *LEIA* C-258 that the OIr. form of this word is *cúae*, cognate with Welsh *cau*, Breton (Vannetais) *keu* 'hollow'. The compound *cocúae* (*com* + *cúae*) 'hollow, empty, concave' is well attested (*DIL* C 285.1–8), including one occurrence in the Milan Glosses (*Thes.* i 328.27). The author may be advising that a horse with a hollow chest is to be avoided.

fothach. *Fothach* is well attested in the meaning 'glanders', a respiratory disease of horses (see p. 211).

etrocht. *Étracht* means 'weakness' (*DIL* E 247.11–15), and is used in relation to a horse in legal commentary at *CIH* i 266.20; iii 787.33 = *AL* iii 182.15. According to a legal gloss (*CIH* vi 2313.3 = Binchy, '*Bretha Déin Chécht*', 40 § 30[4]), it may be also be used of sexual impotence.

crithach. 'Shivering' constitutes an unsoundness in a horse (West, *Black's veterinary dictionary*, 498). It may also indicate disease. As we have seen on p. 202, *Immacallam in dá Thúarad* refers to *crithach* 'shivering' as a disease mainly associated with cattle.

(15) forlim slan soi*mr*i*me* somul. Due to rubbing on the left-hand margin, it is impossible to make out any text before *forlim*, but one or two letters may have been obscured. This uncertainty makes it difficult to understand the context of (15). Unlike the rest of § 2, it seems to refer to good qualities in a horse.

Forlim (*forlaimm*) is vb.n. of *for-ling* 'leaps on, leaps over'. So one could translate *forlim slán* as 'sound leaping', i.e. the horse is a good jumper. *Forl(a)im* is also used in a negative sense of 'leaping-trespass', evidently with the same meaning as *airlim*, e.g. *CIH* i 67.22–3, 28 = *AL* iv 80.22, 29 *ina nairlim* . . . *ina forlim*. But the meaning 'leaping-trespass' does not seem to make sense in our text.

Soimrime is gen. sing. of *immrim(m)* 'riding', with the prefix *so-* 'good'. My translation of *somul* as 'fleet [horse]' is uncertain. This word is also attested at *CIH* vi 2315.20 = Binchy, '*Bretha Dein Chécht*', 44 § 35 *noch(a) dlig somul neich*. Binchy translates 'and is entitled to a well-balanced steed', but notes that his rendering of *somul* as 'well-balanced' is highly doubtful. I suggest that *somul* in both examples refers to speed in a horse. One can compare the following instance in a Life of Saint Findchú: *Gluaisis Finnchua andsin ana shomhulrith charpuit* (Stokes, *Lismore Lives*, 92.3087–8). Stokes translates (p. 240) 'Findchua then drove in his . . . chariot'. Possibly, *somhulrith charpuit* means 'in his swift course by chariot (lit. of chariot)'.

(16) athchuirthe. Read *athchuirthi*, verbal of necessity of *ath-cuirethar* 'returns, restores'. If the horse is found to be seriously defective, it must be returned to the seller with a full refund. See discussion on p. 509 above to Text 1, § 1 (2) *aran athchuir*.

fuillithe. Read *fuillithi* (*fuilnithi*), verbal of neccessity of *fo-lína* 'fills up, makes an extra payment'. The defect may not be sufficiently serious to warrant the return of the horse to the seller. In this case, he must recompense the buyer by making an appropriate compensatory payment.

It is difficult to be certain whether the original had *fuillithi* or *fuilnithi*. The cluster *-ln-* was occasionally being simplified to *-ll-* already in the Old Irish of the Glosses, e.g. *Thes.* i 212.22; 245.24 *giallai* (= *gíallnai*).

Appendix B
Units of measurement

(1) Length

Shorter measurements of length used in early Irish texts are generally based on the human hand (see *ordlach, mér, bas, glacc, dorn*) or foot (see *traig, céim*). An obvious disadvantage of this method is the variation in the size of people's limbs. The texts sometimes specify, therefore, that the measure is that of an average adult male.[1] To judge by our texts, it seems that an average spear-throw was the traditional method of measuring distances of medium length (see *airchor*). Precise measurements of longer distances are rare in our sources, and in legal material the only instance I have noted is the use of the Roman mile of a thousand paces (see *míle chéimenn*) in a Middle Irish commentary. Long distances may be measured very roughly by the audibility of certain sounds: see p. 571 below. Distance may also be measured by reference to a single journey by horse or man. For example, *Ceart Uí Néill* refers to the distance of a horse's run (*rioth eich*).[2] Below, I list the measures of length attested in our sources, with suggested values in inches and feet. It should be stressed that they do not comprise a single integrated system: there is a good deal of overlapping, particularly in measurements based on the handbreadth (*bas, crob, glacc, dorn, airtem*).

Gráinne (= ⅓ inch)

The smallest measurement of length is the cereal-grain (*gráinne*). In a discussion on the fines for wounds of different sizes, the Old Irish law-text *Bretha Déin Chécht* defines the smallest category of wound as *fuil gráin cruithnechta* 'a wound of [the length of] a grain of wheat'. Next is the *fuil dá gráinne cruithnechta* 'a wound of [the length of] two grains of wheat'.[3] The same text states that there are three grains in an inch (*ordlach*),[4] and this equation is repeated in an Old Irish law-text on land-values,[5] and in commentary on *Bretha Étgid*.[6] We can therefore take the early Irish *gráinne*

[1] E.g. *CIH* ii 483.13 = Thurneysen, 'Aus dem irischen Recht I', 355 § 13 *do lāim fir toimsi tēchta* 'of the hand of a man of proper measurement'; cf. *ferglacc* 'man's hand' (Meyer, *Cáin Adamnáin*, 2 § 2); *ferdorn* 'man's fist' (*CIH* iv 1471.4–5 = O'Dav. 218 § 129); *ferchubat* 'man's forearm' (Stokes, *Saltair na Rann*, 84.5756). In literary sources, *ferthraig* 'man's foot' is a common measurement: see *DIL* F 83.61–5.

[2] Dillon, '*Ceart Uí Néill*', 6.59 § 6.

[3] *CIH* vi 2306.25–7 = Binchy, '*Bretha Déin Chécht*', 26 § 5.

[4] *CIH* vi 2309.26 = Binchy, '*Bretha Déin Chécht*', 32 § 15.

[5] *CIH* ii 675.30 = Mac Niocaill, '*Tír cumaile*', 82.16–17.

[6] *CIH* i 299.29 = *AL* iii 334.20. Cf. Carney, 'Ó Cíanáin miscellany', 132–4.

to correspond appoximately to one third of a modern inch. In the imperial system of measures, there are three barleycorns in an inch.[7]

Ordlach (= approx. 1 inch)

Ordlach is a derivative of *ordu* 'thumb',[8] and can be roughly identified with the modern inch (= 2.54 centimetres), i.e. the breadth of an average thumb. In *Bretha Déin Chécht*, the proper inch (*ordlach innraic*) is defined as the *uinge orddan*, 'the thumb inch'.[9] An *ordlach* is also distinguished in legal sources which is the breadth of the little finger (*lútu*) at its lowest joint.[10] In the average hand, this is a slightly shorter measure.

As we have seen in the discussion on the *gráinne* above, the *ordlach* is taken in *Bretha Déin Chécht* to contain three grains of wheat. However, a Middle Irish glossary has the entry: *na trī grāinde adearar a n-ordlach .i. grāinde coirce* 7 *grāinde eorna* 7 *grāinde cruithnecht*[*a*] 'the three grains which are said to be in an inch, i.e. a grain of oats and a grain of barley and a grain of wheat'.[11] This appears to assume that a standard inch is obtained by placing a grain of each type end to end; the result would be a slightly longer inch than the *Bretha Déin Chécht* one.[12] A later text gives an inch of four grains: *ceithri gráine an t-órdlach.*[13] In our sources, the *ordlach* is used as a measure of circumference as well as length. Thus a hen's egg is said to have a circumference of four inches (*.iiii. ordluigh a timcomac*).[14]

Mér (= 1 inch or slightly less)

In the law-texts, measurements are sometimes given in terms of one, two or three fingers. This method is used, for example, when describing the proper dimensions of tools,[15] vessels,[16] and salted meat.[17] In the *Annals of*

[7]Dresner, *Units of measurement*, 249. The same equation is found in medieval Welsh law: see Jenkins, *The law of Hywel Dda*, 316 s.v. *barleycorn*; *GMWL* 192 s.v. *heyd*. Cf. O'Loan, 'A history of early Irish farming (3)', 170 (= O'Loan, 'Livestock in the Brehon laws', 69).

[8]*VKG* ii 55 § 397,5.

[9]*CIH* vi 2309.26 = Binchy, '*Bretha Déin Chécht*', 32 § 15. As Binchy points out in his notes (p. 57), words for 'inch' and 'thumb' are connected in many languages. *Ungae* (*uinge*) is from Latin *uncia* 'twelfth part', and is usually a measure of weight in Irish sources (see p. 585 below), but it is here used of the twelfth part of a foot, i.e. an inch.

[10]E.g. *CIH* v 1781.29 = Thurneysen, 'Aus dem irischen Recht I', 345 § 6. This is glossed *.i. ordluch lūtun* 'i.e. an inch of the little finger' (*CIH* v 1781.29 = *AL* ii 232.22). Cf. *CIH* iii 1108.4 *Ordlach a tighe .i. isin alt īchtarach don lūtain* 'its thickness is an inch i.e. at the lowest joint of the little finger'.

[11]*ACL* i 56 § 279.

[12]See the measurements of cereal-grains given by Binchy, '*Bretha Déin Chécht*', 49.

[13]O'Grady, *Catalogue*, 240 § 15.

[14]*CIH* iv 1521.21 = O'Dav. 440 § 1375; *CIH* iv 1526.9–10 = O'Dav. 461 § 1484.

[15]E.g. *CIH* iii 921.23, 24, 28, 29, 30.

[16]*CIH* iv 1496.40–1 = O'Dav. 330 § 807.

[17]E.g. *CIH* ii 480.25–6 = Thurneysen, 'Aus dem irischen Recht I', 348 § 8.

Tigernach, the dimensions of an enormous salmon caught at Clonmacnois in AD 1113 are given in feet, fists and fingers.[18]

In the Roman system of measurements, the finger (*digitus*) is one sixteenth of a foot, whereas the inch (*uncia*) is one twelfth. There is some evidence in Irish texts of a finger (*mér*) which is different from an inch (*ordlach*) – presumably it is a slightly shorter measurement. For example, the dimensions of a billhook given in a law-text include both the *mér* and the *ordlach*: this suggests that they were not viewed as identical measurements by the author.[19] But the evidence of other law-texts indicates that the finger and the inch could also be regarded as the same. For example, the texts distinguish four different thicknesses of salted meat. The thinnest seems to be the *tinne ordlaig* 'flitch of an inch (in thickness)',[20] then the *tinne dá mér* 'flitch of two fingers',[21] then the *tinne trí mér* 'flitch of three fingers',[22] and finally the *tinne láime* 'flitch of a hand'.[23] The width of the hand is generally taken to be four inches (see under *bas* below), so it looks as if we have an approximate one-inch, two-inch, three-inch, four-inch gradation.

Bas (= 4 inches)

A measure which is occasionally found in our sources is the *bas*, i.e. the palm of the hand, handbreadth. For example, a ninth-century triad gives the three proper handbreadths (*basa téchtai*) as the handbreadth between a person's shoe and his leggings, between his ear and his hair, and between the hem of his tunic and his knee.[24] Legal commentary takes the *bas* to be a four-inch measure.[25] *Bas* is also used as a measure of the amount which can be held in the palm of the hand, e.g. *bas gráin* 'a handful of grain'.[26]

[18] Stokes, 'Annals of Tigernach (3)', 32 s.a. 1113.

[19] *CIH* iii 921.24.

[20] *CIH* iii 778.37 = *CG* 5.110 *tine ordlaighe.*

[21] *CIH* ii 482.4 = Thurneysen, 'Aus dem irischen Recht I', 351 § 9.

[22] *CIH* ii 482.24 = Thurneysen, 'Aus dem irischen Recht I', 353 § 10. The author makes a distinction between this flitch, which is measured at the knuckle (*isin alt medōnach*) and another – slightly larger – three-finger flitch (*CIH* ii 483.1 = Thurneysen, 'Aus dem irischen Recht I', 354 § 11) which is measured at the base of the fingers (*a mbun lāime*).

[23] *CIH* ii 483.13 = Thurneysen, 'Aus dem irischen Recht I', 355 § 13. The glossator explains *.i. bas a tiget a tinne* 'i.e. a handbreadth in the thickness of its flitch'.

[24] Meyer, *Triads*, 28 § 222.

[25] *CIH* i 299.29–30 = *AL* iii 334.20 *.iiii.ri orlaighi i mbais*; cf. O'Grady, *Catalogue*, 240 § 15 *ceithri hórdlaighi an pais* (= *bais*). But a three-inch *bas* is mentioned in Carney, 'Ó Cianáin miscellany', 133–4, and in medieval Welsh law the palmbreadth (*lled y balf*) is likewise three inches (William, *Llyfr Iorwerth*, 59–60 § 90.18–25 = *ALIW*, Venedotian Code, II xvii §§ 5–6). In the imperial system of modern times, the palm is taken to be three inches, whereas the hand is four inches (Dresner, *Units of measurement*, 249).

[26] *CIH* i 307.17 = *AL* iii 382.7.

Glacc (= 4 inches(?))

The word *glacc* is used of the clasping hand, or of the amount of material which can be contained by a handclasp. For example, the early twelfth-century tale *Aislinge Meic Con Glinne* refers to seven handfuls (*glacca*) of edible seaweed.[27] The derivative *imglaice* is quite common in the law-texts,[28] and in *O'Davoren's Glossary* is equated with a fistful (*lán duirn*).[29] *Ferglacc* 'a man's hand' may be used as a measure of length, as in *Cáin Adomnáin*, where a candle is said to be the length of four men's hands (*cainnel cethri ferglac*).[30] It is probably the same as the *bas*, i.e. 4 inches.

A difficult attestation occurs in a description of a sumptuous feast in *Aislinge Meic Con Glinne*, which includes an *ag teóra ferglacc* 'a bullock of three hands'.[31] The context requires a fairly large animal for consumption, so Jackson suggests in his notes the emendation to *ag teóra ferglacc ndéc* 'a bullock of thirteen hands'. However, an argument against this emendation is the fact that the phrase *ag trí nglac* is attested elsewhere, and could refer to some body measurement other than girth. For example, a Life of Saint Mo Laisse of Daiminis refers to the tribute of an *agh trí nglac* to be paid by every minor king to the community of this saint to ward off plague.[32] In the Life of Saint Náile, an *agh trí nglac* is said to measure a *glac* in its horn, a *dorn* in its hoof, and a *bas* in its tail, but these measurements seem pedantic and unrealistic.[33]

Dorn (= 4–6 inches)

The *dorn* 'fist' is commonly used in measurements in our sources, but it is hard to establish its precise length: the *Dictionary of the Irish language* quotes O'Donovan's definition 'a measure of 6 or 4 inches'.[34] Uncertainty about the size of the *dorn* is perhaps not surprising as the width of the adult male fist is particularly variable, depending not only on bone-size but also on the amount of muscle and tissue built up by manual labour. The Old Irish law-text on land-values states that there are six inches in a *dorn*,[35] and this measure is also indicated in commentary on *Bretha Comaithchesa* which takes a *dorn* to be half a foot.[36] Commentary on *Cáin Aicillne* likewise

[27] Jackson, *Aislinge*, 35.1078 = Meyer, *Aislinge*, 89.20.

[28] E.g. *CIH* iii 778.13 = *CG* 3.74; *CIH* ii 482.26 = Thurneysen, 'Aus dem irischen Recht I', 353 § 10; *CIH* ii 483.3 = Thurneysen, 'Aus dem irischen Recht I', 354 § 11.

[29] *CIH* iv 1510.38 = O'Dav. 394 § 1124.

[30] Meyer, *Cáin Adamnáin*, 2 § 2.

[31] Jackson, *Aislinge*, 1.14–15 = Meyer, *Aislinge*, 3.14.

[32] O'Grady, *Silva Gadelica* i 21.11; trans. ii 21.

[33] Plummer, *Miscellanea hagiographica hibernica*, 113 (38).

[34] *DIL* D ii, 360.47–8.

[35] *CIH* ii 675.30–1 = Mac Niocaill, '*Tír cumaile*', 82.17. It is necessary to emend *se ordl*aige *i nordl-* to *sé ordlaige i ndorn*).

[36] *CIH* i 66.25 = *AL* iv 76.21.

equates twenty *dorn* with thirty *bas.*[37] However, it is impossible to make sense of a six-inch *dorn* in the passage on walls and fences in the text of *Bretha Comaithchesa.* Here it is stated that a stone wall or a wattle fence to confine cattle must be twelve fists in height. If we are to take the *dorn* to be six inches, this would mean that the wall or fence is six foot high. Even with modern cattle – which are generally larger than those of the early Irish period – it is not considered necessary to exceed five feet. In my reconstruction in Chapter 11, I take the *dorn* to be only four inches: this gives a fence or wall of four feet in height. See p. 375.

It is also difficult to establish the relationship of the *dorn* to the *crob* in our texts. The word *crob* means 'hand, claw, paw', and one might expect it to be the same length as the *glacc,* which I take to be four inches (see above). However, the law-text *Bretha Nemed Déidenach* describes a cow as having a measurement of thirty hands (*a tomhas trīocha crob*).[38] Comparison with commentary on *Cáin Aicillne* indicates that this measurement refers to the cow's girth. Thus the text of *Cáin Aicillne* speaks of a cow whose girth is twenty fists (*fiche dorn a timcomac*),[39] and an accompanying commentary defines this as *trīcha crobh ina tacmaisi* 'thirty hands in her girth'.[40] If we take the *dorn* to be four inches, the *crob* must then be only 2⅔ inches. If we take the *dorn* to be six inches, the *crob* will be four inches – like the *bas* and perhaps the *glacc* – but the cow will have an impossible girth of 120 inches.[41]

Airtem (= 6 inches (?))

The *airtem* is occasionally used as a unit of measurement.[42] It seems to have been different from the standard *dorn,* as a description of the dimensions of a piece of meat in *Cáin Aicillne* includes both the *airtem* and the *dorn.*[43] A glossator on this text takes the *airtem* to refer to the fist with the thumb pointing upwards, while the *dorn* is the shorter measure of the clenched fist (*máeldorn,* lit. 'bald fist').[44] The O'Davoren gloss on this passage provides no futher help, and simply defines the *airtem* as a *ferdorn* 'male fist'.[45]

37 *CIH* v 1784.35–6 = *AL* ii 252.23.
38 *CIH* iii 1121.23 = E. J. Gwynn, 'Privileges', 28.6.
39 *CIH* ii 483.12; v 1784.32 = Thurneysen, 'Aus dem irischen Recht I', 355 § 13.
40 *CIH* v 1784.34 = *AL* ii 252.21.
41 See p. 31 for a discussion on the girth of early Irish cattle.
42 E.g. *CIH* iii 921.28. See p. 486.
43 *CIH* ii 480.24–5 = Thurneysen, 'Aus dem irischen Recht I', 348 § 8.
44 *CIH* ii 481.4–6 = *AL* ii 240.1–4. The 'bald fist' measure (*moelddyrnfedd*) is also used in Welsh law: see Jenkins, *The law of Hywel Dda,* 272 (note to 129.4).
45 *CIH* iv 1471.4–5 = O'Dav. 218 § 129.

Traig (= 1 foot)

The *traig* 'foot' is very common as a measure of length in legal and other texts. It may be used in one-, two-, or three-dimensional measurements. I have not noted its use in measurements of circumference. Though the length of the adult male foot is quite variable, it has been used as the basis for systems of length-measurements in many societies.[46] Early Irish sources indicate a twelve-inch foot, approximately the same as the modern standard foot of 30.48 centimetres.[47] Legal commentary on *Bretha Étgid* states that there are three *basa* (four-inch measures) in a foot.[48] There is one legal attestation of a shorter measure based on the foot. In the description of a standard fence in *Bretha Comaithchesa* it is specified that the distance between the stakes should be 'the length of a foot as far the joint of the big toe' (*traigh co ruige deil n-ordan*).[49] This must be approximately eight inches.

Cubat (= 1½ feet)

The *cubat* 'cubit' (from Latin *cubitum* (*-us*) 'elbow, forearm, ell, measure of 1½ feet') is used mainly in texts with Biblical themes.[50] In legal commentary it is also used in a sea-measure: see p. 570 below.

Céim (= 2½ feet)

The *céim* 'step' is occasionally used as a measure of length in our sources. As in the case of the modern yard, the *céim* always refers to horizontal measurements. The *traig* 'foot', on the other hand, is used for both vertical and horizontal measurements. In *O'Davoren's Glossary*, the *céim* is identified with the *greis* (from Latin *gressus*), and is stated to be two and a half feet in length (*dā traigh co leth*).[51] In 'A note on O'Davoren's Glossary', 157, E. J. Gwynn suggests that this equation derives from an Old Irish translation of the Latin mensuration tract *Pauca de Mensuris*, which counts 2½ feet (*pedes*) in a *gressus* (= *gradus*). In other texts there is considerable variation as to the length of different types of *céim*. In the gloss quoted above, O'Davoren appears to regard the *céim* 'step', *coiscéim* 'footstep', and *deiscéim* 'step to the right' as being of the same length. However, a glossator on the law-text *Coibnes Uisci Thairidne* states that the proper step (*céim inraic*) is 2½ feet,

[46]See Glare, *Oxford Latin dictionary* s.v. *pes* (9); *OED* s.v. *foot* III (7).

[47]It may have been slightly shorter: compare the Classical Roman foot of approximately 29.6 cm. (Kidson, 'A metrological investigation', 75).

[48]*CIH* i 299.30 = *AL* iii 334.20–21; cf. O'Grady, *Catalogue*, 240 § 15.

[49]*CIH* i 195.32; vi 2133.15–16 = *AL* iv 112.13–14; cf. *CIH* iv 1493.1–2 = O'Dav. 313 § 715; *CIH* i 73.15.

[50]Stokes, *Saltair na Rann*, 35.2449; 36.2463; 37.2527; etc.; Calder, *Auraicept*, 181.2604.

[51]*CIH* iv 1507.29–30 1= O'Dav. 380 § 1048; cf. Carney, 'Ó Cíanáin miscellany', 133 *dā traig colleith i ngreis*, but also *trī traig go leith a ngres.*

whereas the *coiscéim* (*cascéim*) is 1¼ feet.[52] In another source, the *coiscéim* is said to be two feet, half the length of a *pais* (Latin *passus*).[53] A conflicting figure is given in the Old Irish text on land-values where a *deiscéim* is counted as being equivalent to six feet.[54] Much of this uncertainty seems to stem from a confusion as to whether a *céim* consists of a step taken by one foot (Latin *gressus*) or by both feet (Latin *passus*).[55] For a discussion of the use of the *deiscéim* 'step to the right' and *tuaithchéim* 'step to the left' in relation to the *immaire* 'ridge (of ploughland)', see p. 574 below.

Fertach (= 12 feet)

A rare measure of length in our sources is the *fertach* (*pertach*), a borrowing from Latin *pertica* 'perch'. Legal commentary on *Bretha Étgid* takes it to contain 12 feet and to be one twelfth of a *forrach.*[56] Another legal commentary confirms the latter figure: *isi aili dēc na forraighe in phertach* 'the *fertach* is one twelfth of the *forrach*'.[57] Isidore of Seville (*c.*560–636) gives a *pertica* of 10 feet, but the mensuration tract *Pauca de Mensuris* – in a manuscript written at Corbie about the middle of the ninth century – refers to *perticae* of 10, 12, 15 and 17 feet. The larger *perticae* are used to measure poorer land, so that the yields per unit of land are evened out.[58]

Forrach (OIr.: 12 feet (?); Mid.Ir.: 144 feet)

Forrach is a particularly difficult term. In *Cormac's Glossary,* which dates from about AD 900, it is equated with the *fertach* (*pertach*) discussed above. The entry reads: *pertic a pertica .i. forrach tomais tīre* '*pertic* from *pertica,* i.e. a rod(?) for measuring land'.[59] Other Old Irish texts provide evidence in support of this identification of *forrach* as some type of measuring instrument, probably a rod of standard length. A ninth-century triad includes the *forrach* along with two other devices for measurement,[60] and an eighth-century gloss on the Psalms uses the same word to explain the Latin expression *funiculus distributionis* 'cord of division', i.e. a cord or rope used to measure out land. The term *forrach* is also used in a passage in the law-text *Bretha Comaithchesa* which deals with the dimensions of a

[52] *CIH* ii 461.6–7 = Binchy, '*Coibnes Uisci Thairidne*', 70 § 10, gloss[5].

[53] O'Grady, *Catalogue*, 240 § 15.

[54] *CIH* ii 675.31 = Mac Niocaill, '*Tír cumaile*', 82.18. A cow-track (*bóthar*) which provides access to water across another person's land is stated to be a *descéim* in width in *Bretha Nemed Déidenach* (*CIH* iii 1125.18 = E. J. Gwynn, 'Privileges', 33.22).

[55] Dresner, *Units of measurement*, 243, distinguishes an ordinary pace of 5 feet from a military pace of 2½ feet.

[56] *CIH* i 299.30 = *AL* iii 334.21–2 *dā troigid dēc i fertaig, dā fertaigh dēc i forraigh.*

[57] *CIH* i 140.18–19.

[58] Kidson, 'A metrological investigation', 75.

[59] Meyer, *Sanas Cormaic*, 90 § 1055. E. J. Gwynn, 'A note on O'Davoren's Glossary', 158, draws attention to the same correspondence of *forrach* and *pertica* in *CIH* iv 1507.29–30 = O'Dav. 380 § 1048.

[60] Meyer, *Triads*, 18 § 138.

headland (*airchinn*).[61] It is defined as being *tēoru foirge* ⁊ *urcar flescaig* 'three *forrach*-measures and the spear-cast of a youth'. Related legal verse has a slightly different measure: *tommus aircinne cethrai forrgib co n-aurchur flescaich* 'measurement of a headland by four *forrach*s with the spear-cast of a youth'.[62]

The above attestations of *forrach* in Old Irish sources suggest that it is most likely to have been a measure of about 12 feet in length.[63] In Middle Irish legal commentary, on the other hand, it is defined as containing twelve *fertaig*, i.e. 144 feet.[64] An even longer *forrach* is indicated in the table of measures in an Old Irish law-text on land-values.[65] Here it is said that there are 6 feet in a *deiscéim* 'step', 6 steps in an *inntrit*, 6 *inntrit* in a *lait*, and 6 *laiti* in a *forrach*. This gives an impossible *forrach* of 1296 feet. Scholars have referred to the difficulties of this passage,[66] and it is safest to leave it out of consideration. The measures *inntrit* (perhaps from Latin *introitus* 'entrance') and *lait* (Latin *latus* 'side') are otherwise unattested.

The term *forrach* is also used in relation to the marking out of a site. For example, the eighth-century *Additamenta* in the Book of Armagh contains the following passage: *Dulluid iar suidiu Patricc cu Fíacc* ⁊ *durind a locc les* ⁊ *cutsecar* ⁊ *forruim a forrig n-and* 'after this Patrick came to Fíacc and divided out his plot of land for him and blessed it, and fixed his *forrach* there.'[67] In his edition, Bieler quotes a note supplied by D. A. Binchy: '*forrach*, originally a native land-measure (possibly = 144 ft.), but also used in a more specialized meaning as an area of land set aside for a particular purpose (e.g. by a king for holding an assembly)'.[68] *Forrach* is well-attested as an element in place-names.[69]

Airchor (= approx. 60 feet (?))

In modern colloquial usage, a 'stone's throw' is a rough measurement of distance. In early Irish literary sources, we find similar expressions.

[61] *CIH* i 78.9; 200.12 = *AL* iv 138.15. Gaulish *arepennis*, cognate with *airchinn* (*LEIA* A-39), is recorded as a square measure, equivalent to a *semiiugerum* (Columella, *De re rustica* (ed. Forster and Heffner), 5.1.6) or *actus quadratus* (Isidore, *Etymologiarum* (ed. Lindsay), 15.15.4). However, Old Irish *airchinn* is attested as a measure of length only.

[62] *CIH* ii 571.11–12. Binchy's omission of *cethrai forrgib* in his edition ('An archaic legal poem', 159.86–7) does not seem justifiable.

[63] For a Welsh rod of 12 feet, see Rowlands, 'Mesur tir', 272.

[64] *CIH* i 299.30 = *AL* iii 334.21–2.

[65] *CIH* ii 675.30–3 = Mac Niocaill, '*Tír cumaile*', 82.16–20.

[66] *AL* vi (Glossary) 407; E. J. Gwynn, 'A note on O'Davoren's Glossary', 158; Mac Niocaill, '*Tír cumaile*', 84^{10}.

[67] Bieler, *Patrician texts*, 176.

[68] Bieler, *Patrician texts*, 238.

[69] Hogan, *Onomasticon*, 429; Doherty, 'The monastic town in early medieval Ireland', 49–50.

Fled Bricrenn refers to 'the length of a cast from the fort' (*fot n-aurchora ōn cathraig*),[70] and *Togail Bruidne Da Derga* describes how some mysterious birds remained 'the length of a cast' (*fot n-ahurchara*) ahead of Conaire, who was hunting them with a stone and sling.[71] The same expression may be used in relation to a spear-cast, e.g. *focheird a bunsaig rout n-aurchora úad* 'he throws his spear the length of a cast from him'.[72] In legal material, the term *airchor* is used of a fixed unit of length in various contexts. The law-text *Críth Gablach* defines the radius of the *airlise* (see p. 368) as the length of a cast from a dwelling.[73] A similar usage is described in more detail in Middle Irish commentary on the *maigen dígona* 'the area around a house which is under the householder's protection'.[74] In the case of a commoner of *bóaire* rank, this area extends the length of a spear-cast on all sides. In the case of an ordinary lord (*aire déso*), it extends for two spear-casts, in the case of a lord of precedence (*aire tuíseo*) it extends for four spear-casts, and so on in multiples of two up to the king of territories (*rí túath*) whose area extends for sixty-four spear-casts.[75] This commentary even specifies that the spear (*cnairsech*) used should be twelve fists (*dā dorn dēg*) from its iron blade to the end of its handle.

As we have seen in the previous section, *Bretha Comaithchesa* refers to a headland consisting of three *forrach*s and the spear-cast of a youth (*urchar flescaig*). The youth is defined as a *macc bundsaige* 'lad of the spear'.[76] In a discussion on the extent of cattle-trespass, commentary on the same text likewise refers to a distance of eighteen spear-casts of a youth (*.uiii. n-urcora .x. and do ḟleascach*).[77] These passages raise the possibility that the early Irish system of measurement made a distinction between a standard spear-cast (*airchor*) and a shorter cast thrown by a youth (*airchor flescaig*). However, no text distinguishes two different lengths of cast. It seems more likely, therefore, that the youth (*flescach*)[78] is mentioned in the texts merely to give greater precision to the unit of measurement.

[70] *LU* 275.9136 = Windisch, 'Das Fest des Bricriu', 299.11 § 88. Cf. *DIL* E 60.29–32 *ed n-urchair.*

[71] Knott, *Togail Bruidne Da Derga*, 5.138; cf. Ibid. 9.300 = *LU* 209.6787.

[72] Meid, *Táin Bó Fraích*, 2.47–8.

[73] *CIH* ii 564.22–3; iii 780.28–9 = *CG* 9.219–20 *Aurchor snedar cach leth is é córus a airlise* 'a spear-cast which is made in every direction is the proper area of his *airlise*'.

[74] *CIH* iv 1431.32–1432.10; vi 2111.17–25 = *AL* iv 226.1–228.11. For a brief discussion of the *maigen dígona*, see *GEIL* 141.

[75] In the case of a king of higher rank, the *maigen dígona* is reckoned as the area which his livestock reach: see p. 572.

[76] *CIH* i 78.9–10; 200.12–13 = *AL* iv 138.16.

[77] *CIH* i 199.36–7 = *AL* iv 138.4.

[78] The law-text on status *Uraicecht Becc* assigns the *flescach* an honour-price worth one lamb (*CIH* v 1610.17 = *AL* v 84.20).

Noí n-immairi

Another measure of length used occasionally in our sources is *noí n-immairi* 'nine ridges'. Though the *immaire* 'ridge' sometimes has a precise application as a unit of area (see p. 574), *noí n-immairi* seems to be used only as an approximate measure of middle distance similar to an *airchor*. According to Keating, for example, applicants to join the Fian had to defend themselves against spear-casts thrown at a distance of nine ridges.[79] A less common variant of this distance is *secht n-immairi* 'seven ridges'. According to commentary on Broccán's hymn, Saint Brendan never proceeded more than a distance of seven ridges without thinking of God.[80]

Míle chéimenn (= 5,000 feet (?))

The measure of a thousand paces (*míle chéimenn*) is uncommon in our sources, and is doubtless modelled on Latin *mille passus* (or *passuum*) 'mile': see *DIL* M 137.3–13. The only legal attestation is in Middle Irish commentary on the extent of the *maigen dígona* 'the area around a house which is under the householder's protection'.[81] The radius of this area extends for a thousand paces (*míle chéimenn*) from the residence of a saint, bishop, hermit, or recluse. It extends for a radius of two thousand paces (*dá míle*) from a monastic city of refuge (*cathair ataig*).[82] I take the *céim* in *míle chéimenn* to correspond to Latin *passus* '(five-foot) pace', and suggest a mile of five thousand feet.[83] This would be slightly shorter than the modern mile of 5280 feet (= 1.6093 kilometres).

(2) Sea-measures

Noí tonna

At sea, the wave (*tonn*) is used as an approximate measurement of length, and there are many references to the distance of nine waves.[84] The preface to Colmán's hymn in the Liber Hymnorum tells us that a plague cannot reach beyond nine waves (*ní thic teidm tar noí tonna*); Saint Colmán therefore fled with his followers to a coastal island to avoid the plague called the *Buide Conaill*.[85] In a fragment of *Muirbretha* 'sea-judgements', it is stated that what is retrieved beyond nine waves (*dar noí tonnaib*) from the

[79] Keating, *Foras feasa* ii 334.5201.

[80] *Thes.* ii 335.40.

[81] See discussion of this passage by Ó Corráin, 'Irish vernacular law and the Old Testament', 304–5.

[82] *CIH* iv 1432.6–8; vi 2095.24–5 = *AL* iv 228.5–7.

[83] Another text has a mile of six thousand feet (Carney, 'Ó Cíanáin miscellany', 133).

[84] E.g. Macalister, *Lebor Gabála* v 54 § 415; *CIH* iii 916.8; cf. Fenton, *The Northern Isles*, 441.

[85] *Thes.* ii 299.1–16.

shore belongs to the finder.[86] The accompanying commentary equates the distance of nine waves with 200 cubits, i.e. 300 feet.[87]

Muirchrech

Another sea-measure is the *muirchrech*, which seems to refer to the distance at which some small floating object is visible from the shore. For example, *Cáin Adomnáin* describes the fit punishment for a woman who has committed murder or stolen from a church. She is to be abandoned in a boat with one paddle at a distance of a *muirchrech* on an offshore wind; judgement on her is then left to God.[88] In another description of this punishment, the criminal is to be abandoned at sea 'the distance from which a white shield is visible' (*in airet is lēir geilsciath*).[89] In legal commentary on the punishment of a castaway, the equivalent distance is given as 'beyond nine sea-waves' (*tar .ix. tonna mara amach*).[90] A greater distance is indicated in the Old Irish tale of Fergus mac Léti. For the crimes of her son, Dorn daughter of Buide is said to have been set adrift as far as three *muirchrecha* out to sea (*co .iii. mara muircrecha*).[91] The *muirchrech* may also be used as a measure of distance at sea in other contexts. Thus the *Lebor Gabála* describes how the invading sons of Míl raced towards the shore of Ireland. One of them, Ír mac Míled, was a *muirchrech* ahead of his brothers' boats until his oar broke.[92] Máirín Ní Dhonnchadha suggests that *muirchrech* means 'sea-boundary', and quotes O'Rahilly's view that *crech* may be cognate with *crích* 'boundary, end', and would originally have meant something like 'a cutting'.[93]

I have noted no references in early texts to measures of the depth of the sea. In later Irish *feadh* (OIr. *ed*) is used of the fathom, i.e. the length from finger-tip to finger-tip of the outstretched arms = six feet. This unit is used especially in the measurement of lengths of rope, and so of the depth of the sea.[94]

[86] *CIH* i 314.17 = *AL* iii 422.1–2; cf. *CIH* i 60.5, 10, 14 = *AL* v 336.15, 23, 28.

[87] *CIH* i 315.32–3 = *AL* iii 428.5–6.

[88] Meyer, *Cáin Adamnáin*, 30 § 45. See discussion by Máirín Ní Dhonnchadha in her thesis, 'An edition of *Cáin Adomnán*', 240-1.

[89] *CIH* i 109.12; ii 744.33–4; cf. M. Byrne, 'On the punishment of sending adrift', 98.

[90] *CIH* iv 1245.19; 1398.34.

[91] Binchy, 'The saga of Fergus mac Léti', 37 § 2; cf. Binchy, *Scéla Cano*, 5.122 (and note).

[92] *LL* i 48.1533 = Macalister, *Lebor Gabála* v 72 § 434 (this version has *muirrdrēcht*).

[93] T. F. O'Rahilly, '*Creach. Gad. Brad. Slad.*', 168.

[94] R. B. Breatnach, *Seana-chaint na nDéise II*, 182.

(3) SQUARE AND CUBIC AREA

In their descriptions of tools and other objects, the authors of the law-texts sometimes provide full three-dimensional measurements. For example, one text refers to a standard beam of twelve feet by two feet by one foot (*dā traig .x. a fod 7 dā traig a lethed 7 traig a tiged*).[95] Often, however, the author gives a measurement on only one plane, as in the standard description of a vessel as being 'of twelve inches', e.g. *mesair dā ordlach dēc,*[96] *cuad dā ordlach .x.,*[97] *escrad dā orlach dec.*[98] Presumably, the variation in the shape of such vessels was so slight that it was sufficient to specify the height alone: the other dimensions could be deduced from this information. In his description of houses, the author of *Críth Gablach* likewise states simply that a house is 27 feet (*tech .uii. traiged fichet*),[99] or an outhouse is 17 feet (*airchae .uii. traiged ndeec*).[100] Modern archaeological findings suggest that these figures refer to the diameter of the structures (see p. 361).

In the case of small areas of land, measures are generally given in terms of the *traig* 'foot'. For example, a bed for cultivation is said to be twelve feet in length and eight feet in width (*dā traigh .x. fot cach innraid, a hocht a lethat*).[101] To measure larger areas of land, we find a wide variety of usages. In cases where precision is not required, an area may be defined by the audibility of certain sounds. For example, the seventh-century law-text *Bechbretha* defines the extent of a farm's infield or legally prescribed green (*faithche téchtae*) as being 'as far as the sound of a bell or the crowing of a cock reaches'.[102] A more exact method is the definition of an area of land by the amount of cattle which can be grazed on it, though obviously the quality of the land must also be taken into consideration. A clear example is to be found in an entry in the *Annals of Ulster*, which describes the consequences of a storm in Corcu Baiscinn (in the modern Co. Clare). It is recorded that the sea covered over with sand as much land as would support twelve cows (*mēd dā boo deac di thīr*) on the island of Fita.[103] Similarly, Tírechán records the donation in perpetuity to bishop Assicus of 'grazing for a hundred cows with their calves and for twenty oxen'.[104]

[95] *CIH* iii 921.27–8.
[96] *CIH* ii 675.7 = Appendix A, text 1 § 5 (1).
[97] *CIH* iii 779.1 = *CG* 5.116; *CIH* iii 920.36.
[98] *CIH* v 1766.13 = *AL* ii 176.17.
[99] *CIH* ii 563.6; iii 779.26 = *CG* 6.154.
[100] *CIH* ii 564.4–5; iii 780.11–12 = *CG* 8.193.
[101] *CIH* iii 894.14–15.
[102] *CIH* ii 454.13–14 = *BB* 82 § 46 (see notes). Cf. Plummer, *Vitae* i, Introduction p. ciii^{7}. Stokes ('Adamnan's second vision', 440) compares the early German use of the *Glockenklang* 'bell-peal' as a measure of distance.
[103] *AU*^{2} 260 s.a. 803 (recte 804) § 11.
[104] Bieler, *Patrician texts*, 140 § 22 (3); cf. 170 § 5 (4). One can compare the passage in the law-text on joint-herding, which distinguishes the amount grazed by various

Legal commentary also uses grazing by livestock to delimit very roughly the area around his dwelling which is under the protection of an overking (*rí ruirech*).[105] This area extends on every side as far as his pasturage reaches (*an airat rosoich a scor-sidhe for cach leth*).[106] One can compare the idea in another commentary of an area around a farm defined by the extent of grazing (*in airet rosoich cuaird ingelta*).[107]

There are also other ways in which cattle are used to define land-area. For example, in *Críth Gablach* the phrase *tír mbó* 'the land of a cow' refers to an area of land from which one cow is due as rent.[108] This may also be the explanation of a difficult passage in the *Additamenta* in the Book of Armagh.[109] Here it is recorded that the sons of Conlaíd offered to God and Patrick for all time *octo campi pondera, id est uaccas campi octo* lit. 'eight measures of open land, i.e. eight cowlands'.[110] This can hardly refer to land which provides grazing for eight cows, as this would be rather a small area to give to the Church. It is possible, therefore, that this is land for which eight cows would be the annual rent. The commercial value of land may also be expressed in terms of cattle. Thus *tír ndarta* 'the land of a yearling heifer' is used in a law-text of the least valuable parcel of land, i.e. land which is only worth a yearling heifer.[111] In later sources, land-area is regularly defined in terms of the *colpthach* 'two-year-old heifer' (anglicized *colpe, collop*, etc.).[112]

In many medieval systems, arable land is measured by reference to the amount which can be ploughed in a specific time. For example, in medieval England, the amount which an eight-ox team could plough in

---

animals (*CIH* i 192.14–16; ii 577.2–4 = *AL* iv 100.22–4). Two geese graze as much as a sheep, two sheep as much as a *dairt* (yearling heifer), two *dairt*s as much as a *colpthach* (two-year-old heifer), two *colpthach*s as much as a cow, and a cow and a *colpthach* as much as an ox.

[105] *CIH* iv 1432.3–4; vi 2111.24–5 = *AL* iv 226.y-z. This area is the *maigen dígona*: see p. 568 above.

[106] *Scor* means 'unyoking (of horses), herd of horses, area grazed by horses, pasturage': see *DIL* S 102.53–103.13.

[107] *CIH* i 57.17 = *AL* v 328.10–11; cf. (in a different context) *CIH* ii 444.30–1 = *BB* 54 § 8.

[108] *CIH* iii 778.25 = *CG* 4.92. McErlean ('The Irish townland system of landscape organisation', 328) suggests that the post-Norman *baile bó* (angl. *ballyboe*) likewise refers to a unit of land with the annual rent of a cow. One can compare the Scottish Gaelic term *tír unga* 'land of an ounce' (borrowed into Scots as *terunga*), used of the area of land from which rent of an ounce of silver was levied (Bannerman, 'The Scots language and the kin-based society', 9).

[109] Bieler, *Patrician texts*, 170.32–4.

[110] Cf. Sharpe, 'Review of Bieler, *Patrician texts*', 330–1.

[111] *CIH* i 237.12 = *AL* v 466.16. It is contrasted with *tír cumaile* 'land worth a *cumal*': see below.

[112] See discussion of this term on p. 62.

a season was termed the ploughland (Mid.Eng. *ploulond*).[113] The acre (OEng. *æcer*) was the amount which could be ploughed in a day. The Irish evidence in relation to areas of ploughland is complex, and some of the terminology is obscure to me. An important source of information is *O'Davoren's Glossary*, which contains a number of entries dealing with land-area. Gwynn suggests that entries 117, 491, 492, 493, 1048, and 1121 derive from an Old Irish translation of the Latin mensuration tract *Pauca de Mensuris.*[114]

In entry no. 1121, O'Davoren defines Latin *iuger*[*um*] as *lá air* 'a day's ploughing'.[115] This may be the same unit as the *indle*, which is used as a land-measure in the *Additamenta* in the Book of Armagh.[116] Francis John Byrne makes the suggestion that *indle* is a derivative of *indel*(*l*), which can mean 'yoking, harnessing'.[117] An *indle* would thus be the amount of land which could be ploughed at one yoking, i.e. in one day. In entry no. 493, O'Davoren refers to lesser units called in Latin *climata* (nom. plur. of *clima*),[118] which he explains as *damicht*, a compound of *dam* 'ox' and *icht* 'work(?)'.[119] The wording in this gloss is confusing, but the general sense seems to be that the *clima* is a 'piece or division' (*ord*u *nō rand*) of a *iugerum.*[120] Another unit of land which may belong here is the *damaisc*,

[113]The OEng. term is *sullung* from *sulh* 'plough'. In Scotland a term for ploughland was *dabhach* (angl. *davach, davoch*). This is the same word as OIr. *dabach* 'vat, cauldron', and may refer to the area of land which can be sown from a vat full of seed-corn. The use of *dabach* as a land-measure is not attested in Irish sources. See *DIL* D i, 5.16–19; Jackson, *The Gaelic notes in the Book of Deer*, 116–17; Craigie, *A dictionary of the older Scottish tongue* s.v. *davach*; Bannerman, 'The Scots language and the kin-based society', 9.

[114]E. J. Gwynn, 'A note on O'Davoren's Glossary'. For further discussion of this tract, see Kidson, 'A metrological investigation'.

[115]*CIH* iv 1510.33–4 = O'Dav. 393 § 1121 *Iuger .i. lá ferainn nō lā air*, ut est *cenn tuir ingera* (= *iugera*) *.i. in dam oraib nō i lāib n-air. Iugerum* is used as a measure of distance in Tírechán's *Collectanea* (Bieler, *Patrician texts*, 148.35).

[116]Bieler, *Patrician texts*, 170.33; 172.34, 36. The spelling *cach indlea* at 170.33 is difficult to explain: the *-a* ending seems a latinization. The form *-indli* at 172.34, 36 is an acc. plur.

[117]He is quoted by Doherty, 'Some aspects of hagiography', 309[5].

[118]See Du Cange, *Glossarium* s.v. *clima*; Glare, *Oxford Latin dictionary* s.v. *clima* (1).

[119]A commentator at *CIH* i 137.27 explains *damhicht* as *.i. tre icht na ndamh* 'i.e. through the work of the oxen'. It is possible that the later unit, the *gním* 'work' (generally one twelfth of a ploughland), is equivalent to the *damicht*. See Nicholls, 'Gaelic society and economy', 407[3].

[120]*CIH* iv 1484.20 = O'Dav. 276 § 493 *Climata .i. in damicht,* ut est *nō lā air iugera hi climatu .i. ord*u *nō rand.* Following Meyer ('Erläuterungen und Besserungen', 350), Binchy suggests emending *no la air* to *noda-air*, but it seems better to take it to be a gloss on *iugera*, as in no. 1121.

which is attested once in *Bethu Phátraic.*[121] *Damaisc* is clearly also a derivative of *dam*, but the second element is unexplained. Binchy takes *seisrech* in the late Old Irish tale *Scéla Cano meic Gartnáin* to mean 'ploughland', but it could equally in this context have its usual meaning in early texts, 'a team of six [oxen]'.[122] Keating (*Forus feasa* i 112.10) takes a *seisreach* to be an area of 120 acres.

In the Latin system, the unit of area below the *clima* is the *actus*. The smallest *actus* is the *actus minimus*, which is defined as four feet in width and 120 feet in length.[123] O'Davoren cites the definition of the smallest *actus* as a ridge (*immaire*) of four feet in width, but gives no information as to its length.[124] There is also a discussion on the dimension of an *immaire* in a fragment of legal commentary inserted in Rawlinson B 506 (*CIH* i 137.27–30). Here, the width of half an *immaire* is taken to consist of a step to the right (*deiscéim*) and a step to the left (*tuaithchéim*),[125] and an *immaire* is further defined as 'seven sods of the plough' (*.uii. fōit arathair*). A gloss on the same commentary distinguishes a small ridge (*immaire becc*) which is three fifths as large (= wide?) as a large ridge (*immaire mór*). *Immaire* is sometimes used in a general sense of a strip of land. For example, in the preface to Colmán's hymn in the Liber Hymnorum, it is stated that at a time of overpopulation in the reign of the two sons of Áed Sláine, each man in Ireland had thrice nine strips of land (*trí noí n-immaire*): nine strips of moorland (*móin*), nine strips of arable land (*mín*), and nine strips of woodland (*caill*).[126]

Another unit of area which is frequently mentioned in legal commentary is the *tír cumaile* 'land of a *cumal*'. The basic meaning of the word *cumal* is 'female slave', but it is also common as a unit of value equivalent to three or more cows (see p. 592). It is often used in relation to the value of land. For example, the law-text *Bretha Déin Chécht* states that if the bone of a king is broken away from home, he receives land worth seven *cumal*s

[121] Mulchrone, *Bethu Phátraic*, 82.1530.

[122] Binchy, *Scéla Cano*, 1.8 (and note).

[123] Columella, *De re rustica* (ed. Forster and Heffner), 5.1.4; Isidore, *Etymologiarum* (ed. Lindsay), 15.15.4; E. J. Gwynn, 'A note on O'Davoren's Glossary', 157.

[124] *CIH* iv 1470.22–3 = O'Dav. 215 § 117 *Actus .i. tomus ł imaire, ut est actus is lugha a .iiii. do troighaib lethat a lāir .i. imaire is lugha nā int actus aderam 'nar ndegha; ł actus .i. lā air* '*Actus*, i.e. a measure or ridge, as in [the following]: the smallest *actus*, the breadth of its middle is four feet, i.e. the ridge which is smaller than the *actus* which we will speak of later; or *actus* i.e. a day's ploughing'. For the dimensions of the other sizes of *actus*, see Isidore, *Etymologiarum* (ed. Lindsay), 15.15.4–5.

[125] Cf. *CIH* i 98.42 *deisceim ⁊ tuaidcheim* (the width of an access-route across another person's land).

[126] *Thes.* ii 298.7–12. See also *DIL* s.v. *immaire*; Mac Mathúna, 'Some words for "(man-made) ridge" in Irish', 447. For *noí n-immairi* as a rough measure of length, see p. 569 above.

(*tir .uii. cum*al).[127] In *Críth Gablach*, the value of land owned by different classes of person is expressed in the same way; for example, the *bóaire febsa* is expected to own fourteen *cumal*s of land (*tír dá secht cumal*).[128] The use of the word *cumal* as a unit of area is a further development, and uncommon in the Old Irish law-texts.[129] The *tír cumaile* is viewed as a rectangular land-holding with a headland (*airchinn*) and a side (*tóeb*). The side is 12 *forrach*s in length, and the headland is 6 *forrach*s.[130] Our understanding of the dimensions of the *tír cumaile* obviously depends on the interpretation of the length of the *forrach*, discussed above. If we take it to be 12 feet, the *tír cumaile* is 144 feet by 72 feet, i.e. the area of a fair-sized field.[131] This figure is comparable to that given for an *airchenn* in the law-text *Bretha Comaithchesa*.[132] It is defined here as being three *forrach*s and the spear-cast of a youth (*tēoru foirge* ⁊ *urcar flescaig*) in width: this could reasonably be calculated as adding up to about 96 feet (i.e. 3 x 12 + 60 = 96).[133] But the later commentators explicitly calculate the *forrach* as being 144 feet, which gives a much larger *tír cumaile* of 1728 feet by 864 feet. Thus one legal commentary states that the area of the *tír cumaile* is *ocht traigthi .xx.it ar .c.* ⁊ *.iiii. .xx.itt d'.xx.tib do traigtib hi fott tīri cum*aile ⁊ *a leth dia lethett* 'one hundred and twenty-eight feet and four scores of scores of feet in the length of the *tír cumaile* and half that in its width'.[134]

For the military area known as the *trícha cét*, see James Hogan, 'The *tricha cét* and related land-measures'.

(4) Liquid capacity

Measures of liquid capacity in our sources generally refer to drinks, especially milk. However, the smallest measure, *lán blaísce na cnó* 'the fill of a nut-shell', is attested only in relation to blood. As in other languages, terms originally meaning 'container' are sometimes applied to specific measures.

[127] *CIH* vi 2311.8 = Binchy, '*Bretha Dēin Chę́cht*', 36 § 22.
[128] *CIH* ii 563.6; iii 779.26 = *CG* 6.153.
[129] The difficult passage at *CIH* ii 675.30–7 = Mac Niocaill, '*Tír cumaile*', 82.16–26, contains the only Old Irish example which I have noted.
[130] *CIH* ii 675.32–3 = Mac Niocaill, '*Tír cumaile*', 82.19–20; *CIH* i 299.31 = *AL* iii 334.22–3; *CIH* iii 1105.5–6.
[131] See E. MacNeill, 'Ancient Irish law', 286–7².
[132] *CIH* i 78.9; 200.12 = *AL* iv 138.15–16.
[133] For a discussion of the length of a spear-cast, see p. 567 above. Old Irish commentary on *Bretha Comaithchesa* assigns a shorter *airchinn* to the *bóaire*: *tēora foirrge i n-aircenn tīre in bōairech* 'three *forrach*s in the headland of the land of a *bóaire*' (*CIH* iii 907.30).
[134] *CIH* iii 1105.6–7. Binchy reads *.uii. .xx.itt* at 1105.7 but the MS definitely has *.iiii. .xx.itt*. The same figure is doubtless intended at *CIH* iii 1094.30–4 *ocht traighthi .xx.it ara .x.* (emend to *ar .c.*) ⁊ *cethre .xx.it do .xx.ithbh* (= *.xx.tibh*) *traigh a fod tíre cumaile* ⁊ *a leth dia lethat*. At *CIH* iii 1043.10 it is necessary to supply *ar míle* after *troigh* to give a length of 1728 feet.

For example, *cingit* is normally a general term for a goblet, but occasionally is used to signify a precise measure of liquid.[135] The opposite development also takes place: Irish *muide* 'vessel (generally for milk)' is from the Latin term of measurement *modius*, used especially of corn.[136]

Lán blaísce na cnó (= approx. 1/20 fluid ounce)

The smallest liquid measure used in our sources is the half-shell of a hazel-nut. In a discussion on liability for the shedding of small quantities of blood, legal commentaries refer to a measure of blood which is the fill of a nut-shell (*lán blaísce na cnó*).[137] This measure is also called *lán crotbaill na cnó* 'the fill of the skin of the nut',[138] and *lán lestair na cnó* 'the fill of the vessel of the nut', i.e. the half-shell viewed as a container.[139] In one commentary, this measure is said to contain five drops (of blood).[140] Half the shell of a hazel-nut contains approximately 0.05 fluid ounces.

Lán uige circe (= approx. ½ fluid ounce)

The smallest measure of liquid capacity given in the table of measures in commentary on *Bretha Étgid* is the fill of a hen-egg (*lán uige circe*).[141] This is a convenient measure for liquids, as it is of fairly constant size and easily available. As the fill of a nut-shell (*lán blaísce na cnó*) clearly refers to the liquid which can fit in half a nut-shell, it seems likely that the 'fill of a hen-egg' in our sources refers to what is contained in a halved egg-shell.[142] In one non-legal source, this is stated explicitly: *leathlán blaoisge uighe circe* 'the half fill of the shell of a hen-egg'.[143] One can compare the use of the fill of an egg-cup in modern cookery.

As we have seen (p. 104), an Old Irish quotation in *O'Davoren's Glossary* gives the circumference of a hen-egg as four inches, which is about the same as a small pullet or bantam egg.[144] A half egg-shell of this size contains approximately half a fluid ounce, i.e. 14.4 millilitres.

[135] E.g. E. J. Gwynn, *The rule of Tallaght*, 22 § 38 *an tomhas dā ngoirthi cingid* 'the measure which was called a *cingit*'.
[136] *LEIA* M-71 s.v. *muide*.
[137] E.g. *CIH* i 123.24, 26.
[138] *CIH* iii 815.36.
[139] *CIH* iii 815.30, 32; 1022.11, 12.
[140] *CIH* iii 815.22 *.u. bainde besa cnūlestar lān* 'five drops which are the fill of a nut-shell'.
[141] *CIH* i 299.33 = *AL* iii 334.24.
[142] However, Joyce, *A social history of ancient Ireland* ii 375–6, evidently takes this measure to comprise the fill of a whole egg-shell.
[143] See *DIL* B 111.86–112.1.
[144] *CIH* iv 1521.21 = O'Dav. 440 § 1375.

Cern (= approx. 2 fluid ounces)

According to *O'Davoren's Glossary*, there are four fills of a standard-sized egg in a *cern.*[145] This measure is equated with the *selann,*[146] a unit used particularly in monastic diets with regard to a small extra delicacy, usually of butter or honey (see p. 345).

Meisrín (= approx. 6 fluid ounces)

Legal commentary on *Bretha Étgid* states that there is the fill of twelve hen-eggs in a *meisrín,*[147] a term which is a diminutive of *mesar* 'measure', a borrowing from Latin *mensura.*[148] This measure can be equated with the Roman *sextarius* (a sixth of a *congius*), which is defined in a gloss on the *Penitential of Cummean* as being 'the fill of twelve hen-eggs' (*.xii. plenitudo ouorum gall(inaceorum)*).[149] See *sesrae* below.

Sesrae (= approx. 6 fluid ounces)

Latin *sextarius* (a sixth of a *congius*) was borrowed into Irish in the form *sesrae*, which is well attested in literary sources, e.g. *core . . . a ndechotar cetri ficet sesrai do lemlacht* 'a cauldron . . . into which twenty-four *sextarii* of milk went'.[150] See *meisrín* above.

Bochtán (= approx. 6 fluid ounces)

A liquid measure called a *bochtán* is commonly mentioned in penitentials and monastic rules. For example, an Old Irish penitential refers to a *bochtán* of buttermilk, and also to a prime *bochtán* (*prímbochtán*) of good milk.[151] The corresponding passage in the Latin Penitential of Cummean has *sextarius de lacte Romanus* 'a Roman pint of milk'.[152] This confirms the identity of the *bochtán* and the *sesrae* (*sextarius*): see above.

The main law-text on distraint states that a low-ranking defendant is only entitled to have a *bochtán* to drink while in custody.[153] The accompanying gloss explains this measure as *lán eine in boicht* 'the fill of a pauper's cup', and goes on to equate it with the *meisrín*, and to define its capacity as being the fill of twelve hen-eggs.

[145] *CIH* iv 1526.10 = O'Dav. 461 § 1484.
[146] Cf. Meyer, *Sanas Cormaic* 90 § 1051 *Puingin .i. selland imme .i. selland cernæ.*
[147] *CIH* i 299.33 = *AL* iii 334.24.
[148] For *mesar*, see Appendix A, text 1 § 5 (1).
[149] *IP* 112 II § 2.
[150] Stokes, 'The second battle of Moytura', 86 § 89; Gray, *Cath Maige Tuired*, 46 § 89.
[151] E. J. Gwynn, 'An Irish penitential', 140–2 §§ 6–7.
[152] *IP* 114 § 3. Bieler mistakenly translates 'a pint of Roman milk'.
[153] *CIH* ii 363.26; v 1678.1; 1899.19 = *AL* i 106.2.

Foilderb, ollderb (= approx. 3⅗ pints)

Commentary on *Bretha Étgid* states that twelve *meisríns* go to make up an *ollderb.*[154] The form *ollderb* looks like a compound of *oll-* 'great' + *derb* 'pail, vessel'. However, in another legal commentary, what seems to be the same word is spelled with an initial *f-*: *inann an cuadh acon cerd ⁊ an follearbh acon sær, ⁊ inann in meisrín acu leiss* 'the mug is the same to the craftsman as the pail is to the wright, and the small measure is the same to them as well'.[155]

I suggest that the word – both for the vessel and the unit of measurement – is *foilderb,* a compound of *foil* 'ring' + *derb* 'pail'. One can compare the legal gloss which explains *fo*[*i*]*lderb* as 'a vessel with a ring on its side, a bell-shaped container' (*bis ⁊ in foil asa taīb, in cua cluic*).[156] The dat. sing. of the word is found in Triad 75, where the readings of the MSS are *foildeirb, folldeirb, foilleirb.*[157]

Ól (variable)

The evidence on the *ól-*measure is complex. It is likely that the suggested etymology given in *DIL* O 132.5 is correct, and that it is the same word as *ól* (earlier disyllabic *óol*) 'drinking, draught'. However, the citations in the texts indicate a fairly large amount of liquid, much more than could be consumed in a single draught. Thus an Old Irish law-text on the interest (*somoíne*) which must be given in exchange for advances of livestock includes *ian ōil lān de trīb asuib* 'a pail of an *ól-*measure full of three milks', i.e. a pail of fresh milk, a pail of buttermilk, and a pail of cream'.[158] A reference in Broccán's hymn in praise of Saint Brigit indicates an *ól-*measure of similar size.[159] According to an accompanying gloss, an *ól-*measure of mead (*ól meda*) fills a vat (*dabach*), and is part of the entitlements of the king of Leinster. Also of relevance is an entry in the *Annals of Ulster* which describes a monstrous six-legged cow seen on Dalkey Island, which gave three *ól-*measures of milk at each milking.[160] Whatever about the factual basis of this story, it is at least clear that the annalist regarded three *ól-*measures as a fair milk-yield.

Legal commentary on *Bretha Étgid* distinguishes a larger and a smaller *ól-*measure.[161] The smaller measure consists of twelve *ollderbs* (= 43.2 pints),

[154] *CIH* i 299.33 = *AL* iii 334.y-z.
[155] *CIH* v 1571.29–30.
[156] *CIH* ii 371.8–9 = *AL* i 124.5; 134.5.
[157] Meyer, *Triads*, 10 § 75.
[158] *CIH* iii 920.12–13. Cf. 920.32–3 *ian ōil lān di mellit* 'a pail of an *ól-*measure full of hydromel' (see p. 113 above).
[159] *Thes.* ii 347.85.
[160] AU^2 186 s.a. 732 (recte 733) § 11 *Uacca uisa est i nDelggenis Cualann . . . do-omlacht fo trī ōl n-ais caich mbleguin.*
[161] *CIH* i 299.33–5 = *AL* iii 334.25–336.2.

and is called an *ól medach* '*ól* of mead' or an *ól Pátraic* '*ól* of Patrick'.[162] According to a legal gloss, two of these measures go to make up an *ól Féne* '*ól* of the laity'. An *ól Pátraic* is said to be the appropriate amount of drink for twelve clerics or six laymen, presumably at a feast.[163] This would give each cleric a ration of 3.6 pints. The reason for the smaller measure in the case of the clerics is stated to be so that they may not become drunk, and fail to observe the canonical hours.[164]

In general, it seems that the standard *ól*-measure is the *ól Féne.*[165] In *O'Davoren's Glossary* the depth of a vessel which contains an *ól Féne* is given as three fists and three fingers (*trī duirn ⁊ tri mēoir*), i.e. about 15 inches.[166] However, this measurement does not tally with the figures given in the *Bretha Étgid* commentary, unless the vessel has a very wide diameter.

Lestar lulaice

The word *lestar* is used of a wide variety of containers, ranging from a nutshell (see under *lán blaísce na cnó* above) to large vessels for milk or ale.

In both Old and Middle Irish legal commentary, there are a number of references to a liquid measure called *lestar lulaice,* lit. 'a vessel of a milch-cow (*la*[*u*]*lgach*)'. This must be the same as *coire lulaice* 'a cauldron of a milch-cow', which occurs in the law-text *Cáin Aicillne* in the phrase *caire lulaice lan do lemnacht* 'a milch-cow cauldron full of milk'.[167] Thurneysen took this to mean 'a cauldron in which a milch-cow can be cooked'.[168] In the discussion by Charles-Edwards and myself on the occurrences of *lestar lulaice* in glosses and commentary to *Bechbretha,*[169] we follow Thurneysen's translation, and compare the phrase *caire i talla boin co tinniu* 'a cauldron in which there is room for a cow and a flitch (of bacon)' in *Críth Gablach.*[170]

However, this interpretation conflicts with that of the *Cáin Aicillne* glossator, who explains *caire lulaice* as *.i. as fiu lulgaidh* 'i.e. which is worth a milch-cow'.[171] We now believe that the glossator is right, and would

[162] I would guess that a smaller measure is used for mead because of its high alcoholic content. Another commentary (*CIH* v 1571.28–9) distinguishes the *ól Pátraic* and the *ól medach,* but here the discussion seems to be on different types of container rather than on different measures.

[163] *CIH* ii 474.13, 17 = *AL* v 414.1, 7–8 *fuillem cacha lethōile .i. lethōil fēni nō ōl pātraic, sesiur do laechaib nō dā fer dēc do clēirchib* 'the interest of every half *ól,* i.e. half an *ól Féne* or [a full] *ól Pátraic,* six laymen or twelve clerics'.

[164] *CIH* i 299.35–6 = *AL* iii 336.2–4.

[165] E.g. *CIH* ii 483.31, 484.2 = *AL* ii 254.3–4, 16 *lān ueine* (= *n-eine*) *trī n-ol .i. fēine* 'the fill of a container of three *ól*-measures, i.e. of the laity'.

[166] *CIH* iv 1496.40–1 = O'Dav. 330 § 807.

[167] *CIH* ii 483.30 = Thurneysen, 'Aus dem irischen Recht I', 356.1.

[168] Thurneysen, 'Aus dem irischen Recht I', 358.

[169] *BB* 99.

[170] *CIH* ii 567.24; iii 783.31–2 = *CG* 16.405–6.

[171] *CIH* ii 484.2 = *AL* ii 254.15–16.

compare the phrase *lestrae lóge bó* 'vessels worth a cow' in *Críth Gablach*.[172] Another strong argument in favour of the glossator's interpretation is the distinction made in legal commentary between other categories of vessel, e.g. *lestar bó inloíge* 'vessel of an in-calf cow', *lestar samaisce* 'vessel of a three-year-old heifer'. It would make no sense to consign an in-calf cow or a heifer to the cooking-pot. For the use of the *bó inláeg, samaisc,* etc. as units of value, see section (7) below.

One legal commentary gives the weight of the *lestar lulaice* as *dá dírna déc* 'twelve *dírna*' (see p. 586 below) and its dimensions as three fists in the width of its mouth and half that in depth.[173] This seems much too small to tally with the description of the *lestar lulaice* given elsewhere: see next section.

***Lestar colpthaige, lestar dairte,* etc.**

An Old Irish law-text on the lord and client relationship refers to a *lestar lulaice* 'vessel of a milch-cow' full of honey.[174] The text then implies that half this amount of honey is contained in a *lestar colpthaige* 'vessel of a two-year-old heifer', and a third in a *lestar dairte* 'vessel of a yearling heifer'.

Later legal commentary employs a full system of liquid measures based on the value of cattle. In the surviving material, this system is only attested in relation to measures of honey, but it may have also been used in other contexts. Commentary on § 6 of *Bechbretha* 'bee-judgements' discusses the amount of honey which a neighbour is entitled to receive from a bee-keeper in certain exceptional circumstances.[175] If the bees have produced a *lestar lulaice* of honey, the neighbour is entitled to a half-cup (*leithescrae*) of honey,[176] if they have produced a *lestar samaisce,* he is entitled to a third of a cup, if they have produced a *lestar colpthaige,* he is entitled to a quarter of a cup, and if they have produced a *lestar dairte,* he is entitled to a fifth of a cup. The *lestar lulaice* is defined as a vessel which can be lifted (full) by a man as high as his knee, the *lestar samaisce* can be lifted as high as his navel, the *lestar colpthaige* can be lifted as high as his temple, and the *lestar dairte* can be lifted above his head. Commentary at *CIH* vi 2104.32–4 gives an even more complex version of the same system, adding *lestar bo inloigi* 'vessel of an in-calf cow', and *lestar dartadha* 'vessel of a yearling bullock'.

Lestar úanmuilt

O'Davoren's Glossary gives the dimensions of a *lestar úanmuilt* 'vessel of a young wether'.[177] The width of its mouth is a male fist (*ferdorn*) – i.e.

[172] *CIH* ii 566.38; iii 783.5 = *CG* 15.364–5.
[173] *CIH* vi 2093.35–6; cf. 2104.35–6.
[174] *CIH* iii 920.32.
[175] *CIH* ii 445.10–13 = *BB* 52 § 6^{b}.
[176] Legal commentary at *CIH* vi 2094.1–2 takes the width of a cup to be one fist (*dorn*) at the mouth, i.e. one third of the width of a *lestar lulaice.*
[177] *CIH* iv 1491.17–18 = O'Dav. 306 § 680.

between 4 and 6 inches (see p. 564) – and its depth is half its width. Its width is thus the same as that given for the *escrae* 'cup' in commentary at *CIH* vi 2094.1.

Escop

As we have seen above (p. 319), the *escop* is defined in *Cormac's Glossary* as 'a jar for measuring wine among the merchants of the Gauls and Franks'.[178] It also occurs in commentary on *Muirbretha* 'sea-judgements' as a measure of honey: *escup fina no mela* 'a jar of wine or honey'.[179]

(5) Dry capacity

Bas, glacc, dorn (= approx. ¼ pint)

As we have seen above, the human hand is commonly used as a measure of length. What can be contained in the hand is also a convenient measure for small quantities of dry matter, particularly corn, e.g. *bas gráin* 'a handful of grain', *glacc arbae* 'a handful of corn'.

A handful can be taken as roughly equivalent to one quarter of a modern UK pint = 142 millilitres.

Mám (= approx. ½ pint)

Another unit of dry capacity is the *mám*, e.g. *atnaig Ciarán mám don grán 'na ucht* 'Ciarán flings a scoop of the grain into his bosom'.[180] Legal commentary on *Bretha Comaithchesa* states that the fine for rooting-trespass by a piglet is *ceithri máim* 'four scoops [of grain]'.[181] In a note on this passage, O'Donovan suggests that a *mám* is 'as much as can be taken up between the two palms of the hands held together'.[182] This would make it about twice as much as a *bas*.

Airmed

This term is a compound of the preposition *air-* + *med-* 'to measure', and is commonly used as a standard measure of dry matter.[183] For example, a list of items in a food-rent in *Críth Gablach* includes *fidlann airmeide di tharu* 'a wooden container with an *airmed*-measure of dried wheat'.[184] In some glosses, the *airmed* is identified with the *míach* 'bushel',[185] but in general the textual evidence indicates a measure between a *mám* and a *míach*. Thus

[178] Meyer, *Sanas Cormaic*, 45 § 539.
[179] *CIH* i 315.28 = *AL* iii 426.25.
[180] Stokes, *Lismore Lives*, 128.4322–3 .
[181] *CIH* i 191.15 = *AL* iv 98.19. The commentator also gives an alternative fine of half this amount.
[182] *AL* iv 99[5].
[183] Meyer, *Sanas Cormaic*, 45 § 553 *ermed .i. med tomais quia aridas res metitur* '*airmed*, i.e. a measure since it measures dry materials'.
[184] *CIH* iii 779.17 = *CG* 6.139–40. See Binchy's note at pp. 28–9.
[185] *CIH* iii 888.6 *foxal mēch airech .i. airmed* 'taking the bushel of a lord, i.e. an *airmed*'; cf. *CIH* ii 368.26, 28–30 = *AL* i 122.14, 128.5–7.

O'Davoren's Glossary takes an *airmed* to be the fill of a sieve (*críathar*). In this entry, it is stated to be a measure for corn or malt (*meth tomuis arba no bracha*),[186] but there are instances of its use to measure sloes,[187] and butter.[188]

As well as being employed as a unit of measurement, the word *airmed* may also be applied to a container which holds this amount. For example, a list of wooden vessels in *Bretha im Ḟuillemu Gell* includes the *airmed* along with the bucket (*síthal*), kneading-trough (*losat*), sieve (*críathar*), platter (*mías*), mug (*cúad*), etc.[189]

Míach (= approx. 50 pints (?))

The *míach* is a larger measure of dry capacity. It is generally used in relation to grain or malt, but is also attested as a measure of salt, charcoal, acorns, nuts and apples.[190] The law-text *Críth Gablach* states that a *mruigfer* normally sows sixteen such measures of seed-corn each year.[191] According to the same text, a farmer of this rank should have in his house at all times a *míach* of malt, a *míach* of salt, and a *míach* of charcoal.[192] The law-text on base clientship, *Cáin Aicillne*, distinguishes bullocks worth one, two, three or four *míach*s of corn:[193] see p. 588 below.

In the *Ancient Laws of Ireland*, the word *míach* is generally translated 'sack',[194] but this has connotations of the modern jute sack, so it seems better to use the translation 'bushel'. It is uncertain what material would normally have been used to contain a *míach* of grain, malt, salt, etc. Linen would presumably have been too valuable for such purposes, so it is likely that the ordinary material was wickerwork: see p. 244.

It is difficult to arrive at even an approximate figure for a standard early Irish *míach*. In a Middle Irish *Life of Saint Brigit* we find the statement: *ní rothecht acht ænmiach bracha*[195] 'she (Brigit) had only one bushel of malt'.[196] The corresponding passage in the Latin version has *illa non habuit nisi unum*

[186] *CIH* iv 1472.1–2 = O'Dav. 221 § 148.

[187] Meyer, *Aislinge*, 124.20.

[188] *CIH* iii 920.15 *fidlann airmide d'imm* 'a wooden container with an *airmed*-measure of butter'.

[189] *CIH* ii 472.3 = *AL* v 406.16; cf. *CIH* i 242.19 = *AL* v 486.14–15. It is likely that *airmed* means 'measuring-container (for corn)' in Meyer, *Triads*, 18 § 138.

[190] *DIL* s.v. *míach*.

[191] *CIH* ii 564.6–7; iii 780.13–14 = *CG* 8.196–7 *sé mēich deec i talmain.*

[192] *CIH* ii 564.2–3; iii 780.9–10 = *CG* 8.189–90. Cf. *CIH* iii 888.6–7.

[193] *CIH* ii 480.21; v 1782.39 = Thurneysen, 'Aus dem irischen Recht I', 347 § 8 *ag lōige mĕich*, etc.

[194] *AL* vi 566–7 s.v. *míach*.

[195] The Book of Lismore version equates *oencriathar bracha* with *oenmíach bracha* (Stokes, *Lismore Lives*, 41.1357, 1361).

[196] Hogan, *The Latin Lives of the saints*, 72; cf. Ó hAodha, *Bethu Brigte*, 7.191 § 21 *ataa óenmiach mbracha lind.*

modium, so we can take *míach* here to be roughly equivalent to Latin *modius*. A *modius* is generally sixteen *sextarii*. This gives a *míach* of about 50.4 pints = 28.8 litres. This quantity of corn would weigh slightly over 3½ stone (= 22.344 kilograms). This seems a convenient weight which could be carried without strain by an adult farm-worker of average strength. However, if a *míach* of seed-corn weighs 3½ stone, it follows that the *mruigfer* – the most prosperous grade of commoner – sows about 56 stone of seed-corn annually. In modern farming, the sowing-rate for cereals is 10–12 stone per statute acre. In the ridge-sowing system practised by the early Irish (p. 231), the rate must have been considerably less. But even allowing for a rate of 7 stone per acre, it only gives a figure of 8 acres under cereals on the farm of a *mruigfer*. This seems very little for a farmer who is stated to own a full ploughteam.[197]

Seisedach

There are two instances in the annals of *seisedach* 'a sixth' being used as a measure of dry matter. A marginal entry in the *Annals of Ulster* records such an abundance of nuts in the year 1097 that a *sesedach* could be bought for one penny.[198] An entry in the *Annals of the Four Masters* refers to a *seisedhach* of oaten grain being bought in the year 1031 for one penny.[199] In the same entry, *trian* 'a third' is used as a measure of sloes, acorns, and nuts. No doubt this measure was double a *seisedach*.

(6) Weight

The system of weight-measurements used in our sources is essentially the Roman one. This section should be studied in conjunction with the discussion on units of value in the next section, as one of the systems of value is based mainly on two weights of silver: the scruple (*screpul*) and the ounce (*ungae*).

A weighing scales (*med*) was used for the purpose of measuring metal and other goods, and our sources stress that measurements of weight should be accurate. *Bretha Nemed Déidenach* advises 'may you judge iron according to its weight'.[200] Similarly, an eighth-century account in the Book of Armagh *Additamenta* of a land-transaction describes various items of precious metal, and stresses that they were of the regular ancient valuations (*senmesib*) and ancient dimensions (*senairotib*).[201]

[197] *CIH* ii 563.25; iii 780.4 = *CG* 7.181.
[198] *AU*^{2} 532 s.a. 1097 § 8.
[199] *AFM* ii 822.8–13 s.a. 1031. This usage is presumably based on Latin *sextarius* 'sixth (of a *congius*)', which is used as a measure of both dry and liquid capacity: see also p. 577 above.
[200] *CIH* iii 1130.40 = E. J. Gwynn, 'Privileges', 41.23 *meisir iarn iarna thomhus*. Cf. Kelly, *Audacht Morainn*, 12 § 39.
[201] Bieler, *Patrician texts*, 174.9–10.

An entry in *Cormac's Glossary* refers to a *med tomais indile*,[202] which has been taken to mean 'a balance for weighing cattle'.[203] In the legal sources, however, the value of cattle is assessed in accordance with their girth-measurements and age, rather than their weight (see p. 60). I suspect, therefore, that *indile* in this context refers to smaller objects of value.

Atam (= $\frac{1}{576}$ scruple)

The smallest unit of weight given in our sources is the atom (*atam*, Latin *atomus*). According to legal commentary on *Uraicecht Becc*, there are twenty-four atoms in a grain of wheat.[204]

Gráinne cruithnechta (= $\frac{1}{24}$ scruple)

According to legal commentary on *Uraicecht Becc*, a grain of wheat (*gráinne cruithnechta*) weighs one twenty-fourth of a scruple of silver.[205] A grain of wheat is thus one eighth of the weight of a silver penny.[206] An entry in the Leabhar Breac version of *Cormac's Glossary* gives a slightly different equation under the headword *pissire*, which seems to be a term for a device for measuring small weights. Here a pennyweight is said to be the same as seven grains of true wheat.[207]

Screpul

As Cormac points out in his *Glossary*, *screpul* is a borrowing from Latin *scripulus*, a measure of weight which is one twenty-fourth of an *uncia* 'ounce'.[208] An H 3. 18 glossator explains that the *screpul* is named after a stone of this weight found in the Red Sea.[209] In another entry,[210] Cormac equates the scruple with the *puincne*, a unit of weight and value which is believed to be of Germanic origin.[211]

The diminutive *scriplíne* is used in legal commentary on *Cáin Lánamna* of small quantities of dye prepared from the roots of madder (*roid*).[212] In the tale *Erchoitmed Ingine Gulidi* it is used likewise of small quantities of oats.[213]

[202] Meyer, *Sanas Cormaic*, 90 § 1053.
[203] *DIL* M 76.80–81.
[204] *CIH* ii 637.34; v 1594.4 = *AL* v 18.8–9.
[205] *CIH* ii 637.33; v 1594.3–4 = *AL* v 18.7–8.
[206] *CIH* v 1571.22–3 *ocht ngrāinni cruithnechta comtrom na pinginni airgid.* In modern Troy measure, a pennyweight (dwt.) contains twenty-four grains.
[207] Stokes, *Three Irish Glossaries*, 35–6. *Pissire* is presumably from *pís* (Latin *pensum*) with the suffix of agency or instrument -(*a*)*ire*; a *píssire* is thus a 'weigher'.
[208] Meyer, *Sanas Cormaic*, 101 § 1150.
[209] *CIH* iii 1094.39–42.
[210] Meyer, *Sanas Cormaic*, 90 § 1050.
[211] *LEIA* P-16 s.v. *puing.*
[212] *CIH* i 177.14 = *AL* ii 420.5.
[213] Meyer, *Hibernica Minora*, 66 § 9.

Crosach (= 8 scruples (?))

The only attestation of this measure is in the account of the combat of Fer Diad and Cú Chulainn in *Táin Bó Cúailnge.*[214] It occurs in a description of the leaf-brooch of gold (*duileand-dealc óir*) which Queen Medb offered to Fer Diad as part of the bargain if he agreed to fight Cú Chulainn. The brooch is said to weigh *deich fichit unga* ⁊ *deich fichit leathunga* ⁊ *deich fichit crosach* ⁊ *deich fichit cethramthan* '200 ounces and 200 half-ounces and 200 *crosach*s and 200 quarter ounces'. It is implicit that a *crosach* is a measure between a half-ounce and a quarter-ounce, presumably one third of an ounce.

Crosach is most probably a derivative of *cros* 'cross'. One can compare the word *crosóc*, which is applied to a farthing in legal glosses (see p. 597).

Ungae (= 24 scruples)

Cormac makes the correct observation in his glossary that *ungae* is a borrowing from Latin *uncia* 'ounce'.[215] An *uncia* consists of twenty-four *scripuli*, and is one twelfth of a pound (*libra*).[216]

As well as being used as a unit of weight and of value (see next section), *ungae* is also attested as a unit of time. This usage is found in an Old Irish passage on computistics in the Codex Augiensis.[217]

Pún (= 288 scruples)

The Roman pound (*libra* or *pondus*) consists of twelve ounces. In the Old Irish St. Gall glosses, *pondus* is borrowed as *pond*,[218] but in other sources the form is *pún* (later *punt*). As in the Roman system, the Irish *pún* normally contains twelve ounces. For example, in the tale *Scél Mongán*, Mongán is reported as saying: *dobéræ pún findairgit duit fadéin hi fil dí ungi deac* 'you will get a pound of white silver for yourself in which there are twelve ounces'.[219] Similarly, another glossary takes a *pán* (= *pún*) to consist of twelve ounces.[220] However, *Scél Mongán* refers also to a *pún* of gold containing nine ounces.[221]

In modern Troy measure, there are 7000 grains in a pound: this is close to the early Irish figure of 6912 grains of wheat (288 x 24).

[214] *TBC* Rec. I 79.2599; Ó Fiannachta, *Táin Bó Cuailnge*, 46.1626.
[215] Meyer, *Sanas Cormaic*, 112 § 1292.
[216] *IP* 248, note 8.
[217] *Thes.* ii 10.7.
[218] *Thes.* ii 126.24: see *LEIA* P-12 s.v. *pond.*
[219] *LU* 336.11010–11 = Meyer, *The voyage of Bran* i (Appendix) 53.3–4.
[220] *CIH* ii 619.10; iii 1077.6 *Pán .i. pon[d] us .i. dī uinge .x.*
[221] *LU* 336.11015–16 = Meyer, *The voyage of Bran* i (Appendix) 53.8–9 *pun óir . . . i mbíat noí n-ungi.*

Dínnra, dírna (variable)

A short section of legal commentary provides a good deal of information on the system of weights used for various metals, though it is difficult to reconcile all the equivalences which are given. This commentary probably derives from a lost law-text *Bretha Créidini* 'the judgements of Crédine', which dealt with craftsmanship in metal, especially copper. Crédine was a famous coppersmith in early Irish mythology.[222]

The commentary is headed in large letters *dinnra clasaighe*, which seems to mean 'the denarius of the miner', and is likely to be a direct quotation from *Bretha Créidini*.[223] Latin *denarius* means 'consisting of ten' and originally referred to a silver coin containing ten asses. But it is also used in Latin as a unit of weight, and this usage is well attested in Irish.[224] Gwynn was the first to suggest a connection between Latin *denarius* and OIr. *dínnra* (also, with metathesis, *dírna*).[225]

According to this commentary, a miner's denarius (*dīnnra clasaighe*) weighs two ounces; presumably this refers to a weight of copper ore.[226] A craftsman's denarius (*dinnra cerda*) of red copper (*derguma*) weighs six ounces, and is worth a scruple. This is doubtless the same as the coppersmith's denarius (*dírn'a umaidhe*) mentioned in *O'Davoren's Glossary*. A legal gloss also refers to a *dirna umaidi di derguma* 'a coppersmith's denarius of red copper' and assigns it a value of two and a half pence.[227] As a scruple normally equals three pence, this is close to the commentary's equation of a craftsman's denarius of red copper with one scruple.

The commentary goes on to state that white copper (*finnuma*) has twice the value of the same weight of red copper.[228] An ounce of silver is the same weight as an ounce of copper (= red copper?), and red copper has the same value as tin (*sdān*). Finally,[229] four denarii of lead (*.iiii. dīrnna*

[222]See *GEIL* 269 § 16.

[223]*CIH* v 1571.18–24. This phrase is quoted in *O'Davoren's Glossary* (*CIH* iv 1491.15–16 = O'Dav. 306 § 679), with the instruction *fēch fēin bretha creīdhine* 7rl 'look at *Bretha Créidini* etc. yourself'.

[224]E.g. *LU* 12.330, where it is glossed *.i. dīrna ainm tomais mōir* 'i.e. denarius, the name of a large measure'.

[225]E. J. Gwynn, 'Some Irish words', 3. His suggestion is generally accepted: see McManus, 'A chronology', 43.

[226]The Egerton 88 version of *O'Davoren's Glossary* has *dīrna glasaigh* (*CIH* iv 1491.15–16 = O'Dav. 306 § 679), but the correct spelling *clasaighi* is found in Mac Firbhisigh's version.

[227]*CIH* iii 1107.18. The MS has *d-uma*.

[228]For the contrast between white and red copper, cf. Isidore, *Etymologiarum* (ed. Lindsay), 16.20.4, where a distinction is made between copper which is *candidus* 'white, shining' and *fulvus* 'reddish'.

[229]For the discussion in this commentary on the weight of a grain of wheat (*gráinne cruithnechta*), see p. 584 above.

do luaidhi) are said to be equivalent (in value) to one denarius of white copper (*dīnnra finnuma*). The lead, we are told, is used for soldering (*is do luaidhe donīther in tāth*).

In *O'Davoren's Glossary*, the *dírna* is defined as the name of a measure (*ainm tomuis*), but the glossator then states that each *dírna* is not the same, and distinguishes the *galldīrna* 'foreign denarius', the *dīrna clasaighi* 'miner's denarius', the *dīrna umaidhe* 'coppersmith's denarius', and the *dīrna Pātraig* 'Patrick's denarius'.[230] The *dírna clasaighi* and the *dírna umaidhi* are treated in the legal commentary which has been discussed above, but I have found no other references to the *galldírna* or the *dírna Pátraig*.[231]

(7) Units of value

In my *Guide to early Irish law*, I provide a brief account of the units of value employed in the Old Irish law-texts and in later glosses and commentary, and I discuss some of the discrepancies between the figures from different texts.[232] There seems to be no way in which the values from all the known sources can be reconciled in a single system. Some of these problems have been addressed by Marilyn Gerriets in 'Money in early Christian Ireland according to the Irish laws', and by Neil McLeod in 'Interpreting early Irish law: status and currency'. Recently, Thomas Charles-Edwards has produced a synopsis of the evidence in 'A note on Irish units of value'. This is Appendix A of his *Early Irish and Welsh kinship*.

Livestock

Female cattle are the most common units of value in our texts, and are given as fines, rents, bride-prices, tribute, etc. It has been demonstrated in Chapter 1 (p. 58), that the basic unit is the *bó mlicht* or *laulgach* 'milch cow', which is generally taken to be equivalent to 1 *ungae* or ounce (of silver) = 24 scruples = 2 *séts* = ⅓ *cumal*. The milch cow continued as a unit of currency in Ireland long after the Norman invasion. Thus an agreement made in 1456 between Patrick fitzWilliam Duf Haket and Richard fitzThomas Butler refers to a payment of 26s. 8d. and six milch cows or *lelyaghes* (= Ir. *l(a)ulgach*).[233]

Lesser values are commonly given in terms of younger cattle, sheep, hens or fleeces.

[230] *CIH* iv 1491. 15–16 = O'Dav. 306 § 679.
[231] The latter can be compared with the liquid measure called the *ól Pátraic*, discussed on p. 579 above. Gwynn ('Some Irish words', 5–6) compares the *dírna Pátraig* with the *denarius Petri*, and suggests that it refers to a tax levied by Armagh in Patrick's name.
[232] *GEIL* 112–16.
[233] Curtis, *Ormond Deeds* iii 177 § 195.

There is a good deal of variation with regard to the values assigned to the animals below the *colpthach*, but the general pattern indicated by law-texts, glosses and commentaries can be represented as follows:

bó inlóeg 'in-calf cow' = 18 scruples
samaisc 'three-year-old dry heifer' = 12 scruples
colpthach 'two-year-old heifer' = 8 scruples
colpthach fírenn 'two-year-old bullock' = 5⅓–6 scruples(?)[234]
dairt 'yearling heifer' = 4–6 scruples
dartaid 'yearling bullock' = 3–4 scruples[235]
cáera 'sheep' = 2–3 scruples[236]
úan méich 'lamb worth a bushel of grain' = 1–2 scruples
cerc cen rún 'non-broody hen' = ½ scruple
cnaí olla 'fleece of wool' = ½ scruple[237]

Older male livestock and horses also feature occasionally in the currency system. For example, *Bretha Étgid* states that a payment (*fíach*) of half a *cumal* should be made in a single currency, which the glossator explains as 'cattle or horses or silver'. A payment of a *cumal* should be made in two currencies, which the glossator explains as 'cattle and horses, or horses and silver'. A payment of more than one *cumal* should be made in three currencies. In this case, the text states that one third should be cattle, one third horses, and one third silver. A third of the cattle should be oxen, and a third of the horses should be mares.[238]

Míach

As we have seen in section (5) of this Appendix, *míach* 'bushel' is used in our sources as a measure of dry capacity. As a unit of value, *míach* commonly refers to a bushel of grain; for example, *ag lóige méich* is a bullock worth one bushel, *ag lóige dá míach* is a bullock worth two bushels, etc.[239] Charles-Edwards suggests that *míach* as a unit of value refers to a standard measure of malted barley,[240] but this is surely an unneccessary restriction: bushels of other cereals must also have been used as currency. In this connection it is worth noting that the Old Irish law-text *Bretha Nemed Toísech* assigns the same value of one scruple to a bushel of malt (*míach bracha*) and to a bushel of wheat (*míach cruithnechta*).[241]

[234]See discussion on p. 62 above.
[235]See discussion on p. 60 above.
[236]For a discussion of sheep-values, see p. 76 above.
[237]E.g. *CIH* v 1609.16 = *AL* v 80.15–16.
[238]*CIH* iv 1263.15–29 = *AL* iii 150.12–21; 152.3–8.
[239]*CIH* ii 480.21–482.34 = Thurneysen, 'Aus dem irischen Recht I', 347–54 §§ 8–11.
[240]Charles-Edwards, *Early Irish and Welsh kinship*, 478.
[241]*CIH* vi 2230.19–20; cf. *CIH* iii 806.2 (commentary) where a bushel of wheat has the value of one scruple.

This equation of 1 *míach* = 1 scruple is general in legal glosses and commentary, and is also indicated by the above-mentioned passage in *Bretha Nemed Toísech*. However, as Charles-Edwards points out in his study, there is also evidence in the Old Irish material of a higher value for the *míach* in relation to the scruple.[242] It is implicit in a passage on animal-trespass at *CIH* ii 576.7–9 that a cow is worth twelve *míach*s, which gives a *míach* of 2 scruples.

In conclusion, it seems safest to take the *míach* as fluctuating in value between 1 and 2 scruples.

Sét

The basic meaning of this term is 'precious object, jewel, item of value', and it survives in this general sense in Modern Irish *seod*. In the Old Irish law-texts, *sét* is often used in a more precise sense of a unit of value, usually equivalent to 12 scruples or half a milch cow.[243] *Uraicecht Becc* assigns the *ánruth* grade of poet an honour-price of 20 *séts*,[244] which corresponds to the 10 cows which he is given in *Bretha Nemed Déidenach*.[245] Similarly, an Old Irish gloss on *Din Techtugud* equates a *sét* with a three-year-old dry heifer (*samaisc*), which always has the value of 12 scruples.[246] Another gloss equates a *sét* with a *samaisc* or half an ounce (of silver).[247]

However, other texts give different values for the *sét*. In *Coibnes Uisci Thairidne* the *sét* is given the lower value of 10 scruples,[248] while in *Críth Gablach* it has the considerably higher value of four fifths of a milch cow, i.e. about 19 scruples.[249] This variation in the value of the *sét* is reflected in the complex terminology employed in the law-texts. Thus *Bretha Nemed Toísech* distinguishes the *sét indraic* 'standard *sét*' (which equals one *samaisc*) from the *sét gabla* 'forked *sét*'.[250] The difference between these two types of *sét* is difficult to establish, as there is much inconsistency in our sources. In a discussion on the status of poets, *Uraicecht Becc* assigns to the lowest

[242] Charles-Edwards, *Early Irish and Welsh kinship*, 483–4.
[243] I would query Charles-Edwards's claim that the *sét* 'was always primarily a unit of livestock' (*Early Irish and Welsh kinship*, 479).
[244] *CIH* v 1606.18 = *AL* v 68.17.
[245] *CIH* iii 1124.31 = E. J. Gwynn, 'Privileges', 32.22.
[246] *CIH* iii 910.31 *sēt .i. samaisc*. For the value of the *samaisc*, see p. 63.
[247] *CIH* iii 785.6 *leth n-uingi cach .s. nō samaisc*.
[248] *CIH* ii 460.34 = Binchy, '*Coibnes Uisci Thairidne*', 70 § 10 *for sēoit deich screbul*. Charles-Edwards points out that *sét* in this example might not be a unit of exchange, but simply have its basic meaning 'article of value'. See his *Early Irish and Welsh kinship*, 482.
[249] *CIH* iii 778.34 = *CG* 4.105 *ocht mba a turcrec, it ē .x.s. insin*.
[250] *CIH* vi 2227.38. The forked *sét* is probably so called because it commonly consists of two animals, a *dartaid* and a *colpthach*: see *GEIL* 115.

grade, the *fochloc*, an honour-price of one *sét gabla*.[251] The next grade is the *macḟuirmid*, who has an honour-price of three *séts*. A legal gloss takes a *sét gabla* to consist of a *samaisc* and a two-year-old bullock (*colpthach ḟirenn*), which would be approximately 1½ *séts*.[252] This equation would suit the *Uraicecht Becc* example quite well: the *fochloc* would have half the honour-price of the *macḟuirmid*. It would also suit a gloss on the *Díre*-text which takes the *sét gabla* to have the same value as the in-calf cow, which we have seen above to be 18 scruples.[253] However, *Cormac's Glossary* gives a totally different value for the *sét gabla*, taking it to be the smallest *sét*, with a value of less than a *samaisc*.[254]

In Cormac's view, there are three sizes of *sét*. The smallest is the *sét gabla*, which may be equivalent to the combination of a *dartaid ḟirend* 'yearling bullock' and a *colpthach boinend* 'two-year-old heifer'. The intermediate *sét* is the *samaisc* (= 12 scruples), and the largest is the *clitharṡét* 'protective(?) *sét*' or *rígṡét* 'royal *sét*', which is equivalent to a milch cow or a working ox (*lulgach nō dam timchill arathair*), i.e. 24 scruples.[255] The *clitharṡét* (also *sét clithair*) is given the same potential value in legal sources. The law-text on pledge-interests, *Bretha im Ḟuillemu Gell*, states that the value of the *sét clithair* extends up to that of a cow or an ounce (of silver).[256] Similarly, a gloss on *Din Techtugud* identifies a *clitharṡét* as a milch cow.[257] Cormac goes on to explain that a payment of *sét*s should be made in rotation, starting with a *sét gabla*, then a standard *sét*, then a *clitharṡét*, then back to a *sét gabla*, and so on until the end of the payment (*co rría cend na hérci*). This seems too tidy to represent universal practice, but it certainly makes good farming sense that a payment should be made in a variety of cattle, both from the point of view of the payer and the recipient. The arrival of seven milch cows as payment for some offence might not be particularly welcome in a farm which was already producing enough milk. In such cases, a mixture of cattle of different ages might be more acceptable. From the point of view

[251] *CIH* v 1604.12 = *AL* v 58.16; cf. *CIH* vi 2215.22 = L. Breatnach, *Uraicecht na Ríar*, 29.18.

[252] *CIH* iii 785.2–3 *S. gabla .i. samusc ⁊ colpthach ḟirenn ł cētloegh ⁊ colpthach boinen*[*n*].

[253] *CIH* iii 922.20–1 = Thurneysen, *Irisches Recht*, 6 § 6.

[254] Meyer, *Sanas Cormaic*, 19 § 209. A *sét gabla* of less than a standard *sét* is also indicated by an Old Irish passage at *CIH* ii 580.34–8 = *AL* iv 144.29–34. Here the fine for removing ten stakes from a fence is given as five *sét*s, but with the proviso that they are *séuit gabla* (*acht it .s. gabla*). If they were standard *sét*s, the fine would seem disproportionately heavy when compared with the fine for removing lesser numbers of stakes.

[255] The equation of the *clitharṡét* with the in-calf cow (*bó inlaíg*) is mentioned as a possibility ('*alii dicunt*') in *Cormac's Glossary*. This identification is also proposed as an alternative in a gloss on *Din Techtugud* at *CIH* iii 910.30: *Clithar.s. .i. laulgach nō daum timchella arathar nō buō inlaoge.*

[256] *CIH* ii 467.11–12 = *AL* v 392.1–2.

[257] *CIH* i 213.21 *clithear sēt .i. clēithe ar .s.i. loilgech.*

of the payer, the loss of all his milch cows might spell ruin for his whole household, whereas a mixture of cattle could be spared. In the law-texts there are also indications that fines were paid in the manner indicated by Cormac, though not to as rigid a formula. *Bretha Nemed Toísech* refers to a fine of 3 *séts*, which may be paid with a milch cow and a three-year-old dry heifer (*im lulgaig la samaiscc*).[258]

In some texts, reference is made to the way in which different types of valuables make up the amount of *séts* required for a particular fine or payment. *Bretha im Fuillemu Gell* states that the interest payable on a pledged mug belonging to a commoner is three *séts*. This should be paid in the form of a *sét faithche* 'a *sét* of the field', i.e. an animal, a *sét trebe* 'a *sét* of the house', i.e. a domestic vessel', and a *sét dia cholainn fodeissin* 'a *sét* of its own body', i.e. another mug.[259] An extract from the lost law-text on cats, *Catṡlechtae*, refers also to a payment of three *séts*, but in this case it must be paid in the three best *séts* which are in the house to the value of three half-ounces (of silver).[260] It would clearly be unacceptable to pay such a fine in cattle.[261] The law-texts sometimes refer to a *sét accobair* 'a *sét* of desire' in a payment. A glossator suggests that a *sét accobair* can be paid in cattle or gold or silver,[262] so it may be that the payer (or the recipient?) can choose the combination of valuables in such cases.

Cumal

There is evidence of the use of slaves as units of value in many societies. For example, the 'Welsh Canons' (*Canones Wallici*) contain a number of references to the giving of male slaves (*servi*) or female slaves (*ancellae*)[263] in payment for various offences.[264] In one case, the payment consists of a male **or** female slave, so it is clear that a slave of either sex is viewed as being of roughly the same value.[265] According to version A of these canons, the fine for slapping another person's face hard enough to cause

[258] *CIH* vi 2228.1 = *AL* iv 28.16–17.

[259] *CIH* ii 472.31–3 = *AL* v 410.6–8. Similar threefold payments are given elsewhere in this text. For example, the next paragraph refers to an interest payment of three *séts*, comprising a *lucharsēt faithche*, a *lucharsēt treibe* and a *lucharsēt dia cholainn fadeisin*. But what is a *lucharṡét* (or *luchair sét*)? Also, what is an *ec[h]ta sét* at *CIH* ii 465.31 = *AL* v 386.28?

[260] *CIH* v 1550.19–20 *direnur trī .s.aib .i. na trī .s. ata ferra bíte isin teigh co rōsat trī lethuinge.*

[261] For a brief discussion on restrictions in the forms of payment, see *GEIL* 115.

[262] *CIH* ii 476.8 = *AL* v 418.10.

[263] Classical Latin *ancillae*. Latin *ancilla* was borrowed into Irish in the form *aingel*, e.g. *CIH* vi 2218.6. See L. Breatnach, 'Canon law and secular law in early Ireland', 457.

[264] E.g. *IP* 136 §§ 1, 2, 3, 4; 138 §§ 8, 13, etc.

[265] *IP* 138 § 8.

bruising or bleeding is one female slave.[266] In version P the fine for this offence is stated to be one pound of silver (*argenti libra*).[267] This indicates the equation 1 slave = 1 pound of silver.

In Irish sources, there are also references to the use of slaves as currency. For example, the law-text *Bretha Déin Chécht* gives the penalty for wounding a king in the temple of the head as *fergniae dóer* 'an unfree workman'.[268] In another legal source, the sum paid as honour-price (*eneclann*) for the legendary king Conchobur includes fifty male slaves (*.l. moga*) and fifty female slaves (*.l. cumal*).[269] Most commonly, however, the term *cumal* is used in our sources to designate a unit of value rather than a slave. This is implicit in the saga of Fergus mac Léti, which distinguishes the *cumal* as an abstract unit of value – which may be realised in gold, silver or land – from the *duine-chumal* 'human-*cumal*', i.e. an actual female slave.[270]

It is difficult to fix the value of the *cumal* in relation to the other units used in our sources.[271] The legal glosses often refer to a *cumal trí mbó* 'a *cumal* of three cows', and this value is sometimes indicated by the Old Irish law-texts. In *Uraicecht na Ríar*, the honour-price of a chief poet (*ollam*) is given as forty *séts*, which is taken to be the same as seven *cumals*.[272] If we equate 2 *séts* with one milch cow (see above), this gives a *cumal* of nearly three cows. A discussion in *Míadślechtae* on the payment due to bishops for the destruction of an article of value indicates a similar value. If the owner is a celibate bishop, the fine is eight *cumals*; if the owner is a bishop with one wife, the fine is two thirds of that sum, which is given as five *cumals* and two cows.[273] This figure likewise presupposes a *cumal* of roughly three

[266] *IP* 138 § 13.
[267] *IP* 154 § 36.
[268] *CIH* vi 2306.34 = Binchy, '*Bretha Déin Chécht*', 26 § 6. An accompanying gloss takes this male slave to have the value of a *cumal trí mbó*, i.e. a female slave worth three cows.
[269] *CIH* v 1572.30.
[270] Binchy, 'The saga of Fergus mac Léti', 37 § 3. He emends *duine cam-* of the MS to *duine-chumal*.
[271] See discussion at *GEIL* 116; Charles-Edwards, *Early Irish and Welsh kinship*, 482–3.
[272] *CIH* vi 2336.6–12 = L. Breatnach, *Uraicecht na Ríar*, 102 §§ 2–3. See his note on p. 117.
[273] *CIH* ii 589.18, 21–2 = *AL* iv 366.23, 27–8. In *AL* the abbreviation *.s.* is wrongly expanded as *saiges* rather than *sét*.

cows. However, there are also references to *cumal*s of four,[274] five,[275] six[276] and ten cows.[277]

An added complication is the fact that the word *cumal* may sometimes be used in the non-specific sense 'value, payment, fine'.[278] This usage is illustrated in *Bretha im Ḟuillemu Gell*, where a distinction is made between an *ech cumal .x. sét* 'a horse of the value of ten *séts*' and an *ech cumal tríchat sét* 'a horse of the value of thirty *séts*'.[279] Similarly, an entry in the *Annals of Inisfallen* for the year 1093 records how Muirchertach took fifty cows from Ua Flaithbertaig in compensation for the death of the son of Cathal (*i cumalaib meic Cathail*).[280] Finally, *cumal* may also be used of an area of land (p. 574).

Weights of metal

As we have seen on p. 583, the Old Irish law-texts use the Roman system of weights based on the *screpul* 'scruple' (Latin *scripulus*), which is one twenty-fourth of an *ungae* 'ounce' (Latin *uncia*).[281] Values are often expressed in terms of a particular weight of precious metal, which is usually silver. For example, *Bretha im Ḟuillemu Gell* speaks of a pledge-interest reaching the value of an ounce of silver (*lóg n-uinge argit*).[282] One ounce of silver (= 24 scruples) is taken in both text and commentary to be worth a milch cow.[283]

There are few references in our sources to payments being made exclusively in gold, but not infrequently gold is included along with silver. For example, *Bretha Déin Chécht* refers to fines for injury to the face being

[274]E.g. *CIH* vi 2313.21 = Binchy, '*Bretha Déin Chécht*', 40 § 31^2.

[275]E.g. *CIH* i 35.15–16 = *AL* v 240.y.

[276]E.g. *CIH* iii 764.7–8; vi 1968.5. The *taurchrecc* 'fief' which a *mruigḟer* receives from his lord is *dí chumail* 'two *cumal*s' (*CIH* ii 564.11; iii 780.17–18 = *CG* 8.202). This must be more than the twelve cows in the fief of the grade below (*CIH* ii 563.9–10; iii 779.29 = *CG* 6.159).

[277]E.g. *CIH* v 1790.21–3 = Thurneysen, 'Aus dem irischen Recht I', 368 § 29; *CIH* ii 435.33.

[278]See *DIL* C 617.8–28.

[279]*CIH* ii 475.1, 4 = *AL* v 414.22, 24.

[280] *AI* 246 s.a. 1093 § 8.

[281]An exceptional value is given in an unedited passage in the Leabhar Breac (O'Longan transcript, p. 132, col. 2, ll. 2–3), where a royal tribute (*cís rígda*) includes a scruple of gold (*screpall óir*), which is said to be equivalent to one quarter of an ounce of gold (*.i. cethraimthi uinge óir*).

[282]*CIH* ii 464.1–2 = *AL* v 382.2; cf. *CIH* vi 2306.26 = Binchy, '*Bretha Déin Chécht*', 26 § 5; *CIH* vi 2114.2 = Dillon, 'Stories from the law-tracts', 45 § iv; *IP* 170 § 8; Bieler, *Patrician texts*, 174.8–9, 14.

[283]*CIH* ii 467.11–12 = *AL* v 392.1–2; *CIH* iii 845.35 = Appendix A, text 3 § 6; *CIH* vi 1975.4. An exception is to be found at O'Keeffe, '*Cáin Domnaig*', 208 § 30, where the ounce of the Law of Sunday (*unga Chána Domnaig*) is said to be a two-year-old heifer (*colpdach*).

paid in refined gold and refined silver.[284] In his study of units of value,[285] Charles-Edwards points out that the standard ratio between gold and silver in western Europe in the period *c.*700 was twelve to one. If we take an ounce of silver to be equivalent to one (milch) cow, this tallies exactly with a legal passage which equates an ounce of red gold with twelve cows (*uinge dergō[i]r ar dā būaibh dēcc*).[286] Metals other than silver or gold rarely feature in the currency system. As we have seen on p. 588 above, the law-text *Bretha Étgid* states that a payment of more than one *cumal* should be one third cattle, one third horses, and one third silver (*trian airgit*). Of the third of silver, one third may be of copper (*umae*).[287] In another law-text, the fine for illegal satire is said to include a third of tin or copper (*trian di stán ł da huma*).[288] Marilyn Gerriets suggests that payments in metal may sometimes have been made in finished goods, as well as in bullion form.[289]

In legal glosses and commentary, the scruple (*screpul*) is regularly equated with three pence (*trí pinginne*).[290] It is also identified in one gloss with a *dracaim* 'dram' (Latin *drachma*, Greek δραχμή) in the context of the Caesarian tribute (*cís cesarda*).[291] In another gloss,[292] it is equated with a silver shekel (*sical airgid*), Latin *sicolus* or *siculus* (*siclus*).[293] What seems to be a very much higher value is assigned to the shekel (*siclus*) in Irish Canon I, which states that the value of a female slave is twelve capons or thirteen shekels (*.xii. altilia uel .xiii. sicli praetium unius cuiusque ancillae*).[294] As we have seen above, the value of a female slave (*cumal*) normally ranges from about three cows up to ten cows, so this would give a shekel of at least 5½ scruples.[295] The nom. plur. *sicail* is sometimes used in glosses and commentary of the payment which a man makes to a

[284] *CIH* vi 2313.16–17 = Binchy, '*Bretha Déin Chécht*', 40 § 31 *dirirter bronnōr bronnarcat*; cf. *IP* 170 § 2. See discussion of this passage by Ó Corráin, 'Irish law and canon law', 164–5.

[285] Charles-Edwards, *Early Irish and Welsh kinship*, 482.

[286] *CIH* i 149.1. In another gloss (*CIH* vi 2266.39–40) the equation is made between one scruple of red gold and four scruples of silver, but the context here is obscure to me.

[287] *CIH* iv 1263.17–18 = *AL* iii 150.15–16 *trian do anfolam i triun airgit .i. uma indiu* 'a third of alloy(?) in the third of silver, i.e. copper nowadays(?)'.

[288] *CIH* vi 2113.32 = Dillon, 'Stories from the law-tracts', 45 § iv.

[289] Gerriets, 'Money in early Christian Ireland', 333.

[290] E.g. *CIH* iii 961.40; vi 2094.6.

[291] *CIH* iii 1082.39–40. See *DIL* s.vv. *dracmam, dragma.* The Classical Roman *drachma* was one eighth of an *uncia*, i.e. 3 scruples; cf. Isidore, *Etymologiarum* (ed. Lindsay), 16.25.13, 19.

[292] *CIH* iii 961.38–9; vi 2094.4.

[293] See Lewis and Short, *A Latin dictionary*, s.v. *siclus.*

[294] *IP* 160 § 9.

[295] Isidore, *Etymologiarum* (ed. Lindsay), 16.25.18, refers to a shekel among the Latins and Greeks of one quarter of an *uncia*, i.e. 6 scruples.

woman's father after taking her virginity. It is used both in the context of lawful marriage and of rape.[296] Thurneysen has shown that this usage goes back to Deuteronomy 22: 28–9 which states that a man who lies with an unbetrothed maiden must pay her father fifty shekels of silver (*quinquaginta siclos argenti*), and remain with her all his life.[297]

Screpul is often used in a general sense of 'fee, payment, tax, tribute'.[298]

Coinage

Coins have been in circulation in Britain since pre-Roman times. In the second century BC gold coins were minted in southern Britain, based ultimately on Macedonian prototypes. The Roman invasion of the first century AD brought in the Roman system, which used as its basic coin the silver *denarius.* Twenty-five *denarii* made up the gold *solidus,* though this ratio was subsequently reduced.

Roman coins have occasionly been discovered on Irish sites. As we have seen above (p. 586), the Latin word *denarius* was borrowed into Irish in the form *dínnra* (later *dírna*), which was mainly applied to a unit of weight.[299] Latin *obolus* is once attested in legal commentary on *Cáin Lánamna* in the form *abel,* and is taken to be equivalent to one penny.[300]

Archaeology provides no evidence for the presence of Anglo-Saxon coins in Ireland until the end of the eighth century.[301] It has been suggested that the Irish monetary unit known as the *oiffing* (*offaing, affaing*) is to be identified with the silver penny struck by King Offa of Mercia (757–96),[302] though the tenth-century glossator Cormac mac Cuilennáin clearly regarded the *oiffing* as a native coin, and defined it as *screpul nGaoidel* 'the scruple of the Irish'.[303] In *O'Davoren's Glossary* the *af(f)aing* is likewise identified with the scruple: *Afaing .i. screpul . . . .i. uinge ɫ a lōgh .i. iiii. scripuil .xx.it ɫ .iiii. afuingi .xx.it ɫ .iiii. puincni .xx.* '*Afaing* i.e. a scruple . . .

[296] *CIH* ii 502.23 = *AL* ii 342.21; *CIH* ii 370.18 = *AL* i 132.16; *CIH* v 1915.14, 17 = Thurneysen, *Irisches Recht*, 37; *CIH* i 147.32, 35, 36, 38.
[297] *SEIL* 124[1].
[298] *DIL* S 108.39–66.
[299] However, at E. J. Gwynn, 'An Irish penitential', 156 § 10, there is a reference to 'the worth of a *denarius* of private property' (*lóg ndīrnai di sainchrund*).
[300] *CIH* ii 509.26 = *AL* ii 368.24: see *SEIL* 35. However, it is usual in medieval Latin sources for the *obolus* to be worth half a penny (*denarius*), cf. Emanuel, *The Latin texts of the Welsh laws,* 363.30–1. The *obolus* was originally a Greek coin and measure of weight, and was one sixth of a *drachma* (Lewis and Short, *A Latin dictionary* s.v.). In the St. Gall glosses (*Thes.* ii 103.21) it is equated with a half-scruple (*lethscripul*).
[301] Dolley, 'Coinage to 1534: the sign of the times', 816.
[302] See Binchy, *Scéla Cano,* 22 (note to l. 46). For the suggestion that *oiffing* may be a borrowing from a contracted form of Latin *officina* 'workshop', see Bieler and Carney, 'The Lambeth commentary', 52.
[303] Meyer, *Sanas Cormaic,* 90 § 1050.

i.e. an ounce or its worth i.e. twenty-four scruples, or twenty-four *afuingi*, or twenty-four *puincni*'.[304]

In his study 'The earliest Irish coinage' William O'Sullivan concludes that the first Irish coins were minted under the Norse king Sihtric III of Dublin in the late tenth century, and were imitations of the silver pennies of King Ethelred II of England.[305] It is likely that the word *pinginn* (*pinning*, *penginn*) 'penny' came into the Irish language at this time from Old Norse *penningr*,[306] though it is also possible that it was an earlier borrowing from Anglo-Saxon *penning*. There are no attestations of *pinginn* in the Old Irish law-texts, but it is common in Middle Irish legal glosses and commentary, where it is generally equated with one third of a scruple.[307] According to legal commentary on metal values, a silver penny weighs eight grains of wheat (*ocht ngrāinni cruithnechta comtrom na pinginni airgid*).[308]

In an article on the date of some glosses on *Bretha Déin Chécht*, Michael Dolley has remarked that 'before the advent of the Vikings, Ireland was virtually coinless, while even in the later Viking period the use of coin was confined for all practical purposes to the Hiberno-Norsemen, and in the main to the south-eastern quadrant of our island. There was in this period but a single denomination, the penny, and there is no real evidence that the English and Scandinavian practice of halving and quartering was prevalent in Ireland; certainly Hiberno-Norse pennies in Irish hoards are never so treated. It would seem clear, too, that the Anglo-Normans came to an Ireland where even the Ostmen had relapsed into coinlessness. During the lordship of John, a coinage of two denominations, a halfpenny and a farthing apparently, was struck at several mints in Ireland for most of the last decade of the twelfth century, and there is some evidence that the coins did penetrate into parts of the country which remained essentially Gaelic'.[309]

The history of coinage in post-Norman Ireland is obviously closely linked with that of England and Scotland, and a detailed discussion of this topic is outside the scope of this book. Nonetheless, post-Norman coinage cannot be ignored altogether. Our sources from the twelfth to the sixteenth centuries show that both coinage and cattle coexisted as currency during this period. This is clearly illustrated in the glosses to a passage in the law-text *Bretha Déin Chécht* which gives the fines for

[304] *CIH* iv 1467.1–3 = O'Dav. 201 § 25. For *puincne*, see p. 584.

[305] O'Sullivan, 'The earliest Irish coinage', 191; Wallace, 'The English presence in Viking Dublin', 208.

[306] Marstrander, *Bidrag*, 62–3.

[307] Thurneysen, 'Aus dem irischen Recht II', 254.

[308] *CIH* v 1571.22–3.

[309] Dolley, 'The date of some glosses on *Bretha Déin Chécht*', 169–70. See also Dolley, *The Hiberno-Norse coins in the British Museum*; Dolley and O'Sullivan, 'The chronology of the first Anglo-Irish coinage'. For a re-appraisal of Dolley's conclusions, see Kenny, 'The geographical distribution of Irish Viking-age coin hoards', 516–20.

minor wounds to the face of persons of different rank. For example, the glossator states that a high lord is paid a milch cow for a wound between the brows, and an in-calf cow for a wound to the cheek. At the bottom of the social scale, an alien of *inol* rank is paid a *crosóc* for a wound between the brows, and a halfpenny (*lethpinginn*) for a wound to the cheek.[310] The word *crosóc* means '[the coin] with the cross', and Dolley suggests that it refers to the farthing of Edward I.[311] Many coins circulating in Ireland from Norse times onwards were marked with a cross, so this aspect of the identification is inconclusive. However, there is support from another legal gloss, which explains the *crosóc* as one quarter of a penny (*cethrime pinginne*), i.e. a farthing.[312] In commentary on *Uraicecht Becc*, the word for farthing is given as *feoirling*, a borrowing from Anglo-Saxon *feórþling* (later *feórþing*).[313] This commentary also refers to a half farthing (*lethfeoirling*), one seventh of a farthing (*.uii.madh feoirlingi*), and one fourteenth of a farthing (*.uii.madh lethḟeoirlinge*).[314] Commentary on *Bretha Comaithchesa* refers even to one hundredth of a farthing (*dechmadh in dechmaidh d'feorling*).[315] Dolley stresses that such tiny denominations make no sense to the numismatist.[316]

In non-legal sources from the post-Norman period, we likewise find evidence that both cattle and coinage were used as currency. The *Annals of Connacht* record the payment in the year 1400 of 126 cows as *éraic* in a case of accidental death,[317] and there is also annalistic evidence that the use of coinage extended to Gaelic Ireland. The *Annals of Loch Cé* refer to a minting of pence in Ireland by Henry III in the year 1252: *monadh nua d'ordugad do rig Saxan do dhenum an Erinn,* ⁊ *trécedh an airgid do bhí roime*

[310] *CIH* vi 2309.1, 14 = Binchy, '*Bretha Déin Chécht*', 30 § 13[2,11].
[311] Dolley, 'The date of some glosses on *Bretha Déin Chécht*', 170–1. He interprets *crosóc* as 'little-cross-coin', taking *-óc* to be used as a diminutive suffix. However, the suffixes *-óc* and *-án* do not merely have diminutive force. For example, *crossán* can mean 'little cross' (for churning milk), but more often it has the sense of 'one who carries a cross (in a religious procession)', i.e. jester. Similarly, *scolóc* does not mean 'little school', but 'one who is at school, scholar'. I suggest, therefore, that *crosóc* simply means 'the coin with the cross' rather than 'little-cross-coin'. Cf. Russell, *Celtic word-formation*, 109.
[312] *CIH* iv 1207.25–6.
[313] *CIH* ii 556.5–6, 36; cf. *CIH* iv 1233.33, 38. The presence of the *-l-* in the Irish form suggests an early borrowing, but Dolley (quoted above) remarks on the absence of the farthing in pre-Norman Ireland.
[314] One can compare the gloss at *CIH* i 226.29 = *AL* v 436.5, which refers to a *crosóc* and a seventh of a *crosóc* (*.uii.mad crosōigi*). This supports the identification of *crosóc* with the farthing.
[315] *CIH* i 74.5 = *AL* iv 118.14–15.
[316] Dolley, 'On the date of some glosses on *Bretha Déin Chécht*', 169.
[317] *AC* 374 s.a. 1400 § 8.

air sin 'new money was ordered by the king of England to be made in Ireland, and the money which was previously in use was abandoned for it'.[318] There are annalistic references to the groat (*bonn*), which was a coin worth four pence.[319] *Ceart Uí Néill*, a late sixteenth-century text on the rights of Ó Néill, also refers to various types of groat, such as the *bonn chroise caoíle* 'narrow-cross groat'.[320] The mark (*marc, marg*) was a denomination of weight, usually equivalent to eight ounces of silver. The *Annals of Connacht* record the payment of two hundred marks (*dā cēt marc*) in the year 1463 as part of the ransom for Tadhg Ó Conchobhair.[321] The shilling (Old English *scilling*) was used as a money of account in England from at least 900, and was valued at twelve pence since the Norman invasion of 1066. It was generally identified with the *solidus* (see p. 595 above). The first shilling coin was issued by Henry VII in 1503. The reference in the *Annals of Ulster* of the year 1497 to a milch cow worth a shilling (*an loilgech . . . ar sgilling*) belongs therefore to the period before the shilling coin.[322] In the *Annals of Inisfallen* of the year 1315, it is reported that in the Scottish camp (of Edward Bruce) food was so scarce that four quarters of a sheep were worth two shillings sterling (*quatuor quarteria ovis valuerunt duos solidos sterlinquorum*).[323] Under the year 1549, the *Annals of Loch Cé* record the exactions of Ruaidhri Mac Diarmada. Among the payments due to him were twenty shillings rent each year (*fiche sgillinn do chíos gacha bliadhna*) from the descendants of Tadhg, son of Brian Mac Donnchadha, as well as many tributes of cattle from other lords.[324]

Garments

Garments have been used as a form of currency in many societies.[325] In the Old Irish law-texts there are occasional references to this practice. Thus *Bretha Déin Chécht* states that the fine for an injury to the thigh consists of three-year-old dry heifers and linen mantles.[326] In the fine for satirizing King Cernodon of Ulster, another law-text includes a purple cloak worth seven ounces of silver (*brat corcra .uii. nuingi airgit*) and a mantle with its fine

[318] *ALC* i 398 s.a. 1252.
[319] E.g. *AC* 694 s.a. 1536 § 18.
[320] Dillon, '*Ceart Uí Néill*', 6.53 § 5. See also Dolley and Mac Niocaill, 'Some coin-names in *Ceart Uí Néill*'.
[321] *AC* 514 s.a. 1463 § 18. Cf. *AU* iii 422 s.a. 1497 *an mart ar mharg* 'a bullock [could be bought] for a mark'.
[322] *AU* iii 422 s.a. 1497.
[323] *AI* 418 s.a. 1315.
[324] *ALC* ii 354 s.a. 1549.
[325] For example, in some Slavic languages the verb meaning 'to pay' is a derivative of a noun meaning 'linen', e.g. Serbo-Croat *platiti* 'to pay', *platno* 'linen'.
[326] *CIH* vi 2312.32 = Binchy, '*Bretha Dein Chécht*', 40 § 30 *samaiscib dirirther lennaib scēo lennaib līn*. In his notes on p. 61, Binchy suggests omitting the first *lennaib* as dittography. Cf. *CIH* v 1572.29 *in .c. lend* '100 mantles' (in a fine).

hood which is worth twelve scruples (*lēne cona dagculpait bes fīu dā screpul* [*deug*]).[327] In the Book of Armagh *Additamenta,* we likewise find a garment worth half an ounce (*díllat leith ungæ*) included in the price of an area of land.[328]

[327] *CIH* vi 2114.3–4 = Dillon, 'Stories from the law-tracts', 45 § iv; *CIH* iii 1134.18–19 = E. J. Gwynn, 'Privileges', 46.27–8 (*deug* is supplied from this version). Cf. *CIH* vi 2218.14–15.
[328] Bieler, *Patrician texts*, 174.11.

Index 1
Irish index

References are to page-numbers. An 'n' after a page-number indicates that the occurrence of the word is in the footnotes. Unmarked verbal forms are in the 3rd singular present indicative; otherwise person, number and tense are supplied. A hyphen before a verbal form indicates that it is conjunct in the case of a simple verb or prototonic in the case of a compound verb.

The headwords are mainly given in their Old Irish form, and as far as possible in the same spelling as that used in the *Dictionary of the Irish language* (*DIL*). Alternative spellings are frequently supplied in brackets after the headwords, and are cross-referenced for the convenience of the reader.

Within the longer entries, most forms are quoted directly from texts in the *Corpus iuris hibernici* (*CIH*) without normalization. Consequently, there is considerable spelling variation. For instance, *caírech, cáerach, caorach, caírach* and *caíreg* are cited as genitive plurals of **cáera** 'sheep'.

The index includes forms in Texts 1–7 (pp. 503–59) which are discussed in the linguistic notes. In such cases the page-reference is to the discussion, and not to the occurrence(s) of the form in the Texts.

In a few cases, head-words are given in Modern Irish spelling, as there are no attestations from Old or Middle Irish, e.g. **ladhairíneach** 'bovine foul-in-the-foot'; cf. Old Irish *ladrach* 'with defective hoofs (of cattle)'.

The sign * before a form indicates that it is unattested. For example, there is no attestation of the 3sg. pres. indic. **con-fich** 'destroys', but the occurrence of the 3pl. pres. indic. *con-fechad* in Text 2 confirms its existence.

Index 2

Scottish Gaelic and Manx index

References are to page-numbers. An ‘n’ after a page-number indicates that the occurrence of the word is in the footnotes. Scottish Gaelic words are unmarked. Manx words are marked (M).

Index 3

Anglicized forms of Irish and Scottish Gaelic terms

This book is based on Old and Middle Irish texts from the pre-Norman period, but relevant information on farming, hunting, etc. is also supplied by later English sources. I list in this index the anglicized forms of Irish or Scottish Gaelic terms which have beeen quoted. After the headwords I give the Modern Irish or Scottish Gaelic spellings in brackets. The headwords of the latter are marked (S).

References are to page-numbers. An 'n' after a page-number indicates that the occurrence of the word is in the footnotes.

Index 4
Welsh, Breton and Cornish index

References are to page-numbers. An 'n' after a page-number indicates that the occurrence of the word is in the footnotes.

Modern Welsh spellings are unmarked. Middle Welsh spellings are marked (MW), and an Old Welsh spelling is marked (OW). Modern Breton spellings are marked (B), Middle Breton spellings are marked (MB), and Old Breton spellings are marked (OB). Modern Cornish spellings are marked (C), and an Old Cornish spelling is marked (OC).

Index 5
Latin index

References are to page-numbers. An 'n' after a page-number indicates that the occurrence of the word is in the footnotes. The initials of the scientific names for plants and animals are capitalized. The scientific names for cereals are based on the classification used by Zohary and Hopf, *Domestication of plants in the Old World* (2nd ed., Oxford 1993).

The spelling of some Late Latin words is variable: in such cases alternative spellings may be given in brackets after the main entry, e.g. **affrus** (*avrus, aver*) 'draught-animal'.

Index 6
General index

In this index of topics and proper names, references are to page-numbers. An 'n' after a page-number indicates that the reference is to the footnotes.

Bibliography

PUBLISHED SOURCES

Aalen, F. H. A., 'A note on transhumance in the Wicklow mountains', *JRSAI* 93 (1963) 189–90

——'Transhumance in the Wicklow mountains', *Ulster folklife* 10 (1964) 71–84

Abbott, T. K., and E. J. Gwynn, *Catalogue of the Irish manuscripts in the Library of Trinity College, Dublin* (Dublin 1921)

Adshead, S. A. M., *Salt and civilization* (London 1992)

Almqvist, Bo, *Viking ale: studies on folklore contacts between the Northern and the Western worlds* (ed. Éilís Ní Dhuibhne-Almqvist and Séamas Ó Catháin, Aberystwyth 1991)

Amoroso E. C., and P. A. Jewell, 'The exploitation of the milk-ejection reflex by primitive peoples', in *Man and cattle: proceedings of a symposium on domestication at the Royal Anthropological Institute, 24–26 May, 1960* (ed. A. E. Mourant and F. E. Zeuner. Royal Anthropological Institute of Great Britain & Ireland, London 1963) 126–38

Ancient laws of Ireland: see **Hancock, W. N.**

Anderson, A. O., and M. O. Anderson (ed.), *Adomnan's Life of Columba* (London 1961, repr. with new pagination Oxford 1991). I give both paginations and also the foliation of the Schaffhausen manuscript of this text.

An Seabhac (Ó Siochfhradha, Pádraig), 'Triocha-Céad Chorca Dhuibhne, II', 93–136, published as a supplement to *Béaloideas* 8 (1938)

Applebaum, S., 'Roman Britain', in *The agrarian history of England and Wales* I ii (ed. Finberg, Cambridge 1972) 1–277

Armitage, P., B. West and K. Steedman, 'New evidence of the black rat in Roman London', *London Archaeology* 4 (1984) 375–83

Ash, Harrison Boyd (ed.), *Lucius Junius Moderatus Columella on agriculture* i *De re rustica* 1–4 (The Loeb Classical Library, London and Cambridge (Massachusetts) 1941). See **Forster and Heffner**.

Atkinson, Robert (ed.), *The passions and the homilies from the Leabhar Breac: text, translation and glossary* (Todd Lecture series II, Dublin 1887)

——(ed.), *The Yellow Book of Lecan, a collection of pieces (prose and verse) in the Irish language, in part compiled at the end of the fourteenth century.* Photo-lithographic reproduction (Dublin 1896)

Baer, George M. (ed.), *The natural history of rabies* i–ii (Academic Press, New York 1975)

Baillie, M. G. L., 'A horizontal mill of the eighth century A.D. at Drumard, Co. Derry', *Ulster Journal of Archaeology* [third series] 38 (1975) 25–32

——'Dendrochronology: the prospects for dating throughout Ireland', in *Irish antiquity* (ed. Ó Corráin) 1–22

——*Tree-ring dating and archaeology* (London and Canberra, 1982)

Bannerman, John, 'The Scots language and the kin-based society', in *Gaelic and Scots in harmony: proceedings of the Second International Conference on the languages of Scotland* (ed. Derick S. Thomson, Glasgow 1988) 1–19

Bauersfeld, Helmut, 'Die Kriegsaltertümer im Lebor na hUidre', *ZCP* 19 (1933) 294–345

Baumgarten, Rolf, 'The kindred metaphors in 'Bechbretha' and 'Coibnes Uisci Thairidne', *Peritia* 4 (1985) 307–27

Beckett, Colm, *Fealsúnacht Aodha Mhic Dhomhnaill* (Baile Átha Cliath, 1967, repr. 1980)

Bergin, Osborn, (ed.), 'A story of Flann mac Lonáin', *Anecdota* i (ed. Bergin et al.) 45–50

——(ed.), 'What brought the Saxons to Ireland', *Ériu* 7 (1914) 244

——(ed.), 'Irish grammatical tracts', supplements to *Ériu* 8 (1916), 9 (1921–23), 10 (1926–28), 14 (1946), 17 (1955): I. Introductory (pp. 1–36); II. Declension (pp. 37–166); III. Irregular verbs (pp. 167–250); IV. Abstract nouns (pp. 251–7); V. Metrical faults (pp. 259–93)

——'White red-eared cows', *Ériu* 14 (1946) 170

——(ed.), *Irish bardic poetry, texts and translations, together with an introductory lecture.* Compiled and edited by D. Greene and F. Kelly (Dublin 1970, repr. 1984)

Bergin, O., and R. I. Best (ed.), '*Tochmarc Étaíne*', *Ériu* 12 (1938) 137–96

Bergin, O., R. I. Best, K. Meyer, and J. G. O'Keeffe (ed.), *Anecdota from Irish manuscripts* i–v (Halle 1907–13)

Best, R. I. (ed.), 'The adventures of Art son of Conn', *Ériu* 3 (1907) 149–73

——(ed.), 'The settling of the manor of Tara', *Ériu* 4 (1910) 121–72

——(ed.), 'An early monastic grant in the Book of Durrow', *Ériu* 10 (1926–28) 135–42

——(ed.), 'Amairgen son of Ecet Salach', *Irish texts* i (ed. Fraser, Grosjean, and O'Keeffe, London 1931) 32–4

——(ed.), 'Story of Mael Ruain of Tamlacht', *Irish texts* i (ed. Fraser, Grosjean, and O'Keeffe, London 1931) 34–5

——(ed.), 'Some Irish charms', *Ériu* 16 (1952) 27–32

Best, R. I., O. Bergin, M. A. O'Brien, and A. O'Sullivan (ed.), *The Book of Leinster, formerly Lebar na Núachongbála* i–vi (Dublin 1954–83)

Best, R. I., and O. Bergin (ed.), *Lebor na hUidre, Book of the Dun Cow* (Dublin 1929, repr. 1992)

Best, R. I., and Eóin Mac Neill (ed.), *The annals of Inisfallen: reproduced in facsimile from the original manuscript* (Dublin 1933)

Betel, Lisa M., *Land of women: tales of sex and gender from early Ireland* (Ithaca and London 1996)

Bieler, Ludwig (ed.), *The works of St. Patrick* (New York and London, 1953)

——(ed.), *The Irish penitentials*, with an appendix by D. A. Binchy (Scriptores Latini Hiberniae V, Dublin 1963, repr. 1975)

——(ed.). *The Patrician texts in the Book of Armagh*, with a contribution by F. Kelly (Scriptores Latini Hiberniae X, Dublin 1979)

Bieler, Ludwig, and James Carney (ed.), 'The Lambeth commentary', *Ériu* 23 (1972) 1–55

Binchy, Daniel A. (ed. with Myles Dillon), *Studies in early Irish law* (Dublin 1936)

——(ed.), '*Bretha Crólige*', *Ériu* 12 (1938) 1–77

——'Sick-maintenance in Irish law', *Ériu* 12 (1938) 78–134

——'*Aimser Chue*', in *Féil-sgríbhinn Eóin Mhic Néill* (ed. J. Ryan, Dublin 1940) 18–22

——(ed.), *Críth Gablach* (Mediaeval and Modern Irish Series XI, Dublin 1941, repr. 1979)

——(ed.), 'The saga of Fergus mac Léti', *Ériu* 16 (1952) 33–48

——(ed.), '*Coibnes Uisci Thairidne*', *Ériu* 17 (1955) 52–85

——'The fair of Tailtiu and the feast of Tara', *Ériu* 18 (1958) 113–38

——(ed.), 'The Old-Irish table of penitential commutations', *Ériu* 19 (1962) 47–72

——(ed.), *Scéla Cano meic Gartnáin* (Mediaeval and Modern Irish Series XVIII, Dublin 1963, repr. 1975)

——(ed.), '*Bretha Déin Chécht*', *Ériu* 20 (1966) 1–66

——'Varia. III', *Ériu* 20 (1966) 229–37

——(ed.), 'An archaic legal poem', *Celtica* 9 (1971) 152–68

——'Varia hibernica', in *Indo-Celtica* (ed. Pilch and Thurow, Munich 1972) 29–41

——'Distraint in Irish law', *Celtica* 10 (1973) 22–71

——(ed.), 'A text on the forms of distraint', *Celtica* 10 (1973) 72–86

——(ed.), *Corpus iuris hibernici* i–vi (Dublin 1978)

——'Brewing in eighth-century Ireland', in *Studies on early Ireland: essays in honour of M. V. Duignan* (ed. B. G. Scott, Belfast 1981) 3–6

Bischoff, Bernhard, *Latin palaeography: antiquity and the Middle Ages* (Berlin 1979; trans. D. Ó Cróinín and D. Ganz, Cambridge 1990)

Bisseru, B. (ed.), *Rabies* (London 1972)

Bois, D., *Les plantes alimentaires chez tous les peuples et à travers les âges: histoire, utilisation, culture* (Paris 1927)

Bonser, Wilfrid, *The medical background of Anglo-Saxon England: a study in history, psychology and folklore* (London 1963)

Bradley, John, 'A separate-bladed shovel from Moynagh Lough, Co. Meath', *JRSAI* 112 (1982) 117–22

——'The interpretation of Scandinavian settlement in Ireland', in *Settlement and society in medieval Ireland: studies presented to F. X. Martin* (ed. J. Bradley, Boethius Press, Kilkenny 1988) 49–78

Brady, Niall, 'Reconstructing a medieval Irish plough', in *I Jornadas internacionales sobre Tecnologia agraria tradicional*, mai 25–29 1992 (Museo Nacional del Pueblo Español, Madrid 1993) 31–44

——'Labor and agriculture in early medieval Ireland: evidence from the sources', in *The work of work: servitude, slavery, and labor in medieval England* (ed. A. J. Frantzen and D. Moffat, Glasgow 1994) 125–45

Breatnach, Liam (ed.), 'The caldron of poesy', *Ériu* 32 (1981) 45–93

——'On the agent suffix *-e* in Irish', *Ériu* 34 (1983) 194

——'Canon law and secular law in early Ireland: the significance of *Bretha Nemed*', *Peritia* 3 (1984) 439–59

——(ed.), *Uraicecht na Ríar: the poetic grades in early Irish law* (Early Irish Law Series II, Dublin 1987)

——'Lawyers in early Ireland', in *Brehons, serjeants and attorneys: studies in the history of the Irish legal profession* (ed. D. Hogan and W. N. Osborough, Dublin 1990) 1–13

——(ed.), 'The first third of *Bretha Nemed Toísech*', *Ériu* 40 (1989) 1–40

——'On the glossing of early Irish law-texts, fragmentary texts, and some aspects of the laws relating to dogs', in *Celtica Helsingiensia: proceedings from a symposium on Celtic Studies* (ed. A. Ahlqvist, Glyn Welden Banks, R. Latvio, H. Nyberg, T. Sjöblom, Helsinki 1996) 11–20.

Breatnach, Pádraig A., 'The chief's poet', *PRIA* 83 C (1983) 37–79

Breatnach, Risteard B. (ed.), *Seanachaint na nDéise II* (Dublin 1961). See **Sheehan, M**.

Brindley, A. L., J. N. Lanting, and W. G. Mook, 'Radiocarbon dates from Irish fulachta fiadh and other burnt mounds', *Journal of Irish archaeology* 5 (1989–90) 25–33

Bruck, Eberhard F., *Kirchenväter und soziales Erbrecht* (Berlin, Göttingen, Heidelberg 1956)

Butler, Colin G., *The world of the honeybee* (London 1954, revised ed. 1974)

Butler, Richard (ed.), *A treatice of Ireland by John Dymmok* (Dublin 1842), in *Tracts relating to Ireland, printed for the Irish Archaeological Society* ii (1843) 1–90

Byrne, Francis John, 'Tribes and tribalism in early Ireland', *Ériu* 22 (1971) 128–66

——*Irish kings and high-kings* (London 1973, revised ed. 1996)

Byrne, Mary E., '*Airec menman Uraird maic Coisse*', *Anecdota* ii (ed. Bergin et al.) 42–76

——'On the punishment of sending adrift', *Ériu* 11 (1932) 97–102

Calder, George (ed.), *Auraicept na nÉces: the scholars' primer* (Edinburgh 1917, repr. Dublin 1995)

Cameron, Alexander, *Reliquiae celticae: texts, papers and studies in Gaelic literature and philology* i–ii (Inverness 1892–94)

Campbell, John Lorne: see **McDonald, Allan**

Carey, John (ed.), '*Scél Tuáin meic Chairill*', *Ériu* 35 (1984) 93–111

——'Sequence and causation in *Echtra Nerai*', *Ériu* 39 (1988) 67–74

Carmichael, Alexander, *Carmina gadelica: hymns and incantations* i–vi (Edinburgh 1928–71; vols. v–vi ed. Angus Matheson)

Carney, James, (Ó Ceithearnaigh, Séamus) (ed.), *Regimen na sláinte: regimen sanitatis Magnini Mediolanensis* i–iii (Baile Átha Cliath 1942–44)

——(ed.), *The poems of Blathmac son of Cú Brettan, together with the Irish gospel of Thomas and a poem on the Virgin Mary* (Irish Texts Society XLVII, Dublin 1964)

——(ed.), 'The Ó Cianáin miscellany', *Ériu* 21 (1969) 122–47

Caulfield, Séamas, 'The beehive quern in Ireland', *JRSAI* 107 (1977) 104–38

——'The neolithic settlement of North Connaught', in *Landscape archaeology in Ireland* (ed. T. Reeves-Smyth and F. Hamond, British Archaeological Reports: British Series 116, London 1983) 195–215

Champion, Timothy, 'The myth of Iron Age invasions in Ireland', in *Studies on early Ireland* (ed. B. G. Scott, Belfast 1981) 39–44

Charles-Edwards, Thomas M., 'A comparison of Old Irish with medieval Welsh land-law' (Ph.D. thesis, University of Oxford, 1971)

——'Boundaries in Irish law', in *Medieval settlement: continuity and change* (ed. P. H. Sawyer, London 1976) 83–7

——'The Church and settlement', in *Irland und Europa: Die Kirche im Frühmittelalter* (ed. P. Ní Chatháin und M. Richter, Stuttgart 1984) 167–75

——'Críth Gablach and the law of status', *Peritia* 5 (1986) 53–73

——*Early Irish and Welsh kinship* (Oxford 1993)

Charles-Edwards, Thomas, and Fergus Kelly (ed.), *Bechbretha: an Old Irish law-tract on bee-keeping* (Early Irish Law Series I, Dublin 1983)

Charles-Edwards, Thomas, Morfydd E. Owen and D. B. Walters (ed.), *Lawyers and laymen: studies in the history of law presented to Professor Dafydd Jenkins on his seventy-fifth birthday* (Cardiff 1986)

Céitinn, Séathrúin: see **Keating, Geoffrey**

Cleary, R. M., M. F. Hurley and E. A. Twohig (ed.), *Archaeological excavations on the Cork-Dublin gas pipeline 1981–82* (Cork 1987)

Clutton-Brock, Juliet, *A natural history of domesticated animals* (British Museum (Natural History) 1987)

——*Horse power: a history of the horse and the donkey in human societies* (British Museum (Natural History) 1992)

Colgrave, Bertram, and R. A. B. Mynors (ed.), *Bede's ecclesiastical history of the English people* (Oxford 1969)

Comyn, David: see **Keating, Geoffrey**

Connolly, Seán, and Jean-Michel Picard (ed.), 'Cogitosus: Life of Saint Brigit', *JRSAI* 117 (1987) 5–27

Cooney, Gabriel, and Eoin Grogan, *Irish prehistory: a social perspective* (Dublin 1994)

Cosgrove, Art (ed.), *A new history of Ireland* ii (Medieval Ireland 1169–1534) (Oxford 1987)

Craigie, William A., et al. (ed.), *A dictionary of the older Scottish tongue, from the twelfth century to the end of the seventeenth* i- (Chicago and London 1931–73; Aberdeen 1983-)

Crane, Eva, *The archaeology of beekeeping* (London 1983)

Crawford, E. Margaret (ed.), *Famine: the Irish experience 900–1900; subsistence crises and famines in Ireland* (Edinburgh 1989)

Cregeen, Archibald, *A dictionary of the Manks language* (Douglas 1835)

Crigger, Bette-Jane, ' "A man is better than his birth": identity and action in early Irish law', dissertation submitted for the degree of Doctor of Philosophy at the University of Chicago (1991)

Cunliffe, Barry, *Iron Age communities in Britain: an account of England, Scotland and Wales from the seventh century BC until the Roman Conquest* (London 1991)

Curran, Patrick Leonard, *Kerry and Dexter cattle, and other ancient Irish breeds: a history* (Dublin 1990)

Curtis, Edmund, 'Rental of the manor of Lisronagh, 1333, and notes on "betagh" tenure in medieval Ireland', *PRIA* 43 C (1935–37) 41–76

——(ed.), *Calendar of Ormond deeds* (1172–1603) i–vi (Dublin 1932–43)

Danaher, Kevin (Ó Danachair, Caoimhín), 'The flail and other threshing methods', *Journal of the Cork Historical and Archaeological Society* 60 (1955) 6–14

——*The year in Ireland* (Cork 1972)

Davies, Wendy, *An early Welsh microcosm: studies in the Llandaff charters* (London 1978)

——*The Llandaff charters* (Aberystwyth 1979)

——'The Latin charter tradition in western Britain, Brittany and Ireland in the early medieval period', in *Ireland in early medieval Europe: studies in memory of Kathleen Hughes* (ed. D. Whitelock, R. McKitterick, and D. Dumville, Cambridge 1982) 258–80

Davies, Oliver, and David B. Quinn, 'The Irish pipe roll of 14 John, 1211–1212', *Ulster Journal of Archaeology* [third series] 4, supplement, (July 1941) 1–76

Davis, Ralph Henry Carless, *The medieval warhorse: origin, development and redevelopment* (London 1989)

De Bhaldraithe, Tomás, 'Irish names of birds, plants, animals, fishes, and minerals', in *Selections from the Zoilomastix of Philip O'Sullivan Beare* (ed. Thomas J. O'Donnell, Dublin 1960) xxxvii–liv

——(ed.), *Cín Lae Amhlaoibh* (Baile Átha Cliath 1970)

De Brún, Pádraig, *Catalogue of Irish manuscripts in King's Inns Library, Dublin* (Dublin 1972)

De Vries-Edel, D. R., '*Máeláin Muilchi*', *Peritia* 1 (1982) 297–8

Dictionary of the Irish language: see **Quin, E. G.**

Dillon, Myles (ed.), 'Stories from the law-tracts', *Ériu* 11 (1932) 42–65

——(ed.), 'The taboos of the kings of Ireland', *PRIA* 54 C (1951–52) 1–36

——(ed.), 'The story of the finding of Cashel', *Ériu* 16 (1952) 61–73

——(ed.), *Serglige Con Culainn* (Mediaeval and Modern Irish Series XIV, Dublin 1953, repr. 1975)

——(ed.), *Lebor na cert: the book of rights* (Irish Texts Society XLVI, Dublin 1962)

——(ed.), '*Ceart Uí Néill*', *Studia Celtica* 1 (1966) 1–18

——(ed.), *Stories from the Acallam* (Mediaeval and Modern Irish Series XXIII, Dublin 1970, repr. 1984)

Dimock, James, F. (ed.), *Giraldi Cambrensis: Topographia hibernica et Expugnatio hibernica* (Giraldi Cambrensis Opera V, London 1867)

——(ed.), *Giraldi Cambrensis Itinerarium Kambriæ et Descriptio Kambriæ* (Giraldi Cambrensis Opera VI, London 1868)

Dinneen, Patrick S., *Foclóir Gaedhilge agus Béarla: an Irish-English dictionary* (Dublin 1927, repr. 1996)

——see also **Keating, Geoffrey**.

Dobbs, M. E. (ed.), 'On the graves of Leinster men', *ZCP* 24 (1954) 139–53

Doherty, Charles, 'Exchange and trade in early medieval Ireland', *JRSAI* 110 (1980) 67–89

——'Some aspects of hagiography as a source for Irish economic history', *Peritia* 1 (1982) 300–28

——'The basilica in early Ireland', *Peritia* 3 (1984) 303–15

——'The monastic town in early medieval Ireland', in *The comparative history of urban origins in non-Roman Europe* (ed. H. B. Clarke and Anngret Simms, British Archaeological Reports: International Series 255, Oxford 1985) 45–75

Dolley, Michael, *The Hiberno-Norse coins in the British Museum* (London 1966)

——'The date of some glosses on *Bretha Déin Chécht*', *Celtica* 8 (1968) 167–73

——'Coinage to 1534: the sign of the times', in *A new history of Ireland* ii (ed. Art Cosgrove, Oxford 1987) 816–26

Dolley, M., and G. Mac Niocaill, 'Some coin-names in *Ceart Uí Néill*', *Studia Celtica* 2 (1967) 119–24

Dolley, M., and W. O'Sullivan, 'The chronology of the first Anglo-Irish coinage', in *North Munster studies: essays in commemoration of Monsignor Michael Moloney* (ed. Etienne Rynne, Limerick 1967) 437–78

Dottin, Georges, *La langue gauloise: grammaire, textes et glossaire* (Paris 1920)

Down, Kevin, 'Colonial society and economy in the high middle ages', in *A new history of Ireland* ii (ed. Art Cosgrove, Oxford 1987) 439–91

Doyle, Andy, 'Weeds of arable land', supplement to *Irish Farmers' Journal*, 7 May 1994, pp. 45–53

Dresner, Stephen, *Units of measurement: an encyclopaedic dictionary of units both scientific and popular* (Aylesbury, England, 1971)

Du Cange, Charles, *Glossarium mediæ et infimæ Latinitatis* i–x (1883–87, repr. Graz 1954)

Duffy, Seán, 'King John's expedition to Ireland, 1210: the evidence reconsidered', *Irish historical studies* 30/117 (1996) 1–24

Duignan, Michael, 'Irish agriculture in early historic times', *JRSAI* 74 (1944) 124–45

Dumville, David, 'On the dating of the early Breton lawcodes', *Études celtiques* 21 (1984) 207–21

Dunlevy, Mairead, 'A classification of early Irish combs', *PRIA* 88 C (1988) 341–422

Dunn, Peter, *The goatkeeper's veterinary book* (Ipswich 1982, third ed. 1994)

Dwelly, Edward, *Faclair Gaidhlig agus Beurla le dealbhan: the illustrated Gaelic-English dictionary* (Glasgow 1901–11)

Edwards, H. J. (ed.), *Caesar: the Gallic war* (The Loeb Classical Library, London and Cambridge (Massachusetts) 1917, repr. 1963)

Edwards, Nancy, *The archaeology of early medieval Ireland* (London 1990, repr. 1996)

Emanuel, Hywel D. (ed.), *The Latin texts of the Welsh laws* (Cardiff 1967)

Eska, Joseph, 'OIr. *sercol*', *Celtica* 17 (1985) 53–8

Etchingham, Colmán, 'The implications of *paruchia*', *Ériu* 44 (1993) 139–62

Evans, D. Ellis, *Gaulish personal names: a study of some Continental Celtic formations* (Oxford 1967)

Evans, Estyn E., 'Some problems of Irish ethnology: the example of ploughing by the tail', in *Folk and farm: essays in honour of A. T. Lucas* (ed. Ó Danachair, Dublin 1976) 30–39.

Fairclough, H. Rushton (ed.), *Virgil* (1): *Eclogues, Georgics, Aeneid 1–6* (The Loeb Classical Library, Cambridge (Massachusetts) and London 1916, repr. 1978)

Feehan, John, and Grace O'Donovan, *The bogs of Ireland: an introduction to the natural, cultural and industrial heritage of Irish peatlands* (Dublin 1996)

Fenton, Alexander, *The Northern Isles: Orkney and Shetland* (Edinburgh 1978)

Fiachra Éilgeach (ed.), *Scéalta andeas* (Oireachtas 1920, Baile Átha Cliath)

Finberg, H. P. R. (ed.), *The agrarian history of England and Wales* I ii, AD 43–1042 (Cambridge 1972)

Flanagan, Marie Therese, 'Irish and Anglo-Norman warfare in twelfth-century Ireland', in *A military history of Ireland* (ed. Thomas Bartlett and Keith Jeffery, Cambridge 1996) 52–75

Fleuriot, Léon, 'Un fragment en Latin de très anciennes lois bretonnes armoricaines du vi[e] siècle', *Annales de Bretagne* 78 (1971) 601–60

Flower, Robin, *Catalogue of Irish manuscripts in the British Library [formerly British Museum]* ii (London 1926, repr. Dublin 1992)

Forster, E. S., and Edward H. Heffner (ed.), *Lucius Junius Moderatus Columella on agriculture* ii *De re rustica* 5–9 (The Loeb Classical Library, London and Cambridge (Massachusetts) 1954; iii *De re rustica* 10–12, London and Cambridge (Massachusetts) 1955). See **Ash, Harrison Boyd**.

Forsyth, A. A., *British poisonous plants.* Ministry of Agriculture, Fisheries and Food. Bulletin no. 161 (London 1954)

Fox, James B., and Fergus Kelly, 'Capercaillie in Ireland: is the evidence conclusive?', *Irish birdwatching* 4/2 (1995) 20–3

Fox, Peter (ed.), *The Book of Kells, MS 58, Trinity College Library Dublin: commentary* (Luzern 1990)

Fraser, J., P. Grosjean and J. G. O'Keeffe (ed.), *Irish texts* i–v (London 1931–3)

Fraser Darling, F., *Natural history in the Highlands and Islands* (London 1947)

Freeman, A. Martin (ed.), *Annála Connacht: the annals of Connacht* (Dublin 1944, repr. 1983)

Gautier, Achilles, *La domestication: et l'homme créa ses animaux* (Paris 1990)

Geiriadur Prifysgol Cymru: see **Thomas, R. J.**

Geraghty, Siobhán, *Viking Dublin: botanical evidence from Fishamble Street.* Medieval Dublin excavations 1962–81. Series C, vol. 2 (Royal Irish Academy, Dublin 1996)

Gerriets, Marilyn, 'Economy and society: clientship according to the Irish laws', *Cambridge Medieval Celtic Studies* 6 (1983) 43–61

——'Money in early Christian Ireland according to the Irish laws', *Comparative Studies in Society and History* 27 (1985) 323–39

Gilbert, John M., *Hunting and hunting reserves in medieval Scotland* (Edinburgh 1979)

Gilbert, John T., *Facsimiles of National Manuscripts of Ireland, selected and edited under the direction of the Right Hon. Edward Sullivan* i–iv (Dublin 1874–84)

Glare, P. G. W. (ed.), *Oxford Latin dictionary* (Oxford 1982, repr. 1984)

Gowen, Margaret, *Three Irish gas pipelines: new archaeological evidence in Munster* (Dublin 1988)

Graves, Eugene, *The Old Cornish vocabulary* (Ann Arbor, Michigan, 1965)

Gray, Elizabeth A. (ed.), *Cath Maige Tuired: the second battle of Mag Tuired* (Irish Texts Society LII, Dublin 1982)

Greene, David, '*Lafan: llyffant*', *Celtica* 2 (1952) 148–9

——(ed.), *Fingal Rónáin and other stories* (Mediaeval and Modern Irish Series XVI, Dublin 1955, repr. 1993)

——'The Irish war-cry', *Ériu* 22 (1971) 167–73

——'The chariot as described in Irish literature', in *The Iron Age in the Irish Sea Province* (ed. Charles Thomas), Council of British Archaeology Research Report 9 (1972) 59–73

——'OIr. *sopp*, Faroese *soppur*, W. *swp*', *Ériu* 26 (1975) 178–81

——'The influence of Scandinavian on Irish', in *Proceedings of the seventh Viking Congress, Dublin 1973* (ed. B. Almqvist and D. Greene, Royal Irish Academy, Dublin 1976) 75–82

——'The coming of the Celts: the linguistic viewpoint', in *Proceedings of the sixth international congress of Celtic Studies, Galway 1979* (ed. G. Mac Eoin et al., Dublin 1983) 131–7

——'*Cró, crú* and similar words', *Celtica* 15 (1983) 1–9

Groenman-van Waatinge, W., 'Field boundaries in Ireland', in *Irish antiquity* (ed. Ó Corráin) 285–90

Gruffydd, W. J., 'Pangur', *The bulletin of the Board of Celtic Studies* 7 (1935) 4.

Gulick, Charles Burton (ed.), *Athenaeus: the Deinosophists* vol. 2 (The Loeb Classical Library, Cambridge (Massachusetts) and London 1928, repr. 1967)

Gwynn, Aubrey (ed.), *The writings of bishop Patrick 1074–84* (Scriptores Latini Hiberniae I, Dublin 1955)

Gwynn, Edward J., (ed.) *The metrical Dindshenchas* i–v (Dublin 1903–35, repr. 1991)

——'Notes', *Ériu* 3 (1907) 190–3

—— (ed.), 'An Irish penitential', *Ériu* 7 (1914) 121–95

——'A note on O'Davoren's Glossary', *Ériu* 9 (1921–23) 157–8

——'*Miscellanea celtica*', *Hermathena* 20 (1926) 62–72

——(ed.), *The rule of Tallaght.* Second supplemental volume to *Hermathena* 44 (1927)

——(ed.), 'Athirne's mother', *ZCP* 17 (1928) 153–6

——'Some Irish words', *Hermathena* 21 (1931) 1–15

——(ed.), 'An Old-Irish tract on the privileges and responsibilities of poets', *Eriu* 13 (1942) 1–60, 220–36

Gwynn, E. J., and W. J. Purton (ed.), 'The monastery of Tallaght', *PRIA* 29 C (1911) 115–80

Gwynn, Lucius (ed.), 'The Life of St. Lasair', *Ériu* 5 (1911) 73–109

——(ed.), '*De Ṡíl Chonairi Móir*', *Ériu* 6 (1912) 130–43

Hamlin, Ann, 'The archaeology of the Irish church in the eighth century', *Peritia* 4 (1985) 279–99

Hamlin, Ann, and Chris Lynn (ed.), *Pieces of the past: archaeological excavations by the Department of the Environment for Northern Ireland 1970–1986* (Belfast 1988)

Hamp, Eric P. '*Imbolc, Óimelc*', *Studia Celtica* 14–15 (1979–80) 106–13

Hancock, W. Neilson, Thaddeus O'Mahony, A. G. Richey, William M. Hennessy and Robert Atkinson (ed.), John O'Donovan and Eugene O'Curry (text and trans.), *Ancient laws of Ireland* i–vi (Dublin 1865–1901)
Harbison, Peter, 'The Old Irish "chariot"', *Antiquity* 45 (1971) 171–7
——*The high crosses of Ireland: an iconographical and photographic survey* i–iii (Bonn 1992)
Hardiman, James (ed.), *Ancient Irish deeds and writings* (Royal Irish Academy Transactions XV, 3–95, 1826)
Harrison, S. G., G. B. Masefield, Michael Wallis, and B. E. Nicholson, *The Oxford book of food plants* (Oxford 1969)
Heist, W. W. (ed.), *Vitae sanctorum Hiberniae.* Subsidia hagiographica XXVIII (Bruxelles 1965)
Hemon, Roparz, *Geriadur istorel ar Brezhoneg: dictionnaire historique du Breton* (Brest 1958, Rennes 1959–79)
Hencken, Hugh O'Neill, 'Ballinderry crannóg no. 1', *PRIA* 43 C (1935–37) 103–239
——'Ballinderry crannóg no. 2', *PRIA* 47 C (1941–42) 1–76
——'Lagore crannóg: an Irish royal residence of the 7th to 10th centuries A.D.', *PRIA* 53 C (1950–51) 1–247
Henderson, George, *From Durrow to Kells: the Insular Gospel-books 650–800* (London 1987)
Henderson, Isabel, *The Picts* (London 1967)
Hennessy, William M. (ed.), *Chronicum Scotorum* (London 1866)
——(ed.), *The annals of Loch Cé* i–ii (London 1871)
Hennessy, William M., and Bartholomew MacCarthy (ed.), *Annála Uladh: annals of Ulster* i–iv (Dublin 1887–1901)
Henry, Françoise, 'Remarks on the decoration of three Irish psalters', *PRIA* 61 C (1960–61) 23–40
Henry, Françoise, and G. L. Marsh-Micheli, 'A century of Irish illumination (1070–1170)', *PRIA* 62 C (1961–63) 101–66
Henry, P. L. 'A linguistic survey of Ireland: preliminary report', *Lochlann* 1 (1958) 49–208
Herbert, Máire, 'Charter material from Kells', in *The Book of Kells* (ed. O'Mahony) 60–77
Herity, Michael, and George Eogan, *Ireland in prehistory* (London 1977, repr. 1996)
Herren, Michael W. (ed.), *The Hisperica Famina: I. The A-text* (Toronto 1974)
Hillman, G., 'Evidence for malting spelt', in *Excavations at Catsgore* (ed. Leech, Bristol 1982) 137–41
Hogan, Edmund (ed.), *Cath Ruis na Ríg for Bóinn* (Todd Lecture Series IV, Dublin 1892)
——*The Latin lives of the saints* (Todd Lecture Series V, Dublin 1894)
——*Onomasticon Goedelicum* (Dublin 1910, repr. 1993)

Hogan, James, 'The *tricha cét* and related land measures', *PRIA* 38 C (1929) 148–235

Holm, Poul, 'The slave trade of Dublin, ninth to twelfth centuries', *Peritia* 5 (1986) 317–45

Hooper, William Davis, and Harrison Boyd Ash, *Marcus Porcius Cato on agriculture; Marcus Terentius Varro on agriculture* (The Loeb Classical Library, Cambridge (Massachusetts) and London 1934, repr. 1979)

Howlett, David R. (ed.), *The book of letters of Saint Patrick the bishop* (Dublin 1994)

Hughes, Kathleen, *The Church in early Irish society* (London 1966)

Hull, Vernam (ed.), '*Aided Meidbe*: the violent death of Medb', *Speculum: a journal of mediaeval studies* 13 (1938) 52–61

——(ed.), 'Conall Corc and the Corco Luigde', *Publications of the Modern Language Association of America* 62 (1947) 887–909

——(ed.), *Longes mac n-Uislenn: the exile of the sons of Uisliu* (New York 1949)

——(ed.), '*Bretha im Gatta*', *ZCP* 25 (1956) 211–25

——(ed.), '*Cáin Domnaig*', *Ériu* 20 (1966) 151–77

——(ed.), '*Noínden Ulad*', *Celtica* 8 (1968) 1–42

——(ed.), '*Apgitir Chrábaid*: the alphabet of piety', *Celtica* 8 (1968) 44–89

Hurry, J. B., *The woad plant and its dye* (Oxford 1930)

Jackson, Kenneth (ed.), *Cath Maighe Léna* (Mediaeval and Modern Irish Series IX, Dublin 1938, repr. 1990)

——*Language and history in early Britain* (Edinburgh 1953, repr. Dublin 1994)

——(ed.), *The Gaelic notes in the Book of Deer* (Cambridge 1972)

——(ed.), *Aislinge Meic Con Glinne* (Dublin 1990)

James, Montague R. (ed.), *The apocryphal New Testament* (Oxford 1924, repr. 1955)

Jenkins, Dafydd (ed.), *The law of Hywel Dda* (Gomer Press, Llandysul, Wales, 1986)

Johns, C. H. W. (ed.), *The oldest code of laws in the world: the code of laws promulgated by Hammurabi, king of Babylon B.C. 2285–2242* (Edinburgh 1903)

Jones, W. H. S. (ed.), *Pliny, Natural History, books 20–23* (The Loeb Classical Library, London and Cambridge (Massachusetts) 1951, repr. 1961)

Joyce, Patrick Weston, *The origin and history of Irish names of places* i (Dublin 1869); ii (Dublin 1875); iii (Dublin 1913)

——*A social history of ancient Ireland* i–ii (Dublin 1920)

Kavanagh, Rhoda M., 'The horse in Viking Ireland', in *Settlement and society in medieval Ireland: studies presented to F. X. Martin* (ed. John Bradley, Boethius Press, Kilkenny 1988) 89–121

Keating, Geoffrey (Céitinn, Séathrúin) *Foras feasa ar Éirinn: the history of Ireland* i–iv (ed. David Comyn and Patrick S. Dinneen, Irish Texts Society IV, VIII–IX, XV, London 1902 for 1901, 1908 for 1905 and 1906, 1914 for 1913)

Kelly, Fergus (ed.), 'A poem in praise of Columb Cille', *Ériu* 24 (1973) 1–34

——(ed.), '*Tiughraind Bhécáin*', *Ériu* 26 (1975) 66–98

——'The Old Irish tree-list', *Celtica* 11 (1976) 107–24

——(ed.) *Audacht Morainn* (Dublin 1976)

——(ed.) 'An Old-Irish text on court procedure', *Peritia* 5 (1986) 74–106

——*A guide to early Irish law* (Early Irish Law Series III, Dublin 1988, repr. 1991)

Kelly, Patricia, 'Dialekte im Altirischen?', in *Innsbrucker Beiträge zur Kulturwissenschaft* 50 (ed. Wolfgang Meid, Hermann Ölberg und Hans Schmeja, Innsbruck 1982) 85–9

Kenny, Michael, 'The geographical distribution of Irish Viking-age coin hoards', *PRIA* 87 C (1987) 507–25

Kidson, Peter, 'A metrological investigation', *Journal of the Warburg and Courtauld Institutes* 53 (1990) 71–97

Knott, Eleanor, '*Bó thúir*', *Ériu* 7 (1914) 26

——(ed.), *Togail Bruidne Da Derga* (Medieval and Modern Irish Series VIII, Dublin 1936, repr. 1975)

——(ed.), 'A poem of prophecies', *Ériu* 18 (1958) 55–84

Lambert, Pierre-Yves, 'Les gloses bibliques de Jean Scot: l'élément vieil-irlandais', *Études celtiques* 22 (1985) 205–24

——*La langue gauloise* (Paris 1994)

Langdon, John, *Horses, oxen and technological innovation: the use of draught animals in English farming from 1066 to 1500* (Cambridge 1986)

Larson, Laurence M. (ed.), *The earliest Norwegian laws, being the Gulathing law and the Frostathing law* (New York 1935)

Le Duc, Gwenaël, 'Notes sur les anciennes lois bretonnes (*Canones Wallici*)', in *Bretagne et pays celtiques: mélanges offerts à la mémoire de Léon Fleuriot* (ed. Le Menn and Le Moing, Rennes 1992) 101–9

Leech, Roger (ed.), *Excavations at Catsgore 1970–73: a Romano-British village* (Bristol 1982)

Lehmann, Ruth (ed.), *Fled Dúin na nGéd* (Mediaeval and Modern Irish Series XXI, Dublin 1964)

Leman, A. D., et al., *Diseases of swine* (Iowa State University Press 1958, seventh ed. 1992)

Lewis, Charlton T., and Charles Short, *A Latin dictionary* (Oxford 1879, repr. 1966)

Lewis, Timothy, *A glossary of mediaeval Welsh law, based upon the Black Book of Chirk* (Manchester 1913)

Liebermann, F. (ed.), *Die Gesetze der Angelsachsen* i–iii (Halle 1903–16, repr. Aalen 1960)

Lindeman, Fredrik Otto, 'Archaic Irish *ferba fíra*: a speculative note', *Ériu* 41 (1990) 123–5

Lindsay, W. M. (ed.), *Isidori Hispalensis episcopi etymologiarum sive originum* i–ii (Oxford 1911)

Linnard, William, 'The nine huntings: a re-examination of *Y Naw Helwriaeth*', *The bulletin of the Board of Celtic Studies* 31 (1984) 119–32

Lockie, J. D., and D. Stephen, 'Eagles, lambs, and lamb management in Lewis', *Journal of animal ecology* 28 (1959) 43–50

Logan, Patrick, *Making the cure: a look at Irish folk medicine* (Dublin 1972)

Loth, Joseph, 'Les noms et les variétés du froment chez les Celtes insulaires', *RC* 41 (1924) 193–203

Lucas, A. T., 'The horizontal mill in Ireland', *JRSAI* 83 (1953) 1–36

——'Nettles and charlock as famine food', *Breifne: Journal of Cumann Seanchais Bhreifne* 1 (1959) 137–46

——'Irish food before the potato', *Gwerin* 3 (1960–62) 8–43

——'The sacred trees of Ireland', *Journal of the Cork Historical and Archaeological Society* 68 (1963) 16–54

——'Washing and bathing in ancient Ireland', *JRSAI* 95 (1965) 65–114

——'Sea sand and shells as manure', in *Studies in folk life: essays in honour of Iorwerth C. Peate* (ed. Geraint Jenkins, London 1969) 183–203

——'Irish ploughing practices (1)', *Tools and tillage* (Copenhagen) 2/1 (1972) 52–62; 'Irish ploughing practices (2)', *Tools and tillage* 2/2 (1973) 67–83; 'Irish ploughing practices (3)', *Tools and tillage* 2/3 (1974) 149–60; 'Irish ploughing practices (4)', *Tools and tillage* 2/4 (1975) 195–210

——'Toghers or causeways: some evidence from archaeological, literary, historical and place-name sources', *PRIA* 85 C (1985) 37–60

——*Cattle in ancient Ireland* (Kilkenny 1989)

Luce, A. A., et al. (ed.), *Evangeliorum quattuor Codex Durmachensis* i–ii (Urs Graf-Verlag; Olten, Lausanne and Freiburg 1960)

Lynn, Chris, 'Houses and related outbuildings in early Christian Ireland' (Ph.D. thesis, University College, Dublin, 1986)

——'The excavation of Rathmullan, a raised rath and motte in County Down', *Ulster Journal of Archaeology* [third series] 44–45 (1981–82) 65–171

——'Navan fort: a draft summary account of D. M. Waterman's excavations', *Emania: bulletin of the Navan research group* 1 (1986) 11–19

——'Deer Park farms, Glenarm, Co. Antrim', *Archaeology Ireland* 1/1 (1987) 11–15

——'Houses in rural Ireland, AD 500–1000', *Ulster Journal of Archaeology* [third series] 57 (1994) 81–94

Mabey, Richard, *Food for free* (London 1972, repr. 1992)

Mac Airt, Seán (ed.), *The annals of Inisfallen* (Dublin 1951, repr. 1988)
——'Lexicographical notes', *Celtica* 3 (1956) 262–71

Mac Airt, Seán, and Gearóid Mac Niocaill (ed.), *The annals of Ulster (to A.D. 1131)* i (Dublin 1983)

Macalister, Robert Alexander Stewart (ed.), *Lebor Gabála Érenn: the book of the taking of Ireland* i–v (Irish Texts Society XXXIV, XXXV, XXXIX, XLI, XLIV, Dublin 1938–56)
——(ed.), *The Book of Uí Maine.* Collotype facsimile (Dublin 1941)

Mac an Bhaird, Alan (ed.), 'Tadhg mac Céin and the badgers', *Ériu* 31 (1980) 150–5

MacArthur, William P., 'The identification of some pestilences recorded in the Irish annals', *Irish historical studies* 6/23 (1949) 169–88

Mac Cana, Proinsias, *Celtic mythology* (London 1970)
——*The learned tales of medieval Ireland* (Dublin 1980)
——'Ir. *buaball*, W. *bual* "drinking horn"', *Ériu* 44 (1993) 81–93

McCone, Kim, 'Werewolves, cyclopes, *díberga*, and *fianna*: juvenile delinquency in early Ireland', *Cambridge Medieval Celtic Studies* 12 (1986) 1–22
——'A tale of two ditties: poet and satirist in *Cath Maige Tuired*', in *Sages, saints, and storytellers: Celtic studies in honour of Professor James Carney* (ed. D. Ó Corráin, L. Breatnach, and K. McCone, Maynooth 1989) 122–43
——*Pagan past and Christian present in early Irish literature* (Maynooth 1990)

Mac Con Iomaire, Séamus, *Cladaigh Chonamara* (Baile Átha Cliath 1938)

MacCormack, Geoffrey, 'On thing-liability (*Sachhaftung*) in early law', *The Irish Jurist* [new series] 19 (1984) 322–49

McCormick, Finbar, 'Dairying and beef production in early Christian Ireland: the faunal evidence', in *Landscape archaeology in Ireland* (ed. T. Reeves-Smyth and F. Hamond, British Archaeological Reports: British Series 116, London 1983) 253–67
——'Stockrearing in early Christian Ireland' (Ph.D. thesis, Queen's University, Belfast, 1987)
——'The domesticated cat in early Christian and medieval Ireland', in *Keimelia: studies in medieval archaeology and history in memory of Tom Delaney* (ed. G. Mac Niocaill and P. Wallace, Galway 1988) 218–28
——'The animal bones', in *Three Irish gas pipelines: new archaeological evidence in Munster* (M. Gowen, Dublin 1988) 182–4

McDonald, Allan (ed. J. L. Campbell), *Gaelic words and expressions from South Uist and Eriskay* (1st ed. Dublin 1958, 2nd ed. with supplement Oxford 1972, repr. Dublin 1991)

Mac Eoin, Gearóid (ed.), 'A poem by Airbertach mac Cosse', *Ériu* 20 (1966) 112–39
——'The early Irish vocabulary of mills and milling', in *Studies on early Ireland: essays in honour of M. V. Duignan* (ed. B. G. Scott, Belfast 1981) 13–19

——'The death of the boys in the mill', *Celtica* 15 (1983) 60–4

Mac Eoin, Gearóid, Anders Ahlqvist and Donncha Ó hAodha (ed.), *Proceedings of the sixth international congress of Celtic Studies, Galway, 6–13 July 1979* (Dublin 1983)

McErlean, Thomas, 'The Irish townland system of landscape organisation', in *Landscape archaeology in Ireland* (ed. T. Reeves-Smyth and F. Hamond, British Archaeological Reports: British Series 116, London 1983) 315–39

McKenna, Lambert (ed.), *Aithdioghluim dána* i–ii (Irish Texts Society XXXVII, XL, Dublin 1939, 1940)

McLaughlin, Róisín, 'Aspects of early Irish satire' (M.Litt. thesis, Trinity College, Dublin, 1989)

McLeod, Neil, 'Interpreting early Irish law: status and currency (1)', *ZCP* 41 (1986) 46–65; 'Interpreting early Irish law: status and currency (2)', *ZCP* 42 (1987) 41–115

——*Early Irish contract law* (Sydney Series in Celtic Studies I, Sydney [1992])

McManus, Damian, 'A chronology of the Latin loan-words in early Irish', *Ériu* 34 (1983) 21–71

——'On final syllables in the Latin loan-words in early Irish', *Ériu* 35 (1984) 137–62

——'Irish letter-names and their kennings', *Ériu* 39 (1988) 127–68

——*A guide to Ogam* (Maynooth 1991)

Mac Mathúna, Liam, 'Some words for "(man-made) ridge" in Irish: *fu(i)th(a)irbe*; *immaire*; *indra, indrad*', *Bwletin y Bwrdd Gwybodau Celtaidd* : *Bulletin of the Board of Celtic Studies* 26 (1976) 445–9

Mac Neill, Eóin (John), (ed.), *Duanaire Finn* i (Early Irish Texts Society VII, London 1908). See also **Murphy, Gerard**

——'Ancient Irish law: the law of status or franchise', *PRIA* 36 C (1923) 265–316

——'The Vita Tripartita of St. Patrick', *Ériu* 11 (1932) 1–41

Mac Néill, Máire, *The festival of Lughnasa* (Oxford 1962)

Mac Niocaill, Gearóid, *Notitiae as Leabhar Cheanannais 1033–1161* (Baile Átha Cliath 1961)

——'The origins of the betagh', *The Irish Jurist* [new series] 1/2 (1966) 292–8

——'Register of St. Saviour's Chantry, Waterford', *Analecta Hibernica* 23 (1966) 135–222

——'A propos du vocabulaire social irlandais du bas moyen âge', *Études celtiques* 12 (1968–71) 512–46

——'Jetsam, treasure trove and the lord's share in medieval Ireland', *The Irish Jurist* [new series] 6 (1971) 103–10

——(ed.) '*Tír Cumaile*', *Ériu* 22 (1971) 81–6

——'Investment in early Irish agriculture', in *Studies on early Ireland: essays in honour of M. V. Duignan* (ed. B. G. Scott, Belfast 1981) 7–9

——'Christian influences in early Irish law', in *Irland und Europa: Die Kirche im Frühmittelalter* (ed. P. Ní Chatháin and M. Richter, Stuttgart 1984) 151–6

——'The Irish "charters"', in *The Book of Kells: commentary* (ed. Peter Fox, Luzern 1990) 153–65

Mahaffy, John Pentland (ed.), 'Two early tours in Ireland', *Hermathena* 18 (1914) 1–16

Mahon, Bríd, 'Oatmeal as food', *Sinsear: the folklore journal* 2 (1980) 123–7

——'Traditional dyestuffs in Ireland', in *Gold under the furze: studies in folk tradition presented to Caoimhín Ó Danachair* (ed. A. Gailey and D. Ó hÓgáin, Dublin 1983) 115–28

Malone, Kemp, 'Bonnyclabber', *Celtica* 5 (1960) 142

Marchant, E. C. (ed.), Xenephon, vol. VII, *Scripta minora* (Cambridge (Massachusetts) and London 1925, repr. 1984)

Marstrander, Carl J. S., *Bidrag til det norske sprogs historie i Irland* (Kristiania 1915)

——'Review of J. Vendryes, *Lexique étymologique de l'irlandais ancien: A*', *Lochlann* 2 (1962) 196–226

Martin, Martin (ed. Donald MacLeod), *A description of the Western Islands of Scotland, circa 1695* (Stirling 1934)

Meehan, Bernard, *The Book of Kells: an illustrated introduction to the manuscript in Trinity College Dublin* (London 1994)

Meid, Wolfgang, *Táin Bó Fraích* (Mediaeval and Modern Irish Series XXII, Dublin 1967, revised ed. 1974)

Méniel, Patrice, *Chasse et élevage chez les Gaulois* (Paris 1987)

——*Les sacrifices d'animaux chez les Gaulois* (Paris 1992)

Meroney, Howard, 'Irish in Old English charms', *Speculum* 20 (1945) 172–82

——'Studies in early Irish satire', *Journal of Celtic Studies* 1 (1949–50) 199–226

Meyer, Kuno, (ed.) '*Macgnímartha Find*', *RC* 5 (1881–83) 195–204

——(ed.), 'Die Geschichte von Philipp und Alexander von Macedonien aus dem Lebar Brecc, mit deutscher Uebersetzung und mit Excerpten aus dem Book of Ballymote', in *Irische Texte* II ii (ed. Stokes and Windisch, Leipzig 1887) 1–108

——(ed.), 'The oldest version of *Tochmarc Emire*', *RC* 11 (1890) 433–57

——(ed.), *Aislinge Meic Conglinne: the vision of MacConglinne, a Middle-Irish wonder tale* (London 1892)

——(ed.), *Hibernica minora, being a fragment of an Old-Irish treatise on the Psalter* (Anecdota Oxoniensia, Oxford 1894)

——*The voyage of Bran son of Febal to the Land of the Living* (London 1895, repr. Llannerch, Wales, 1994)

——(ed.), 'The death of Finn mac Cumaill', *ZCP* 1 (1897) 462–5
——(ed.), 'Zwiegespräch zwischen König Gúaire von Aidne und seinem Bruder Marbán, dem Einsiedler', *ZCP* 3 (1901) 455–7
——(ed.), 'The death of Conla', *Ériu* 1 (1904) 113–21
——(ed.), *Cáin Adamnáin: an Old-Irish treatise on the law of Adamnan.* Anecdota Oxoniensia. Mediaeval and Modern Series, Part XII (Oxford 1905)
——(ed.), *The triads of Ireland* (Todd Lecture Series XIII, Dublin 1906)
——(ed.), *The death tales of the Ulster heroes* (Todd Lecture Series XIV, Dublin 1906, repr. 1993)
——(ed.), 'The expulsion of the Déssi', *Ériu* 3 (1907) 135–42
——(ed.), 'König Guaire und Oennu maccu Laigse', *ACL* iii (1907) 1–2
——(ed.), '*Tucait indarba na nDéssi*', *Anecdota* i (ed. Bergin et al.) 15–24
——(ed.), *Tecosca Cormaic: the instructions of King Cormac Mac Airt* (Todd Lecture Series XV, Dublin 1909)
——(ed.), *Fianaigecht, being a collection of hitherto inedited Irish poems and tales relating to Finn and his Fiana* (Todd Lecture Series XVI, Dublin 1910; repr. 1993)
——'Irish mirabilia in the Norse "Speculum regale"', *Ériu* 4 (1910) 1–16
——(ed.), 'Conall Corc and the Corco Luigde', *Anecdota* iii (ed. Bergin et al.) 57–63
——(ed.), *Betha Colmáin maic Lúacháin: Life of Colmán son of Lúachan* (Todd Lecture Series XVII, Dublin 1911)
——(ed.), *Sanas Cormaic: an Old-Irish glossary compiled by Cormac Úa Cuilennáin*, *Anecdota* iv (ed. Bergin et al.) 1–128 (repr. Llannerch, Wales, 1994)
——'Aed Dub mac Colmáin, bishop-abbot of Kildare', *ZCP* 9 (1913) 458–60
——(ed.), *Über die älteste irische Dichtung* i–ii. Abhandlungen der königlich preussischen Akademie der Wissenschaften. Jahrgang 1913. Phil.-Hist. Classe. Nr. 6, Nr. 10 (Berlin 1913–14)
——(ed.), 'The guesting of Athirne', *Ériu* 7 (1914) 1–9
——'Erläuterungen und Besserungen zu irischen Texten', *ZCP* 10 (1915) 349–57
——'Tätowierung bei den Iren', *ZCP* 10 (1915) 400–1
——'Zur keltischen Wortkunde. VII', *Sitzungsberichte der königlich preussischen Akademie der Wissenschaften.* Sitzung der phil.-hist. Klasse (Berlin 1917) 624–53
——'Cormacs Glossar nach der Handschrift des Buches der Ui Maine', *Sitzungsberichte der preussischen Akademie der Wissenschaften.* Sitzung der phil.-hist. Klasse (Berlin 1919) 290–319
Mitchell, G. F. (Frank), 'The relative ages of archaeological objects recently found in bogs in Ireland', *PRIA* 50 C (1945) 1–20
——*The Irish landscape* (London 1976)

——*The Shell guide to reading the Irish landscape* (incorporating *The Irish landscape*) (Dublin 1986)

——*Archaeology and environment in early Dublin*, with contributions by Camilla A. Dickson and J. H. Dickson. Medieval Dublin excavations 1962–81. Series C, vol. 1 (Royal Irish Academy, Dublin 1987)

Mitchell, Frank and Michael Ryan, *Reading the Irish landscape* (Dublin 1997). This is a revision of *The Shell guide to reading the Irish landscape*, and was published when the present work was in page-proofs: it was therefore too late to add references to it in the footnotes.

Mitchell, John, and Arthur Rook, *Botanical dermatology: plants and plant products injurious to the skin* (Vancouver 1979)

Monk, Mick, 'Post-Roman drying kilns and the problem of function: a preliminary statement', in *Irish antiquity: essays and studies presented to Professor M. J. O'Kelly* (ed. D. Ó Corráin, Cork 1981) 216–30

——'Evidence from macroscopic plant remains for crop husbandry in prehistoric and early historic Ireland: a review', *Journal of Irish Archaeology* 3 (1985–86) 31–6

——'Charred seed and plant remains', in *Archaeological excavations on the Cork-Dublin gas pipeline 1981–82* (ed. Cleary et al., Cork 1987) 98–9

——'Excavations at Lisleagh Ringfort, North County Cork', *Archaeology Ireland* 2/2 (1988) 57–60

——'Archaeobotanical study of samples from pipeline sites', in *Three Irish gas pipelines: new archaeological evidence in Munster* (M. Gowen, Dublin 1988) 185–91

——'The archaeobotanical evidence for field crop plants in early historic Ireland', in *New light on early farming: recent developments in palaeoethnobotany* (ed. Jane M. Renfrew, Edinburgh 1991) 315–28

Monsen, Erling (ed.), *Heimskringla or the Lives of the Norse kings by Snorre Sturlason* (Cambridge 1932)

Moore, Fionnbarr, 'Ireland's oldest bridge – at Clonmacnoise', *Archaeology Ireland* 10/4 (1996) 24–7

Morgan, Hiram, 'Lawes of Ireland: a tract by Sir John Davies', *The Irish Jurist* [new series] 28–30 (1993–95) 307–13

Moritz, L. A., *Grain-mills and flour in Classical antiquity* (Oxford 1958)

Morley, Henry (ed.), *Ireland under Elizabeth and James the First: described by Edmund Spenser, by Sir John Davies, and by Fynes Moryson* (London 1890)

Mulchrone, Kathleen, (ed.), *Bethu Phátraic: the Tripartite Life of Patrick* (Dublin 1939)

Murphy, Denis (ed.), *The annals of Clonmacnoise, translated into English A.D. 1627 by Conell Mageoghagan* (Dublin 1896, repr. Llannerch, Wales, 1993)

Murphy, Gerard (ed.), *Duanaire Finn* ii (Irish Texts Society XXVIII, London 1933); *Duanaire Finn* iii (Irish Texts Society XLIII, Dublin 1953). See also **MacNeill, Eóin**

——(ed.), *Early Irish lyrics, eighth to twelfth century* (Oxford 1956, repr. Dublin 1996)

Mytum, Harold, *The origins of early Christian Ireland* (London and New York 1992)

Nettlau, Max (ed.), 'The fragment of the *Táin Bó Cuailnge* in MS. Egerton 93, (1)', *RC* 14 (1893) 254–66; 'The fragment of the *Táin Bó Cuailnge* in MS. Egerton 93, (2)', *RC* 15 (1894) 62–78, 198–208

Ní Chatháin, Próinséas, 'Swineherds, seers and druids', *Studia Celtica* 14–15 (1979–80) 200–11

Nicholls, Kenneth, *Land, law and society in sixteenth-century Ireland.* O'Donnell Lecture 1976, National University of Ireland [Dublin]

——'Anglo-French Ireland and after', *Peritia* 1 (1982) 370–403

——'Gaelic society and economy in the high middle ages', in *A new history of Ireland* ii (ed. Art Cosgrove, Oxford 1987) 397–438

Ní Dhonnchadha, Máirín, 'An edition of *Cáin Adomnán*' (Ph.D. thesis, University College, Cork, 1991)

——'The *Lex Innocentium*: Adomnán's law for women, clerics and youths, 697 A.D.', in *Chattel, servant, or citizen: women's status in Church, state and society* (ed. Mary O'Dowd and Sabine Wichert, Belfast 1995) 58–69

Ní Shéaghdha, Nessa (ed.), *Agallamh na Seanórach* i–iii (Baile Átha Cliath 1942–45)

——(ed.), *Tóruigheacht Dhiarmada agus Ghráinne: the pursuit of Diarmaid and Gráinne* (Irish Texts Society XLVIII, Dublin 1967)

Norman, E. R., and J. K. S. St Joseph, *The early development of Irish society: the evidence of aerial photography* (Cambridge 1969)

O'Brien, George, *The economic history of Ireland in the seventeenth century* (Dublin 1919)

O'Brien, Michael A., '*Scéla Cano*', *Ériu* 11 (1932) 86–8

——'Ir. *spel*, Welsh *paladur* "scythe"', *Ériu* 11 (1932) 159–60

——'*Eo cétnama*', *Ériu* 12 (1938) 244

——'*Uchtlach* "lapful". *Littiu* "porridge"', *Celtica* 2 (1954) 353

——(ed.), *Corpus genealogiarum Hiberniae* (Dublin 1962, repr. 1976)

——(ed. R. Baumgarten), 'Old Irish personal names': Rhys Lecture Notes, 1957, *Celtica* 10 (1973) 211–36

O'Brien, William, *Mount Gabriel: Bronze Age mining in Ireland* (Galway 1994)

Ó Buachalla, Liam, 'Leinster tribute feud', *Journal of the Cork Historical and Archaeological Society* 66 (1961) 13–25

Ó Cathasaigh, Tomás (ed.), *The heroic biography of Cormac mac Airt* (Dublin 1977)

Ó Ceithearnaigh, Séamus: see **Carney, James**

Ó Corráin, Donnchadh, *Ireland before the Normans* (Dublin 1972)

——(ed.), *Irish antiquity: essays and studies presented to Professor M. J. O'Kelly* (Cork 1981, repr. Dublin 1994)

——'Some legal references to fences and fencing in Early Historic Ireland', in *Landscape archaeology in Ireland* (ed. T. Reeves-Smyth and F. Hamond, British Archaeological Reports: British Series 116, London 1983) 247–51

——'Irish law and canon law', in *Irland und Europa: Die Kirche im Frühmittelalter* (ed. P. Ní Chatháin and M. Richter, Stuttgart 1984) 157–66

——'Marriage in early Ireland', in *Marriage in Ireland* (ed. Art Cosgrove, Dublin 1985) 5–24

——'Irish vernacular law and the Old Testament', in *Irland und die Christenheit: Bibelstudien und Mission* (ed. P. Ní Chatháin und M. Richter, Stuttgart 1987) 284–307

——'Early Irish hermit poetry?', in *Sages, saints, and storytellers: Celtic studies in honour of Professor James Carney* (ed. D. Ó Corráin, L. Breatnach, and K. McCone, Maynooth 1989) 251–67

——'Women and the law in early Ireland', in *Chattel, servant, or citizen: women's status in Church, state and society* (ed. Mary O'Dowd and Sabine Wichert, Belfast 1995) 45–57

Ó Corráin, D., L. Breatnach, and A. Breen, 'The laws of the Irish', *Peritia* 3 (1984) 382–438

Ó Cuív, Brian (ed.), 'Fragments of two mediaeval treatises on horses', *Celtica* 2 (1952) 30–63

——*Cnósach focal ó Bhaile Bhúirne i gCunndae Chorcaí.* Mícheál Ó Briain (1866–1942) a bhailig. (Baile Átha Cliath 1947)

——'Review of *Études celtiques*, vol. IX, fascicule 2', *Éigse* 10/4 (1963) 337–9

——(ed.), 'Three Middle Irish poems', *Éigse* 16/1 (1975) 1–17

——(ed.), 'Fragments of Irish medieval treatises on horses', *Celtica* 17 (1985) 113–22

O Daly, Máirín (ed.), 'A poem on the Airgialla', *Ériu* 16 (1952) 179–88

——(ed.), '*Lānellach tigi rīch 7 ruirech*', *Ériu* 19 (1962) 81–6

——(ed.), *Cath Maige Mucrama* (Irish Texts Society L, Dublin 1975)

Ó Danachair, Caoimhín: see **Danaher, Kevin**

O'Donnell, Thomas J. (ed.), *Selections from the Zoilomastix of Philip O'Sullivan Beare* (Dublin 1960)

O'Donoghue, Tadhg (ed.), 'Advice to a prince', *Ériu* 9 (1921–23) 43–54

O'Donovan, John (ed.), *Annála ríoghachta Éireann: annals of the kingdom of Ireland by the four masters, from the earliest period to the year 1616* i–vii (Dublin 1848–51)

O'Donovan, John (ed. W. Stokes), *Sanas Chormaic: Cormac's Glossary* (Calcutta 1868)

O'Dowd, Anne, *Meitheal: a study of co-operative labour in rural Ireland* (Dublin 1981)

Ó Drisceoil, Diarmuid A., 'Burnt mounds: cooking or bathing?', *Antiquity* 62 (1988) 671–80

Ó Dubhda, Seán; 'Foclóir agus téarmaí feirmeoireachta, 7c.', *Béaloideas* 13 (1944 for 1943) 3–39

Ó Fiannachta, Pádraig (ed.), *Táin Bó Cuailnge* (Dublin 1966)

O'Grady, Standish Hayes (ed.), *Silva gadelica: the colloquy of the ancients* i–ii (London 1892)

——*Catalogue of Irish manuscripts in the British Library* [*formerly British Museum*] i (London 1926, repr. Dublin 1992). See also **Flower, Robin**

Ó hAodha, Donncha (ed.), *Bethu Brigte* (Dublin 1978)

Ó hEochaidh, Seán, 'Seanchas iascaireachta agus farraige', *Béaloideas* 33 (1967 for 1965) 1–96

——'Seanchas na caorach', *Béaloideas* 37–38 (1973 for 1969–70) 131–209

——'Seanchas éanlaithe Iar-Uladh', *Béaloideas* 37–38 (1973 for 1969–70) 210–337

O'Keeffe, James G. (ed.), 'Colman mac Duach and Guaire', *Ériu* 1 (1904) 43–8

——(ed.), 'The rule of Patrick', *Ériu* 1 (1904) 216–24

——(ed.), '*Cáin Domnaig*', *Ériu* 2 (1905) 189–214

——(ed.), 'A poem on the day of judgment', *Ériu* 3 (1907) 29–33

——(ed.), '*Dál Caladbuig* and reciprocal services between the kings of Cashel and various Munster states', *Irish texts* i (ed. Fraser, Grosjean, and O'Keeffe, London 1931) 19–21

——(ed.), *Buile Ṡuibhne* (Mediaeval and Modern Irish Series I, Dublin 1931; repr. 1975)

O'Kelleher, A., and G. Schoepperle (ed.), *Betha Colaim Chille: Life of Columcille* (University of Illinois 1918, repr. Dublin 1994)

O'Kelly, Michael J., 'Excavations and experiments in ancient Irish cooking-places', *JRSAI* 84 (1954) 105–55

——'Church Island near Valencia, Co. Kerry', *PRIA* 59 C (1957–59) 57–136

——'Two ring-forts at Garryduff, Co. Cork', *PRIA* 63 C (1962–64) 17–126

——*Early Ireland: an introduction to Irish prehistory* (Cambridge 1989)

O'Loan, John, 'Livestock in the Brehon laws', *The agricultural history review* 7/2 (1959) 65–74

——'A history of early Irish farming (1)', *Journal of the Department of Agriculture* 60 (1963) 178–219; 'A history of early Irish farming (2)', *Journal of the Department of Agriculture* 61 (1964) 242–84; 'A history of early Irish farming (3)', *Journal of the Department of Agriculture* 62 (1965) 131–97

Ó Lochlainn, Colm, 'Roadways in ancient Ireland', in *Féil-sgríbhinn Eóin Mhic Néill* (ed. J. Ryan, Dublin 1940) 465–74

O'Longan, Joseph, *Leabhar Breac: the Speckled Book, otherwise styled Leabhar Mór Dúna Doighre: the Great Book of Dún Doighre.* Lithographic reproduction of O'Longan's transcript, with a preface by Samuel Ferguson (Dublin 1872–76)

O'Mahony, Felicity (ed.), *The Book of Kells: proceedings of a conference at Trinity College Dublin, 6–9 September 1992* (Dublin 1994)
Ó Máille, Tomás, *An béal beo* (Baile Átha Cliath 1936)
O'Meara, John J. (ed.), *Giraldus Cambrensis (Gerald of Wales): the history and topography of Ireland (Topographia Hiberniae)* (Dundalgan Press 1951, revised ed. Dolmen Press 1982)
Ó Murchadha, Diarmuid, '*Sódh* i logainmneacha', *Éigse* 28 (1994–5) 129–34
O'Neill, Joseph (ed.), '*Cath Boinde*', *Ériu* 2 (1905) 173–85
——(ed.), 'The rule of Ailbe of Emly', *Ériu* 3 (1907) 92–115
O'Neill, Timothy, *Merchants and mariners in medieval Ireland* (Dublin 1987)
O'Rahilly, Cecile (ed.), *The Stowe version of Táin Bó Cuailnge* (Dublin 1961, repr. 1978)
——(ed.), *Táin Bó Cúalnge from the Book of Leinster* (Dublin 1967, repr. 1984)
——(ed.), *Tain Bó Cúailnge*, Recension I (Dublin 1976)
——'The bleeding of living cattle', *Celtica* 12 (1977) 185–8
O'Rahilly, Thomas F. (ed.), *A miscellany of Irish proverbs* (Dublin 1922)
——'*Sp-* in Gaelic', *Scottish Gaelic Studies* 2 (1927) 24–9
——'*Spoch. Foich.*', *Scottish Gaelic Studies* 3 (1931) 63–4
——'Mid. Ir. *lága, láige*; *lágan, láigen*', *Ériu* 13 (1942) 152–3
——'*Creach. Gad. Brad. Slad.*', *Ériu* 13 (1942) 168–70
——'Sc. *fàl, fàladair*; W. *pal, p(a)ladur*', *Ériu* 13 (1942) 190–1
——'Ir. *aobh, aoibheall*, etc. W. *ufel, uwel.* Gaul. *Esus*', *Ériu* 14 (1946) 1–6
——'On the origin of the names *Érainn* and *Ériu*', *Ériu* 14 (1946) 7–28
——*Early Irish history and mythology* (Dublin 1946, repr. 1984)
——'*Rámha. ruamh. rómhar*', *Celtica* 1 (1950) 363–5
Ó Riain, Pádraig, 'The *crech ríg* or "royal prey"', *Éigse* 15/1 (1973) 24–30
——(ed.), *Corpus genealogiarum sanctorum Hiberniae* (Dublin 1985)
——'A misunderstood annal: a hitherto unnoticed *cáin*', *Celtica* 21 (1990) 561–6
Ó Ríordáin, Seán P., 'Lough Gur excavations: Neolithic and Bronze Age houses on Knockadoon', *PRIA* 56 C (1953–54) 297–459
Osborough, W. N., 'Roman law in Ireland', *The Irish Jurist* [new series] 25–26 (1990–92) 212–68
Ó Sé, Mícheál, 'Old Irish cheeses and other milk products', *Journal of the Cork Historical and Archaeological Society* 53 (1948) 82–7
——'Old Irish buttermaking', *Journal of the Cork Historical and Archaeological Society* 54 (1949) 61–7
——*The history of native Irish cattle* (= *Agricultural bulletin* no. 7) (Cork 1954)
O'Shea, Michael: see **Ó Sé, Mícheál**
Ó Siochfhradha, Pádraig: see **An Seabhac**
Ó Súilleabháin, Pádraig, 'Nótaí ar thrí fhocal ó na hAnnála', *Éigse* 15/1 (1973) 22–3

O'Sullivan, Aidan, 'Intertidal survey on the Fergus estuary and the Shannon estuary', *Discovery Programme reports* 1 (Royal Irish Academy, Dublin 1993) 61–8

——'Harvesting the waters: fishweirs in early Ireland', *Archaeology Ireland* 8/1 (1994) 10–12

——'Intertidal archaeological survey and estuarine wetlands in North Munster, Ireland', in *Archaeology in the Severn Estuary 1994* (ed. Martin Bell, University of Wales, Lampeter 1995) 145–55

O'Sullivan, William, 'The earliest Irish coinage', *JRSAI* 79 (1949) 190–235

Otway-Ruthven, A. J., 'The organization of Anglo-Irish agriculture in the Middle Ages', *JRSAI* 81 (1951) 1–13

——*A history of medieval Ireland*, with an introduction by Kathleen Hughes (London 1968)

Owen, Aneurin (ed.), *Ancient laws and institutes of Wales* i–ii (London 1841)

Patterson, Nerys, 'Material and symbolic exchange in early Irish clientship', in *Proceedings of the Harvard Celtic Colloquium* 1 (ed. James Doan and Cornelius Buttimer, Harvard 1981) 53–61

——*Cattle-lords and clansmen: kinship and rank in Early Ireland* (New York and London 1991; revised ed. *Cattle-lords and clansmen: the social structure of Early Ireland*, University of Notre Dame 1994)

Payne, Ffransis G., *Yr aradr gymreig* (Caerdydd, 1954)

——'The Welsh plough team to 1600', in *Studies in folk life: essays in honour of Iorwerth C. Peate* (ed. Geraint Jenkins, London 1969) 235–52

Pearson, A. I. (ed.), 'A medieval glossary', *Ériu* 13 (1942) 61–87

Pedersen, Holger, *Vergleichende Grammatik der keltischen Sprachen* i–ii (Göttingen 1908–13)

Peterson, William and Maurice Hutton (ed.), *Tacitus: Dialogus, Agricola, Germania* (The Loeb Classical Library, London and Cambridge (Massachusetts) 1914, repr. 1963)

Petrie, George, 'On the history and antiquities of Tara hill', *The Transactions of the Royal Irish Academy* 18 (1839). Antiquities III, 25–232

——'The ecclesiastical architecture of Ireland, anterior to the Anglo-Norman invasion, comprising an essay on the origin and uses of the round towers of Ireland' (second ed., Dublin 1845)

Piggott, Stuart (ed.), *The agrarian history of England and Wales* I i, Prehistory (Cambridge 1981)

Pilch, Herbert, and Joachim Thurow (ed.), *Indo-Celtica: Gedächtnisschrift für Alf Sommerfelt* (Munich 1972)

Plummer, Charles (ed.), '*Betha Farannáin*', *Anecdota* iii (ed. Bergin et al.) 1–7

——(ed.), *Vitae sanctorum Hiberniae* i–ii (Oxford 1910, repr. Dublin 1996)

——(ed.), *Bethada náem nÉrenn: Lives of Irish saints* i–ii (Oxford 1922, repr. Dublin 1996)

——(ed.), *Miscellanea hagiographica hibernica.* Subsidia hagiographica XV (Bruxelles 1925)

——'Notes on some passages in the Brehon laws IV', *Ériu* 10 (1926–28) 113–29

Pokorny, Julius, *Indogermanisches etymologisches Wörterbuch* i–ii (Bern 1959–69)

Price, Liam, 'The origin of the word *betagius*', *Ériu* 20 (1966) 185–90

Quentel, Paul, 'Goémon et goémoniers de deux communes du Léon, Kerlouan et Brignogan', *Études celtiques* 9 (1960) 190–240, 549–69

Quin, E. G., et al. (ed.), (*Contributions to a*) *Dictionary of the Irish language, based mainly on Old and Middle Irish materials* (Dublin, Royal Irish Academy, 1913–76; repr. in compact ed. 1983)

Quin, E. G. (ed.), '*Truagh truagh an mhuc*', *Hermathena* 101 (1965) 27–37

Quin, E. G., and T. W. Freeman, 'Some Irish topographical terms', *Irish Geography* (*Bulletin of the Geographical Society of Ireland*) 1/4 (1947) 85–9; 1/5 (1948) 151–5

Rackham, H., *Pliny, Natural History, books 8–11* (The Loeb Classical Library, London and Cambridge (Massachusetts) 1940, repr. 1956); *Pliny, Natural History, books 17–19* (The Loeb Classical Library, London and Cambridge (Massachusetts) 1950, repr. 1961)

Rackham, James, '*Rattus rattus*: the introduction of the black rat into Britain', *Antiquity* 53 (1979) 112–20

Rackham, Oliver, *Trees and woodland in the British landscape* (London 1976, revised ed. 1990, repr. 1995)

——*The history of the countryside* (London 1986, repr. 1993)

Radner, Joan Newlon (ed.), *Fragmentary annals of Ireland* (Dublin 1978)

Raftery, Barry, *La Tène in Ireland: problems of origin and chronology* (Marburg 1984)

——*Trackways through time* (Rush, Co. Dublin, 1990)

——'The Celtic Iron Age in Ireland: problems of origin', *Emania: bulletin of the Navan research group* 9 (1991) 28–32

——*Pagan Celtic Ireland: the enigma of the Irish Iron Age* (London 1994)

——'The conundrum of Irish Iron Age pottery', in *Sites and sights of the Iron Age: essays on fieldwork and museum research presented to Ian Mathieson Stead* (ed. Barry Raftery with V. Megaw and V. Rigby, Oxford 1995) 149–56

——(ed.), *Trackway excavations in the Mountdillon bogs, Co. Longford, 1985–1991.* Irish Archaeological Wetland Unit Transactions: vol. 3 (Dublin 1996)

Reaney, P. H., *The origin of English place-names* (London 1960)

Rees, Sian, *Agricultural implements in prehistoric and Roman Britain* i–ii (Oxford 1979)

——*Ancient agricultural implements* (Aylesbury, England, 1981)

Reeves-Smyth, Terence, and Fred Hamond (ed.), *Landscape archaeology in Ireland.* British Archaeological Reports: British Series 116 (London 1983)

Rivers, Theodore John, *Laws of the Salian and Ripuarian Franks* (New York 1986)

Risk, Henry, 'French loan-words in Irish (1)', *Études celtiques* 12 (1968–71) 585–655; 'French loan-words in Irish (2)', *Études celtiques* 14 (1974–75) 67–98

Roider, Ulrike, *De chophur in dá muccida.* Innsbrucker Beiträge zur Sprachwissenschaft 28 (Innsbruck 1979)

Ross, Anne, 'Esus et les trois "grues"', *Études celtiques* 9 (1960) 405–38

Rowlands, Eurys, 'Mesur tir: land measurement', *Studia Celtica* 14–15 (1979–80) 270–84

Russell, Paul, 'The Celtic preverb **uss* and related matters', *Ériu* 39 (1988) 95–126

——'The sounds of a silence: the growth of Cormac's Glossary', *Cambridge Medieval Celtic Studies* 15 (1988) 1–30

——*Celtic word-formation: the velar suffixes* (Dublin 1990)

——'Brittonic words in Irish glossaries', in *Hispano-Gallo-Brittonica: essays in honour of Professor D. Ellis Evans on the occasion of his sixty-fifth birthday* (ed. J. Eska, R. Geraint Gruffydd, and N. Jacobs, Cardiff 1995) 166–82

Ryan, John, *Irish monasticism: origins and early development* (Dublin 1931, repr. 1993)

——(ed.), *Féil-sgríbhinn Eóin Mhic Néill .i. essays and studies presented to Professor Eoin MacNeill* (Dublin 1940, repr. 1995)

Ryan, Kathleen, 'Holes and flaws in medieval Irish manuscripts', *Peritia* 6–7 (1987–88) 243–64

Ryder, M. L., 'Livestock', in *The agrarian history of England and Wales* I i (ed. Piggott, Cambridge 1981) 301–410

——*Sheep and man* (London 1983)

Rynne, Colin, 'The introduction of the vertical watermill into Ireland: some recent archaeological evidence', *Medieval Archaeology* 32 (1989) 21–31

——'Milling in the 7th century: Europe's earliest tide mills', *Archaeology Ireland* 6/2 (1992) 22–4

——'Early medieval horizontal-wheeled mill penstocks from Co. Cork', *Journal of the Cork Historical and Archaeological Society* 97 (1992) 54–68

Sayers, William, 'Old Irish *fert* "tie-pole", *fertas* "swingletree", and the seeress Fedelm', *Études celtiques* 21 (1984) 171–83

——'A cut above: ration and station in an Irish king's hall', *Food and foodways* 4/2 (1990) 89–110

——'Textual notes on descriptions of the Old Irish chariot and team', *Studia Celtica Japonica* 4 (1991) 15–35

Scharff, R. F., 'On the breeds of dogs peculiar to Ireland and their origin', *The Irish naturalist* 33 (1924) 77–95

Schmitz, Herm. Jos. (ed.), *Die Bussbücher und die Bussdisciplin der Kirche* i; *Die Bussbücher und das kanonische Bussverfahren* ii (1883–98, repr. Graz 1958)

Schoepperle, G.: see **O'Kelleher, A.**

Schröder, Franz Rolf, 'Ein altirischer Krönungsritus und das indogermanische Rossopfer', *ZCP* 16 (1927) 310–12

Scott, A. B., and F. X. Martin (ed.), *Expugnatio hibernica: the conquest of Ireland, by Giraldus Cambrensis* (Dublin 1978)

Scott, B. G. (ed.), *Studies on early Ireland: essays in honour of M. V. Duignan* (Belfast 1981)

——'Some conflicts and correspondences of evidence in the study of Irish archaeology and language', in *Studies on early Ireland* (ed. B. G. Scott, Belfast 1981) 115–19

Selmer, Carl (ed.), *Navigatio Sancti Brendani abbatis* (Notre Dame, Indiana 1959, repr. Dublin 1989)

Sexton, Mary Regina, 'Cereals and cereal foodstuffs in early historic Ireland' (M.A. thesis, University College, Cork, 1993)

Sharpe, Richard, 'Review of *The Patrician texts in the Book of Armagh* (ed. Bieler, Dublin 1979)', *Éigse* 18/2 (1981) 329–32

Sheehan, Michael, *Sean-chaint na nDéise: the idiom of living Irish* (Dublin 1906; 2nd ed. 1944). See **Breatnach, R. B.**

Sherman, Bob, and Carol Bowen, 'The Green gardening and cooking guide' (London 1990)

Simington, Robert S. (ed.), *The Civil survey, A.D. 1654–1656.* County of Tipperary i–ii (Dublin 1931–34)

Simms, Katharine, 'Nomadry in medieval Ireland: the origins of the creaght or *caoraigheacht*', *Peritia* 5 (1980) 379–91

——*From kings to warlords: the changing political structure of Gaelic Ireland in the later Middle Ages* (Woodbridge, Suffolk, 1987)

Singer, Charles, E. J. Holmyard, and A. R. Hall, *A history of technology* i (Oxford 1954)

Sjoestedt, Marie-Louise, '*Forbuis Droma Damhghaire: le siège de Druim Damhghaire*', *RC* 43 (1926) 1–123; *RC* 44 (1927) 157–86

Small, John (ed.), *The image of Irelande, with a discouerie of woodkarne, by John Derricke, 1581* (Edinburgh 1883)

Smith, Roland M. (ed.), 'The *Senbriathra Fithail* and related texts', *RC* 45 (1928) 1–92

——'Fithal and Flann Fína', *RC* 47 (1930) 30–8

——'Further light on the *Finnsruth Fithail*', *RC* 48 (1931) 325–31

——(ed.), 'The advice to Doidin', *Ériu* 11 (1932) 66–85

Smyth, Alfred P., *Celtic Leinster: towards an historical geography of early Irish civilization AD 500–1600* (Dublin 1982)

Spufford, Peter, *Handbook of medieval exchange* (London 1986)

Stacey, Robin Chapman, '*Berrad Airechta*: an Old Irish tract on suretyship', in *Lawyers and laymen* (ed. Charles-Edwards, Owen, and Walters, Cardiff 1986) 210–33

Stokes, Whitley (ed.), *Irish glosses: a mediæval tract on Latin declension* (The Irish Archæological and Celtic Society, Dublin 1860)

——*Three Irish glossaries* (London 1862)

——(ed.), *Sanas Chormaic: Cormac's Glossary* (Calcutta 1868): see **O'Donovan, John**

——(ed.), 'On the Calendar of Oengus', *Transactions of the Royal Irish Academy.* Irish Manuscript Series I. Dublin 1880)

——(ed.), *The Saltair na Rann, a collection of early Middle Irish poems* (Oxford 1883)

——(ed.), 'Find and the phantoms', *RC* 7 (1886) 289–307

——(ed.), 'On the *materia medica* of the mediaeval Irish', *RC* 9 (1888) 224–44

——(ed.), 'The voyage of Mael Dúin (1)', *RC* 9 (1888) 447–95; 'The voyage of Mael Dúin (2)', *RC* 10 (1889) 50–95

——(ed.), *Lives of saints from the Book of Lismore* (Oxford 1890)

——(ed.), 'On the Bodleian fragment of Cormac's Glossary', *Transactions of the Philological Society* (1891–94) 149–206 (separate pagination: 1–58)

——(ed.), 'The second battle of Moytura', *RC* 12 (1891) 52–130

——(ed.), 'Life of S. Féchín of Fore', *RC* 12 (1891) 318–53

——(ed.), 'Adamnan's second vision', *RC* 12 (1891) 420–43

——(ed.), '*Scél na Fír Flatha, Echtra Cormaic i Tir Tairngiri ocus Ceart Claidib Cormaic*', in *Irische Texte* III i (ed. Stokes and Windisch, Leipzig 1891) 185–229

——(ed.), 'The *Bórama*', *RC* 13 (1892) 32–124

——(ed.), 'The battle of Mag Mucrime', *RC* 13 (1892) 426–74

——(ed.), 'The voyage of the Húi Corra', *RC* 14 (1893) 22–69

——(ed.), 'The prose tales in the Rennes Dindṡenchas (1)', *RC* 15 (1894) 272–336, 418–84

——(ed.), 'The prose tales in the Rennes Dindṡenchas (2)', *RC* 16 (1895) 31–83, 284–312; 135–67 (1st supplement), 269–83 (2nd supplement)

——(ed.), 'The annals of Tigernach (1)', *RC* 16 (1895) 374–419; 'The annals of Tigernach (2)', *RC* 17 (1896) 6–33, 116–263, 337–420; 'The annals of Tigernach (3)', *RC* 18 (1897) 9–58, 150–303, 374–91 (repr. in two volumes, Llannerch, Wales, 1993)

——(ed.), '*Cóir anmann* (fitness of names)', in *Irische Texte* III ii (ed. Stokes and Windisch, Leipzig 1897) 285–444

——(ed.), 'The Bodleian Amra Choluimb chille', *RC* 20 (1899) 30–55, 132–83, 248–89, 400–37

——(ed.), 'O'Mulconry's Glossary', *ACL* i (1900) 232–324

——(ed.), *Acallamh na Senórach*, in *Irische Texte* IV i (ed. Stokes and Windisch, Leipzig 1900)
——(ed.), 'The destruction of Dá Derga's hostel', *RC* 22 (1901) 9–61, 165–215, 282–329, 390–437
——(ed.), 'The death of Crimthann son of Fidach, and the adventures of the sons of Eochaid Muigmedón', *RC* 24 (1903) 172–207
——(ed.), 'O'Davoren's Glossary', *ACL* ii (1904) 197–504
——(ed.), *Félire Óengusso Céli Dé: the martyrology of Oengus the Culdee* (London 1905, repr. Dublin 1984)
——(ed.), 'The colloquy of the two sages', *RC* 26 (1905) 4–64
——(ed.), 'The adventure of St. Columba's clerics', *RC* 26 (1905) 130–70
——(ed.), 'The birth and life of St. Moling', *RC* 27 (1906) 257–312
Stokes, Whitley, and Kuno Meyer (ed.), *Archiv für celtische Lexikographie* (*ACL*) i–iii (Halle 1900–1907)
Stokes, Whitley, and John Strachan (ed.), *Thesaurus palaeohibernicus* i–ii (Cambridge 1901–3, repr. Dublin 1987)
Stokes, Whitley and Ernst Windisch (ed.), *Irische Texte mit Übersetzungen und Wörterbuch* II–IV (Leipzig 1884–1909)
Stout, Matthew, 'Ringforts in the South-West Midlands of Ireland', *PRIA* 91 C (1991) 201–43
——*The Irish ringfort* (Dublin 1997)
Strachan, John (ed.), *Stories from the Táin* (Dublin 1908, third ed. revised by O. Bergin 1944)
Sveinsson, Einar Ól. (ed.), *Brennu-Njáls Saga* (Reykjavík 1954)
Swan, Leo, 'The early Christian ecclesiastical sites of County Westmeath', in *Settlement and society in medieval Ireland: studies presented to F. X. Martin* (ed. John Bradley, Boethius Press, Kilkenny 1988) 3–31
Thomas, R. J., et al. (ed.), *Geiriadur Prifysgol Cymru: a dictionary of the Welsh language* (Caerdydd 1950-)
Thompson, D'Arcy Wentworth (ed.), *The works of Aristotle: Historia animalium* (Oxford 1910)
Thomson, Derick S. (ed.), *Branwen uerch Lyr* (Mediaeval and Modern Welsh Series II, Dublin 1961, repr. 1986)
Thomson, R. L. (ed.), *Pwyll Pendeuic Dyuet* (Mediaeval and Modern Welsh Series I, Dublin 1957, repr. 1986)
Thornton, David, 'Locusts in Ireland? A problem in the Welsh and Frankish annals', *Cambrian Medieval Celtic Studies* 31 (1996) 37–53
Thurneysen, Rudolf (ed.), 'Mittelirische Verslehren', in *Irische Texte* III i (ed. Stokes and Windisch, Leipzig 1891) 1–182
——(ed.), 'Táin Bó Cúailghni nach H.2.17', *ZCP* 8 (1912) 525–54
——(ed.), 'Morands Fürstenspiegel', *ZCP* 11 (1917) 56–106
——'Ir. *marbu* "ich töte"', *ZCP* 13 (1921) 106
——(ed.), '*Tochmarc Ailbe*: "das Werben um Ailbe"', *ZCP* 13 (1921) 251–82

——'Ir. *fotha(e)* und *foth*', *ZCP* 13 (1921) 300–3

——'Aus dem irischen Recht I' [1. Das Unfrei-Lehen], *ZCP* 14 (1923) 335–94

——'Aus dem irischen Recht II' [2. Das Frei-Lehen; 3. Das Fasten beim Pfändungsverfahren], *ZCP* 15 (1925) 238–76

——'Aus dem irischen Recht III' [4. Die falschen Urteilssprüche Caratnia's; 5. Zur Überlieferung und zur Ausgabe der Texte über das Unfrei-Lehen und das Frei-Lehen], *ZCP* 15 (1925) 302–76

——(ed.), *Cóic Conara Fugill: die fünf Wege zum Urteil.* Abhandlungen der preussischen Akademie der Wissenschaften. Jahrgang 1925. Phil.-Hist. Klasse Nr. 7 (Berlin 1926)

——'Aus dem irischen Recht IV' [6. Zu den bisherigen Ausgaben der irischen Rechtstexte], *ZCP* 16 (1927) 167–230

——Review of George Calder (ed.), *Togail na Tebe*, *ZCP* 17 (1927) 420–1

——(ed.), *Die Bürgschaft im irischen Recht.* Abhandlungen der preussischen Akademie der Wissenschaften. Jahrgang 1928. Phil.-Hist. Klasse. Nr. 2 (Berlin 1928)

——'MIr. *melle*', *ZCP* 18 (1930) 105–6

——(ed.), *Irisches Recht.* I. Díre. Ein altirischer Rechtstext, II. Zu den unteren Ständen in Irland. Abhandlungen der preussischen Akademie der Wissenschaften. Jahrgang 1931. Phil.-Hist. Klasse. Nr. 2 (Berlin 1931)

——(ed.), 'Die Flöte von Mac Dīchoeme', *ZCP* 19 (1933) 117–24

——'Altirisch *canait* 'Sängerin', *ZCP* 19 (1933) 189–90

——(ed.), *Scéla Mucce Meic Dathó* (Mediaeval and Modern Irish Series VI, Dublin 1935, repr. 1986)

——(ed.), '*Cáin Lánamna*', in *Studies in early Irish law* (ed. Binchy, Dublin 1936) 1–80

——'Singularisches Verb bei dualischem Subjekt im Irischen', *ZCP* 20 (1936) 369–71

——'Air. *nis* "sind nicht"', *ZCP* 20 (1936) 371–2

——'*Sochor*', in *Féil-sgríbhinn Eóin Mhic Néill* (ed. J. Ryan, Dublin 1940) 158–9

——'*Rodb-*', *ZCP* 22 (1941) 32–7

——*A grammar of Old Irish* (Dublin 1946, repr. 1981)

Tierney, J. J. (ed.), *Dicuili liber de mensura orbis terrae*, with contributions by L. Bieler (Scriptores Latini Hiberniae VI, Dublin 1967)

Todd, James Henthorn (ed.), *The Irish version of the Historia Britonum of Nennius* (Dublin 1848)

——(ed.), *Cogadh Gaedhel re Gallaibh: the war of the Gaedhil with the Gaill* (London 1867)

Toussaint-Samat, Maguelonne, *A history of food*, trans. from French by Anthea Bell (Blackwell, Cambridge USA / Oxford UK 1992)

Tymoczko, Maria, 'Animal imagery in *Loinges Mac nUislenn*', *Studia Celtica* 20–21 (1985–86) 145–66

Ucko, Peter J., and G. W. Dimbleby, *The domestication and exploitation of plants and animals* (London 1969)

Uí Bheirn, Úna M., *Cnuasach focal as Teileann* (Baile Átha Cliath 1989)

Ussher, Richard J., and Robert Warren, *The birds of Ireland* (London 1900)

Van Hamel, A. G. (ed.), *Compert Chon Culainn and other stories* (Mediaeval and Modern Irish Series III, Dublin 1933, repr. 1978)

——(ed.) *Immrama* (Mediaeval and Modern Irish Series X, Dublin 1941)

Van Wijngaarden-Bakker, Louise H., 'The animals remains from the Beaker settlement at Newgrange, Co. Meath: first report', *PRIA* 74 C (1974) 313–83

——'The animals remains from the Beaker settlement at Newgrange, Co. Meath: final report', *PRIA* 86 C (1986) 17–112

Vendryes, Joseph, '*Imbolc*', *RC* 41 (1924) 241–4

——(ed.), *Airne Fíngein* (Mediaeval and Modern Irish Series XV, Dublin 1953, repr. 1975)

Vendryes, J., E. Bachellery and P.-Y. Lambert, *Lexique étymologique de l'irlandais ancien* (Dublin and Paris 1959-)

Von Bunau, Gunther, 'Fallow versus red', letter in *The Irish Farmers' Journal*, 31 December 1994, p. 7

Waddell, John, 'The question of the Celticization of Ireland', *Emania: bulletin of the Navan research group* 9 (1991) 5–16

Wade-Evans, A. W. (ed.), *Welsh medieval law, being a text of the laws of Howel the Good* (Oxford 1909)

Wagner, Heinrich, *Linguistic atlas and survey of Irish dialects* i–iv (Dublin 1958–69, repr. 1981–82)

Wallace, Patrick F., 'The English presence in Viking Dublin', in *Anglo-Saxon monetary history: essays in memory of Michael Dolley* (ed. M. A. S. Blackburn, Leicester 1986) 201–21

Wasserschleben, Hermann, (ed.), *Die irische Kanonensammlung* (Leipzig 1885, repr. Aalen 1966)

Watkins, Calvert, 'Irish *milchobur*', *Ériu* 19 (1962) 114–16

——'Preliminaries to a historical and comparative analysis of the syntax of the Old Irish verb', *Celtica* 6 (1963) 1–49

——'Indo-European metrics and archaic Irish verse', *Celtica* 6 (1963) 194–249

——'*In essar dam do á*', *Ériu* 29 (1978) 161–5

Watson, J. Carmichael (ed.), *Mesca Ulad* (Mediaeval and Modern Irish Series XIII, Dublin 1941, repr. 1983)

Watson, William J., *The history of the Celtic place-names of Scotland* (Edinburgh 1926)

Webb, David A., *An Irish Flora* (Dundalk 1943, seventh ed. revised by Declan Doogue and John Parnell, Dublin 1996)

Weir, David A., 'A palynological study of landscape and agricultural development in County Louth from the second millennium BC to the first millennium AD', in *Discovery Programme reports* 2 (Royal Irish Academy, Dublin 1995) 77–126

Went, Arthur E. J., 'The Galway fishery: an account of the modes of fishing together with notes on other matters connected with the fishery', *PRIA* 49 C (1943–44) 187–219

——'Irish fishing weirs', *JRSAI* 76 (1946) 176–94

——'The pursuit of salmon in Ireland', *PRIA* 63 C (1962–64) 191–244

——'Giraldus Cambrensis' notes on Irish fish', *Irish naturalists' journal* 9/9 (1949) 221–4

West, Geoffrey P., (ed.), *Black's veterinary dictionary* (18th ed., London 1995)

Westropp, Thomas Johnson, 'The forests of the counties of the Lower Shannon valley', *PRIA* 27 C (1908-09) 270–300

White, K. D., *Agricultural implements of the Roman world* (Cambridge 1967)

White, Lynn, *Medieval technology and social change* (Oxford 1962, repr. 1980)

White, Newport B., (ed.), *The Red Book of Ormond* (Dublin 1932)

White, Newport J. D., *St. Patrick: his writings and life* (London 1920)

Wiliam, Aled Rhys, (ed.), *Llyfr Iorwerth: a critical text of the Venedotian Code of medieval Welsh law* (Cardiff 1960)

Williams ab Ithel, John (ed.), *Annales Cambriæ* (London 1860)

Williams, J. E. Caerwyn, 'Welsh *drythyll, trythyll*; Irish *drettel, treitell*', *Celtica* 15 (1983) 150–7

Williams, Nicholas J. A., 'Some Irish plant names', in *Sages, saints, and storytellers: Celtic studies in honour of Professor James Carney* (ed. D. Ó Corráin, L. Breatnach, and K. McCone, Maynooth 1989) 449–62

——*Díolaim luibheanna* (Baile Átha Cliath 1993)

Williams, Stephen J., a J. Enoch Powell (ed.), *Cyfreithiau Hywel Dda yn ôl Llyfr Blegywryd (Dull Dyfed)* (Caerdydd 1942)

Wilson, J., 'The Scandinavian origin of the hornless cattle of the British Isles', *Scientific Proceedings of the Royal Dublin Society* 12 (15) (1909) 145–64

Windisch, Ernst (ed.), *Irische Texte mit Wörterbuch* (Leipzig 1880)

——(ed.), 'Das Fest des Bricriu', in *Irische Texte mit Wörterbuch* (ed. Windisch, Leipzig 1880) 235–311

——(ed.), '*De chophur in dá muccida*', in *Irische Texte* III i (ed. Stokes and Windisch, Leipzig 1891) 230–78

——(ed.), '*Tochmarc Ferbe*', in *Irische Texte* III ii (ed. Stokes and Windisch, Leipzig 1897) 445–556

Wooding, Jonathan M., *Communication and commerce along the western sealanes AD 400–800* (British Archaeological Reports: International Series 654, Oxford 1996)

——'Cargoes in trade along the western seaboard', in *External contacts and the economy of Late Roman and Post-Roman Britain* (ed. K. R. Dark, Woodbridge, England, 1997) 67–82

Wulff, Winifred (ed.), *Rosa anglica* (Irish Texts Society XXV, London 1929)

Zohary, Daniel, and Maria Hopf, *Domestication of plants in the Old World* (Oxford 1988, second ed. 1993)

Unpublished sources

(1) Glosses in the Trinity College Dublin manuscript 1337 (H 3. 18), p. 607

(2) Fragment of the law-text *Bretha Étgid* in the Royal Irish Academy manuscript 670 (D v 2), ff. 1–32 (pencilled foliation). I am grateful to William O'Sullivan for bringing the existence of this fragment to my attention.

(3) Versions of *Sanas Cormaic* (Cormac's Glossary) in the Trinity College Dublin manuscript 1317 (H 2. 15B), pp. 13–37 and pp. 79–102

(4) Royal Irish Academy manuscript 461 (24 B 2)

(5) Royal Irish Academy manuscript 462 (24 M 34)

(6) Royal Irish Academy manuscript 465 (23 N 20)

(7) 'A note of such kinds of rents and duties as the lordes and ffreholders in Desmound did customablie use to paye to the earle of Clancarty' in Lambeth Palace Library, Carew MS 625 ff. 28–41. I acknowledge the use of a transcript made by the late Myles Dillon.

Addenda and corrigenda

ADDENDA AND CORRIGENDA

P. 42. For the explanation of *airthend* as 'creeping bent-grass' I am indebted to Seosamh Ó Cuaig (*Foinse,* 3 Bealtaine 1998) and to Máire Nic Mhaoláin (*Foinse,* 10 Bealtaine 1998).

P. 55. In the first edition I suggested erroneously that the tallow candle tapered from the width of an inch to the width of a grain. In fact *Cáin Aicillne* simply states that the base of the rush – where it is cut – must be 'as thick as the grain by which an inch is measured' (*cutruma graine frisi tomister ordlach*). This is taken by the glossator to be a grain of barley, i.e. one third of an inch: see pp. 560–1.

P. 195. Michael Baillie suggests (*A slice through time* (Batsford 1995), 102) that these phenomena were associated with the eruption of the Icelandic volcano Hekla in AD 1222. The livestock may have suffered flourine poisoning.

P. 290. *Tuillem línchuir* is more likely to refer to 'the earnings of a medical bag', i.e. a physician's fee. See Binchy, 'A text on the forms of distraint', 83.

P. 325. In 'Aus dem irischen Recht I', 357, Thurneysen translated *cen fothana na ruba ann* as 'ohne Dünnliches und ohne Schädigendes darin' (without wateriness or without anything noxious in it). However Liam Breatnach points out that it makes better sense to take *ruba* to be for *rubai,* 3 sg. consuetudinal pres. of *atá,* i.e. 'that it (the cheese) should not be there'. The general sense of the passage is unaffected.

P. 403. John Carey suggests that there may be an earlier example (perhaps early eleventh century) of a poet seeking a gift of land from a king at *LL* i 218.6497.

P. 488. The *clúasa* 'ears' of the foreign axe are probably the spurs or wings on the shaft-holes of Viking axe-heads, see Cormac Bourke, 'Antiquities from the River Blackwater', *Ulster Journal of Archaeology* (forthcoming).